Young Researchers 2006

WATER AND ENVIRONMENTAL MANAGEMENT SERIES

Young Researchers 2006

Edited by Richard Stuetz and Lim Teik-Thye

IWA Publishing

Published by IWA Publishing, Alliance House, 12 Caxton Street, London SW1H 0QS, UK
Telephone: +44 (0) 20 7654 5500; Fax: +44 (0) 20 7654 5555; Email: publications@iwap.co.uk
Web: www.iwapublishing.com

First published 2006

British Library Cataloguing in Publication Data
A CIP catalogue record for this book is available from the British Library

Library of Congress Cataloging- in-Publication Data
A catalog record for this book is available from the Library of Congress

Disclaimer

ISBN13: 9781843395133
ISSN: 1476-1785

Contents

Modelling

Water quality and wastewater treatment

Preface

The water industry over the past 15 years has undergone significant change in term of advances in treatment technologies, increasing regulation, changing managerial structures and a greater focus on customer service. These changes have been unfortunately been accompanied by an aging of the industry's labour force as a results of industry downsizing and the lack of career opportunities for younger staff and new graduates. The impending loss of industry knowledge as older industry leaders retire is therefore a looming threat to the industry.

To meet this challenge the water industry has recognised the importance that new staff, particularly young water professional must play in developing innovative technological solutions for the treatment of polluted water, gas, solid waste, soil and sediments as well as providing an environment that promotes better customer relations, more sustainable solutions, and an enhanced commercial focus.

To assist in the development of new skills (*up-skilling*) and improved *multi-skilling* among young water scientists and engineers, the IWA has engaged in the development of strategies to enhance young professionals' engagement within the Association, and to raise awareness of the needs of this key stakeholder group with industry, academia and government members. Through its Young Water Professional program, IWA is providing increased networking opportunities for young water professionals through dedicated national and international conferences, developing web-based medium for the disseminations of young professional activities and promoting researcher exchange and internships opportunities to enhance mobility for young professionals.

As part of the IWA role to promote networking opportunities and career development for young researchers and professionals, IWA established the Young Researchers Conference (YRC) in 2002 as a biennial international forum for young water professionals to discuss current and future environmental concerns and network with peers from different research institutions, disciplines and countries. Moreover, the IWA Young Researchers Conferences aim to address issues related to the development of careers in the water sector. This forum has grown from its initial conference at Cranfield University, UK, in 2002 that was made up of 27 presentations from 19 countries to the 2nd YRC at Wageningen University, The Netherlands in 2004 that attracted over 130 submissions from which 52 papers were selected for oral and poster presentations.

The selection of papers in this special issue of WEMS represents papers that were presented at the 3rd Young Researchers Conference (YRC06) held on the 24–26 May 2006 at Nanyang Technological University, Singapore. The 58 papers from 23 countries were selected from 238 submissions and illustrate various aspects of water and wastewater engineering, technology and management. The conference opening address was presented by Dr. Yaacob Ibrahim, Minister for the Environment and Water Resources, Singapore and sponsored by the Public Utilities Board, Singapore, Agilent Technologies Singapore (Sales) Pte Ltd, Pan Asian Water Solutions Lte, PerkinElmer Singapore Pte Ltd, Olympus Singapore Pte Ltd, and Applied Biosystems Asia Pte Ltd.

The conference organisers thank Nanyang Technological University, Public Utilities Board, Singapore, the Singapore Water Association, the Economic Development Board, Singapore, delegates from the Water Utilities Forum and members of the local organising committee for their support.

Assoc Prof Richard M Stuetz
The University of New South Wales

Dr Lim Teik Thye
Nanyang Technological University

April 2006

Wastewater treatment

Microbial enhancement of granulation in sequencing batch reactors

Wang Xiao Hui* and Volodymyr Ivanov**

School of Civil and Environmental Engineering, Nanyang Technological University, 50 Nanyang Ave., Singapore 639798 (E-mail: *wang0048@ntu.edu.sg*; CVIvanov@ntu.edu.sg)

Abstract Bio-granulation is a new technology that can be used in municipal and industrial aerobic wastewater treatment. However, the formation of microbial granules from activated sludge requires several weeks, whereas activated sludge system can be started up in several days. To overcome it, aim of this research was to enhance aerobic granulation by using fast aggregated, environmentally safe microbial strains as inoculum. Pure cultures were obtained by enrichment of aggregated cultures and direct isolation form formed microbial granules. The isolated microbial strains were tested for their self-aggregation index, and the strains with AI higher than 50% were selected for strain identification by 16s rRNA gene analysis. Based on the information of identified strains in previous publications, *Pseudomonas veronii* and *Pantoea agglomerans* were selected as starter cultures to start up microbial granulation in SBR because they were not pathogenic bacteria. Aerobic granulation process in sequencing batch reactors (SBRs) with these two strains showed that the addition of strains with higher aggregation abilities diminished the duration of granulation process from several weeks to 3 and 8 days, respectively. This evaluation of granule formation time was based on value of SVI, which is lower than 70 mL g^{-1} of the typical value of matured granules. From day 8, the COD removal efficiencies for both control and experiment reactors were stable at 85–98%.
Keywords Microbial granule, microbial aggregation, starter culture

Introduction

Bio-granulation is a new technology, which can be used in aerobic treatment of municipal and industrial wastewaters. Microbial granules have some advantages over activated sludge such as fast settling, higher resistance to toxic substance (Jiang *et al.* 2004; Tay *et al.* 2004). It has diverse microbial community and structure, and can be used to remove particles (Ivanov *et al.* 2004). Owing to its diverse microbial community and presence of aerobic layer, anoxic layer and anaerobic layer in a single microbial granule, aerobic granular sludge has been able to perform aerobic and anaerobic processes (Ivanov *et al.* 2005a).

The development of aerobic granulation technology previously concentrated on how to form microbial granules. Different process parameters were manipulated in finding out the conditions under which microbial granules could be formed. It was found that a short settling time of less than 20 min, an aeration rate in terms of superficial up-flow air velocities higher than 1.2 cm s^{-1}, and a short cycling time within the range 3–6 h were necessary conditions for successful cultivation of microbial granules aerobically (Tay *et al.* 2001; Tay 2002a,b; Qin *et al.* 2004). Carbon source has no obvious effects on formation of microbial granules but significantly affected microbial community composition and structure of formed granules (Moy *et al.* 2002; Liu and Tay 2004; Zheng *et al.* 2005).

Recently the focus of the research in aerobic granulation has shifted to its scale-up from lab scale to pilot scale and finally to its industry application. Both de Bruin *et al.* (2004a,b) and Tay *et al.* (2005a) showed promising results of pilot research on aerobic granular sludge technology. Tay *et al.* used stored granules as initial inoculums with synthetic wastewater, whereas de Bruin *et al.* used activated sludge as inoculum and municipal wastewater as feeding at WWTP.

The development of aerobic technology for wastewater treatment was an effort for a compact and simple design of bioreactor. The advantages of aerobic granulation technology are mainly based on compactness of the installation (de Kreuk *et al.* 2004). It is because the compact structure of the microbial granules makes it possible to obtain high biomass concentrations in these systems that the volume load of these reactors can be high. The good settleability of the granules makes settling tanks unnecessary (de Bruin et al. 2004a,b; de Kreuk *et al.* 2004). Compared with conventional activated sludge, however, the start-up period of aerobic granulation was very long. It took several weeks for the formation of aerobic granules from activated sludge, whereas activated sludge system can start up within several days. As the formation of microbial granules was considered a selection process, and the rate of such microbial selection in a mixed sample will depend on both the starting microbial community and the selection pressure applied to the system such as short settling time, higher hydraulic shear force, etc. Therefore, by analogy with other wastewater treatment systems, formation of the granules can be enhanced by selected microbial cultures (Beun *et al.* 1999; Ivanov *et al.* 2005a,b). Beun *et al.* (1999) used different inoculums as seed sludge. One bioreactor took 30 days to reach steady state, while the other only took 10 days. This study suggested that inoculum is one of the important factor that determines granulation start up. Ivanov *et al.* (2005a) applied enrichment culture with higher cell-surface hydrophobicity as initial inoculum to form granules in sequencing batch reactors (SBRs). Mechanically stronger granules, which were suitable for the reactors with mechanical stirring, were formed for several days after start of the cultivation. The application of mixture of *Pseudomonas veronii*, *Klebsiella pneumoniae* and activated sludge as initial inoculums showed that microbial granules formed after 8 days cultivation in SBRs (Ivanov *et al.*, 2005b). After application of acetate-fed granules as the seeds for the development of phenol degrading granules stable phenol-degrading granules were developed within one week (Tay *et al.* 2005b).

The aim of this study was to select environmentally safe microbial strains with high aggregation ability, and testing these microbial strains to shorten start up of aerobic granular sludge system. To achieve it, pure microbial cultures were isolated from enrichment culture of aggregated cell and from microbial granules formed from activated sludge. Then, these cultures were screened by their self-aggregation ability and bio-safety assessment. Only environmentally safe microbial strains with high aggregation ability were used as inoculum for start up of aerobic granulation in SBRs.

Methods

Selection of enrichment culture to facilitate formation of microbial aggregates

The microbial granules were collected from different laboratory SBRs and were dispersed by the beating of 200 mg (wet weight) of granules in 2 mL tube without glass beads for 3 min at 420 strokes per min in Mini-Beadbeater (Biospec, Bartlesville, OK, USA). One milliltre of this suspension was diluted in 9 mL of 0.1 M phosphate buffer solution (PBS), pH 7.2, in 15 mL conical tube and stayed for 10 min. Then, 1 mL of cell suspension and aggregates from the bottom of the conical tube was transferred into fresh medium and incubated for 48 h at 25°C on shaker with 150 rpm. The medium composition was as follows: tryptone, 5.0 g L^{-1}; dextrose 1.0 g L^{-1}; yeast extract, 2.5 g L^{-1}; antifungal agent (cycloheximide), 50 mg L^{-1}. Starting from the second transfer, the microbial aggregates were collected and washed in PBS five times using 15 mL conical tubes. Aggregates were treated in beater for 3 min before washing to remove weakly attached and suspended cells. Then, the procedure followed the same steps as described above. The total number of transfers was 13.

Formation of granules

Control was performed in one lab-scale column type reactor R1, a sequencing batch reactor (SBR) to cultivate the granules from activated sludge. The reactor has a working volume of 2.5 L, the internal diameter of 50 mm and a filling height of 1200 mm. Reactor R1 was inoculated with 100 ml of activated sludge from local municipal wastewater treatment plant with concentration of mixed liquor volatile suspended solids (MLVSS) 3500 mg L^{-1}. The initial concentration of biomass was 0.14 g dry biomass L^{-1}. The reactor was operated as sequencing batch reactor with a cycle of 3 h, including 5 min of influent filling, 169 min of aeration, 1 min settling and 5 min of effluent discharge. Effluent was discharged from the middle port of the reactor with a volumetric exchange ratio of 50%, giving a hydraulic retention time (HRT) of 6 h. Air was supplied through a dispenser at the reactor bottom with airflow rate of 2.5 L min^{-1}. Temperature in the reactor was ambient, varying from 25°C to 32°C. Synthetic wastewater was composed of glucose 2 g L^{-1}, $(NH_4)_2SO_4$, 0.2 g L^{-1}; Nutrient Broth (Difco, Detroit, MI, USA), 0.15 g L^{-1}; KH_2PO_4, 0.1 g L^{-1}, $MgSO_4 .7H_2O$, 0.04 g L^{-1}; $CaCl_2.2H_2O$, 0.04 g L^{-1}; $FeSO_4.7H_2O$, 0.04 g $L^{-1}$and trace element solution, 1 ml L^{-1}. The trace element solution contained (g L^{-1}): $ZnSO_4.7H_2O$, 0.2; H_3BO_3, 0.6; $MnCl_2.4H2O$, 0.06; $CoCl_2.6H2O$, 0.4; $CuSO_4.4H_2O$, 0.02 and $NaMoO_4.2H_2O$, 0.06. Chemical oxygen demand (COD) loading of reactors was 8.5g COD L^{-1} day^{-1}.

Isolation of pure cultures

The microbial aggregates of the enrichment culture after 13th transfer, and the granules produced in control reactor R1 after 9 days of cultivation were separately dispersed in phosphate buffer solution (PBS) by beating 200 mg (wet weight) of aggregates in 2 mL tube for 3 min at 420 strokes per min using beater. The suspensions were diluted in PBS to 10^{-4}, 10^{-5} and 10^{-6} times and 0.1 mL aliquots of these suspensions were spread on Petri dishes with Nutrient Agar (Difco, Detroit, MI, USA) and cultivated for 24 h at 27°C. Pure cultures were transferred to fresh Nutrient Broth (Difco, Detroit, MI, USA), cultivated for 16 h and studied for self-aggregation using aggregation index described by (Malik *et al.* 2003).

16S rRNA gene sequencing

A whole-cell direct lysis polymerase chain reaction (PCR) amplification method was used to amplify the 16S rRNA gene of isolates, as described previously (Maszenan *et al.* 2000). The nearly full-length 16S rRNA gene was amplified by PCR with forward primer Eubac27F and reverse primer Universal 1492R1 (Proligo, Paris, France) (Lane 1991). The PCR product was purified with a PCR purification kit (Wizard® SV Gel and PCR Clean-Up System, Promega, Madison, WI, USA) according to manufacturer's instructions and stored at –20°C before sequencing. The 16S rRNA gene sequence of the isolates was determined by using an ABI model 310A DNA sequencer (Applied Biosystems/Perkin-Elmer) and an ABI PRISM BigDye Terminator Ready Reaction cycle sequencing kit (version 3.0; Applied Biosystems/Perkin-Elmer). The partial sequences were assembled by using BioEdit software and analyzed with BLAST and other algorithms as described previously (Tay *et al.* 2002).

Formation of granules from pure selected culture

Two lab-scale column type SBRs, R2, R3, were used to produce granules from selected pure culture of *Pseudomonas veronii* strain B and *Pantoea agglomerans* respectively. The reactor setup and synthetic wastewater composition was the same as described above for reactor R1. Experimental reactors R2 and R3 were inoculated with 100 mL cell suspension of *P. veronii* strain B and *Pantoea agglomerans* respectively. Optical density of this suspension was pre-adjusted to 1.2; initial concentrations of biomass in R2 and R3 were 32 mg dry biomass L^{-1}.

Measurements

Chemical oxygen demand (COD), mixed liquor suspended solids (MLSS), and sludge volume index (SVI) were determined by standard methods of wastewater examination (APHA 1998). Size of the granules was measured by a laser-particle size analysis system MasterSizer Series 2000 (Malvern Instruments, Worcestershire, UK). The aggregation ability of pure cultures was determined by aggregation index (AI), which shows the percentage of aggregated and fast settling cells. AI was determined as relative decrease of optical density of bacterial suspension at 660 nm after slow centrifugation at 650*g* for 2 min as described by (Malik *et al.* 2003).

Results and discussion

The purpose of this research was to select aggregate-associated bacterial cultures from microbial granules and to examine their ability to accelerate formation of granules during wastewater treatment.

Selection of enrichment culture with high self-aggregation ability

One way to achieve this goal was to was to select self-formed microcolonies after destruction of granules, to separate micro-colonies/micro-aggregates by fast settling, and then to grow them in fresh medium. By repeating this selection procedure, aggregate-forming microbial culture could be enriched, and microbial strains with high aggregation ability can be isolated. Aggregation index of cells in enrichment culture increased from 4% to 34% after 13 cycles of selection transfers to liquid medium (Fig. 1). Enrichment culture produced after 13 cycles of selection was used because there were no significant changes of cell aggregation ability after 13 cycles (Fig. 1b). During batch cultivation of enrichment culture after 13 transfers, cell aggregation started only in during stationary phase (Fig. 1a). This enrichment culture left a clear supernatant after 10 min settling (sedimentation) of microbial biomass. The supernatant turbidity after 10 min settling was 1946 nephelometric turbidity units (NTU) at the beginning of selection process and 318 NTU after 13 transfers. Both the increased AI and decreased supernatant turbidity after settling of biomass indicated the good aggregation ability of enrichment culture.

Formation of microbial granules from activated sludge

Another way to select bio-granule forming microbial strains was direct microbial pure culture isolation from microbial aggregates, which were collected during transition period from activated sludge flocs to microbial aggregates. The hypos sis behind of it was that aerobic granulation was considered a selection process due to high selection pressure imposed to aerobic granular SBR, including high aeration rate, short settling period, and relatively short hydraulic retention time which related to wash out frequencies, etc (Liu and Tay 2004). Under such selection pressure, the light and dispersed sludge is washed out and relatively strong aggregates are retained in the reactor. After operation for 70–80 cycles when activated sludge started to turn into microbial aggregates and small granules, it was believed that the strong aggregated cells were enriched and dominated in bioreactors. Therefore, microbial granules were cultivated from conventional activated sludge in one SBR for 9 days, in the preparation of microbial aggregates for microbial pure culture isolation.

Dense microbial aggregates and granules were formed from the flocs of activated sludge during cultivation in GSBR (Fig. 1c). It was conventionally considered that mean size of the particles which are granules was 500 μm. In this case, the granules were formed from the flocs of activated sludge for 9 days of cultivation in GSBR (Fig. 1d). SVI values were 154 mL g^{-1} at the beginning of cultivation and 106 mL g^{-1} after 9 days of cultivation. The particle mean size was 76 ± 15 μm at the beginning of cultivation and 745 ± 197 μm after 9 days cultivation.

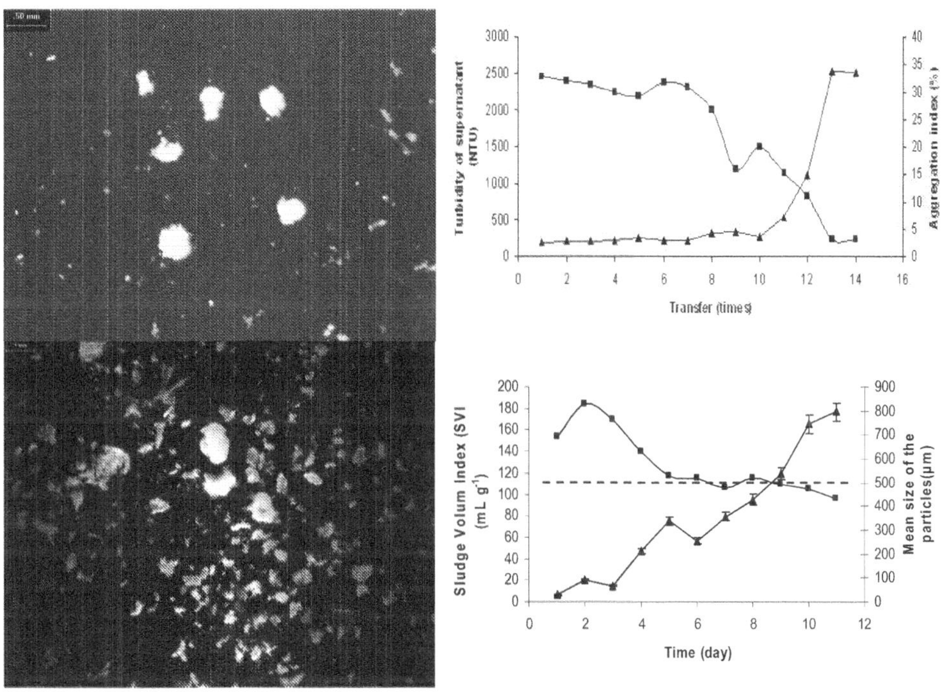

Figure 1 Formation of cell aggregates in enrichment culture and formation of microbial granules from activated sludge. (a) Cell aggregates formed after 13 times transfer in enrichment culture, scale bar is 0.5 mm; (b) aggregation index and supernatant turbidity change during enrichment culture transfers; ▲: Aggregation Index; ■: Turbidity of supernatant after 10 min settling. (c) granules formed from activated sludge after 9 days of cultivation. Scale bar is 1 mm. (d) mean size of the particles (▲) and SVI (■) changes during formation of granules from activated sludge. Dotted line shows 500 μm, mean size of the particles conventionally considered as granules.

Isolation, screening and identification of pure cultures

In total, five pure cultures were isolated from enrichment culture after 13 time transfers, and 10 pure cultures were isolated from the formed microbial granules after 9 days of cultivation. After testing of their self-aggregation index, four microbial strains from enrichment culture and five microbial strains from granules with aggregation index higher than 50% were selected for further study and 16s rRNA gene sequences analysis. The results of identification are shown in Table 1.

The goal of wastewater treatment is the protection of the environment in a manner commensurate with public health, economic, social and political concerns. Therefore, not only environmental problems but also public health and economics must be considered in biological wastewater treatment. Using a safe starter culture for fast and enhanced granulation, the wastewater treatment could be operated with high specific degrading efficiency and reduced risk to public health.

Table 3.1 Taxonomic, phylogenetic and aggregation characterization of isolates

Strain	Morphology	Gram stain	Closest relatives	Taxonomic affiliation(s)	16S rRNA gene sequences idendity (%)	No. of bases used to establish identity	Self aggregation index
E01	Cocci	-	*Pantoea sp.P101* [AF394538]	γ-proteobacteria	99.6	1405	69
E02	Cocci	-	*Pseudomonas veronii* AY267192	γ-proteobacteria	99.3	1298	50
E03	Rod	+	*Bacillus cerus*	Bacillales	98	Partial	51
G01	Rod	-	*Klebsiella pneumoniae* AY292866.1	γ-proteobacteria	99.4	1338	65
G02	Cocci	-	*Pseudomonas veronii* AY512619.1	γ-proteobacteria	99.9	1408	51
G03	Rod	-	*Aeromonas caviae* gi:38692	γ-proteobacteria	99.7	1503	67
G04	Cocci	-	*Enterobacter agglomerans* AF157688	γ-proteobacteria	99.8	1349	72
G05	Cocci	-	*Pantoea agglomerans* AF199029	γ-proteobacteria	98.7	1491	84

There were many human pathogens among bacterial strains isolated from the granules and having high aggregation ability. Strain EN03, *Bacillus cereus* was a Gram-positive, spore forming bacteria. It has been recognized as an agent of food poisoning since 1955, and known to produce an emetic toxin causing vomiting, and at least three different enterotoxins, which cause diarrhea (Prub *et al.* 1999). Strain G01, close to *Klebsiella pneumoniae,* and strain G03, close to *Aeromonas caviae*, may be human pathogens. *Klebsiella pneumoniae* is an agent causing pneumonia (Bodmann 2005). The strains of *Aeromonas caviae* seem to be associated with hospitalized cases of diarrhea, exhibiting multidrug resistance (Sinha *et al.* 2004) and should be considered a possible etiologic agent of contact lens-associated keratitis (Pinna *et al.* 2004). It seems that bacterial pathogens are often present in the microbial aggregates and biofilms used for wastewater treatment (Hu *et al.* 2003). Species of strains E01, G04 and G05, *Pantoea agglomerans,* the previous name of *Enterobacter agglomerans* (Edwards *et al.* 2004)**,** is listed in Biohazard group 1. This strain is known as plant pathogenic bacteria, but it is prevalent in the general environment and usually relatively benign. Cells of *Pseudomonas veronii* are Gram-negative rods, motile by means of a single polar flagellum (Elomari *et al.* 1996). *Pseudomonas veronii* is known as species, which has not been implicated in human infection, except one case of intestinal inflammatory pseudotumour (Cheuk *et al.* 2000). In this study, *Pantoea agglomerans* and *Pseudomonas veronii* were used as initial inoculums to start up aerobic granulation in R2 and R3, respectively.

Granulation process with *P. veronii* and *Pantoea agglomerans* as starter culture

The morphological changes of microbial aggregates produced from cells of *P. veronii* and *Pantoea agglomerans* in R2, R3 respectively are shown in Fig. 3. After use of cells of *P. veronii and Pantoea agglomerans* as starter cultures the granules (particles with size bigger 500 μm) were observed after 3 days of cultivation in GSBR (Fig. 3 b2 and c2). Granules produced from starter culture *P. veronii* str. B were more compact, SVI was 70 mL g^{-1} (Fig. 2a) than those produced from *Pantoea agglomerans* with SVI 102 mL g^{-1} and those from activated sludge with SVI 106 mL g^{-1} (Fig.2.a). The removal efficiency of chemical oxygen demand (COD) and accumulation of in GSBR were similar when the granules were produced either from activated sludge flocs or starter culture of *P. veronii* or *P. agglomerans* (Fig. 2b). It means that major advantage of the using of *P. veronii* as the starter culture is fast formation of compact granules but there are no advantages in biodegradation and growth rates over granules produced from conventional activated sludge.

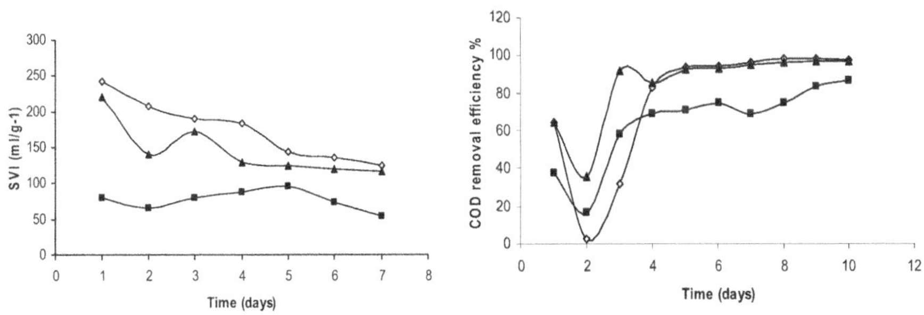

Figure 2 (a) SVI change during granule formation; (b) COD removal efficiency change during granule formation. ◊: control in R1; ■: experiment in reactor R2, inoculated with *Pantoea agglomerans*; ▲: experiment in reactor R3, inoculated with, *Pseudomonas veronii*.

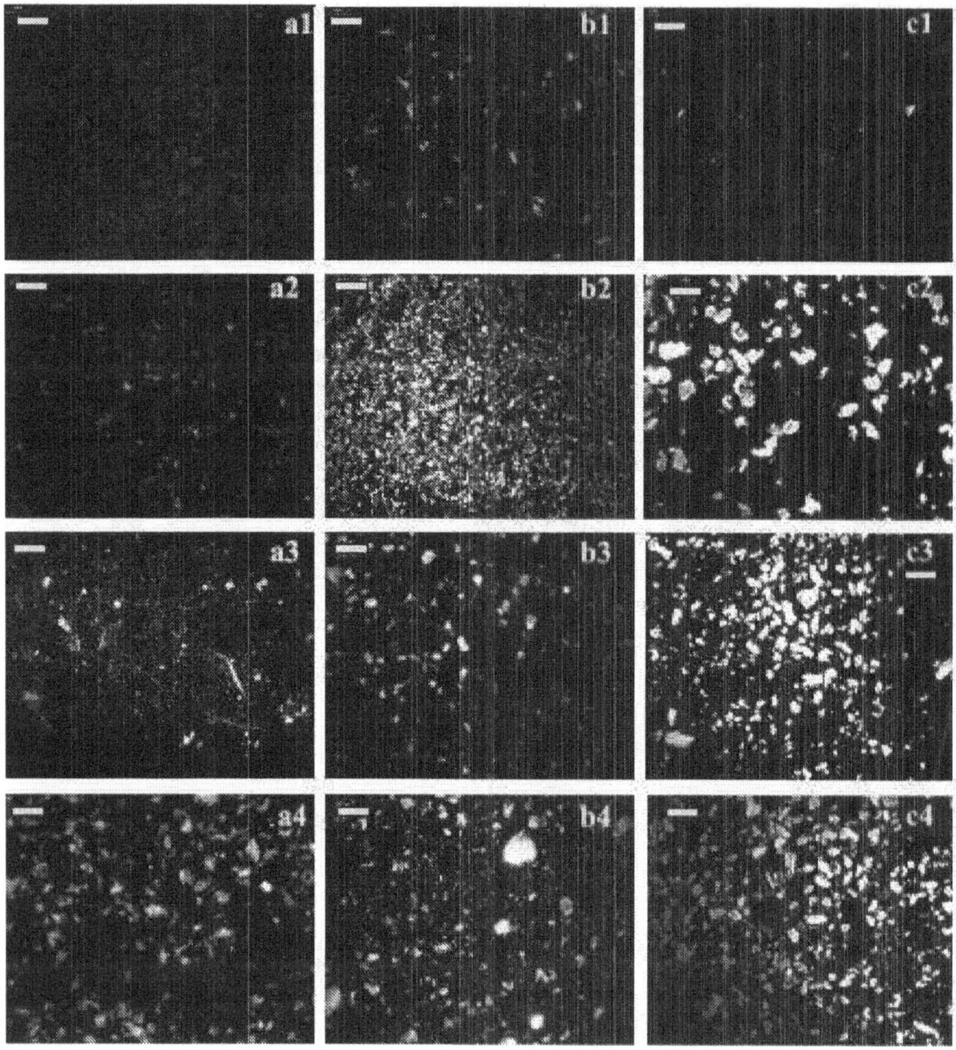

Figure 3 Morphology change during aerobic granulation. (a1): R1, activated sludge – day 1; (a2):R1, activated sludge – day3; (a3): R1, activated sludge – day5; (a4):R1, activated sludge – day7; (b1): R2, *Pantoea agglomerans* – day 1; (b2):R2, *Pantoea agglomerans* – day3; (b3): R2, *Pantoea agglomerans* – day5; (b4):R2, *Pantoea agglomerans* – day7; (c1): R3, *Pseudomonas* veronii – day 1; (c2):R3, *Pseudomonas veronii* – day3; (c3): R3, *Pseudomonas veronii* – day5; (c4):R3, *Pseudomonas veronii* – day7.

Bioaugmentation of activated sludge systems with specialized bacteria strains (microbial seeds), has been practiced since the 1960s. Their application in wastewater treatment was originally the efforts to solve operational problems, such as shock loading in treatment plant (Limbergen *et al.* 1998). Bioaugmentation could be a powerful tool and cost effective method to improve several aspects in wastewater treatment process such as improved flocculation and degradation of recalcitrant compounds (Limbergen *et al.* 1998). A stable enhanced biological phosphorus removal (EBPR) in bioreactor was installed within 15 days using bioaugmentation of sludge by the phosphorus-accumulating organism (Dabert *et al.*

2005). In our experiments, environmental conditions and process parameters in both control and experimental reactors were same except the addition of two selected strains with high aggregation ability into experimental reactors respectively. The formation of granules from cells of *P. veronii* was for three days instead of 9 days when from the flocs of activated sludge. Experimental data demonstrated that this addition significantly diminished time of granule formation and facilitate formation of dense granule with low SVI.

Conclusion

Pseudomonas veronii could be used for further large-scale trials as a starter culture for facilitated formation of microbial granules in aerobic wastewater treatment. However, survival of introduced strains in the system is a common problem in bioaugmentation applications. Therefore, a selected strain of *Pseudomonas veronii* cannot be considered as universal strain for all cases, where microbial granulation is required. The feasibility of microbial granulation with this strain must be tested for every specific process.

References

APHA (1998) *Standard Methods for the Examination of Water and Wastewater*, 19th edn. American Public Health Association, Washington, DC, USA.

Beun JJ, Hendriks A, van Loosdrecht MCM, Morgenroth E, Wilderer PA, Heijnen JJ (1999) Aerobic granulation in a sequencing batch reactor. *Wat. Res.* **33**: 2283–2290.

Bodmann KF (2005) Current guidelines for the treatment of severe pneumonia and sepsis. *Chemotherapy* **51**: 227–233.

Cheuk W, Woo PC, Yuen KY, Yu PH, Chan JK (2000) Intestinal inflammatory pseudotumour with regional lymph node involvement: identification of a new bacterium as the aetiological agent. *J. Pathol.* **192**: 289–292.

Dabert P, Delgenès J-P, Godon J-J (2005) Monitoring the impact of bioaugmentation on the start up of biological phosphorus removal in a laboratory scale activated sludge ecosystem. *Appl. Microbiol. Biotechnol.* **66**: 575–588.

de Bruin L, de Kreuk, MK, van der Roest HFR, Uijterlinde, C, van Loosdrecht, MCM (2004) Aerobic granular sludge technology: an alternative to activated sludge? *Wat. Sci. Technol.* **49**: 1–7.

de Bruin LMM, van der Roest HF, de Kreuk MK, Van Loosdrecht MCM (2004) Promising results pilot research aerobic granular sludge technology at WWTP Ede. In: Bathe S, de Kreuk M, McSwain B, Schwarzenbeck N (eds) *Aerobic Granular Sludge*. IWA Publishing, London, pp 135–142.

de Kreuk M, de Bruin LMM, Van Loosdrecht MCM (2004) Aerobic granular sludge-from idear to pilot plant. In: Bathe S, de Kreuk M, McSwain B, Schwarzenbeck N (eds) *Aerobic Granular Sludge*. IWA Publishing, London, pp 111–124.

Edwards J, Baque E, Tang J (2004) Proposal to acknowledge Beijerinck as the original author of the species *Pantoea agglomerans*. Request for an Opinion. *Int. J. Syst. Evol. Microbiol.* **54**: 24–37.

Elomari M, Coroler L, Hoste B, Gillis M, Izard D, Leclerc H (1996) DNA relatedness among Pseudomonas strains isolated from natural mineral waters and proposal of *Pseudomonas veronii* sp. nov. *Int. J. Syst. Bacteriol.* **46**: 1138–1144.

Hu JY Hu, JY; Fan, Y; Lin, Y-H; Zhang, H-B; Ong, SL; Dong, N; Xu, J-L; Ng, WJ; Zhang, L-H (2003) Microbial diversity and prevalence of virulent pathogens in biofilms developed in a water reclamation system. *Res. Microbiol.* **154**: 623–629.

Ivanov V *et al.* (2005a) Microstructural optimization of wastewater treatment by aerobic granular sludge. In: Bathe S, Kreuk Md, McSwain B, Schwarzenbeck N (eds) *Aerobic Granular Sludge*. IWA Publishing, London, pp 43–52.

Ivanov V, Tay J-H, Tay ST-L, Jiang H-L (2004) Removal of micro-particles by microbial granules used for aerobic wastewater treatment. *Wat. Sci. Technol* **50**: 147–154.

Ivanov V, Wang XH, Tay STL, Tay JH (2005b) Bioaugmentation and enhanced formation of microbial granules used in aerobic wastewater treatment. *Appl. Microbiol. Biotechnol.* **70**: 374–381.

Jiang HL, Tay JH, Maszenan AM, Tay STL (2004) Bacterial diversity and function of aerobic granules engineered in a sequencing batch reactor for phenol degradation. *Appl. Environ. Microbiol.* **70**: 6767–6775.

Lane DJ (1991) 16S/23S rRNA sequencing. In: Stackebrandt E, Goodfellow M (eds) *Nucleic Acid Techniques in Bacterial Systematics*. John Wiley & Sons, Ltd., Chichester, England, pp 115–175.

Limbergen HV, Top EM, Verstraete W (1998) Bioaugmentation in activated sludge: current features and future perspectives. *Appl Microbiol. Biotechnol.* **50**: 16–23.

Liu Y, Tay J-H (2004) State of the art of biogranulation technology for wastewater treatment. *Biotechnology Advances* **22**: 533–563.

Malik A et al. (2003) Coaggregation among nonflocculating bacteria isolated from activated sludge. *Appl. Environ. Microbiol.* **69**: 6056–6063.

Maszenan AM et al. (2000) Three isolates of novel polyphosphateaccumulating Gram-positive cocci, obtained from activated sludge, belong to a new genus, *Tetrasphaera* gen. nov., and description of two new species, *Tetrasphaera japonica* sp. nov. and *Tetrasphaera australiensis* sp. nov. *Int. J. System. Evol. Microbiol.* **50**: 593–603.

Moy BYP, Tay JH, Toh SK, Liu Y, Tay STL (2002) High organic loading influences the physical characteristics of aerobic sludge granules. *Letts Appl. Microbiol.* **34**: 407–412.

Pinna A, Sechi LA, Zanetti S, Usai D, Carta F (2004) *Aeromonas caviae* keratitis associated with contact lens wear. *Ophthalmology* **111**: 348–351.

Prub BM, Dietrich R, Nibler B, Martlbauer E, Scherer S (1999) The hemolytic enterotoxin HBL is broadly distributed among species of the *Bacillus cereus* group. *Appl. Environ. Microbiol.* **65**: 5436–5442.

Qin L, Tay J-H, Liu Y (2004) Selection pressure is a driving force of aerobic granulation in sequencing batch reactors. *Process Biochemistry* 39: 579–584.

Sinha S et al. (2004) Prevalence, serotype distribution, antibiotic susceptibility and genetic profiles of mesophilic *Aeromonas* species isolated from hospitalized diarrhoeal cases in Kolkata. *Ind. J. Med. Microbiol.* 53: 527–534.

Tay JH, Jiang HL, Tay STL (2004) High-rate biodegradation of phenol by aerobically grown microbial granules. *J. Environmental Engineering – ASCE* **130**: 1415–1423.

Tay J-H, Liu Q-S, Liu Y (2001) The effects of shear force on the formation, structure and metablism of aerobic granules. *Appl. Microbiol. Biotechnol.* **57**: 227–233.

Tay J-H, Liu Q-S, Liu Y, Show K-Y, Ivanov V, Tay ST-L (2005a) A comparative study of aerobic granulation in pilot- and laboratory-scale SBRs. In: Bathe S, Kreuk Md, McSwain B, Schwarzenbeck N (eds) *Aerobic Granular Sludge*. IWA Publishing, London, pp 125–134.

Tay JH, Yang, S.F., Liu, Y. (2002) Hydraulic selection pressure-induced nitrifying granulation in sequencing batch reactors. *Appl. Microbiol. Biotechnol.* **59**: 332–337.

Tay ST-L, Ivanov V, Yi S, Zhuang W-Q, Tay J-H (2002) Presence of anaerobic bacteroides in aerobically grown microbial granules. *Microbial Ecol.* 44: 278–285.

Tay STL, Moy BYP, Jiang HL, Tay JH (2005b) Rapid cultivation of stable aerobic phenol-degrading granules using acetate-fed granules as microbial seed. *J. Biotechnol.* **115**: 387–395.

Zheng YM, Yu HQ, Sheng GP (2005) Physical and chemical characteristics of granular activated sludge from a sequencing batch airlift reactor. *Process Biochemistry* **40**: 645–650.

Fouling control mechanisms in anaerobic bioreactors for the treatment of industrial wastewater

G. James*, G. Leslie and F. Zha***

* Memcor Australia, 1 MEMTEC Parkway South Windsor, NSW, 2756, Australia
(E-mail: *gerin.james@siemens.com*; *fufang.zha@siemens.com*)
** UNESCO Centre for Membrane Science and Technology, School of Chemical Engineering and Industrial Chemistry, University of New South Wales, 2052, Sydney, Australia
(E-mail: *g.leslie@unsw.edu*)

Abstract Membrane bioreactors (MBRs) have been used successfully in aerobic biological wastewater treatment to solve the perennial problem of effective solids-liquid separations. Adapting the MBR concept to anaerobic systems, via an Anaerobic Membrane Bioreactor (An-MBR), has the potential to similarly enhance the performance of anaerobic processes by allowing the complete retention of slow growing anaerobic bacteria within the reactor. Anaerobic processes are generally required to treat industrial wastewater containing high Chemical Oxygen Demand (COD). In this study an An-MBR, configured with membranes that use different mechanisms for fouling control, was operated on a synthetic industrial wastewater. Although the bioreactor achieved only modest removal of COD (average reduction 26%), it was apparent that the fouling control techniques used in aerobic systems could be readily adapted to control the accumulation of anaerobic biomass on the membrane surface. And maintain a minimum flux of 16 $L/m^2/h$, which is 60% above the industry standard for industrial waste water treatment using MBR's. The higher flux achieved by the jet mixed module compared to the direct injection module was found to relate to the effectiveness of the bubbles generated by jet mixing which achieved greater and more random fibre displacement per bubble than the small bubble direct injection module. This phenomenon was explained by developing the concept of fibre "expansivity" to attempt to describe the random movement of fibres in the bundle.

Keywords Anaerobic membrane bioreactors; industrial wastewater treatment; high COD wastewater

Introduction

Anaerobic treatment of wastewater has many advantages over other biological and physical/chemical treatment processes including an ability to deal with high strength waste, high removal efficiency for biodegradable and recalcitrant materials, methane generation, no oxygen requirements, reduction of pathogens, reduced solids production and production of stabilised solid waste (Speece, 1996). It is expected that anaerobic waste treatment will become more popular as environmental restrictions on effluent quality, biomass disposal and energy consumption criteria are tightened (Rajeshwan *et al.*, 2000).

The purpose of this paper is to investigate possible synergistic benefits of combining anaerobic treatment with membrane separation as a means of overcoming limitations inherent in conventional anaerobic systems. These limitations derive from the slow accumulation of biomass and growth rate of anaerobic microorganisms which require the use of uneconomical reactor designs and compromise the efficiency of downstream separation processes. In addition, the sensitivity of bacteria to chemical, nutrient, flow or temperature variations in the feed necessitate complex process monitoring and control and significant upstream buffering capacity (Speece, 1996).

Our approach is to exploit the ability of membranes processes to increase biomass retention to simultaneously improved process stability and improve downstream separation efficiency (Rajeshwan *et al.*, 2000) Advances in aerobic membrane bioreactors (MBRs) and anaerobic membrane bioreactors (An-MBR) over the past 15 years afford the opportunity for the enhancement and improvement of conventional anaerobic systems for various wastewater treatment applications, particularly those dealing with high concentrations of chemical oxygen demand (Hernandez *et al.*, 2002).

The higher biomass concentration required for anaerobic treatment and in particular industrial wastes which often contain more refractory organics, mainly owing to a lack of nutrients or an unfavourable carbon to nitrogen ratio (Speece, 1996; Fuchs *et al.*, 2003) is a significant issue that must be addressed in the application of membrane technology to membrane processes.

Submerged membrane systems use gas bubbles to perform cleaning. Typically the membranes are suspended above the gas diffuser to maximise solids removal. The gas liquid mixture generates a two-phase flow which scours the membrane creating surface shear. When the bubble train (or gas slug) moves across the vertical hollow fibre membrane surface, part of the mass transfer boundary layer is physically removed (Bellara *et al.*, 1996). Anaerobic systems employing both membranes with intermittent (or cyclic) or cyclic bubbling were developed in the mid-1990s (Takagki *et al.*, 1997); however, the bubbling method has not been optimised, particularly for membrane modules with variable degrees of fibre tension (or slack).

In modules with fibres of sufficient slack, two-phase flow also generates lateral fibre movement. This fibre movement enables mechanical cleaning increasing cake solids removal (Cui *et al.*, 2003).

In this study, we operated a pilot An-MBR reactor and determined the conditions required to achieve stable membrane performance through fouling control in a high Mixed Liquor Suspended Solids (MLSS) anaerobic environment containing 25–55g/L treatment (aerobic 5–15g/L). For this study, two different two-phase flow configurations were used to control membrane fouling: direct injection and jet mixing.

Our previous studies using a lab-scale reactor (5 L) processed a synthetic high COD organic wastewater (12,000 COD mg/L) demonstrated a COD removal of up to 65% at a HRT of 44hrs. The basic tenet of this study is to investigate the possibility of utilizing the 2–2.5 LPM of biogas (methane content greater than 70%) to improve the operating performance of the membrane systems.

Methods

The pilot anaerobic reactor was constructed with the support of Memcor Australia at their Membrane Bioreactor trial site. Seed biomass was obtained from the anoxic tank of the local sewage treatment plant and concentrated to 30 g/L. The reactor operated on synthetic high strength industrial waste (~15,000 mg/L COD) delivered from a storage tank (2400 L) at a rate of 20L/hr and detained in mixed (200 rpm) a pre-acidification tank (160 L) for 13 h.

The mixed liquor was processed in an up-flow style anaerobic reactor (440L) with a gas headspace of 60 L, and a HRT of 33 h. The total system HRT was 45 h. The reactor was insulated and temperature maintained at 35 °C, by an external water bath circulating water through an immersed heating coil.

Higher biomass retention was achieved by diverting the mixed liquor into a membrane tank (5L) containing two PVdF hollow fibre membrane modules ($0.5\ m^2$) (Memcor Australia) with a nominal pore size of 0.1 μm.

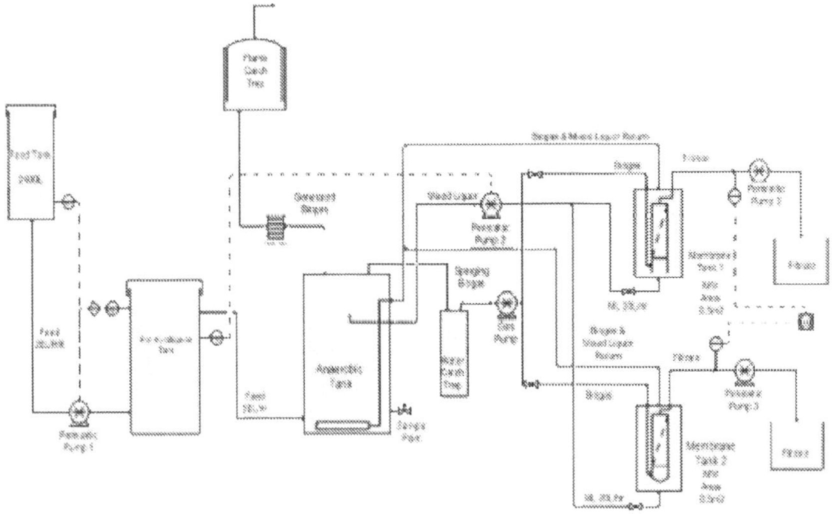

Figure 1 Experimental set up of the An-MBR system.

Membrane fouling was controlled by establishing a two-phase flow (gas and liquid) that scoured the accumulated biomass from the membrane. One module was configured using the patented Memjet configuration that uses external mixing of gas and liquid to establish a well-developed two-phase flow, while the second module used direct injection of gas and the base of the membrane fibres.

Biogas generated by the anaerobic system was used to establish two-phase flow. Biogas generated passes through a flame trap and gas flow meter before discharged to the atmosphere.

The membrane modules operate in a dead-end suction mode (Peristaltic Pump, Cole-Parmer) with an imposed suction flux of 10 LMH, filtrate flow rate was measured daily, while the filtrate was collected and the average flow rate (and thus flux) determined. Fouling was determined by measuring trans-membrane pressure using a pressure transducer (–50 to 100 kPa) connected to a data logger.

Filtrate tests were carried out using a HACH DR2000 using he appropriate HACH test method. MLSS was measured using a test based on AS

Flux was calculated using the following equation.

Flux (LMH or $l/m^2.h$) = filtrate flow rate (Q) (m^3/hr)/membrane area (m^2)

Permeability at temperature T (°C) was calculated using the following equation:

Permeability (LMH/Bar or $l/m^2.h.bar$) = Flux (LMH)/TMP (bar)

To allow for comparison between modules the permeability was corrected to 20°C.

Results and discussion

After a biomass accumulation period of 6 weeks, initial operation of the reactor showed that significant flux reduction (and increase in TMP) occurred in 7 days. Under constant imposed filtration the flux decreased lead to an increase in HRT lead to increased COD removal (Figure 2).

To achieve more consistent operation a fibre relaxation step was added to the processes, this resulted in consistent flux for the three days followed by a reversible increase in the transmembrane pressure. To restore membrane permeability recovery cleans using sodium hypochlorite, 2000 ppm was performed.

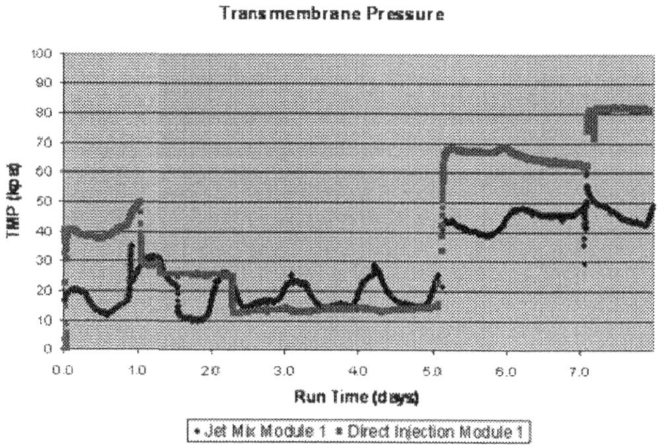

Figure 2 Initial TMP of jet mixed and direct injection modules.

Each module started up with similar TMP, 6 kPa and permeability above 200 LMH/Bar; once steady state conditions were established the Memjet® system was far superior to that of the direct biogas injection system. A significantly higher flux 16.5 LMH vs. 14.2 LMH was also recorded for the Memjet® system.

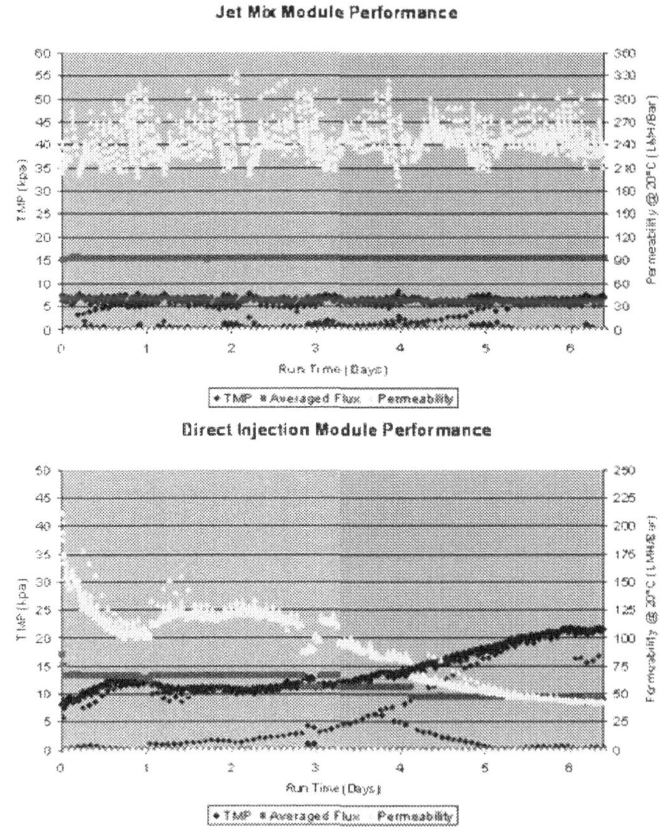

Figure 3 Start up permeability of jet mixed and direct injection modules.

The start-up permeability of the jet mixing module averaged 240 LMH/Bar for the first six days of operation (Figure 3). TMP was steady at 6–7 kPa and the averaged flux was 16 LMH. The biogas jet mixed module was observed to produce a turbulent regime inside the membrane tank. A circulation pattern was found to establish itself with the ellipsoidal biogas bubbles flowing up between the fibres before reaching the top of the module. Here some bubbles and mixed liquor were pushed against the wall and flowed to the bottom of the membrane tank.

The direct injection module started up with a TMP similar to that of the jet mixed module (Figure 3), in the first day the TMP increased to 11 kPa where it stayed for a day before increasing at a rate of 3 kPa per day until day 6 reached 22 kPa. Start up permeability (211 LMH/Bar) was lower, due to the lower average flux (13.3 LMH) and the rise in TMP.

The biogas bubbles observed in the direct injection module had a very different flow pattern to that of the jet mixed module. There were fewer large ellipsoidal bubbles than found in the jet mixed membrane tank; the bubbles were more spherical in shape flowing straight up through the fibres and out of the tank, without inducing circulation within the tank. A separation layer of denser biomass at the base of the direct injection membrane tank was also observed at times.

As seen in Figure 3 the direct injection module's permeability decreased rapidly, it stabilised below 50 LMH/bar and continued at this level until the process upset. The jet mixed module was found to maintain a stable permeability of greater than 100 LMH/bar for 30 days with no maintenance cleans until a process upset caused by power failure caused a large rise in TMP.

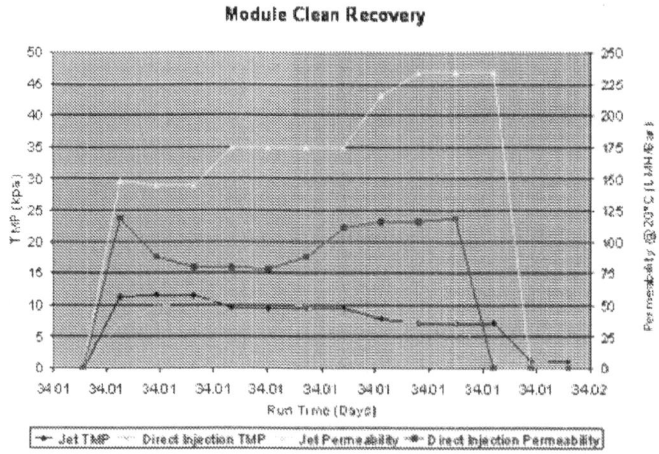

Figure 4 Jet mixed and direct injection module TMP and permeability recovery after clean.

Chemical cleaning

After the process upset the flux of both modules declined and a chemical clean of the membranes was conducted. Each membrane tank was drained and the reactor isolated. Each tank was rinsed with clean water and then filled with a 2750 ppm sodium hypochlorite solution.

During the 2 h soak over half the available chlorine was consumed indicating the fouled state of both membranes. The clean recovered the jet mixed module to close to start up performance: flux (20LMH), TMP (10 kPa) and permeability (225 LMH/bar).

Large amounts of solids were removed after cleaning the more highly fouled direct injection module. While the severe surface fouling was removed the pore plugging fouling and inorganic fouling was not. Thus recovered permeability (116 LMH/Bar) was well below start up permeability (211 LMH/Bar).

Effect of extended relaxation period

To optimise membrane performance an extended fibre relaxation protocol was instigated on day 10 had a positive impact upon permeability. Filtration was set for thirteen (standard being fourteen) 81 min periods followed by twelve 22 min (standard: thirteen 21 min periods) and one 125 minute relaxation period (standard: one 33 minute relaxation period).

As the experiment continued, the jet mixed module displayed less change in permeability and TMP within each filtration cycle and the average permeability increased (Figure 5).

This was a short-term test and is inconclusive as whether this behaviour is part of the long-term fouling trend of the module or that an extended relaxation period is of benefit. Any decreases in TMP with the resulting increase in membrane permeability and extend membrane life must be compared to the cost of lost filtration time. However, in an anaerobic system with longer hydraulic retention times, it may be possible to design longer relaxation periods; however, research is focused on developing high-rate anaerobic processes (Rejeshwari *et al.*, 2000).

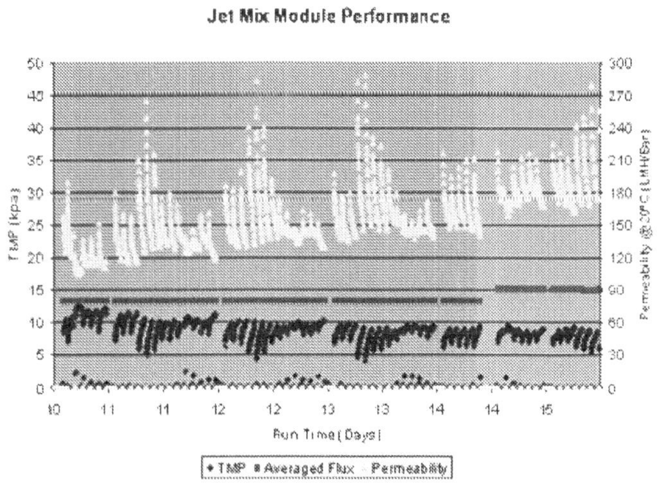

Figure 5 Performance of jet mixed module with extended relaxation period.

COD removal

The COD removal was found to vary greatly from greater than 65% initially to 30%, a decline in process efficiency then lead to removal efficacies of below 10% before the system returned to 30% removal. Previous work had shown that with a feed of 1000 mg/L 40–60% removal could be sustained. The variability of these results reflects the sensitive nature of anaerobic processes to operation outside optimum parameters, caused by the anaerobic bacteria. The lower than expected COD removal was also due to the increased COD loading. Previous work by the author at Memcor Australia, had investigated an organic loading rate (OLR) of 5–6 g COD/L/d, this work has investigated a loading of 8–10 g COD/L/d.

Energy generation

Biogas production was non-constant as it was affected by upsets in process and changes in system pressure. Instead, biogas production can be estimated from the average COD removal rate (3.3 g COD/L) and the average daily filtrate flow rate (255.2 L/day). Multiplying these values, it can be seen that an average of 842.2 g COD was consumed per day. Since the average composition of biogas was 79%, 665.3 g COD were converted to methane per day.

Speece (1996) states that 12 × 106 BTU of energy is produced for every 1000 kg COD converted to methane. This equates to an energy output of 8.4 MJ/day or 2.3 kWh/day for the pilot anaerobic membrane bioreactor operating at 26% COD removal efficiency. This is equivalent to the energy required to produce 1000L of desalinated water.

Bubbling experiments

To try and understand more exactly the behaviour of each module a series of video recorded bubbling experiments were performed on both module types (experimental Wicaksana *et al.*, 2005). They were immersed in a water filled Perspex tank and the gas flowrate varied at 5L/h and 20 L/h in 5 l/h increments. To more fully approximate the anaerobic environment a glycerol/water mixture with a viscosity of 10 cp (equivalent to 20 mg/L mixed liquor solids) was used for a second set of experiments.

Analysis of the results was very complex previous work on single fibres (Wicaksana *et al.*, 2005) and nine fibre modules (Yeo and Fane, 2005) with single a bubble size had gathered data such as rate of oscillation and degree of displacement.

In a 0.5 m^2 module containing over 300 fibres there are complex interactions, resulting it being to difficult to determine the movement of individual fibres. The fibres do not oscillate in a controlled manner, interaction with other fibres changes the direction of movement and stops full oscillation occurring. Bubbles of different sizes are formed and break up as they interact with the modules' fibres.

The efficiency of bubbling to reduce fouling will depend on the number of fibres that are displaced per bubble, the pattern and random nature of the fibres movement at particular flow rate, the viscosity and fibre slackness. A properly gas scoured module will have well separated fibres which pulse independently of each other. While a poorly scoured module will have only some outer fibres moving with the inner fibres are very close together and oscillate in the same direction with the same displacement.

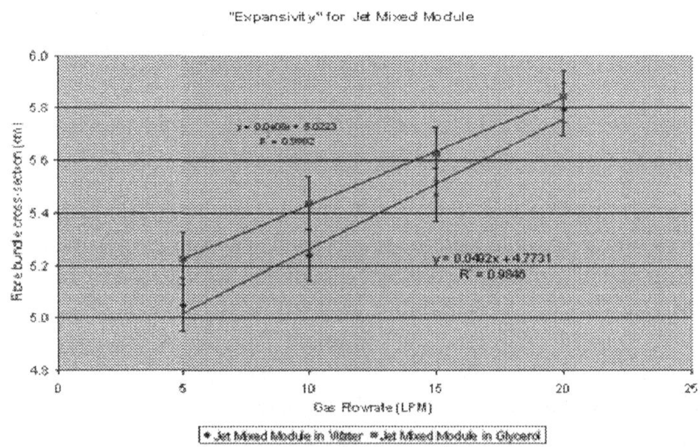

Figure 6 Relationship between module cross section and aeration rate for the jet mixed module.

To encompass the broad range of fibre behaviour observed and reconcile it with efficient gas scouring, the concept of 'expansivity' is introduced. This proposes that in a well-aerated module all fibres will move independently of each other at the maximum displacement allowed by the surrounding fibres with the rate of oscillation being dependent upon the rate of bubbling. It is proposed that a measure of how well fibre movement is induced, at a particular flow rate, viscosity or degree of fibre slack, can be derived from the cross section of the fibre bundle, at the mid-point of the module (Figure 6).

Using the expansivity principle a strong linear relationship was found between the fibre cross-section and gas flow rate for the jet mixed module in both the low viscosity (water) and the high viscosity (glycerol/water) solutions (Figure 6). No relationship was found between gas flow rate and the cross-section of the direct injection module, suggesting that the observed low permeability is related to poor fibre movement generated.

Conclusion

These projects showed it is possible to develop an operating An-MBR with stable permeability, through the instigation of a fibre relaxation protocol. The use of jet mixing technology was found to significantly enhance membrane performance with the mixed module operating at a flux (16 LMH) above those usually obtained for industrial wastewater (10 LMH).

The higher flux achieved by the jet mixed module compared to the direct injection module was found to relate to the effectiveness of the coarse jet bubbles generated by jet mixing which achieved larger and more random fibre displacement per bubble than the small bubble direct injection module.

Module performance was stable at biomass concentration between 25–30 g/L. Even with the occurrence of sever process interruptions causing increases in MLSS the jet mixed modules flux was higher than predicted by literature for aerobic systems. Leading to the conclusion that fouling in anaerobic systems is influenced by additional factors. It can be seen that simple fibre models do not adequately describe factors such as bubble efficiency or fibre interaction, which are dependent upon the gas injection and dispersion system.

The usefulness of fibre 'expansivity' in a module as a measure of effective fouling control has yet to be fully established and needs further work to test and develop the concept more rigorously. However, it was found useful by the authors to understand the performance of jet mixing and direct injection technology.

At a HRT of 65 h partial degradation (26%) was achieved, thus the energy generation of the system being 2.3 kWh per day. Further is being conducted with the OLR reduced to less than 2.5 g/L/d with the target of a COD removal rate greater than 90%.

Acknowledgements

This work was only possible with the support of Memcor Australia, the UNESCO centre for membrane science and technology as well as Hawkesbury City Council.

References

Bellara S.R.; Cui Z.F, Pepper D.S. (1996) Gas sparging to enhance permeate flux in ultra-filtration using hollow fibre membranes. *Journal of Membrane Science*, **121**(2), 175–184.

Cui, Z.F., Chang, S. and Fane A.G. (2003) The use of gas bubbling to enhance membrane processes. *Journal of Membrane Science*, **221**(1), 1–35.

Fuchs, W., Binder, H.., Mavrias, G. and Braun, R. (2003) Anaerobic treatment of wastewater with high organic content using a stirred tank reactor coupled with a membrane filtration unit. *Water Research*, **37**, 902–908.

Hernandez, A.E., Belalcazar, L.C., Rodriguez, M.S. and Giraldo E. (2002) Retention of granular sludge at hydraulic loading rates in an anaerobic membrane bioreactor with immersed filtration. *Water Science and Technology*, **45**(10), 169–174.

Rajeshwan, K.V., Balacrishnan, M., Kansal, A., Lata, K. and Kishore, V.V.N. (2000) State of the art of anaerobic digestion technology for industrial wastewater treatment. *Renewable and Sustainable Energy Reviews*, **4**, 135–156.

Speece R.E. (1996) *Anaerobic Biotechnology for Industrial Wastewater Treatment*. Nashville, TN: Archae Press.

Takagki, S., Yoshika, S. and Katsuhiko, Y. (1997) A method for intermittently aerating sewage. Japanese Patent 09192688.

Wicaksana, F. Fane, A.G. and Chen, V. (2005) The relationship between critical flux and fibre movement induced by bubbling in a submerged hollow fibre system. *Water Science and Technology*, **51**(6–7), 115–122.

Yeo, A. and Fane, A. G. (2005) Performance of individual fibres in a submerged hollow fibre bundle *Water Science and Technology*, **51**(6–7), 165–172.

Rapid evaluation of carbon sources for biological nutrient removal

P. Kampas*, S. A. Parsons*, P. Pearce, S. Ledoux**, P. Vale***, J. Churchley*** and E. Cartmell***

* School of Water Sciences, Cranfield University, Cranfield, Bedfordshire MK43 0AL, UK (E-mail: *p.kampas.2003@cranfield.ac.uk*)
** Research & Technology, Thames Water, Spencer House, Manor Farm Road, Reading, Berkshire RG2 0JN, UK (E-mail: *pete.pearce@thameswater.co.uk*)
*** Technology & Development, Severn Trent Water Ltd, Avon House, St Martins Road, Coventry CV3 6PR, UK (E-mail: *peter.vale@severntrent.co.uk*)

Abstract Biological nutrient removal (BNR) is a complex process that is strongly affected by wastewater characteristics with the lack of available carbon to be a major issue for a successful BNR. The need of a reliable predictive parameter is essential for testing not only the wastewater characteristics but also different carbon sources that can possibly be applied. This study examines the application of disintegrated sludge on phosphorus release and denitrification process. Batch laboratory scale tests were conducted to measure the phosphorus release and the reduction of nitrates under anaerobic conditions. The phosphorus (P) release test (P test) was developed to become more suitable for examining the application of the external carbon source by minimizing the storage time between sampling and experiment. Similarly the denitrification test (N test) was modified to asses the impact of the addition of extra carbon by adding the solid free fraction of disintegrated sludge which contains the organic matter that can be utilized. According to these tests the comparison of the results with and without addition of carbon showed that there was higher phosphorus release (460 %) and higher denitrification rates (244 %) when the carbon was added.

Keywords Biological nutrient removal, carbon source, denitrification rate, disintegrated sludge, phosphorus release, predictive test, wastewater

Introduction

Biological nutrient removal (BNR) is a well documented wastewater treatment process that has been implemented in many countries (Pavan *et al.*, 1998 and Jonsson *et al.*, 1996). BNR processes have increasingly gained more importance and become more widely applied over recent years. The key for successful BNR is the availability of carbon in wastewater that has to be assimilated during the phosphorus release and denitrification processes. The most suitable carbon source for BNR has been found to be volatile fatty acids (VFA) and readily biodegradable chemical oxygen demand (RBCOD) (Abu-ghararah and Randall, 1991; Randall *et al.*, 1997).

The need of a predictive parameter for examining the suitability of wastewaters for BNR became essential the recent years, when more standard activated sludge plants have been converted to BNR plants. The most popular predictive parameters that have been used were the carbon to phosphorus and carbon to nitrogen ratios. The carbon can be represented either as biological oxygen demand (BOD), COD or VFA. It was reported that for a successful BNR process a ratio of COD: P above 50 (Carlsson *et al.*, 1996) or BOD: P. above 20 (Randall *et al.*, 1992; Cooper *et al.*, 1995) was necessary. Another ratio used for examining the wastewater carbon availability was the VFA potential to phosphorus, which has to be above 14-20 in order to be suitable for BNR (Abu-gararah and Randall, 1991; Johnsson et al., 1996; Curto 2001). Similarly, for nitrogen the higher the carbon to nitrogen ratio (COD: N) in wastewater the more likely the wastewater is able to give good nitrogen removal (Henze, 1991; Kujawa and Klapwick, 1999; Henze *et al.*, 2002).

Unfortunately, these parameters proved to be unreliable for indicating BNR potential as in many cases although, the initial indications showed successful BNR, the performance was poor or the opposite (German *et al.*, 1998; Llabres *et al.*, 1999; Avendano 2003). Therefore, further tests were developed, which instead of recycling on the stoichiometric ratios to predict phosphorus and nitrogen removal, batch laboratory scale tests were used, such as the phosphorus release test (Park *et al.*, 2001; Vale *et al.*, 2005) for phosphorus and the nitrate utilization rate (NUR) test for nitrogen (Kujawa and Klapwijk 1999). The basic idea of the phosphorus test is to monitor the phosphorus release under anaerobic conditions, with and without acetate addition. High phosphorus release (~30 mg l^{-1}) without any acetate addition and similar to the release with acetate, can indicate high uptake under aerobic conditions and hence efficient phosphorus removal. The reduction of nitrates was observed in NUR tests with the fast and complete denitrification to be a good indicator for a satisfactory nitrogen removal.

All the above predictive parameters and tests were used to examine the potential of wastewater for successful BNR and how the results correlate with the wastewater characteristics. On the other hand, the aim of this study was to investigate the quality and suitability of an external carbon source for improving BNR performances. The modifications and the overall development of these tests in order to make them appropriate for assessing the impact of an external carbon source to BNR process are presented in this paper.

Materials and methods

Grab samples of wastewater were collected and used in this study from four different wastewater treatment plants (WTPs) (Table 1). Sites 3 and 4 are successful BNR sites with the domestic wastewater mixed with industrial effluent rich in organic matter (~500 mg l^{-1} VFA), while sites 1 and 2 are standard activated plants that have to be converted to BNR the next few years. The wastewater from sites 3 and 4 was domestic as was collected before the inputs of industrial sources. Moreover, for the completion of phosphorus release tests bio-P sludge (return activated sludge, RAS) was sampled from a successful BNR site (site 4). Initially, the wastewater was stored at 4 °C for one day and RAS stored at 4 °C and aerated. The experiment was then modified and a minimum storage time (2–3 hours) for RAS was applied under aerobic conditions. The external carbon source tested in this research was mechanically disintegrated sludge (surplus activated sludge, SAS) with carbon one and carbon two appeared in the following graphs to correspond to the sludge, disintegrated by the same equipment but on different retention times and hence different carbon release (Kampas *et al.*, in preparation). This carbon source was also compared with industrial wastewater (brewery effluent) when the phosphorus release test was conducted for site 3.

Table 1 Collection of samples and number of test completions

	Site 1	Site 2	Site 3	Site 4
BNR mode			✓	✓
Non BNR mode	✓	✓		
Settled sewage	✓	✓	✓	✓
RAS				✓
SAS	✓	✓	✓	✓
Phosphorus release test	1	2	2	1
Denitrification test	-	2	1	4

Phosphorus release test

The phosphorus release test consisted of four plastic vessels with 1L of wastewater and 1L of return activated sludge (RAS) that were kept under anaerobic conditions for 2 hours. During the test there was no inhibition by the presence of nitrates as their concentration was always very low (<1 mg l^{-1}) (Kuba *et al.*, 1994). The first vessel was the control one without any amendment. The second and third were operated with the addition of external carbon source (disintegrated sludge) and the fourth with the addition of sodium acetate. In every reactor the same amount of carbon was added, as all the carbon sources were matched in terms of VFA or SCOD (25 mg COD l^{-1}) and with a limit in volume for disintegrated sludge (100 ml). When 100 ml of the added sludge contained less than 25 mg COD l^{-1}, the match of carbon for the three reactors was on the minimum concentration of COD. All vessels were sealed and a nitrogen blanket was over the solution, which was mixed by magnetic stirrers. Moreover, all the vessels were inside a water bath with a heater and circulation pump to keep the temperature constant (20 °C). During the experiment pH and dissolved oxygen (DO) were observed taking measurements every 30 minutes. The experimental rig is demonstrated in Figure 1. Samples were taken every 30 minutes, filtered immediately with a syringe filter (0.45 μm) and the concentration of phosphates (PO_4) was measured.

Denitrification test

The above experimental rig was also used for the denitrification test with the difference here that 2L of wastewater was used in the reaction vessels. Again, the first vessel was unamended with carbon, the second and third with the addition of external carbon and the fourth one with sodium acetate. The carbon was matched again in terms of VFA or SCOD (25 mg l^{-1}) and with the limit of 100 ml as well. Moreover, the same amount of nitrate (30 mg l^{-1} N-NO_3) in the form of KNO_3 was added in all reactors. Samples were taken every 30 minutes, filtered immediately with syringe filter (0.45 μm) and the concentrations of nitrate (NO_3) and nitrite (NO_2) were measured. The volatile suspended solids (VSS) were also measured from the solution of each vessel.

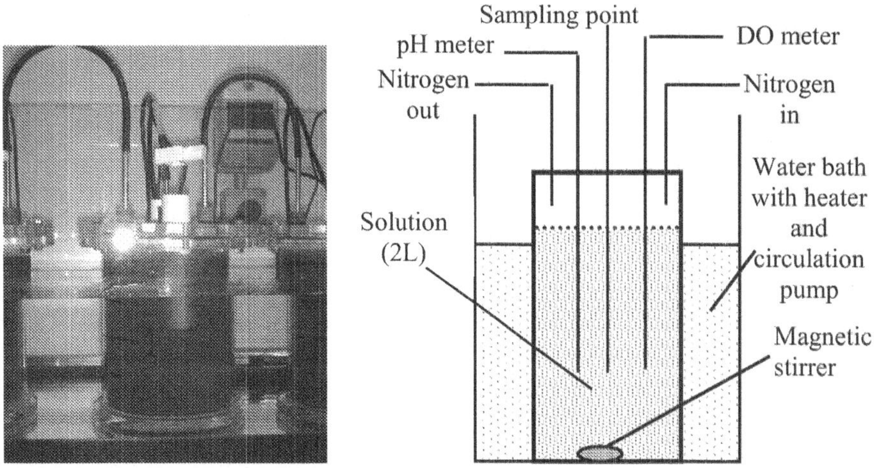

Figure 1 A picture and a schematic of the experimental rig.

Analytical methods

The concentration of soluble COD, VFA, total solids (TS), VSS, and PO_4 was measured according to APHA Standard Methods (1998). Analysis of nitrates (NO_3) and nitrites (NO_2) from the denitrification tests was carried out using ion exchange chromatography (IEC) system, (Dionex, DX500 series, UK).

Results and discussion

Phosphorus release test

In phosphorus release test (P test) the release of phosphorus when the biomass (RAS) was using the carbon that was available only in wastewater, was compared to the release, when external carbon sources were available. Moreover, the extra carbon was compared to acetate, which as mentioned above is considered the most suitable for P release. Our expectations were to receive a higher release in the reactor with the added carbon, equivalent to the release in acetate reactor. Unfortunately, not only we did not observe that, but also instead of seeing a release under anaerobic conditions, phosphorus was taken up in one of the reactors despite the fact that soluble carbon was available (Figure 2, site 2). Failure in the phosphorus release process has been observed in laboratory tests and full-scale operation. In our case the conditions were not anoxic, where it is possible phosphorus uptake to take place (Kern-Jespersen and Henze 1993), but anaerobic. Other possible reasons are the presence of glycogen accumulating organisms (GAOs), which compete for the carbon with polyphosphate accumulating organisms (PAOs), responsible for phosphorus release (Cech and Hartman 1993) and the shortage of potassium and other elements (Brdjanovic *et al.*, 1998). Taking into account that a normal domestic sewage contains sufficient amount of the required inorganic elements and the WTP that RAS was taken from was operated under normal and usual conditions (without any GAO interference), was assumed that the awkward results from phosphorus release process were due to the storage of RAS for 24–48 hours under aeration before the experiment. Moreover, the higher phosphorus release observed in the control reactor than the reactor with the added carbon (Figure 2, site 1) can be characterized as secondary phosphorus release, which takes place after prolonged aeration and lack of available carbon (Metcalf and Eddy 2003). For all these reasons it was decided that the time between sampling and experiment had to be minimized.

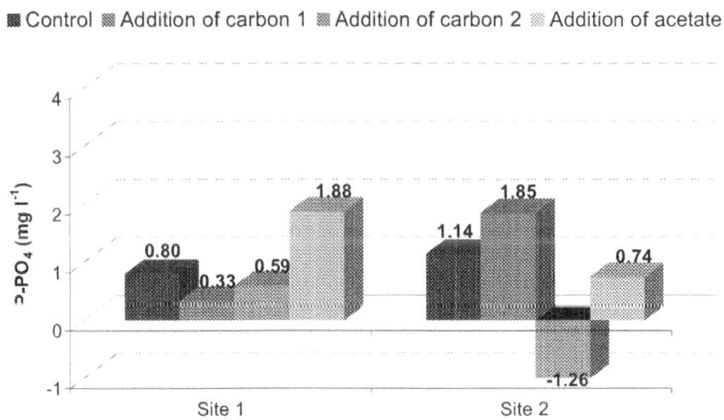

Figure 2 Phosphorus release when the test was conducted with RAS being stored 24-48 h after sampling from site 1 and site 2.

The next step was to compare P releases from two tests with the same wastewater and RAS and the storage time of RAS varying only, expecting to see different amount of phosphorus to be released. Indeed, the amount of phosphorus released was much higher when "fresh" RAS was used (Figure 3). However, the results were following the same trend and as expected the control reactor gave the lowest release. The examined external carbon source was able to be taken up by the PAOs releasing more phosphates than the control, but still lower than the reactor with acetate addition.

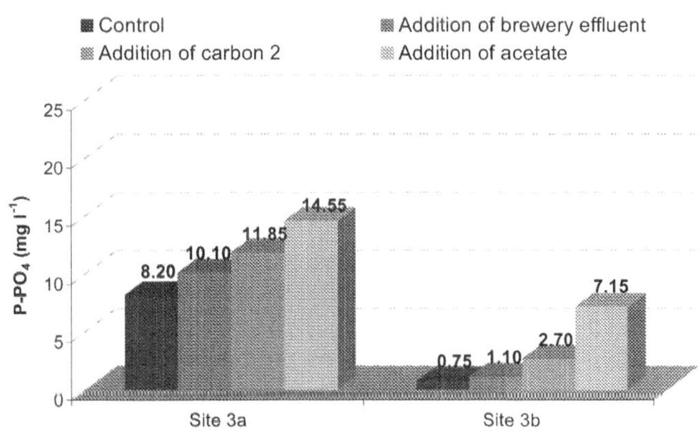

Figure 3 Results from phosphorus release test from site 3. a: test immediately after sampling, b: RAS stored for 24-48 h after sampling.

The amended P test (minimum RAS storage) was conducted again for sites 2 and 4, testing the same carbon sources and comparing them with acetate by adding the same amount in terms of VFA or SCOD investigating which is the most preferable to the microorganisms (Figure 4). The test for site 2 took place with the external carbon to be matched in terms of VFA (25 mg COD l^{-1}) and for site 4 in terms of SCOD (25mg COD l^{-1}). In first case the SCOD concentration of the acetate was 25 mg COD l^{-1} and for the disintegrated sludge was 214 and 442 mg l^{-1}. A comparison between the results presented in Figure 4 shows primarily the different wastewater quality in these two WTPs and also the preference of PAOs bacteria to the external carbon (disintegrated sludge) rather than the acetate in the test with wastewater from site 2, despite that the carbon was matched in terms of VFA. This is a strong indication that VFA is not the only carbon that can be utilized by the bacteria, but also a proportion of SCOD produced from sludge by the disintegration.

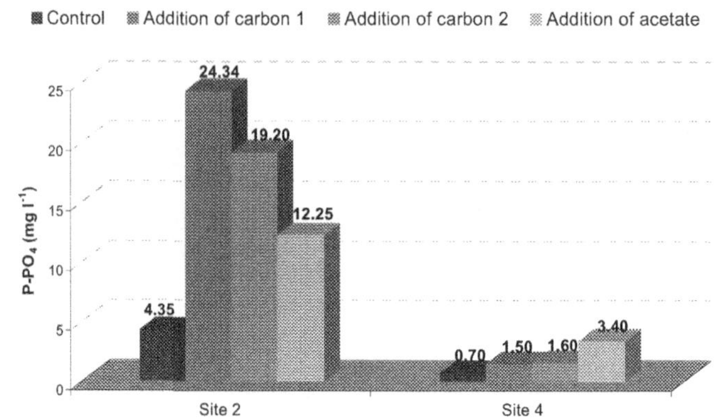

Figure 4 Phosphorus release when test was conducted immediately after sampling.

Denitrification test

The denitrification tests (N test) were carried out to examine whether the external carbon source could be assimilated by the denitrifying bacteria. It was assumed that the N test could take place without any biomass but using only the bacteria that are available in wastewater. Initially the extra carbon, which was always matched in terms of VFA or SCOD (25 mg COD l^{-1}), was added in the form of sludge resulting in rapid denitrification in these reactors much faster than the control and acetate reactor (Figure 5). That happened because of the addition of extra biomass. It was assumed that the disintegrated sludge contained enough active microorganisms, which with the presence of organic carbon gave immediate denitrification.

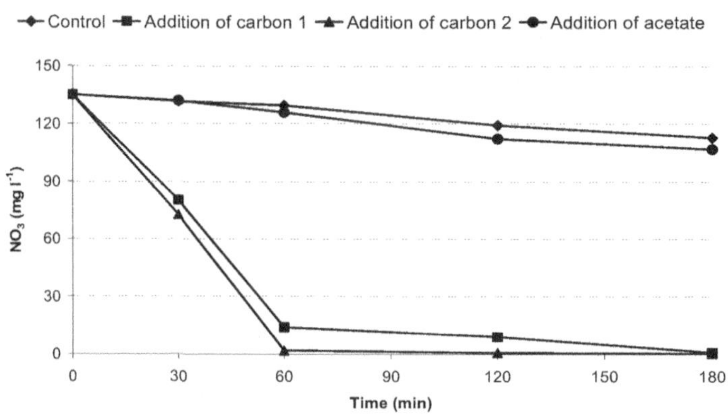

Figure 5 Reduction of nitrates against time when the additional carbon was sludge (site 4).

The N test was modified and the solid free fraction of disintegrated sludge was applied. In this case the VSS concentration was equal in all reactors (~100 mg l^{-1}) and the reduction of nitrate was almost similar (Figure 6). The denitrification rates as mg N-NO_3 reduced in the first hour of experiment per g of VSS available could reveal the importance of the external carbon source (Figure 7). Comparing the denitrification rates produced from the modified N test and with different wastewaters the application of the examined carbon source was always beneficial to denitrification process with those reactors to give higher rates than the control and most of the times higher than the acetate reactor. These rates were higher than the rates reported in literature using a range of carbon sources, such as primary sludge hydrolysate (4.1 mg NO_3-N g^{-1} VSS h^{-1}) (Isaacs and Henze 1995), methanol (3 mg NO_3-N g^{-1} VSS h^{-1}) (Nyberg *et al.*, 1996), ethanol (10 mg NO_3-N g^{-1} VSS h^{-1}) (Nyberg *et al.*, 1996), acetic acid (7.3 mg NO_3-N g^{-1} MLSS h^{-1}) (Naidoo *et al.*, 1998), fermented solid waste (2.5 mg NO_3-N g^{-1} VSS h^{-1}) (Pavan *et al.*, 1998), fermented swine wastes (7.3 mg NO_3-N g^{-1} VSS h^{-1}) (Lee *et al.*, 1997) and septic sludge (2.4 mg NO_3-N g^{-1} VSS h^{-1}) (Morling 2001). That may be due to the different solid concentration on the substrate used, where the extra carbon source was added, as in all the above cases biomass was used instead of wastewater and also due to the different initial concentration of NO_3. Similar rates were reported when wine distillery effluent (35.2 mg NO_3-N g^{-1} VSS h^{-1}) (Bernet *et al.*, 1996) and mixture of VFAs (31.9 mg NO_3-N g^{-1} SS h^{-1}) (Fass *et al.*, 1994) were the carbon source. At this point has to be mentioned that we do not expect to see these rates in full scale denitrification process as the conditions will be completely different. There will not be the same initial high concentration of nitrates and the substrate will be mixed liquors suspended solids, (MLSS) and not wastewater only. However, the difference in denitrification rates when the examined carbon sources were added, the ability to compare them to the rates produced from the addition of acetate (an excellent carbon source) and the overall applicability of the test, made us conclude that the N test is an efficient tool for evaluating external carbon sources for denitrification.

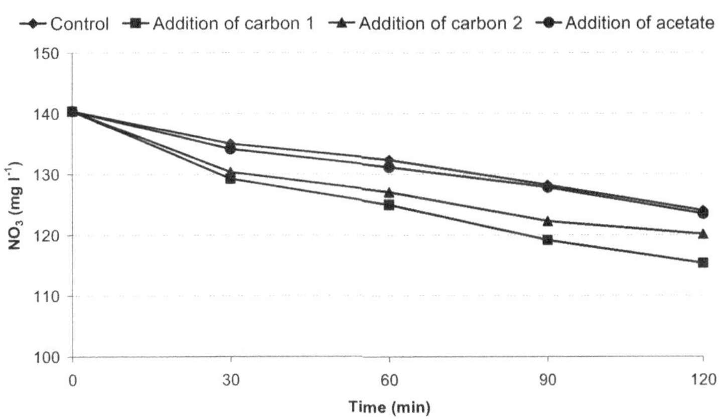

Figure 6 Reduction of nitrates against time when additional carbon was solid free fraction of sludge and the substrate was wastewater (site 2).

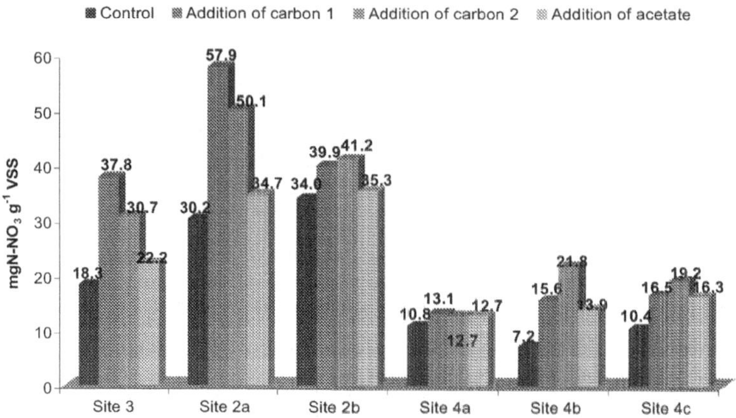

Figure 7 Denitrification rates from the N test conducted for three different WTP. (a, b and c represent different sampling times).

Conclusions and recommendations

- The phosphorus release test described here can be used for comparing carbon source suitability of enhanced biological phosphorus removal (EBPR), but the time between the sampling and test should be kept as short as possible. The higher phosphorus release in the reactor with the added carbon than the control one, is a strong indication that the carbon source tested is likely to be able to improve EBPR.
- The denitrification test described here can be used for carbon source comparison when the substrate used contains a similar amount of active bacteria measured as VSS. The denitrification rates produced indicate whether the examined carbon can improve denitrification.
- Overall the two tests were modified respective to the needs of this study and were able to deliver satisfactory results. However, there are some issues that have to be further examined, which for the P test could be the amount of carbon taken up simultaneously with the phosphorus release in order to be sure that there is no secondary release. Moreover, the test has to be conducted with the addition of solid free fraction of disintegrated sludge in order to ensure that the phosphorus being released is coming from the substrate (RAS) and not from the added sludge. For the N test future research will investigate the performance in the presence of those bacteria responsible for denitrification. In addition, a combination of the two tests by changing the anaerobic conditions after two hours to anoxic by adding a certain amount of nitrates would indicate the performance of phosphorus release, phosphorus uptake under anoxic conditions and denitrification process.

Acknowledgements

We acknowledge Severn Trent Water Ltd and Thames Water for their financial support.

References

Abu-ghararah, Z. H., Randall, C. W. (1991). The effect of organic compounds on biological phosphorus removal. *Water Science and Technology* **23** (4–6), 585–594.

APHA (1998). *Standard methods for the examination of water and wastewater.* 20th edition. American Public Health Association, American Water Works Association and Water Environment Federation, Washington, USA.

Avendano, E. S. (2003). Wastewater characterization and testing for enhanced biological phosphorus removal. MSc thesis, Cranfield University, Beds, UK.

Bernet, N., Habouzit, F. & Moletta, R. (1996). Use of an industrial effluent as carbon source for denitrification of a high-strength wastewater. *Applied Microbiology and Biotechnology* **46**, 92–97.

Brdjanovic, D., Slamet, A., van Loosdrect, M. C. M., Hooijmans, C. M., Alaerts, G. J. & Heijnen, J. J. (1998). Impact of excessive aeration on biological phosphorus removal from wastewater. *Water Research* **32** (1), 200–208.

Carlsson, H., Aspegren, H. & Hilmer, A. (1996). Interactions between wastewater quality and phosphorus release in the anaerobic reactor of the EBPR process. *Water Research* **30** (6), 1517–1527.

Cech, J. S. & Hartman, P. (1993). Competitions between polyphosphate and polysaccharide accumulating bacteria in enhanced biological phosphate removal systems. *Water Research* **27** (7) 1219–1225.

Cooper, P., Upton, J. E., Smith, M. & Churchley, J. (1995). Biological nutrient removal: design snags, operational problems and costs. *Journal of Institution of Water and Environmental Management* **9** (1), 7–18.

Curto, P. R. (2001). Determining the potential for enhanced biological phosphorus removal based on wastewater characteristics. MSc thesis, Cranfield University, Beds, UK.

Fass, S., Ganaye, V., Urbain, V., Manem, J. & Block, J. C. (1994). Volatile fatty acids as organic carbon sources in denitrification. *Environmental Technology* **15,** 459–467.

German, C.-R., Oscar, G.-B. & Simon, G.-M. (1998). Wastewater fermentation and nutrient removal in sequencing batch reactors. *Water Science and Technology* **38** (1), 255–264.

Henze, M. (1991).Capabilities of biological nitrogen removal process from wastewater. *Water Science and Technology* **23**, 669–679.

Henze, M., Harremoes, P., Jansen, J. C. & Arvin, E. (2002). *Wastewater treatment; Biological and Chemical Processes*, 3rd edition. Springer.

Isaacs, S. H. & Henze, M. (1995). Controlled carbon source addition to an alternating nitrification–denitrification wastewater treatment process including biological P removal. *Water Resources* **29** (1), 77–89.

Jonsson, K., Johansson, P., Christensson, M., Lee, N., Lie, E. & Welander, T. (1996). Operational factors affecting enhanced biological phosphorus removal at the waste water treatment plant in Helsingborg, Sweden. *Water Science and Technology* **34** (1–2), 67–74.

Kampas, P., Parsons, S. A., Pearce, P., Ledoux, S., Vale, P., Churchley, J. & Cartmell, E. (2006). Mechanical sludge disintegration for the production of carbon source for biological nutrient removal. *Water Research.* (In preparation).

Kern-Jespersen, J. P. & Henze, M. (1993). Biological phosphorus uptake under anoxic and anaerobic conditions. *Water Research* **27**, 617–624.

Kuba , T., Wachtmeister, A., van Loosdrecht, M. C. M. & Heinjen, J. J. (1994). Effect of nitrate oon phosphorus release in biological phosphorus removal systems. *Water Science and Technology* **30** (6), 263–269.

Kujawa, K. & Klapwijk, B. (1999). A method to estimate denitrification potential for predenitrification systems using NUR batch test. *Water Research* **33** (10), 2291–2300.

Lee, S. I., Park, J.-O., Ko, K.-B. & Koopman, B. (1997). Effect of fermented swine wastes on biological nutrient removal in sequencing batch reactors. *Water Research* **31** (7), 1807–1812.

Llabres, P., Pavan, P., Battistoni, P., Cecchi, F. and Mata-Alvarez, J. (1999). The use of organic fraction of municipal solid waste hydrolysis products for biological nutrient removal in wastewater treatment plants, *Water Research* **33** (1), 214–222.

Metcalf & Eddy (2003). *Wastewater Engineering – Treatment and Reuse*, 4th edition, revised by Tchobanoglous, G., Burton, F. L. & Stensel, H. D. McGraw Hill.

Morling, S. (2001).Performance of an SBR-plant for advanced nutrient removal, using septic sludge as a carbon source. *Water Science and Technology* **43** (3), 131–138.

Naidoo, V., Urbain, V. & Buckley, C. A. (1998). Characterization of wastewater and activated sludge from European municipal wastewater treatment plants using the NUR test. *Water Science and Technology* **38** (1), 303–310.

Nyberg, U., Andersson, B. & Aspegren, H. (1996). Long-term experiences with external carbon sources for nitrogen removal. *Water Science and Technology* **33** (12), 109–116.

Pavan, P., Battistoni, P., Traverso, P., Musacco, A. & Cecchi, F. (1998). Effect of addition of Anaerobic fermented OFMW (organic fraction of municipal solid waste) on biological nutrient removal (BNR) process: preliminary results. *Water Science and Technology* **38** (1), 327-334.

Randall, W. C., Barnard, L. J. & Stensel, H. D. (1992). Design and retrofit of wastewater treatment plants for biological nutrient removal. *Water Quality Management Library*, vol. 5.

Randall, A. A., Benefield, L. D. & Hill, W. E. (1997). Induction of phosphorus removal in an enhanced biological phosphorus removal bacterial population. *Water Research* **31** (11), 2869–2877.

Park, J. K., Whang, L. M., Wang, J. C. & Novotny, G. (2001). A biological phosphorus removal potential test for wastewaters. *Water environment Research* 73 (3), 374–382.

Pavan, P., Battistoni, P., Traverso, P., Musacco, A. & Cecchi, F. (1998). Effect of addition of Anaerobic fermented OFMW (organic fraction of municipal solid waste) on biological nutrient removal (BNR) process: preliminary results. *Water Science and Technology* **38** (1), 327–334.

Vale, P., Avendano, S., and Parsons, S. A. (2005). Wastewater characterisation and testing for enhanced biological phosphorus removal. In *Proceedings: IWA Specialized Conference Nutrient Management in Wastewater Treatment Processes and Recycles Streams*. Krakow, 19–21 September 2005, pp 783–790.

Symbiotic algal bacterial wastewater treatment: characteristics of biomass and their relation to the process performance

M. Medina and U. Neis

University of Technology Hamburg Harburg, Institute of Wastewater Management and Water Protection. Eissendorfer Strasse 42, D–21073 Hamburg, Germany
(E-mail: *marcela.medina@tuhh.de*)

Abstract Algae incorporation to biomass is important in an innovative wastewater treatment that exploits the symbiosis between bacterial activated biomass and microalgae (*Chlorella vulgaris* sp. Hamburg), since it allows a good and easy algae separation by means of clarification. The symbiotic biomass was characterized during the first 6 weeks after its incubation, being this study, up to our knowledge, the first in its class. Measured characteristics were correlated with the quality criteria of the proposed treatment, namely settleability and algae concentration in the supernatant and biomass. The properties of symbiotic biomass are in the same ranges as those reported for conventional biomass. Calcium concentration, hydrophobicity and filament index seem to affect the performance of the symbiotic wastewater treatment.

Keywords Bioflocculation, *Chlorella vulgaris*, floc characterization, settleability, symbiotic algae bacteria biomass

Introduction

An innovative wastewater treatment which integrates a normal activated sludge with a microalgae population (*Chlorella vulgaris* sp. Hamburg) was implemented in a previous project with the aim to improve the performance of conventional waste stabilization ponds through the incorporation of algae on the biomass (Gutzeit *et al.*, 2005). The foreseen complementarity of photosynthetic algal O_2 and bacterial CO_2 took place, rendering an effective removal of organic matter, nitrogen and phosphorus from the influent with no need of aeration. The project also demonstrated that microalgae removal is possible via sedimentation, since they were incorporated into the activated sludge, yielding a clean and algae-free effluent. In this way, the floc formation step turns to be a key step for the success of the proposed innovation.

Although a crucial step in the activated sludge process, bioflocculation still remains a not full understood aspect. The floc structure has been characterized as a net of exocellular polymeric substances (EPS) where bacteria, bacteria colonies, inorganic particles are enmeshed (Nielsen, 2002). The relevance of bivalent cations as linking agents of negative charged groups of EPS in the floc structure has been also acknowledged (Urbain *et al.*, 1993; Sanin and Vesilind, 2000; Sobeck and Higgins, 2002). Filamentous microorganisms in moderated proportions contribute as well to the structure, providing "backbone" structure

(Jenkins *et al.*, 2004) and filament index, among other properties, was considered as an important parameter for the floc stability under shear stress (Wilén *et al.*, 2003)

Wastewater treatment process parameters like inflow composition (Bura *et al.*, 1998; Sponza, 2003), sludge retention time (SRT) (Liao *et al.*, 2002) and aerobic activity (Wilén *et al.*, 2004) can affect floc characteristics. Given the complexity and dynamics of the biomass system, this research field now tends to identify species in situ and correlate them to usual studied characteristics (Nielsen *et al.*, 2004).

In the case of symbiotic flocs, algae surface properties should be also considered for assessing their incorporation to the sludge. It is already known that green algae are able to form biofilms and flocs, in account to their own EPS production (Watanabe *et al.*, 2005) and their proneness to attach to rough/hydrophobic surfaces or existing biofilms (Sekar *et al.*, 2004). The symbiotic sludge project considered *Chlorella vulgaris* sp. Hamburg as the most appropriated alga species due to its abilities to grow either autotrophically or heterotrophically under wastewater conditions and to form flocs (Gutzeit *et al*, 2005). It was found that symbiotic floc stability was reached within the first 4 weeks since the incubation of conventional biomass with algae (Gutzeit, 2005). The success of the proposed treatment can be easy assessed by the settleability of the biomass, determined by the sludge volume index (SVI), as well as by the algae concentration in the system phases (biomass and supernatant), measured as chlorophyll a concentrations.

The present study assessed the physicochemical characteristics of this symbiotic biomass and their effects on the process performance of the treatment. Up to our knowledge, this is the first characterization study of symbiotic algae – bacteria biomass for wastewater treatment.

Materials and methods

1l of activated sludge from the aeration basin of the Seevetal wastewater treatment plant (100,000 pp) was mixed with 4l of *Chlorella vulgaris* sp. Hamburg culture in steady growth phase (Institute of Botany, University of Hamburg), in $0.03m^3$ stirred, 12 hours/day illuminated (2,000 μmol s^{-1} m^{-2}), thermoregulated reactors. They were operated in sequence batch reactor (SBR) mode and fed with synthetic wastewater, as referred by Gutzeit *et al.* (2005). Reactors with symbiotic sludge were operated under two different food-to-microorganism ratios (F/M = 0.15 and 0.56 mg COD l^{-1} (mg VSS $l^{-1})^{-1}$ d^{-1}) at constant hydraulic retention time (HRT = 4 d). Samples were taken twice a week during the first 4–6 weeks after incubation, to explore the stabilization period. Figure 1 illustrates the laboratory set up.

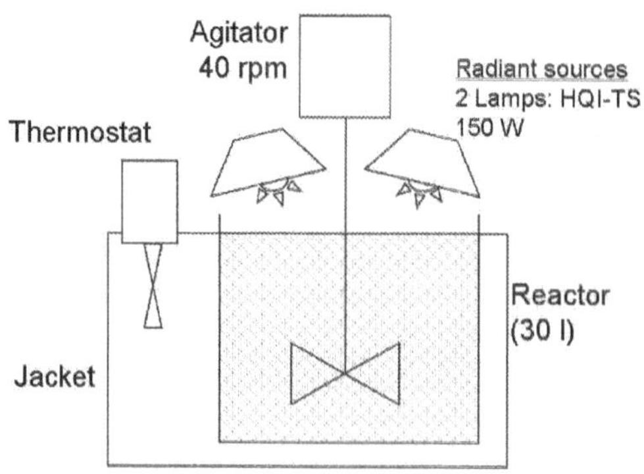

Figure 1 Schema of the laboratory set up.

Suspended solids, SVI (sludge volume index), EPS content and composition (proteins and carbohydrates), hydrophobicity, electrophoretic mobility, calcium content, filament index, and chlorophyll α content in the supernatant and in the sludge, as indicator of algae presence, were measured in order to characterize the symbiotic system and its biomass. SVI was selected as criterion to assess the settleablity of symbiotic sludge and chlorophyll a concentration in the supernatant to evaluate effectiveness of algae incorporation to biomass. Chlorophyll a concentration in the sludge was chosen to estimate the algae activity in the flocs. The three criteria can be considered as indicators of the treatment's success.

Chlorophyll a was measured according to the German standard 38412, part 16 (DIN, 1985), based on ethanol (90%) extraction.

Suspended solids (SS), volatile suspended solids (VSS) and Sludge volume index (SVI) were measured according to the Standard Methods. (APHA, 1985) Chemical oxygen demand (COD) was photometrically measured after 120 minutes oxidation time at 148 C with cuvette tests (Lange, Düsseldorf, Germany).

EPS were extracted with the formaldehyde / NaOH method (Liu and Fang, 2002). A 500 ml sample was washed twice with buffer solution (pH = 7.0) and centrifuged at 2,000*g*, 4C. 6 μl/ml formaldehyde (36,7%) were added, mixed and kept at 4C; 1 h later 100 μl/ml NaOH 1M were added and kept 1 h at 4C. The treated sludge was centrifuged at 20,000 G, 4C, and the supernatant was dialyzed against demineralized water at room temperature (Spectrapor membrane tube, molecular weight cut-off = 3,500, Spectrum Medical Industries, Los Angeles, USA). EPS content is determined putting a determined volume in a drying dish previously weighted and drying until constant weight at 105 °C.

Carbohydrate content of EPS was estimated as glucose by the method proposed by (Dubois *et al.*, 1956), and protein was measured as bovine serum by the method suggested by Lowry, as described by (Frolund *et al.*, 1996).

Hydrophobicity was estimated according to a modified MATH (Microbial adhesion to hydrocarbons) method (Guellil *et al.*, 1998). The sludge was sonicated at 50W for 30 sec in an ice bath and centrifuged at 5,700 G, 4C for 15 min. Optical density at 600 nm (spectrophotometer V-550, Jasco, Tokyo, Japan) was adjusted to 0.3 with $NaNO_3$ 1%, 0.4 ml hexadecane ml_{sample}^{-1} was added and mixed for 2 min with a vortex mixer; left quiet for 15 min and optical density of aqueous phase taken with a Pasteur pipette was measured again. Hydrophobicity was calculated after Equation 1:

$$Hydrophobicity = (1 - \frac{Abs_1}{Abs_0}) * 100 \qquad (1)$$

where Abs_0: initial optical density at 600 nm, Abs_1: final optical density at 600 nm.

Electrophoretic mobility was measured with a Zeta-Meter (Zeta-Meter Inc, New York, USA).

Calcium content was estimated with atomic adsorption spectroscopy, washing the sludge twice with demineralized water; afterwards, samples were lyophilized and treated with HNO_3; measurements were done in a PE-Optima DV OES with ICP spectrometer.

Filament index was determined by light microscopy (Olympus BX 41, Olympus Optical Co. Ltd., Tokyo, Japan), following the subjective scoring of filament abundance suggested by Jenkins *et al.* (2004).

Experimental data were evaluated with analysis of variance ANOVA (p = 95%) and Pearson's correlation coefficients (p = 95%).

Results and discussion

General characterization of symbiotic sludge

Table 1 summarizes the measured properties of sludge during all experimental runs and compares them with those reported in the literature for conventional sludge (excepting the chlorophyll a concentration in the sludge)

Table 1 Characteristics of the symbiotic sludge

Characteristic	Range	Referenced value	Source
Sludge volume index (SVI) ml g SS^{-1}	61.4–117.5		
Extracellular polymeric substances (EPS)			
Concentration			
mg EPS g VSS^{-1}	60.4–217.4	164.9–417	
Protein[a] fraction %	16.3–53.3	33.1–58.3	Liu and Fang. 2002
Carbohydrate[b] fraction %	5.9–16.5	24.5–11.5	Frolund *et al., 1996*
Ratio protein : carbohydrate	3.02–4.78	1.35–5.06	
Surface properties			
Zeta potential mV	–29.7 to –17.1	–24.9	Chao and Keinath, 1979
Hydrophobicity %	30–69	18–21	Boyette *et al.*, 2001
Calcium concentration meq Ca l^{-1}	3.01–12.88	4.99–6.96	Murthy and Novak, 1998
Chlorophyll a content in the sludge (as Chl a g oSS^{-1})	4.7–16.3	16.5 –18.1	Gutzeit, 2005

a: As bovine serum; b: as glucose.

It can be appreciated that the SVI, a variable of common use to evaluate and control the operation of activated sludge processes, moved in a range which is still under the acceptable values of performance (< 150 ml gSS^{-1}). Obtained values for EPS concentrations and protein and carbohydrate fractions are in the ranges given in the literature for conventional activated sludge. Zeta potential of symbiotic sludge agrees with the average value obtained by Chao and Keinath (1979). Hydrophobicity values are definitively higher than those reported by Boyette *et al.* (2001). Since the measurement of this property in conventional activated biomass yielded similar results as those from symbiotic one (Medina and Neis, in preparation a), differences with the referred values could be attributable to the experimental procedure. The obtained range of calcium concentration is wider than that referred by Murthy and Novak (1998), but still in the same order of magnitude. Comparison between concentration values of activated and symbiotic biomass did not show any significant difference (Medina and Neis, in preparation a); the variation could be due to the wastewater characteristics and / or wastewater treatment plant operation conditions where the biomass for the experiments was taken. Chlorophyll a content in the sludge was in the same range reported by Gutzeit (2005).

Correlations of biomass properties and process quality criteria

Results of the quality criteria (sludge settleability and algae distribution between aqueous and biomass phases) were correlated with the measured sludge characteristics. Pearson's coefficients obtained for those correlations are listed in Table 2.

Table 2 Pearson's coefficients for the correlations between quality criteria and biomass properties

Property	Sludge volume index (SVI)	Chl a in supernatant	Chl a in sludge
EPS:			
Concentration	0.23 (54)	0.13 (53)	0.02 (54)
% Protein	0.25 (53)	7×10^{-3} (53)	0.26 (53)
% Carbohydrate	0.11 (39)	0.12 (37)	0.23 (42)
Protein : Carb.	0.14 (42)	0.08 (42)	0.04 (43)
Surface properties:			
Hydrophobicity	0.36 (17)	0.46 (33)*	0.17 (35)
E. mobility	0.02 (31)	0.17 (33)	0.08 (32)
Ca concentration in sludge	0.37 (32)*	0.26 (31)	0.51 (35)*
Filament index	0.05 (45)	0.22 (45)	0.41 (38)*

Values in parentheses are the number of observations. Marked coefficients (*) are significant for $p = 95\%$.

Correlations for settleability

The influence of EPS concentration and their different fractions on the SVI has been discussed in other studies. While some researchers found a proportional interaction between the EPS concentration and SVI (Urbain *et al.*, 1993), other authors proposed that the interaction is not significant enough to be considered (Liao *et al.*, 2001). Similar analyses have been made for the influence of the protein and carbohydrate fractions of EPS and their ratio on the SVI. Martinez *et al.* (2000) recognized a tendency of SVI to increase with the protein fraction of EPS but showed no interaction with the carbohydrate fraction, while Jobbagy *et al.* (2002) found a clear effect of the carbohydrate fraction on the augment of SVI.

Surface properties of the symbiotic biomass did not show either any significant correlation with its settleability, contrary to the observations of Forster (1968) who found a linear relation between electrophoretic mobility and SVI. Mikkelsen and Keiding (2002) found a direct correlation between EPS concentration and Z potential –derived from electrophoretic mobility-, but settleability was not considered in their study.

Although the correlation between filament index and SVI is wide known -abundance of these microorganisms can originate filamentous bulking and foaming (Jenkins *et al.*, 2004) -, the current results indicate that it is not significant. It should be noted that this variable remained in the range 1–3 and no negative effect in the SVI was noticed.

The only significant Pearson's coefficient found for settleability in this study was that of SVI and calcium concentration in the sludge, illustrated in Figure 2. Although the role of calcium as a linking element in the biomass net has been acknowledged (Urbain *et al.*, 1993; Sanin and Vesilind, 2000; Sobeck and Higgins, 2002), no reference concerning the direct correlation between these two variables was found. Nevertheless, the SVI values remained under the established criterion for good settleability ($\leq$ 150 mg l^{-1}) for the whole range of evaluated calcium. That makes difficult to assess if the correlation applies for SVI values $\geq$ 150 mg l^{-1}.

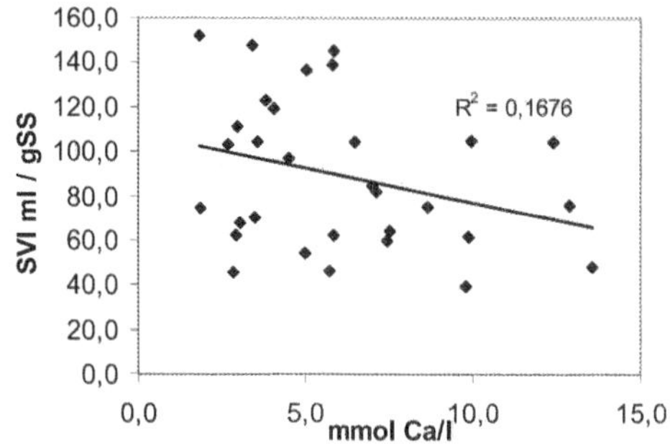

Figure 2 Correlation between calcium concentration in the sludge and SVI.

Correlations for free-swimming algae

Although the symbiotic biomass showed to have lower EPS concentrations and higher carbohydrate fraction when compared with conventional biomass (Medina and Neis, in preparation a), the obtained results indicate that distribution of microalgae in the system (their concentration in the aqueous phase and in the biomass) is not sensible to the EPS concentration nor to the different fractions in the biomass.

The only significant correlation for the quality criterion "free swimming algae", measured as Chl a $ml_{supernatant}^{-1}$, was that with the hydrophobicity of biomass used for the symbiosis, illustrated in Figure 3a). More than a correlation in the evaluated range, results suggest that there is a minimum value of hydrophobicity (around 20%) required for the beginning of symbiosis and floc formation. When biomass hydrophobicity is higher than this threshold, correlation turns inexistent but Chl a concentration in supernatant keeps under 3 mg Chl a l^{-1}. Although statistically not significant, Figure 3b) shows the data of Chl a in the sludge and hydrophobicity. There it is possible to recognize the tendency of low hydrophobicity values ($\leq$ 20%) to yield just low concentrations of Chl a in the sludge, which is consequent with the results for high Chl a in the supernatant. The evaluation of algae attachment to biomass types with different hydrophobicities suggested that this surface characteristic is relevant only during the light periods, indicating a possible relevant role of algae surface properties in hydrophobic interactions when photosynthesizing (Medina and Neis, submitted).

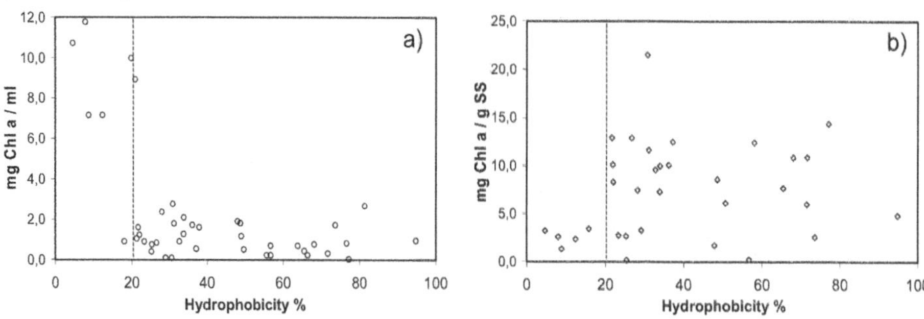

Figure 3 Correlation between Chl a distribution in the system and hydrophobicity of biomass. a) Chl a in the supernatant b) Chl a in the sludge.

Correlations for algae incorporated to biomass

As mentioned above, algae activity in the biomass, evaluated with the concentration of Chl a in the sludge, was not affected by the EPS concentration nor by the component fractions, as shown by the coefficients in Table 1.

The significant correlations for the algae incorporated to biomass were found with the filament index and calcium concentration in the biomass. They are illustrated in Figure 4. As already discussed, calcium ions are an important linking component of biomass, but in this case it seems that these ions have a negative effect on the algae incorporation. Calcium concentration values measured for algal biomass without incubation of conventional biomass (0.04 – 0.21 mmol Ca $l_{biomass}^{-1}$) were even under the concentration range obtained for the symbiotic biomass (2.22 – 12.41 mmol Ca $l_{biomass}^{-1}$). Floc stability tests performed for different concentrations of EDTA demonstrated that algae in the biomass were not sensitive to that effect when stabilized (Medina and Neis, in preparation b). Moreover, some studies have found that calcium ions tend to bind more proteins than carbohydrates (Urbain *et al.*, 1993; Dignac *et al.*, 1998) and measurement of EPS components for algae in steady growth phase yielded values of protein to carbohydrate ratio (P:C) < 1 (data not showed). It could be suggested that algae not only do not use calcium ions to incorporate to the biomass, but given the composition of their own EPS, the bivalent ions could have a negative effect for their incorporation in the floc net.

In the comparison with conventional activated biomass, the parameter of filament index was one of the significant differences between them (Medina and Neis, in preparation a). Filamentous microorganisms have been acknowledged as a constituent part of flocs, providing them with a "backbone" structure (Nielsen, 2002; Jenkins *et al.*, 2004). This parameter has been also considered an important floc macrostructure property that governs floc stability (Wilén *et al.*, 2003). This parameter showed to have values one unit higher than for conventional biomass without detriment of the settleability quality. It seems that algae could prefer filamentous microorganisms to attach to biomass, but further research would be needed to explain exactly the nature of this association. Up to now just few *Chlorella* symbionts and their adhesion mechanisms have been identified (Watanabe *et al.*, 2005). None of them corresponds to identified filamentous microorganisms.

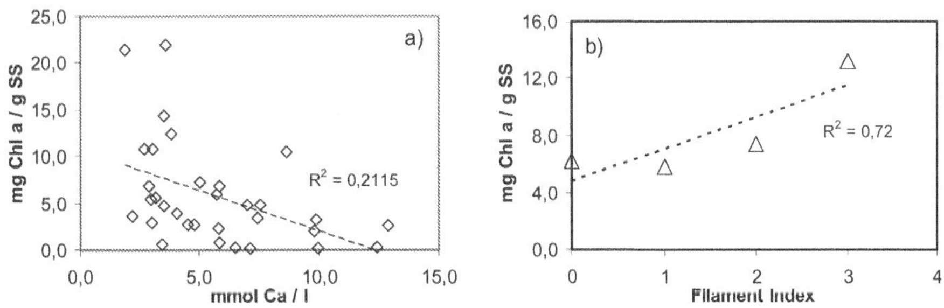

Figure 4 Correlations between quality criteria of symbiotic biomass treatment and calcium. a) with settleability (SVI) b) with algae activity in the biomass (as Chl a mg SS^{-1}).

Conclusions

Symbiotic biomass characteristics are between the ranges referred in the literature for conventional activated sludge. Common characteristics of symbiotic biomass with the literature references are similar EPS concentrations, protein and carbohydrates as the main EPS fractions and negative charge of flocs.

The results of this study indicate that the evaluation of calcium concentration, hydrophobicity and filament index of the microbial biomass component is pertinent for the implementation of symbiotic biomass. These were the biomass properties which reported an effect on the quality parameters of the symbiotic system. Calcium concentration showed to affect settleability and algae incorporation to the biomass. The first effect remains uncertain, since obtained SVI values remained under the threshold value of 150 mg l^{-1}. The interaction between calcium concentration and Chl a in the biomass hinted that calcium could have a negative effect on the algae incorporation, presumably due to the richer fraction of carbohydrates in the algae EPS and the tendency of calcium to bind protein rather than carbohydrate molecules.

The significant correlation resultant between hydrophobicity and Chl a in the supernatant indicates the relevance of this biomass surface property as a precondition for the algae incorporation: a minimum value of 20% was determined in these results for algae incorporation to take place.

Filament index, whose values kept under 3 during the experiments, showed to have a strong positive effect on algae incorporation, probably due to affinities between surface characteristics of both types of microorganisms.

Even though the evaluation of the mentioned bacterial biomass properties could be useful for the implementation of the symbiotic wastewater treatment, such evaluations may not be available in the places where this technology is intended. Nevertheless, this knowledge could be pertinent for the process designers, since it indicates the main characteristics of local bacteria biomass that should be considered. This would save resources in the biomass characterization, in case to be affordable, since fewer tests should be done, or it points specific local conditions that could affect local biomass and would have an effect on the symbiosis properties (e.g. hardness of drinking water, characteristics of wastewater, foam and bulking in places where bacteria are taken, a.o.)

Acknowledgements

This research was possible thanks to a scholarship from the IPSWat Program of the German Federal Ministry of Education and Research. Special thanks to Dr. Dieter Lorch, Institute of Botany, University of Hamburg, for the *Chlorella vulgaris* cultures.

References

APHA (1985) *Standard Methods for the Examination of Water and Wastewater*, 16th edn, American Public Health Association/American Water Works Association/Water Environment Federation, Washington DC, USA.

Boyette, S. M., Lovett, J. M., Gaboda, W. G. and Soares, J. A. (2001). Cell surface and exopolymer characterization of laboratory stabilized activated sludge from a beverage bottling plant. *Wat Sci Technol*, **43**(6), 175–184.

Bura, R., Cheung, M., Liao, B., Finlayson, J., Lee, B. C., Droppo, I. G., Leppard, G. G. and Liss, N. (1998). Composition of extracellular polymeric substances in the activated sludge floc matrix. *Wat Sci Technol*, **37**(4–5), 325–333.

Chao, A. C. and Keinath, T. M. (1979). Influence of process loading intensity on sludge clarification and thickening characteristics. *Wat Res*, **13**, 1213–1223.

Deutsches Institut für Normung (DIN) (1985). Deutsche Norm 38 412, Teil 16. Bestimmung des Chlorophyll-a-Gehalts von Oberflächenwasser Beuth Verlag GmbH, Berlin.

Dignac, M. F., Urbain, V., Rybacki, D., Bruchet, A., Snidaro, D. and Scribe, P. (1998). Chemical description of extracellular polymers: Implication on activated sludge floc structure. *Wat Sci Technol*, **38**,(8–9), 45–53.

Dubois, M., Gilles, K. A., Hamilton, J. K., Rebers, P. A. and Smith, F. (1956). Colorimetric method for determination of sugars and related substances. *Analytical Chemistry*, **28**, 350–356.

Forster, C. F. (1968). The surface of activated sludge particles in relation to their settling characteristics. *Wat Res*, **2**, 767–776.
Frolund, B., Palmgren, R., Keiding, K. and Nielsen, P. H. (1996). Extraction of extracellular polymers from activated sludge using a cation exchange resin. *Wat Res*, **30**(8), 1749–1758.
Guellil, A., Block, J. C. and Urbain, V. (1998). Adaptation of MATH for measuring act. sludge hydrophobicity. *Wat Sci Technol*, **37**,(4–5), 359–362.
Gutzeit, G. (2005). Entwicklung und Modellierung eines neuartigen Abwasserreinigungsverfahrens mit symbiotischer Algen-Bakterien-Biomasse. Doktor Ingenieur Dissertation, Technische Universität Hamburg-Harburg.
Gutzeit, G., Lorch, D., Weber, A., Engels, M. and Neis, U. (2005). Bioflocculent algal-bacterial biomass improves low-cost wastewater treatment. *Wat Sci Technol*, **52**(12), 9–18.
Jenkins, D., Richard, M. G., & Daigger, G. T. (2004). *Manual on the causes and control of activated sludge bulking, foaming, and other solids separation problems*, 3rd edn. Lewis Publishers, Boca Raton, a.o.
Jobbagy, A., Literathy, B. and Tardy, G. (2002). Implementation of glycogen accumulating bacteria in treating nutrient-deficient wastewater. *Wat Sci Technol*, **46**(1–2), 185–190.
Liao, B., Allen, D. G., Droppo, I. G., Leppard, G. G. and Liss, N. (2001). Surface properties of sludge and their role in bioflocculation and settleability. *Wat Res*, 35,(2), 339-350.
Liao, B., Allen, D. G., Leppard, G. G., Droppo, I. G. and Liss, N. (2002). Interparticle interactions affecting the stability of sludge flocs. *J Colloid and Interface Science*, **249**, 372–380.
Liu, H. and Fang, H. H. P. (2002). Extraction of extracellular polymeric substances of sludge. *J Biotechnol*, **95**, 249–256.
Martinez, F., Favela-Torres, E. and Gomez, J. (2000). Oscillations of expoloymeric compostion and sludge volume index in nitrifying flocs. *Appl Biochem Biotechnol*, **87**(3), 177–188.
Medina, M. and Neis, U. (in preparation, a). Symbiotic algal bacterial wastewater treatment: Comparison between symbiotic and conventional activated biomass.
Medina, M. and Neis, U. (in preparation, b). Symbiotic algal bacterial wastewater treatment: Interparticle effects on the stability of symbiotic biomass.
Medina, M. and Neis, U. (submitted). Symbiotic algal bacterial wastewater treatment: Role of the sludge hydrophobicity on algae incorporation.
Mikkelsen, L. H. and Keiding, K. (2002). Physico-chemical characteristics of full scale sewage sludge with implications to dewatering. *Wat Res*, **36**, 2451–2462.
Murthy, S. N. and Novak, J. T. (1998). Effects of potassium ion on sludge settling, dewatering and effluent properties. *Wat Sci Technol*, **37**(4–5), 317–324.
Nielsen, P. H. (2002). Activated sludge – the floc. In *Encyclopedia of Environmental Microbiology*, G. Bitton, ed., Wiley, New York, a.o., pp 54–61.
Nielsen, P. H., Thomsen, J. R. and Nielsen, J. L. (2004). Bacterial composition of activated sludge – importance for floc and sludge properties. *Wat Sci Technol*, **49**(10), 51–58.
Sanin, F. D. and Vesilind, P. A. (2000). Bioflocculation of activated sludge: the role of calcium ions and extracellular polymers. *Env Technol*, **21**, 1405–1412.
Sekar, R., Venugopalan, V. P., Satpathy, K. K., Nair, K. V. K. and Rao, V. N. R. (2004). Laboratory studies on adhesion of microalgae to hard substrates. *Hydrobiologia*, **512**, 109–116.
Sobeck, D. C. and Higgins, M. J. (2002). Examination of three theories for mechanisms of cation-induced bioflocculation. *Wat Res*, **36**, 527–538.
Sponza, D. T. (2003). Investigation of EPS and physicochemical properties of different activated sludge flocs under steady-state conditions. *Enzyme and Microbial Tech*, **32**, 375–385.
Urbain, V., Block, J. C. and Manem, J. (1993). Bioflocculation in activated sludge. An analytic approach. *Wat Res*, 27(5), 829–838.
Watanabe, K., Noritaka, T., Aoyagi, H., Hanada, S., Watanabe, Y., Ohmura, N., Saiki, H. and Tanaka, H. (2005). Symbiotic association in Chlorella culture. FEMS Microbiology Ecology, 51, 187–196.
Wilén, B. M., Jin, B. and Lant, P. A. (2003). Impacts of structural characteristics on activated sludge stability. *Wat Res*, **37**, 3632–3645.
Wilén, B. M., Keiding, K. and Nielsen, P. H. (2004). Flocculation of activated sludge flocs by stimulation of the aerobic biological activity. *Wat Res*, **38**, 3909–3919.

Four years of pollutant removal performance of two-stage inverse vertical subsurface flow constructed wetlands (IVSSFCW)

J. Yin*, Y.B. Cui*, K. Zhao**, Ch. Dong*, Zh.Sh. Liu**, L.H. Zhao* and B.Y. Zhu*

* Department of Environmental Engineering, Jilin Architectural and Civil Engineering Institute, 130021 China (E-mail: *cui.yubo@163.com*)
** School of Municipal and Environmental Engineering, Harbin Institute of Technology, 150090 China (E-mail: *zhaoke326@tom.com*)

Abstract Pollutant removal performance of two-stage inverse vertical subsurface flow constructed wetlands (IVSSFCW) changes with running time, and pollutant removal mechanisms are different under different running times and running modes. This paper investigates the performance variation of two-stage IVSSFCW wastewater treatment systems for a four year running period. The results show that plants and microorganisms in IVSSFCW required an adjustment period: in the first-stage constructed wetland, COD removal rate tended to increase with running year, while the accumulated plant residue had a significant effect on COD removal during the system start up; batch feed and drain could strengthen re-aeration and improve ammonia nitrogen (NH_4-N) removal; vertical down-flow CW was more than 90% of NH_4-N, which had no significant relationship with batch or continuous running mode; phosphorus removal performance was stable and remained at more than 90% of removal rate.

Keywords Inverse vertical subsurface flow constructed wetlands (IVSSFCW), nutrients removal, wastewater treatment

Introduction

Constructed wetlands are one kind of semi-natural wastewater treatment technology, which can meet the objective of water purification under the role of plants, microorganisms and media using the Sun's energy (Yin and Shen, 1995; Koottaatep *et al.*, 2001). This technology has been developed for more than 30 years to remove BOD, COD, TN and TP in municipal wastewater and to treat rain wastewater, sludge and leachate.

Wetland treatment of wastewater relies largely on biological and biochemical processes (McNevin *et al.*, 2000; Mitchell and McNevin, 2001), so oxygen availability greatly affects the system treatment efficiency. Recent research showed that batch flow system can improve aerated degradation process (McBride and Tanner, 2000), while Burgon *et al.* (1995) got the conclusion of no treatment efficiency improvement by batch fill-and-drain operations. The purpose of this research is to investigate constructed wetland long-term running performance under the operations of batch flow and continuous flow to find optimum operation conditions for wastewater treatment.

Method

The experimental unit consists of two-stage constructed wetlands (Figure 1). The first wetland is an inverse vertical subsurface flow (IVSSF) constructed wetland (CW), which is named by the upward plug-flow style of wastewater; it measured 3.34 m in length, 0.94 m in width, and 0.56 m in depth. The substratum height was 0.38 m, including a 0.1 m layer of crushed limestone (2–7 cm diameter), a 0.08 m layer of volcano cinder (0.5–1 cm diameter), a 0.06 m layer of volcano cinder (0.1–0.5 cm diameter), and a 0.14 m layer of silver sand (0.1–0.3 cm diameter) from the bottom to the top. The wetland was planted with reeds (*Phragmites*) 0.1 m deep in the silver sand layer. The sampling ports were set at 0.8 m, 1.6 m and 2.4 m from the inlet along the flow direction in the first wetland. The sampling deep was 0.22 m.

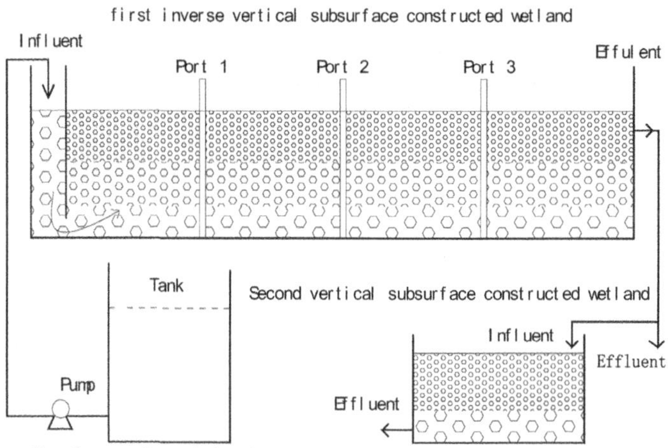

Figure 1 Schematic of experimental setup.

The second wetland is a vertical subsurface flow constructed wetland, which measured 1.25 m in length, 0.92 m in width, and 0.52 m in depth. Wastewater flows downwards. The substratum height was 0.38 m, including a 0.1 m layer of crushed limestone (2–7 cm diameter) and a 0.28 m layer of silver sand (0.1–0.3 cm diameter) from the bottom to the top. The wetland was planted with reeds (*Phragmites*) 0.1 m deep in the silver sand layer. The sampling ports were set at the middle of the wetland.

Synthetic domestic wastewater was prepared daily with $C_6H_{12}O_6$, NH_4Cl and KH_2PO_4. The characteristics of the wastewater were: COD 100–400 mg L^{-1}, NH_4-N 10–50 mg L^{-1}, PO_4-P 1.2–2.3 mg L^{-1}, pH 5–8. The system was run in batch flow in 2001 and 2002, and run in continuous flow in 2003 and 2004. Operation conditions are shown in Table 1.

Results and discussion

Performance variation of first-stage constructed wetland

Organic matter removal

Figure 2 shows COD removal variation by month during 2001–2004. Year 2001 was the first year of the system start-up, starting in July. From the trend of COD removal variation, we can see that COD removals are relatively lower, ranging from 36% to 60%, and have a trend from decrease to increase. Even though batch fill-and-drain can

Table 1 Two-stage constructed wetlands operation conditions

	Year	Month	April	May	June	July	August	September	October
First Stage CW	2001	HRT(d)					3	3	3
		Period(fill/drain)					12h/12h	6h/6h	6h/6h
	2002	HRT(d)	6	3	3	3			
		Period(fill/drain)	12h/12h	6h/6h	3h/9h	3h/9h			
	2003	HRT(d)	6	4	3	2	2	1	1
		Period(fill/drain)				Continuous flow			
	2004	HRT(d)	3	3	3	2	1.5	1.5	1.5
		Period(fill/drain)				Continuous flow			
Second Stage CW	2001	HRT(d)					1.25	1.25	1.25
		Period(fill/drain)					12h/12h	12h/12h	12h/12h
	2002	HRT(d)	0.54	0.54	0.54	0.54			
		Period(fill/drain)	6h/6h	6h/6h	6h/6h	6h/6h			
	2003	HRT(d)	1.25	1	0.75	0.75	0.75	0.625	0.625
		Period(fill/drain)				Continuous flow			
	2004	HRT(d)	1.5	1.5	1.5	1.5	0.94	0.94	0.94
		Period(fill/drain)				Continuous flow			

re-aeration, the main COD removal mechanisms are filter and adsorption, microorganism growth went through the process of generation, adaptation and regeneration during the early days. Biochemical processes became the main removal mechanism due to plant growth and adaptive microorganism generation during later months.

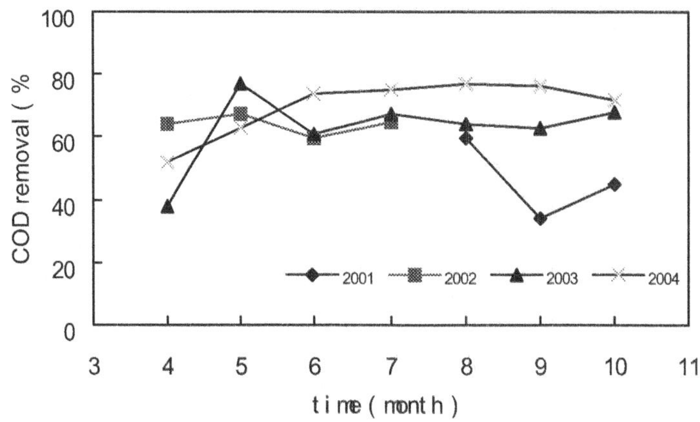

Figure 2 Monthly COD removals variation.

COD removal remained stable, ranging from 60% to 70% during the second year and the third year, while in the fourth year COD removal increased to more than 70%. So COD removal increased with running time, because of plant and microorganism adjustment. Operation style was the main reason for COD removal staying at the same level in 2002 and 2003.

It is worth noting that COD removal at the first start-up month was relatively higher during 2001 and 2002, while lower than that of later months in 2003 and 2004. The main reason was organic residue development, including plant residue, roots and dead microorganisms. These began to degrade during the system start-up, thus causeing COD removal.

Four years' organic loading rates were 7.36–11.27 gCOD/(m^2.d) (2001), 5.52–11.7 g COD/(m^2.d) (2002), 3.34–32.53 g COD/(m^2.d) (2003) and 8.06–27 g COD/(m^2.d) (2004), a trend of increase with running years. COD removals also increased, which showed the processes of microorganism adaptation, adjustment and regeneration and the role of plant growth indirectly.

Ammonia nitrogen removal

Figure 3 shows NH_4-N removal variation with month from 2001 to 2004. As shown in Figure 3, COD removal of 72% in July 2003 is a bizarre point. Because constructed wetlands have a delay effect, when the inflow NH_4-N loading rate increased, the corresponding outflow NH_4-N concentration increase needed to spend a period, lower NH_4-N concentration and higher loading caused high NH_4-N removal.

NH_4-N removals were higher, up to 50%, due to re-aeration under batch fill-and-drain operations during the first two years. Whereas in 2003 and 2004 NH_4-N removals decreased greatly due to shortage of re-aeration under continuous flow operation. During the early months of the fourth running year, NH_4-N removals were negative.

Four years' NH_4-N loading rates were 0.6–0.72 g NH_4-N/(m^2.d) (2001), 0.25–0.94 g NH_4-N/(m^2.d) (2002), 0.42–5.46 g NH_4-N/(m^2.d) (2003) and 1.32–2.32 g NH_4-N/(m^2.d) (2004).

The system NH_4-N removal decrease could be caused by organic loading increase, such as NH_4-N removal decreasing greatly with loading increase in August and September 2003.

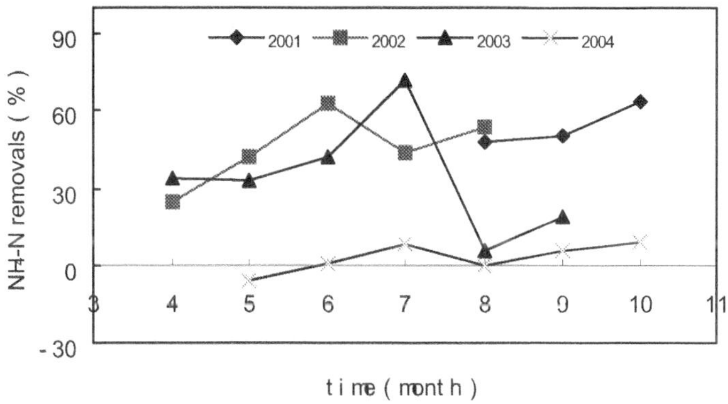

Figure 3 Monthly NH_4-N removals variation.

Combined with four years' NH_4-N removal variation characteristics, we can conclude that batch fill-and-drain can improve NH_4-N removal, NH_4-N loading has negative effect on the removal, plant residue may cause NH_4-N release and further affect NH_4-N removal with running time. Loading rate has negative effect on the removal.

Performance variation of second stage constructed wetland

Organic matter removal

Figure 4 shows COD removal variation with month from 2001 to 2004. COD loading rates were 3.25–8.17 g COD/(m^2.d) (2001), 5.87–8.7 g COD/(m^2.d) (2002), 2.88–7.65 g COD/(m^2.d) (2003) and 2.33–4.82 g COD/(m^2.d) (2004).

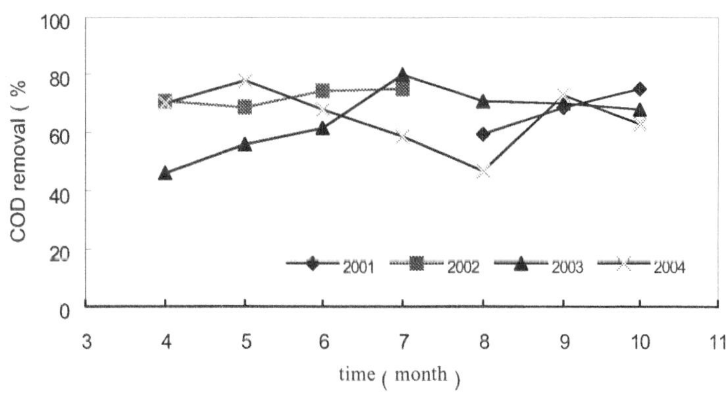

Figure 4 Monthly COD removal rates variation.

As shown in Figure 4, COD removals are 58–76% in 2001, 68–80% in 2002, 46–80% in 2003 and 47–80% in 2004. The operation style of the second-stage wetland is vertical subsurface flow. From the theoretical point of view, batch flow and continuous flow while draining at the down part of CW are beneficial to re-aeration, which can improve organic matter degradation. In the first three years, organic loading rates and COD removal kept at the same level. In 2004, organic loading rates were relatively lower, but discharge height changed three times: 0.1 m, 0.2 m and 0.3 m from the bottom in June, July and September, individually; i.e., aerobic time became shorter and anaerobic time became longer. From the curve trend of 2004, we can see that COD removal began to decrease from June, while COD removal increase in September may have a relationship with adaptive microorganism growth and re-generation.

Ammonia nitrogen removal

Figure 5 shows NH_4-N removal variation with month from 2001 to 2004. NH_4-N loading rates are 0.28–0.34 g NH_4-N/(m^2.d) (2001), 0.68–0.83 g NH_4-N/(m^2.d) (2002), 0.61–3.23 g NH_4-N/(m^2.d) (2003), 0.93–1.6 g NH_4-N/(m^2.d) (2004).

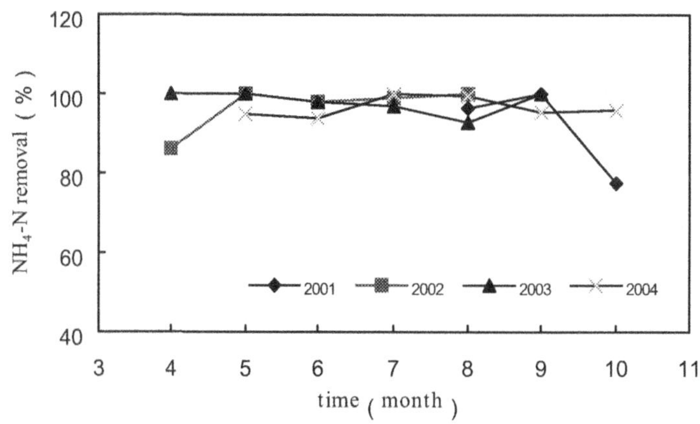

Figure 5 Monthly NH_4-N removal rates variation.

As shown in Figure 5, NH_4-N removals ranged from 90% to 100% beside lower removal in October 2001 owing to the effect of temperature decrease. The second-stage wetland ran in batch style in the first two years, in continuous style in the third year while draining at the bottom, and in continuous style in the fourth year while draining at several heights, which also kept an unsaturated region in the wetland. These are beneficial for generating an aerobic environment, which is beneficial to nitrification. At the same time, local anaerobic environments and anaerobic environments caused by flooded conditions are beneficial to denitrification. Thus the second stage wetland can keep high NH_4-N removals.

Phosphorus removal

Figure 6 shows TP removals in 2003 and 2004. The first-stage wetland TP loading rates are 0.052–0.179 g TP/(m^2.d) (2003) and 0.074–0.16 g TP/(m^2.d) (2004), the second-stage wetland TP loading rates are 0.041–0.061 g TP/(m^2.d) (2003) and 0.068–0.11 g TP/(m^2.d) (2004).

The first-stage wetland removed 55% of TP in 2003 and 2004, and the second-stage wetland removed 92–100% of TP. TP loading rate increased, TP removals also increased. So TP loading rates had little effect on TP removal in the range of experimental loading rates. The main mechanisms of TP removal are absorption, settlement, and plant and microorganism intake; and especially limestone used for the media in this study played a key role in TP removal. Temperature and running time have no obvious effect.

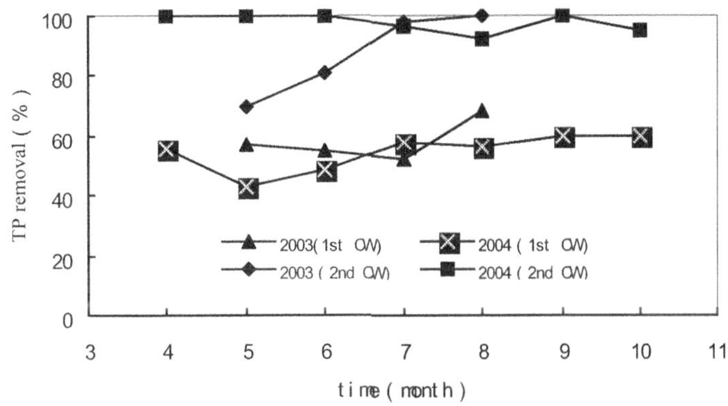

Figure 6 Monthly TP removal variation.

Conclusions

Plants and microorganisms in the constructed wetland require a process to adapt, adjust and grow. In the first wetland, COD removal had a trend of increasing with running years. But organic residue can accumulate in the system year after year: this matter will degrade during the period of system start-up, which can affect COD removal.

Batch fill and drain can strengthen NH_4-N removal, while NH_4-N loading rate has a negative effect on the removal. Plant residue may generate NH_4-N release, and affect NH_4-N removal. Vertical subsurface flow can get higher NH_4-N removals of more than 90%, with no obvious relationship with operation style. TP removals are higher, up to 100%.

References

Burgon P. S., Reddy K. R., Debusk T. A. (1995). Performance of subsurface flow wetlands with batch load and continuous-flow condition. *Water Environment Research*, **67**(5): 855–862.

Koottaatep T., Polprasert C., Koanh N.T., Heinss U., Montangero A. (2001). Strauss M. Septage dewatering in vertical-flow constructed wetlands located in the tropics. *Water Sci. Technol.*, **44**(2–3): 181–188.

McBride G.B., Tanner C.C. (2000). Modelling biofilm nitrogen transformation in constructed wetland mesocosms with fluctuating water levels. *Ecol. Engng*, **14**: 93–106.

McNevin D., Harrison M., King A., David K., Mitchell C. (2000). Towards an integrated performance model for subsurface flow constructed wetlands. *Environ. Sci. Health.*, **A35**(8): 1415–1429.

Mitchell C., McNevin D. (2001). Alternative analysis of BOD removal in subsurface flow constructed wetlands employing Monod kinetics. *Water Res.*, **35**(5): 1295–1303.

Yin H., Shen W. (1995). Using reed beds for winter operation of wetland treatment system for waster water. *Wat. Sci. Technol.* **32**(3): 111–117.

Temperature as a factor influencing the Anammox process performance

B. Szatkowska and E. Plaza

Department of Land and Water Resources Engineering, Royal Institute of Technology, S–100 44, Sweden (E-mail: *beata@kth.se*; *elap@kth.se*)

Abstract A new technology combining partial nitritation and Anammox processes for removal of nitrogen from highly concentrated streams is widely studied at present. To better understand relationships between factors influencing the process, a laboratory-scale plant (17l) working as a moving-bed biofilm reactor with Kaldnes rings was constructed over 4 years ago. In these studies the influence of temperature on the partial nitritation/Anammox process was considered, both for one-stage and two-stage system. Together with the pilot plant studies, a series of batch tests were carried out. It occurred that a sudden decrease in temperature affected Anammox nitrogen removal rates. A linear relation between nitrogen removal rates and temperature was found. Similar correlation was established for a single stage process with the partial nitritation/Anammox. However, the results indicated that after an adaptation period the Anammox bacterial culture adapted to lower temperature conditions. An addition of artificial NO_2-N solution speeded up the nitrogen removal reaction. Therefore, the lack of NO_2-N occurred to be the bottleneck for the simultaneous partial nitritation/Anammox process. The progress of reaction was suppressed with the presence of allylthiourea (ATU) – the inhibitor of *Nitrosomonas* bacteria. Nitrifiers were inhibited and no nitrite was produced. Therefore, the Anammox reaction could not take place.

Keywords Ammonium removal; Anammox; nitritation; temperature

Introduction

Application of traditional biological treatment methods with nitrification/denitrification processes for nitrogen removal from highly concentrated streams is expensive. Thus, a better alternative is implementation of a partial nitritation/Anammox process with decreased oxygen demand and without external carbon addition. In that novel process, the first step is well-known biological conversion while Anammox still requires a better understanding and increased knowledge. Many research groups have studied the influence of different factors on the Anammox process performance and have found that temperature can be an important factor. Researchers from The Netherlands reported temperature conditions of 20–43 °C (with an optimum at 40 °C) as suitable for the Anammox process (van Dongen *et al.*, 2001). Other researchers dealing with mathematical models (Hao *et al.*, 2002) reported that the Anammox process can work with the efficiency of 88% under the temperature of 20 °C combined with ammonium surface load (ASL) equal to 0.5 g $NH4^+$-N m^{-2} d^{-1}, in a biofilm of at least 0.7 mm thickness and under the dissolved oxygen (DO) concentration in the bulk liquid at the level of 0.3 g O_2 m^{-3}. Egli *et al.*, (2001) studying Anammox bacteria in Switzerland in a rotating biological contactor found that the highest Anammox activity could be obtained for the temperature of 37 °C, while no reaction was noticed at 45 °C. In that study it was also demonstrated that the Anammox activity at 11°C was approximately 24% of that at 37 °C. In Spain, the Anammox was tested in an SBR and in a gas-lift reactor, and the temperature parameter was set as 35 and 30 °C, respectively (Dapena-Mora *et al.*, 2004a,b). All these values of optimal temperature found in wastewater treatment systems deviate from the optimal temperature of the Anammox bacteria growing in the marine sediments. Dalsgaard and Thamdrup (2002) give the temperature of 15°C as suitable for Anammox bacteria present in sediments. Rysgaard *et al.* (2004) reported similar value of 12 °C.

The effect of temperature on nitrification process is better known. Especially the knowledge concerning the nitrification in systems with suspended biomass was widely studied. Different level of knowledge concerning temperature is for biofilm systems. Here other temperature-dependent factors such as substrate diffusion, transport and dissolved oxygen concentration determine the nitrification process performance in biofilms (Fdz-Polanco *et al.*, 1994; Zhu and Chen, 2002) what make it more complex.

To study the partial nitritation/Anammox process, a laboratory-scale pilot plant was build and operated for more than four years at the Department of Land and Water Resources Engineering, Royal Institute of Technology (KTH), Sweden. The bacteria responsible for nitrogen removal in the Anammox process in the laboratory-scale pilot plant were recognized as *Brocadia anammoxidans*. To evaluate temperature as a factor influencing the activity of the biofilm culture combining nitrifiers and Anammox bacteria, a series of batch tests were carried out.

Methodology

Laboratory-scale plant

The laboratory-scale plant consisted of two reactors (with a volume of 7.1 and 9.8 l) in series filled with Kaldnes rings as the carrier material for fixed biofilm growth. The pilot was supplied with a supernatant from dewatering of the digested sludge from Bromma Wastewater Treatment Plant (WWTP). The WWTP receives municipal wastewater from central and western part of the Stockholm region. To obtain the required concentration of influent ammonium, the supernatant was diluted with tap water. Each reactor was equipped with a heater to keep temperature stable and with a mixer to assure proper mixing and adequate oxygen conditions. During the most of the period the pH level was corrected with a continuous dosage of Na_2CO_3 solution to the first reactor (to keep the pH value around 8.2). Later on the correction was stopped to evaluate system's ability to utilise nitrogen without the pH adjustment. In the pilot plant experiment, different operational strategies were applied. Two separate stages consisting of partial nitritation (reactor 1) and Anammox (reactor 2) and one-stage completely autotrophic nitrogen removal over nitrite similar to CANON configuration (in both reactors) were tested. Detailed description can be found in earlier publications (Płaza *et al.*, 2002; Szatkowska *et al.*, 2004a,b).

Batch tests

Series 1

The series was run when the partial nitritation and Anammox process took place in two separate reactors. Tests were carried out in three parallel bottles of 1 litre working volume with the Kaldnes carriers taken from the second reactor (Anammox) of the continuously working pilot plant. An $NaNO_2$ solution was added to obtain a required concentration of

NO_2-N. To keep temperature on different levels during one batch test, each bottle was purposefully located in a laboratory. The bottles with temperatures kept at 25 or 26 °C were placed in a constant room. For the lowest (20 or 23 °C) and the highest (30 or 35 °C) temperatures bottles were placed in a water bath. Every batch test lasted 8 hours. Every hour a 10 ml sample was taken from a bottle by a syringe. Then the sample was filtrated with a pre-filter and a 0.45 μm filter, and analysed for NH_4-N, NO_2-N and NO_3-N. Moreover, the pH value, dissolved oxygen concentration in the bulk liquid and conductivity were measured.

Series 2

The series was run parallel to the experiment decreasing the process temperature in the plant when both partial nitritation and Anammox simultaneously took place in both reactors. Temperature in the batch corresponded to the temperature in the pilot plant reactors. Tests were carried out in two parallel bottles. Bottle one was filled with the medium taken from reactor 1 and bottle two with the medium from reactor 2. The temperature was kept stable due to placing the bottles in the water bath. The difference in nitrogen removal rates under temperature of around 31, 28 and 26 °C was tested (three tests at each temperature). Each of the batch tests lasted 5 hours. Samples were filtrated and analysed for NH_4-N, NO_2-N and NO_3-N. Dissolved oxygen concentration in the bulk liquid, the pH value and conductivity were measured as well.

Series 3

The last series was carried out after a four-month adaptation period of the pilot plant to the temperature of 20 °C. The tests were performed during two following days. On the first day, three bottles with the medium from reactor 1 were used. On the second one also three bottles with the medium from the reactor 2 were prepared. In bottle one, it was intended to simulate the reaction present in the reactor. In bottle two, a NO_2-N solution was added to verify the effect of an additional amount of nitrite on reaction rates. In bottle three an allylthiourea (ATU) solution was dosed to inhibit nitrifiers. The concentration of ATU in batch bottle was equal to 5 g m^{-3} according to Surmacz-Górska *et al.* (1996). The rest of batch test methodology was kept like in previous tests.

Results and discussion

Laboratory-scale plant

The plant has been operated for more than 4 years. The period described present results gathered during three years of the experiment when different strategies were tested. The first series of batch were carried out in the period from 3 July to 8 September 2003 when typical partial nitritation (reactor 1) and Anammox (reactor 2) processes took place in separate rectors (Figure 1). At that time, the plant was operated at the average temperature of 31.6±0.5 °C. Dissolved oxygen (DO) concentration amounted to 0.40±0.13 and 0.28±0.09 g O_2 m^{-3} for reactors 1 and 2, respectively. The pH value was equal to 8.1. The total inorganic nitrogen removal efficiency for reactor 1 was equal to 17%. Mainly the conversion of half of ammonium to nitrite proceeded. The nitrogen removal efficiency for reactor 2 was as much as 78% due to the established Anammox reaction. The second series of batch tests were run during the period from 8 June to 4 August 2005. Then, the plant was operated in such a way that partial nitritation and Anammox processes were simultaneously carried out in both reactors. During the period, the temperature parameter was decreasing step-by-step from 31 down to 20 °C. The nitrogen removal efficiency in the system during this time was equal to 75% (46 and 29% for reactor 1 and 2, respectively). The last four months of the presented period concerns operation of the plant at temperature of around 20 °C. In that case, a very small decrease in the efficiency of the second reactor (about 4%) was observed, while the efficiency of reactor one stayed at the same level. At the end of this period the third series of batch tests was run.

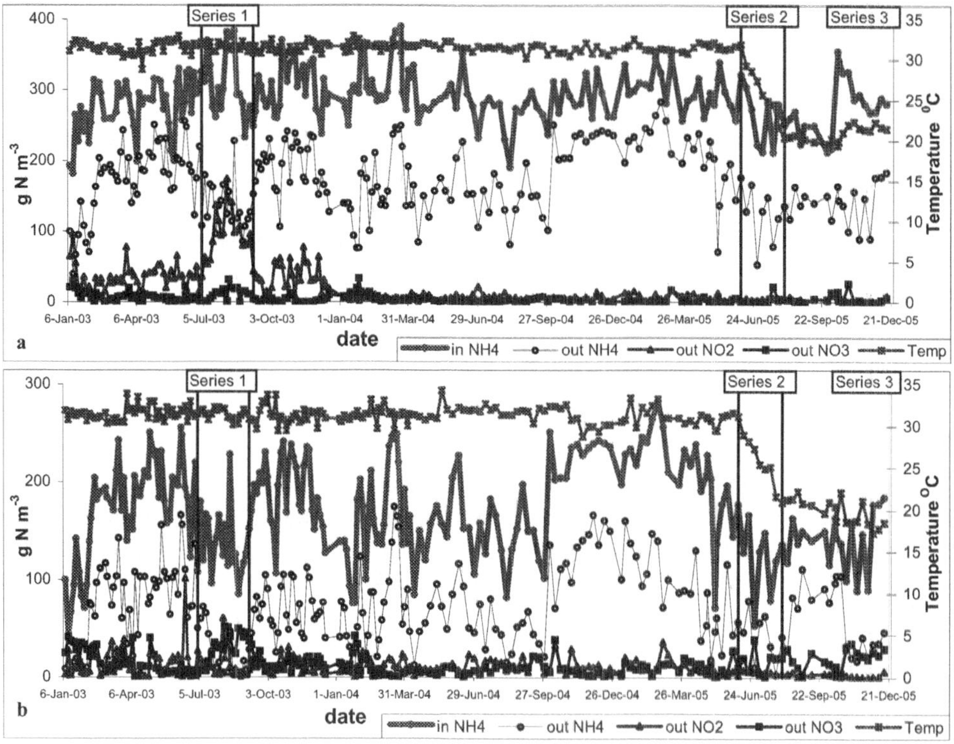

Figure 1 Nitrogen conversions in: a. reactor 1; b. reactor 2.

Batch tests series 1

The first series of batch tests was performed in order to evaluate the influence of temperature on the Anammox reaction rate. By evaluation of results of nitrogen variation from all batch tests series 1, it was possible to calculate the average removal rates (Figure 2a).

The decrease in the temperature caused a decrease in nitrite removal rates and a linear correlation was found with high correlation coefficient (Figure 2b). At the temperature of 20°C, the nitrite removal rate constituted only half of the ammonium removal rate. A stepwise decrease in the temperature during the batch test series performance caused also a decrease in values of inorganic nitrogen removal rates.

The temperature of 25–26 °C appeared to be a sensitive point where the relationship between ammonium and nitrite removal rates was altered. Below this temperature nitrite removal rates were lower than ammonium removal rates, while above this temperature nitrite removal rates were higher than ammonium. The nitrite-to-ammonium removal rate ratio close to the stoichiometrical value for the Anammox reaction was obtained for the temperature of 26 °C and amounted to 1.30.

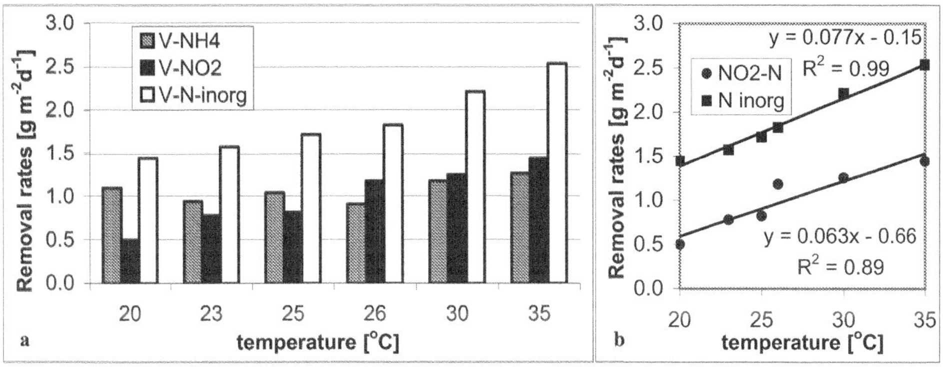

Figure 2 (a) The average values of nitrogen removal rate for batch test series 1; (b) Correlation between temperature and nitrogen removal rates.

Total inorganic nitrogen removal rates were increasing parallel with the increase in temperature. Generally, it can be said that sudden changes in the temperature influence nitrite removal rate to a large extend (variation from 0.5 g N $m^{-2}d^{-1}$ in the temperature of 20 °C up to 1.45 g N $m^{-2}d^{-1}$ at the temperature of 35 °C) (see Figure 2a). Ammonium removal rates varied in the smaller range of 0.91–1.27 g N $m^{-2}d^{-1}$. The highest values of ammonium and nitrite removal rates during short-term experiments were obtained for the temperature of 35 °C. The decrease in temperature from 35 to 20 °C diminished the inorganic nitrogen removal rate by 43%.

Batch tests series 2

The second series was run to find out the influence of temperature on partial nitritation/Anammox processes that simultaneously took place in each reactor. The temperature in the plant was decreasing in three weeks' time intervals. In each interval three batch tests were run. The conditions of batch tests performance are presented in Table 1.

Table 1 Physical conditions of batch tests performance

	Reactor 1			Reactor 2		
Batch test no	pH	DO	T	pH	DO	T
1	8.00	0.21	31.0	7.91	0.20	31.0
2	7.87	0.29	30.8	7.62	0.27	31.0
3	8.19	0.19	31.6	8.09	0.23	31.6
4	8.12	0.23	28.9	7.68	0.25	28.9
5	8.16	0.27	27.4	8.05	0.29	27.4
6	7.91	0.23	27.3	7.52	0.26	27.3
7	8.07	0.39	26.7	7.92	0.34	26.6
8	8.05	0.38	25.8	7.82	0.32	25.9
9	7.71	0.39	24.5	7.40	0.41	24.6

For these tests, the averages of ammonium and total inorganic nitrogen removal rates were calculated for the medium taken from both reactors of the pilot plant. Linear relationships between temperature values and ammonium and inorganic nitrogen removal rates with high correlation coefficients were found for both reactors. The decrease in temperature from 31 to 27.8 °C in reactor 1 diminished the ammonium removal rate by 30%, while the decrease to 25.6 °C caused the decrease in the ammonium nitrogen removal rate of about 43%. The average ammonium nitrogen removal rate for reactor 2 at the temperature of 27.9 °C was estimated as 42.4% of the average rate obtained for the test at the temperature of 31.2 °C, while as 34.2% for the temperature of 25.7 °C. In general, the removal rates for the first reactor were higher (see Figure 3).

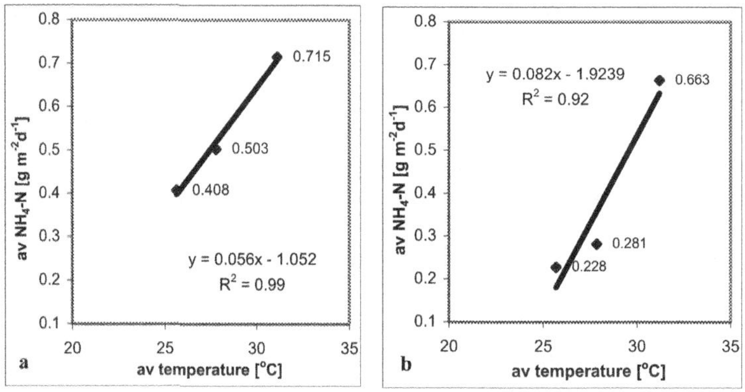

Figure 3 Relation between removal rates and temperature in: (a) reactor 1; (b) reactor 2 (av=average).

Batch tests series 3

The series was run after the period of four-month adaptation of the system to the temperature of around 20 °C. It occurred that the bacterial culture adapted to new conditions and the nitrogen removal rate increased compared to tests that were run in the previous series. For the bottle where the partial nitritation/Anammox process was established ammonium nitrogen removal rates were equal to 0.51 and 0.26 g N m^{-2} d^{-1} for reactor 1 and 2, respectively. The value for reactor 1 corresponds to the value of removal rate at temperature of 28 °C from the previous series 2 (Figure 3a), while the value of removal rate for the second reactor was equal to that obtained at a temperature of around 24 °C (Figure 3b).

In case where the NO_2-N solution was added, the nitrogen removal rate was as much as 1.72 and 2.40 g N $m^{-2}d^{-1}$ (Figure 1.4b) for reactor 1 and 2, respectively. The addition of nitrite significantly speeded up the whole reaction. It means that the Anammox would be able to utilise more ammonium and nitrite but due to the lack of produced nitrite during the nitritation step the reaction rate is lower. Ammonium and nitrite removal rates for reactor 1 were equal to 0.86 and 1.51 g N $m^{-2}d^{-1}$ that gave the value of nitrite-to-ammonium ratio 1.76. Nitrate production amounted to 37.3%. In reactor 2, the ammonium removal rate amounted to 1.17 g N m^{-2} d^{-1}, while the nitrite removal rate was equal to 1.55 g N $m^{-2}d^{-1}$. It resulted in the nitrite-to-ammonium ratio equal to 1.33. At the same time 13.7% of nitrate was produced.

Higher values of nitrogen removal rates for the partial nitritation/Anammox process in reactor 1 compared with the removal rates for reactor 2 could be explained by process diversity in the pilot plant in previous studies. Then, both processes were run as two separate steps. The partial nitritation was performed in reactor 1 and therefore more nitrifiers than Anammox bacteria are still present in this reactor. Such situation assures higher production of nitrite and higher nitrogen removal rates. The other explanation could be that partial nitritation can be less sensitive to temperature changes than Anammox bacteria.

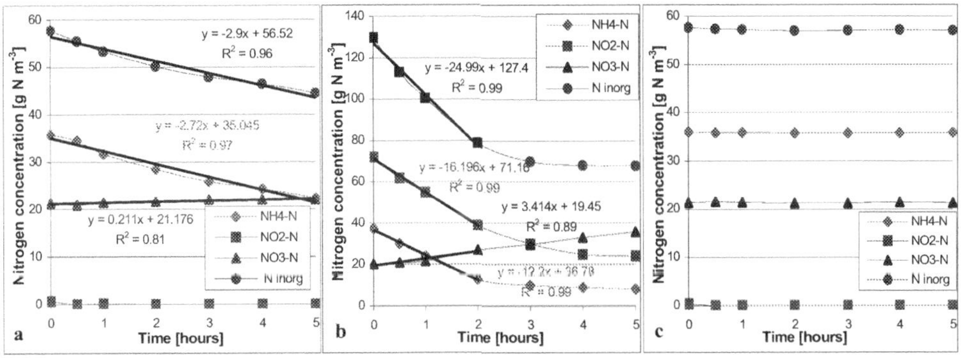

Figure 4 Nitrogen conversions during batch tests (Kaldnes rings from reactor 2): (a) one-stage partial nitritation/Anammox; (b) with NO_2-N addition; (c) with ATU addition.

To confirm that the reaction, which is taking place in both reactors of pilot plant, is a reaction of simultaneous partial nitritation and Anammox, the ATU solution was used. It inhibited the first step of the whole process what resulted in no nitrite production and no reaction was observed in this bottle (Figure 4c).

Conclusions

Conducted studies demonstrated that the impact of temperature on Anammox or partial nitritation/Anammox processes is significant in the case of a sudden decrease in temperature. Then, the negative impact was registered and resulted in the decrease in nitrogen removal rates. If a longer-term adaptation period to lower temperatures is assured, a bacterial culture is able to utilise nitrogen at higher removal rates. It is in agreement with natural conditions where at low temperatures of marine sediments the Anammox reaction still proceeds.

In the single stage partial nitritation/Anammox process, higher values of ammonium removal rates were registered in reactor 1 where previously mainly the partial nitritation took place. It resulted in higher amounts of nitrifiers present in the first reactor. It could also result from the difference in the influence of temperature on partial nitritation and Anammox process. It seems to be that the partial nitritation is less sensitive than the Anammox bacteria to lower temperatures.

For simultaneous partial nitritation/Anammox processes the bottleneck in stable performance occurred to be the lack of produced nitrite. The addition of nitrite caused an increase of the reaction rates.

The presence of simultaneous partial nitritation/Anammox processes was confirmed by the usage of the ATU solution in the test run. That selective inhibitor of *Nitrosomonas* bacteria caused the inhibition of nitrite production. Therefore, the second step, Anammox process that requires both ammonium and nitrite, was stopped.

In a full-scale application, the importance of the temperature must be considered to find a compromise between high removal rates and the cost of heating.

Acknowledgement

We thank Monica Löwén and Master Students Dorota Sosnowska and Agnieszka Comber for all the help during experiments and Bengt Hultman for valuable discussions. Financial support from J. Gust. Richert Foundation and Swedish Institute is acknowledged.

References

Dalsgaard T. and Thamdrup B. (2002). Factors controlling anaerobic ammonium oxidation with nitrite in marine sediments. *Applied and Environmental Microbiology*, **68**(8), 3802–3808.

Dapena-Mora A., Van Hulle S. WH., Campos J. L., Méndez R., Vanrolleghem P. A. and Jetten M. (2004a). Enrichment of Anammox biomass from municipal activated sludge: experimental and modelling results. *J. Chem. Technol. Biotechnol.*, **79**, 1421–1428.

Dapena-Mora A., Campos J. L., Mosquera-Corral A., Jetten M. S. M. and Méndez R. (2004b). Stability of the Anammox process in a gas-lift reactor and a SBR. *Journal of Biotech.*, **110**, 159–170.

Egli K., Fanger U., Alvarez P. J. J., Siegrist H., van der Meer J. R. and Zehnder A. J. B. (2001). Enrichment and characterization of an Anammox bacterium from a rotating biological contactor treating ammonium-rich leachate. *Arch. Microbiol.*, **175**, 198–207.

Fdz-Polanco F., Villaverde S. and García P. A. (1994). Temperature effect on nitrifying bacteria activity in biofilters: activation and free ammonia inhibition. *Wat. Sci. Tech.* 30(11), 121–130.

Hao X, Heijnen J. J, and Van Loosdrecht M. C. (2002). Model-based evaluation of temperature and inflow variations on a partial nitrification-ANAMMOX biofilm process. *Water Res.* **36**(19), 4839-49.

Płaza E., Trela J. and Hultman B. (2002) Treatment of ammonium-rich waste streams with low content of organic matter. In: *Proceedings of the "Enviro 2002 IWA World Water Congress"*. Melbourne, 7–12 April 2002.

Surmacz–Górska J., Gernaey K., Demuynck C., Vanrolleghem P. and Verstreate W. (1996). Nitrification monitoring in activated sludge by oxygen uptake rate (OUR) measurements. *Wat. Res.*, **30**(5), 1228 – 1236.

Szatkowska B., Płaza E. and Trela J. (2004a). Nitrogen removal rates in the deammonification process: batch experiments. In: Proceedings of the "*4th IWA World Water Congress and Exhibition*", Marrakech, 19–24 September 2004.

Szatkowska B. (2004b). *Treatment of ammonium-rich wastewater by partial nitritation/Anammox in a biofilm system.* Royal Institute of Technology, Stockholm. TRITA-LWR.LIC 2023 (licentiate thesis).

Zhu, S. and Chen, S., (2002). The impact of temperature on nitrification rate of fixed film biofilters. *Aquaculture Eng.*, **26**: 221–237.

Van Dongen L. G. J. M., Jetten M. S. M. and van Loosdrecht, M. C. M. (2001). *The combined Sharon/Anammox Process, A sustainable method for N-removal from sludge water.* STOWA Report, IWA Publishing, London, UK.

Optimisation of parameters for efficient decolourisation of textile dye by white rot fungus

Awantika Dixit and Rashmi Sanghi

302, Southern Laboratories, Facility for Ecological and Analytical Testing, Indian Institute of Technology, Kanpur - 20 8016, Uttar Pradesh, India (E-mail: *awantika@iitk.ac.in*; *rsanghi@iitk.ac.in*)

Abstract The environmental problems related to industrial wastewater are becoming more and more complex with the increasing diversity of industrial products. Public demands for colour-free discharges to receiving waters have made decolourisation of a variety of industrial wastes a top priority. The textile dyeing wastewater pollutants are generally characterised by high concentrations of colour, COD, BOD, TOC (total organic carbon) and dissolved solids. However, more important is the presence of toxic dyes and heavy metals, which are persistent in nature. Many commonly used dyes break down into toxic amines, which are carcinogenic in nature and cause health hazards to humans. Azo-reactive dyes are the largest class of water-soluble synthetic dyes, with the greatest variety of colours and structures and are generally resistant to aerobic biodegradation processes. Under anaerobic bacterial reduction, they can form potentially carcinogenic compounds. Microbial decolourisation processes offer a complete cleanup of pollutants in a natural way as it reduces the colour components to carbon dioxide, ammonia and water by initiating cleavage of the bonds in the dyes rather than creating possible toxic fragments of dyes. Wood-rotting fungi have interesting properties in that they are capable of even degrading lignin, which is a polymeric structure with a lot of aromatic rings and thus can be proposed as a less expensive and less environmentally intrusive alternative. This paper reports the optimisation of parameters for better and efficient decolourisation of the difficult-to-decolourise reactive dye Remazol Brilliant Violet 5R (RV) using the pure strain of white rot fungus *Coriolus versicolor.* It was possible to substantially increase the dye decolourising activity of the fungus by carefully selecting the operational conditions such as media composition, glucose concentration, nitrogen source, initial dye concentration and sustainability of the fungus in continuous seven cycles of 7 days each. Since fungi decolourise dyes in secondary metabolism and dyes cannot be used as the primary carbon source for fungi growth, a primary carbon source such as glucose is necessary in fungal dye decolourisation. In the absence of glucose, almost all fungi have no decolourisation capability. Hence, an attempt is made to study the effect of glucose in basal medium whereby the presence of glucose enhances the rate of decolourisation and a concentration of 5.0 g/l was found to be sufficient to achieve the maximum decolourisation percentage.

Keywords *Coriolus versicolor*; decolourisation; degradation; Remazol Brilliant Violet

Introduction

Substantial volumes of water and chemicals are consumed during the wet processing of textiles. Wastewaters from these processes are the main route of environmental contamination. The textile dyeing wastewater (Ganesh *et al.*, 1994) pollutants are generally characterised by high concentrations of colour, COD, BOD, TOC (total organic carbon) and dissolved solids. But more important is the presence of toxic dyes and heavy metals, which are persistent in nature. Textile industry wastewater contains a mixture of different dyestuffs that are usually aromatic in structure with diverse functionalities and are difficult to degrade. The discharge of those coloured wastewaters into rivers and lakes leads to a reduction of sunlight penetration in natural water bodies, which in turn decreases both photosynthetic activity and dissolved oxygen concentration. This will create anaerobic condition thereby killing aerobic marine organisms. Colour removal from wastewater is as important as the removal of soluble colourless organic substances, which contribute the major function of chemical oxygen demand (Banat *et al.*, 1996). Dyes are designed to be resistant to light, water and oxidising agents and are difficult to degrade once released in the aquatic system (Nigam *et al.*, 2000). The decolourisation of industrial dyes is possible with conventional physico-chemical methods like adsorption, precipitation, chemical degradation, photodegradation but they are highly expensive which limits their application. Biodegradation by different microorganisms appears to be another possibility as these are environmental friendly and cost competent alternatives. In aerobic conditions, decolourisation generally occurs by adsorption of dyestuffs on bacteria rather than oxidation in aerobic systems. Some anaerobic microorganisms can biodegrade dyestuffs by azo-reductase activity. However, the effluent from biodegradation of dyestuff could be toxic (Brown and Devoito 1993, Chung and Stevens, 1993). These problems limit the full-scale application of microbial decolourisation. Several biotechnological approaches have been suggested by researchers as of potential interest towards combating this pollution source in an ecoefficient manner, including the use of bacteria or fungi, often in combination with physicochemical processes (Willmott *et al.*, 1998, Robinson *et al.*, 2001, Borchert and Libra, 2001, Beydilli *et al.*, 1998, Zissi and Lyberatos, 2001). By far the single class of microorganisms, most efficient in breaking down synthetic dyes is the white-rot fungi (WRF). These constitute a diverse eco-physiological group comprising mostly basidiomycetous fungi capable of extensive aerobic lignin depolymerisation and mineralisation. This property is based on the WRF's capacity to produce one or more extracellular lignin-modifying enzymes (LME), which, thanks to their lack of substrate specificity, are also capable of degrading a wide range of xenobiotics. White rot fungi (WRF) can degrade a wide variety of recalcitrant compounds like lignin by their extracellular lignolytic enzyme system. The enzymes which allow WRF to degrade lignin, also allows them to degrade a wide range of hazardous xenobiotics (Pointing, 2001). WRF offer significant advantages as the ligninolytic enzymes produced by them are substrate nonspecific, therefore can degrade wide variety of recalcitrant compounds, especially complex aromatic pollutants like all the way to carbon dioxide. WRF do not require preconditioning to particular pollutants, because enzyme secretion depends on nutrient limitation, either nitrogen or carbon rather than the presence of a pollutant. In addition, because the enzymes are extracellular, the substrate diffusion limitation into the cell, generally observed in bacteria, is not encountered and they enable the WRF to tolerate higher concentration of pollutants such as cyanide. Besides, the fungi can even degrade very insoluble chemicals and hazardous environmental pollutants.

Since fungi decolourise dyes (Cripps *et al.*, 1990) in secondary metabolism and dyes cannot be utilised as the primary carbon source for fungi growth, a primary carbon source such as glucose is necessary in fungal dye decolourisation. Increasing the glucose concentration in the medium improves fungal decolourisation (Zhang *et al.*, 1999). In the absence of glucose, almost all fungi have no decolourisation capability (Borchert *et al.*, 2001).

The present work aims to optimise the conditions for enhancing the decolourisation of a difficult to degrade reactive dye using *Coriolus versicolor* and investigates its potential to degrade the dye in repeated batch kinetic experiments.

Material and Methods

Chemicals

All chemicals used were of analytical reagent grade. The reactive Remazol Brilliant Violet dye (RBV), a vinyl sulphone dye, was purchased from Colour-chem Ltd, Mumbai. It is a commercial dye and was used without further purification.

Growth Media

Two types of media compositions were used in this study namely *Media 1* (M1) and *Media 2* (M2). *M1* contained 10 g/l glucose and 5 g/l malt extract. *M2* contained 5 g/l glucose, 2 g/l KH_2PO_4, 1 g/l peptone, 0.5 g/l $MgSO_4.7H_2O$ and 0.099 g/l $CaCl_2$.

Growth and Maintenance of the fungus

Coriolus versicolor was obtained from Institute of Microbial Technology, Chandigarh. The strain was maintained at 4°C on malt agar slants. The growth medium used contained 10 g/l glucose and 5 g/l malt extract. The media was always autoclaved and cooled before use and the initial pH was 5.6. All growth experiments were performed at room temperature i.e., 25–30°C.

Decolourisation studies

The Batch kinetic reactors

The batch reactor consisted of 500 ml Erlenmeyer flasks fitted with cork and the glass tubing. The air was purged into the reactors by an aquarium pump through glass tubing. The aqua pump was connected to the flasks via another Erlenmeyer flask that contained 10 N H_2SO_4 solution to prevent contamination. The volume of the media taken in the reactor was 400 ml. The temperature was 25°C and the unadjusted pH of the media was 5.4 which reduced down to 3.2 during the course of decolourisation. The reactor was operated in repeated or sequential batch mode. The media was replaced with fresh media containing dye when the old media was decolourised almost completely, the biomass being the same through out the kinetic studies. The samples were daily taken out, centrifuged, the clarity of the samples were assessed visually and then the supernatant was pipette out and analysed for different parameter.

Experimental Analysis

The decolourisation of Remazol Brilliant violet was measured at λ_{max} (560 nm) using UV-Vis Spectrophotometer (Perkin-Elmer, lambda 40). The standard curve was obtained for different dye concentrations. The decolourisation was monitored by percentage reduction in the absorbance at 560 nm. The absorbance spectrum was monitored daily for abiotic control, but it did not show any change. The COD measurements were done according to the standard

APHA procedure (American Public Health Association, 1998), using HACH COD digestor. The pH of the reactors was monitored with the digital pH meter (MK VI Systronics). The glucose concentration is determined using the dinitrosalicylate method for glucose determination as described by Plummer (1987).

Results and discussion

Comparative study with different media MI and M2 in sequential batch kinetic reactors

The decolourising capabilities of two medium M1 and M2 were compared in sequential batch reactors. The sequential studies were done for seven cycles of 7 days each. At every cycle when the old media was decolourised fully, each media was replaced with fresh media containing dye, the biomass being the same throughout the kinetic studies.

The results showed that in the early stages of the studies, i.e., in first four cycles media MI has better decolourising capability than M2. Even the decolourisation rate is also high, but gradually this efficiency is decreased and M2 seems to be supersede M1 as far as the longevity of the decolourisation capability of the fungus is compared (Figure 1). The COD reduction also showed the corresponding trend in the reduction of COD. This indicated that the presence of nutrient supplement in M2 makes the fungus culture healthy and capable of sustaining even through adverse nutrient condition. Since M1 was found to be better then M2 in the first four cycles, all further experiments were conducted with M1 only.

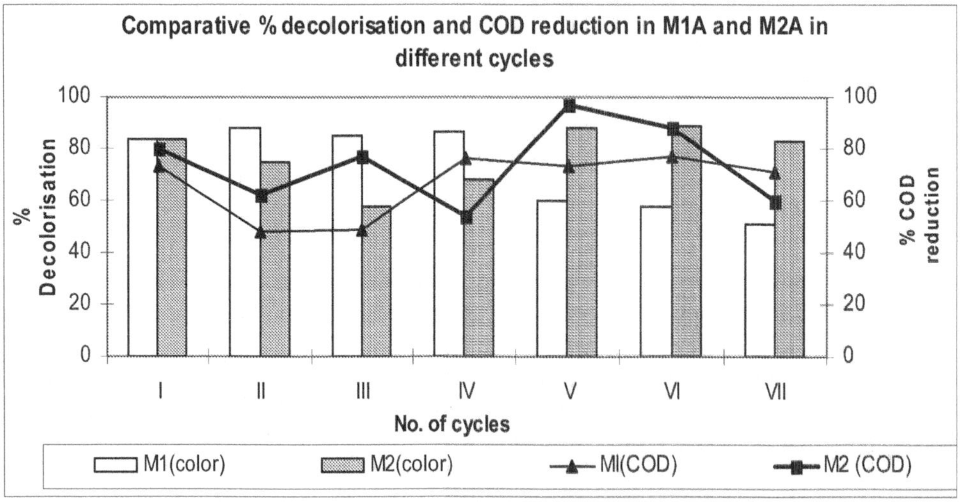

Figure 1 The comparative decolourisation of RBV in M1 and M2.

Role of carbon source

Using M1, three types of reactor were set up, first, the reactor in which the media was replaced at the beginning of each cycle with the fresh spike of dye (MIA). Second was the reactor in which only glucose supplement was given in the beginning of the cycle besides the new spike of dye (M1B) and the third, in which only the fresh spike of dye was given in each cycle and it was not supplied with either new media or the glucose supplement (M1C). Seven such cycles were run for 7 days each.

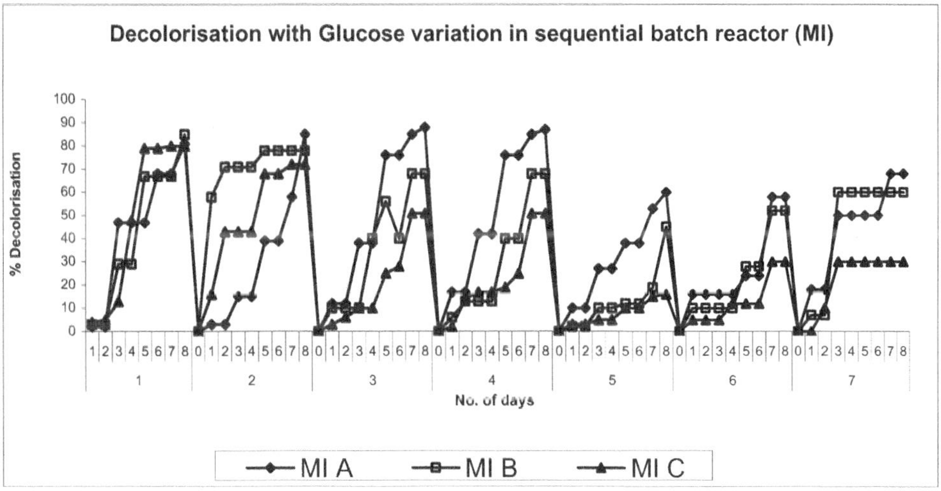

Figure 2 Graph showing comparative decolourisation in MIA, MIB and MIC.

The results showed that the decolourisation in MIA and MIB is comparable, indicating the vital role of glucose supplement in the decolourisation of the dye by the fungus. Infact, MIB seems to be more efficient in the last two cycles. While in the case of the reactor MIC, the decolourisation decreased gradually and almost ceased, as it was not supplied with any nutrient supplement. This comparative study indicates the importance of the carbon source in the decolourisation of the dye, as revealed from the results. It is the key factor that determines not only the extent but also the rate of decolourisation. Moreover the results also indicates that just by optimising the carbon source we can enhance the capability of the fungus, *C. versicolor* , to decolourise the dye for longer time duration and for many cycles as done in the above study. There was corresponding changes observed in the reduction of COD as well.

Studies with glucose variation

Earlier reports on decolourisation studies (Radha *et al*., 2005) indicated the necessity of glucose to enhance the decolourisation process. Hence, an attempt is made to study the effect of glucose in basal medium on the decolourisation and for this, reactors were set up that ran for 8 days. The initial rates of colour reduction significantly improved, as the concentration of glucose was increased from 1 to 5 g/l. The results of the studies show that the presence of glucose enhances the rate of decolourisation and a concentration of 5.0 g/l was found to be sufficient to achieve the maximum decolourisation percentage (Figure 3) of almost 95%. As the concentration of glucose was increased from the optimum level, the decolourisation rate decreased which may be due to the metabolic pathways, where there is more prevalent growth of the fungus rather than the utilisation of the glucose by the fungus for the decolourisation of the dyes (Radha *et al*., 2005).

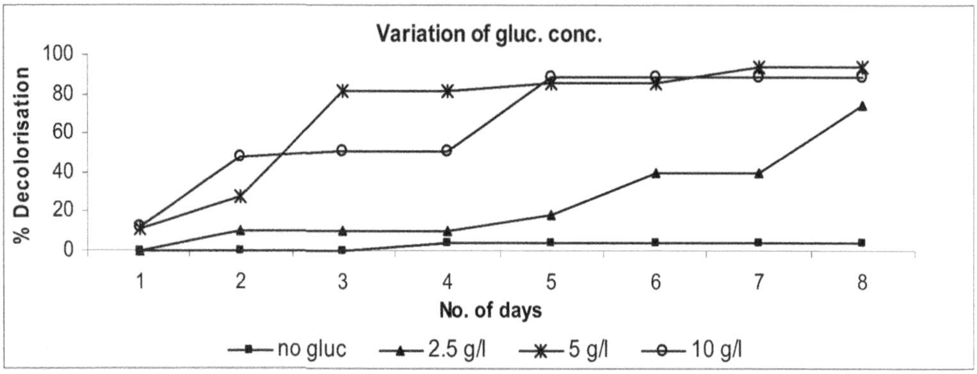

Figure 3 Graph showing the decolourisation using different glucose concentration at optimum temperature.

Effect of nitrogen

Similarly, an optimum concentration of nitrogen for the enhancement of decolourisation of dyes was studied using urea and peptone as nitrogen source keeping the nitrogen content in equal amount. The study showed that peptone is a better alternative over urea as it gave the 95% decolourisation whereas the urea could decolourise only 15% of the dye (graph not given).

Effect of temperature

To study the effect of temperature, experiments were set up, in which the reactors were run at 25°C and 35°C,under the controlled conditions. The results shows, that the temperature also has a vital role to play in the process of decolourisation. As seen in Figure 4, the decolourisation was 82% on third day, 86% on fifth day and 94% on the seventh day, while in the experiment done at 35°C, the maximum decolourisation that could be achieved was 20% which was observed on the fifth day but there was no further reduction in the colour was found even on the fourteenth day, i.e. the last day of the set up. Thus for all the studies conducted 25°C was taken as the operational temperature.

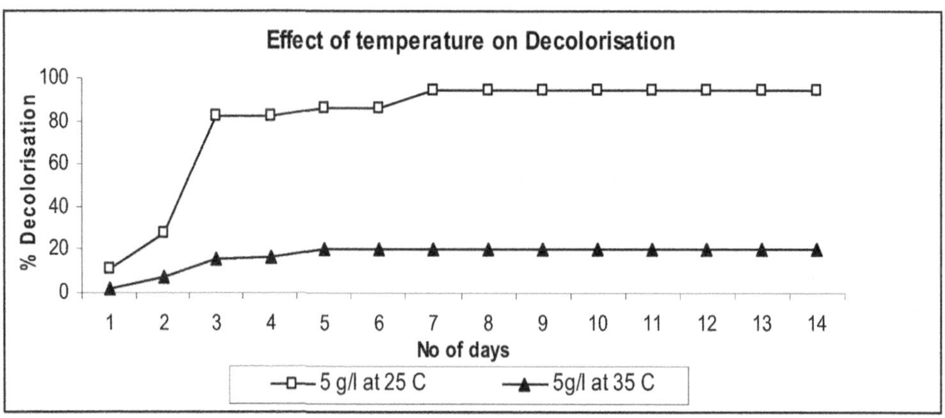

Figure 4 Graph showing the effect of temperature on the decolourisation.

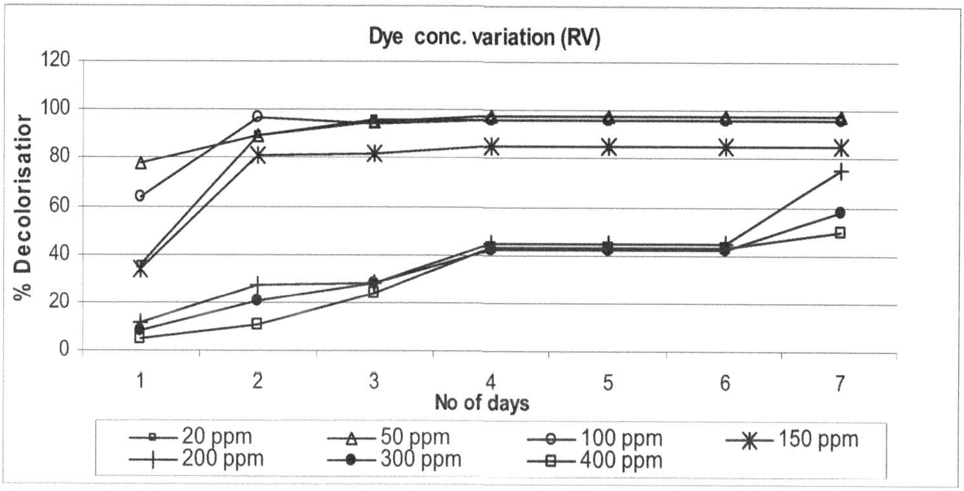

Figure 5 Graph showing the effect of dye concentration on the decolourisation.

Effect of initial dye concentration

Dye concentration plays an important role in deciding the extent and rate of decolourisation. As in the case of RBV dye, concentrations 20, 50, 100 ppm showed almost the same pattern of decolourisation since the beginning of the experiment and reached to almost 95% decolourisation on third day of the set-up and up to 98–100% on the seventh day. Dye concentration of 150 ppm seems to be showing a similar pattern but the extent of decolourisation was a little less as compared to earlier cases, though it could remove up to 80% dye on third day. Beyond 150 ppm the decolourisation rate and the percentage dye removal was found to be slow and less. The dye concentration above 150 mg/l starts becoming toxic for the fungus, which leads to the reduced efficiency of the fungus to decolourise the dye. Thus for all the experiments reported above dye concentration of 100 mg/l was taken.

Spectral Changes

As shown in Figure 6a,b, spectral changes were observed during the course of decolourisation indicating that there is degradation occurring simultaneously leading to decolourisation. The spectra shows the shift in the characteristic peak of the dye, from 560 nm to 520 nm on third day, 516 nm on fourth day indicating the formation of some intermediate (Figure 6a), which is then cleaved by auto mechanism of the fungi, as on the fifth day no peak was observed.

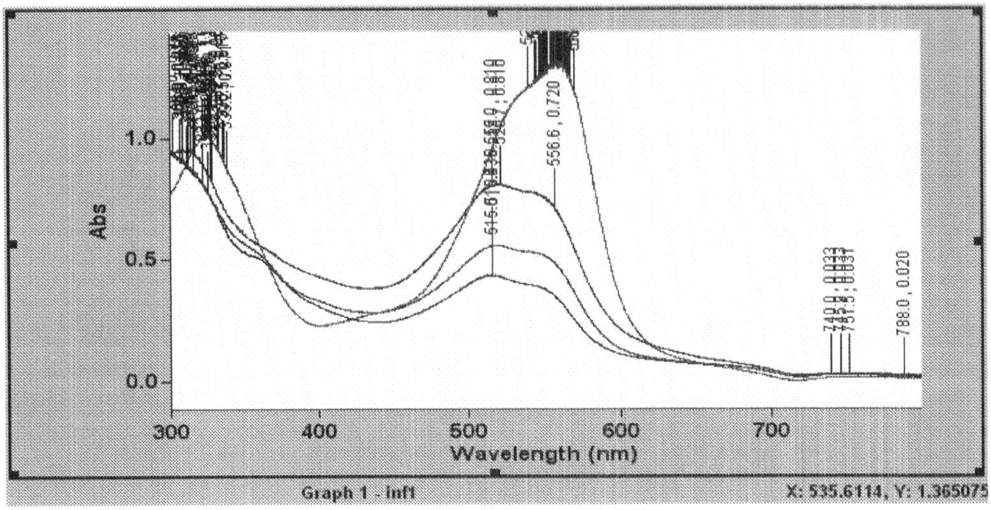

Figure 6 (a) Spectrum showing the peak shift.

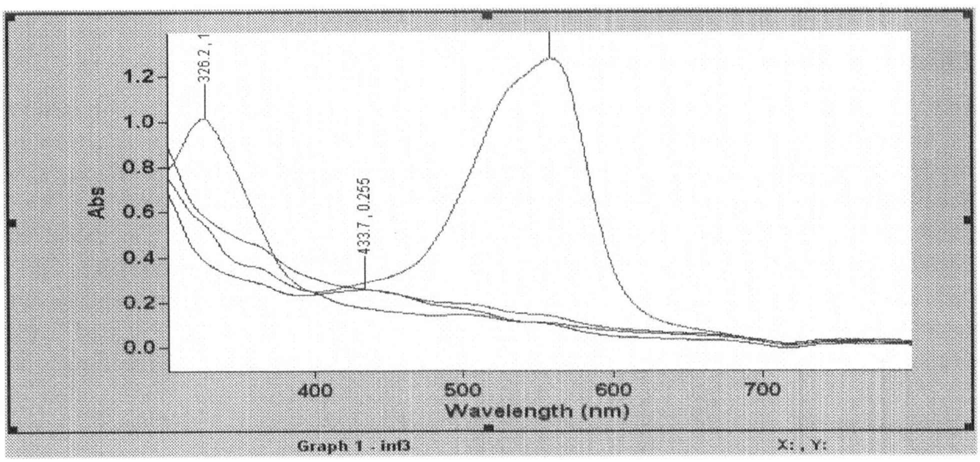

Figure 6 (b) Spectrum showing the cleavage of the peak.

Conclusions

The studies conducted shows that by optimising the operational parameter, the efficiency of the white rot fungus *Coriolus versicolor* to degrade the dye RBV can be much enhanced, using the same biomass. The vital role of the carbon source and temperature has also been discussed in the decolourisation process. The concentration of the dye also matters as far as the rate and extent of dye removal is concerned. These studies show that there lies a greater potential in this fungus to degrade and decolourise the reactive azo dyes.

References

American Public Health Association (1998) *Standard Methods for the Examination of Water and Wastewater*, 20th edn, American Public Health Association/American Water Works Association/Water Environment Federation, Washington DC, USA.

Banat I.M., Nigam P., Singh D. and Merchant R. (1996). Microbial decolourisation of textile dye containing effluent: a review: *Bioresource Technology*, **58**, 217–227.

Beydilli M.I, Pavlostathis S.G and Tincher W.C. (1998). Decolourisation and toxicity screening of selected reactive azo dyes under methanogenic conditions. *Water Science Technology*, **38**, 225–232.

Borchert M. and Libra J.A. (2001). Decolourisation of reactive dyes by the white rot fungus *Trametes versicolor* in sequencing batch reactors. *Biotechnology Bioengineering*, **75**, 313–321.

Brown M.A. and Devoito S.C. (1993). Predicting azo dye toxicity. Critical Review, *Environment Science and Technology*, **23**, 249–324.

Chung K.T. and Stevens S. (1993).Decolourisation of azo dyes by environmental micro-organism and helminthes, *Environmental Toxicology and Chemistry*, **12**(11), 2121–2132.

Cripps C., Bumpus J. A. and Aust S. D. (1990). Biodegradation of azo and heterocyclic dyes by *P. chrysosporium, Applied and Environmental Microbiology*, **56**(4), 1114–1118.

Ganesh, R., Boardman G.D. and Michelsen D. (1994). Fate of azo dyes in sludges, *Water Research*, **28**(6), 1367–1376.

Nigam P., Armon G., Banat I.M., Singh D. and Merchant R. (2000). Physical removal of textile dyes from effluents and solid state fermentation of dye adsorbed agricultural residues, *Bioresource Technology*, **72**, 219–226.

Plummer D.T. (1987). *An introduction to practical biochemistry*, (3rd Edn).

Pointing S.B. (2001) *Exploiting the versatile ligninolytic system of white rot fungi. In: Pointing SB, Hyde KD (Eds) Bio exploitation of filamentous fungi. Fungal diversity research series 6.* Fungal Diversity press. The univ of Hong kong, Hongkong, 253–290.

Radha K.V, Regupathi I., Arunagiri A. and Murugesan T. (2005). Decolourisation studies of synthetic dyes using *Phanerochaete chrysosporium* and their kinetics, **40**(10), 3337–3345.

Robinson T., McMullan G., Marchant R. and Nigam P. (2001). Remediation of dyes in textile effluent: a critical review on current treatment technologies with a proposed alternative. *Bioresource Technology*, **77**, 247–255.

Willmott N., Guthrie J. and Nelson G. (1998). The biotechnology approaches to colour removal from textile effluent, *Journal of Society of Dyers and colourage,* **114**, 38–41.

Zhang F.M., Knapp J.S. and Tapley K.N. (1999). Decolourisation of cotton bleaching effluent with white rot fungus, *Water Res*earch, **33** (4) 919–928.

Zissi U. and Lyberatos G., (2001) Partial degradation of *p*-aminoazobenzene by a defined mixed culture of *Bacillus subtilis* and *Stenotrophomonas maltophilia. Biotechnology Bioengineering*, **72**, 49–54.

Influence of sulphanilamide in wastewater on Anammox process performance

A. Bąkowska*, E. Płaza and K. Mędrzycka***

* Department of Fats and Detergents Technology, Chemical Faculty, Gdansk University of Technology, Narutowicza 11/12, 80 952 Gdansk, Poland (E-mail: *anioa@chem.pg.gda.pl*; *krystyna@chem.pg.gda.pl*)

** Department of Land and Water Resources Engineering, Royal Institute of Technology, 100 44 Stockholm, Sweden (E-mail: *elap@kth.se*)

Abstract The influence of selected sulphanilamide (SA) on the Anammox process was investigated. The short-term (14 hours) and long-term (90 day) exposure of SA on mixed Anammox culture were carried out. The batch-test performances result in the N-removal efficiency decrease along with the increase of the SA concentration. The N-removal efficiency at the SA concentration: 10, 20, 40, 80, 100 and 1000 mg SA/dm^3 was lower than in blank test by about 10%, 20%, 23% 28%, 40% and 46%, respectively. The long-term exposure study were conducted in continuous-flow bioreactors. The test concentrations of SA were: 0 (blank), 20 and 1000 mg SA/dm^3. The results confirmed that SA inhibits the Anammox process. The higher the SA concentration was, the stronger inhibition of microorganisms' activity was observed. Thus, we conclude that sulphanilamide can cause problems during treatment of wastewater using Anammox process.

Keywords Anammox; nitrogen removal; sulphonamide

Introduction

ANaerobic AMMonium OXidation (Anammox) bacteria were discovered in late 1990s and have been used in development and application of a new method for sustainable nitrogen removal from wastewater – the Anammox process. The Anammox process consists in simultaneous nitrite and ammonium removal under anaerobic condition with no external carbon source supply. Ammonium and nitrite are directly converted to the gaseous nitrogen, which is the Anammox process final product. The process is described by equation 1 (Van de Graaf *et al.*, 1996; Strous *et al.*, 1998).

$$NH_4^+ + 1.32\,NO_2^- + 0.066\,HCO_3^- + 0.13\,H^+ \rightarrow$$
$$1.02\,N_2 + 0.26\,NO_3^- + 2.03\,H_2O + 0.066\,CH_2O_{0.5}N_{0.15} \qquad (1)$$

The partial nitritation coupled with the Anammox process can be applied for high-strength ammonium wastewater treatment. The method can be used for treatment of high-ammonium concentrated streams generated during dewatering of digested sludge or during landfilling the waste. It also could be used for treatment of many industrial side streams (Keller *et al.*, 1997; Banas *et al.*, 1999; Hippen *et al.*, 2001; Carrera *et al.*, 2003)

However, the Anammox bacteria appeared to be very sensitive for the process conditions and the maintenance of the stable Anammox process performance is very crucial problem. Thus, the researches on Anammox inhibition were carried out. The inhibition tests were focused on factors affecting the process such as: dissolved oxygen concentration (DO), temperature, and pH and substrates and products concentrations. The reversible inhibition influence of the oxygen on the Anammox process was reported and oxygen concentration in the range 0.2–0.4 mg O_2/dm^3 was found to be the optimal DO concentration for the process performance (Szatkowska *et al.*, 2004). In the temperature range of 6–43°C (optimum at 37°C) the Anammox bacteria are active. The pH level for the optimum temperature is in the range from 6.5 to 8.5 (Egli *et al.*, 2001). In the case of proper Anammox process performance the increase of initial pH value by at least one unit is observed (Gut *et al.*, 2005). The nitrite nitrogen concentration exceeding 70 mg/dm^3 irreversibly inhibits the process (Van Dongen *et al.*, 2001). The nitrite to ammonium ratio (NAR) in the Anammox process is 1.3 (Equation 1); however, another NAR value (1.43) as more suitable for the process has been reported by Wyffels *et al.* (2003).

Wastewater and groundwater may contain many different chemical substances, which can inhibit microorganisms' growth and cause various troubles during wastewater treatment. These harmful substances, depending on their physical and chemical properties, can be not readily biodegradable and remain toxic during the entire wastewater treatment process (Halling-Sørensen *et al.*, 2000). Examples of such compounds are sulphonamides that were found in wastewater treatment plant effluents and surface water (Hartig *et al.*, 1999; Hirsch *et al.*, 1999). Sulphonamides adsorb weakly to both to soil and activated sludge (Huang *et al.*, 2001). Sulphonamides were reported as not readily biodegradable substances (Ingerslev and Halling-Sørensen 2000; Huang *et al.*, 2001).

There were no reports on investigation describing how pharmaceuticals contained in wastewaters could influence the Anammox process. The choice of sulphanilamide (SA) was based on the results from previous study, where the compound was found to be resistant to biodegradation. Thus, sulphanilamide remaining unchanged in wastewater is likely to cause problems during wastewater treatment (Mędrzycka *et al.*, 2004). The main aim of this study was to investigate the sulphanilamide influence on the Anammox process during short-term (14 hour) and long-term (90 day) exposure to different concentrations of SA. The wastewater selected for investigation was high-strength ammonium supernatant from dewatering of the sludge.

Methods

Chemicals

For the study there were used chemicals such as sulphanilamide (SA), sodium nitrite ($NaNO_2$) and ammonium chloride (NH_4Cl).The chemical structure and some properties of sulphanilamide are shown in Table 1.

Table 1 Characteristics of the sulphanilamide (SA) used in the study

Nomenclature	4-amino-benzenesulphonamide
Molecular weight	172.21 g/mol
Molecular formula	$C_6H_8N_2O_2S$
Structural formula	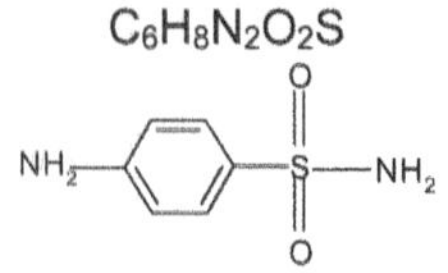
Solubility	7.5 g/dm^3
pH	5.8–6.1 (0.5% solution, 20°C)
Form	White powder

SA was purchased from Sigma-Aldrich Co. All chemicals were analytical reagent grade (at least 99% purity) and were obtained from commercial sources.

Source of Anammox anaerobic culture

Kaldnes rings with well-established Anammox bacterial culture and supernatant from dewatering of digested sludge were taken from the half-technical scale pilot plant of deammonification process (partial nitritation/Anammox system) at Himmerfjärden WWTP in Grödinge (Sweden). Kaldnes rings ensured support for growth of the Anammox bacteria as well as hold them up in the system preventing the bacteria washing out. The supernatant was decanted, filtrated and stored in cooling room before use. The average total nitrogen concentration in the supernatant during storage was about 364 mg N/dm^3. Characteristics of the supernatant is shown in Table 2.

Table 2 Characteristics of supernatant

	NH_4^+-N [mg/dm^3]	NO_2^--N [mg/dm^3]	NO_3^--N [mg/dm^3]	COD [mg O_2/dm^3]
Average	346.2	3.2	14.1	297.8
Standard deviation	12.3	1.8	5.8	1.7

The nitrite to ammonium ratio in the supernatant was too low (NAR=0.012) in relation to requirements (NAR=1.3) therefore the NO_2^--N supply was needed.

Batch tests

To estimate influence of sulphanilamide on the Anammox process efficiency the series of batch tests were run in seven bottles (0.5 l working volume) in the closed reaction system. The test bottles equipped with magnetic stirrers were filled in 50 % of their volume with the Kaldnes rings. The parameters: DO, pH, temperature were kept stable at the optimum level. The initial concentrations of NH_4^+-N and NO_2^--N were set for each test. Required initial nitrogen forms concentration (about 150 mg NH_4^+-N/dm^3 and about 65 mg NO_2^--N/dm^3) were obtained by addition of NH_4Cl and $NaNO_2$ solutions to the liquor taken from the Anammox reactor of pilot plant at Himmerfjärden WWTP. The tests were run for 14 hours and there was investigated the effect of differ sulphanilamide concentrations: 0, 10, 20, 40, 80, 100 and 1000 mg SA/dm^3.

Long-term exposure study

The long-term exposure study was carried out within 90 days to estimate long-term influence of sulphanilamide on Anammox process. The experimental set-up was consisted of three reactors of one litre working volume. Each reactor was equipped with mixers to keep homogeneity of reactor volume. A water bath kept temperature in the range of 30–33°C. In order to avoid oxygen diffusion from the air all reactors were covered. Each reactor was filled in 50% of working volume with Kaldnes rings carrying Anammox bacterial culture. The reactors were operated simultaneously and were continuously fed with sulphanilamide – enriched feeding wastewater. The feeding wastewater was prepared using the supernatant as a source of NH_4^+-N and $NaNO_2$ as a source of NO_2^--N. To obtain stoichiometric, initial concentration of 100 mg NH_4^+-N/dm^3 and 130 mg/dm^3 NO_2^--N the supernatant was diluted with tap water and further, the required volume of $NaNO_2$ solution was added. After first 65 days of the process run the initial concentrations of ammonium and nitrite forms were increased up to 150 mg NH_4^+-N/dm^3 and 195 mg NO_2^--N/dm^3. The change was made in order to investigate the process performance at higher ammonium and nitrite concentrations, usually observed in supernatant. The effect of two concentrations of sulphanilamide (20 mg SA/dm^3 and 1000 mg SA/dm^3) was investigated during the whole operating period. The blank reactor was fed with sulphonamide-free feeding wastewater.

Analytical procedures

Samples were collected twice a week both, from inlet and outlet of each reactor during the long-exposure study and once per an hour during batch test performance. The samples were analysed for ammonium nitrogen, nitrite and nitrate nitrogen as well as total nitrogen with TECATOR – AQUATEC 5400 ANALYZER (flow injection system based on Vis-spectrophotometry). The process parameters such as pH, DO and conductivity were measured once per every day in long-term experiment and once per one hour during batch tests.

Results and discussion

This paper reports results obtained during the batch test study and the three-month exposure of the Anammox bacteria to the presence of sulphanilamide.

In the batch-test study the process efficiency during short-exposure of the mixed Anammox culture to sulphanilamide was estimated. The N-removal efficiency was calculated as a percent of the total inorganic nitrogen removal. The Figure 1 demonstrates the results obtained during 14 hour exposure of Anammox bacteria to different sulphanilamide concentrations, respectively. During batch tests the temperature in reactors was kept at 31.5°C. The pH level was not corrected and it was ranging from 7.43 to 7.48. DO level was in the range from 0.27 to 0.35 mg O_2/dm^3.

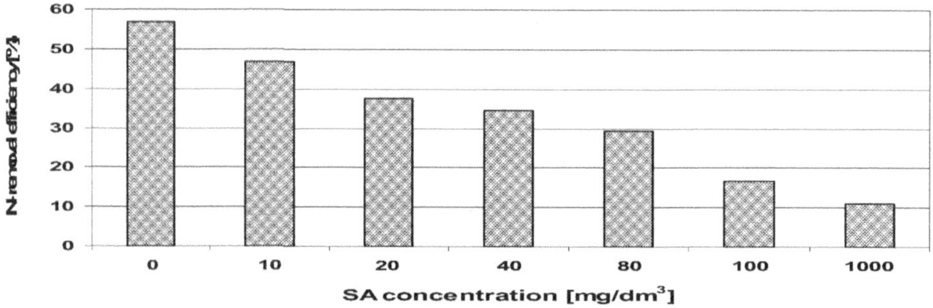

Figure 1 N-removal efficiency in 14-hour batch tests.

The decrease in microorganism activity was observed along with the increase of SA concentration. The higher was the SA concentration, the lower N-removal efficiency. The N-removal efficiency in presence of 20 mg SA/dm^3 and in presence of 1000 mg SA/dm^3 was about 37% and 11%, while in the blank reactor it was about 57%.

The long-term exposure study was divided into two periods: first, 65 day period; second, 25 day period. The process was being operated under condition demonstrated in Tables 3 and 4. The oxygen-limited conditions (0.2–0.4 mg O_2/dm^3) and constant temperature level (30–31°C) were kept for the proper process performance. The hydraulic retention time (HRT) was about 2.1 days during the experiment performance.

Table 3 Operational parameters and experimental results for the first period of long-term exposure study (65 days)

Parameter	Reactor 1 Blank	Reactor 2 20 mg SA/dm^3	Reactor 3 1000 mg SA/dm^3
Influent inorganic N [mg/dm^3]	235.6±21.9	228.9±24.4	236.4±21.0
Effluent inorganic N [mg/dm^3]	96.9±19.1	99.0±15.3	142.7±31.7
Dissolved oxygen [mg/dm^3]	0.29±0.09	0.33±0.09	0.33±0.06
pH influent	7.95±0.24	7.99±0.15	8.02±0.17
pH effluent	8.1±0.24	7.87±0.22	7.68±0.31

Table 4 Operational parameters and experimental results for the second period of long-term exposure study (25 days)

Parameter	Reactor 1 Blank	Reactor 2 20 mg SA/dm^3	Reactor 3 1000 mg SA/dm^3
Influent inorganic N [mg/dm^3]	367.7±11.8	357.9±8.9	364.7±10.9
Effluent inorganic N [mg/dm^3]	62.6±9.95	73.7±15.5	240.0±45.5
Dissolved oxygen [mg/dm^3]	0.31±0.04	0.33±0.03	0.35±0.05
pH influent	7.76±0.19	7.75±0.08	7.86±0.14
pH effluent	8.29±0.21	7.63±0.30	7.33±0.20

The pH-increasing tendency was observed only in blank reactor while in the presence of SA, pH slightly decreased (0.1–0.53). One of the factors that confirm proper Anammox process performance is the increase of pH during process run (Gut *et al.*, 2005). As it can be seen from the tables 3 and 4 it takes place only in the reactor 1 (no sulphanilamide), while in reactors 2 and 3 pH decreases. This may suggest that Anammox process has been disturbed by sulphanilamide. However, the results of nitrogenous compounds conversion can give sufficient proof that sulphanilamide inhibits Anammox process. In Figures 2–4 the results from long-term experiments are shown. The NH_4^+-N, NO_2^--N, NO_3^--N concentrations both in the influent and in the effluent from the reactors are presented.

The initial total nitrogen (N_{tot}) concentration in all reactors was much lower than in feeding wastewaters. Thus, the observed increase of total nitrogen concentration within first six days occurs due to the gradual enriching of the liquor (taken from the Anammox reactor with Kaldnes rings) with feeding wastewater.

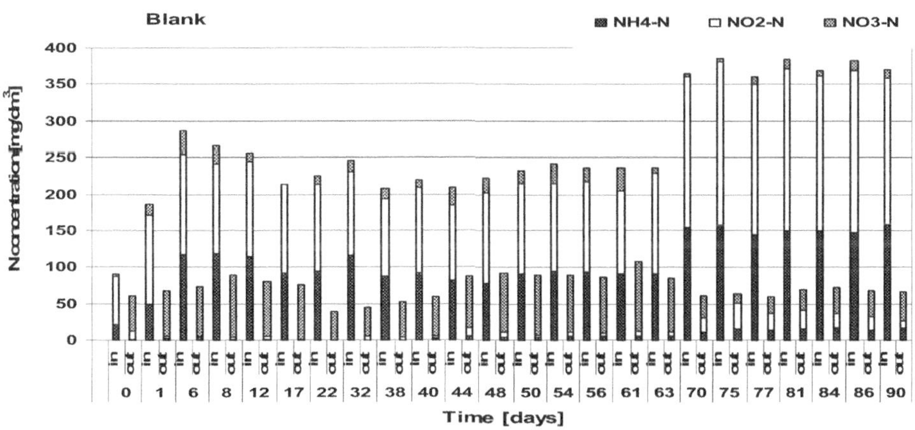

Figure 2 Profiles of nitrogen forms in reactor 1 during Anammox process performance (without SA addition).

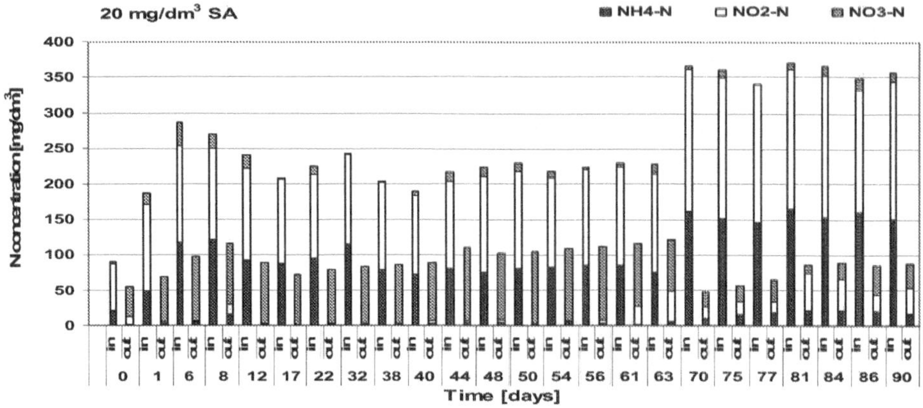

Figure 3 Profiles of nitrogen forms in reactor 2 during Anammox process.

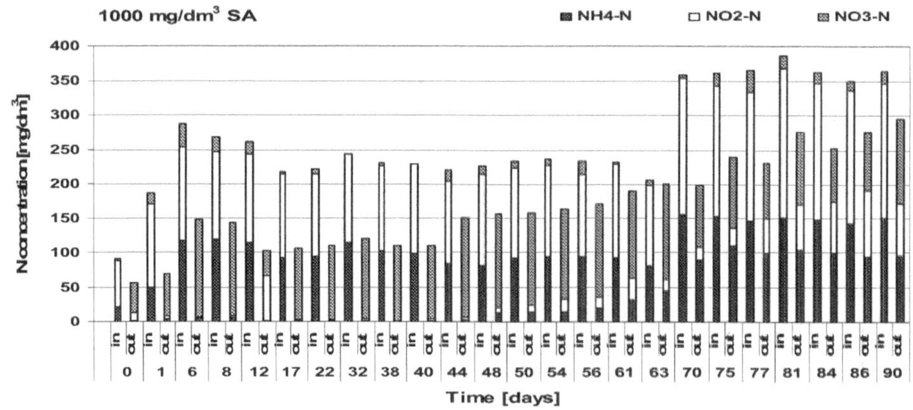

Figure 4 Profiles of nitrogen forms in reactor 3 during Anammox process.

Generally, one can state that fluctuations of the total nitrogen concentration between 6 and 63 days (Figures 2–4) may be caused by sorption and desorption processes on the biofilm surface. The changes of N-removal efficiency (percent of total inorganic N-removal) during long-term exposure study are presented in Figure 5.

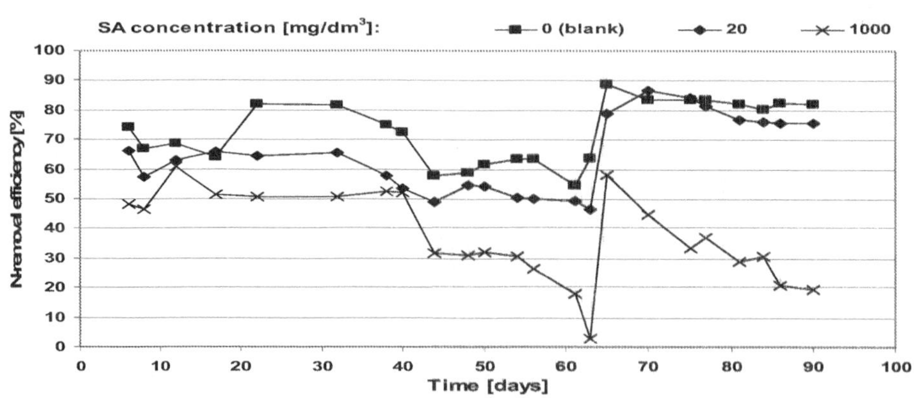

Figure 5 N-removal efficiency during long-term exposure study.

During the first period the Anammox culture was fed with feeding wastewater containing 100 mg NH_4^+-N/dm^3 and 130 mg NO_2^--N/dm^3 and the total initial concentrations in all reactors were fluctuated around about 230 mg N_{Tot}/dm^3 (Table 3). The nitrite to ammonium ratio in all reactors was about 1.3. In the first period the almost complete disappearing of NH_4^+-N and NO_2^--N is observed. As a consequence the total nitrogen concentration decreases and simultaneously nitrates (NO_3^--N) are produced. In reactor 1 the nitrate production had fixed about 22% of the total nitrogen and in the presence of sulphanilamide it had fixed about 34% and 46% of N_{tot} (reactor 2 and 3, respectively). According to the Anammox process reaction (Equation 1) the calculated theoretical nitrate production should not exceed 11.2% of initial nitrogen amount. Basing on these results one can state that beside the Anammox process, in all reactors also nitrification reaction proceeds.

During the first period the N-removal efficiency in reactor 1 and reactor 2 were fluctuated (Figure 5). Only in the reactor 3 the process efficiency has rapidly dropped after 40th day of run (up to 3 % removal). The average N-removal efficiency was about 67% in reactor 1, and in reactors 2 and 3 it was about 56% and 38%, respectively (Figure 5).

After the increase of NH_4^+-N and NO_2^--N concentrations (during the second period the N_{tot} in all reactors was about 360 mg/dm^3) the N-removal efficiency increased rapidly in all reactors (Figure 5). Within this period the effluents' composition from reactor 1 (blank) and reactor 2 was similar (Figures 2 and 3). The effluent from reactor 1 contained about 13 mg NH_4^+-N/dm^3 and about 19.4 mg NO_2^--N/dm^3 and the effluent from reactor 2 contained 16 mg NH_4^+-N/dm^3 and 26 mg NH_4^+-N/dm^3. In the effluents from reactor 3 the ammonium nitrogen concentration was about 100 mg NH_4^+-N/dm^3, whereas the nitrite concentration about 96 mg NO_2^--N/dm^3. The NO_3^--N production exceeded 11.2% only in reactor 3 (1000 mg SA/dm^3) and it was about 22%. In both reactor 1 and second reactor the nitrate production consumed about 6% of the total nitrogen. During the second period ammonium and nitrite forms were removed with efficiency higher than during first period. The Anammox process remained stable in reactor 1 and in reactor 2 during the whole 25 day period (Figure 5). It can be said that during long-term exposure to 20 mg SA/dm^3 the Anammox process was slightly disturbed. N-removal efficiency obtained in reactor 2 was lower by about 14% than in reactor 1 (blank). However, in the presence of 1000 mg SA/dm^3 (reactor 3), the Anammox process was accompanied by nitrification process.

Conclusions

The results obtained from batch tests and long-term exposure experiments confirm hypothesis that sulphanilamide inhibits the Anammox process. The higher the SA concentration, the greater inhibition effect was observed. It has been stated that significant decrease of nitrogen removal is not an immediate response to sulphanilamide presence and depending on SA concentration it can occur after time-lag. However, microorganisms may acclimate to SA presence in the process's environment and along with time the inhibition effect can be reduced. Hence, further research on Anammox and nitrification processes in the presence of sulphanilamide should be continued.

Acknowledgements

Great appreciation goes to the Swedish Institute (Grant No. 01882/2004 382), SYVAB AB, VA-forsk and Royal Institute of Technology in Stockholm for financial support. We are also grateful for Beata Szatkowska, Luiza Gut, Maja Długołęcka, Monica Löwen and Jan Bosander for assistance during this study.

References

Banas J., Płaza E.,Styka W. and Trela J. (1999). SBR technology used for advanced combined municipal and tannery wastewater treatment with high receiving water standards. *Wat. Sci. Technol.*, **40** (4–5), 451–458.

Carrera J., Baeza J. A., Vicet T. and Lafuente J. (2003). Biological nitrogen removal of high-strength ammonium industrial wastewater with two sludge system. *Wat. Res.*, **37** (17), 4211–4221.

Egli K., Fanger U., Alvarez P. J. J., Siegrist H., van der Meer J. R. and Zehnder A. J. B. (2001). Enrichment and characterization of an Anammox bacterium from a rotating biological contactor treating ammonium-rich leachate. Arch. Microbiol., **175**, 198–207.

Gut L., Płaza E., Trela J., Hultman B. and Bosander J. (2003). Combined partial nitritation/Anammox system for treatment of digester supernatant. In: *Proceedings of the IWA Specialized Conference "Nutrient Management in Wastewater Treatment Processes and Recycle Streams",* 19–21 September 2005, Cracow, Poland.

Halling-Sørensen B., Nors Nielsen S., Lanzky P. F., Ingerslev F., Holten Lützhøft, H.-C. and Jørgensen S.E. (1998). Occurrence, fate and effects of pharmaceutical substances in the Environment – a review. *Chemosphere*, **36**, 357–393.

Hartig C., Storm T. and Jekel M. (1999). Detection and identification of sulphonamide drugs in municipal wastewater by liquid chromatography coupled with electrospray ionization tandem mass spectrometry. *Journal of Chromatography A*, **854**, 163–173.

Hippen A., Johansson P., Beier M., Rosenwinkel K. H. and Seyfried C. F. (2001). Six years' practical experience with aerobic/anoxic deammonification in biofilm systems. *Wat. Sci. Technol.*, **44** (2–3), 39–48.

Hirsch R., Ternes T., Haberer K. and Kratz K.-L. (1999). Occurrence of antibiotics in the aquatic environment. *Sci. Total Environ.*, **225**, 109–118.

Huang C.-H., Renew J. E., Pinkston K. and Sedlak D. L. (2001) Occurrence and fate of antibiotics compounds in municipal wastewater and animal waste. In: *Proceedings of the* "74th *Water Environment Federation Annual Conference and Exposition (WEFTEC.)", 13–17 October 2001, Atlanta, USA.*

Ingerslev F. and Halling-Sørensen B. (2000). Biodegradability properties of sulfonamides in activated sludge. *Environmental Toxicology and Chemistry*, **19**, 2467–2473.

Keller J., Subramaniam K., Gösswein J. and Greenfield P.F. (1997). Nutrient removal from industrial wastewater using single tank sequencing batch reactors. *Wat. Sci. Tech.*, **35** (6), 137–144.

Mędrzycka K., Tomczak-Wandzel R., Bąkowska A. and Cytawa S. (2004). Nitrogen monitoring in atmospheric deposition and its removal from municipal and industrial wastewater. In: *Pathways of pollutants and mitigation strategies of their impact on the ecosystems.* Serie: Monografie Komitetu Inżynierii Środowiska PAN, vol. 27, Lublin 2004.

Strous M., Heinen J. J., Kuenen J. G. and Jetten M. S. M. (1998). The sequencing batch reactor as a powerful tool for the study of slowly growing anaerobic ammonium-oxidizing microorganisms. *Appl. Microbiol. Biotechnol.*, **50**, 589–596.

Szatkowska B., Plaza E. and Trela J. (2004). Nitrogen removal rates in the deammonification process: batch experiments. In: *Proceedings of the "IWA World Water Congress and exhibition",* 19–24 September 2004, Marrakech.

Van de Graaf A. A., De Bruijn P., Robertson L. A., Jetten M. S. M. and Keunen J. G. (1996). Autotrophic growth of anaerobic ammonium- oxidising microorganisms in fluidised bed reactor. *Microbiology,* **142**, 2187–2196.

Van Dongen U., Jetten M. S. M. and Van Loosdrecht M. C. M. (2001). The SHARON® -Anammox® process for treatment of ammonium rich wastewater. *Wat. Sci. Technol.*, **44** (1), 153–160.

Wyffels S., Pynaert K., Boeckx, P., Vestraete W. and Van Cleemput O. (2003). Identification and quantification of nitrogen removal contactor by ^{15}N tracer techniques. *Wat. Res.*, **37**, 1252–1259.

Modification of filtration characteristics by PAC addition during dead-end ultrafiltration of WWTP effluent

A.N. Janssen*, M. Thomas, J. de Koning* and J.H.J.M van der Graaf***

* Department of Watermanagement, section Sanitary Engineering, Delft University of Technology, PO Box 5048, 2600 GA Delft, The Netherlands (E-mail: *A.N.Janssen@tudelft.nl*)
** Chemical Engineering and Environment Department, Institute National des Sciences Appliquées (INSA), 31077 Toulouse cedex 4, France

Abstract Membrane filtration is one of the advanced treatment techniques to polish wastewater treatment plant (WWTP) effluent. Membrane fouling by extracellular polymeric substances (EPS), humic substances and colour are one of the major limiting factors of the process. Combination powdered activated carbon (PAC) with ultrafiltration seems to be a promising technique to remove foulants which increase the filterability and reversibility of membrane fouling. This paper focus on the filterability measured as the Specific Ultrafiltration Resistance (SUR). The influence of PAC in terms of SUR decrease was tested of two types of PAC with different particle size. A significant difference was presented between both types of PAC. This differencet can be related to the difference in particle size. Experiments to investigate the potential of PAC on filterability and foulants show high removal of proteins, colours and humic substances. Also large decreases of SUR were observed.
Keywords Adsorption, dead-end ultrafiltration; fouling, PAC; SUR; WWTP effluent

Introduction

One of the techniques to upgrade wastewater treatment plant (WWTP) effluent to improve water quality is membrane filtration. To separate suspended solids, bacteria and viruses ultrafiltration membranes are used. During the filtration process some constituents deposit on the membrane surface or in the membrane matrix causing (so-called) membrane fouling. Membrane fouling reduces the filtration performance of WWTP effluent significantly. To obtain long-term stable ultrafiltration filtration characteristics and foulants of WWTP effluent have to be taken into account.

To measure the filterability of WWTP effluent, Roorda (2004) developed a parameter called specific ultrafiltration resistance (SUR). This parameter provides useful information regarding the filterability and can be measured within a short time (30 minutes) and with a simple filtration set-up which is shown in Figure 1.

During filtration of WWTP effluent, soluble substances seem to have a rather large impact on membrane perfomance (Te Poele, 2006). These compounds, determined as extracellular polymeric substances (EPS), humic substances and colour form irreversible fouling.

Filterability and reversibility of foulants are strongly related (Roorda, 2004). Good filterability goes hand in hand with a good reversibility. Therefore, removing or transforming the foulants before ultrafiltration will probably increase the initial filterability and

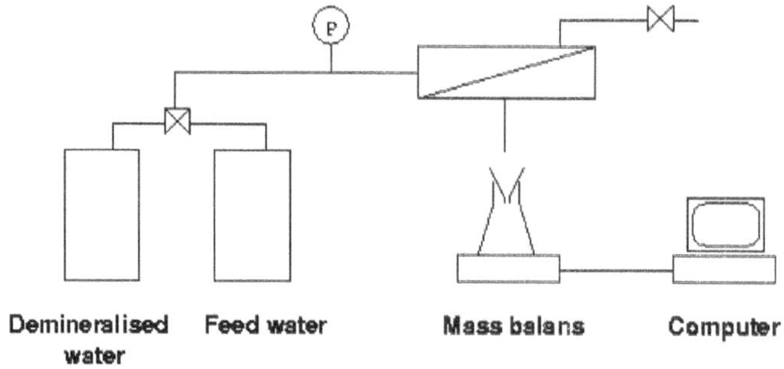

Figure 1 Schematic drawing of the laboratory test set-up.

reversibility. During previous research several pre-treatment techniques (coagulation, sandfiltration and multi-media filtration) were investigated but only a moderate improvement of the initial filterability was reached. Therefore other pre-treatment techniques have to be investigated.

Powdered activated carbon (PAC) is often combined with ultrafiltration and microfiltration to alleviate membrane fouling and improve the effectiveness of the removal of contaminants (Shon *et al.*, 2004; Mozia *et al.*, 2005; Lee *et al.*, 2005). However, the effect of PAC on membrane fouling is disputable. Both, reduction of membrane fouling and significantly decreases of flux by PAC addition are presented. Therefore, there is a need to perform further study on the effect of PAC addition on filtration characteristics during dead-end ultrafiltration of WWTP effluent.

The experiments and results described in this paper aim to investigate the potential of PAC to modify the SUR. Firstly one PAC was selected and secondly the potential of the selected PAC to increase the initial filterability of WWTP effluent was investigated.

Methods

WWTP effluent

Effluent for the experiments was taken from the WWTP Berkel and WWTP Horstermeer. The WWTP Berkel is designed for 27,000 population equivalents. The wastewater is treated in an oxidation ditch and after final sedimentation the effluent is discharged into the surface water. The WWTP Horstermeer is designed for 200,000 population equivalents. The household wastewater is also treated in an oxidation ditch. In Table 1 some effluent quality parameters are presented

Table 1 Effluent water quality at WWTP Berkel (Roorda, 2004) and WWTP Horstermeer (Menkveld *et al.*, 2006)

	COD (mg/L)	BOD (mg/L)	Suspended solids (mg SS/L)	Turbidity (NTU)	P-total (mg P/L)	N-total (mg N/L)
Wwtp Berkel	38.8	2.9	5.5	—	1.6	7.3
Wwtp Horstermeer	30	—	—	4.2	0.7	10.3

Powdered activated carbon

Two types of PAC were selected for the experiments in consultation with Norit. PAC-A and PAC-B correspondent respectively with NORIT SAE SUPER and NORIT SA UF. Table 2 shows the most important properties of the two types of PAC, given by Norit.

Table 2 Properties of the powdered activated carbon used in the experiments

	PAC-A	PAC-B
Activation method	Steam	Steam
Iodine number	1150	1100
Methylene blue adsorption (g/g)	0.28	0.24
Phenol adsorption (w%)	—	5
Total surface area (BET) (m^2/g)	1300	1200
Apparent density (kg/m^3)	425	160
Particle size > 150 μm (w%)	3	—
Particle size D_{50} (μm)	15	7
Ash content (w%)	—	10
pH	Akaline	Alkaline
Moisture (w%)	—	2

Experiments

All experiments were performed at the laboratory of Sanitary Engineering of the Delft University of Technology. The experiments were performed on the same day as sampling, or after storage for a maximum of one day in a refrigerator (5 °C). The experimental procedures were as follows:

PAC selection

Effluent samples were taken from the WWTP Berkel and WWTP Horstermeer. To remove algae or large particles effluent was sieved (400 μm) during sampling. Two litres of (sieved) WWTP effluent was stirred with PAC for 1 h at 100 rpm in a 2 L vessel. To separate PAC and WWTP effluent a paper filter (Schleichel&Schuell 589^2) was used. The SUR was measured for two times of the filtrate after paper filtration. Also the raw WWTP effluent passes the paper filter before SUR measurement. During the SUR measurement the temperature of WWTP effluent was kept at 20 °C. Foulants were measured after filtration the paper filtrate through 0.45 μm (Sartorius, Cellulose Acetate Filter). The applied concentrations of PAC were 0, 10, 25 and 50 mg/L.

Potential of PAC

To investigate the potential of the selected PAC to improve SUR two experiments were performed with effluent of WWTP Berkel on two different days. The experimental procedure was exactly the same as the procedure of the PAC selection experiment. Only the applied concentration of PAC differed. During the experiments the applied concentrations of PAC were 0, 50, 100 and 200 mg/L for the first experiment and 0, 100, 200, 300, 400 and 500 mg/L for the second experiment.

SUR measurement

To measure the filterability of WWTP effluent Roorda (2004) developed a parameter called Specific Ultrafiltration Resistance (SUR). The SUR is calculated from a filtration curve (t/V versus V) that is measured within 30 minutes of filtration over an ultrafiltration membrane at a constant temperature (~ 20 °C) and a transmembrane pressure (TMP) of 0.5 bar. The SUR values for effluent are within the range of 5×10^{12} m^{-2} for effluent with a high filterability and 30×10^{12} m^{-2} for effluent with a low filterability (Roorda, 2004).

The membranes used for all experiments were capillary ultrafiltration membranes with an internal diameter of 0.8 mm, a MWCO of 50–80 kDa and prepared of PES/PVP (type: UFC M5 ID 0.8 mm, X-Flow). New membrane modules, provided a membrane area of about 10×10^{-4} m^2, were soaked for 30 minutes in NaOCl (400 ppm) and flushed with demineralised water before a filtration experiment. After each filtration experiment the membrane module was flushed with demineralised water and cleaned with Divos 120CL (1.25 w%) during 10 minutes at 40 °C. After cleaning the clean water flux (CWF) was measured before starting a filtration experiment. At the beginning of each experiment a similar CWF is required for valid SUR values.

Analyses

Extracellular polymeric substances

Wwtp effluent contains mainly non-bound or soluble EPS. Therefore, proteins and polysaccharides can be directly analysed in order to characterise EPS (Te Poele, 2006).

The method of Rosenberger (2003) was modified to measure proteins in WWTP-effluent. The method is based on a colour reaction (Lowry *et al.,* 1951). The adsorption of the formed colour is measured at 750 nm in a 4 cm glass cuvet by the photo spectrometer (Milton Roy Spectromic 401). The amount of proteins is expressed in mg/L.

Polysaccharides are also measured by a modified method of Rosenberger (2003). The method (Dubois *et al.,* 1956) is based on a colour labelling. The adsorption of the formed colour is measured at 487 nm in a 4 cm glass cuvet by the photo spectrometer (Milton Roy Spectromic 401). The amount of polysaccharides expressed in mg/L.

Humic substances

Humic substances are measured by UV-absorption at 254 nm. UV-absorption is measured at 254 nm in a 1 cm quartz glass cuvet by the UV-VIS spectrophotometer Perkin-emler Lambda 16. The amount of humic substances is expressed as m^{-1}.

Colour

To determine the colour the NEN 1056 11.2 method was used. The absorption of colour at 455 nm in a 4 cm quartz glass cuvet is measured by the spectrophotometer Milton Roy Spectromic 401. The amount of colour is expressed as mg Pt/L.

Results and discussion

PAC selection

In Table 3 and Figure 2 the results of the experiments for the selection of one type of PAC are presented. In Table 3 the absolute SUR values are given of WWTP Berkel and WWTP Horstermeer. A different SUR value between both WWTP's for the untreated effluent (PAC dose 0 mg/L) can be observed. This difference is in accordance with results of previous researchers. Roorda (2004) presented SUR values in the range $5 \times 10^{12} - 30 \times 10^{12}$ m^{-2} for effluent of various WWTPs in The Netherlands.

Also a small difference in SUR can be noticed between the untreated effluent for PAC-A and PAC-B of one WWTP. These small differences are a matter of the experimental procedure. Experiments with PAC-A were performed on the same day of sampling. Experiments with PAC-B were performed after one day of storage the sample at 5 °C. However, when the variation of the SUR values is taken into account these differences can be neglected.

From Table 3, a decrease of SUR with increasing dosage of PAC-A and PAC-B can be observed for effluent of WWTP Berkel. The decrease of SUR by PAC-A is less than PAC-B. Effluent of WWTP Horstermeer shows the same trend. Especially, when a PAC dosage of 50 mg/L is applied the difference is significant.

Table 3 Absolute SUR values of effluent of WWTP Berkel and WWTP Horstermeer after addition of PAC-A and PAC-B

	WWTP Berkel		WWTP Horstermeer	
PAC (mg/L)	PAC-A SUR (10^{12} m^{-2})	PAC-B SUR (10^{12} m^{-2})	PAC-A SUR (10^{12} m^{-2})	PAC-B SUR (10^{12} m^{-2})
0	13.6 ± 1.1	14.3 ± 0.3	10.9 ± 0.7	11.3 ± 0.5
10	12.6 ± 0.6	12.5 ± 0.7	11.2 ± 0.1	10.1 ± 0.1
25	12.4 ± 0.1	11.8 ± 0.5	11.5 ± 1.9	10.0 ± 0.5
50	11.1[a]	8.6 ± 0.2	10.1 ± 0.1	7.8 ± 0.1

[a] Only one SUR measurement.

In Figure 2, a graphical presentation of the relative SUR values is given. With PAC-A a maximal SUR decrease of 18% and 8 % is obtained for respectively WWTP Berkel and WWTP Horstermeer. Applying PAC-B presents a significant difference. The maximal SUR decrease by PAC-B is 40% and 31% for respectively WWTP Berkel and WWTP Horstermeer. PAC dosages of 10 mg/L and 25 mg/L present almost the same trend. Applying PAC-B resulted in significantly lower SUR values than PAC-A. Therefore, PAC-B is selected to investigate the potential of PAC.

The difference in SUR decrease by PAC-A and PAC-B can be explained by the difference in particle size of the carbon particles (Table 2). Compared to PAC-A, PAC-B has a significantly smaller particle size which results in a larger outer total surface area and faster adsorption kinetics (Matsui *et al.*, 2004).

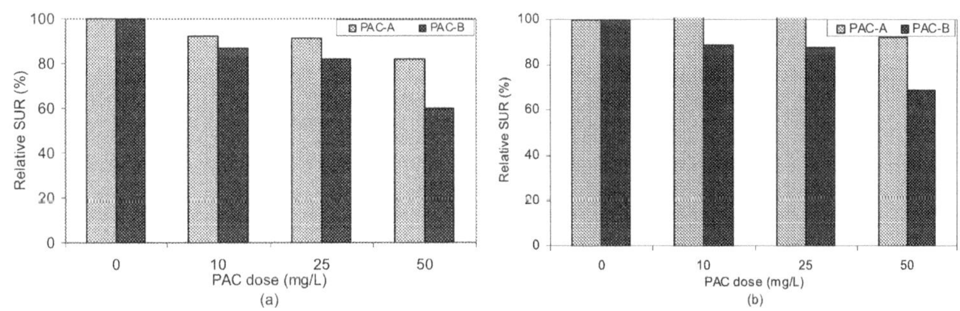

Figure 2 Relative SUR values of effluent of WWTP Berkel (a) and WWTP Horstermeer (b) after addition of PAC-A and PAC-BPotential of PAC.

Modification of SUR

The potential of PAC-B to modify the SUR was determined during two experiments with different samples (I and II). Table 4 shows the absolute SUR values and Figure 3 presents relative SUR values of the experiments. The SUR values of raw effluent (PAC dose 0 mg/L) shows a large difference between both experiments. This large variation was also found by Roorda (2004) and Te Poele *et al.* (2004). Roorda (2004) presented variations in the range of $6 \times 10^{12} - 40 \times 10^{12}$ m^{-2} for effluent of WWTP Berkel.

In spite of the different SUR values of raw effluent, both experiments show a significant decrease of SUR. Strongest decrease of SUR can be observed after adding 50 mg/L (I) and 100 mg/L (II). Increasing the PAC-B dose still results in a decrease of SUR but less. On the other side, the potential of PAC-B in terms of maximal SUR decrease was not obtained during both experiments. No stabilization of SUR with increasing PAC-B dose can be noticed from Table 4.

Table 4 Absolute SUR values of effluent of WWTP Berkel after addition of PAC-A

Experiment	I	II
PAC (mg/L)	SUR (10^{12} m^{-2})	SUR (10^{12} m^{-2})
0	11.3 ± 0.7	31.2 ± 1.1
50	6.6 ± 0.5	nd
100	5.6[a]	20.0 ± 0.5
200	4.1 ± 0.7	13.6 ± 0.9
300	nd[b]	13.8
400	nd	8.7 ± 1.2
500	nd	5.9

[a] Only one SUR measurement; [b] not determined.

In Figure 3 the relative SUR values of Table 4 are presented. Roughly, the same trend can be observed for both experiments. An initial fast decrease of SUR followed by a more gradual and smaller decrease of SUR. Therefore, it seems that the adsorption process is limited by external mass transfer of the adsorbate substances (foulants) from the bulk to PAC-B and subsequently by the internal mass transfer within the carbon particles. Higher PAC dosages (larger surface area) did not present even higher SUR decreases. However, this is only the case when a relation between concentration of foulants and SUR is assumed.

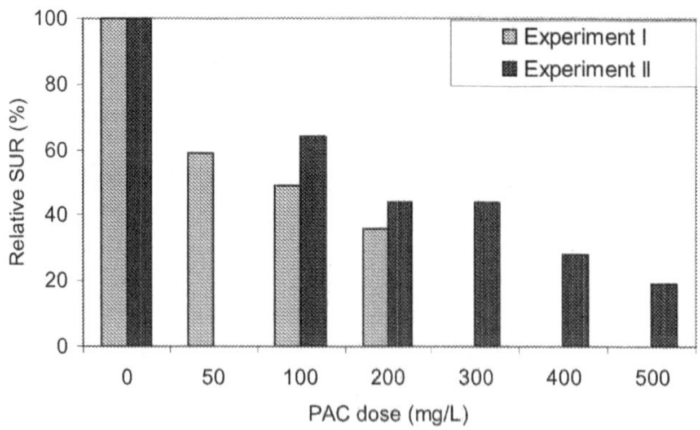

Figure 3 Relative SUR value versus concentration of PAC-B (0–200 mg/L (I) and 0–500 mg/L (II)).

Modification of foulants

In Figure 4 the changes in concentration of EPS (mg/L) are presented. The EPS concentrations present the same trend as the relative SUR values (Figure 3). During the initial phase (PAC dose of 0–100 mg/L) a fast decrease of EPS can be observed for both experiments. During the following phase (PAC dose of 200–500 mg/L) EPS concentrations still decrease but less significant than in the initial phase.

Figure 4 shows a difference between the removal of polysaccharides and proteins. Proteins can be removed almost totally instead of polysaccharides. However, with higher dosages of PAC the removal of polysaccharides improve.

Adsorption of proteins on membrane surface was indicated as one of the major fouling mechanisms during ultrafiltration of WWTP effluent (Te Poele, 2006). Therefore, adsorption of proteins on activated carbon will probably prevent adsorption of proteins on membrane surface.

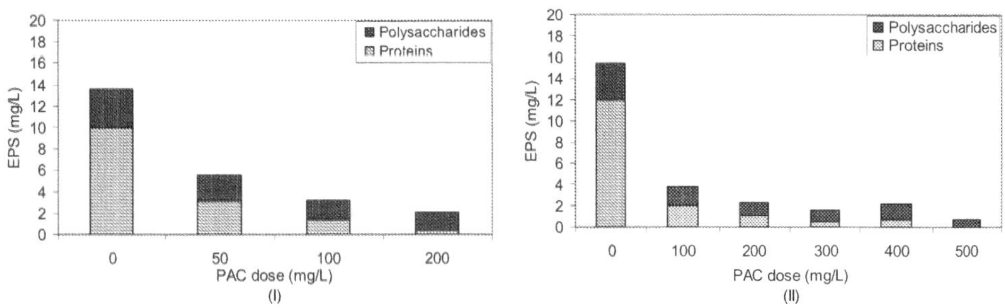

Figure 4 EPS concentrations versus concentration of PAC-A (0–200 mg/L (I) and 0-500 mg/L (II)).

Figure 5 outlines the effect of PAC addition on the concentrations of humic substances (m^{-1}) and colour (mg Pt/L). Like the previous presented figures, a fast decrease of both compounds can be observed for less dosage. PAC dosages of more than 100 mg PAC-A/L removes all the colour and PAC dosages of more than 200 mg PAC-A/L removes significant amounts of humic substances. This is in accordance with results of other research. Tomaszewska and Mozia (2005) presented a removal of 96% of colour and 89% decrease in humic substances concentration for PAC dose of 50 and 100 mg/L in a combined PAC/UF system. When the UF process was conducted without PAC addition only 60% of colour and 40% of humic substances were removed.

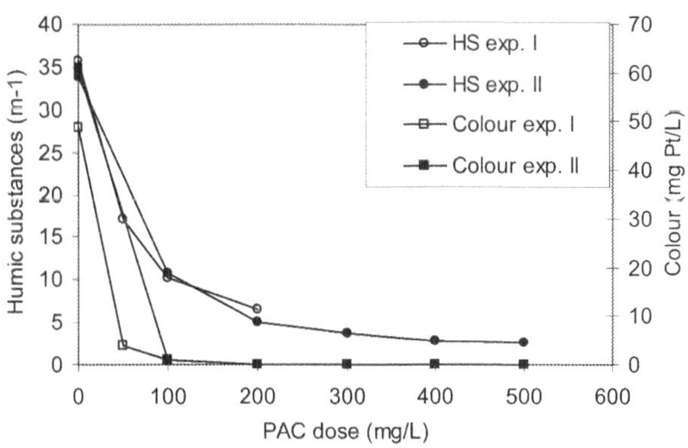

Figure 5 Concentrations of humic substances (HS) and colour versus concentration of PAC-A (0–200 mg/L (I) and 0-500 mg/L (II)).

Conclusions

The experiments with different types of PAC show different results in terms of SUR decrease. Applying doses of 50 mg PAC/L present most significant difference in advantage of PAC-B. Probably, the difference can be explained by difference in particle size.

Maximum decrease of SUR was not obtained during the experiment to investigate the potential of PAC. However, it seems that the relative SUR decrease is related to the initial (raw effluent) SUR value. Higher initial SUR values means higher doses of PAC to decrease SUR.

Foulants seems to be well adsorbed by PAC. Especially, proteins, humic substances and colour are removed fast instead of polysaccharides. Probably, PAC will increase the reversibility due to the adsorption of EPS and humic substances.

The actual relation between PAC addition and SUR seems to be in some way uneconomic; only with high dosages of PAC the SUR reaches attractive low values. Therefore, more attention should be given to optimize the efficiency of the PAC addition for example by combining with the addition of coagulants.

Acknowledgements

This work is financially supported by Evides Industrie, Veolia Water and Witteveen+Bos.

References

Dubois, M., Gilles, K.A., Hamilton, J.K., Rebers, P.A. and Smith F. (1956). Colourmetric method for determination of sugars and related substances. *Analytical Chemistry*, **28**, 350–356.

Lee, C.W., Bae, S.D., Han, S.W. and Kang, L.S. (2005) Application of UF hybrid processes for reuse of secondary effluent. In Proceedings of the IWA Specialty Conference 'Wastewater Reclamation & Reuse for Sustainability', Jeju, Korea, November 8–11, 2005

Lowry, O.H., Rosebrough, N.J., Farr, A.L. and Randall, R.J. (1951) Protein measurement with the Folin phenol reagent. *Journal of Biological Chemistry*, **193**, 265–275.

Matsui, Y., Murase, R., Sanogawa, T., Aoki, N., Mima, S., Inoue, T. and Matsushita, T. (2004). Micro-ground powdered activated carbon for effective removal of natural organic matter during water treatment. *Water Science and Technology: Water Supply*, **51**(6–7), 249–256.

Mozia, S., Tomaszewska, M., and Morawski, A.W. (2005). Studies on the effect of humic acids and phenol on adsorption–ultrafiltration process performance. *Water Research*, **39**(2–3), 501–509.

Menkveld, H.W.H., Miska, V. and Gorter, K. (2006). *Op weg naar EKW, praktijkonderzoek RWZI Horstermeer naar de aanvullende verwijdering van nutriënten, zware metalen en prioritaire stiffen,* Voortgangsrapportage 2, Witteveen+Bos, Deventer, The Netherlands.

Roorda J.H. (2004). *Filtration characteristics in dead-end ultrafiltration of WWTP-effluent*. PhD thesis, Department of Sanitary Engineering, Technical University of Delft, Delft, The Netherlands.

Rosenberger, S. (2003) Charkterisierung von belebtem Schlamm in Membranbelebungsreaktoren zur Abwasserreiniging. PhD thesis. Technical University of Berlin, Berlin, Germany.

Shon, H.K., Vigneswaran, S., Kim, I.S., Cho, J. and Ngo, H.H. (2004). Effect of pretreatment on the fouling of membranes: application in biologically treated sewage effluent. *Journal of Membrane Science*, **234**(1–2), 111–120.

Te Poele S. (2006) *Foulants in ultrafiltration of WWTP effluent*. PhD thesis, Department of Sanitary Engineering, Technical University of Delft, Delft, The Netherlands.

Te Poele, S., Roorda, J.H. and Van der Graaf, J.H.J.M. (2004). Influence of the size of membrane foulants on the filterability of WWTP-effluent. *Water Science and Technology*, **50**(12), 111–118.

Tomaszewska, M. and Mozia, S. (2002). Removal of organic matter from water by PAC/UF system. *Water Research*, **36**(16), 4137–4143.

Dechlorination kinetics of carbon tetrachloride and chloroform with nano-scale Pd/Fe

Jing Feng and Teik-Thye Lim

School of Civil and Environmental Engineering, Nanyang Technological University, 50 Nanyang Avenue, Singapore 639798, Singapore (E-mail: *fengjing@pmail.ntu.edu.sg*; *cttlim@ntu.edu.sg*)

Abstract In this study, nano-scale 0.2% and 1% Pd/Fe particles were synthesized and the ability of these particles to reduce carbon tetrachloride (CTC) and chloroform (CF) was observed. BET surface area of the nano-scale particles was around 26.5 $m^2\ g^{-1}$. The reduction of CTC and CF with the Pd/Fe particles followed pseudo-first-order kinetics. The reduction rates were higher for the highly chlorinated compounds than the lightly chlorinated ones. Meanwhile, comparing the 0.2% Pd/Fe and the 1.0% Pd/Fe particles, the latter exhibited higher reduction rates of CTC and CF. Coupled hydrogenolysis-elimination process was the important mechanism for complete dehalogenation of the chlorinated methanes. Employing the noble metal in the Fe particles could also show some influence to the reduction pathways of CTC and CF.

Keywords Carbon tetrachloride; chloroform; nano-scale particles; Pd/Fe

Introduction

Chlorinated organic compounds are widely used as solvents in industry and synthetic intermediates for plastic and herbicides. Since these compounds are recalcitrant, they have been found to accumulate in the surface waters, sedimentations and groundwater. Abiotic reductive transformation is one promising method for removal of chlorinated organic compounds (Matheson and Tratnyek, 1994; Arnold and Roberts, 1998; Mu *et al.*, 2004; Doong and Lai, 2005). Reduction with zero-valent Fe (ZVI) and catalytic dehalogenation with noble metal represent two promising approaches.

Although ZVI technology has been proved to be an effective treatment method for chlorinated organic solvents, there still exist some challenges to apply this technology, such as low reduction rate of lightly chlorinated compounds and incomplete dechlorination, etc (Wang and Zhang, 1997; Cheng and Wu, 2000). Usually the improvements are primarily related to two aspects, including decreasing the particle size and alloying with another metal. Since the reaction has a direct relationship to the available metal surface area, decreasing the metal particle size will enhance the reduction rate even with a lower metal loading. Nano-scale Fe particles have been synthesized by many researchers to accelerate the reaction rates (Wang and Zhang, 1997; Choe *et al.*, 2001; Lowry and Johnson, 2004; Nurmi *et al.*, 2005). The higher reactivity of the nano-scale particles as compared with their micro-scale counterparts has been attributed to the greater density and higher intrinsic reactivity of their surface reactive sites (Klabunde *et al.*, 1996; Nurmi *et al.*, 2005). Alloying a noble metal, such as Au, Ag and Pd, etc., with ZVI particles can be another possible approach to improve the performance of zero-valent Fe (Wang and Zhang, 1997; Li and Klabunde, 1998; Schrick *et al.*, 2002; Liu *et al.*, 2005). Nano-scale bimetallic particles can not only increase the reduction rate of the target compound, but also can change the reduction pathways and decrease the chlorinated intermediate byproducts. Among all the noble metals, Pd has been proved the most effective (Heinrichs *et al.*, 1997; Nutt *et al.*, 2005).

Thus, in this study, nano-scale Pd/Fe particles with different Pd contents were synthesized in the laboratory. The influence of Pd content to the reduction rate of carbon tetrachloride (CTC) and chloroform (CF) were investigated. Reduction kinetics and pathways of CTC and CF with the Pd/Fe particles were also proposed.

Methodology

All aqueous solutions were prepared with ultra-pure water (18.2 MΩ cm). All chemicals were purchased without further purification: CTC (BDH, UK), CF (Merck, Germany), methylene chloride (DCM) (J. T. Baker, NJ), palladium acetate ($Pd(C_2H_3O_2)_2$) (Merck, Germany), acetone (Fisher scientific, NJ), sodium chloride (Lancaster, UK), sodium borohydride (Lancaster, UK), ferrous sulfate ($FeSO_4{\cdot}7H_2O$) (Reachim, Russia), ferric nitrate ($Fe(NO_3)_3{\cdot}9H_2O$) (Sigma Aldrich, USA), mercuric thiocyanate ($Hg(SCN)_2$) (Merck, Germany).

Syntheses of the nano-scale Fe particles were carried out with sodium borohydride ($NaBH_4$) as the reductant in the batch solution. The process involved the following reaction:

$$2Fe^{2+} + 2H_2O + BH_4^- \rightarrow 2Fe\downarrow + BO_2^- + 4H^+ + 2H_2\uparrow \quad (1)$$

After separating the synthesized nano-scale Fe from the solution, the particles was washed with ultra-pure water (>300ml) and re-suspended in the acetone solution of palladium acetate. To compare the performance of the nano-scale Pd/Fe particles with different amounts of Pd coating, 0.2% w/w Pd/Fe and 1% w/w Pd/Fe were synthesized. Pd would be reduced by the nano-scale Fe, and formed a coating layer on the Fe surface, as depicted by the following equation:

$$2Pd^{2+} + Fe \rightarrow Pd\downarrow + Fe^{2+} \quad (2)$$

When the color of the solution changed from orange to achromaticity, the black particles could be filtered through 1.2-μm glass fiber membrane. The separated particles were dried in the freeze-drier over night before use.

Brunauer-Emmett-Teller (BET) surface area analysis of the nano-scale particles were performed with Quanta Chrome autosorb automated gas sorption system. Transmission electron microscope (TEM) images were obtained using JEOL electron microscope (JEM–2010) operated at an acceleration voltage of 300 kV. The surface element composition of Pd/Fe particles was analyzed with scan electron microscope (SEM, JSM-6360) combined with energy-dispersive X-ray spectrometry (EDX, JED–2003 X-ray analyzer). X-ray diffraction (XRD) analysis of mineralogical characteristics of the lab synthesized nanoscale particles were carried out with Bruker Dt8 advanced XRD diffractometer.

Batch reduction experiments without headspace were carried out with 40 mL TFE-sealed screwed cap bottle. Reduction reactions of CTC and CF were studied using a metal loading of 2.5 g l^{-1}. Chlorinated methanes were introduced to the system via a methanol spike resulting in an initial concentration of 20 mg l^{-1}. At each time interval, 5 ml aqueous sample was injected into purge and trap (Tekmar Dohrmann solatek 3100 sample concentrator) and then analyzed by gas chromatography (GC, HP5890) with a HP-5-mega column (30 m long × 0.53 mm ID with 1.5 μm film thickness) and flame ionization detector. Another 5 ml solution was withdrawn for analysis of the concentration of chloride anion using Flow Injection Analysis Colorimetry with ion chromatography (FIA/IC, Lachat, QuickChem 8000).

Results and discussion

Characterization of nano-scale particles

Due to the magnetic effect between the small particles, the nano-scale Pd/Fe particles tended to aggregate together forming chain structures (Figure not shown). The average diameter of the particles was around 50 nm. The BET surface area of the Pd/Fe particles was 26.5 m^2g^{-1}. The peaks identified on XRD patterns belonged to body-centered cubic Fe. The diffraction peaks of Pd were not observed due to the small percentage of Pd content. Based on the SEM-EDX elemental mapping of the Pd/Fe particles, Pd was observed to distribute evenly on the surface of the Fe particles in the region of the SEM images (Figure not shown). Because Pd was mainly distributed on the surface layer, the surface content of Pd was apparently higher than the Pd theoretical content. For the 1% Pd/Fe, over 8% surface mass was covered by Pd. Nutt *et al.* (2005) also observed similar results of the nano-scale Pd/Au particles. The existence of O in the SEM-EDX elemental mapping indicated that small fraction of surface Fe might be oxidized to Fe_2O_3 and small amount of the body-centered cubic Fe might be covered by a thin discontinuous oxide film. Once these particles contacted water, the oxide film could be reduced rapidly through autoreduction reaction (Ritter *et al.*, 2002):

$$Fe_2O_3 + Fe + 6H^+ \rightarrow 3Fe^{2+} + 3H_2O \tag{3}$$

Transformation of CTC and CF with Pd/Fe

The experimental results of CTC with the 0.2% Pd/Fe are depicted in Figure 1. Complete reduction of CTC was achieved within 40 min. CF and DCM were the two main chlorinated intermediate byproducts. CF concentration reached a maximum at 20 min, and reduced to 0.6 mg l^{-1} at 6 h. With the disappearance of CTC and CF, the DCM concentration increased steadily to 1.2 mg l^{-1} which accounted for around 11% of CTC lost. DCM concentration approached a stable level at the end of the 6 h reaction. Methane was identified in the headspace of the vial as the non-chlorinated end reduction product of CTC degradation. The chlorine mass balance achieved was around 99% at the end of the experiment.

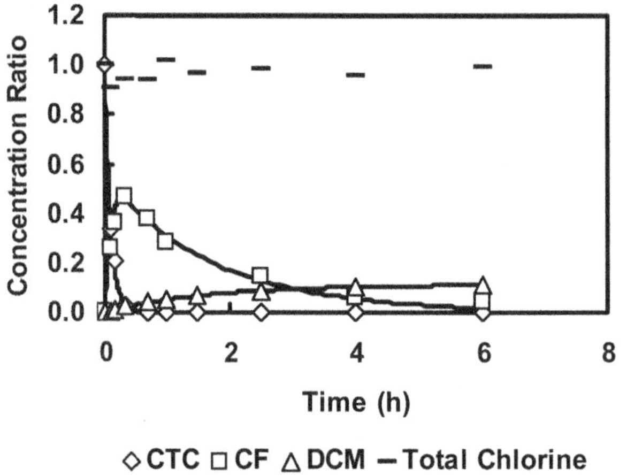

Figure 1 Transformation results and simulation curves of CTC with 2.5 g l^{-1} nano-scale 0.2% Pd/Fe.

CF reduction with the 0.2% Pd/Fe showed a slower degradation rate as compared to that of CTC, as presented in Figure 2. It took around 10 h to reduce the initial CF concentration to 0.1 mg l^{-1}. The ultimate yield of DCM was 3.6 mg l^{-1} (25% of initial CF concentration). The chlorine mass balance in the solution was over 90% after 10 h. To confirm the persistency of DCM with the 0.2% Pd/Fe, a separate experiment was performed with DCM as the parent compound. DCM concentration was found rather stable in the system for 25 h reduction.

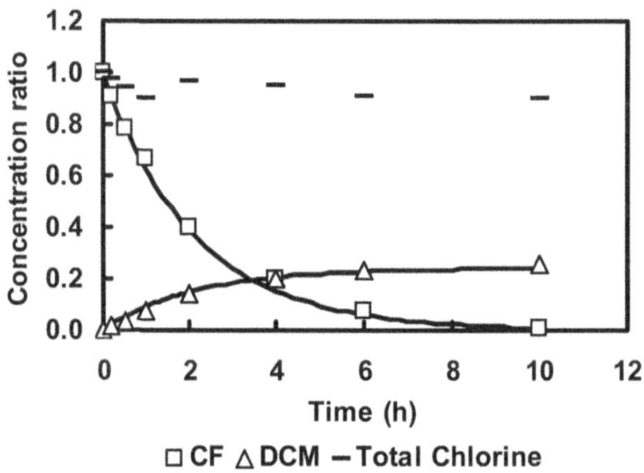

Figure 2 Transformation results and simulation curves of CF with 2.5 g l^{-1} nano-scale 0.2% Pd/Fe.

Transformations of CTC and CF with the 1% Pd/Fe particles were also observed (Figure not shown). In general, the reductions of CTC and CF with the 1% Pd/Fe showed similar trends as those with the 0.2% Pd/Fe. Compared to the reductions with the 0.2% Pd/Fe, the reduction rates of CTC and CF, both as the parent compounds or intermediate by-products, were apparently increased. However, DCM was still the main chlorinated intermediate byproduct which accumulated in the aqueous phase. Methane was the primary end product during the reduction.

Kinetics of transformation and influence of Pd content

The reduction kinetics of the chlorinated methanes with the nano-scale particles followed pseudo-first-order kinetics model. Thus, the reduction rate of CT and CF with the nano-scale particles can be expressed by the following equation (Matheson and Tratnyek, 1994; Johnson *et al.*, 1996; Wüst *et al.* 1999):

$$\frac{dC}{dt} = -k_{obs}C = -k_{SA}\rho_{\alpha}C \quad (4)$$

where C is the concentration of parent compound (mg l^{-1}), k_{obs} is the pseudo-first-order rate constant (h^{-1}), k_{SA} is the specific rate reaction rate constant (l h^{-1} m^{-2}), and ρ_{α} is the surface area concentration of metal in solution (m^2 l^{-1}). The derived reduction rate constants of chlorinated methanes with the nano-scale particles are shown in Table 1.

Table 1 The values of derived rate constants for CTC and CF

Particles		k_{obs} (h^{-1})	k_{SA} (l h^{-1} m^{-2})	$t_{1/2}$ (h)	R^2
0.2% Pd/Fe	CTC	10.64	1.61×10^{-1}	6.51×10^{-2}	0.988
	CF	4.67×10^{-1}	7.05×10^{-3}	1.48	0.991
1% Pd/Fe	CTC	56.04	8.46×10^{-1}	1.24×10^{-2}	0.997
	CF	8.91×10^{-1}	1.34×10^{-2}	7.78×10^{-1}	0.975
Fe[a]	CTC	3.96	5.99×10^{-2}	1.75×10^{-1}	0.998
	CF	3.39×10^{-2}	5.13×10^{-4}	20.44	0.963

[a]The data are from Feng and Lim (2005).

The contributions of Pd to CTC and CF reduction mainly included galvanic cell effect and catalytic dechlorination. Pd, with a lower hydrogen over potential compared to Fe, could form a galvanic cell with Fe which could increase the reduction rate of Fe metal and reduce the accumulation of insoluble Fe oxide that could block Fe surface reactive sites (Wang and Zhang, 1997; Nutt *et al.*, 2005). Pd also served as a catalyst for the catalytic dechlorination of chlorinated organic compounds with hydrogen gas which was produced from water reduction with the Fe particles (Lowry and Reinhard, 2001). Increasing the Pd amount would also enhance the degradation rates of chlorinated methanes. However, Pd, as a noble metal, could not serve as an electron donor directly. Thus, if the Fe surface was completely covered by Pd, the reduction rate of the target compounds would decrease sharply due to the absence of the electron donor in the system.

Reduction pathways of CTC and CF

The reduction of chlorinated methanes followed coupled sequential and parallel pathways. Since mono-chloromethane was not observed during the reduction of CTC and CF and the recalcitrance of DCM in the presence of the Pd/Fe particles, the reduction pathways can be simplified to Figure 3.

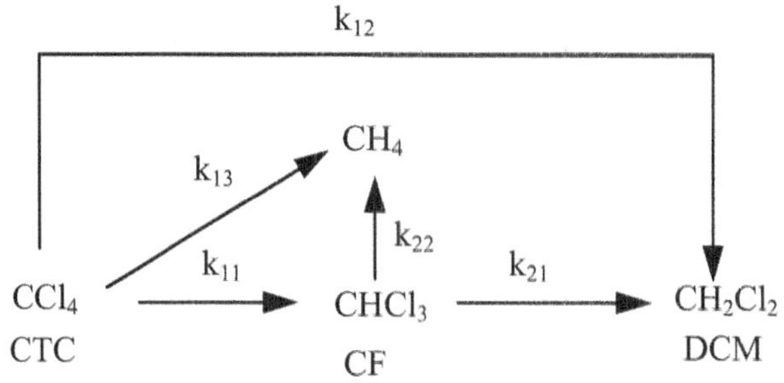

Figure 3 Hypothesized reduction pathways for chlorinated methanes and the annotated rate constants.

Based on the pseudo-first-order kinetics, the generalized rate equation can be expressed as follows:

$$\frac{dC_i}{dt} = -\left(\sum_{j=1}^{4-i} k_{ij}\right) C_i \tag{5}$$

where k value is the rate constant (h^{-1}), i is the counter for the types of parent compounds (e.g.,1 for CTC, 2 for CF) and j is the counter for the hypothesized pathways as annotated in Figure 3. For CF or DCM produced as intermediate products, the generalized rate equation for their formations/degradations can be expressed as follow:

$$\frac{dC_m}{dt} = \sum_{i=1}^{m-1} k_{i,m-i} C_i - \left(\sum_{j=1}^{4-m} k_{mj}\right) C_m \tag{6}$$

where m is the counter for the type of intermediate products (e.g., 2 for CF). On the basis of the chlorinated intermediate products distribution, each rate constant (k) can be calculated. The prediction error sum of square (PRESS) was used as the criterion to assess the goodness of fit of the chosen k values to the experimental results. The set of k values which gave the lowest PRESS were chosen.

Table 2 Best fitted rate constants of CTC and CF with nano-scale particles

Parent Compounds	ZVI	k_{11} (h^{-1})	k_{12} (h^{-1})	k_{13} (h^{-1})	k_{21} (h^{-1})	k_{22} (h^{-1})
	Fe [a]	3.65	0.00	0.31	0.01	0.02
CTC	0.2% Pd/Fe	6.04	0.07	4.53	0.12	0.48
	1% Pd/Fe	37.98	0.00	18.06	0.34	1.54
	Fe [a]	-	-	-	0.01	0.02
CF	0.2% Pd/Fe	-	-	-	0.12	0.35
	1% Pd/Fe	-	-	-	0.19	0.70

[a] The data are from Feng and Lim (2005).

The best-fitted rate constants are presented in Table 2. The main reduction pathway of CTC was hydrogenolysis pathway to CF (k_{11}). However, in the presence of the Pd/Fe bimetallic particles, the significance of reductive elimination pathway of CTC to methane (k_{13}) increased. The main reduction pathway of CF was through reductive elimination pathway to methane (k_{22}). The ratios of k_{22} to k_{21} with the Pd/Fe particles were higher than that of the Fe particles which depicted that complete reduction pathway of CF was more dominant with the Pd/Fe particles. Thus, Pd/Fe particles could change the reduction pathways of chlorinated methanes to decrease the amount of recalcitrant DCM produced.

Conclusions

The nano-scale 0.2% and 1% Pd/Fe particles were synthesized in this study. Batch reduction experiments were carried out to investigate the performance of the lab-synthesized nano-scale particles. The average diameter of the nano-scale particles was around 50 nm. Small amount of the Pd/Fe particles were covered by a thin discontinuous layer of Fe_2O_3, and Pd distributed evenly on the surface of the particles. In the presence of the 0.2% Pd/Fe, the reduction rate constant of CTC was over two times higher than that of the Fe particles. Meanwhile, the rate increase was even higher for that of CF. Pd could not only enhance the reduction rate of chlorinated methanes, but also change the reduction pathways to avoid the formation of recalcitrant chlorinated intermediate byproduct. Thus, the nano-scale Pd/Fe particles will be favorable for environmental applications.

References

Arnold, W. A., Roberts, A. L. (1998). Pathways of chlorinated ethylene and chlorinated acetylene reaction with Zn(0). *Environmental Science and Technology*, **32**, 3017–3025.

Cheng, S.-F., Wu, S.-C. (2000). The enhancement methods for the degradation of TCE by zero-valent metals. *Chemosphere*, **41**, 1263–1270.

Choe, S., Lee, S.-H., Chang, Y.-Y., Hwang, K.-Y., Khim, J. (2001). Rapid reductive destruction of hazardous organic compounds by nanoscale Fe^0. *Chemosphere*, **42**, 367–372.

Doong, R. A., Lai, Y. J. (2005). Dechlorination of tetrachloroethylene by palladized iron in the presence of humic acid. *Water Research*, **39** (11), 2309–2318.

Feng, J., Lim, T. T. (2005). Pathways and kinetics of carbon tetrachloride and chloroform reductions by nano-scale Fe and Fe/Ni particles: comparison with commercial micro-scale Fe and Zn. *Chemosphere*, **59**, 1267–1277

Heinrichs, B., Delhez, P., Schoebrechts, J.-P., Pirard, J.-P. (1997). Palladium–silver sol–gel catalysts for selective hydrodechlorination of 1,2-dichloroethane into ethylene. *Journal of Catalyst*, **172**, 322–335.

Johnson, T. L., Scherer, M. M., Tratnyek, P. G. (1996). Kinetics of halogenated organic compound reduction by iron metal, *Environmental Science and Technology*, **30**, 2634–2640.

Klabunde, K. J., Stark, J., Koper, O., Mohs, C., Park, D. G., Decker, S., Jiang, Y., Lagadic, I., Zhang, D. J. (1996). Nanocrystals as stoichiometric reagents with unique surface chemistry. *Journal of Physical Chemistry*, **100**, 12142–12153.

Li, W., Klabunde, K. J. (1998). Ultrafine zinc and nickel, palladium, silver coated zinc particles used for reductive dehalogenation of chlorinated ethylenes in aqueous solution. *Croatica Chemica Acta*, **71**, 853–872.

Liu, Y., Majetich, S. A., Tilton, R. D., Sholl, D. S., Lowry, G. V. (2005). TCE dechlorination rates, pathways and efficiency of nano-scale iron particles with different properties. *Environmental Science and Technology*, **39**, 1338–1345.

Lowry, G. V., Johnson, K. M. (2004). Congener-specific dechlorination of dissolved PCBs by microscale and nanoscale zerovalent iron in a water/methanol solution. *Environmental Science and Technology*, **38**, 5208–5216.

Lowry, G. V., Reinhard, M. (2001). Pd-catalyzed TCE dechlorination in water: effect of $[H_2](aq)$ and H_2-utilizing competitive solutes on the TCE dechlorination rate and product distribution. *Environmental Science and Technology*, **35**, 696–702.

Matheson, L. J., Tratnyek, P.G. (1994). Reductive dehalogenation of chlorinated methanes by iron metal, *Environmental Science and Technology*, **28**, 2045–2053.

Mu, Y., Yu, H.-Q., Zheng, J.-C., Zhang, S.-J., Sheng, G.-P. (2004). Reductive degradation of nitrobenzene in aqueous solution by zero-valent iron. *Chemosphere*, **54**(7), 789–794.

Nurmi, J. T., Tratnyek, P. G., Sarathy, V., Baer, D. R., Amonette, J. E., Pecher, K., *et al.* (2005). Characterization and properties of metallic iron nanoparticles: spectroscopy, electrochemistry, and kinetics. *Environmental Science and Technology*, **39**, 1221–1230.

Nutt, M. O., Hughes, J. B., Wong, M. S. (2005). Designing Pd-on-Au bimetallic nano-particle catalysts for trichloroethene hydrodechlorination. *Environmental Science and Technology*, **39**, 1346–1353.

Ritter, K., Odziemkowski, M. S., Gillham, R. W. (2002). An in situ study of the role of surface films on granular iron in the permeable iron wall technology. *Journal of Contaminant Hydrology*, **55**, 87–111.

Schrick, B., Blough, J. L., Jones, A.D., Mallouk, T. E. (2002). Hydrodechlorination of trichloroethylene to hydrocarbons using bimetallic nickel–iron nanoparticles. *Chemistry of Materials*, **14**, 5140–5147.

Wang, C. B., Zhang, W. X. (1997). Synthesizing nano-scale iron particles fro rapid and complete dechlorination of TCE and PCBs. *Environmental Science and Technology*, **31**, 2154–2156.

Wüst, W. F., Kőber, R., Schlicker, O., Dahmke, A. (1999). Combined zero- and first-order kinetic model of the reduction of TCE and cis-DCE with commercial iron. *Environmental Science and Technology*, **33**, 4304–4309.

Removal of persistent organic compounds from wastewater using iron-reducing bacteria

J.W. Lim, V. Ivanov and Y.H. Gin

School of Civil and Environmental Engineering, Nanyang Technological University, 50 Nanyang Avenue, Singapore 639798 (E-mail: *s8122278i@ntu.edu.sg*; *cvivanov@ntu.edu.sg*; *cyhgin@ntu.edu.sg*)

Abstract The widespread usage of pharmaceuticals and chemicals has led to an increasing amount of persistent organic compounds (POCs) being released into wastewater effluent. These POCs may remain in wastewater, even after aerobic treatment and anaerobic digestion of activated sludge suspension. Therefore, the effect of additional oxidant, Fe(III), and iron-reducing bacteria (IRB) on degradation of such POCs as ibuprofen, naproxen, 17α-ethynylestradiol (EE2), 17β-estradiol (E2) , and estriol (E3) was investigated. Obligate anaerobic and oxygen-tolerant enrichment and pure cultures of IRB, which were able to degrade targeted pharmaceuticals and steroids using $Fe(OH)_3$ and iron ore as oxidants, were isolated and tested. Biodegradation of these POCs and accumulation of their metabolites were analysed using liquid chromatography–mass spectrometry (LC–MS). Removal efficiencies for ibuprofen and naproxen were 31% and 25%, respectively. EE2 remained resistant to biodegradation by IRB while natural estrogens such as E2 and E3 were removed by more than 90%. The removal efficiency depends probably on physicochemical properties and chemical structures of POCs. Introduction of ferric oxide or hydroxide as oxidant stimulated degradation of POCs, pharmaceuticals and steroids, due to activity of iron-reducing bacteria.

Keywords Estrone; estriol; 17β-estradiol; 17α-ethynylestradiol; ibuprofen; naproxen; iron-reducing bacteria; persistent organic compounds

Introduction

During the past few decades, the growth of the chemical and pharmaceutical industry has accelerated the release of xenobiotics into the environment via discharge of wastewater effluent. These compounds are generally recalcitrant to biodegradation and have been classified as persistent organic compounds (POCs). The POCs reach the sewage system commonly through excretion after administration and direct disposal of surplus drugs. For example, the oral contraceptive pill contains between 30 and 50 μg of 17α-ethynylestradiol (EE2) per pill. However, there is a concern that POCs found in wastewater persist both aerobic and anaerobic wastewater treatments and result in residual concentrations, which might pose potential adverse human health effects. Studies revealed that despite the efficient elimination of 96–99% of ibuprofen in some wastewater treatment plants in Switzerland, the substance was still detected in sewage effluents and surface waters (Buser *et al.* 1999). For EE2, inactive conjugates of this hormone with sulphuric or glucuronic acids may even be reconverted to free steroids by bacteria in the environment (Ternes *et al.*, 1999; Baronti *et al.*, 2000). Thus, more effective measures to enhance conventional treatment processes are needed to remove the pharmaceutical POCs.

Based on the abundant literature on biodegradability of toxic pollutants in wastewater treatment plants, the trend is now inclined towards the identification of microorganisms responsible for removal of POCs. For example, studies by Vader *et al.* (2000) showed that nitrifying activated sludge could degrade EE2 completely within 6 days. Hypothetically, anaerobic microorganisms, which would be able for anaerobic oxidation of the mentioned POCs by such acceptors of electrons as SO_4^{2-}, Fe^{3+} and NO_3^- could be useful in biodegradation. However, only Fe^{3+} could be suitable for the large scale practice of wastewater treatment because sulphate will be transformed to toxic and bad smelling H_2S, and nitrate is too expensive to be used. Therefore, Fe^{3+} and iron-reducing bacteria (IRB) were tested in this research as the factors enhancing biodegradation of organic substances persisting aerobic and anaerobic municipal wastewater treatment. IRB were found to be an ecologically important group of microorganisms. Iron reduction was responsible for 20-80% of the organic carbon oxidation in some anoxic biotopes, such as coastal, river and aquifer sediments (Lovley, 1991; Nealson and Saffarini, 1994).

Iron-reducing bacteria in anaerobic digesters use Fe(III), which can be derived from cheap iron oxides and hydroxides, as an electron acceptor, and reduce them to Fe(II) (Ivanov *et al.*, 2002, 2004):

$$2Fe^{3+} + [2H] \rightarrow 2Fe^{2+} + 2H^+$$

IRB oxidize different organic substances under anaerobic condition using Fe^{3+} (Lovely, 2000; Francis *et al.*, 2000). Nielson (1996) and Ivanov *et al.* (2002, 2004) concluded that IRB are present in anaerobic reactors as part of a microbial consortium. Therefore, they can be used as a source of IRB for the biodegradation of POCs. Two pure cultures of IRB, *Stenotrophomonas maltophilia* BK and *Brachymonas denitrificans* MK, identified by 16S rRNA gene sequencing, were isolated from anaerobic sewage sludge of wastewater treatment plant using return liquor as carbon and energy source and iron hydroxide as oxidant (Ivanov *et al.*, 2005). Ivanov *et al.* (2005) further showed the ability of *Stenotrophomonas maltophilia* BK to reduce Fe(III) using xenobiotics such as diphenylamine, m-cresol, 2,4-dichlorphenol and p-phenylphenol.

Based on these findings, it was assumed that the concentration of the targeted POCs in return liquor after aerobic and anaerobic treatment of municipal wastewater could be diminished by stimulation of microbial activity of IRB. The kinetics of biodegradation of targeted pharmaceuticals such as ibuprofen (Ibu), naproxen (Nap) and steroids estrone, estriol, 17β-estradiol, 17α-ethynylestradiol by the enrichment and pure cultures of IRB were studied. The pharmaceuticals Ibu and Nap were chosen since they are commonly used as anti-inflammatory drugs while EE2 constituted a major component of synthetic estrogens. Biodegradation of natural estrogens such as 17β-estradiol (E2) and estriol (E3) using IRB were also investigated.

Methods and materials

Growth of enrichment culture of IRB

An enrichment culture of IRB was selected using return liquor under anaerobic conditions. The enrichment culture was proportioned as follows, all in millilitres: return liquor after anaerobic digester of activated sludge suspension, 28.25; mixture of 2 ppm ibuprofen, naproxen and EE2, 1.75 to give final concentration of 50 ppb of each compound in the mixture; anaerobic sludge suspension, 20; suspension of 2.0 g/L $Fe(OH)_3$, 20. Individual samples were sparged with N_2 gas before mixing to eliminate or minimize dissolved oxygen concentration to create anaerobic conditions pivotal to IRB growth. $Fe(OH)_3$, returned liquor and anaerobic sludge were then mixed using an anaerobic chamber. The resulting mixture was incubated in 100 ml serum bottles and sealed with rubber stoppers. Finally, the samples were placed on a shaker (Shimadzu Corporation, Kyoto, Japan) at 120 rpm at ambient temperature that varied from 27°C to 32°C before further analysis on days 1, 3, 6, 9, 12 and 15. Six samples were prepared. Each sample made up a total volume of 70 ml and the remaining 30 ml capacity in each bottle was used to accommodate gaseous products during the course of cultivation. An additional six samples were needed as controls where anaerobic sludge was replaced with distilled water. Indigenous IRB from anaerobic sludge were used as they would have adapted to conditions in the sludge. Biodegradation of the natural estrogens under facultative conditions was studied by repeating the experiment using dewatered sludge and spiking 100 ppb of E2 and E3.

Determination of Fe(II) concentration by phenanthroline method

A phenanthroline method was used to determine the concentration of total ferrous iron produced from the reduction of Fe(III). Measurements were carried out with reference to APHA (1998) and as performed by Ivanov *et al.* (2005). The two main chemical reagents used included phenanthroline solution and ammonium acetate buffer solution. Phenanthroline solution was produced from dissolving 100 mg 1,10-phenanthroline monohydrate in 100 ml water followed by 2 drops concentrated HCl. Ammonium acetate buffer solution consisted 250 g $NH_4C_2H_3O_2$ in 150 ml of water. Aliquots of 5 ml of the sample was obtained by syringe from the serum bottles and then acidified by 1 mol/l HCl in volume ratio 1:1. The acidified sample was incubated for 25 min at 25°C to extract Fe(II). Addition of HCl was necessary to prevent oxidation of ferrous iron to ferric iron. Aliquots of 5 ml of acidified sample was immediately withdrawn and 2 ml of phenanthroline solution and 1 ml of $NH_4C_2H_3O_2$ solution were added. Sample was then subjected to vigorous shaking. Subsequently, water was added as a buffer to make up a total volume of 10 ml and colour intensity was measured within 5–10 min. The colour intensity after reaction with the phenanthroline solution was measured by spectrophotometer UV–1201 V (Shimadzu Corporation, Kyoto, Japan) at a predetermined wavelength of 510 nm. A calibration curve was prepared from serially diluted Fe(II) solution, 1000 mg l^{-1}, (0.355 g of $FeCl_2 \cdot 4H_2O$ in 100 ml of 0.1 mol l^{-1} HCl). Concentration of Fe(II) in standard were, in mg l^{-1}: 1, 5, 10, 25, 100, 1000. The absorbance values of the samples were compared against the calibrated curve to determine the total ferrous content in each sample.

Isolation of IRB from enrichment culture

To isolate species of IRB, responsible for iron reduction, the enrichment culture was used after 15 days of cultivation. The source of Fe(III) for isolation was ferric pyrophosphate (FePPi) since FePPi was a soluble source of iron and had been successfully used to isolate IRB such as *Geovibrio ferrireducens* and *Geothrix fermentans*. To prepare solid medium for IRB isolation, 13 g of Bactoagar (Difco Laboratories, Detroit, USA) and 0.2 L of 5.0 g l^{-1} Fe^{3+} were added to 0.3 l of returned liquor and the medium was sterilized for 15 min at 121°C. The solid medium was then evenly distributed on the Petri dish. The enrichment culture was serially diluted 10-fold with distilled water and 100 μl of resulting preparation was spread onto the solid medium. Colonies appeared on plates with 10^{-3} dilution after incubation for 5 days. A total of 6 colonies with different shapes and sizes were transferred to 2.0 ml microcentrifuge tubes containing 0.4 ml of $Fe(OH)_3$ and 1.0 ml of returned liquor. The tubes were kept in the anaerobic incubator under room temperature of 25–31°C for 5 days to test iron reduction by phenanthroline assay. Pure cultures of IRB from the tube with the greatest color change were further used in research.

Liquid–liquid extraction (LLE) for the detection of POCs in wastewater

LLE was carried out to determine the biodegradation of targeted analytes over a period of 15 days. A total of five standards, two blanks, three controls and three spiked samples for each sampling day were required. Standards were prepared using returned liquor (RL) as the liquid medium with stocks of the analytes at concentrations of 1, 10, 25, 50 and 100 ppb. Samples retrieved were filtered through 0.45 μm filter papers. Each 0.5 ml sample was extracted with 0.5 ml dichloromethane (DCM) for 3 minutes in 2 ml centrifuge tubes. After extraction, the tubes were centrifuged for 3 minutes and top aqueous layers were subsequently removed. The bottom DCM extract was transferred into new 2 ml centrifuge tubes and rinsed with a small volume of DCM for quantitative transfer. The solvents were then evaporated to dryness. Dried samples were reconstituted with 200 μl of 50/50 acetonitrile/H_20 containing 10 ppb E1d4 for samples containing estrogens and 10 ppb Fenoprop for pharmaceuticals as internal standard. Finally, reconstituted samples were vortexed for 15 minutes at 10,000 rpm before being transferred to a HPLC vial for analysis with LCMSMS. The LC/MS/MS system used in this study is a high performance liquid chromatography (HPLC) (Shimadzu, Singapore) coupled to mass spectrometry and tandem mass spectrometry (AB Applied Biosystems, Singapore) that offers good sensitivity and selectivity in the trace analysis of environmental pollutants.

Results and discussion

Fe(III) reduction by enrichment culture

The Fe(III) reduction ability of the enrichment culture was tested using $Fe(OH)_3$ (Figure 1). Disappearance of yellow color of $Fe(OH)_3$ and formation of precipitates such as ferrous salts that gave the sample a pale green color were indicators of Fe(III) reduction. Approximately 13.5% of the 2 g/L Fe(III) were converted to Fe(II) but concentration of Fe(II) is expected to continue, increasing exponentially from 15 to 21 days. Maximum rate of Fe(III) reduction occurred during days 12 to 15. The data can be used in the design of pharmaceutical removals in wastewater to optimize the reduction ability of the IRB.

It was noted that not all Fe(III) added was able to be reduced to Fe(II). This may be due to ferric iron forming soluble chelates with organic acids of return liquor or oxidation of a portion of ferrous to ferric upon contact with air during sampling.

Biodegradation of POCs by enrichment culture of IRB

Reduction in concentrations of the spiked POCs was determined from the difference between recovery concentrations in aqueous samples of the enrichment culture and control. Biodegradation was significant after day 6 of cultivation. Removal efficiencies of 31% and 25% were achieved for ibuprofen and naproxen, respectively, while EE2 remained resistant to biodegradation by IRB (Figure 2). Assuming the POCs obeyed first-order kinetics, maximum degradation rates of 0.12 d^{-1}, 0.09 d^{-1}, 0.05 d^{-1} were obtained for ibuprofen, naproxen and EE2, respectively. Both ibuprofen and naproxen are polar carboxylic acids and are more likely to be chelated with Fe(III) particles. Biodegradation of EE2 includes oxygenation and cleavage of aromatic ring and can be performed, probably, only under aerobic conditions. Ternes *et al.* (1999) confirmed that batch cultivation with activated sludge under aerobic conditions resulted in approximately 20% removal of initial EE2 concentration of 1 μg L^{-1}. Removal of POCs may, in part, depend on their physiochemical properties such as water solubility, affinity to surfaces, and molecular weight. Thus, EE2 was found to be the most persistent among the POCs investigated and this may be attributed to its low solubility.

Based on LC-MS/MS results from the control samples of 6–15 days of IRB cultivation, recovery of EE2 concentration ranged from 1 to 3 μg/L, which is in contrast with recovery concentrations of 8–25 μg/L for both ibuprofen and naproxen. This is consistent with the fact that EE2 possess a three to seven times higher absorption coefficient than ibuprofen and naproxen. Therefore, most of the EE2 may have been removed via sorption onto solid matrices such as the anaerobic sludge and particles of $Fe(OH)_3$ that were subsequently retained on filter membranes before being subjected to LCMSMS analysis. Thus, a smaller amount of EE2 may be present in the aqueous phase of the enrichment and pure cultures to undergo oxidation, contributing to its lower percent of biodegradation by IRB. It was further observed that the recovery concentrations of ibuprofen and naproxen in the aqueous phase from control samples decreased with time. This agrees with findings that indicate an increase in water sludge partition coefficient, k_p, for compounds with one carboxyl group such as ibuprofen and naproxen as pH is lowered. Urase and Kikuta (2005) found k_p values for pharmaceuticals with 1 carboxyl groups increasing 2.5–3.0 times higher when pH was lowered from 6.7 to 5.6. Hence ibuprofen and naproxen may more likely be subjected to removal by absorption and sedimentation from the aqueous phase as production of H^+ ions accompanies Fe(III) reduction and ferrous ions hydrolysis. EE2, however, experienced insignificant changes at lowering pH in its k_p value since it does not contain a carboxyl functional group and thus its concentration in the aqueous phase remained unchanged.

Biodegradation of naproxen, after 15 days of cultivation, produced a metabolite (Figure 3), which is presumed to be desmethyl-naproxen since this ether cleavage is well documented for mammalian metabolism (Sidelmann *et al.*, 2001) and transformation by fungi (Zhong *et al.*, 2003). It was observed that the removal of naproxen on day 15 was less than on day 12. This indicated false positive negative results were obtained and can be attributed to matrix ionization suppression effects since return liquor constitutes a complex matrix.

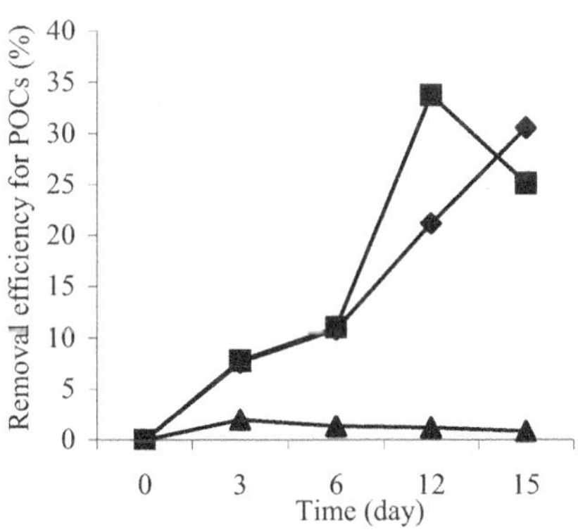

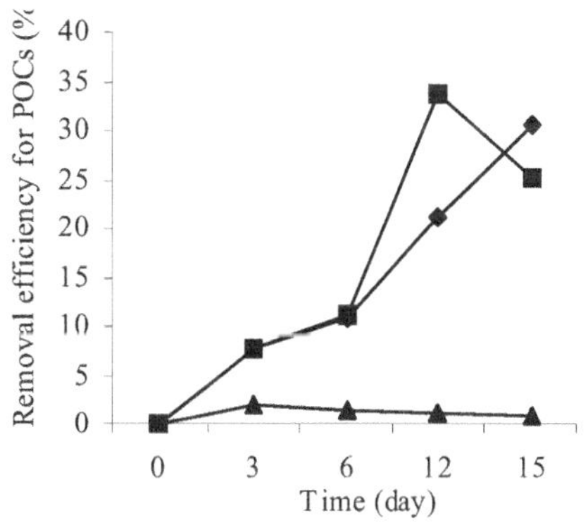

Figure 1 (left) Changes of total ferrous concentration in enrichment culture (♦) and control without Fe^{3+} (■). All data are averages of triplicates.

Figure 2 (right) Per cent removal of ibu (♦), nap (■) and EE2 (▲) over 15 days using $Fe(OH)_3$ as Fe(III) source.

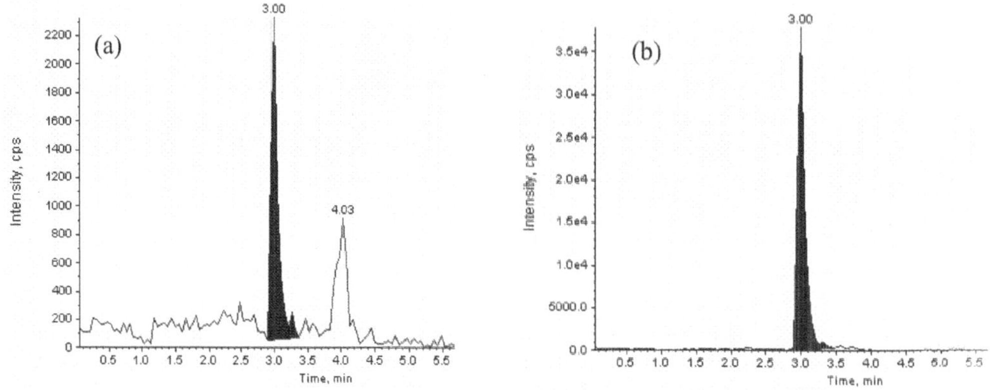

Figure 3 MRM chromatograms of (a) naproxen (RT: 3.0) with possible formation of metabolite (RT: 4.0) after 15 days of cultivation, and (b) naproxen in control sample.

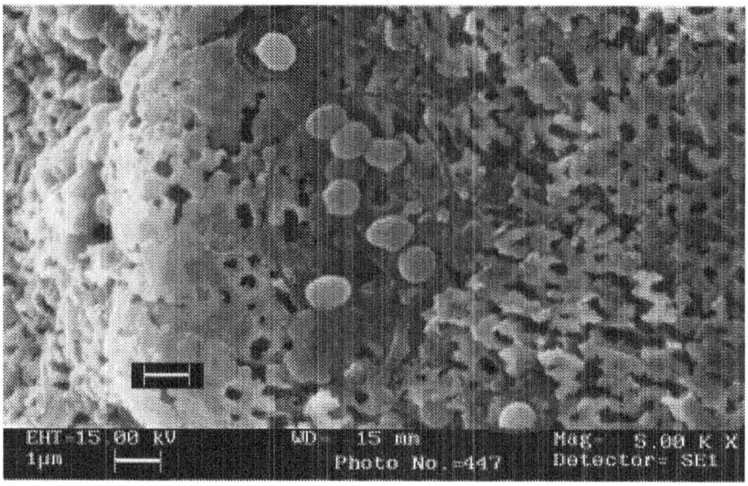

Figure 4 SEM view of the isolated colonies at magnification 5000×; scale bar, 1 µm.

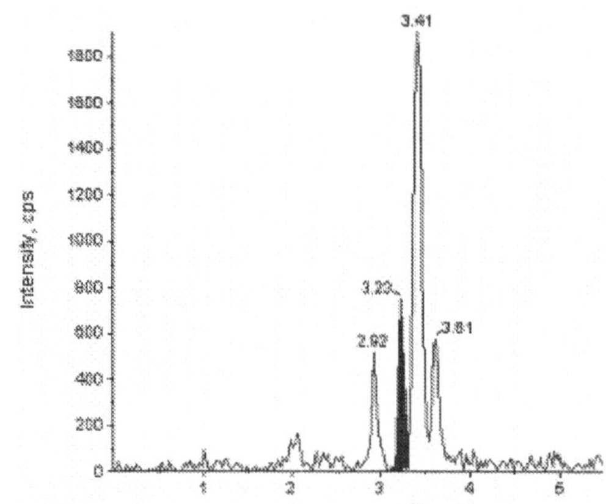

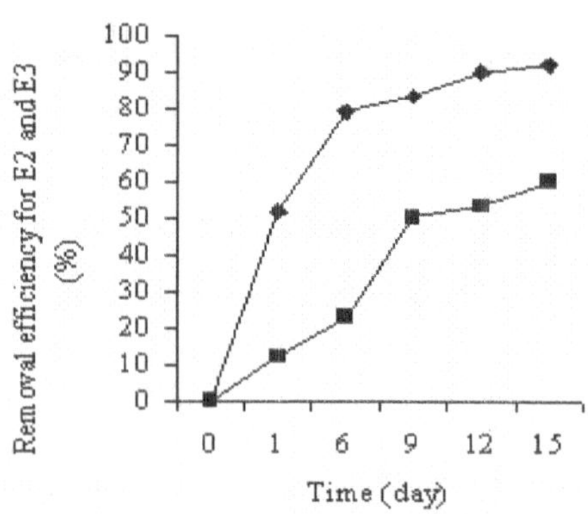

Figure 5 (left) MRM chromatography of E2 (RT: 3.23) being converted to E1(RT: 3.41) and 2 possible metabolites from sample of day 6.

Figure 6 (right) Percent removal of E2 (♦) and E3 (■) over a period of 15 days under facultative conditions.

Isolation of pure IRB culture

Colonies appeared on plates with 10^{-3} dilution after incubation for 5 days. Cells were examined using the SEM (Figure 4). Larger colonies present may be due to cells aggregation. Some of the white spots may presumably be white precipitates of vivianite, $Fe_3(PO_4)_2$. Five colonies with different shapes and sizes were isolated. The cultivation was repeated using an inoculum of the pure culture (5% v/v), which was added to 38 ml of returned liquor and 28.5 ml of $Fe(OH)_3$. A sample on day 15 indicated Fe(II) production to be approximately 195 mg l^{-1}. This corresponds to 72% of Fe(III) reduction. Consequently, biodegradation of ibuprofen and naproxen was observed and removal efficiencies were 24% and 19%, respectively.

Biodegradation of natural estrogens using enrichment culture of IRB

Natural estrogens have slightly higher solubilities and smaller partition coefficient values as compared with synthetic steroids such as EE2. This suggests that E2 and E3 can better dissolve in the aqueous phase for biodegradation by IRB and are thus chosen as POCs to be investigated. Natural estrogens have been suggested as the major compounds responsible for endocrine disruption in wildlife (Chang *et al.*, 1996) and E2 was reported to have the most significant contribution to the endocrine disruption potential in treated water (Yasuoko *et al.*, 1999). The experiments were conducted under anoxic conditions with Fe(III) since Ying *et al.* (2003) found degradation of E2 under anaerobic conditions to be slow, with a half-life of 107 days.

E2 achieved more than 80% removal on day 6 and near complete removal after day 12 cultivation with $Fe(OH)_3$ and IRB while E3 in the aqueous phase was reduced by more than 55% (Figure 6). The solubility of EE2 was approximately 3-fold less than those of the natural estrogens and this fact is believed to contribute to the increased resistance of EE2 to biodegradation as compared to E2 and E3. It was also noted that large fraction of E2 was quantitatively converted to E1, as shown in Figure 5. This is partially consistent with findings by Lee and Liu (2002) who found 88% of E2 to be aerobically biodegraded within a day while more than 50% of E2 was removed after 7 days under anaerobic conditions. Thus, it appears that E1 is likely one of the common key intermediates involved in the degradation of E2 by sewage organisms. A metabolite was further observed to have elucidated at retention time of 3.6 min and this can be tentatively identified as a lactone since it is well known that lactones can be easily formed by dehydratation of γ-hydroxy acids and further oxidation of E1 coupled with cleavage at C-17 produces unstable hydroxy acid. Also, the early eluting peak at a retention time of 2.92 min may indicate the production of a hydrophilic organic compound from further hydroxylation of E2 and E1.

Conclusion

EE2 showed to be the most persistent POC for the biodegradation by IRB among the targeted compounds. In contrast, E2 was biodegraded using IRB by more than 90%. The difference in removal efficiencies between pharmaceuticals and estrogens demonstrated that chemical structures of the POCs and levels of oxygenation may also be important factors to consider. Removal of POCs, in this study, took into consideration only those that were available for biodegradation in the aqueous phase while those removed by absorption onto Fe(III) particles were not accounted for. Overall, addition of Fe(III) sources and IRB proved to be proficient in the removal of POCs under anaerobic conditions.

References

APHA (1998). *Standard Methods for the Examination of Water and Wastewater*, 20th edn, American Public Health Association/American Water Works Association/Water Environment Federation, Washington DC, USA.

Baronti C., Curini, R., D'Ascenzo G., Di Corcia A., Gentili A. and Samperi R. (2000). Monitoring natural and synthetic estrogens at activated treatment plants and in receiving river water. *Environmental Science and Technology*, **34**, 5059–5066.

Buser, H.R., Poiger, T. and Muller, M.D. (1999). Occurrence and environmental behavior of the chiral pharmaceutical drug ibuprofen in surface waters and in wastewater. *Environmental Science and Technology*, **33**, 2529–2535.

Chang, L.T., Yu, N.W., Hsu, C.Y. and Liu, H.W. (1996). Gonadal transformation in male *Rana catesbeiana* tadpoles intraperitoneally implanted with estradiol capsules. *General Comp. Endocrinol.*, **102**, 299–306.

Ivanov, V., Stabnikova, E., Stabnikov, V., Kim, I.S. and Zubair, A. (2002). Influence of iron compounds on the treatment of fat-containing wastewaters. *Applied Biochemistry and Microbiology*, **38**, 255–258.

Ivanov V., Tay S. T.-L., Wang J.-Y., Stabnikova O., Stabnikov V., Xing Z. and Tay J.-H., (2004). Improvement of sludge quality by iron-reducing bacteria. *Residuals Science and Technology*, **1**, 165–168.

Ivanov, V.N., Stabnikov, E.V., Zhuang, W.Q., Tay, J.H. and Tay, S.T.L. (2005). Phosphate removal from the returned liquor of municipal wastewater treatment plant using iron-reducing bacteria. *Journal of Applied Microbiology*, **98**, 1152–1161.

Lee, H.B. and Liu, D. (2002). Degradation of 17β-estradiol and its metabolites by sewage bacteria. *Journal of Water, Air, and Soil Pollution*, **134**, 353–368.

Lovley, D.R. (1991). Dissimilatory Fe(III) and Mn(IV) reduction. *Microbiology Review*, **55**, 259–287.

Nealson, K.H. and Saffarini, D. (1994). Iron and manganese anaerobic respiration – environmental significance, physiology and regulation. *Annual Review on Microbiology*, **48**, 311–343.

Nielsen, P.H. (1996). The significance of microbial Fe(III) reduction in the activated sludge process. *Water Science and Technology*, **34**, 129–136.

Sidelmann, U.G., Bjornsdottir, I., Shockcor, J.P., Hansen, S.H., Lindon, J.C. and Nicholson, J.K. (2001). Directly coupled HPLC-NMR and HPLC-MS approaches for the rapid characterisation of drug metabolites in urine: application to the human metabolism of naproxen. *Journal of Pharmaceutical and Biochemical Analysis*, **24**, 569–579.

Ternes, T.A., Stumpf, M., Mueller, J., Haberer, K., Wilken, R.D. and Servos, M. (1999). Behavior and occurrence of estrogens in municipal sewage treatment plants. *Science of the Total Environment*, **225**, 81–90.

Urase. T. and Kikuta, T. (2005). Separate estimation of adsorption and degradation of pharmaceutical substances and estrogens in activated sludge in acidic operational condition. *Water Research*, **39**, 1289–1300.

Vader, J.S., Ginkel, C.G. van, Stokman, F.M., Sperling, G.M., Jong, J. de, Boer, W. de, Graff, J.S. de, Most, M. vander, and Stokkman, P.G.W. (2000). Degradation of ethinly estradiol by nitrifying activated sludge. *Chemosphere*, **41**, 1239–1243.

Yakou, Y., Takahashi, A., Higashitani, T. and Tanaka, H. (1999). Measurement of estrogen-like activity in wastewater using recombinant yeast. *Environmental Engineering Research*, **36**, 199–208.

Ying, G., Kookana, R.S. and Dillon, P. (2003). Sorption and degradation of selected five endocrine disrupting chemicals in aquifier material. *Water Research*, **37**, 3785–3791.

Zhong, D.F., Sun, L., Ei, L. and Huang, H.-H. (2003). Microbial transformation of naproxen by *Cunninghamella* species. *Acta Pharmacology Sinica*, **24**, 442–447.

Comparative study of formation and characterization of aerobic granules in SBR and SABR for treatment of 4-chloroaniline-containing organic wastewater

Xu Xiangyang*, Zhu Liang, Luo Weiguo, Zhang Nini and Li Jun**

Department of Environmental Engineering, Zhejiang University, Hangzhou 310029, China (E-mail: *xuxy@zju.edu.cn*)

Abstract The formation and characterization of aerobic granules in three column-type sequencing batch reactors (R1, R2 and R3, which had different geometrical configuration) for treatment of 4-chloroaniline(4-CA)-containing organic wastewater were investigated in this paper. Granular sludges were observed since 15 days after start-up in R2 and R3 which had high H/D ratio. Since then and within the subsequent 75 days, the granulation of aerobic sludge was apparently developed by decreased settling time and gradual increased 4-CA concentration to about 400 $mg \cdot L^{-1}$ in R1 to R3. Although the performance of aerobic sludges in three reactors was declined with further increase in 4-CA concentration beyond 450 $mg \cdot L^{-1}$, the removal efficiency of 4-CA and the performance of granular sludge came back within 10 days. The granulation of aerobic sludges tended to be mature in all reactors that were continuously operated in a 12 hour cycle time with a settling time of 5 min, superficial air velocity of 2.4 $cm \cdot s^{-1}$, COD loading rate of 2.4~3.6 $kg \cdot m^{-3} \cdot d^{-1}$ and 4-CA loading rate of around 800 $g \cdot m^{-3} \cdot d^{-1}$, and the removal efficiencies of COD, TN and 4-CA were maintained above 93%, 70% and 99.9% respectively. Mature aerobic granules in R1 to R3 featured with average diameter of 0.78, 1.68 and 1.25 mm, minimal settling velocity of 20.5, 70.1 and 66.6 $m \cdot h^{-1}$, SVI of 38, 31 and 31 $mL \cdot g^{-1}$, SOUR of 18.4, 26.4 and 30.2 $mgDO \cdot gVSS^{-1} \cdot h^{-1}$, specific 4-CA degradation rates of 0.14, 0.21 and 0.27 $g \cdot gVSS^{-1} \cdot d^{-1}$, and PN/PS of 8.2, 10.8 and 13.7 $mg \cdot mg^{-1}$, respectively. The aerobic granules in R2 and R3 had larger granule size, denser granular structure, higher specific 4-CA degradation rate and SOUR than the granular sludge of R1. This study demonstrated that the advantages of the aerobic sludge granular technology to handle high-strength toxic organic wastewater, and reactor with a high H/D ratio and an internal-circulation favoured the granulation and stabilization of aerobic sludge.

Keywords 4-chloroaniline(4-CA); aerobic granule; organic wastewater; sequential air-lift bioreactor(SABR); SBR

Introduction

Bio-granules are compact and dense microbial aggregates packed with different bacterial species formed through self-immobilization of microorganisms. Compared to the conventional floc-sludge, bio-granules have a excellent settling property for enabling a high biomass retention, and withstand high-strength wastewater and its shock loading. Many wastewater treatment plants already apply anaerobic granulation technology, the feasibility and efficiency of UASB reactors and their various modifications for removing biodegradable organic matter have been successfully demonstrated from municipal and industrial wastewater (Hulshoff and Lettinga, 1983; Schmidt and Ahring, 1996). However, anaerobic granulation technology has some drawbacks: these include the need for a long start-up period, a relatively high operation temperature, unsuitability for the removal of nutrients (N and P) and for low-strength organic wastewater.

To overcome those weaknesses, Mishima and Nakamura (1991) first found aerobic granular sludge in a continuous aerobic up-flow sludge blanket reactor. Since then and within the subsequent 10 years, aerobic granulation has been reported by many researchers in treating wastewater containing organics, nitrogen and phosphorus in column-type upflow reactors (Morgenroth *et al.*, 1997; Beun *et al.*, 1999; Etterer and Wilderer, 2001; Tay *et al.*, 2001a,b, 2002; Moy *et al.*, 2002; Liu *et al.*, 2004). Almost all aerobic granules have been cultivated on simple and relatively benign substrates such as glucose, acetate and alcohol, and not much is known about the ability of these granules for high-rate degradation of toxic chemicals, only Tay *et al.* (2004) reported high-rate biodegradation of phenol by aerobic granules in recent years.

(Chloro)-anilines, as an important intermediate in the production of a very wide range of synthetic organic chemicals and polymers including polyurethanes, rubber additives, dyes, pharmaceuticals, pesticides and herbicides, are highly toxic and readily absorbed through the skin in dangerous amounts and fatal if swallowed or if their vapours are inhaled. While not as easily biodegraded as aniline and phenol, (chloro-)anilines are none the less biodegradable and its biodegradation has been reported by several authors (O'Neill *et al.*, 2000; Boon *et al.*, 2002, 2003; Jiang *et al.*, 2002; Tay *et al.*, 2004). The main objective in the current study was to investigate the feasibility of the aerobic sludge granular technology to handle high-strength 4-chloroaniline-containing organic wastewater in SBR and sequencing airlift bioreactor (SABR) that had different geometrical configurations, and would be helpful in the development of aerobic granule-based bioreactor for high-strength toxic organic wastewater.

Methods

Experimental design and setup

Experiments were performed in three column-type sequencing batch reactors, which had different geometrical configurations, designated R1, R2 and R3. R1 and R2 were conventional SBRs with different H/D (R1 of 14 cm in diameter and 3 in H/D, R2 of 8 cm in diameter and 12 in H/D), and R3 was SABR, the internal diameter of the down-comer in R3 was 8 cm, and the riser was 64 cm in height and 5 cm in internal diameter. The configuration of SABR (R3) is shown in Figure 1.

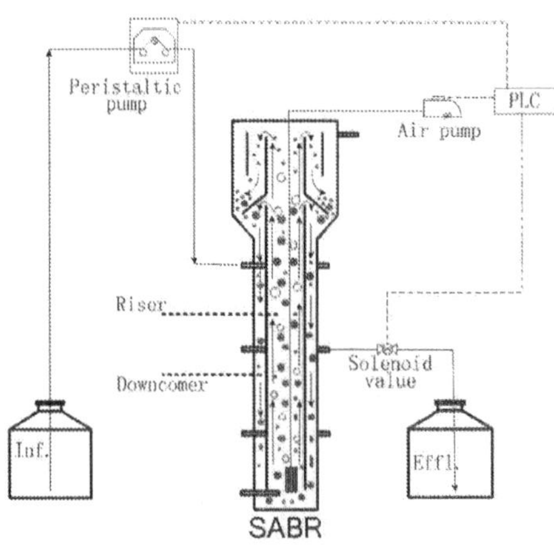

Figure 1 Schematic diagram of sequencing airlift bioreactor (SABR).

Three reactors were all operated sequentially in 12 h cycle with a working volume of 5 l in a temperature-controlled room at 25±2°C. Effluent was discharged from the relevant port of the column reactor at a volumetric exchange ratio of 70%, the hydraulic residence time (HRT) was 17.2 hrs. Fine air bubbles for aeration were supplied through a dispenser at the reactors bottom at a superficial air velocity of 2.4 cm·s^{-1}. pH was controlled at 6.7±0.2 with the addition of $NaHCO_3$ in the influent.

Seed sludge

Reactors R1–R3 were inoculated by 2.5 l of fresh activated sludge (equivalent to 7.8–8.0 gSS·L^{-1}) taken from Hangzhou municipal wastewater treatment plant. This gave an initial reactor biomass concentration of about 4 g L^{-1}. The activated sludge had the following characteristics: volatile suspended solid/suspended solid (VSS/SS) of about 55%, sludge volume index (SVI) of about 120 ml·g^{-1}, specific oxygen utilization rate (SOUR) of around 8.0 mgDO·gVSS^{-1}·h^{-1}.

Medium composition

Three reactors were fed with 4-chloroaniline(4-CA) as target contamination and a fresh synthetic wastewater whose composition was as follows (concentration in mg·l^{-1}): glucose, 240–870 (260–930 as COD); CH_3COONa, 360–1290 (240–870 as COD); yeast, 0.0006; NH_4Cl, 120 (30 as N); KH_2PO_4, 44 (10 as P); $K_2HPO_4{\cdot}3H_2O$, 59; $CaCl_2{\cdot}2H_2O$, 23; $MgSO_4{\cdot}7H_2O$, 25; $FeSO_4{\cdot}7H_2O$, 20; $NaHCO_3$, 200 (120 as $CaCO_3$) and trace mineral solution, 1.0 ml. The composition of the trace mineral solution was based on that by Tay *et al.* (2001a,b).

Experimental scheme

Experiments were conducted in three phases; an overview of the experiments is given in Table.1. The aim of Phases I and II was to develop granulation at an increased 4-CA loading rate of about 800 g·m^{-3}·d^{-1}. Because the reactors were deteriorated by higher 4-CA concentration of 450 mg·L^{-1} at the end of Phase II, in Phase III three reactors had undergone a renewed period which 4-CA loading was achieved by increasing the influent 4-CA concentration from 180 to 400 mg·l^{-1}, and the granulation of aerobic sludge finally tended to be mature in the three reactors.

Table 1 Overview of the operational conditions for granulation of aerobic sludge

Operational parameter \ Period	I	II	III
Cycle time /h	12	12	12
Influent /min	20	20	20
Aeration /min	865	870	870
Settling /min	10	5	5
Effluent withdrawal /min	5	5	5
HRT /h	17.2	17.2	17.2
Influent COD /mg·l^{-1}	500–1200	1000–1800	700–1800
Influent 4-CA /mg·l^{-1}	0–10	10-450	180-400
Superficial air velocity /cm·s^{-1}	2.4	2.4	2.4

Analytical methods

Wastewater samples in the reactors were periodically analyzed for pH, chemical organic carbon(COD), sludge volume index (SVI), average biomass density, minimal settling velocity, Specific oxygen utilization rate (SOUR), SS/VSS using standard methods (APHA, 1998).

To measure 4-CA, samples were centrifuged for 10 min at 10,000 rpm, filtered using 0.22 μm filter paper and analyzed with high performance liquid chromatography, which consisted of a Waters1525 binary HPLC pump, a Waters2487 dual λ absorbance detector, and a Waters717plus autosampler (Waters, Milford, MA, USA). A ZORBAX Eclipse SB-C18 column (4.6 mm × 150 mm I.D., 5 μm particle size) (Agilent, USA) was used for separation. Reversed-phase separations were performed using a mobile phase of water–methanol (30:70, v/v) flowing at 1.0 $mL \cdot min^{-1}$. The detection wavelength was set at 240 nm.

Granule strength was determined according to the method of Ghangrekar *et al.* (1996), it was expressed as an integrity coefficient (IC, %), in which the ratio of residual volatile suspended solid (VSS) after a defined amount of agitation to total VSS of the intact granules prior to agitation.

Extracellular polymers (EPS) was recovered by first centrifuging granules (2,500*g*, 10 min) to a final concentration of about 5 $g \cdot L^{-1}$ in a 20 mL volume of phosphate buffer (pH 7.6); Then sample was placed in an ice bath and sonicated over four 30 s periods (interrupted by a short 10 s resting period) using a sonicator operating at 50 W, 20 KHz; EPS were recovered by centrifuging (14,000g, 20 min) at 4°C. A phenol-sulfuric acid method was used to quantify polysaccharides with glucose as the standard (Dubois *et al.*, 1956), and proteins were measured as described previously (Bradford, 1976), with bovine serum albumin (BSA) as the standard.

The morphology and size of granular sludge was obtained by an image analysis system (DMLB+QCOLite, Leica Cambridge Instruments). For observation with scanning electron microscopy (SEM), the specimen was first fixed with 2.5% glutaraldehyde in phosphate buffer (pH 7.0) for more than 4 hrs and washed three times in the phosphate buffer, then post-fixed with 1% OsO_4 for 1 hour and washed three times in the phosphate buffer (pH 7.0). Fixed granule was dehydrated by a graded series of ethanol (50%, 70%, 80%, 90%, 95% and 100%) for about 20 mins at each step, transferred to the mixture of alcohol and iso-amyl acetate (v:v = 1:1) for about 30 min, then transferred to pure iso-amyl acetate for about 1 h. In the end, the specimen was coated with gold–palladium after dehydrated in Hitachi Model HCP-2 critical point dryer with liquid CO_2, then observed in Philips Model XL30 ESEM.

Results and discussion

Granulation of aerobic sludge

The process of sludge granulation could be divided into three different stages, i.e., start-up, aerobic granule formation and maturation, based on the changes of sludge morphology, minimal settling velocity, SVI and 4-CA-degrading performance. The main results of the three respective reactors treating the inhibitory substrate 4-CA were presented in Table 2 and Figures 2–6.

Table 2 Characteristics of aerobic granular sludges in different operation stages

Periods	Reactor	MLSS ($g{\cdot}L^{-1}$)	Effluent SS ($g{\cdot}L^{-1}$)	SVI ($mL{\cdot}g^{-1}$)	SRT d^{-1}	Minimal settling velocity ($m{\cdot}h^{-1}$)	SOUR ($mg{\cdot}g^{-1}{\cdot}h^{-1}$)
Seed sludge	All	~7.5	–	~120	–	< 1	~8.0
I Start-up	R1	1.29–2.69	1.03–1.57	83–208	2.4	2.1	26.3–29.1
	R2	1.76–3.16	1.05–1.45	80–128	2.2	8.2	13.4–26.1
	R3	1.44–2.61	0.99–1.83	72–187	1.7	8.1	26.1–26.5
II Loading increase and granule formation	R1	2.91–5.73	1.61–2.37	81–213	3.1	8.6	13.1–38.3
	R2	3.15–10.65	1.21–1.62	26–75	6.3	16.7	10.6–30.6
	R3	2.65–10.10	1.11–1.95	25–94	3.7	15.5	12.2-34.5
III Reactor resteadying and granule maturation	R1	3.28–10.39	0.85–2.87	38–141	6.3	20.5	12.2-27.2
	R2	3.66–9.32	0.63–2.37	31–128	3.9	70.1	20.5-26.4
	R3	5.42–9.76	0.50–1.88	31–129	3.9	66.6	15.7-30.2

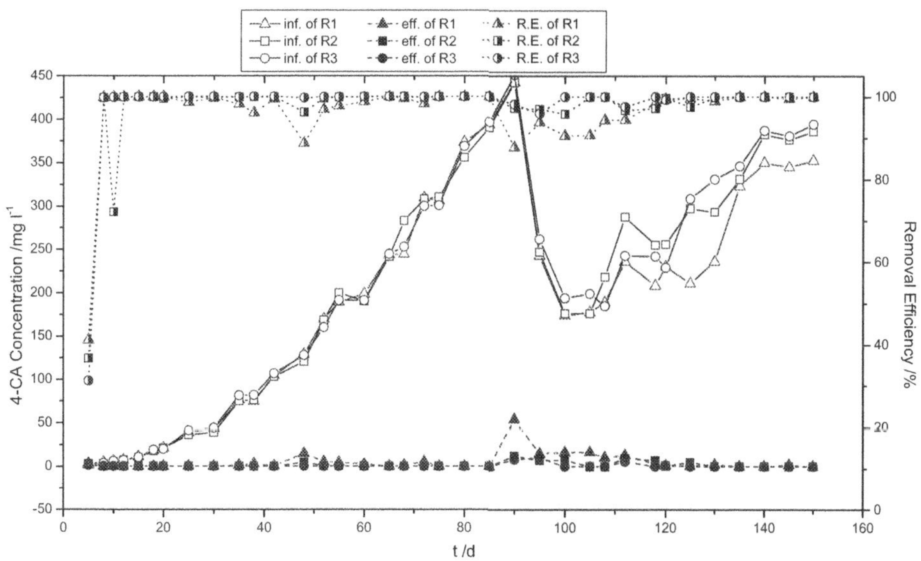

Figure 2 Removal of 4-CA during the whole aerobic sludge granulation of R1 to R3.

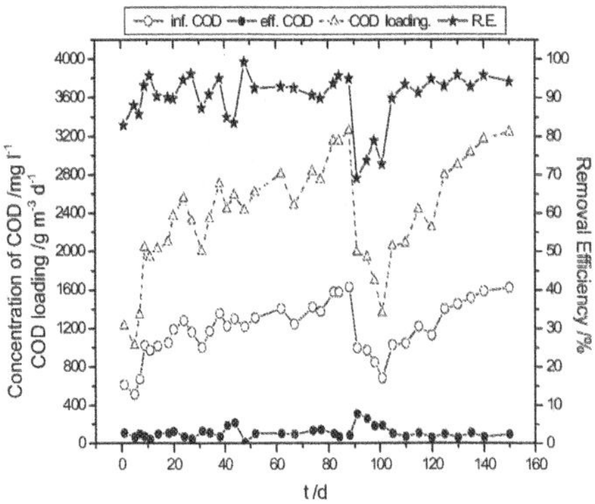

Figure 3 COD loading rate and removal performance of R1 during sludge granulation.

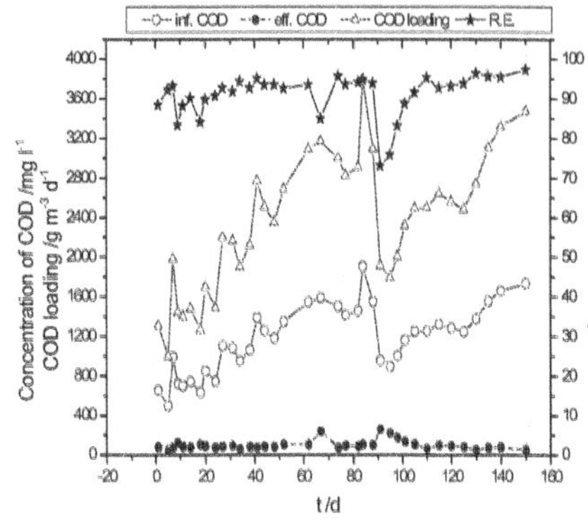

Figure 4 COD loading rate and removal performance of R2 during sludge granulation.

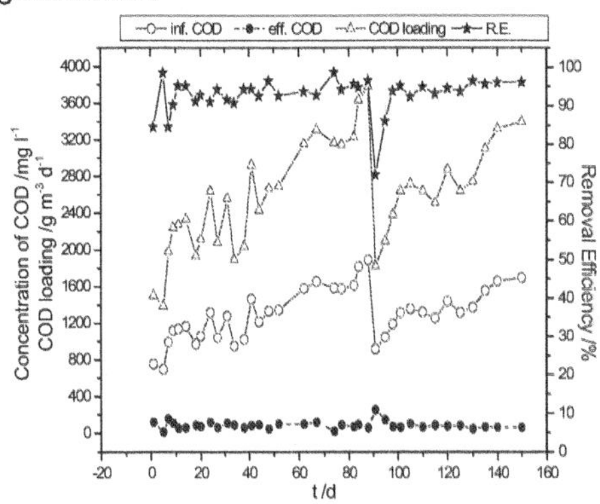

Figure 5 COD loading rate and removal performance of R3 during sludge granulation.

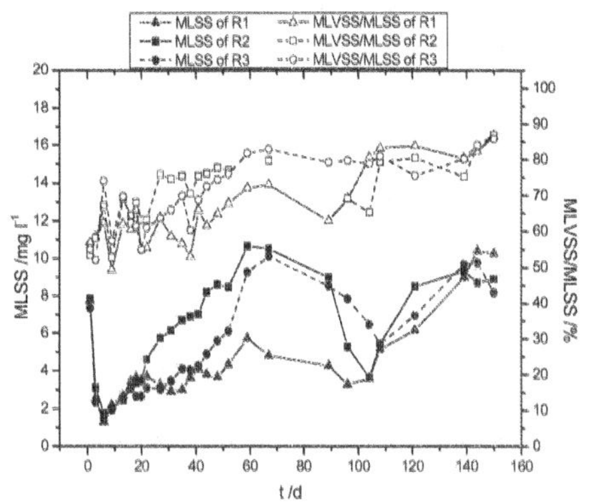

Figure 6 Variation of VSS/SS during sludge granulation in R1 to R3.

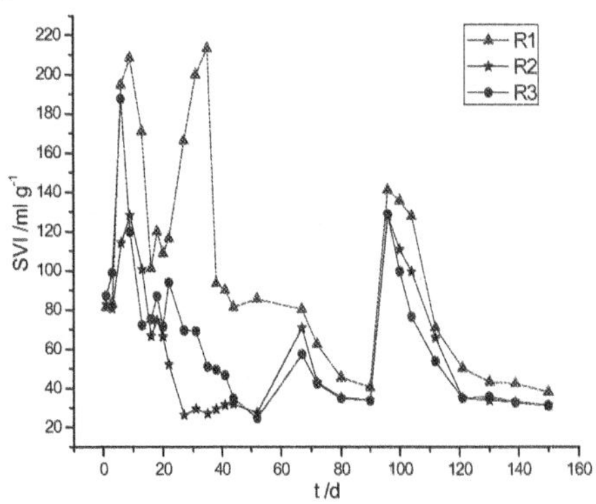

Figure 7 Variation of SVI during sludge granulation in R1 to R3.

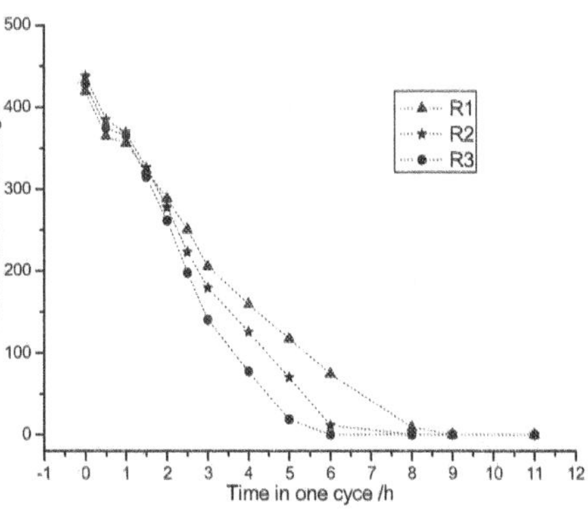

Figure 8 Profiles of 4-CA removal performances in one cycle.

Phase I: start-up of the reactors

Three reactors were started up in 12 h cycle time with settling time of 10 min, superficial air velocity of 2.4 cm·s^{-1}, COD loading rate of 1.0–2.4 kg·m^{-3}·d^{-1}. A shorter settling period was applied to preferentially select for the growth of fast settling microbial aggregates and wash out the sludge with a poor settleability during the initial phase. After Day 6, three reactors were operated at a 4-CA loading rate of 20 g m^{-3} d^{-1}, 4-CA removal efficiency gradually reached 99% (Figure 2).

Highly floc-like sludge dominated in three reactors over several days, small granules first appeared in R2 and R3 on Day 15, by which time the mean biomass sizes were around 0.32 and 0.38 mm, and the SVI value had decreased to less than 100 mL·g^{-1} (Figure 7). At the end of Phase I, the biomass concentration stabilized at around 3 g·L^{-1}, the minimal settling velocity of sludges in R1 to R3 increased from less than 1 m·h^{-1} to 2.1, 8.2 and 8.1 m h^{-1}, respectively (Table 2).

Phase II: loading increase and granule formation

Since Day 16, three reactors were operated at settling time of 5 min, COD loading rates of 1.0~3.6 kg·m^{-3}·d^{-1} and an increased 4-CA loading rate from 0.04 to 0.8 kg m^{-3} d^{-1}. Compared with increased 4-CA influent concentration, 4-CA removal efficiency gradually reached 99.9% beyond Day 50 in R2 and R3, and 4-CA concentration in the effluent both decreased to below 0.01 mg L^{-1}. After Day 75, 4-CA removal efficiency of R1 reached 99.9% too. During the next 10 days, the rapid increase of 4-CA concentration to 400 mg·L^{-1} did not diminish the ability of the aerobic granules to remove 4-CA, 4-CA and COD concentrations in the effluent of three reactors were all below 0.01 and 100 mg·L^{-1}, 4-CA and COD removal efficiency were higher than 99.9% and 90% (Figures 2–5).

Most of floc-like sludge was continuously washed out along with some major hydraulic selection pressures (e.g., hydrodynamic shear force, settling time) and increased organic loading, then new exiguous granulation occurred. By Day 35, the aerobic granules had a more or less "steady state" average diameter of 0.60 and 0.64 mm in R2 and R3, the SVI value had declined to less than 50 mL·g^{-1} (Figure 7), indicating that the granules eventually displaced the activated sludge flocs to become the dominant form of biomass in these two reactors. The rapid transition from activated floc-sludge to granular sludge showed that the chosen operating strategy was successful. Since then and within the subsequent 50 days, the increase in 4-CA loading rate in R1–R3 was accompanied by the biomass concentration increased, and reached at 4.27, 8.98 and 8.53 g L^{-1} (Figure 6), and the mean granule sizes also to 0.86, 1.57 and 1.19 mm, respectively.

While the three reactors were operated at a 4-CA concentration of maximum approximately 450 mg·L^{-1} on Day 88, 4-CA removal efficiency suddenly declined to 87.7%, 95.9% and 96.0%, and COD removal efficiency declined to 69.2%, 73.1% and 77.5%, respectively (Figures 2–5). The SVI value had all increased from about 50 ml·g^{-1} to higher than 120 ml g^{-1} in three reactors (Figure 7). The rapid transition from granular sludge to activated sludge flocs showed that three reactors were deteriorated by higher 4-CA concentration.

Phase III: reactor re-steadying and granule maturation

The self-immobilization of microbial cells into granule could improve the ability of cells inside the granules to achieve good structural properties and tolerate high toxic concentration. Although the 4-CA degradation performances of aerobic sludge in three reactors were declined at 4-CA concentration beyond 450 mg·L^{-1}, three reactors were operated at 4-CA loading rate from 0.5 to 0.8 $kg \cdot m^{-3} \cdot d^{-1}$ in phase III, to evaluate the performance of the cultivated aerobic granule again to high 4-CA concentration of around 400 $mg \cdot L^{-1}$.

After 4-CA concentration was reduced to about 180 $mg \cdot L^{-1}$, 4-CA and COD removal efficiency in R2 and R3 both reached around 99% and 90% within 10 days [Fig.2~Fig.5], subsequently the performance of R3 reactor came back too. After Day 140, three reactors tended to be steady, the removal efficiencies of 4-CA, COD and TN was maintained above 99.9%, 93%, 70%, respectively.

At the end of the experiment, the SVI values of 38, 31 and 31 $mL \cdot g^{-1}$, the biomass concentrations of 10.26, 8.91 and 8.18 $g \cdot L^{-1}$, and the net biomass yield of 0.68, 0.67, 0.39 $gVSS \cdot gCA^{-1}$ were finally maintained in R1 to R3. Furthermore, the increase in 4-CA loading rate did not lead to a decrease in granule activity, as represented by SOUR. On the contrary, a higher SOUR values of 18.4, 26.4 and 30.2 $mgDO \cdot gVSS^{-1} \cdot h^{-1}$ were achieved for the granules sampled at the end of Phase III. The granulation of aerobic sludge tended to be mature, and aerobic granules successfully removed most of the 4-CA fed into the three reactors.

Physical characteristics of aerobic granules

Granule size and strength

Granule size is a direct parameter in moulding the physical performance and characteristics of granular sludge, and plays a significant role in the limitation of mass transport and diffusion. Compared with other reported aerobic and anaerobic granules (Schmidt and Ahring, 1996; Beun *et al.*, 2000, 2002; Liu and Tay, 2004), the aerobic granules cultivated in our experiment had smaller size, the average mature granule diameters in R1–R3 were 0.78, 1.68 and 1.25 mm, respectively. The physical strength of aerobic granules in R1 to R3 was 97.5±0.5%, 98.5±0.4% and 99±0.4%, respectively. Results validated that the aerobic granules with smaller sizes tend to be more compact compared with larger aerobic granules (Toh *et al.*, 2003).

SVI and settling velocity

After 15 day s of operation of R1–R3, the SVI decreased from around 120 $ml \cdot g^{-1}$ to 101.0, 66.6 and 75.3 $ml \cdot g^{-1}$, respectively. Along with the formation and maturation of aerobic granule, the SVI in R1 to R3 gradually decreased to only 38, 31 and 31 $mL \cdot g^{-1}$ at the termination of phase III.

The average settling velocity of cultivated aerobic granules was 20.0, 70.1 and 66.6 $m\ h^{-1}$. Schmit and Ahring (2001) reported that the settling velocity of anaerobic granule ranged between 18 and 100 $m \cdot h^{-1}$, the settling velocity of the aerobic granules cultured in this study was in this range. At the same time, the higher settling velocity of aerobic granule was corresponding to the fact that the water content in granular sludge decreased. The water content of cultivated aerobic granules decreased from about 98.5% to 97.2%, 96.3% and 96.1%. The specific gravity of the aerobic granules was 1.058, 1.088 and 1.101 in R1–R3.

It was clearly observed that the cultivated aerobic granule completely settled down within 10 s, whereas the floc-like sludge was still suspended even after 1 min, suggested that the mature granule had excellent settling capacity. Furthermore, the granules formed in R2 and R3 exhibited higher sedimentation ability than R1, indicating that the granules could be effectively retained in the reactors with higher H/D under perfectly mixed-flow conditions at a high hydrodynamic shear force.

Extracellular polymer substances (EPS)

Extracellular polymer substances (EPS) has been shown to be a rich matrix of polymers, including proteins (PN), polysaccharides (PS), glycoproteins, nucleic acids, phospholipids, and humic acids. Many researches demonstrated that EPS were a major component of cell flocs and bio-films, so they are hypothesized to play a central role in all types of biofilm formation, including flocculation and granulation.

Tay *et al.* (2001a,b) hypothesized PS content to contribute more to the granule structure and stability than the protein content, it reported a PS/PN ratio of 15. However, the PN content was much higher than the PS content in the current experiment, the PN values of aerobic sludges in R1 to R3 increased from 4.2 mg·gVSS^{-1} of inoculated sludge to 27.6, 28.0 and 30.9 mg·gVSS^{-1}, and the PS values increased from 1.2 mg·gVSS^{-1} to 3.35, 2.29 and 2.25 mg·gVSS^{-1} , and the PN/PS ratios increased from 3.5 mg·mg^{-1} to 8.2, 10.8 and 13.7 mg·mg^{-1}, respectively.

In fact, this preferential production of PN over PS in EPS has also been observed in other biofilms exposed to toxic organic compounds (McSwain *et al.*, 2005). Possible explanations for the elevated production of PN include induction of heat shock-like proteins as a defence mechanism against potentially toxic compounds. At the same time, sludge hydrophobicity increased with protein content of the EPS, and an increased hydrophobicity generally led to a better flocculation.

Aerobic granule structure

SEM showed that the 4-CA-degrading granules at the end of Phase III had a smooth outer surface with a compact and regular appearance (Figure 9a,c,e). A close examination of granule surface revealed the presence of a large diversity of microbial morphotypes, including bacterial rods-like and cocci-like (Figure 9b,d,f). These microorganisms were embedded in extracellular polymeric matrix. The granule structure suggested that the granular sludge should be less vulnerable than floc-like sludge to the toxic pollutants in wastewater. Many large cavities (3–6 μm in diameter) could be observed in the outer surface of granule, which can be used to enhance the transport of oxygen and nutrients into the inner cores of the granule, and to facilitate the expulsion of toxic metabolic by-products away from the microbial consortia residing in the cores of the granule. Compared with aerobic granule in R1 and R2, the granules in R3 had smoother outer surface and more compact structure.

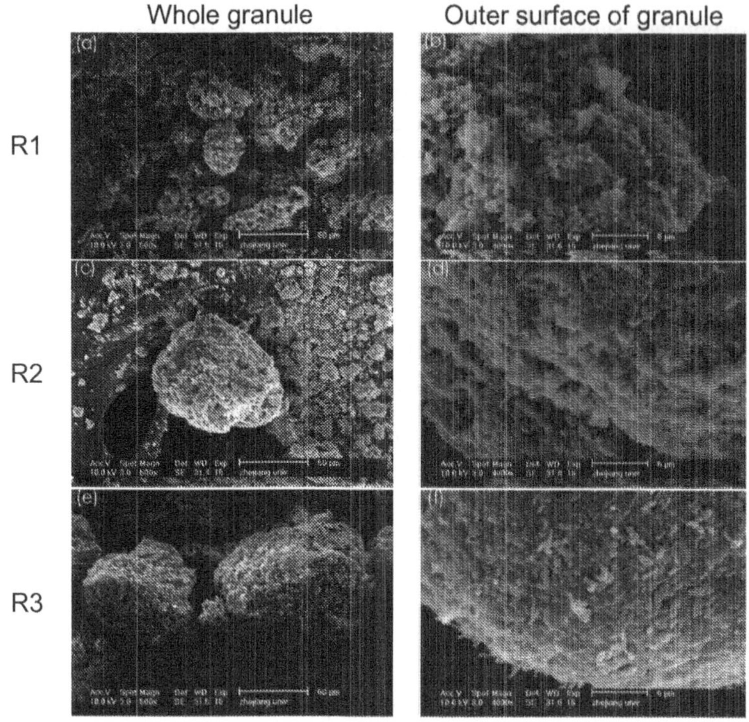

Figure 9 SEM images of aerobic granule in R1–R3.

4-CA-degrading performance for aerobic granule

While chloranilines removal has been carried out for many years by conventional continuous activated sludge systems, these systems were sensitive to fluctuation in the organic loading, e.g., high 3-CA concentration exceeding 250 $mg \cdot l^{-1}$ have been reported to cause system failure (Boon *et al.*, 2003). Furthermore, Boon *et al.* (2002) used immobilized cells to treat a 3-CA contaminated wastewater in a lab-scale semi-continuous activated sludge system, this simple immobilization procedure was effective to maintain a 3-CA-degrading population within the activated sludge community, but the reactor failed to effectively degrade 3-CA because of the decrease of 3-CA metabolizing bacterium (*Comamonas testosteroni* strain I2).

Therefore, the survival and activity of special metabolizing bacterium in bioreactor is critical to obtain successful biodegradation of non- or slowly degradable pollutants. In the current study, the aerobic granules were successfully cultivated in the reactors, and able to sustain the maximum 4-CA loading rate of 0.8 $kg \cdot m^{-3} \cdot d^{-1}$. 4-CA concentration was measured during one cycle in the three reactors at the end of Phase III (Figure 8). While influent 4-CA concentration exceeded 400 $mg \cdot L^{-1}$ in three reactors, only 6 h was required for complete removal of 4-CA in R2 and R3, and it took nearly 8 h for complete removal of 4-CA in R1. The specific 4-CA degradation rates were still around 0.14, 0.21 and 0.27 $gCA \cdot gVSS^{-1} \cdot d^{-1}$ in R1–R3. Although the underlying mechanism for 4-CA tolerance in the aerobic granule has not been pinpointed, and may be the result of several synergistic mechanisms, it is clear from this study that the formation of dense, compact granules could facilitate survival of micro-organisms in the bioreactor.

Influence of reactor structure

Reactor configuration has an impact on the flow pattern of liquid and microbial aggregates in the bioreactor. In almost all reported, aerobic granules were produced in column-type upflow reactors. Column-type upflow reactor and completely mixed tank reactor (CMTR) have very different hydrodynamic behaviors in terms of the interaction between flow and microbial aggregates. In a column-type upflow reactor, a high ratio of reactor height to diameter (H/D) can ensure a longer circular flow trajectory which in turn provides a relatively homogenous circular flow and localizes vortexing along the reactor's axis, and can improve selection of microbial aggregates by the difference in settling velocity. Under such circumstance, microbial aggregates are constantly subject to varying localized hydrodynamic shear force, upflow trajectories and random collisions, only granules with regular shape and size occasionally form instead of irregular flocs.

Compared to conventional column-type upflow reactor, airlift reactor, e.g., SABR used in the current study, consists of two concentric tubes, a steady circulation of liquid is achieved inside the system when gas is fed into the riser. The inherent benefits of the airlift reactor are attributed to their high oxygen transfer efficiency, excellent mixing and substrate transfer capabilities.

As our data showed, the ability of R2 and R3 to cultivate and retain well-settling granules played a significant role in enhancing the overall ability of the reactors to remove 4-CA, and the aerobic granules formed in R2 and R3 had larger granule size, denser granular structure, higher specific 4-CA degradation rate. The granule specific gravity, strength, SOUR and specific 4-CA degradation rate were all highest in R3, at 1.101, 99%, 30.2 mgDO$\cdot$gVSS$^{-1}\cdot$h^{-1} and 0.27 gCA$\cdot$gVSS$^{-1}\cdot$d^{-1}, respectively.

Results demonstrated that the reactors with a high ratio of reactor height to diameter (H/D) and internal-circulation favoured the granulation of aerobic sludges, and affected the characteristics of aerobic granules. In a SBR process, the excellent settling characteristics of the biomass allow a short standstill time for settling and a more efficient use of reactor volume. A high H/D ratio and the absence of an external settler result in a reactor with a small footprint. However, high shear force is an environmental selection pressure that is known to be essential for aerobic sludge granulation, and the application of a high shear force is generally associated with a high energy requirement. If conditions can be created in appropriate reactor to stimulate the growth of those microorganisms responsible for cell aggregation, aerobic granules can conceivably be formed more economically in the presence of weaker environmental selection pressures.

Conclusions

1. Aerobic granules capable of handling high 4-CA loading can be successfully cultivated in the SBR or SABR by seeding activated sludge.

2. Aerobic granules were observed since 15 days after start-up in R2 and R3 which had high H/D ratio. After 140 days of operation, the granulation of aerobic sludges tended to be mature while the removal efficiencies of COD, TN and 4-CA were maintained above 93%, 70% and 99.9% respectively, when three reactors were continuously operated with COD loading rate of 2.4~3.6 kg$\cdot$m$^{-3}\cdot$d^{-1} and 4-CA loading rate of around 0.8 kg$\cdot$m$^{-3}\cdot$d^{-1}.

3. Mature aerobic granules in R1–R3 had an average diameter of 0.78, 1.68 and 1.25 mm, minimal settling velocity of 20.5, 70.1 and 66.6 m$\cdot$h^{-1}, SVI of 38, 31 and 31 mL$\cdot$g^{-1}, SOUR of 18.4, 26.4 and 30.2 mgDO$\cdot$gVSS$^{-1}\cdot$h^{-1}, specific 4-CA degradation rates of 0.14, 0.21 and 0.27 g$\cdot$gVSS$^{-1}\cdot$d^{-1}, and PN/PS of 8.2, 10.8 and 13.7 mg$\cdot$mg^{-1}, respectively.

4. The aerobic granules formed in R2 and R3 had larger granule size, denser granular structure, higher specific 4-CA degradation rate and SOUR than the granules of R1. This study demonstrated that the advantages of the aerobic sludge granular technology to handle high-strength toxic organic wastewater, and the reactor with a high H/D ratio and internal-circulation favoured the granulation and stabilization of aerobic sludge.

Acknowledgement

Foundation item: National Natural Science Foundation of China(No.50378082); Key Scientific Research Items of Zhejiang Province(No. 2004C2302).

References

American Public Health Association (APHA). (1998). *Standard methods for the examination of water and wastewater*. Washington, D.C.

Beun, J.J., van Loosdrecht, M.C.M., Heijnen, J.J. (2000). Aerobic granulation. *Wat. Sci. Technol.*, 41: 41–48.

Beun, J.J., Loosdrecht, M.C.M., Heijnen, J.J. (2002). Aerobic granulation in a sequencing batch airlift reactor. *Wat. Res.*, **36**: 702–712.

Boon, N., De Gelder, L.D., Lievens, H., *et al.* (2002). Bioaugmenting bioreactors for the continuous removal of 3-chloroaniline by a slow release. *Environ. Sci. Technol.*, **36**: 4698–4704.

Boon, N., Top, E.M, Verstraete, W., *et al.* (2003). Bioaugmenting as a tool to protect the structure and function of an activated-sludge microbial community against a 3-chloroaniline shock load. *Appl. Environ. Microbiol.* **69**: 1511–1520.

Ghangrekar, M.M., Asolekar, S.R., Ranganathan, K.R., Jodhi, S.G. (1996). Experience with UASB reactor start-up under different operating conditions. *Wat. Sci. Technol.*, **34**: 421–428.

Hulshoff Pol, L.W., Lettinga, G. (1983). Granulation in UASB reactor. *Wat. Sci. Technol.*, **15**(2): 293–304.

Liu, Y., Tay Joo-Hwa. (2004). State of the art of biogranulation technology for wastewater treatment. *Biotechnol. Adv.*, **22**: 533–563.

Liu, Y., Yang S.-F., Tay J.-H. (2004). Improved stability of aerobic granules by selecting slow-growing nitrifying bacteria. *J. Biotechnol*, **108**: 161–169.

Liu, Y., Tay, J.H. (2002). The essential role of hydrodynamic shear force in the formation of biofilm and granular sludge. *Wat. Res.*, **36**: 1653–1665.

McSwain, B.S., Irvine, R.L., Hausner, M., *et al.* (2005). Composition and distribution of extracellular polymeric substances in aerobic flocs and granular sludge. *Appl. Environ. Microbiol.*, **71**(2): 1051–1057.

Mishima, K., Nakamura, M. (1991). Self-immobilization of aerobic activated sludge – a pilot study of the aerobic upflow sludge blanket process in municipal sewage treatment. *Wat. Sci. Technol.*, **23**: 981–990.

Morgenroth, E., Sherden, T., Van Loosdrecht, M. *et al.* (1997). Aerobic granular sludge in a sequencing batch reactor. *Wat. Res.*, 31(12): 3191–3194.

O'Neill, F.J., Bromley-Challenor, K.C.A., Greenwood, R.J., *et al.* (2000). Bacterial growth on aniline: Implication for the biotreatment of industrial wastewater. *Wat. Res.*, **34**(18): 4397–4409.

Schmidt, J.E., Ahring, B.K. (1996). Granular sludge formation in upflow anaerobic sludge blanket (UASB) reactors. *Biotech.Bioengng*, **49**(3): 229–245.

Tay, J.H., Liu, Q.S., Liu, Y. (2001a). Microscopic observation of aerobic granulation in sequential aerobic sludge blanket reactor. *J. Appl. Microbiol.*, **91**(1): 168–175.

Tay, J.H., Liu, Q.S., Liu, Y. (2001b). The effects of shear force on the formation, structure and metabolism of aerobic granules grown. *Appl. Microbiol. Biotechnol.*, 57: 227–233.

Tay, J.H., Liu, Q.S., Liu, Y. (2002). Characteristics of aerobic granules grown on glucose and acetate in sequential aerobic sludge blanket reactors. *Environ. Technol.*, **23**: 931–936.

Tsuneda, S., Nagano, T., Hoshino, T., *et al.* (2003). Characterization of nitrifying granules produced in an aerobic upflow fluidized bed reactor. *Wat. Res.*, **37**: 4965–4973.

Wang, Q., Du, G., Chen, J. (2004). Aerobic granular sludge cultivated under the selective pressure as a driving force. *Process Biochem.*, **39**: 557–563.

Watanabe, K., Miyashita, M., Harayama, S. (2000). Starvation improves survival of bacteria introduced into activated sludge. *Appl. Environ. Microbiol.*, **66**(9): 3905–3910.

Yang S.-F., Tay J.-H., Liu, Y. (2003). A novel granular sludge sequencing batch reactor for removal of organic and nitrogen from wastewater. *J Biotechnol.*, 106: 77–86.

Zhu, J., Wilderer, P.A. (2003). Effects of extended idle condition on structure and activity of granular activated sludge. *Wat. Res.*, **37**: 2013–2018.

Determination of odorous volatile organic compounds at wastewater treatment plants in Finland

J. Nykänen and A. Veijanen

Department of Biological and Environmental Science, University of Jyväskylä, P.O. Box 35, FI-40014 University of Jyväskylä, Finland (E-mail: *jenykane@cc.jyu.fi*; *veijanen@bytl.jyu.fi*)

Abstract Odour and VOC (volatile organic compound) emissions can be a problem for wastewater treatment plants because of the odour nuisance to the neighbourhood. Plant workers also are frequently exposed to the odours and VOCs. Although the single VOCs generally do not exceed their harmful concentrations, several VOCs simultaneously may cause symptoms such as nausea, headache and hypersensitivity effects. Odour complaints are also becoming more common. In this study concentrations of VOCs, ammonia and hydrogen sulphide were measured at the municipal wastewater and the pulp mill effluent treatment plants. Samples were taken from different process stages from municipal wastewater treatment plant. In the pulp mill effluent plant VOC samples were taken from the activated sludge pond (aeration pond) and from the office building in the vicinity of treatment plant. Carboxylic acids, sulphides, aldehydes and aromatics were the main odour causes in the municipal wastewater treatment plants. At the pulp mill effluent treatment plant the most odorous compounds were different kind of sulphides, thiophenes, monoterpenes and sesquiterpenes. Some chlorinated compounds were present at both the municipal and pulp mill effluent plants.

Keywords Odour; pulp mill effluent; VOCs; wastewater

Introduction

Odour abatement is a major issue for many municipal or industrial waste water treatment plants. Investments in odour control and abatement are made and a lot of improvement concerning odour emissions has already been done in Finland. Still people working at different kind of wastewater treatment plants or living in the neighbourhood of them are exposed to odour causing volatile organic compounds (VOCs) originating from different wastewater treatment processes. Occurrence of odour and odorous compounds may arouse a concern that these compounds might present a risk to the health. People are also more aware of their rights and more ready to complain about the unpleasant odour. In Finland, uniform legislation concerning odour emissions does not exist at the present moment but odour monitoring or odour measurements are included into the environmental permits of companies or plants by environmental authorities.

Prevailing air quality and odour at the working site is a result of dozens of compounds mixed together and the synergetic effect of many VOCs is not well known. Some studies show that even at low concentrations VOCs may cause symptoms such as nausea, headache and hypersensitivity effects (Mølhave *et al* 1991). However, in many cases, moderately offensive odours may cause symptoms via annoyance-mediation. It is suggested that behavioural sensitisation or stress related mechanisms are involved in many symptoms claimed to be caused by offensive odours (Neutra *et al.* 1991). Recent study has also shown that there is not necessarily a direct association between odour perceptibility and potential for inducing health effects (Rosenkranz *et al.* 2003).

The aim of the study was to determine the levels of odour causing or in other way harmful VOCs in municipal and pulp mill wastewater treatment plants. The most crucial sites in treatment processes regarding to odour and VOCs were determined. Also hydrogen sulphide and ammonia concentrations were measured.

Methods

Sampling

Air samples were collected into the adsorption tubes filled with Tenax GR adsorption resin by using an air pump. Samples were taken at municipal wastewater treatment plants from the screening room and sludge dewatering room. Studied plants were of moderate size in Finnish scale. Plant A treated about 15 million m^3 of effluent per year (2004) and plant B about 3 million m^3 per year (2003). At plant A, incoming wastewater was mainly from households but also in some extent from the paper mill nearby. About 8% of all incoming water in plant B originated from industrial sources. Activated sludge process was in use at both of the plants.

At the pulp mill, effluent treatment plant samples were taken from the activated sludge pond and from the office building situated at immediate vicinity of the treatment plant. Studied pulp mill wastewater treatment plant handles typically 45 000 m^3 of effluent per day. About 75% of all incoming water is from pulping process, 23% is from turpentine manufacturing process and the rest (about 2%) from polymer manufacturing. Sanitary waste water from the pulp mill is not treated in this process but is led to the city municipal wastewater treatment plant.

Analytical methods

In the laboratory the VOC samples were analysed using a thermal desorption / gas chromatograph / mass spectrometer device (Tekmar Purge & Trap 3000 / Agilent 6890+ / 5973N MSD), connected with simultaneous sniffing. A double column system was used so that the other column led to the mass spectrometer and the other to a glass funnel. The separated compounds were sniffed in the funnel. Intensity of every odour coming to the funnel was evaluated by four staged scale from weak (stage one) to very strong (stage four).

Desorption temperature was 230 °C and it was held for ten minutes. During this time desorbed compounds were concentrated in the cold trap (–120 °C). Helium (99.995% purity) was used as the carrier gas. The GC oven temperature was held at 40 °C for 2 min, increased at 5 °C/min to 150 °C and then at 15 °C/min to 250 °C. The final temperature was held for 5 min. Concentrations of single VOCs were calculated according to external standard including 25 different volatile organic compounds (Tolvanen *et al.* 2005).

Perkin Elmer Portable gas chromatograph (GC/PID) was used for in-situ measurements of hydrogen sulphide (H_2S), methyl mercaptane (MeSH), dimethyl sulphide (DMS) and dimethyl disulphide (DMDS). Results of GC/MS and GC/PID were compared to verify the reliability of the analysers.

Results and discussion

Municipal wastewater treatment

Typical compounds found at the air of municipal and industrial waste water treatment plants are organic compounds containing sulphur, nitrogen and oxygen. These compounds are usually amines, carboxylic acids, esters, etc. (Gostelow *et al.* 2001, Gostelow & Parsons 2000, Islam Nurul *et al.* 1998). These compounds have typically very low threshold odour concentrations (TOC) so they do not have to appear in high concentrations to cause odour problems. Odorous inorganic compounds at the waste water treatment plants are hydrogen sulphide, (H_2S, TOC: 0.0001 mg/m^3), and ammonia, (NH_3, TOC: 0.03 mg/m^3) (O'Neill & Phillips, 1992). In many cases, sludge treatment is considered as the most odorous process stage in the municipal wastewater treatment. Some studies show that dimethyl trisulphide (DMTS) might be a great contributor to the overall odour in sludge processing (Winter *et al.* 2000).

Air samples from municipal wastewater treatment plant in summer 2004 and 2005 consisted of hundreds of different compounds from which several dozens contributed to the overall odour. Carboxylic acids, sulphides, aldehydes, aromatics (many of them used as solvents in industrial processes) and chlorinated compounds were discovered in our study. Some aromatics like styrene (odour: solvent, glue) and toluene (odour: glue, paint, rubber) were even determined as the main compounds at certain measurement points. Typical fecal odour causing compounds such as indole and scatole were not detected. Differences in VOC concentrations between process stages were significant. Concentrations of the VOCs in the municipal wastewater treatment and their threshold odour limits in literature are shown in Table 1.

Table 1 Concentrations (µg/m^3) of odorous and/or harmful compounds at municipal wastewater treatment stages and threshold odour concentrations from literature (O'Neill and Phillips,1992; Ruth, 1986; Williams, 1995)

Compound	Trash rake room		Sludge dewatering		Odour threshold	Odour description
	Plant A	Plant B	Plant A	Plant B		
H_2S	<140*	<140*	<140*	<140*	0.1	rotten egg
MeSH	<140*	<140*	170	<140*	0.04	decayed
DMS	0.9	7.7	19.6	103.8	0.3	decayed
DMDS	10.2	16.2	29.6	32.5	0.1	decayed
DMTS	nd	3.4	24.9	9.60	6.2	sewer
acetic acid	78.7	nd	44.4	nd	250	acetic
butanoic acid	1.5	nd	6.1	nd	0.4	rancid
hexanal	8.0	13.5	7.6	4.40	28	green
toluene	107.8	51.4	59.3	709.8	80	solvent
styrene	18.1	nd	200.6	nd	73	solvent
1,2-dichloroethane	151.2	nd	87.9	nd	-	solvent
limonene	56.0	32.7	69.8	13.1	10	lemon

*Below detection limit (140 µg/m^3); nd: not detected.

Pulp mill effluent

During the kraft pulping of wood a large number of low molecular weight and volatile compounds such as sulphur compounds are formed from lignin and polysaccharides and many kind of extractives like terpenes and sesquiterpenes are released from wood tissue structures. Odours originating from pulp mill effluents are primarily due to generation of volatile organic sulphur compounds such as hydrogen sulphide, methyl mercaptan, dimethyl sulphide, and dimethyl disulphide (TRS-compounds) (Hagen & Hartung 1997, Niemelä,

2001). They can be transported even for dozens or hundreds of kilometres along the water stream or in the air (Brownlee, *et al.* 1995).

Another potential odour causing group of VOCs is terpenes that are also detected in the air plumes from kraft pulp mills (Strömvall and Petersson, 1993). Terpenes are common in pulp extraction products and process waters (Niemelä, 2001) and they are drifted to the waste water ponds from the pulping process water streams.

Volatile organochlorine compounds, for example chloroform and 2,5-dichlorotiophene, are formed during kraft bleaching process when ClO_2 is used as a bleaching agent (Juuti *et al.* 1996). Volatile organochlorines are not considered much as the odour causes but are known to be harmful to the environment and human health.

At the pulp mill effluent treatment plant the most odorous compounds in our study were different kind of sulphides and thiophenes (odours: decayed cabbage, onion, garlic). It was also found that monoterpenes (odour: coniferous, turpentine) and sesquiterpenes (earthy, musty) were very important in overall odour. Results from VOC analysis at pulp mill effluent treatment and in the office building are shown in Table 2.

Table 2 Odorous compounds detected at the pulp mill effluent plant and the odour threshold concentrations found in literature (O'Neill and Phillips 1992, Ruth, 1986). Concentrations are in µg/m^3.

Compound	Aeration pond	Office building	Odour threshold	Odour description
H_2S	<140*	<140*	0.1	rotten egg
DMS	438.6	8.10	0.3	decayed
DMDS	20.4	14.4	0.1	decayed
thiophene	18.9	nd	78	garlic
2-methylthiophene	28.4	nd	-	onion, garlic
hexanal	14.7	5.60	28	green
toluene	102.6	14.3	80	solvent
styrene	24.7	4.60	73	solvent
chloroform	249.6	nd	-	sharp
2,5-dichlorothiophene	73.6	nd	-	-
alpha-pinene	1794	437.2	16	coniferous
3-carene	748.4	221.3	-	lemon
p-cymene	388.9	33.9	4.0	pungent
limonene	2639	106.5	10	lemon
sesquiterpenes (total sum)	4830	236.2	-	earthy

*Below detection limit (140 µg/m^3); nd: not detected.

Conclusions

Odour at both the municipal treatment plants was caused by many sulphur compounds, especially hydrogen sulphide. Other odorous compounds at plant A were also acetic and butanoic acid and hexanal. Acetic acid and hexanal caused distinct odour in sniffing analysis although their published TOCs (threshold odour concentration) were not exceeded. Noticeable is also that toluene and styrene considered as quite harmful compounds to human health occurred in relatively high concentrations.

Odour at the pulp mill effluent plant was due to different kind of compounds that were present at quite high concentrations (Table 2). Published TOC for DMS is 0.3 µg/m^3 (O'Neill and Phillips, 1992). Our results at maximum for DMS in the pulp mill effluent exceeded the TOC by thousand times. It was also found out that alpha-pinene exceeded its published TOC by thousand times.

Chlorinated organic compounds were found both at municipal and pulp mill effluent plants. 1,2-dichloroethane was detected in the municipal wastewater treatment plant A at quite high concentrations. Chloroform was the main chlorinated compound found in the air of pulp mill effluent aeration pond.

Odorous compounds found in our study have typically very low threshold odour concentrations so they do not necessarily appear in high concentrations to cause odour problems. Therefore it is very difficult to trace them from samples that consist of hundreds of compounds and the most odorous compounds usually are not the largest peaks in chromatograms. Sniffing technique combined with GC/MS is very important and helpful method in these problematic cases.

References

Brownlee, B.G., Kenefick, S.L., MacInnis, G.A., Hrudey, S.E. (1995). Characterization of odorous compounds from bleached kraft pulp mill effluent. *Water Science and Technology*, **31**(11), 35–40.

Gostelow, P. & Parsons, S. A. (2000). Sewage treatment works odour measurement. *Water Science and Technology*, **41**(6), 33–40.

Gostelow, P., Parsons, S. A., Stuetz, R. M. (2001). Odour measurement for sewage treatment works. *Water Research*, **35**(3), 579–597.

Hagen, C. E., Hartung, R. W. (1997). New chemical treatment method controls waste water system odour. *Pulp & Paper*, **71**(11).

Islam Nurul, A. K. M., Hanaki K. and Matsuo T. (1998). Fate of dissolved odorous compounds in sewage treatment. *Water Science and Technology*, **38**(3), 337–344.

Juuti, S., Vartiainen, T., Joutsenoja, P., Ruuskanen, J., (1996). Volatile organochlorine compounds formend in the bleaching of pulp with ClO_2. *Chemosphere*, **33**, 437–448.

Mølhave, L., Grönkjär Jensen, J., Larsen, S. (1991). Subjective reactions to volatile organic compounds as air pollutants. *Atmospheric Environment*, 25A(7), 1283-1293.

Neutra, R., Lipscomb, J., Satin, K. and Schusterman, D. (1991). Hypotheses to explain the higher symptom rates observed around hazardous waste sites. *Environmental Health Perspectives*, **94**, 31–34.

Niemelä, K. (2001). Identification of novel volatile sulphur and nitrogen compounds in kraft pulping process streams. In *Proc. 11th International symposium on Wood and Pulping Chemistry*, Nice, France, 2001, vol. 1, pp. 125–128.

O'Neill, D. H., Phillips, V. R. (1992). A review of the control of odour nuisance from livestock buildings: Part 3, properties of the odorous substances which have been identified in livestock wastes or in the air around them. *J. Agric. Enging Res.* **53**, 23–50.

Rosenkranz, H., S., Cunningham, A. R. (2003). Environmental odours and health hazards. *The Science of the Total Environment*, **313**, 15–24.

Ruth, J. H. (1986). Odour thresholds and irritation levels of several chemical substances: a review. *Am. Ind. Hyg. Assoc. J.*, **47**.

Strömvall, A.-M., Petersson, G. (1993). Photo-oxidant-forming monoterpenes in air plumes from kraft pulp mill industries. *Environmental Pollution*, **79**, 219–223.

Tolvanen, O., Nykänen, J., Nivukoski, U., Himanen, M., Veijanen, A. and Hänninen, K. (2005). Occupational hygiene in a Finnish drum composting plant, *Waste Management*, **25**, 427–433.

Williams, T. O. (1995). Odours and VOC emission control methods. *BioCycle*, **36**(5).

Winter, P. and Duckham, S., C. (2000). Analysis of volatile odour compounds in digested sewage sludge and aged sewage sludge cake. *Water Science and Technology*, 41(6), 73–80.

Identification of functional groups of denitrifiers in activated sludge

M. Hagman*, J. L. Nielsen, P. H. Nielsen** and J. la C. Jansen***

* Dept. of Water and Environmental Engineering, Lund Institute of Technology, P.O. Box 118 SE - 221 00 Lund, Sweden (E-mail: *marinette.hagman@vateknik.lth.se*)
** Dept. of Biotechnology, Chemistry and Environmental Engineering, Aalborg University, Sohngaardsholmsvej 57, DK-9000, Aalborg, Denmark

Abstract The process performance of activated sludge is dependent on the distribution of the present microbial community. The internal composition of heterotrophic biomass is very much related to the carbon content in the wastewater, of any external carbon addition and on the plant operation. Oxygen uptake rate (OUR) and nitrate uptake rate (NUR) measurements were performed with addition of different carbon sources alone and with mixtures to sludge from two full-scale activated sludge plants, one without and one with addition of an external carbon source. Addition of a mixture of carbon sources increase the process rates compared to addition of only one carbon source in both aerobic and anoxic environments for both sludges. For sludge from the plant without addition of external carbon source the increases varied from time to time between the sum of the rates of the individual carbon sources and insignificant increase. For the sludge from the plant with addition of external carbon, the increases were always in between that of each of the individual carbon sources and the sum of the rates. For practical use of mixed carbon sources for increased denitrification rates further information about the microbial populations and the carbon uptake involved in the aerobic and anoxic conversion of the carbon could be very useful.

Keywords Acetate; activated sludge; denitrification; ethanol; methanol; oxygen uptake rate

Introduction

The population of heterotrophic biomass in activated sludge is determined by many factors. Most important are the composition of the organic matter in the wastewater and the operation of the treatment plant. The impact on the microbial functional groups will be especially significant when the operation includes additions of an external carbon source. The process rate for the aerobic degradation of heterotrophic biomass is determined by the oxygen uptake rate (OUR) and the process rate for the denitrifying heterotrophic bacteria can be found by measuring the nitrate uptake rate (NUR). By measuring these rates, the process for degradation of organic matter can be analysed. Further on, the fractions of denitrifiers in activated sludge is calculated from the proportion of the maximum OUR and the NUR. The proportion of denitrifying bacteria in activated sludge is found to vary significantly from one type of activated sludge to another (Orhon *et al.*, 1996).

When an external carbon source is added to activated sludge, adaptation is normally achieved, resulting in a growth of a specific denitrifying functional group of bacteria. The activated sludge will then contain one specialised microbial functional group concentrating on degrading one special carbon source and another mixed group of organisms will be adapted to the organic substrates in the wastewater. This is not very optimal from an operational point of view, since a significant part of the denitrifiers may be unable to utilise the added external carbon with a much lower nitrate reduction rate as the result. In order to reach an increased activity, an addition of a mixed carbon source could be a more optimal solution since more than one group of bacteria can turn active.

Material and method

Experimental strategy

In order to experience the impact of a single carbon source versus a mixture of carbon sources as external carbon addition to activated sludge, OUR and NUR was measured at four separate occasions. Activated sludge was collected at Källby wastewater treatment plant (WWTP), Lund, Sweden and Ryaverket WWTP in Gothenburg, Sweden. Källby WWTP is a pre-denitrification plant operated without addition of external carbon source. Ryaverket WWTP is a high loaded activated sludge plant where the denitrification process takes place in an activated sludge process and the nitrification in a plastic media trickling filter.

The first two experiments were made with activated sludge collected at Källby WWTP with six months interval, performed in August and in January the following year. The third and fourth experiments were performed to compare the process rates for activated sludge with and without adaptation to an external carbon source. The third experiment was made with activated sludge from Källby WWTP and the fourth with activated sludge from Ryaverket WWTP. In all occasions OUR- and NUR experiments were performed in parallel with addition of single carbon sources and mixtures of those to the same activated sludge.

Carbon sources chosen for the experiments were; acetate, methanol, ethanol, mixtures of acetate and methanol together with mixtures of acetate and ethanol. Acetate was chosen since it is known as a very easily biodegradable substrate present in high amount in municipal wastewater (Henze *et al.* 2002), methanol since it is often used and has shown to be a strong selector for specific denitrifiers (Christensson *et al.*, 1994; Nyberg *et al.*, 1992) and ethanol also commonly used and applied as external carbon source at Ryaverket WWTP.

Oxygen and nitrate uptake rate measurements

The oxygen uptake rate and nitrate uptake rate experiments were performed in 1 litre batch reactors at 20°C. Buffer and nutrient were added to the activated sludge in order to keep the pH stable and not to cause any limitations of essential nutrients. In reactors for OUR measurements, ATU was added in order not to include any oxygen related to nitrification during the experiment. More details about OUR and NUR experimental procedures and performances are further described in Kristensen *et al.*, (1992), Roš (1993) and Spanjers *et al.*, (1998).

The oxygen uptake rate was obtained by measuring the decrease of oxygen during five minutes every ten minute. The decrease in oxygen is found to be linear and by calculating the slope of that curve, the result gives the oxygen uptake rate. By taking the volatile suspended solids (VSS) into account, the specific OUR for a particular activated sludge can be obtained.

To establish the nitrate uptake rate in activated sludge, the nitrate concentration in batch reactors was measured. Nitrate was added in excess and nitrogen gas was bubbled into the suspension, to keep an anoxic environment in the reactors during the experiments. The decrease in nitrate concentration was further estimated when additions of different carbon sources were made, one carbon source or a mixture of two was added to each reactor. Since the decrease in nitrate concentration is linear, the slope of that curve gives the nitrate uptake rate. By taking the volatile suspended solids (VSS) into account, the specific NUR for the particular activated sludge can be obtained.

Chemical analysis

Suspended solids (SS) and volatile suspended solids (VSS) were analysed according to standard methods by filtration through fibreglass filters (European Standard SS-EN 872 (1996)). Nitrate-nitrogen was analysed with the spectrophotometric method from HACH LANGE (LCK 339).

Results

The NUR and the OUR were measured in activated sludge from Källby WWTP for acetate, methanol and a mixture of acetate and methanol as shown in Figure 1.

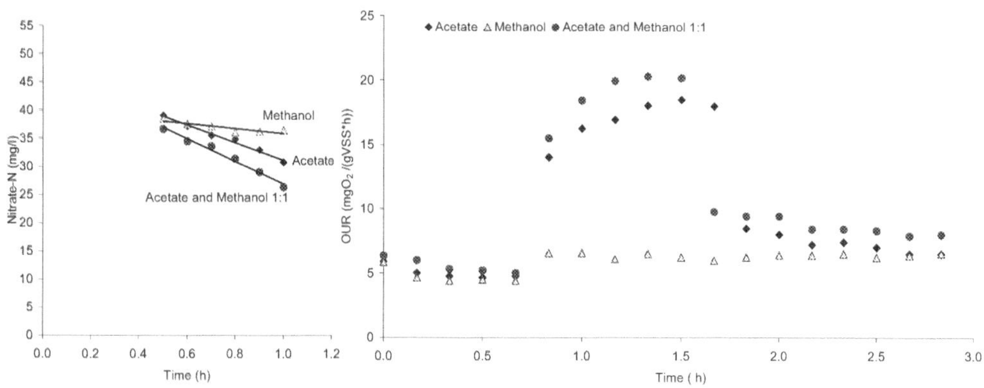

Figure 1 NUR and OUR for activated sludge from Källby WWTP with additions of acetate, methanol and a mixture of acetate and methanol 1:1.

Figure 1 clearly shows that additions of a mixture of acetate and methanol to activated sludge, increases the process rates in both anoxic and aerobic environment compared to additions with single carbon sources. The increase approximately corresponded to the sum of the rates for the two individual carbon sources as long as both were present (Table 1) (for the OUR, the endogenous respiration is subtracted).

Table 1 Maximum NUR and OUR with additions of different carbon sources to activated sludge from Källby WWTP

Carbon source	NUR ($mgNO_3^-$-N/(gVSS*h))	Maximum OUR (mgO_2/(gVSS*h))	Maximum OUR (mgO_2/(gVSS*h)) (endogenous respiration is subtracted)
Acetate	4.5	17.8	13.0
Methanol	1.3	6.4	1.6
Acetate+Methanol 1:1	5.8	19.7	14.9

The experiment with activated sludge from Källby WWTP was repeated six months later. This time there was no indication of any increase of the process rates for neither oxygen nor nitrate reduction with addition of a mixed carbon source as seen in Figure 2. The denitrification rate for the mixture of acetate and methanol reached about the same rate as for acetate alone. Similar results presented in Table 2, turned out for the OUR. The aerobic respiration rate for the mixture was close to that for acetate.

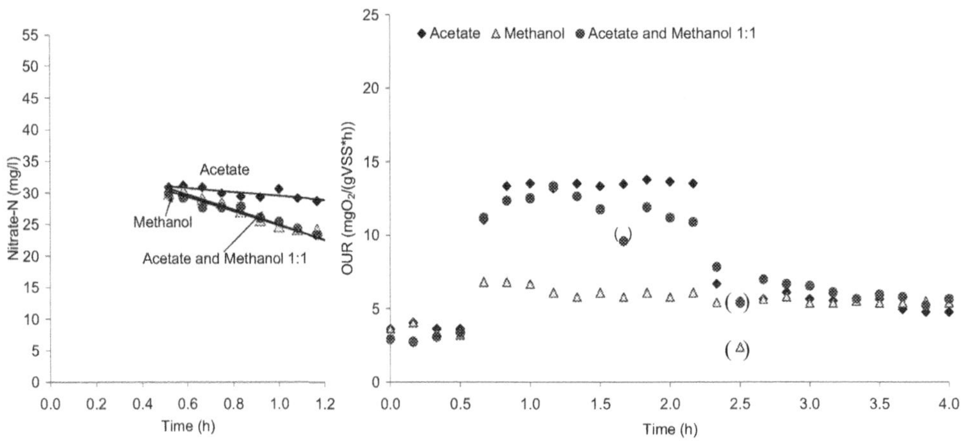

Figure 2 NUR and OUR for activated sludge from Källby WWTP with additions of acetate, methanol and a mixture of acetate and methanol 1:1 (measured value in brackets indicate error with the oxygen probe).

Table 2 NUR and OUR for activated sludge from Källby WWTP with additions of different carbon sources together

Carbon source	NUR ($mgNO_3^-$-N/(gVSS*h))	Maximum OUR (mgO_2/(gVSS*h))	Maximum OUR (mgO_2/(gVSS*h)) (endogenous respiration is subtracted)
Acetate	3.8	13.4	9.7
Methanol	1.1	6.6	2.9
Acetate + Methanol 1:1	3.6	12.3	8.6

In Table 3 the results of an identical experiment with activated sludge form Källby WWTP are presented, but this time performed with additions of acetate, ethanol and a mixture of acetate and ethanol. An increase in process rates for the activated sludge was observed with additions of a mixed carbon source during both anoxic and aerobic degradation. Even though, the increase with the mixture of acetate and ethanol does not correspond to the sum of the two individual carbon sources, it seems as if the increase in NUR and OUR are of the same magnitude.

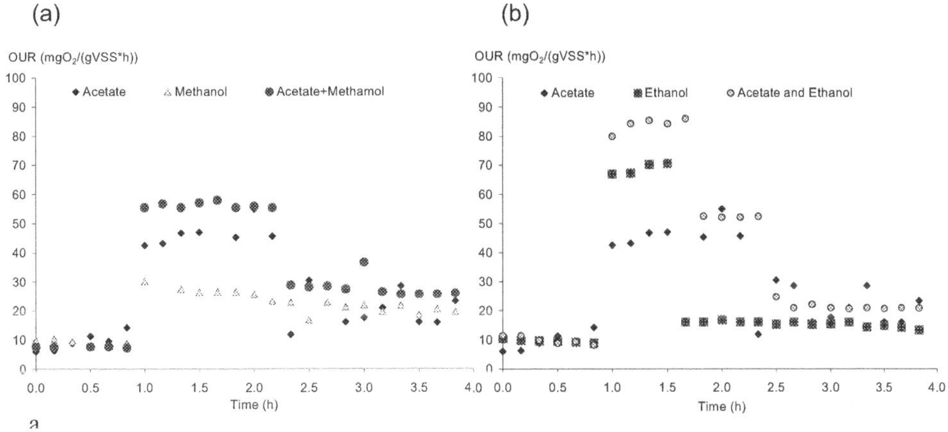

Figure 3 OUR for activated sludge from Ryaverket with additions of: (a) acetate, methanol and a mixture of acetate and methanol 1:1; (b) acetate, ethanol and a mixture of acetate and ethanol 1:1.

Table 3 Maximum NUR and OUR for activated sludge from Källby WWTP with additions of different carbon sources

Carbon source	**NUR ($mgNO_3^-$-N/(gVSS*h))**	**Maximum OUR (mgO_2/(gVSS*h))**	**Maximum OUR (mgO_2/(gVSS*h)) (endogenous respiration is subtracted)**
Acetate	3.4	17.2	12.0
Ethanol	3.3	10.4	5.2
Acetate+Ethanol 1:1	4.6	18.7	13.5

In the fourth experiment, activated sludge from Ryaverket in Gothenburg, adapted to ethanol, was tested with the same procedure as sludge from Källby WWTP. In Figure 3, a distinct increase in oxygen consumption is registered for mixtures of acetate and methanol and mixtures of acetate and ethanol compared with the single carbon addition. The increased OUR does not correspond to the sum of the activity for the two individual single carbon sources. Figure 3a,b illustrates that the OUR first increases above the rates for the two single carbon sources, after a while it decreases to the same rate as for acetate and eventually reaches the initial level, probably as a result of subsequent degradation of ethanol and acetate.

According to the NUR measurements a mixture of acetate and methanol increased the denitrification very minor compared to acetate alone as seen in Table 4. For addition of acetate mixed with ethanol a minor decrease in NUR was observed.

Table 4 Maximum NUR and OUR for activated sludge from Ryaverket with additions of different carbon sources

Carbon source	NUR ($mgNO_3^-$-N/(gVSS*h))	Maximum OUR (mgO_2/(gVSS*h))	Maximum OUR (mgO_2/(gVSS*h)) (endogenous respiration is subtracted)
Acetate	13.3	45.0	36.1
Methanol	5.4	27.5	18.6
Ethanol	15.2	72.2	63.3
Acetate + Methanol 1:1	14.0	59.4	50.5
Acetate+Ethanol 1:1	14.1	84.0	75.1

Discussion

Additions of methanol, acetate and mixtures of acetate and methanol were made at three different occasions to activated sludge from Källby and Ryaverket and three different results were obtained. In activated sludge from Källby an increased activity for both anoxic and aerobic consumption was observed in the first experiment. The increase approximately corresponded to the sum of the two individual carbon sources. This phenomenon supports the idea that there are two microbial functional groups present in the activated sludge and that each group utilises different carbon sources simultaneously. If this is the case, an improved denitrification would be possible by using a mixture of carbon sources instead of single substance, in order to increase the process rate.

The second experiment was in contradiction with the first experiment, since no increases in process rates were found when a mixture of acetate and methanol was added. The observed uptake rate for nitrate and oxygen were about the same for the mixture of acetate and methanol as for acetate. It could indicate the presence of only one group of bacteria where all bacteria prefers acetate. Preliminary results from MAR analyses also show very low uptake of methanol under both aerobic and anoxic conditions. It is though clear that further microbiological analysis is needed in order to be able to establish any correlations between the process rates, the microbial population distribution and carbon preferences. With the combined use of MAR and FISH (MAR-FISH) it is possible to detect cells with a positive uptake of a single radioactively labelled substrate, and subsequently determine cell identity using a wide range of already published fluorescently labelled oligonucleotide probes. (Nielsen *et al.*, 2003; Nielsen and Nielsen, 2005). With this approach a phylogenetically characterisation is possible of the populations involved in the degradation of a specific carbon source under aerobic and nitrate reducing conditions. Further more, it is possible to show substrate preferences by the involved microbial consortia, and thereby allow prediction of the benefit by adding single or mixtures of carbon sources to increase the denitrification capabilities.

The experiments show that activated sludge from the same plant not always reaches higher process rates with addition of mixed carbon source. It may be depending on what time of the year the experiments are performed. In this case high increases in process rates for mixed carbon addition were identified during the warmer season of the year.

Experiments with activated sludge from Ryaverket generally showed much higher process rates, which is due to the operation of the plant. The high load and addition of ethanol as external carbon source for denitrification, impacts the processes and microbial population very much. The large increase in OUR for the mixture of acetate and methanol can depend on two major groups of bacteria, one using acetate and the other using methanol as their first choice. Only a minor increase in NUR for the same mixture indicates that the bacteria that can reduce methanol with oxygen, can not do so with nitrate. It probably depends on the presence of a very specialised denitrifying population of bacteria as a result of the addition of ethanol as external carbon source at the plant.

Mixtures of acetate and ethanol were added to adapted activated sludge from Ryaverket. As expected, the highest process rates were found for additions of ethanol, which depends on the presence of a specialised group of bacteria developed due to the ethanol addition. When a mixture of acetate and ethanol was added to the sludge, the OUR increased above the level for that of ethanol. It demonstrates the presence of two separate functional groups of bacteria, where one uses acetate and the other uses ethanol. However since the increase do not correspond to the sum of the rates for each carbon source a more complicated pattern of functional groups present. In the anoxic environment the denitrification rate was in between the rates for acetate and ethanol respectively. This could be explained by the presence of functional groups of bacteria all able to denitrify with both acetate and ethanol, but when exposed to both in a mixture some will prefer acetate and some will prefer ethanol. As a result the NUR will be in between the rates from the carbon sources.

In activated sludge from Källby only a small increase in OUR and NUR is observed when a mixture of acetate and ethanol is added supporting the idea of functional groups able to degrade only one of the carbon sources or both.

The experiments shows that rate measurements such as OUR and NUR gives valuable information about utilisation of different external carbon sources in activated sludge but also that more information is needed about the microbial populations involved in the aerobic and anoxic conversion of the carbon if the information should be used for improved denitrification with mixtures of carbon sources.

Conclusion

A mixture of carbon sources added to activated sludge results in a higher heterotrophic process rates in both anoxic and aerobic environment compared to a single carbon source.

In sludge without adaptation to external carbon sources the increase in process rate for an added mixture of two carbon sources corresponds to the sum of the process rate for the two single carbon sources, indicating that the degradation activity is caused by two different microbial functional groups each reducing one carbon source. In some experiments the increase in process rate for mixed carbon sources was minor or not observed at all, which probably indicate that two functional groups of bacteria only exist in the sludge during some periods.

Activated sludge adapted to a certain carbon source will be specialised in utilising that particular one with a significantly higher NUR due to the chosen carbon source as a result. By adapting activated sludge to a mixture of different carbon sources it would be possible to optimize the denitrification process even further.

Extended microbiological analyses are needed to improve interpretation of the OUR and NUR measurements. By using the combination of MAR-FISH it might be possible to identify the population distribution and degradation pattern in activated sludge during additions of mixed carbon sources compared to single substrate.

Acknowledgement

The results presented in this paper are experiments based on activated sludge from the municipal WWTP in Lund, Källbyverket and Ryaverket in Gothenburg, both in Sweden. The staff at the two treatment plants are thanked for providing sludge for the examinations.

References

Christensson, M., Lie E. and Welander, T. (1994). A comparison between ethanol and methanol as carbon sources for denitrification. *Wat. Sci. Tech.*, **30**(6) 83–90.

European Standard (1996). SS-EN 872.

Henze M., Harremoës, P., Jansen, J. la Cour and Arvin E. (2002). *Wastewater treatment – Biological and chemical processes*, 3rd edition. Springer-Verlag. Berlin Heidelberg. ISBN: 3-540-42228-5.

Kristensen, H. G., Elberg Jørgensen, P. and Henze, M. (1992). Characterization of functional microorganism groups and substrate in activated sludge and wastewater by AUR, NUR and OUR. *Wat. Sci. Tech.*, **25**(6) 43–57.

Nielsen, J. L., Christensen, D., Kloppenborg, M., and Nielsen, P. H. (2003). Quantification of cell-specific substrate uptake by probe-defined bacteria under *in situ* conditions by microautoradiography and flourecence *in situ* hybridization. *Env. Microbiol.*, **5**(3), 202–211.

Nielsen, J. L. and Nielsen, P. H. (2005). Advances in Microscopy Microautoradiography of Single Cells. *Meth. Enzymol.*, **397**.

Nyberg, U., Aspegren, H., Andersson, B., Jansen J., la C. and Villadsen I., S. (1992). Full-scale application of nitrogen removal with methanol as carbon scource. *Wat. Sci. Tech.*, **26**(5–6) 1077–1086.

Orhon D., Sözen S. and Artan N. (1996). The effect of heterotrophic yield on the assessment of the correction factor for anoxic growth. *Wat. Sci. Tech.*, **34**(5–6), 67–74.

Roš M. (1993). *Respirometry of activated sludge*. Technomic Publishing Company, Inc. U.S.A. ISBN No. 1-56676-029-1.

Spanjers H., Vanrolleghem P.A., Olsson G. And Dold P.L. (1998). *Respirometry in control of the activated sludge process: Principles*. (*Scientific and Technical Report* no.7.) IAWQ, London. ISBN: 1-900222-04-3.

Preparation and evaluation of triacetate cellulose and activated carbon submicron particles composite membranes for the treatment of arsenic contaminated water

L.P. Terrazas-Bandala*, L. Manjarrez-Nevárez*,, J.C. Duran-Alvarez**, O. Ayala-Valenzuela*, G. González-Sánchez*, L. Ballinas-Casarrubias****

* Advanced Materials Research Center (CIMAV, S.C.) Miguel de Cervantes 120 Compl. Ind. Chih. 31109, Chihuahua, Chihuahua, México (E-mail: *piroshka.terrazas@cimav.edu.mx*)
** Chemistry Sciences Faculty, Universidad Autónoma de Chihuahua, Circuito Universitario S/N, Chihuahua, Chihuahua, México (E-mail: *mballinas@uach.mx*)

Abstract Submicronic activated carbon particles for membrane preparation were obtained by a solvation method using THF and characterized by laser diffraction technique and Transmission electronic microscopy. Results showed that for a 72 h solvation process a submicronic distribution was obtained. Composite membranes of triacetate cellulose and activated carbon particles were prepared by the evaporation process at different temperature and relative humidity conditions in a three factorial design (35–55°C and 30–70% RH). Membranes were observed microscopically by SEM and evaluated mechanically by DMA analysis. Results demonstrated that materials totally dispersed have the highest rupture stress. Selected material, based on its mechanical behaviour and structural conformation, were tested on a continue flow cell in order to remove arsenic at different concentration (100–500 ppb) from synthetic contaminated water, in a pressure driven membrane process (10–15 bar). Arsenic removal was totally performed using totally dispersed carbon composite membranes (1% carbon loading, 35°C and 70% RH).

Keywords: Arsenic; membranes; polymer composites; water treatment

Introduction

Arsenic is a common naturally occurring element found in the earth's crust. Natural geochemical contamination through mineral leaching is the primary contributor of dissolved arsenic in groundwater around the world. In recent years arsenic poisoning due to its presence in contaminated groundwater has become one of the major environmental concerns in global scale. Specifically, arsenic presence at high concentrations in groundwater has been a serious problem in the northwest region of Mexico. In 1994 the National Water Commission made a hydrologic study founding arsenic concentrations from 0.05 to 0.5 mg^{-1} in well water (CNA, 1996).

The United States Environmental Protection Agency (US EPA) set since 1976, a maximum contaminant level (MCL) of 50 $\mu g\ L^{-1}$ for As. In 1998, they conducted a risk assessment for As in drinking water and proposed that the MCL must be set at 10 $\mu g\ L^{-1}$ and complied by January, 2006 (US EPA, 2001).

According USEPA estimation, the newly promulgated 10 μg L^{-1} arsenic MCL in drinking water would require corrective action for over four thousand water supply systems, serving an approximate population of 20 million. A vast majority of these systems use groundwater (US EPA, 2002). So that, the lowering of the As-MCL along with better understanding of the harmful effects of arsenic have been catalysts for an increased focus on developing new As treatment technologies

Arsenic cannot be destroyed; it can only be transformed into different forms or combined with other elements to be converted into insoluble compounds. The conventional physical chemical processes used for arsenic removal can be classified based on the principles involved: (a) precipitation, (b) ion exchange, (c) adsorption, (d) membrane technology.

Precipitation is a process in which anions are combined with cations forming insoluble compounds. However, to effectively decrease arsenic to acceptable levels by this process requires a large amount of chemicals creating high quantities of volumetric sludge. This is not easy to treat further or dispose of directly. Arsenic (III) sulfide, calcium arsenate, or ferric arsenate were the most common precipitation methods of arsenic removal, but it has been shown that all of these materials are unstable under certain conditions and therefore not suitable for directly disposal (Robins *et al.*, 2001).

Ion exchange is the process by exchanging arsenic anions from chloride or other anions at active sites bound to a resin. This process has the disadvantage of releasing noxious chemical reagents used in the resin regeneration into the environment (Chen *et al.*, 2002)

The technology of adsorption is a process by using materials that have a strong affinity for dissolved arsenic. Arsenic is attracted to the sorption site on the adsorbent's surface, and is removed from water. Activated carbon is by far the widely used adsorbent. Nevertheless arsenic removal efficiencies are low and carbon is not selective.

Membrane process removes arsenic by allowing water to pass through special filter material, which physically retains the impurities of water. Membrane technologies remove more than just the arsenic, they also could effectively remove portions of all of the dissolved minerals, and even prevent the microorganisms to pass through the membrane to diminish the harmful diseases. The membrane itself does not accumulate arsenic, so disposal of used membranes would be simple, maintenance and operation requirements are minimal in some process, and no chemicals need to be added (Shih, 2005). Pressure driven membrane filtrations are classified into two categories, high-pressure membranes, such as reverse osmosis and nanofiltration; and low-pressure membranes such as microfiltration and ultrafiltration. For arsenic removal, high efficiencies (up to 90%) are found when high-pressure membranes are used. Even though, these process are limited by energy and materials costs, specifically in reverse osmosis (Yamamoto *et al.*, 2000).

One of the mayor issues of study in the membrane technology field is the development of new materials for membrane implementation. In the state of the art, hybrid materials composed by organic polymers and different inorganic solids like zeolites, activated carbon, and alumina, have been used for membrane applications. They show high performance in pressure driven processes due to the synergistic interaction of its components, offering unique physicochemical properties (Esteves *et al.*, 2005; Min *et al.*, 2003).

The broad objective of this study is to prepare and evaluate a new hybrid polymeric-inorganic material that can be used in membrane processes with excellent mechanical strength and attrition resistance properties. Cellulose triacetate was used as polymer material and activated carbon as submicronic particles added. Membranes were tested in their capability to remove arsenic from synthetic polluted water.

Methods

Membrane Preparation

Granular activated carbon (Carbochem LQ 1000, 1058 m^2/g surface area, 5.75 Å average pore size) was triturated by a steel balls mill (8000 Mixer/Mill, Spex Certiprep) for 10 minutes. Then, particles were dissolved in THF (Merck 99%) with agitation and controlled temperature (25 °C) for submicron particle generation as reported elsewhere (Terrazas, 2006 submitted). Particles were stabilized on ethanol and were sonicated before its analysis by laser diffraction technique (Mastersizer MS2000, Malvern Instruments) and transmission electron microscopy (TEM Philips CM–200). Membranes were prepared by the evaporation method in a closed chamber (Shell Lab) at controlled temperature and relative humidity. Cellulose triacetate (Sigma-Aldrich) and activated carbon particles (3.3% w/v, 0-3% w/w respectively) were dissolved into methylene chloride (Sigma-Aldrich) separately and agitated for 24 h at controlled temperature (25 °C). Solutions were mixed and poured into a glass plate and left for evaporation for 2 8 h. Membranes were preserved on distilled water for further evaluation.

Membranes produced were fractured in liquid nitrogen and then treated further with gold in a covering system (Denton Desk-II Gatan) previous to be characterized by scanning electron microscopy (SEM JEOL JSM5800-LV).

Membrane Characterization

Mechanical behaviour

The dynamic mechanical properties of the materials were studied with a DMA-RSA3 TA Instrument analyzer, applying tensile stress. The rupture strain was measured in three independent membrane samples and results as a function of stress were obtained (25 °C and extension rate of 0.005 mm/s). Sample dimensions were 8 mm length and 0.384 mm^2 of transversal area.

Arsenic permeability

Membranes were tested for arsenic permeability in a continuous flow cell SEPA (Osmonics). Arsenic synthetic solutions of 100, 300 and 500 ppb were used. Arsenic concentration was measured by hydride generation atomic absorption spectrometry (AAnalyst 800.Perkin Elmer) in permeate and re circulated streams.

Results and discussion

Membrane Preparation

Membranes were prepared using carbon obtained by the solvation method. Carbon was milled and characterized by laser diffraction technique. Specific surface area found was of 1.75 $m^2\ g^{-1}$. Particle size distribution was: d (0.1) = 1.47 μm, d (0.5) = 6.793 μm, d (0.9) = 22.194 μm. Dispersant used was ethanol, with a refractive index of 1.32. After solvation, carbon was analyzed by TEM analysis. Results obtained are shown in figure 1. Images were interpreted using IFME® program (Torras, 2003). IFME is a software created specifically to treat and analyze images obtained with SEM and TEM (although it can also be used in other applications). IFME calculates average particle diameter d (0.5) as 1.6 μm, reducing in 77% their raw size.

Membranes were synthesized using a three factorial experimental design. Carbon loading, relative humidity and temperature were varied in the range of 0–3%; 10–70% RH and 35–55 °C respectively. Preliminary results showed that carbon dispersion is the principal technical limitation (Terrazas, 2006 submitted).

Nevertheless, after carbon and polymer dilution into methylene chloride separately for 24 hours, previous to membrane preparation, dispersion was attained in materials obtained at high humidity as it is shown in Figure 2.

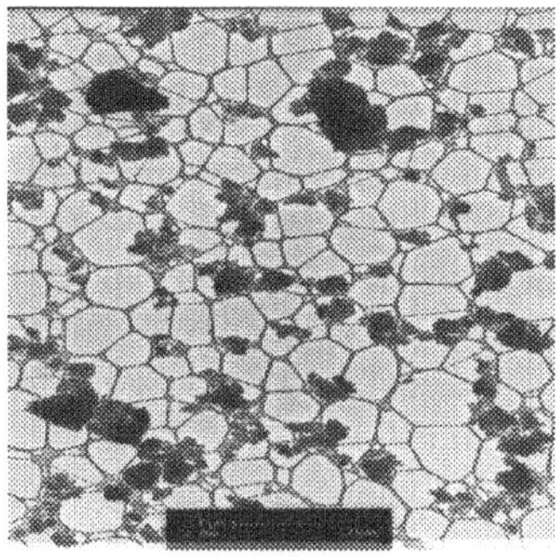

Figure 1 TEM Image of submicron particles of activated carbon, obtained at 72 h of solvation with tetrahydrofuran. Magn 2000×.

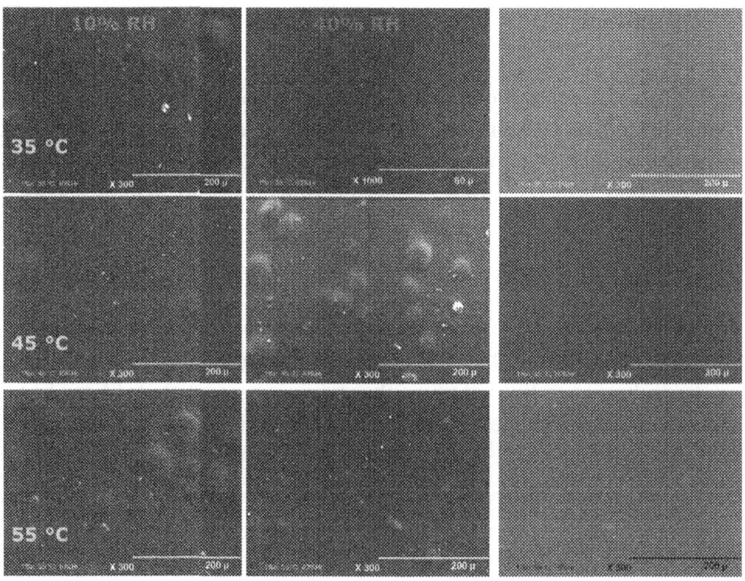

Figure 2 SEM top surface of polymer composite 1% activated carbon loading, obtained at HR 10, 40 and 70%; and T 35, 45 and 55 °C. Magn 300×.

Membrane characterization

Membranes were tested by DMA for its mechanical behavior. Results obtained of rupture stress are shown in Table 1.

Table 1 Rupture stress of membranes tested by DMA

%Carbon Loading	Temperature °C	% Relative Humidity	Rupture stress (dyne cm^2)
0	35	10	2.8
0	55	10	1.7
0	35	70	2.0
0	55	70	2.2
1	35	10	2.3
1	55	10	2.0
1	35	70	4.0
1	55	70	1.5
3	35	10	1.8
3	55	10	1.0
3	35	70	3.0
3	55	70	1.2

As can be observed, membranes prepared at 35 °C and at high humidity are the more resistant of the series tested. In these conditions, evaporation rate is the slowest so that material obtained is more homogenous (Strawhecker *et al.*, 2001). From morphological evidence by SEM analysis, the best materials are those in which carbon is totally dispersed.

Membranes selected for arsenic removal, were the ones located at the extremes of the experimental design performed during their preparation. These were obtained with a cross section of 0.0266 m^2 to be tested in a continuous flux cell. Results obtained with different arsenic synthetic solutions are summarized in Table 2.

Table 2 Arsenic removal for membranes obtained at 70% RH and operated at 14 bar

%Carbon Loading	Temperature °C	As in feed solution (μg L^{-1})	Arsenic removal (%)
0	35	100	56.0
0	35	300	60.0
0	35	500	68.0
1	35	100	80.0
1	35	300	85.0
1	35	500	100
0	55	100	58.0
0	55	300	48.0
0	55	500	10.0
1	55	100	35.0
1	55	300	3.4
1	55	500	3.3

Arsenic was totally removed using carbon composite membrane at 1% loading obtained at 35°C. This membrane is also the one with the highest resistance and dispersion. Removal using membrane obtained at 55°C was not attained, as it was expected due to is homogeneity scarcity. Triacetate cellulose membranes evaluated showed removal efficiencies in the same order of the ones reported earlier in a reverse osmosis (RO) process (Clifford *et al.*, 1986).

Arsenic removal using this membrane technology has been reported with efficiencies up to 90%, which are pH and valence dependent (Amy *et al.*, 1998; Kang *et al.*, 2000). Carbon addition strongly modifies bulk material performance, in one case increasing arsenic permeability ($T = 55°C$) and in the other increasing its rejection ($T = 35°C$).

RO membranes contain extremely small pores (< 0.001 μm) (Schneiter and Middlebrooks, 1983). Composite materials obtained in this study are totally dense, as it was shown in Figure 2. The transport of the solvent is accomplished through the free volume of the material with the need of high driven pressure been operated. So that, fluxes obtained with the best material are of a RO process (0.05 L m^{-2} h^{-1} bar^{-1}).

Conclusions

Composite membranes made of submicronic activated carbon particles and triacetate cellulose membranes were optimal for arsenic removal in a reverse osmosis process with efficiencies up to 100%. Homogeneous composite membranes present dense morphology and high rupture stress. Composite membranes with the highest resistance can operate at high pressures, allowing the highest arsenic rejection.

Acknowledgment

This research was supported by CONACYT México project number J44189-Q, and by PROMEP program.

References

Amy, G. L., Edwards, M., Benjamin, M., Carlson, J. Chwirka, Brandhuber, P., McNeill, L., Vagliasindi, F. (1998). Draft Report, AWWARF.

Comisión Nacional del Agua (CNA) (1996). *Estudio Hidrológico, Hidrogeoquímico y de la Incidencia de Arsénico, Flúor y Hierro en las Zonas Acuíferas de Delicias-Meoqui y Jiménez-Camargo*, Chih. Gerencia de Aguas Subterráneas.

Chen, A. S. C., Fields K. A., Sorg, T. J., Wang, L. L. (2002), Field evaluation of As removal by conventional plants. *J. Am. Waterworks Assoc.* **94** (9), 64–77.

Clifford, D., Subramonian, S., Sorg, T. (1986). *Environ. Sci. Technol.*, **20**, 1072–1080.

Esteves A. C., Barros-Timmonsa A. M., Martinsb J.A, Zhangb W., Cruz-Pintoa J. and Trindade T., (2005). Crystallization behaviour of new poly (tetramethyleneterephthalamide) nanocomposites containing SiO_2 fillers with distinct morphologies. *Composites: Part B* **36**, 51–59.

Kang, M., Kawasaki, M., Tamada, S., Kamei, T., Magara, Y. (2000). *Desalination*, **131**, 293–298.

Min, J.-S., Kiyozumi, Y. and Itoh, N., A. (2003). Sealant-free preparation technique for high-temperature use of a composite zeolite membrane. *Ind. Eng. Chem. Res.* **42**, 80–84.

Robins, R. G., Nishimura, T., Singh, P. (2001). BUET-UNU International Workshop on Technologies for Arsenic Removal from Drinking Water, Dhaka, Bangladesh, pp. 59–68.

Schneiter, R. W., Middlebrooks, E. J. (1983). *Environ. Intern.* **9** (4) 289–291.

Shih, M. (2005). An overview of arsenic removal by pressure-driven membrane processes. *Desalination*, **172**, 85–97.

Strawhecker, K., Kumar, S., Douglas, J. Alamgir, K., The critical role of solvent evaporation on the roughness of spin-cast polymer films. (2001). *Macromolecules*, **34**, 4669–4672.

Terrazas, L. P. (2006). Structural and performance variation of activated carbon-polymer films. *Polymers for Advanced Technologies*. (Submitted.)

Torras C. (2003). IFME ®. Registered software 02/2003/3395, Spain.

US EPA (2001). Federal Register 66, 6975-7066.

US EPA-542-S-02-002 June 2002. Proven alternatives for above ground treatment of arsenic in ground water, www.epa.gov/tio/tsp.

Yamamoto, J.I., Kitawaki, H., Nakao, S., Sugawara, Rahman, M. M., Rahman, M. H. (2000). Desalination, **131**, 307–314.

Automation of an anaerobic/aerobic SBR for toxic wastewater treatment using ORP

A. Vargas, D. González, A. Estival and G. Buitrón

Environmental Bioprocesses Dept. Engineering Institute – UNAM, Circuito Interior s/n Ciudad Universitaria, 04510 Coyoacán, D.F., Mexico (E-mail: *avargasc@iingen.unam.mx*)

Abstract. The present paper reports the experimental implementation of an automatic control strategy for the robust operation of a sequencing batch biological reactor (SBR) used to treat toxic wastewater, specifically synthetic wastewater with p-nitrophenol as model substrate and propionic acid as co-substrate. The SBR had an initial anaerobic reaction subphase where the toxic compound was reduced and a subsequent aerobic subphase where the intermediate amine was oxidized and mineralized. Using only oxidation-reduction potential (ORP) on-line measurements the control algorithm was able to establish the optimal duration of both the anaerobic and the aerobic subphases. On the other hand, monitoring of the biodegradation dynamics was accomplished using UV-Visible spectrophotometry and a novel concentration estimation technique.

Keywords Anaerobic/aerobic biodegradation; oxidation–reduction potential; SBR; spectrophotometry

Introduction

Many industrial wastewaters are difficult to treat by conventional methods, for example with activated sludge systems. However, the sequencing batch reactor (SBR) has proven to be a viable alternative for this type of wastewaters (Melgoza and Buitrón 2001). SBR systems are periodic in nature with usually five predefined phases: fill, react, settle, draw, and idle. While the durations of the fill, settle, draw, and idle phases are fixed depending on parameters that do not change over time, such as physical characteristics of the reactor, making the reaction phase variable may allow for greater flexibility in accommodating varying effluent conditions and microorganism evolution and adaptation. Besides the use of an effective treatment process, strategies and proper control variables are required to eliminate the need for frequent measurements using expensive equipment, and also to increase the efficiency of the process.

On the other hand, for some effluents of an industrial origin, it is convenient to allow two conditions for reaction, depending on the oxygen concentration within the reactor. Such is the case when some nitroaromatic compounds are present. In this case, in an initial anaerobic or anoxic subphase, the original substrate is reduced to some other compound that is easier to oxidize and mineralize in a subsequent aerobic subphase (Melgoza and Buitrón 2001). This may even be accomplished in the same tank using only one consortium of (facultative) microorganisms.

The present work aims at controlling both the durations of the anaerobic and the aerobic subphases dynamically using the on-line measured oxidation-reduction potential (ORP) as control variable. This has already been successfully implemented controlling only the duration of the anaerobic subphase (Buitrón *et al.* 2003). Here modifications of this algorithm and an additional ORP-based control for the aerobic subphase are implemented and reported. The model compound used was p-nitrophenol (PNP), which is reduced to p-aminophenol (PAP) under anaerobic conditions, and then this produced PAP is mineralized during the aerobic subphase.

It has been widely reported that regarding the problem of nitrogen removal, which requires anoxic and aerobic environments, a so-called "knee" appears on the ORP signal upon removal of oxidized nitrogen during the anoxic phase (Holman and Wareham 2003), *i.e.* the ORP signal undergoes a sudden change in its rate or velocity. Having observed this knee also for PNP biodegradation, it has also been used successfully to establish the end of the anaerobic subphase (Buitrón *et al.* 2003). On the other hand, under a significant supply of oxygen, it has been observed that the ORP signal closely follows the dissolved oxygen concentration signal. The oxygen uptake rate is known to be an indicator of the microbial activity (Spanjers *et al.* 1996). When there is no more substrate present in the medium, activity diminishes and thus dissolved oxygen concentration tends to its saturation value; the ORP signal then also tends to a constant value. Both indicators on the ORP signal are used here to control the duration of the anaerobic and aerobic reaction subphases automatically using a proprietary algorithm (Martínez *et al.* 2005).

Furthermore, this work reports the use of a novel spectrophotometric technique called nonlinear spectral deconvolution (NSD) (Vargas and Buitrón 2006) during experimentation, in order to assess that the control algorithm did indeed work as expected. The technique allows the quick measurement of the present concentrations of some compounds in the mixed liquor while the reaction is taking place. Therefore, using these estimations it is possible to see practically in real time the biodegradation dynamics. That is, in this particular case, both the time evolution of the transformation of PNP to PAP and later the biodegradation of PAP can be monitored practically on-line.

Materials and methods

Reactor

A single pilot double-walled temperature-controlled (27°C) reactor with 8 L of working volume was used. That is, both the aerobic and anaerobic reactions took place in the same tank and the (facultative) biomass was able to survive and work in both environments. The exchange volume was 6 L. Peristaltic pumps (Cole-Parmer) were used to fill with synthetic wastewater and to draw the supernatant at the end of the reaction. Both the pumps and a mixer were controlled by a computer through a data acquisition card and peripherals (PCI-6025E, National Instruments) and software coded in *LabView* (National Instruments) through a proprietary algorithm called *BioReC*. During the aerobic subphase, air was diffused through the bottom of the reactor regulating the airflow at 2.5 L/min using a digital mass flow controller (Cole-Parmer). The pH was monitored and kept within 7±0.2. The reactor operated as an SBR allowing 6 min to fill, 40 min to settle, 6 min to draw, and 5 min of idle time, while the reaction times were variable and dynamically set by the control algorithm.

Biomass

The reactor was inoculated with sludge from a municipal wastewater treatment plant and acclimated according to the variable time procedure described by Melgoza *et al.* (2000).

Nutrient enriched synthetic wastewater containing 35 mg/L of p-nitrophenol (PNP) and propionic acid (PA) in a molar PNP:PA relationship of 1:20 was used. The nutrients were added according to AFNOR (1985).

Control algorithm

Measurement of the oxidation-reduction potential signal was accomplished through a commercial industrial Pt-Ag/AgCl electrode and transmitter (+GF+ Signet). The preamplified voltage signal was acquired through a data acquisition card (PCI-6025E, National Instruments) and read by *LabView* software (National Instruments), where the control algorithm was implemented as an operating mode called *AnA-VTC* of the *BioReC* program. The algorithm to control the operation of the reactor and the control algorithm based on the ORP signal measurement is described in detail by Martínez *et al.* (2005). Nevertheless, its basic aspects are as follows.

The derivative of the filtered ORP signal, dubbed dORP, is numerically calculated using a nonlinear filter. This signal is then used to detect when a knee in the ORP signal occurs, since usually a sudden change in the velocity of the ORP signal will cause the dORP signal to become more negative. A negative peak is then sought, such that a minimum of the dORP signal indicates that the end of the so-called knee is about to come. The algorithm then waits for dORP to become a fraction of its minimum value to end the anaerobic reaction phase. It then starts aeration and therefore the ORP signal rises. During the aerobic reaction it remains almost constant, until a sudden increase up to positive values indicates complete mineralization. The ORP signal then remains at an almost constant (positive) value. The algorithm again uses the dORP signal, waiting for a sudden increase and posterior tendency towards zero of this signal to end the aerobic reaction and start sedimentation.

Analysis and concentrations estimation

The total (TSS) and volatile suspended solids (VSS) were quantified according to *Standard Methods* (1992). PNP and PAP concentrations were not monitored using a standard technique. Instead, samples were drawn periodically from the reactor and passed through a UV/Vis spectrophotometer and the nonlinear spectral deconvolution (NSD) method was used to estimate the concentrations (Vargas and Buitrón 2006). This technique is a modification of the spectral deconvolution (SD) technique proposed by Constant *et al.* (1996). The idea is based on the Lambert-Beer law by assuming that the measured absorbance spectrum $s(\lambda)$, where λ is the wavelength of a sample with n compounds will be a weighted sum of n individual base spectra, one for each of the constituents of the sample. In this case, one may propose that the measured spectrum is a linear combination of the base spectra of the compounds (Escalas *et al.* 2003), i.e.

$$s(\lambda) = \sum_{i=1}^{n} \beta_i \cdot \alpha_i(\lambda), \tag{1}$$

which expressed in vector notation becomes

$$s = \sum_{i=1}^{n} \alpha_i \beta_i = [\alpha_1 \quad \dots \quad \alpha_n] \cdot \begin{bmatrix} \beta_1 \\ \vdots \\ \beta_n \end{bmatrix} = \mathbf{A} \cdot \boldsymbol{\beta}. \tag{2}$$

Ideally, if the Lambert-Beer law holds exactly, the coefficient vector $\boldsymbol{\beta}$ will equal the concentration vector $\mathbf{x}$ and thus SD finds this vector by simple least squares techniques. Nonlinear spectral deconvolution adds some level of flexibility and considers a quadratic nonlinear relationship defined by the matrix $\mathbf{M}$:

$$\boldsymbol{\beta} = \varphi(\mathbf{x}) = \mathbf{x} + \mathbf{M} \cdot \xi(\mathbf{x})$$
$$\xi(\mathbf{x}) = \begin{bmatrix} x_1 x_2 & \dots & x_1 x_n & x_2 x_3 & \dots & x_{n-1} x_n \end{bmatrix}^T. \tag{3}$$

The matrix $\mathbf{A}$ of base spectra and the nonlinearity matrix $\mathbf{M}$ are obtained through a previous calibration with samples of known concentrations. To obtain an estimate $\hat{\mathbf{x}}$ of the concentrations vector in the sample given a measured spectrum s, a weighted least squares procedure is first used to obtain $\boldsymbol{\beta}$, using a covariance matrix that is also obtained during calibration to account for variability at different wavelengths. This estimate of $\boldsymbol{\beta}$ is then used to solve a nonlinear constrained optimization problem based on minimizing an error cost function using equation (4). This estimation problem can be solved very efficiently using numerical solvers.

For the bioreactor studied, samples of 10 mL were taken at different times during both the anaerobic and aerobic phases of the reaction, centrifuged at 3000 rpm (Sol-Bat J600) for 3 min and filtered through a 1.6 μm microfiber filter (Millex). The filtered samples were then measured by manually filling the corresponding quartz cell in the spectrophotometer. This measurement was taken in the 210–480 nm range. Distilled water was used as the reference sample. After measurement, the samples and the centrifuged sludge were returned to the reactor. The whole procedure lasted approximately 4 min. For each SBR cycle the spectrophotometer was calibrated and baseline corrected only once. Estimation of the concentrations, based on the measured spectral data, was done in MATLAB (The Mathworks) using a specifically designed toolbox called *ConSpect*.

Results

In figure 1, the results for one representative cycle are shown. The filtered ORP signal is shown, as well as its time derivative dORP/dt (this is the numerically computed result). Also, the estimated PNP and PAP concentrations time evolution are shown, which were computed using the spectrophotometric data. It is interesting to notice that the ORP knee around t=300 min concides with the complete reduction of PNP to PAP. This is more evident with the dORP signal, where a minimum occurs. The control algorithm has detected such a knee and some (security) time later turns on the aeration, such that the aerobic subphase begins. Notice then the sudden increase in ORP values only due to the change in the environment. Then ORP oscilates around some negative values during PAP oxidation, until a sharp increase signals complete mineralization, which coincides with PAP concentration values near zero. This is again detected by the control algorithm correctly and therefore this subphase ends to begin sedimentation, where the ORP signal again decreases to negative values (aeration has been turned off and the reamining oxygen is consumed due to endogenous respiration).

Figure 2, on the other hand, shows the time evolution of the spectrograms for the samples taken from the representative cycle of figure 1. Notice how PNP has a distinctive peak around 390 nm, which as the anaerobic reaction evolves diminishes and the two distinctive peaks of PAP at 220 nm and 325 nm begin to appear. By the end of the anaerobic reaction only the spectrogram of PAP is distinctive. During the aerobic reaction this compound is biodegraded and thus the spectrogram becomes that of the effluent.

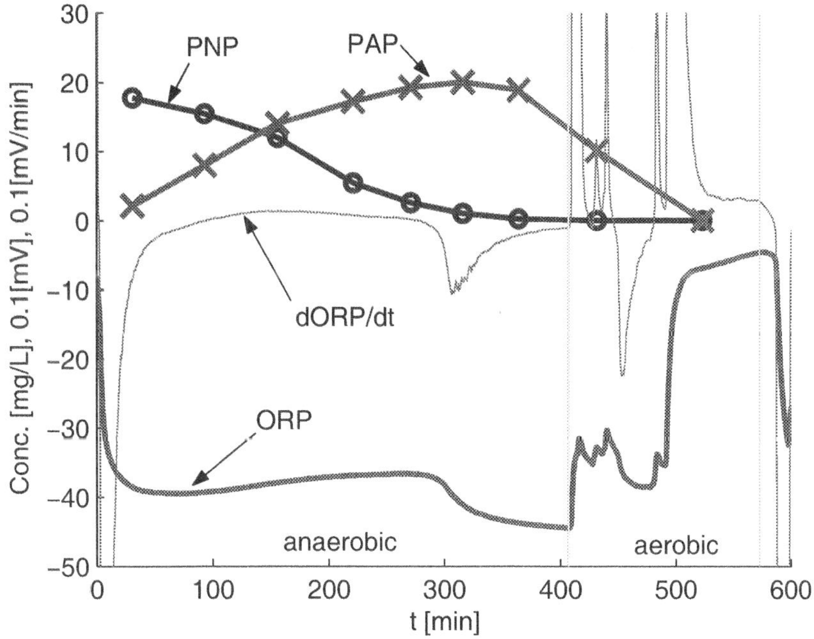

Figure 1 A typical reaction phase in the anaerobic/aerobic SBR showing the ORP signal evolution, its time derivative dORP/dt, and the PNP and PAP concentrations estimation using the NSD technique.

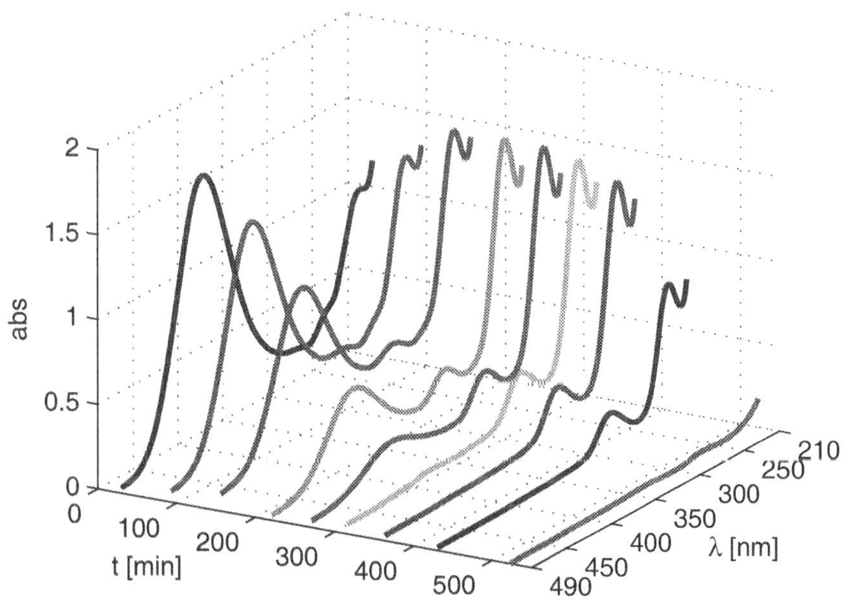

Figure 2 Time evolution of the spectrograms during a typical reaction phase.

Figure 3 is presented to show that the ORP signal is indeed a suitable control signal, since its evolution is rather repetitive for consecutive cycles. In this figure also the noisy measured signal is shown together with the filtered one, in order to illustrate that signal conditioning is essential for correct operation of the algorithm. Otherwise it would be impossible to detect the curve characteristics that are sought.

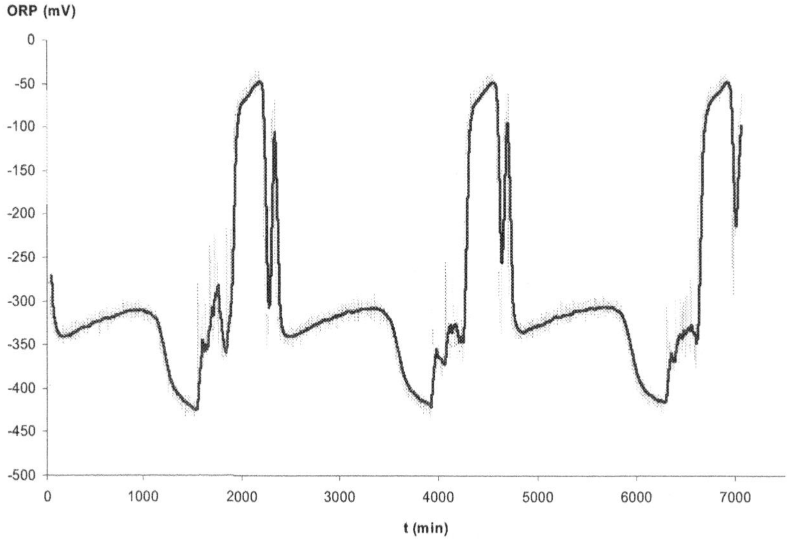

Figure 3 Several consecutive cycles, showing the noisy and the filtered ORP signal.

Finally, Figure 4 shows the time evolution of TSS and VSS some days after acclimation with the control strategy working. It is clearly seen that VSS remained around 2600 mg/L and thus was rather stable.

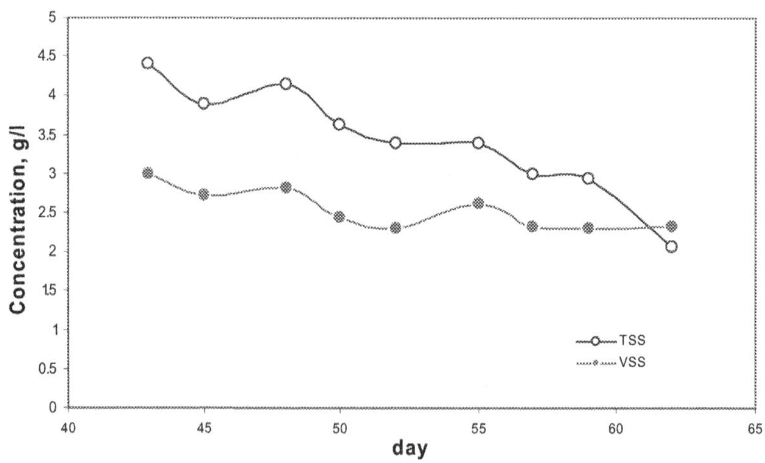

Figure 4 Total (TSS) and volatile suspended solids in the bioreactor after acclimation.

Conclusions

The present work has presented the results of the application of a control algorithm to dynamically set the duration of both the anaerobic and aerobic subphases of an SBR operated with two alternating enviornments on the same tank. Such an algorithm is based on the on-line measurement of the ORP signal and on detecting some of its salient features, namely the appearance of a so-called knee for the first (anaerobic) subphase, and stabilization of the signal after a sudden increase for the second (aerobic) subphase. Furthermore, it has also been shown the effectiveness of a novel spectrophotometric technique to assess the time evolution of the concentration of some compounds of interest during the reaction, in practically real time. The model compound used was p-nitrophenol, but current investigations have also shown the possibility of using the same control algorithm for more complex toxic compounds, such as some azo dyes.

Acknowledgements

This project was financially supported by DGAPA-UNAM project IN104805. The second author also thanks CONACYT for financial support as a scholarship.

References

Buitrón, G., Betancur, M.J., Moreno, G., and Moreno, J.A. (2003) Oxidation-reduction potential as a control variable for the anaerobic stage during anaerobic–aerobic p-nitrophenol degradation. *Biotechnol. Prog.* **19**, 1822–1827.

Escalas, A., Droguet, M., Guadayol, J., and Caixach, J. (2003). Estimating DOC regime in a wastewater treatment plant by UV deconvolution, *Wat. Res.*, **37**, 2627–2635.

Holman, J.B. and Wareham, D.G. (2003) Oxidation-reduction potential as a monitoring tool in a low dissolved oxygen wastewater treatment process, *J. Environ. Eng. Div.* **129**(1), 52–58.

Martínez K.M., Vargas, A., and Buitrón G. (2005) Using ORP as control variable for an anaerobic/aerobic SBR for nitrophenol biodegradation, *3rd IWA Leading Edge Conference*, Sapporo, Japan, July 2005.

Melgoza, R.M., Chew, M., and Buitrón, G. (2000) Start-up of a sequential anaerobic-aerobic batch reactor for the mineralization of PNP. *Wat. Sci. Tech.* **42**(5/6), 289–292.

Melgoza R.M. and Buitrón, G. (2001) Degradation of p-nitrophenol in a batch biofilter under sequential anaerobic/aerobic environments, *Wat. Sci. Tech.* **44**(4),151–158.

Spanjers, H., Vanrolleghem, P, Olsson G., and Dold, P. (1996) Respirometry in control of the activated sludge process. *Wat. Sci. Tech.* **36**(3/4), 117–126.

Standard Methods for the Examination of Water and Wastewater (1992), 18th ed., APHA/AWWA/WEF, Washington DC, USA.

Vargas, A. and Buitrón, G. (2006). On-line concentration measurements in wastewater using nonlinear deconvolution and partial least squares of spectrophotometric data, *Wat. Sci. Tech.* (In press.)

Effect of cationic polymer on biomass granulation in UASB reactors treating domestic sewage

Puspendu Bhunia and M.M. Ghangrekar

Department of Civil Engineering, Indian Institute of Technology, Kharagpur – 721 302, India
(E-mail: *puspendu@civil.iitkgp.ernet.in*; *ghangrekar@civil.iitkgp.ernet.in*)

Abstract This paper examines the effect of cationic polymer additives on granulation in lab-scale UASB reactors treating synthetic sewage (COD 650 mg/l). Under identical conditions, two reactors were operated in parallel at different organic loading rate: reactor R1 without any polymer additives and reactor R2 with addition of polyacrylamide in the inoculum. Polymer dose was decided based upon the results of Jar test and settling test. Under the same operational and environmental conditions, for inoculum having SS concentration greater than 120 g/l and VSS/SS ratio less than 0.3, granulation was obtained in both the reactor operated at organic loading rate (OLR) 3.87 kg COD/m^3.d. However, addition of polymer to such inoculum sludge was found to deteriorate reactor performance by means of increasing the start-up time and decreasing the COD removal efficiency. In addition, polymer additives were found to reduce the formation of larger diameter granules. Thus, addition of polymer does not show any favourable effect in sludge granulation when inoculum itself is very thick. Better granulation of the sludge can be obtained with such inoculum if proper operating conditions are maintained in the reactor during start-up.

Keywords UASB, granules, polymer dose, minimum size of granules

Introduction

Upflow anaerobic sludge blanket (UASB) reactors are becoming more popular for sewage treatment after the knowledge gained during the operation of several municipal anaerobic plants all over the world. Important parameters affecting the treatment efficiency of UASB reactors include granulation in the reactor, the characteristics of wastewater to be treated, selection of proper inoculum, the influence of nutrients and several other physical and environmental factors. Among these parameters, the biomass granulation is the most crucial parameter to enhance the efficiency, especially for the treatment of low strength wastewater, such as sewage, and granulation itself depends upon several parameters.

Although the application of UASB reactor for treatment of sewage at large scale has been applied in several tropical countries, the process is unable to produce higher COD removal efficiency in single-stage secondary treatment compared with aerobic processes (Elmitwalli *et al.*, 2002; Singh and Viraraghavan, 1998). In the literature, the UASB reactor is reported to give 60–75% COD removal efficiency while treating sewage, at organic loading rates (OLR) between 0.4 and 3.0 kg COD/m^3.d and temperature range 15–30°C.

The success of the UASB reactor depends on the formation of active and settleable sludge granules, which results from microbial self-immobilization and, subsequently, aggregate formation and growth. Apart from improving the settleability of biomass leading to effective bacterial retention in the reactor, granulation also improves the physiological conditions and helps in the formation of syntrophic association in the anaerobic system. In engineering practice, we are interested in biomass granulation because of two reasons. First, it is a natural process under favorable conditions (owing to the tendency of bacteria to self-immobilize), hence, eliminating the need of any media support for bacterial attachment; and second, due to its better retention ability inside the reactor, which in turn favours high efficiency. Formation of stable granules is essential for successful operation of the reactor (Show *et al.*, 2004), and it is reported to increase the COD removal efficiency from 60–80% (with flocculent sludge) to greater than 90% (Jeong *et al.*, 2005).

It is reported that external additives in the form of multivalent cations such as, iron and aluminium enhance granulation in high strength wastewater (Yu *et al.*, 2001; Liu and Tay, 2004). In recent past, the development of granular biomass in UASB reactor using polymer addition is evaluated by many researchers (Show *et al.*, 2004; Jeong *et al.*, 2005; Wang *et al.*, 2004). The time required for the formation of biogranules due to addition of polymer varies from 5 min (Jeong *et al.*, 2005) to 50 days (Show *et al.*, 2004). Although, studies have been carried out by many researchers to evaluate the effects of polymer dose on biomass granulation, the polymer dose has been determined on an arbitrary basis by evaluating the reactors performance under a series of different polymer dose. Against this context, the present study seeks to determine the optimum polymer dose, and to compare the characteristics of the sludge developed and performance of the UASB reactor, treating synthetic sewage, without and with polymer addition. It was hypothesized that, if the operating conditions are maintained under favorable conditions, for certain inoculum sludge, granulation can be obtained in UASB reactor without any polymer addition.

Materials and methods

Two identical reactors (R1 and R2), made of Plexiglas having effective volume 12.57 l, internal diameter 100 mm, and effective height 1.6 m were used in the study. The reactors were operated at same operational and environmental conditions with reactor R1 without any polymer addition and R2 with polymer addition in the inoculum, as presented in Table 1.

Table 1 Experimental design (mean ± std. deviation)

Reactor No.	Polymer addition	OLR ($kgCOD/m^3.d$)	SLR (kgCOD/kg VSS.d)	Flow rate (l/d)	Influent COD (mg/l)	Temp. (°C)
R1	No	3.87 ±0.16	0.251	75.4	647 ± 27.37	28.31±0.78
R2	Yes	3.87 ±0.16	0.251	75.4	647 ± 27.37	28.31±0.78

Synthetic feed with sucrose as carbon source was used in the study. All the nutrients required were added in appropriate proportion with the trace metals. The reactors were seeded with sludge obtained from bottom of a septic tank. The seed sludge was flocculent in nature with suspended solids (SS) of 120.4 g/l and volatile suspended solids (VSS) to SS ratio of 0.26. Reactor R1 was inoculated without polymer addition and reactor R2 was inoculated with a cationic polymer (Polyacrylamide containing DADMAC, Ondeo Nalco India Ltd.) dose of 0.083 ml/g of inoculum SS. The required volume of polymer was first diluted to 100 ml with distilled water and after adding with inoculum, mixing between polymer and inoculum was carried out by mechanical stirrer.

Determination of optimum polymer dose

The septic tank sludge was first diluted two times with tap water and jar test was carried out with 300 ml of the diluted sludge sample containing different polymer doses of 0–0.503 ml/g SS. During the jar test, the rapid mixing was carried out for 2 minutes at 150 rpm and flocculation for 15 minutes at 30 rpm. The samples from the jar test apparatus was then transferred and diluted to 1000 ml into a one litre measuring cylinder for determination of SVI. Against different polymer dose SVI was plotted, and the range of doses corresponding to the minimum SVI was considered as optimum range of doses. The cross verification of the dose was done by performing settling test. Settling test was thought to be another appropriate test to evaluate optimum dose. In this test, 150 ml of the diluted sludge sample with different polymer dose was taken in a closed conical flask assembly. The sludge sample was kept on a platform shaker at 200 rpm for first 5 minutes and then at 50–80 rpm for the next 15 minutes. The sludge sample was then transferred into a 150 ml graduated glass cylinder. After 1 minute settling, SS of the supernatant was determined. The range of doses corresponding to the minimum SS in the supernatant was considered as optimum range.

Sampling and analytical techniques

Influent and effluent samples were analyzed for pH, settled COD, and alkalinity every day, for SS, VSS twice in a week, and BOD_5 and volatile fatty acids (VFA) once in a week. Sludge samples form the reactors were analyzed for SS, VSS, settling velocity, strength of granules, sludge volume index (SVI), specific gravity, and specific methanogenic activity (SMA) after 60–65 days of operation. COD, pH, Alkalinity, volatile fatty acid (VFA) in the effluent sample, SS and VSS in the effluent and sludge, and SVI of the sludge were determined as per APHA (1998). Specific methanogenic activity (SMA) was carried out according to the method proposed by Ghangrekar *et al.* (1996). Measurement of settling velocity followed the method used by Ghangrekar *et al.* (2005). A glass column of 7.6 cm diameter and height 1.32 m filled with tap water was used for determination of average settling velocity of sludge. The test procedure described by Ghangrekar *et al.* (1996) was used for determination of strength of granules. Scanning electron microscopy (SEM) was carried out according to Singh *et al.*, (1996).

Fractional distribution of different size of granules

Settling velocity of sludge settled at the bottom of the settling column after fixed time intervals (0.5, 1, 2, 3, 5, 7, 15, 30, and 60 min) was considered to determine corresponding size (d_p) of the biomass fractions, using Stokes and Newton's law of particle settling theory. Average specific gravity of the sludge was carried out according to the APHA *et al.* (1998), and considered for the determination of granules diameter using settling theory. From the percentage mass fractions of the settled sludge at different time intervals, diameter of granules present in sludge was calculated.

Results and discussion

Optimum polymer dose

The agglomeration between two bio-particles can be understood by a physico-chemical approach by considering that bacteria behave as inert particles (Hulshoff Pol *et al.*, 2004). The interaction between two particles depends on the Van der Waals force of attraction and

Coulomb's law of electro-static repulsion. The main function of polymer additives is to neutralize the surface charge of the biomass and thereby eliminating the electro-static repulsive force between the two approaching particles, leading to the decrease in SVI, as the volume occupied by the sludge will be reduced.

Figure 1 shows the variation of SVI with polymer dose. SVI will be lowest when surface charges of bio-particles will be neutralized and hence, indicating optimum dose. At polymer dose in the range 0.05 ml/g of SS to 0.15 ml/g of SS, the SVI was found to be lowest (about 4.0 ml/g of SS). Hence, a polymer dose in this range was considered the optimum dose.

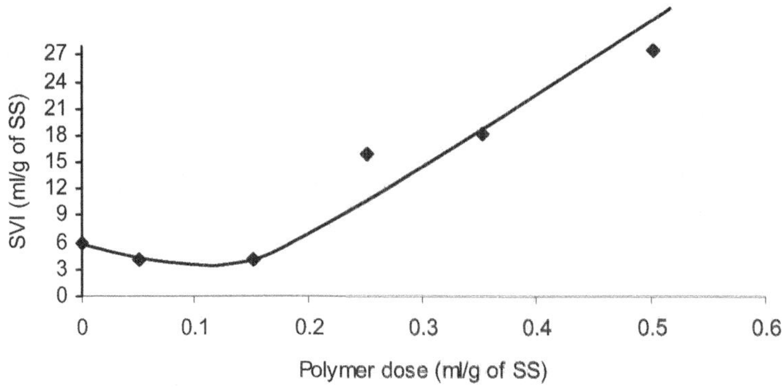

Figure 1 Variation of SVI with polymer dose.

As the range of polymer dose obtained by jar test was very large, in addition to SVI determination, settling test was carried out. Supernatant SS concentration observed after settling was plotted against polymer doses as shown in figure 2. In absence of electro-static repulsive force, the SS of the supernatant will be minimal at optimum polymer dose. A range of polymer dose (0.062 ml/g of SS to 0.0933 ml/g of SS) is observed from figure 2, where SS of the supernatant lies at the lower range of 0.62 g/l to 0.50 g/l. Therefore, from the consideration of charge neutralization of bacterial surface observed by SVI values and settling test at different polymer dose, 0.06 ml/g of SS to 0.09 ml/g of SS has been considered the optimum polymer dose. Hence, in the present study, a polymer dose of 0.083 ml/g of inoculum SS was selected to study effect of polymer dose by comparing the performance of the reactor under identical conditions with and without polymer addition.

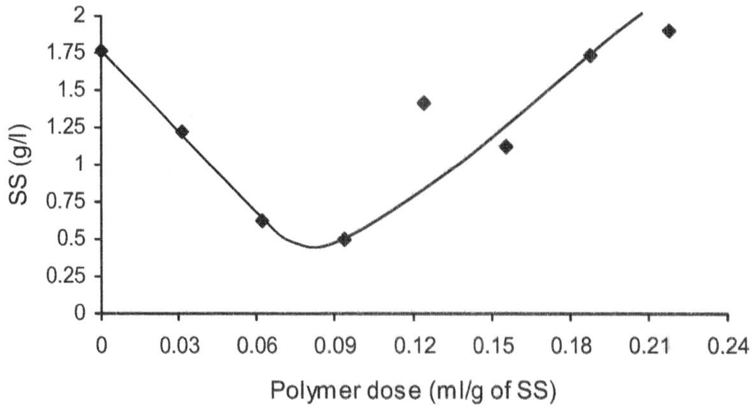

Figure 2 Variation of supernatant SS with polymer dose.

Reactor start-up, steady state and performance

Steady state condition in R1 and R2 was achieved after 29 days and 38 days of operation, respectively. An average COD removal efficiency of 92.07% and 87.5% was observed for R1 and R2, respectively for the operational period of 60 days under steady state condition. Significant differences in COD and BOD_5 removal efficiency were observed in R1 and R2 after reaching steady state condition (Table 2), indicating the improper functioning of the micro-organisms due to addition of polymer with inoculum sludge. Although, cationic polymer is reported to favour the reactors performance (Show *et al.*, 2004; Jeong *et al.*, 2005; Wang *et al.*, 2004), by enhancing the biomass granulation, however, in the present study deterioration in removal efficiency was observed with polymer addition. This could be because of the high SS content of inoculum sludge, where further addition of polymer made it highly compact in nature, which probably has reduced the substrate diffusion inside the sludge.

Table 2 COD, and BOD_5, and removal efficiency (mean ± std. deviation) at steady state

Parameters	R1 (without polymer additives)	R2 (with polymer additives)
Influent COD (mg/l)	638.41± 49.45	638.41± 49.45
Effluent COD (mg/l)	55.43 ± 14.97	81.30 ± 6.82
%COD removal efficiency	92.07 ± 1.87	87.5 ± 1.23
Influent BOD_5 (mg/l)	288.4 ± 11.84	288.4 ± 11.84
Effluent BOD_5 (mg/l)	14.42 ± 4.29	25.1 ± 3.91
%BOD_5 removal efficiency	94.93 ± 1.38	91.29 ± 1.36

Biomass granulation

Granule size distribution was obtained through Stokes and Newton's law of settling theory by performing settling velocity and specific gravity test. Faster rate of formation of granules was observed in R1 than in R2. Although formation of biogranules was observed in both the cases but reactor with polymer additives (R2) took 43 days for granulation whereas reactor without polymer additives (R1) took 22 days. Granules thus obtained in both the reactors also showed differences in settleability character. The average settling velocity of sludge in reactor R1 and R2 was 42.48 m/h and 35.68 m/h, respectively.

In spite of the polymer addition, this is supposed to favour granulation, the figure 3 shows that the percentage of granules having an average diameter in the range of 3.03 mm to 0.45 mm was more in R1 (without polymer) in comparison of R2 (with polymer). Whereas, particles having an average diameter of 0.34 mm to 0.11 mm were more in R2 (with polymer) in comparison to R1 (without polymer). High percentage of granules with size less than 0.34 mm may not be favorable in reactor operation, as settling of these particles is supposed to be predominantly governed by viscosity of the surrounding fluid medium. As the settling of these bio-particles does not governed by mass of the particles, therefore any turbulence produced by the up flow velocity of the influent wastewater and/or the biogas may lead to wash out of bio-particles from the reactor. Based upon the Stokes and Newton's law of particle settling theory and considering the inertia force to be predominant over the viscosity of the surrounding medium, the minimum diameter of bio-particles to be considered as minimum size of granules is 0.34 mm for specific gravity of gross sludge as 1.035. With this consideration, better granulation was obtained without polymer addition.

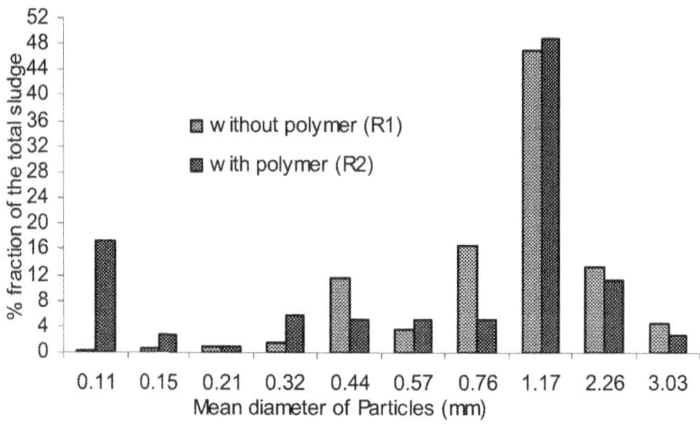

Figure 3 Distribution of different diameter granules in R1 and R2.

Although addition of cationic polymer is reported to enhance the granulation process and thereby reducing the start-up time (Show *et al.*, 2004; Jeong *et al.*, 2005; Wang *et al.*, 2004), results obtained at polymer dose adopted in the present study contradict in many aspects. Thus from the present study, it appears that polymer addition is not always advisable, particularly for a thick inoculum having SS greater than 100 mg/l and VSS/SS ratio less than 0.3. For such an inoculum, addition of polymer does not enhance the biomass granulation; rather it disturbs the proper contact between the micro-organisms and influent wastewater probably by increasing the compactness of the bed and hence reduces the overall COD removal efficiency of the reactor.

Physical properties of the sludge

Significant differences in characteristics of sludge were observed with or without polymer addition. The physical properties of the sludge observed at steady state are presented in table 3. Based upon the Stokes and Newton's law of particle settling theory and settling column analysis, the fraction of granules (considering minimum diameter of granules $\geq$ 0.34 mm) as mass of the total sludge, developed in R1 and R2 was 98.34% and 84%, respectively. On account of the production of higher percentage of bigger size granules (in the range of 3.03 mm to 0.34 mm), higher settling velocity of the sludge was observed in R1 (without polymer additives) in comparison to R2 (with polymer additives).

Table 3 Physical properties of sludge developed in R1 and R2.

Parameters	R1 (without polymer)	R2 (with polymer)
Settling velocity (m/h)	42.48	35.68
SMA (kg CH_4 COD/kg VSS.d)	0.502	0.587
Specific gravity	1.035	1.035
Integrity coefficient (%)	19.73	5.49
SVI	16.27	13.39
VSS/SS ratio	0.54	0.56

Although, higher percentage of granules was observed in R1 than R2, however, integrity coefficient was significantly less when polymer was added than without polymer addition, indicating higher strength of the granules. This can be explained as the increase in microbial density of individual granules due to the addition of cationic polymer. The positively charged polymer may form a bridge among negatively charged bacterial cells through electro-static charge attraction. The resulting chain-like structure may lead to a complex network structure for enhanced bacterial aggregation, which shows better withstanding ability against high abrasion and shear force.

Marginal improvement in methanogenic activity of sludge was observed due to polymer addition. The minimum SMA of 0.3 kg CH_4 COD/kg VSS.d is reported as the activity of granular sludge (MacLeod *et al.*, 1990; Fang and Chui, 1993). Hence, apart from the settling velocity, SMA also supports the formation of granules in R1 and R2. SEM photographs (Figure 4) of granules revealed that the mixed populations with some inert materials, which supports the idea that granulation may be initiated by non-specific ionic bonds formed between positively charged inert material and the negatively charged surfaces of anaerobic bacteria. This implies that the addition of cationic polymer did not result in a difference in predominant microorganisms.

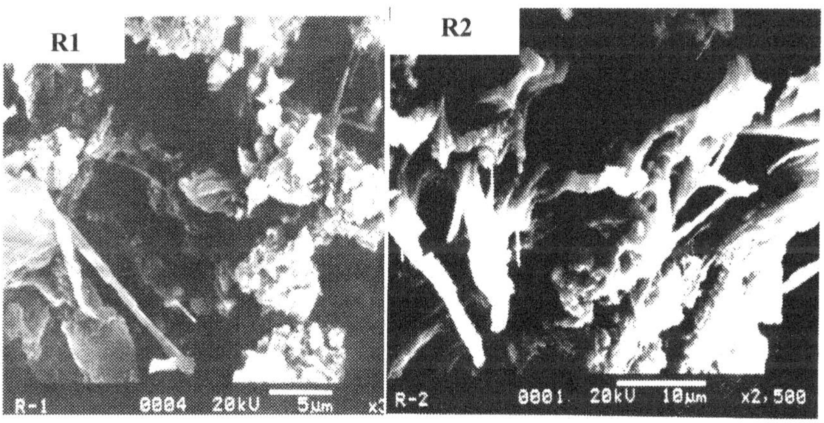

Figure 4 SEM photographs of granules in R1 and R2.

The results of the present study contradict the beneficial effects of polymer addition on performance of reactor and biomass granulation. However, it showed better methanogenic activity and strength of granules when compared against the granules obtained without any polymer additives (R1), under the same operational and environmental conditions. Studies, where a beneficial effect of cationic polymer on granulation was claimed, sludge having SS of 20–80 g/l and VSS/SS ratio greater than 0.35 was used as inoculum (Show *et al.*, 2004; Wang *et al.*, 2004; Jeong *et al.*, 2004). The cells of these types of inoculum are dispersed in nature and very much susceptible to wash out from the reactor. Therefore, under such thin inoculum sludge, addition of polymer helps in enhancing granulation by reducing wash out of biomass from the reactors. For such thin inoculum, mixing between the sludge and wastewater is not a critical factor even after compaction offered by adding polymer. An inoculum with SS greater than 120 g/l and VSS/SS ratio less than 0.3 is compact in nature without any polymer additives and contains higher percentage of inert matters. Compaction of such thick types of inoculum are good enough to prevent the excessive wash out of sludge from the reactors as well as maintaining a proper mixing between the inoculum sludge and influent wastewater. Addition of polymer with this sludge proved to have detrimental effects on reactor performance by making it highly compact in nature.

Conclusions

As indicated in jar test and subsequent SVI determination under different polymer dose, even at optimum dose, improvement in SVI was marginal as compared to SVI of inoculum sludge without polymer addition.

Based upon the results of the present study, it can be concluded that addition of polymer not always helps in granulation for the treatment of low strength wastewater. Instead, for certain inoculum sludge, it deteriorates the reactors performance. Anaerobic biomass granulation for the treatments of low strength domestic wastewater basically depends on the selection of proper organic loading rate (OLR) and types of inoculum used. For thin sludge (SS < 100 g/l and VSS/SS >0.35), addition of polymer is reported to have enhancing effects on biomass granulation. However, for thick sludge (SS > 120 g/l and VSS/SS < 0.3) addition of polymer does not have any significant enhancing effects on biomass granulation rather it reduces the formation of higher diameter as well higher percentage (by mass) of granules. The result of the present study reveals that for such inoculum sludge, addition of polymer is not necessary for biomass granulation and better reactors performance.

References

APHA, AWWA, WPCF (1998). *Standard Methods for Examination of water and wastewater, 20th ed.* American Public Health Association, Washington, DC.

Elmitwalli, A.T., Oahn, T.L.K., Zeeman, G., Lettinga, G. (2002). Treatment of domestic sewage in a two-step anaerobic filter/anaerobic hybrid system at low temperature. *Water Research*, **36**, 2225–2232.

Fang, H.H.P., Chui, H.K. (1993). Maximum COD loading capacity in UASB reactors at 37°C. *J. Environ. Engng Div. ASCE.*, **119** (1), 103–119.

Ghangrekar, M.M., Asolekar, S.R., Joshi, S.G. (2005). Characteristics of sludge developed under different loading conditions during UASB reactor start-up and granulation. *Water Research*, **39** (6), 1123–1133.

Ghangrekar, M.M., Asolekar, R.S., Ranganathan, R.K., Joshi, G.S. (1996). Experience with UASB reactor start-up under different operating conditions. *Water Science and Technology*, **34**, 421–428.

Hulshoff Pol, W.L., Lopes, C.I.S., Lettinga, G., Lens, L.N.P. (2004). Anaerobic sludge granulation, *Water Research*, **38**, 1376–1389.

Jeong, S. H/, Kim, H. Y., Yeom, H. S., Song, K. B., Lee, S. (2005). Facilitated UASB granule formation using organic–inorganic hybrid polymers. *Process Biochemistry*, **40** (1), 89–94.

Liu, Y., Tay, H.J. (2004). State of art of biogranulation technology for wastewater treatment. *Biotechnology Advances*, **22**, 533–563.

MacLeod, F.A., Guiot, S.R., Costerton, J.W. (1990). Layered structure of bacterial aggregates produced in an UASB and filter reactor. *Appl. Environ. Microbiol.*, **50** (6), 1598–1607.

Show, Y. K., Wang, Y., Foong, F. S., & Tay, H. J. (2004). Accelerated start-up and enhanced granulation in upflow anaerobic sludge blanket reactors. *Water Research*, **38**, 2293–2304.

Singh, S.K., Viraraghavan, T. (1998). Start-up and operation of UASB reactors at 20°C for municipal wastewater treatment. *Journal of Fermentation and Bioengineering*, **85**, 609–614.

Singh, S.K., Harada, H., Viraraghavan, T.B. (1996). Low strength wastewater treatment by a UASB reactor. *Bioresource Technology*, **55**, 187–194.

Wang, Y, Show, Y. K., Tay, H. J., Sim, H. K. (2004). Effects of cationic polymer on start-up and granulation in upflow anaerobic sludge blanket reactors. *Journal of Chemical Technology and Biotechnology*, **79**, 219–228.

Yu, Q.H., Tay, H.J., Fang, P.H.H. (2001). The roles of calcium in sludge granulation during UASB reactor start-up. *Water Research*, **35** (4), 1052–1060.

Effects of organic loads on organic matter removal in a moving bed biofilm reactor

Ahmet Aygun, Bilgehan Nas, Tuba Ertugrul and Ali Berktay

Department of Environmental Engineering, Selcuk University, 42075, Konya, Turkey
(E-mail: *bnas@selcuk.edu.tr*)

Abstract A moving bed biofilm reactor (MBBR), where biomass is attached to small carriers elements that move freely along with the water in the reactor, has been tested for organic matter removal at five various organic loads. A lab-scale reactor with volume of 2L was constructed and fed continuously with synthetic wastewater. Oxidation reactor was filled with the kaldnes biomedia K1 to 50% of the volume of empty reactor. The HRT in the oxidation tank and in the settler was adjusted 8 and 4 hours, respectively. A start-up period of about 4 weeks for biofilm growth on the carrier was followed by 10 weeks testing period. By changing the wastewater flow, the operation of the system was adjusted, one after the other, to five various organic loading rates: 6,12,24,48 and 96 g COD/m^2.d. By increasing organic loads, organic removal efficiencies were decreased and totCOD removal efficiencies were calculated as 95%, 94%, 91%, 69% and 45%, respectively. In the MBRR reactor biofilm reached an average concentration of 3.28 kg TSS/m^3 in the highest organic load. The ratio between the TSS production and the totCOD removal resulted equal to 0.38 kg TSS/kg totCOD while influent totCOD 1000 mg/l.

Keywords Biofilm; moving bed biofilm reactor; organic carbon removal; organic load; wastewater

Introduction

The patented moving bed biofilm reactor (MBBR) has been developed in Scandinavia in the late 1980s by the Norwegian company Kaldnes MiljØteknologi (KMT), in cooperation with the SINTEF research organization. Currently more than 300 plants, using the MBBR process for treatment of industrial or municipal wastewater, are in operation worldwide (Wessman *et al.*, 2004). It has been a huge success worldwide for treatment of wastewaters (Rusten *et al.*, 2005).

The Kaldnes MBBR is a completely mixed continuously operated biofilm reactor where the biomass is grown on small carrier elements that move along with the water in the reactor. The interest for MBBR is justified by the following advantages: high sludge age, no sludge recycling, less bulking troubles. In the case of upgrade of an existing plant, the MBBR solution does not increase head loss significantly, allows a continuous feed, does not present clogging problems and does not require backwashing. MBBRs allow applying higher loads compared to those applied to the conventional activated sludge, in case the specific surface is sufficiently high (Rusten *et al.*, 2005; Andreottola *et al.*, 2002, 2003).

The biofilm carrier elements are made of polyethylene (density 0.92–0.96 g/cm^3) and shaped like small cylinders with a cross inside the cylinder and longitudinal fins on the outside. Two basic sizes – K1 and K2 – are available. To keep the biofilm carriers in the reactor, a sieve is placed at the outlet of the reactor. The filling ratio and the specific area of the biofilm carriers are the two main design parameters. The filling ratio ranges from 30–70% of the total reactor volume. It is recommended that the filling fractions should be below 70% so as to be able to move the carriers freely (Ødegaard, 2000).

MBBRs have been already used in the treatment of dairy wastewater (Rusten *et al.*, 1992; Andreottola *et al.* 2002), forest industry wastewater (Dalentoft and Thulin, 1997), cheese factory wastewater (Rusten *et al.*, 1996), potato chips wastewater (Ødegaard *et al.* 1994), newsprint mill wastewater (Broch-Due *et al.*, 1997), thermo-mechanical pulping whitewater (Jahren *et al.*, 2002), municipal wastewater (Ødegaard *et al.* 1993; Orantes and González-Martinez, 2002), and for nitrification (Rusten *et al.*, 1995; Welander *et al.*, 1997) and denitrification (Rusten *et al.*, 1995; Aspegren *et al.*, 1998; Maurer *et al.*, 2001; Welander and Mattiasson, 2003)

MBBR is a good process for upgrading current wastewater treatment systems. Many authors mentioned examples of new plant construction or of existing plants upgrading with satisfactory results in relation to the removal both organic matter and nutrients (Ødegaard *et al.* 1993; Hem *et al.* 1994; Orantes *et al.* 2002; Daude and Stephenson 2003; Andreottola *et al.* 2003). Andreottola *et al.* (2000), found that hydraulic retention time (HRT) affected the COD removal in MBBR processes and suggested that the HRT should be higher than 5 h. Another study performed by Andreottola *et al.* (2003), they evaluated the application of MBRR system for the upgrading of an overloaded MWWTP. A preliminary experimentation at pilot-scale was carried out in order to evaluate the efficiency of the system in the removal of organic matter using a high loaded MBBR. KMT plastic media was applied with a filling ratio of 50%. The applied surface loads ranged between 5.4 and 32 gtotCOD/ m^2.d

The aim of this research is to investigate the removal of organic carbon in synthetic wastewater by using the Kaldnes biomedia K1, which is used in the patented Kaldnes Moving Bed™ biofilm process, in various high organic loads (6 to 96 g totCOD/ m^2.d).

Methods

Carriers

The Kaldnes K1 biofilm carrier elements are made of polyethylene and shape like small cylinders (a nominal diameter of 9.1 mm and a nominal length of 7,2 mm) with a cross inside the cylinder and longitudinal fins on the outside. The Kaldnes carriers had a specific gravity of 0.96 and a specific biofilm surface area of 500 m^2 per m^3 bulk volume of carriers. With a filling ratio of 50%, the available surface area (referred to the reactor volume) was 250 m^2/m^3 (considering only the internal surface of cylinders, as biofilm does not grow on the external surface because of collisions among the elements).

Lab-scale reactor and wastewater

In this study, the system consisted of an oxidation reactor (volume equal to 2 l) followed by a final settler (volume equal to 1.2 l) according to the scheme in Figure 1. Oxidation reactor was filled with the Kaldnes biomedia K1 to 50% of the volume of empty reactor. No sludge recycle was implemented. Membrane diffusers were adopted for oxygen supply and mixing.

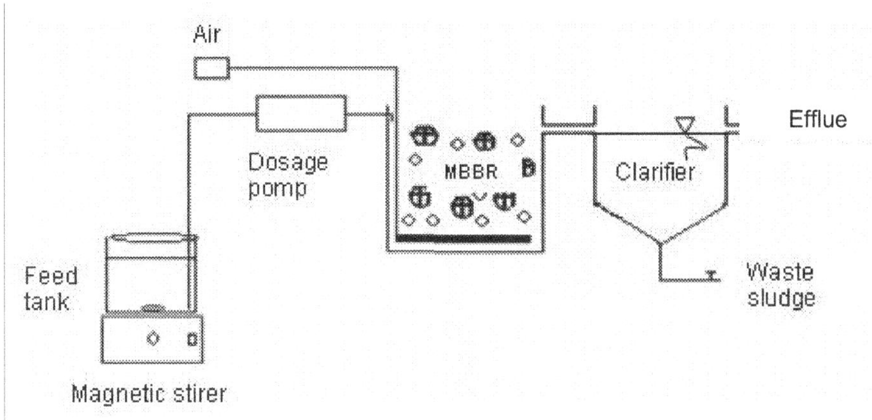

Figure 1 Simplified flow-sheet of the lab-scale MBBR.

Activated sludge were obtained from a local municipal WWTP. The MBBR was fed with a dosage pump. The theoretical hydraulic retention time (HRT) in the oxidation tank and in the settler was adjusted 8 hours and 4 hours, respectively. The settled sludge was removed from bottom of settler.

Synthetic wastewater was used to provide a source of carbon, nitrogen, phosphorus and trace metals required for biomass growth. It was prepared daily with deionized water to reach a total chemical oxygen demand (totCOD) of 500, 1000, 2000, 4000 and 8000 mg/l. Synthetic wastewater composed of glucose (258 mg/l), sodium acetate (471 mg/l), NH_4Cl (95.5 mg/l), KH_2PO_4 (22 mg/l), $NaHCO_3$ (295 mg/l), $MgSO_4.7\ H_2O$ (100 mg/l) and as trace element $CaCl_2.2\ H_2O$ (50 mg/l), $FeCl_3.\ 6\ H_2O$ (50 mg/l) were used for 500 mg/l COD. Other concentrations were prepared according to this composition. The calculated COD/N/P ratio of the synthetic wastewater is 100/5/1.

A start-up period of about 4 weeks for biofilm growth on the carrier was followed by 10 weeks testing period. Steady-state conditions were resumed for a minimum of one week before the next trial commenced. The dissolved oxygen (DO) concentrations in the MBRR ranged from 0.3 to 3.1 mgO_2/l depending on the influent organic load. The temperature and pH in the reactor varied in the range of 18.4 to 23.6 °C, 6.72 to 7.88, respectively. Table 1 shows operational conditions for the reactor during all the experimental periods.

Table 1 Operational conditions of MBBR reactor

Influent totCOD (mg/l)	500	1000	2000	4000	8000
Volumetric load (kg totCOD/ m^3.d)	1.5	3	6	12	24
Surface load (g totCOD/ m^2.d)	6	12	24	48	96
TSS (kg/m^3)	2.34	2.5	2.76	3.23	3.28
VSS (kg/m^3)	1.73	1.89	2.10	2.46	2.42
Air (l/min)	0.5	0.5	1.8	1.8	2.3
Dissolved oxygen (mg/l)	3.1	2.8	2.5	0.8	0.3
Operation (days)	15	10	14	13	10

Analyses

To measure the performance of the lab-scale MBBR, samples were taken from the biofilm reactor and the final effluent. Total COD, filtered COD (on membrane of 0.45 μm porosity), TSS, VSS, N-NO_3, N-NH_4, DO and pH were measured on samples every day.

Closed reflux colorimetric method (Method 5220 D) was used for COD analysis as specified in the Standard Methods (APHA, AWWA, WEF, 1998). TSS and VSS concentrations referred to the fixed biomass in the MBBR were also measured. TSS analysis of fixed biomass was obtained detaching the biomass from 10 KMT elements, diluting the biomass in a known amount of demineralized water, referring the values to the surface of an element and taking into account the number of elements per litre (Andreottola *et al.* 2000; Andreottola *et al.* 2003). N-NO_3 and N-NH_4 were analyzed with Orion 710A advanced ion selective meter with Method 4500-NO_3-D nitrate and 4500-NH_4-D ammonia-selective electrode method (APHA, AWWA, WEF, 1998). DO and pH measurements were done by using the WTW Multiparameter 340i.

Results and discussion

Experimental work in the laboratory was carried out in order to evaluate the efficiency of the system in the removal of organic matter using a high loaded MBBR. During the experimental of five different organic loads were applied to the reactor. Surface organic load was between 6 and 96 g totCOD/m^2.d (corresponding to the range 1.5–24 kg totCOD/m^3.d).

Effluent totCOD and filtered COD concentrations and COD removal rates versus time are shown in Figure 2. At the lowest organic load (Figure 2a.), average effluent totCOD concentration was 25 mg/l and average totCOD removal efficiency was 95%. At 12, 24, 48 and 96 gCOD/m^2.d organic loads average effluent totCOD concentrations were 51, 172, 1253, 4381 mg/l respectively, while the average totCOD removal efficiencies were 94%, 91%, 69%, and 45%.

Average effluent filtered COD concentrations were 16, 38, 135, 1070, and 3786 mg/l respectively, while the average filtered COD removal efficiencies were 97%, 96%, 93%, 73%, and 53%. Figure 3 shows effluent totCOD of MBBR at the five different organic loads. As shown in the Figure 3, in general with increasing organic loads, COD removal efficiencies were decreased.

The DO was measured on the reactor every day with 2 hours interval. The average DO levels in the MBBR ranged from 0.3 to 3.1 mgO_2/l depending on the influent organic load. Although the amount of air increased to 1.8 and 2.3 l/min, DO levels decreased to average 0.8 mg/l and 0.3 mg/l when 48 and 96 g totCOD/ m^2.d surface organic loads applied, respectively. Low DO concentrations could be affected COD removal efficiency at 48 and 96 g totCOD/ m^2.d surface organic load.

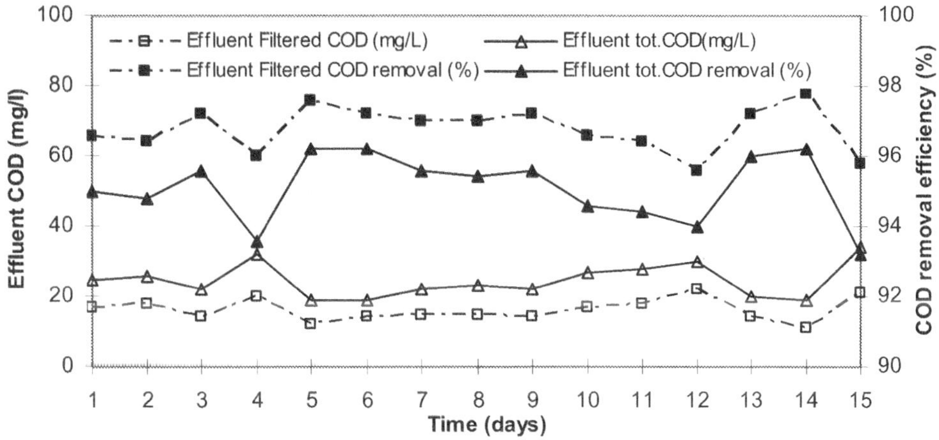
Effluent Filtered COD (mg/L)
Effluent tot.COD(mg/L)
Effluent Filtered COD removal (%)
Effluent tot.COD removal (%)
Effluent COD (mg/l)
COD removal efficiency (%)
Time (days)

(a)

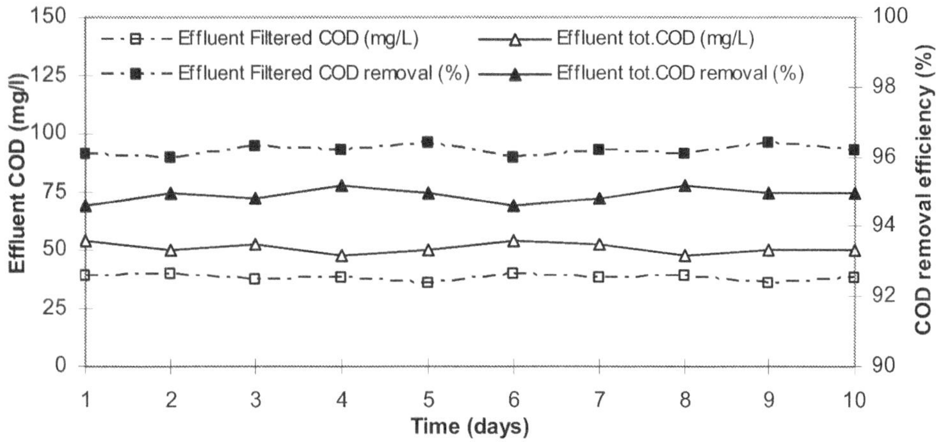
Effluent Filtered COD (mg/L)
Effluent tot.COD (mg/L)
Effluent Filtered COD removal (%)
Effluent tot.COD removal (%)
Effluent COD (mg/l)
COD removal efficiency (%)
Time (days)

(b)

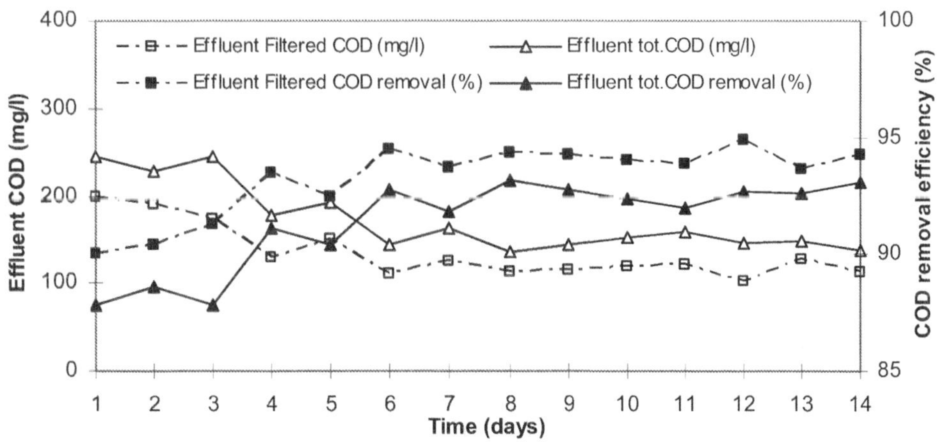
Effluent Filtered COD (mg/l)
Effluent tot.COD (mg/l)
Effluent Filtered COD removal (%)
Effluent tot.COD removal (%)
Effluent COD (mg/l)
COD removal efficiency (%)
Time (days)

(c)

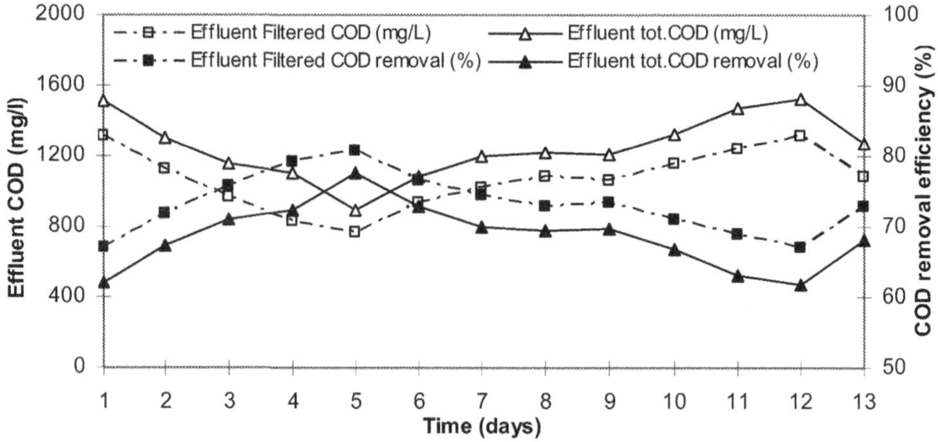

(d)

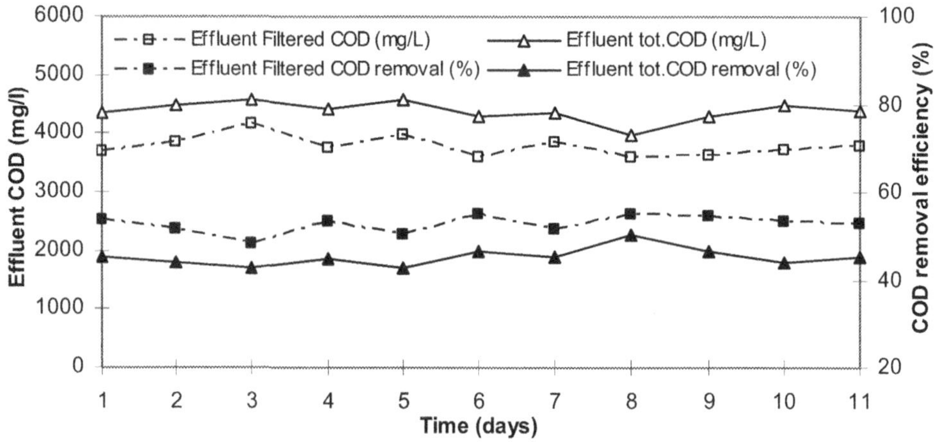

(e)

Figure 2 Effluent totCOD, filtered COD and COD removal efficiencies a) 6 gCOD /m^2.d organic load b) 12 gCOD /m^2.d organic load c) 24 gCOD /m^2.d organic load d) 48 gCOD /m^2.d organic load e) 96 gCOD /m^2.d organic load.

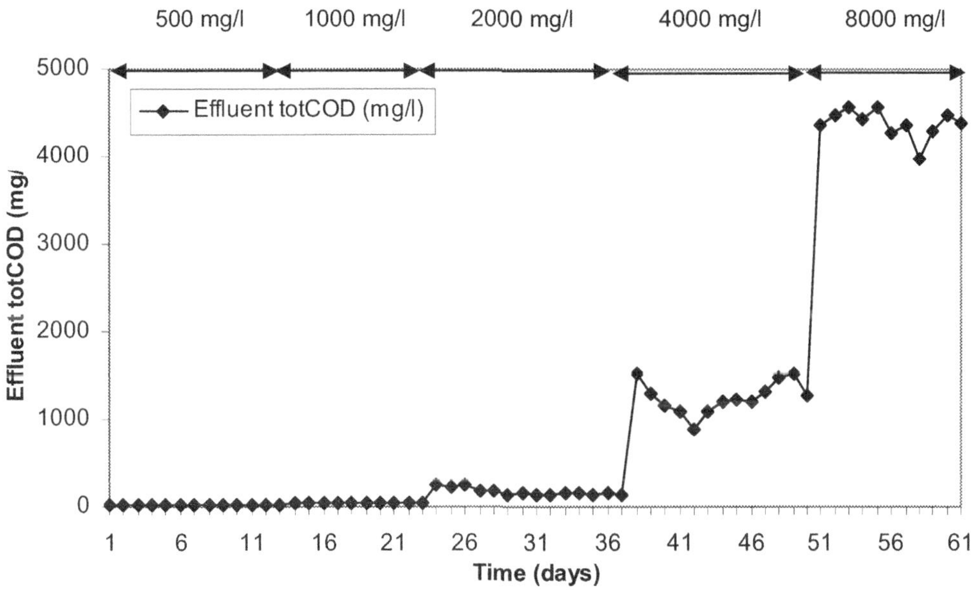

Figure 3 Effluent totCOD of MBBR at the various organic loads.

Figure 4 shows the removal efficiency versus volumetric and surface applied loads. By applying a surface organic load between 6 and 24 g totCOD/m^2.d (corresponding to the range 1.5–6 kg totCOD/m^3.d) it was possible to obtain the highest removal efficiency of the totCOD in the MBRR. However, removal efficiency of the totCOD was significantly decreased to 45% when 96 g totCOD/ m^2.d surface organic load applied.

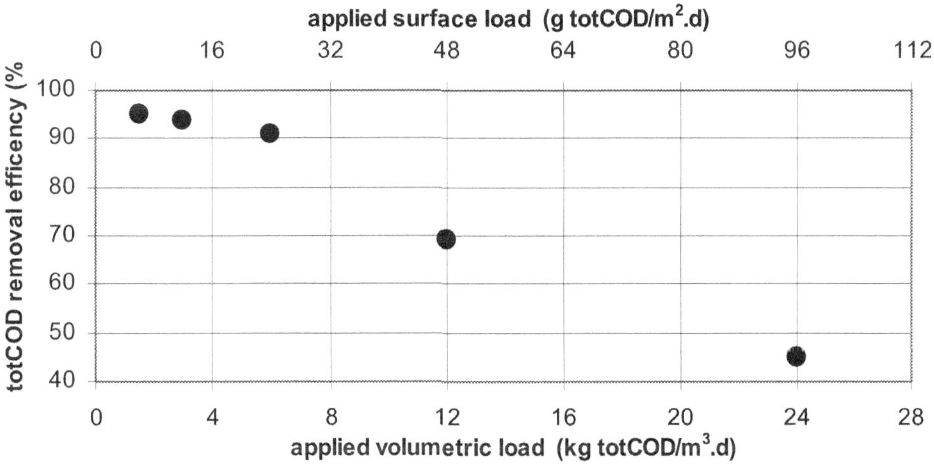

Figure 4 Total COD removal efficiency versus applied organic load.

The amount of biomass attached to the carrier material increased as the organic load increased (Figure 5). In the MBBR reactor biofilm reached an average concentration between 2.34 and 3.28 kgTSS/m^3 at various organic loads. The ratio of VSS/TSS came to between 69% and 80% in the carriers.

The ratio of the amount of biomass produced to the amount of substrate consumed (g biomass/g substrate) is defined as the biomass yield. The observed solids yield accounts for the actual sludge production that would be measured for the system. The sludge production (or solid yield) in the MBBR was calculated by measuring the sludge volume extracted from the settler and the TSS concentration in the sludge, in a period with an average totCOD concentration in the influent and in the effluent of 1000 and 172 mg/l respectively and a flowrate of 6 l/d. Sludge production resulted equal to 1.9 g TSS/d in correspondence of totCOD removal in the plant of about 4.97 g totCOD/d. Therefore, the ratio between the TSS production and the totCOD removal resulted equal to 0.38 kg TSS/kg totCOD. Observed yields, calculated as average for the each test period and resulted equal to 0.12, 0.38, 0.37, 0.42, and 0.56 kg TSS/kg totCOD, for influent totCOD 500, 1000, 2000, 4000, and 8000 mg/l, respectively. Therefore, solid yields increased with the organic loads increased.

Rusten *et al.* (1994) and Andreottola *et al.* (2003) found the same value of 0.36 kg TSS/kg totCOD for a full-scale plant designed for nitrification and postdenitrification and upgraded a small activated sludge plant, respectively.

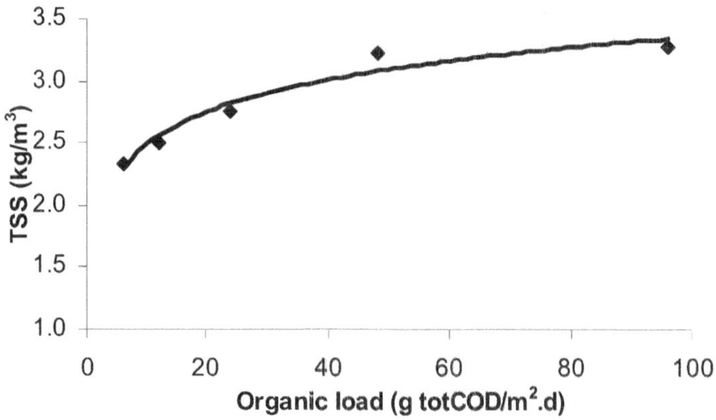

Figure 5 Solid attached to the biofilm carrier.

Conclusion

There are several advantages of the Kaldnes MBBR process compared with the conventional activated sludge process. It could be possible to operate the MBBR at very high organic loads with obtaining high COD removal, for instance at organic load up to 24 kg totCOD/m^3.d applied in this study.

Various organic loads applied to MBBR process studies showed that MBBR process can be operated organic loads up to 6 kg totCOD/ m^3.d (corresponding to 2000 mgCOD/l) providing a removal efficiency of 91% with HRT of 8 h. In concentrations higher than 2000 mgCOD/l, COD removal efficiencies significantly decreased down to 45% at 8000 mgCOD/l. Observed yields, calculated as average for the each test period and resulted equal to 0.12, 0.38, 0.37, 0.42, and 0.56 kg TSS/kg totCOD, for influent totCOD 500, 1000, 2000, 4000, and 8000 mg/l, respectively.

MBBR process could be an efficient alternative process to the conventional biological treatment processes for organic matter removal at high organic loads or of existing plants upgrading with satisfactory results

Acknowledgments

This study is supported by Selcuk University Scientific Research Fund with project No: 2004/085. The authors wish to thank Kaldnes MiljØteknologi AS for supplying the KMT-biofilm carriers.

References

Andreottola, G., Foladori, P., Ragazzi M. (2000). Upgrading of a small wastewater treatment plant in a cold climate region using a moving bed biofilm reactor (MBBR) system. *Water Science and Technology*, **41**(1), 177–185.

Andreottola, G., Foladori, Ragazzi M., Villa, R. (2002). Dairy wastewater treatment in a moving bed biofilm reactor. *Water Science and Technology*, **45**(12), 321–328.

Andreottola, G., Foladori, P. Gatti, G., Nardelli P., Pettena M., Ragazzi M. (2003). Upgrading of a small overloaded activated sludge plant using a MBBR system. *Journal of Environmental Science and Health Part A – Toxic/Hazardous Substances & Environmental Engineering*, **A38**(10), 2317–2328.

Standard Methods for the Examination of Water and Wastewater (1998). 20th edn, American Public Health Association/American Water Works Association/Water Environment Federation, Washington DC, USA.

Aspegren, H., Nyberg, U., Andersson, B., Gotthardsson, S., Jansen, J. (1998). Post denitrification in a moving bed biofilm reactor process. *Water Science and Technology*, **38**(1), 31–38.

Broch-Due, A., Andersen, R., Opheim, B. (1997). Treatment of integrated newsprint mill wastewater in moving bed biofilm reactors. *Water Science and Technology*, **35**(2–3), 173–180.

Daude, D., Stephenson, T. (2003). Moving bed biofilm reactors: a small-scale treatment solution. *Water Science and Techonology*, **48**(11–12), 251–257.

Jahren, S.J., Rintala, J.A., Ødegaard, H. (2002). Aerobic moving bed biofilm reactor treating thermomechanical pulping whitewater under thermophilic conditions. *Water Research,* **36**, 1067–1075.

Maurer, M., Fux, C., Graff, M., Siegrist, H. (2001). Moving-bed biological treatment (MBBT) of municipal wastewater: denitrification. *Water Science and Technology*, **43**(11), 337–344.

Orantes, J.C., González-Martinez, S. (2002). *A new low-cost biofilm carrier for the treatment of municipal wastewater in a moving bed reactor*. 5th specialized conference on small water and wastewater treatment systems, Istanbul, Turkey.

Ødegaard, H., Rusten, B., Badin, H. (1993). Small wastewater treatment plants based on moving bed biofilm reactors. *Water Science and Technology*, **28**(10), 351–359.

Ødegaard, H., Rusten, B., Westrum, T. (1994). A new moving bed biofilm reactor – applications and results. *Water Science and Technology*, **29**(10–11), 157–165.

Rusten, B., Ødegaard, H., Lundar, A. (1992). Treatment of dairy wastewater in a novel moving bed biofilm reactor. *Water Science and Technology,* **26**(3–4), 703–711.

Rusten, B., Siljudalen, J.G., Nordeidet, B. (1994). Upgrading to nitrogen removal with KMT moving-bed biofilm process. *Water Science and Technology*, **29**(12), 185–195.

Rusten, B., Hem, L.J., Ødegaard, H. (1995). Nitrification of municipal wastewater in moving-bed biofilm reactors. *Wat. Env. Res.*, **67**(1), 75–86.

Rusten, B., Siljudalen, J.G., Strand, H. (1996). Upgrading of a biological–chemical treatment plant for cheese factory wastewater. *Water Science and Technology*, **34**(11), 41–49.

Rusten, B., Eikebrokk, B., Ulgenes,Y., Lygren, E. (2005). Design and operations of the Kaldnes moving bed biofilm reactors. *Aquacultural Engineering,* In Press.

Welander, U., Henrysson, T., Welander, T. (1997). Nitrification of landfill leachate using suspended-carrier biofilm technology. *Water Science and Technology*, **31**(9), 2351–2355.

Welander, U., Mattiasson, B. (2003). Denitrification at low temperatures using a suspended carrier biofilm process. *Water Research*, **37**, 2394–2398.

Wessman, F.G., Yuegen, E.Y., Zheng, Q., Welander, T., Rusten, B. (2004). Increasing the capacity for treatment of chemical plant wastewater by replacing existing suspanded carrier madia with Kaldnes moving bed media at a plant in Singapore. *Water Science and Technology*, **49**(11–12), 199–205.

Microbial biodegradation of dimethyl phthalate by phthalic acid aerobic granules in sequencing batch reactors

Ping Zeng*, Stephen Tiong-Lee Tay*and Joo-Hwa Tay**

* Environmental Engineering Research Centre, School of Civil and Environmental Engineering, Nanyang Technological University, 50 Nanyang Avenue, Singapore 639798 (E-mail: *zengping@ntu.edu.sg*, *cltay@ntu.edu.sg*)
** Innovation Centre, Block2, Unit 237, Nanyang Technological University, 18 Nanyang Drive, Singapore 637723 (E-mail: *jhtay@ntu.edu.sg*)

Abstract Biological treatment of a dimethyl phthalate (DMP) wastewater was evaluated in sequencing batch reactor (SBR) using phthalic acid (PA) granules as bioseeds. This strategy achieved success with 100% DMP removal efficiency after one day's operation and the reactor reached stabilization in short period. The kinetic study of DMP degradation by phthalic acid (PA) aerobic granules showed that no intermediates accumulation in the degradation process. The structure of aerobic granules was not affected by the shift of substrate from PA to DMP. The study widens our current knowledge on aerobic granules' application.

Keywords aerobic granules; phthalic acid; DMP; biodegradation; kinetic

Introduction

Phthalate esters (PAEs) are anthropogenic compounds widely used as plasticizers in polyvinyl resin, cellulosic and polyurethane polymers for the manufacturing of building materials, home furnishings, transportation, clothing, and packaging of food and medical products (Seymour and Carraher, 1993). Dimethyl phthalate (DMP) is typically used in cellulose-ester based plastics, such as cellulose acetate and butyrate. To maintain the flexibility of plastics, plasticizer is not bound covalently to the plastic resin. Migration of phthalate esters into the environment during manufacturing, use and disposal has caused serious concerns (Bauer and Herrman, 1997). Some of PAEs and their degradation intermediates are suspected to be mutagens (Kozumbo *et al.*, 1982) and carcinogens (Huff and Kluwe, 1984), even at very low concentrations they might interfer with reproductive systems and behavior in humans and wildlife, through disturbance of the endocrine system (Colon *et al.*, 2000; Petrovilc *et al.*, 2001). As a result, US Environmental Protection Agency (US EPA, 1992 and update) classified phthalate esters as a top priority pollutant for risk assessment, mandating the reduction and control of phthalate pollution.

Since the rates of abiotic degradation are very slow, microbial degradation is considered one of the major routes for the environmental degradation of PAEs (Staples *et al.*, 1997, Scholz *et al.*, 1997). However, some of phthalates are refractory, especially those with high molecular weight and complex structure. Conventional activated sludge process cannot effectively treat recalcitrant phthalate wastewater (Huang *et al.*, 1994).

As one of strategies for recalcitrant compound degradation, bioaugmentation offers the benefits in high populations of degrading microorganisms and highly active species to maintain the high efficiency of the system. However, there are reports of bioaugmentation failure and the real efficiency of the procedure remains controversial (Boon *et al.*, 2000; Stephenson and Stephenson, 1992; Tchelet *et al.*, 1999). Many reasons could lead to the failure of bioaugmentation are numerous and complex. Undoubtedly the initial selection of bioseeds is one of the most important factors (Van Der Gast *et al.*, 2002). Besides high removal ability of the recalcitrant compound, good survive and high retention of bioseeds in ecosystem is also the desired properties of bioseeds.

Aerobic granulation, a new form of cell immobilization in biological wastewater treatment, has shown high efficiency in various wastewater treatments (Tay *et al.*, 2001, Arrojo *et al.*, 2004, Dulekgurgen *et al.*, 2003, McSwain *et al.*, 2005). Aerobic granulation can provide large quantity of biomass for the refractory compounds. In addition, aggregation of activated sludge cells into aerobic granules helps establish syntrophic relationships and create a more favourable microenvironment for microorganism under unfavourable conditions. Because of these characteristics, aerobic granules can be selected as bioseeds to treat refractory industrial wastewaters.

It is generally accepted that PA is one of the degradation intermediates of PAEs. The accumulation of PA might hinder the complete mineralization of PAEs. However, addition of PA degrading microorganisms could enhance the degradation of PAEs (Fan *et al.* 2001). The PA-degrading microorganisms existed in phthalic acid aerobic granules (hereafter phthalic acid aerobic granules was defined as PA granules) could enhance the degradation of PAEs by degrading the intermediates. At the same time, the granular sludge could ensure the high retention of PA-degrading bacteria in the reactor. Therefore PA granules were selected as bioseeds for the degradation of PAEs.

The DMP, which is one of PAEs, was selected as the model of PAEs. PA aerobic granules were used as bioseeds for DMP degradation. The main objective of this study was to evaluate the strategy of using aerobic granules as bioseeds for bioaugmentation.

Materials and methods

Inoculum

The matured PA granules which were obtained from a sequencing batch reactor (SBR) were used as seed sludge. They were cultivated from acclimated activated sludge. PA was fed as sole carbon and energy source with a synthetic wastewater with the following composition (l^{-1}): 0.8 g NH_4Cl, 0.045 g K_2HPO_4, 0.03 g $CaCl_2{\cdot}H_2O$, 0.025 g $MgSO_4{\cdot}7H_2O$, 0.02 g $FeSO_4{\cdot}7H_2O$ and micronutrients, which had been described previously (Moy *et al.*, 2002). Organic loading rate was 3 kg COD m^{-3} d^{-1}.

Experiment design and set-up

Experiments were performed in a column-type SBR with a working volume of 2.4 l and a diameter of 5 cm. Reactors were operated sequentially in 4-h cycles (4 min of influent filling, 228 min of aeration, 5 min of settling and 4 min of effluent withdrawal) with a hydraulic residence time (HRT) of 8 h. The reactors were housed in a temperature-controlled room at 25°C. Effluent was discharged at a volumetric exchange ratio of 50%. Fine air bubbles for aeration were supplied through a dispenser at the reactor bottom at an airflow rate of 3.5 l min^{-1} and air flow was controlled by a gas flow meter. DMP was fed as sole carbon source with influent DMP concentration of 577 mg l^{-1}. The other components were same as that feeding to PA granules.

Analytical methods

Suspended solids (SS), volatile suspended solids (VSS), sludge volume index (SVI) and dissolved oxygen (DO) was measured periodically according to the standard methods (APHA, 1995). To measure soluble total organic carbon (TOC), samples were filtered using 0.45 μm filter paper and analyzed with a TOC analyzer (TOC-Vcsh, Shimadzu). The concentrations of DMP and PA are determined by high-pressure liquid chromatography (HPLC). Liquid samples were filtered through a 0.22 μm filter. Separation of the DMP was obtained by using an Inertsil C8–3 (5 μm, 150 × 4.6mmI.D.). A mixture of acetonitrile and water in a 70/30 ratio was used as a carrier. The applied flow rate was 0.3 ml min^{-1}. The separated components were detected by a UV detector (Spectroflow 773) at a wavelength of 275 nm. Typical retention time for DMP and PA was 5.3 min and 4.7 min respectively (Fan *et al.*, 2001; Kleerebezem *et al.*, 1999). Each determination was performed in triplicate.

Granule morphology and size were determined with an image analysis system (Quantimet 500 Image Analyser, Leica Cambridge Instruments). Microscopic study of aerobic granular samples was conducted with with a scanning electron microscope (Stereoscan, 420, Leica Cambridge Instruments), as described previously (Moy *et al.*, 2002).

Results and discussion

Performance of phthalic acid granules feeding with DMP

Aerobic granules were first cultivated in sequencing batch reactors with phthalic acid as sole carbon source at a loading rate of 3.0 kg COD m^{-3} d^{-1}. DMP was then added to the reactor at the same loading rate when phthalic acid aerobic granules matured. The reactor operation was not affected by the feeding of new substrate. The biomass concentration increased when the feeding substrate changed from PA to DMP (day 0). 15 days later, the reactor reached steady state with a steady biomass concentration of 6.0 g l^{-1} (Figure 1). The aerobic granules maintained their good settling ability when the carbon source changed into DMP. The initial SVI value of PA granules was 41 ml g^{-1}. It decreased slightly in the first 10 days operation. Then it levelled off at approximately 30 ml g^{-1} SS when reactor reached steady state from day 15 on. The mean biomass size increased from 0.37 (day 0) to 1.00 mm (day 4) and levelled off at approximately 1.10 mm (Figure 2).

The PA granules successfully removed 98.9% DMP fed into the reactor on day 0. The DMP removal efficiency was 100% in the whole process from day 0 on (Figure 1). The TOC removal efficiency was 45% on day 0. And it increased rapidly to 90% on day 4. Then it levelled off at 98% at steady state (29 days onwards) (Figure 1).

The PA granules acclimated quickly to the DMP, reached steady state only 2 week after DMP was introduced. The granules exhibited good settling ability with good biomass retention and good metabolic activity, as evidenced by the low SVI values, stable biomass concentrations and high removal of DMP (Figure 1). No lag phase was observed during initial stage.

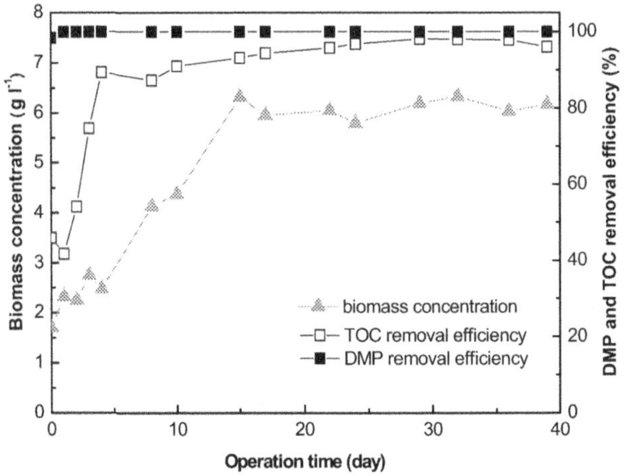

Figure 1 Profiles of biomass concentration and DMP, TOC removal efficiency.

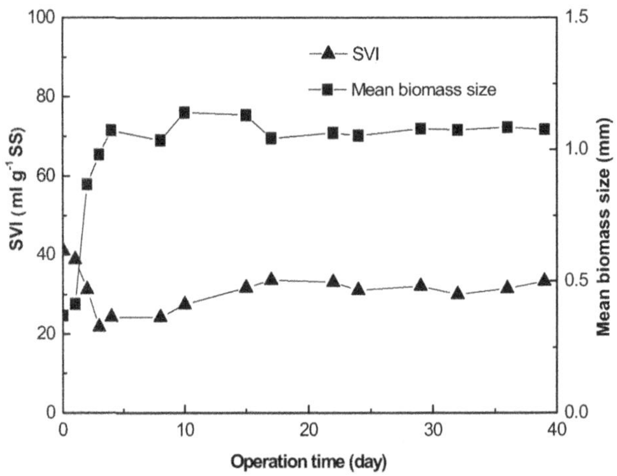

Figure 2 Profiles of sludge volume index (SVI) and mean biomass size of DMP granules.

Kinetics of DMP degradation by PA granules

Concentrations of DMP and PA were monitored during a single cycle time in the reactor. The monitoring was conducted at both the beginning of the operation (day 0) and at the end of experiment (day 39). At beginning of operation, it took 180 minutes for completely removal of DMP. As the intermediate of DMP degradation, PA appeared at 5 minutes with concentration of 23 mg l^{-1}, and quickly exhausted in 5 minutes. At 15 minutes, the PA accumulated to 8 mg l^{-1} and decreased to 1.6 mg l^{-1} at 20 minutes. At 40 minutes, it accumulated to 9 mg l^{-1} and was completely degraded at 60 minutes (Figure 3A). At the end of operation, 40 minutes was required to completely removal of DMP. PA appeared at 5 minutes with concentration of 15 mg l^{-1}. It was degraded in 5 minutes and no accumulation happened again (Figure 3B).

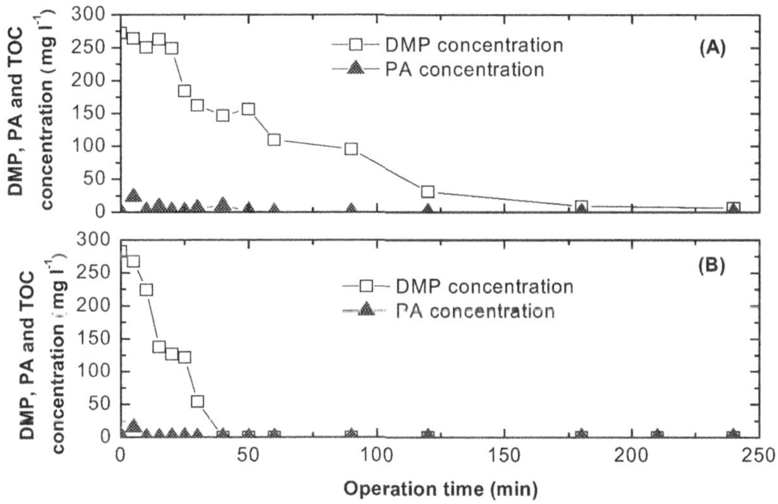

Figure 3 DMP and PA Concentration variation in one cycle time (A) at beginning of operation (day 0) (B) at the end of operation (day 39).

As the volumetric exchange ratio for the reactor is 50% and effluent DMP concentration was insignificant compared to influent DMP concentration. It merits mentioning the DMP concentrations in the reactor at time zero were approximately half of the influent DMP concentrations. The results indicate that the DMP degradation ability improved greatly. There is no PA accumulation during a cycle time both at beginning and at the end of operation.

It is well known that the aerobic degradation of DMP involves several steps of transformation before the substrate becomes fully mineralized (Staples *et al.*, 1997). It starts with the hydrolysis of DMP by esterase to monomethyl phthalate (MMP) and phthalic acid followed by a complete mineralization of phthalic acid to CO_2 and water. If the intermediates can not be degraded completely, they would hider the overall DMP mineralization. At the same time, the accumulation of intermediates, especially PA, would decrease pH of the medium in reactor, which would in turn influence the activity of DMP-degrading enzymes (Fan *et al.*, 2001). In the PA granules, the PA-degrading microorganisms could contribute to the degradation of PA and a favourable environment was maintained at a suitable level for the microorganisms and the degrading enzymes. Thus the bioaugmentation of PA granules would enhance the overall DMP degradation.

The structure of aerobic granules

The aerobic granules kept compact structure after the feeding substrate changed from PA to DMP. SEM images revealed that both the PA granules and the DMP aerobic granules cultivated from PA granules (hereafter DMP aerobic granules were defined as DMP granules) had cauliflower out appearance (Figure 4a,c). Bacterial cells were tightly clustered together and embedded within extracellular polymeric matrix. These results indicate that both granules had compact structures no matter what carbon source was used for cultivation.

In PA granules, non-filamentous bacteria such as rods and cocci dominated the microbial community residing in the granules (Figure 4b). In DMP granules, filamentous bacteria appeared in the microbial community besides rods and cocci (Figure 4d). Energy-rich substrates are known to support the proliferation of filamentous bacteria in activated sludge, anaerobic and denitrifying granular sludge (Chudoba *et al.*, 1985; Thaveesri *et al.*, 1995; Van Der Hoek, 1988). These results agree with our findings.

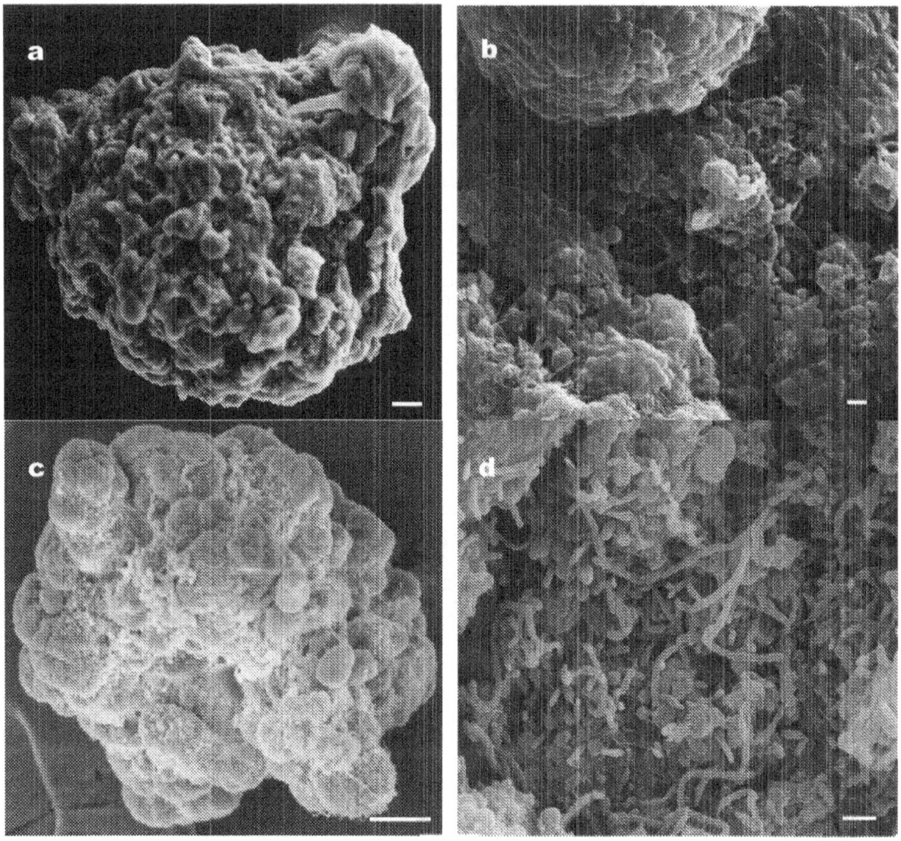

Figure 4 Scanning electron micrographs of aerobic granules, (a) Phthalic acid aerobic granules (bar: 20 μm); (b) Surface of phthalic acid aerobic granules (bar: 1 μm); (c) DMP aerobic granules (bar: 30 μm) (d): Surface of DMP aerobic granules (bar: 3 μm).

Conclusions

PA granules used as the bioseeds for DMP degradation could offer the benefits on both fast start-up and enhancement degradation of DMP. The compact structure of aerobic granules did not affected by the changing of carbon source. The PA-degrading microorganisms in PA granules play a key role in overall DMP mineralization and thus contribute to the steady operation of the reactor. The study widens our current knowledge on aerobic granules' application.

References

APHA (1995). *Standard methods for the examination of water and wastewater*, 19th edn. Washington, DC, USA: American Public Health Association.

Arrojo, B., Mosquera-Corral, A., Garrido, J. M., & Mendez, R. (2004). Aerobic granulation with industrial wastewater in sequencing batch reactors. *Wat. Res.*, **38**(14–15), 3389–3399.

Bauer, M. J., & Herrmann, R. (1997). Estimation of the environmental contamination by phthalic acid esters leaching from household wastes. *Sci. Total Environ.*, **208**(1–2), 49–57.

Boon, N., Goris, J., De Vos, P., Verstraete, W., & Top, E. M. (2000). Bioaugmentation of activated sludge by an indigenous 3-chloroaniline-degrading *Comamonas testosteroni* strain, I2gfp. *Appl. Environ. Microbiol.*, **66**(7), 2906–2913.

Chudoba, J., Cesh, J. S., Frakac, J., & Grau, P. (1985). Control of activated sludge filamentous bulking VI Formulation of basic principles. *Wat. Res.*, 19, 1389–1406.

Colon, I., Caro, D., Bourdony, C.J., & Rosario, O. (2000). Identification of phthalate esters in the serum of young Puerto Rican girls with premature breast development. *Environmental Health Perspectives*, **108**, 895–900.

Dulekgurgen, E., Ovez, S., Artan, N., & Orhon, D. (2003). Enhanced biological phosphate removal by granular sludge in a sequenicng reactor. *Biotechnol. Lett.*, **25**, 687–693.

Fan, Y. Z., Cheng S. P., & Gu J. D. (2001). Degradation of phthalic acid and dimethyl phthalate ester by enrichment cultures of aerobic bacteria. Paper presented at the Conference proceedings of IWA Asia environmental technology 2001, Singapore.

Huang, X., Zhang, X. J., & Fu, Q. (1994). Assessment for biodegradability of organics in municipal wastewater and controlling measure for its refractory organics. *Journal of Environmental Science* (in Chinese), **15**, 5–19.

Huff, J. E., & Kluwe, W. M. (1984). Phthalate esters carcinogenicity in F344/N rates and B6C3 F mice. *Progress in Clinical Biology Research*, **141**, 137–154.

Kleerebezem, R., Pol, L. W. H., & Lettinga, G. (1999). Anaerobic degradation of phthalate isomers by methanogenic consortia. *Appl. Environ. Microbiol.*, **65**(3), 1152–1160.

Kozumbo, W.J; Kroll, R., & Robin, R.J. (1982). Assessment of mutagenicity of phthalate esters. *Environmental Health Perspectives*, **45**, 103–110.

Mcswain, B. S., Irvine, R. L., Hausner, M., & Wilderer, P. A. (2005). Composition and distribution of extracellular polymeric substances in aerobic flocs and granular sludge. *Appl. Environ. Microbiol.*, **71**(2), 1051–1057.

Moy, B. Y. P., Tay, J. H., Toh, S. K., Liu, Y., & Tay, S. T. L. (2002). High organic loading influences the physical characteristics of aerobic sludge granules. *Lett. Appl. Microbiol.*, **34**(6), 407–412.

Petrovilc, M., Eljarrat, E., Liopez, M.J., & Barcelio, D. (2001). Analysis and environmental levels of endocrine-disrupting compounds in freshwater sediments. *Trends in Analytical Chemistry*, **20**, 637–648.

Scholz, N., Diefenbach, R., Rademacher, I. , & Linneman, D. . (1997). Biodegradation of DEHP, DBP, and DINP: poorly water soluble and widely used phthalate plasticizers. *Bulletin of Environmental Contamination and Toxicology*, **58**, 527–534.

Seymour, R. B., & Carraher, C. R. (1993). *Polymer Chemistry–An Introduction.* New York: Marcel Dekker.

Staples, C. A., Peterson, D. R., Parkerton, T. F., & Adams, W. J. (1997). The environmental fate of phthalate esters: A literature review. *Chemosphere*, **35**(4), 667–749.

Stephenson, D., & Stephenson, T. (1992). Bioaugmentation for enhancing biological wastewater treatment. *Biotech. Adv.*, **10**, 549–559.

Tay, J. H., Liu, Q. S., & Liu, Y. (2001). Microscopic observation of aerobic granulation in sequential aerobic sludge blanket reactor. *J. Appl. Microbiol.*, **91**(1), 168–175.

Tchelet, R., Meckenstock, R., Steinle, P., & Van Der Meer, J. R. (1999). Population dynamics of an introduced bacterium degrading chlorinated benzenes in a soil column and in sewage sludge. *Biodegradation*, **10**(2), 113–125.

Thaveesri, J., Liessens, B., & Verstraete, W. (1995). Granular sludge growth under different reactor liquid surface tensions in lab-scale upflow anaerobic sludge blanket reactors treating waste-water from sugar-beet processing. *Appl. Microbiol. Biotechnol.*, **43**(6), 1122–1127.

US EPA. (1992 and update). Code of Federal Regulations, 40 CFR, Part 136.

Van Der Gast, C. J., Knowles, C. J., Starkey, M., & Thompson, I. P. (2002). Selection of microbial consortia for treating metal-working fluids. *J. Ind. Microbiol. Biotechnol.*, **29**(1), 20–27.

Van Der Hoek, J. P. (1988). Granulation of denitrifying sludge. In G. Lettinga, A. J. B. Zehnder, J. T. C. Grotenhuis & L. W. Hulshoff Pol (eds.), *Granular Anaerobic Sludge* (pp. 203–210). Wageningen: Pudoc.

Algal–bacterial system for the treatment of hydrocarbon-rich wastewaters

Anal Chavan and Suparna Mukherji

Centre for Environmental Science and Engineering, Indian Institute of Technology Bombay, Mumbai - 400 076, India (E-mail: *anal12@iitb.ac.in*; *mitras@iitb.ac.in*)

Abstract Association of algae with oil-degrading bacteria has been reported to facilitate the bioremediation of oil polluted marine environments. This work reports batch studies to illustrate the effect of algae on bacterial degradation of oil. A *Burkholderia cepacia* culture (designated as ES1) isolated from Arabian Sea sediments was reported to utilize the aliphatic fraction of diesel oil as a sole source of carbon and energy. A detailed study on media formulation for obtaining good growth of both the bacterial and algal culture revealed three primary aspects pertaining to media formulation: salt concentration, source of phosphate and trace elements. Both the cultures were acclimatized to 0.5% salt concentration. Higher concentration of trace elements (present in the algal medium) inhibited the growth of bacteria. The presence of phosphate at high concentration in the bacterial medium inhibited algal growth. The algal culture was found to tolerate n-hexadecane but was incapable of utilizing this hydrocarbon as a source of carbon and energy. 33% n-hexadecane biodegradation (higher than in the bacterial system) was observed in batch algal-bacterial system demonstrating the indirect role of algal culture in hydrocarbon degradation. This study demonstrates the feasibility of using algal-bacterial system for the treatment of hydrocarbon containing wastewaters in bioreactors.

Keywords *Burkholderia cepacia*; biodegradation; marine algae; media optimization; n-hexadecane

Introduction

Petrochemical and petroleum refinery industries generate large volume of wastewaters containing high concentration of hydrocarbons, which are difficult to biodegrade. Although physico-chemical treatment and conventional biological treatment have been reported, these methods have largely not been accepted due to the high cost incurred per unit volume of effluent treated and generation of chemical and biological sludge that pose disposal problems. In addition, oil-degrading cultures are difficult to settle and cause problems in conventional biological treatment. Thus, there is need for development of a technology, which will not only degrade the hydrocarbons but also improve the settleability of the sludge. Algal bacterial system has been reported to be a techno-economically feasible treatment scheme for nutrient rich industrial wastewaters (John and Bokil, 1979). In addition, it was found to degrade wastewaters containing aliphatic and aromatic hydrocarbons (Semple *et al.*, 1999) and pig farm and palm-oil mill wastewaters (Aziz and Ng, 1992). In these studies, it was observed that treatment efficiency is more compared with the bacterial system alone. This work reports the utilization of symbiotic association of an oil-degrading bacterial culture with an algal culture for the treatment of hydrocarbon containing wastewaters. Hydrocarbon tolerance of the algal culture, suitable media conditions for coexistence of the bacterial and algal cultures and degradation of n-hexadecane (n-C_{16}) in algal-bacterial batch system is reported in this work.

Materials and methods

A *Burkholderia cepacia* culture isolated from Arabian Sea sediments (designated as ES1), was demonstrated to grow on and utilize the aliphatic fraction of diesel oil as sole source of carbon and energy (Mukherji *et al.*, 2004). This culture earlier reported as *Pseudomonas* based on biochemical tests was later identified as *Burkholderia cepacia* based on 16s rDNA analysis (Bangalore Genei, India). The algal culture, utilized for the study was obtained from the surface of rocks covered by algal mats in Kurusadai Island, Tamil Nadu, India.

The algal culture was originally grown in ASN III media (recommended media for marine algae) containing 2.5% salt. For the algal growth studies, a method was developed for breaking the clumps of algal culture with glass beads so as to form an uniform suspension for inoculating in multiple test tubes (with 5 ml of media and 500 μl culture) which were incubated under continuous illumination of 1500 lux at 30°C. Algal growth was studied by determining the concentration of chlorophyll-a over time (MacKinney, 1941). Methanol was used as solvent for chlorophyll-a extraction, which was estimated as follows: Chl-a (mg/L) = (16.5 A_{665}–8.3 A_{650}) (mg/L)*extract volume (L)/volume of sample (L). Tolerance of algal culture to hydrocarbon was tested by supplementing the ASN III media with 1% (v/v) n-hexadecane.

The bacterial growth studies were set up in 500 ml flasks with 100 ml media and 1% (v/v) n-hexadecane and growth studies were conducted as described by Mukherji *et al.* (2004). Bacterial growth was measured as absorbance at 600 nm measured using visible spectrophotometer (Spectronic 20, Genesys) which was converted to MPN count (N and N_0 represent MPN/ml at time t and $t = 0$, respectively) for diesel degraders based on a predetermined correlation. The MPN values were obtained as suggested by Biswas *et al.* (2005). The flasks were incubated in a rotary shaker at 30°C and 120 rpm.

ASN III media had a significantly different composition than that used for the growth of *B. cepacia* (ES1 media). The media composition of ES1 media, ASN III and Al-Hasan's media (used in algal–bacterial study) is shown in Table 1. *B. cepacia* and algal culture was acclimatized to 0.5% salt concentration by subsequent transfers. The media optimization studies were performed by growing each culture individually in media of varying composition and following the culture growth profiles.

The n-hexadecane (n-C_{16}) degradation rate study was carried out in presence of both algae and bacteria using the optimized media, in multiple batch system (with 1% v/v n-hexadecane in 100 ml optimized media), where samples were analyzed at various time over a period of 15 days. Residual n-C_{16} was extracted using hexane. Extraction of n-hexadecane associated with the bacterial and algal cell surface was carried out by centrifugation (15,500 rpm, 45 min), and by washing the cell pellet with hexane (using vortexing, ultra-sonication and centrifugation). Three serial extractions with hexane (10:1 ratio) were used for extracting the aqueous phase. The combined extract was then analyzed in a Gas Chromatograph (Agilent 6890N) equipped with HP5 capillary column, FID detector and nitrogen gas as the carrier gas. 2 μl of sample was injected in a split-less mode. The oven temperature programme used was as follows: initial temperature 100°C, with hold time of 1 minute, ramping at 25°C per minute up to 250°C, with hold time of 3 minute, with total run time of 10 minutes. The injector temperature was set at 100°C while the detector temperature was set at 300°C. For quantitative estimation, a standard curve (correlation coefficient, 0.99) was generated using n-hexadecane concentration range of 5–20 μg/mL. In all standards and samples, 5 μg/mL of n-octadecane was added as an internal standard. To determine the amount of biomass in the system, cultures as well as n-hexadecane were added to multiple test tubes in similar proportions as in the flasks. Total volatile suspended solids (VSS) concentration was measured over a period of time as per *Standard Methods for Examination for Water and Wastewater* (APHA, 1998).

Table 1 Composition of media

Mediums	**Algal medium**	**Bacterial medium**	**Algal–bacterial medium**
Ingredients	**ASN III**	**Mukherji *et al.*, 2004**	**Al-Hasan *et al.*, 1994**
Di sodium EDTA	0.5	–	1
Citric acid	3	–	–
$NaNO_3$	750	–	850
NH_4Cl	–	850	–
KH_2PO_4	–	170	560
K_2HPO_4	20	435	–
Na_2HPO_4	–	668	860
$MgSO_4.7H_2O$	3500	22.5	370
$MgCl_2.6H_2O$	2000	–	–
$CaCl_2.2H_2O$	500	27.5	7
KCl	500	–	–
K_2SO_4	–	–	170
NaCl	25000	–	–
Na_2CO_3 (H_2O)	20	–	–
Ferric ammonium citrate (FAC)	3	–	–
$FeCl_3$	–	0.25	–
Fe (III) EDTA	–	–	4
H_3BO_3	2.860	0.0620	0.560
$MnCl_2.4H_2O$	1.810	0.2807	–
$MnSO_4.H_2O$	–	–	1.780
$ZnCl_2$	–	0.1363	–
$ZnSO_4.7H_2O$	0.220	–	2.320
$CuSO_4.5H_2O$	0.080	–	1.000
$CuCl_2.2H_2O$	–	0.0392	–
$Na_2MoO_4.2H_2O$	0.390	0.0254	0.390
$NiCl_2.6H_2O$	–	0.0130	0.004
$FeCl_2.4H_2O$	–	0.7016	–
$FeSO_4.\ 7H_2O$	–	–	0.400
Na_2SO_4	–	0.1420	–
$CoCl_2.6H_2O$	–	0.0382	–
$Co(NO_3)_2.6H_2O$	0.049	–	–

Note: All values are in mg/L.

Results and discussion

Algal culture characteristics and acclimatization

The algal culture was found to be a consortium of three cultures, out of which two were filamentous (of varying thickness), whereas, the third one was identified as *Chrococcus*, as shown in Figure 1a. To acclimatize the culture to low salt concentration, subsequent transfers were done using lower salt concentration in successive transfers, i.e. 2.5%, 2%, 1.5%, 1% and 0.5%. The algal culture was acclimatized to 2% and then to 1.5% within a period of 1 month, whereas acclimatization from 1.5% to 1% and 1% to 0.5% could only be achieved over a period of 2 months. Simultaneously, the algal culture was also acclimatized to 1% (v/v) n-hexadecane. During acclimatization of the algal culture to low salt concentration, microscopic observation revealed that there was a change in the diversity of microorganisms at the lower salt concentration and in the presence of n-C_{16}. It was found that the thinner filamentous algae were getting predominant (Figure 1b) over a period of time and the growth of thick filamentous culture and *Chrococcus* was inhibited.

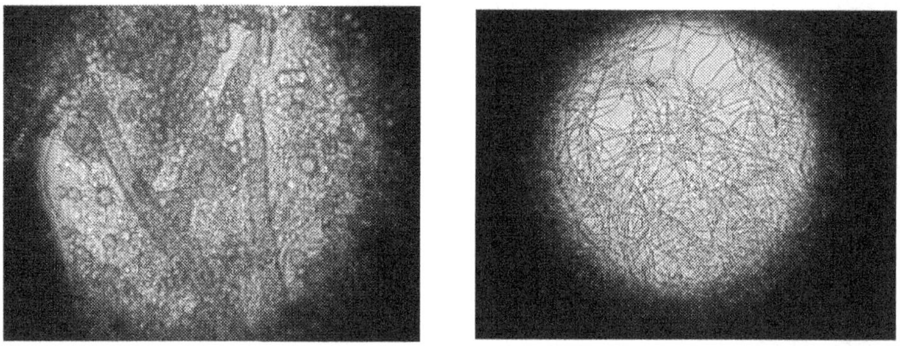

(a) (b)

Figure 1 Microscopic examination of algal culture grown in ASN III medium with (a) 2.5% salt and (b) 0.5% salt.

An algal growth study was conducted after acclimatizing the culture to 1.5% salt. Beyond a lag phase of about 4 days, a rapid increase in chlorophyll-a concentration up to a maximum concentration of 6.36 mg/l at the end of log phase at 10 days was observed (baseline, depicted in Figure 2a). After that, a long stationary phase was observed. Further studies were carried out to observe tolerance of the algal culture to hydrocarbons, using n-C_{16} (1% v/v) as a model compound. Growth of algal culture was observed in ASN III medium with 1.5% NaCl concentration, with n-C_{16} both under continuous illumination condition as well as under dark condition. In the presence of n-C_{16} under continuous illumination, beyond a 6 day lag, there was a rapid rise in chlorophyll-a concentration with maximum concentration of 5.42 mg/l at the end of 11th day, indicating almost comparable growth as depicted in Figure 2a. In contrast, under dark conditions, it was found that growth of the algal culture was negligible with maximum chlorophyll-a concentration of 0.86 mg/l at the end of 11th day, which was followed by decrease in chlorophyll-a concentration. Thus, it can be concluded that this algal culture is capable of tolerating 1% n-C_{16} without manifesting any toxic effects. However, the algal culture is not capable of utilizing n-C_{16} as a carbon source. Microscopic examination of the algal culture in presence of n-C_{16} at 1.5% salt concentration revealed presence of all the three cultures (Figure 2b), although, at 0.5% salt concentration, the thin filamentous culture was found to predominate.

Media optimization studies

In Figures 3 and 4, depicting growth profiles of bacteria and algae in various media formulations, the 'baseline' in bacterial and algal profiles indicates growth in ES1 media and ASN III media with 1.5% NaCl, respectively. The baseline profiles were obtained simultaneously whenever growth profiles of algae and bacteria in new media formulations were obtained. *B. cepacia* could not be grown in ASN III media even after the exclusion of 2.5% salt (Figure 3a) and the algal culture could not be grown in the ES1 media even after addition of 2.5% salt. In addition, while *B. cepacia* was able to grow in Al-Hasan's media (algal-bacterial media), the algal culture could not grow in this media as depicted in Figure 3b). Thus, media optimization studies were carried out to allow growth of both the *B. cepacia* as well as the algal culture in the same media and to grow both the cultures with minimum concentration of inorganic components. It was found that the media recommended for marine algae (ASN III) contains higher concentrations of NaCl, trace elements, as well as other inorganic constituents as compared to the ES1 media, while the ES1 media contains higher concentration of phosphate salts as compared to the ASN III medium.

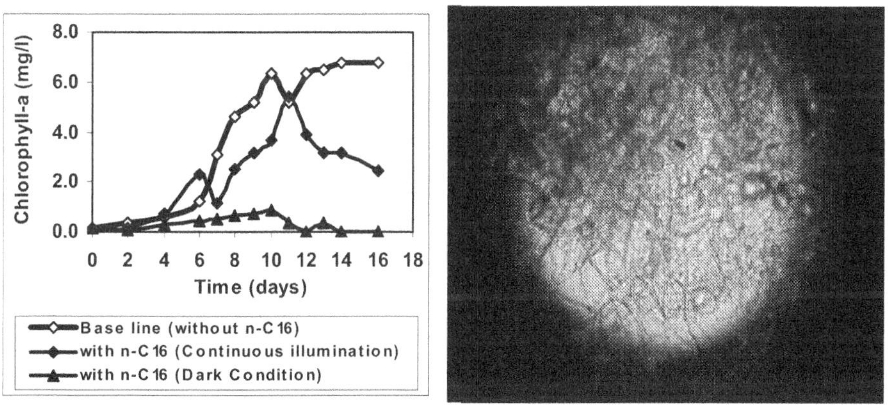

(a) (b)

Figure 2 (a) Growth of algal culture in ASN III medium (1.5% NaCl), in absence and presence of 1% n-hexadecane and (b) Microscopic examination of algal culture in presence of n-hexadecane.

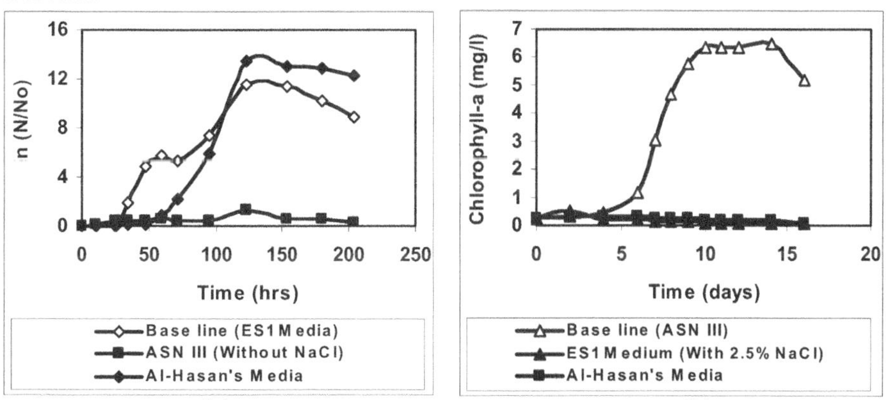

(a) (b)

Figure 3 Effect of: (a) ASN III media (Without 2.5% NaCl) on ES1 culture and (b) ES1 mineral medium (with NaCl) on growth of algal culture.

When growth study for *B. cepacia* was carried out by replacing the trace elements of ES1 media with trace elements of ASN III media, it was found that there was negligible growth of *B. cepacia*. Thus, it was concluded that high concentrations of trace elements present in ASN III medium was toxic to *B. cepacia* and cannot be chosen for the optimized media. This sensitivity of *B. cepacia* to high concentrations of trace elements was also reported by Mukherji *et al.* (2004). When the algal culture was grown with trace elemental solution (TES) of ES1 media, growth of algal culture was observed, although growth was comparatively less than in the ASN III media. Since lower concentration of trace elements could support growth of the algal culture, TES composition as in the ES1 media was chosen for the optimized media. When KCl, Na_2CO_3 and citric acid, at the concentrations present in ASN III media were added to the ES1 media, no adverse effect was observed on growth of *B. cepacia*. However, absence of these components resulted in negligible growth of the algal culture; thus, these components were retained in the optimized media. For the components, $MgSO_4$ and $CaCl_2$, good growth of both *B. cepacia* as well as the algal culture was obtained at concentrations present in Al-Hasan's media, thus, these concentrations were chosen for the optimized media. No adverse effect of absence of $MgCl_2$ was observed for algal culture and this component was, thus, eliminated. Ferric ammonium citrate (3 mg/l) was found to be good source of iron for both the cultures. EDTA at 0.5 mg/l was added as chelating agent and depicted no adverse effect on *B. cepacia* growth at this low concentration. Effects of these components are as shown in Table 2.

Table 2 Effect of various media components on growth of *B. cepacia* and algal culture

Inorganic components	*B. cepacia* growth	Algal growth
TES	– (TES of ASN III)	+ (TES of ES1 Media)
Citric acid	++ (3)	– (0)
KCl	++ (500)	– (0)
Na_2CO_3 (H_2O)	++ (20)	– (0)
$MgSO_4.7H_2O$	++ (370)	++ (370)
$MgCl_2.6H_2O$	++ (2000)	+ (0)
$CaCl_2.2H_2O$	++ (7)	+ (7)
$CaCl_2.2H_2O$	+ (500)	++ (500)
Iron source	++ FAC (3)	+ $FeCl_3$ (0.25)
Di-sodium EDTA	++ (0.5)	+ (0)

Note: Values in parentheses represents concentration in mg/L. ++ indicates good growth of cultures, comparable to the baseline mineral media, + indicates low growth of cultures in comparison to baseline media but no inhibition and – indicates growth inhibition.

Growth of *B. cepacia* was carried out with increased $NaNO_3$ concentration from 37.5 mg/l to 750 mg/l with K_2HPO_4 (20 mg/l) as the only source of phosphorus as per ASN III medium, N: P ratio varied from 5:1 to 100:1. It was observed that increased N: P ratio of 5:1 resulted in increased growth of *B. cepacia*, but 100:1 adversely affected growth of *B. cepacia* (Figure 4a). Thus, it was concluded that K_2HPO_4 alone as a source of phosphorus cannot support the growth of *B. cepacia*. When growth of the algal culture was carried out by including the ES1 media phosphate buffer containing high concentration of phosphate, negligible growth of algae was observed (Figure 4a). Thus, it was found that algal culture was not able to tolerate high concentrations of phosphate in the medium and ES1 phosphate buffer was not suitable for the optimized media. Subsequently, phosphate buffer of ES1 media was added in fixed proportions of 0.05 X, (where X is ES1 Phosphate Buffer: K_2HPO_4: 435 mg/l, KH_2PO_4:170 mg/l and Na_2HPO_4. $7H_2O$: 668 mg/l) to both the mediums.

It was observed that even at a very low concentration of phosphate buffer (0.05 X), good growth of *B. cepacia* and algae was obtained (Figure 4b), thus, 0.05 X phosphate buffer was incorporated in the optimized media.

The optimum medium formulation for algal-bacterial system was identified as (concentration in mg/L): NaCl (5000), $NaNO_3$ (750), $MgSO_4.7H_2O$ (370), $CaCl_2.2H_2O$ (7), and ferric ammonium citrate (3), KH_2PO_4 (8.5), K_2HPO_4 (21.8), $Na_2HPO_4.7H_2O$ (33.4), EDTA (0.5), citric acid (3), KCl (500), Na_2CO_3 (20) and trace element solution same as in the bacterial medium (Mukherji *et al.*, 2004).

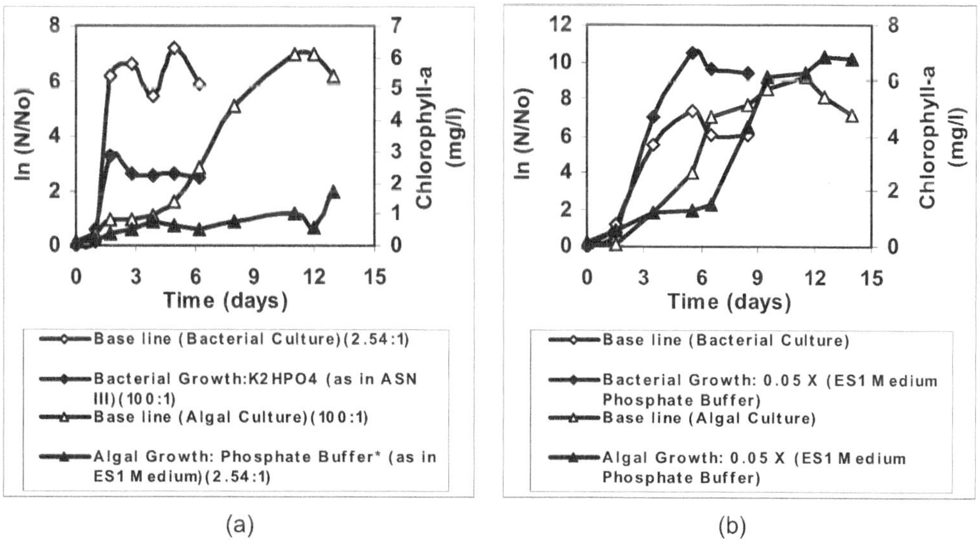

Figure 4 Effect of: (a) K_2HPO_4 alone as phosphorus source on bacterial culture and ES1 medium phosphate buffer on algal culture and (b) low concentration of phosphate buffer on bacterial and algal culture.

Biodegradation rate studies

In batch system, rapid degradation of n-hexadecane was found to occur over the first four days after inoculation, with maximum degradation of 40% over and above controls, after which biodegradation was found to stabilize. Extraction losses and abiotic losses over a period of 15 days observed in the system were found to be same, i.e. 13.33% each. Decreases in the concentration of n-hexadecane with time in control and experimental batches are shown in Figure 5a. The total mass balance of the system at 15 days is shown in Figure 5b. Based on visual observations, good algal growth was observed only beyond 7–8 days of inoculation. Estimation of chlorophyll-a with methanol extraction was not possible in this system due to presence of the bacterial cells growing in suspension. Total n-hexadecane removal achieved in the system was 59.5%, over and above controls. Biodegradation observed in the system was 33%, which was found to be higher than in bacterial system alone with biodegradation of 19% (Mukherji and Mohapatra, 2005). Thus, although the algal culture did not directly utilize n-hexadecane, it played an indirect role and facilitated bacterial degradation of n-hexadecane. Volatile suspended solids (VSS) concentration of bacterial and algal culture at time $t = 0$ is 20 and 50 mg/l, respectively. Total volatile suspended solids concentration in the system was determined over a period of time (Figure 5a) and maximum VSS attained at 15 days was 1300 mg/l.

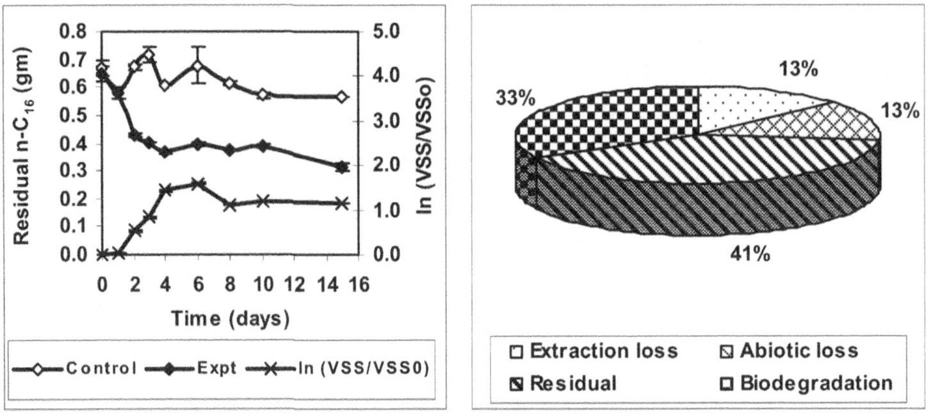

(a) (b)

Figure 5 (a) Degradation of 1% n-hexadecane in algal-bacterial system and (b) mass balance of the system at 15 days.

Conclusions

The above study established that, it is possible to grow both the marine algae and the bacteria in the same medium after systematic media optimization studies are conducted. The algal culture was found to be tolerant to n-hexadecane but incapable of utilizing this hydrocarbon as a source of carbon and energy. Thirty-three per cent n-hexadecane biodegradation (higher than in the bacterial system) was observed in batch algal–bacterial system demonstrating the indirect role of algal culture in hydrocarbon degradation. This study demonstrates the feasibility of treating hydrocarbon-rich wastewaters using algal–bacterial system in bioreactors.

References

APHA (1998). *Standard Methods for Examination of Water and Wastewater*, 20th edn, APHA/AWWA/WEF, Washington DC, USA.

Al-Hasan, R.H., Sorkhoh, N.A., Al-Bader, D. and Radwan, S.S. (1994). Utilization of hydrocarbons by cyanobacteria from microbial mats on oily coasts of the Gulf. *Applied Microbiology and Biotechnology*, **41**(6), 615–619.

Aziz, M. A. and Ng, W. J. (1992). Feasibility of wastewater treatment using activated algal process. *Bioresource Technology*, **40**(3), 205–208.

Biswas, S., Chaudhari, S.K. and Mukherji, S. (2005). Microbial uptake of diesel oil sorbed on soil and oil spill clean-up sorbents. *Journal of Chemical Technology and Biotechnology*, **80**(5), 587-593.

Cohen, Y. (2002). Bioremediation of oil by marine microbial mats, *International Microbiology*, **5**(4), 189–193.

John, U.L. and Bokil, S.D. (1979). Flocculating algal bacterial system: a new method of wastewater treatment. *Indian Journal of Environmental Health*, **21**(5), 1-9.

MacKinney, G. (1941). Absorption of light by chlorophyll solutions. *Journal of Biological Chemistry*, **140**(2), 315–322.

Mukherji, S., Jagadevan, S., Mohapatra, G. and Vijay, A. (2004). Biodegradation of diesel oil by an Arabian Sea sediment culture isolated from vicinity of an oil field. *Bioresource Technology*, **95**(3), 281–286.

Mukherji, S. and Mohapatra, G. (2005). Surface hydrophobicity and oil uptake by naturally occurring bacterial cultures. *IUMS – Microbes in Changing World*, San Francisco, CA, USA, 23–28 July.

Semple, K.T., Cain, R.B. and Schmidt, S. (1999). Biodegradation of aromatic compounds by microalgae, *FEMS Microbiology Letters*, **170**(2), 291–300.

Water policy and management

Safeguarding Canada's drinking water: defining and assessing the federal role

A. Holdway

Institute of Intergovernmental Relations, Queen's University, Kingston, Ontario, K7M 1B6, Canada (E-mail: *holdwaya@post.queensu.ca*)

Abstract Drinking water quality has received increased national attention in Canada since the contamination of the municipal water supply in Walkerton, Ontario, with *E. coli* O157:H7 in May 2000. This disastrous event, in which seven people died and thousands became ill, shook Canadians' confidence in the safety of their drinking water and demonstrated that Canada needs a national approach on drinking water safety with greater coordination and accountability at all levels of government. The present text outlines some of the problems encountered in the federal government's responsibilities for drinking water. Documentary analysis and interviews with key governmental and non-governmental officials demonstrate that the diffuse division of responsibilities for drinking water among government departments, a stalled *Federal Water Policy*, and limited funding have contributed to weak coordination of federal activities on drinking water and have impeded efforts to reverse the situation. The result has been a failure to meet some of the basic federal responsibilities in safeguarding drinking water, putting the health of millions of Canadians at risk.

Keywords Division of powers; drinking water quality; federalism; intergovernmental relations

Introduction

The analysis outlined in this text is part of a larger study aimed at identifying the gaps and overlaps in the protection of drinking water quality in selected jurisdictions in Canada and making suggestions on intergovernmental mechanisms that could be put in place to better ensure the safety of drinking water in Canada. In addition to the mechanisms for coordination at the federal level, the study focuses on the provinces of British Columbia, Alberta, and Ontario, and their relations with the federal government and with municipalities within their borders. Alberta and British Columbia were chosen because of their excellent and poor records, respectively, of waterborne disease outbreaks in recent decades. Ontario was chosen because it was the province in which the Walkerton *E. coli* outbreak occurred, and thus begs the question of how much the provincial drinking water regime has improved since 2000. The research focuses on both the overall provincial regimes for drinking water protection and case studies within each province: urban water systems in Vancouver, British Columbia, and Calgary, Alberta; rural and small municipal water systems in Alberta and Ontario; and water systems on First Nations reservations in British Columbia and Ontario. The present text examines some of the weaknesses in the federal government's approach to its responsibilities for safeguarding drinking water.

Constitutional basis

As with many other issues in Canada, jurisdictional responsibility over drinking water is divided between various orders of government. The division of powers is largely determined by the *Constitution Act, 1867*, though this document does not provide a clear indication as to which order of government should have jurisdiction – indeed, the constitution does not even mention water. Section 109 of the constitution gives the provinces ownership of resources such as "lands, mines, and minerals," and this has been interpreted as giving the provinces proprietary rights to surface waters and groundwater within their provincial boundaries, except those occurring on federal lands within the province. (The constitutional amendment of 1930 that gave Alberta, Saskatchewan, and Manitoba "a position of equality with the other Provinces of Confederation with respect to the administration and control of [their] natural resources" does make reference to water.) S. 92 of the constitution gives the provinces legislative powers in areas such as the "management and sale of . . . public lands" (s. 92(5)), "property and civil rights" (s. 92(13)), "local works and undertakings" (s. 92(10)), "municipal institutions" (s. 92(8)), and "generally all matters of a purely local and private nature" (s. 92(16)). Thus the provinces are effectively the water managers, having the day-to-day responsibility for drinking water. Some responsibility is delegated by provincial governments to municipal governments, such as the operation of treatment plants.

Though the constitution does not clearly identify a federal role in drinking water protection, a number of powers can be inferred. Since the constitution gives the federal Parliament jurisdiction over lands such as First Nation reserves (s. 91(24)), military bases (s. 91(7)), and national parks (s. 91(1A)), drinking water on these lands is a federal responsibility. The federal government is also responsible for interprovincial and international transportation (s. 91(13), 92(10)(a)-(b)), so it is responsible for drinking water onboard common carriers such as airplanes, ships, and trains. Through its jurisdiction over fisheries (s. 91(12)), the federal government is responsible for protecting fish habitat from pollution, giving the federal government some power over waters to which the provinces hold the proprietary rights. The federal government also has indirect powers that can be used to justify involvement in drinking water matters. These include the statistics power (s. 91(6)), used as a basis for gathering water quality data; the spending power (e.g., s. 91(1A)), by which the federal government can influence provincial water management through conditional grants; and criminal law powers (s. 91(27)), which, according to the Supreme Court of Canada decision in *R. v. Hydro-Québec* (1997), can be used to regulate pollution. Two more controversial powers that are available to the federal government are the declaratory power and the "residual" power. The declaratory power in s. 92(10)(c) gives the federal Parliament the authority to exercise jurisdiction over works within a province by declaring them "to be for the general advantage of Canada or for the advantage of two or more of the provinces." This is a power that is rarely used, making it of limited practical use in the area of drinking water protection. Equally controversial is the statement in the preamble to s. 91 that gives the federal Parliament the power "to make laws for the peace, order, and good government of Canada" (often referred to as the "POGG" power). POGG is termed a "residual" power because it gives the federal Parliament the authority to legislate on any matter not assigned to either order of government elsewhere in the constitution.

Further, the federal and provincial governments have a concurrent responsibility for agriculture (s. 95). Though not mentioned specifically in the constitution, the two orders of government have been interpreted also to have a shared responsibility over health through the provinces' responsibilities for hospitals (s. 92(7)), property and civil rights (s. 92(13)), and matters of a purely local or private nature (s. 92(16)) and the federal government's spending powers (s. 91(1A)) and responsibility for criminal law (s. 91(27)).

One of the greatest areas of uncertainty in the division of powers over drinking water is in authority over interjurisdictional waters. The prevailing assumption for many years was that each province held virtual sovereignty over waters within its boundaries. As long as federal jurisdiction over fisheries was respected, an upstream province could thus pollute without regard to the consequences for a downstream province. This belief was challenged in the Supreme Court of Canada in the case of *Interprovincial Co-operatives* v. *The Queen*, [1976] 1 S.C.R. 477. Interprovincial Co-operatives was licensed in Ontario and Saskatchewan to release effluents containing mercury that found their way downstream to Manitoba, where they contaminated fish and forced the closure of a commercial fishery. Manitoba sued Interprovincial Co-operatives, raising the question of whether a government has the authority to legislate on the quality of water entering its jurisdiction. The central question was whether the wrongful act was committed in the jurisdiction where it took place or where its consequences were felt. The judges were divided 4-3, with the majority ruling that since "property and civil rights" is a provincial head of power under s. 92(13), Manitoba could not make laws affecting the civil right of upstream provinces to pollute in downstream provinces. The minority, however, held that neither could Ontario and Saskatchewan interfere with the rights of Manitobans to clean water. This view was not contradicted by the majority, adding to the confusion over what future decisions the court might make on interprovincial water issues. Some interpreted the ruling as recognizing a federal responsibility in this area under POGG, by which authority federal legislation in this area would have to be justified (Pearse *et al.*, 1985, p. 66). Without such legislation, it is unclear how the court will deal with such problems in the future. Clarification is thus needed at the political level.

Methods

As mentioned above, the present analysis is part of a larger study on intergovernmental relations in the protection of drinking water quality in Canada. A detailed documentary analysis for the jurisdictions under study considered some 200 pertinent government documents, including statutes, regulations, bills, policy papers, guidelines, reports, transcripts from government debates, and documents from government inquiries. Some 200 pertinent documents from outside government were also analyzed, including journal articles, conference proceedings, and reports from academics, industry associations, and other non-governmental organizations. Additionally, approximately 450 newspaper articles, magazine articles, and press releases were reviewed.

The documentary analysis was supplemented by semi-structured interviews with civil servants in various government departments in each jurisdiction under study, as well as with heads of industry associations and non governmental organizations. Thirteen interviews were conducted in person and by telephone, of which seven were with civil servants. The documentary analysis was used to identify most interviewees; others were identified by other study participants. The interviews explored the mechanisms used to coordinate drinking water programs and policies in each jurisdiction, both within the government and with other orders of government, and alternative mechanisms that might be employed to improve such coordination.

Results and discussion

Defining the role of the federal government

The federal government has attempted to define its role in the protection of drinking water quality through a small number of statutes and regulations. The provinces, having proprietary rights over water within their boundaries, have the primary authority to legislate on drinking water matters. The federal government, however, has stated on a small number of occasions an intention to introduce, or at least consider, federal legislation on drinking water protection – for example, in 1984 (Grover and Zussman, 1985, p. 9), 1991 (Interdepartmental Committee on Water, 1994, p. 15), and 2001 (Office of the Auditor General of Canada, 2001, 1.3.4.11) – but such legislation has never come before Parliament for a vote. In the absence of such legislation, drinking water receives little mention in federal legislation. The *Canada Water Act*, for example, the federal government's central piece of legislation on water issues, does not mention drinking water. In fact, only one active federal statute, the *Canada Labour Code*, refers to the quality of potable water. Regulations under several other statutes touch briefly on drinking water quality, including the *Canada Agricultural Products Act*, the *Canada Transportation Act*, and the *Canada Shipping Act*. The *Pest Control Products Act* of 2002 also touches on drinking water quality, though this statute is not yet in force. The *Potable Water Regulations for Common Carriers* under the *Department of Health Act*, which define federal inspection procedures on ships, trains, and airplanes travelling between provinces and internationally, are the most direct reference to a federal role in drinking water. A few other statutes have some bearing on drinking water without mentioning it by name. Through the *Prairie Farm Rehabilitation Act*, the federal government has established drinking water supplies for farms and small towns in the prairies, something it chooses to do through agreements with the provinces. The *Canadian Environmental Protection Act (CEPA)* mandates the federal government to control the release of toxic substances, pollutants, and wastes, many of which may end up in drinking water sources. (The Supreme Court decision in *R. v. Hydro-Québec* (1997) that upheld the federal government's right to legislate "the protection of a clean environment" means the federal government could use CEPA more aggressively to regulate water quality, though it appears reluctant to do so.) More tangentially, the *Fisheries Act* allows the federal government to prohibit the pollution of fish habitat, though court decisions have suggested that this does not allow much scope for federal drinking water regulations. The federal government may also be able to exert influence on freshwater management through the *Canadian Environmental Assessment Act*. The *Food and Drugs Act* regulates the quality of water that is used in food, drinks, and pre-packaged ice sold in Canada. It is not, however, applied to drinking water sold by municipalities.

The lack of precision in defining the federal role in safeguarding drinking water has been evident for some time. When the federal government called an Inquiry on Federal Water Policy in January 1984 "to identify and describe emerging water issues in Canada, including their interjurisdictional dimensions" (Inquiry on Federal Water Policy, 1985, p. 4), the stated theme of its *Final Report* was "the need to clarify the federal role in water management in Canada" (Pearse *et al.*, 1985, p. 182). Noting that "even senior public servants admit they are sometimes uncertain about the scope of their responsibilities" (Pearse *et al.*, 1985, p. 182), the authors concluded that "cooperation among governments, although critical, was not enough to sort out the confusion. The business of the federal government had to be clarified, communicated, and consistently acted upon" (Inquiry on Federal Water Policy, 1985, p. 38). The Inquiry seemed to augur well for greater clarity in the federal role in drinking water, but after collecting "more information about water management in Canada than has ever before been assembled" (Pearse *et al.*, 1985, p. 3) and producing 55 recommendations for government, the final report of more than 200 pages devoted only *two* pages to drinking water quality.

This may be part of the reason for the lack of focus on drinking water quality in the government's subsequent *Federal Water Policy*, published in 1987. The government's response to the Inquiry, the *Federal Water Policy* contained only one, generic policy statement on drinking water:

> The federal government is committed to ensuring safe drinking water within areas under its jurisdiction and to promoting and encouraging a consistent approach to protection and improvement of the nation's drinking water by provinces, territories, and local governments (Environment Canada, 1987, p. 17).

In addition to lacking specificity, however, the *Federal Water Policy* came to suffer from a lack of coordination. Environment Canada had intended to implement the *Policy* through its Inland Waters Directorate, but the Directorate was broken up in 1993 and its staff dispersed throughout the Ministry. The *Federal Water Policy* called for an annual report on its implementation, but only two progress reports were issued, in 1990 and 1994. A draft discussion document produced in 1998, *Towards a Federal Freshwater Strategy*, was intended to lead to an update of the *Federal Water Policy*, but this did not happen. A second discussion document, *Fresh Water: A Federal Discussion Document*, was prepared in 2000, but its status remains unclear. Furthermore, over the course of nearly 20 years, the government has never explicitly allocated responsibilities or funds to carry out the actions set out in the *Federal Water Policy* (OAG, 2001, 1.3.7.6).

Responsibilities for water within the federal bureaucracy are highly dispersed, with at least 19 departments and agencies involved (OAG, 2005a, 4.56), and in the case of drinking water, "responsibility is even more diffuse" (Saunders, 1988, p. 42). Health Canada has the largest role, with responsibility to monitor and investigate waterborne diseases, monitor drinking water quality on federal lands and on board common carriers, and regulate water used in food and drink. Health Canada also serves as the Secretariat for the Federal-Provincial-Territorial Subcommittee on Drinking Water (DWS), which produces national drinking water guidelines. Environment Canada also has a seat on DWS and serves as the Technical Secretariat for the Water Quality Task Group of the Canadian Council of Ministers of the Environment. Together, DWS and the Water Quality Task Group produced *From Source to Tap: The Multi-barrier Approach to Safe Drinking Water* (2002), a document prepared for a general audience that describes the concept of a multi-barrier approach to drinking water protection, as well as an in-depth guidance document for drinking water system owners and operators (2004). Environment Canada's interest in drinking water is generally more peripheral, however, as most of its water-related activities – for example, administering the *Canada Water Act* – focus on protecting water quality in general. Other departments involved in drinking water include Indian and Northern Affairs Canada, which shares responsibility with Health Canada for drinking water on First Nations lands; Agriculture Canada, which has been involved in the construction of water supply systems for farms and farm-based communities in the prairies under the *Prairie Farm Rehabilitation Act*; Infrastructure Canada, which has supported improvements to municipal water systems through, for example, the Infrastructure Canada Program and the Canada Strategic Infrastructure Fund; and Natural Resources Canada, which helped to create the National Ad Hoc Committee on Groundwater and its *Canadian Framework for Collaboration on*

Groundwater (2003). The federal government has thus chosen a collaborative approach in many of its drinking water activities, engaging the provinces as active partners (this will be examined in more detail in the final report on this study), but the coordination of its disparate activities at the federal level is wanting in a number of ways.

Weaknesses in federal coordination

Of the many ways in which the federal government is involved in the area of drinking water quality, a few examples highlight some of the weaknesses in coordination and follow-through. One of these is the government's main attempt to ensure communication and cooperation among federal departments involved in water, the Interdepartmental Committee on Water (ICW). Consisting of officials from some two dozen departments and agencies involved in water issues and chaired by Environment Canada, the ICW was initially intended to be an interim committee, formed in 1968 when the government was considering the Canada Water Bill (later to become the *Canada Water Act*). The ICW's terms of reference were

> to consider and approve all federal government water programs until a permanent mechanism was established, and that a committee of ministers should formulate a permanent mechanism for resolving interdepartmental conflicts on water programs (Mactavish, 1985, p. 61).

The Committee played a significant role for the first few years, but after the *Canada Water Act* was passed in 1970, the status of the ICW was not reconfirmed and the committee of ministers was never formed. The ICW eventually created revised terms of reference for itself and continued to meet, limiting itself to exchanging information and dealing with routine matters (Mactavish, 1985, p. 62). Over the years it became clear that, far from being asked to "approve all federal government water programs," the ICW was given projects for review "only as a token gesture" (Saunders, 1988, p. 45). The Committee's loss of stature and influence has been underlined by its membership, which, once consisting of senior officials, is now typically made up of more junior representatives. The resulting "organizational silos and turf protection" mean that "anything that requires action across departments is particularly vulnerable to collapse" (OAG, 2005a, p. 12). Clearly, the ICW, with its lack of clear responsibilities and authority, cannot perform the important task of coordinating the federal government's efforts on water in general, and drinking water in particular.

A second set of problems arises with the *Guidelines for Canadian Drinking Water Quality*. Health Canada leads the Federal-Provincial-Territorial Subcommittee on Drinking Water (DWS) in producing these national guidelines, which establish "recommended limits for substances and conditions that affect the quality of drinking water." The *Guidelines* are "intended to apply to all drinking water supplies, public and private" (DWS, 1996, p. 7), but because water is largely a provincial responsibility, they are not legally enforceable unless adopted by the provinces. Health Canada's role in DWS is to provide coordination and scientific expertise, keeping abreast of the latest health data relating to existing guidelines and leading the development of new guidelines (DWS, 1999, pp. 8-9). One of the problems in this coordination of DWS is the amount of time taken to develop or revise guidelines. Health Canada says this process should take two to three years (OAG, 2005a, 4.19). In practice, however, Canada's Commissioner of the Environment and Sustainable Development has found that the process can take far longer. The review of the guideline for aluminum took four years, for example, that for *E. coli* took seven years, and trichloroethylene took 12 years, despite being declared a priority substance in 1993. The guideline for arsenic, after eight years, is still only at the public consultation stage (OAG, 2005a, 4.19). Goff Jenkins, a former longtime member of DWS told the inquiry into the *E. coli* outbreak in Walkerton that this slow pace is due to a lack of funding for research at Health Canada (O'Connor, 2002, p. 154). This lack of resources has created a great backlog of substances requiring study. There are currently no guidelines, for example, for viruses or protozoa, even though these can be responsible for fatal illnesses, and even though the cause of one of Canada's latest high-profile outbreaks of waterborne disease was the protozoan *Cryptosporidium*. It has been ten years since the latest edition of the *Guidelines* reported that these microorganisms were "being reviewed" (DWS, 1996, p. 14). The Commissioner of the Environment and Sustainable Development reports that the creation of guidelines for existing pesticides, a frequent contaminant of drinking water, cannot keep pace with the registration and use of new pesticides (OAG, 2005a, 4.25). Jenkins says there is also a "desperate need" (Perkel, 2001) for guidelines on disinfection byproducts, some of which are known to be carcinogenic. The current backlog of some 50 guidelines could take 10 years to deal with at the current pace (OAG, 2005a, p. 6), and the situation is unlikely to improve. Despite a commitment from the federal government to accelerate this work in its 2002 Speech from the Throne, DWS's budget has decreased by 20 percent between 2001 and 2005 (OAG, 2005a, 4.29).

Another equally serious problem is the federal government's lack of commitment to the very guidelines it helps to create. Although the provinces have the main responsibility for managing water in Canada, millions of people are affected by federal responsibilities for drinking water at military bases, national parks, and other federal facilities; on passenger trains, ships, and airplanes travelling between provinces or internationally; and in First Nations communities. The *2005 Report of the Commissioner of the Environment and Sustainable Development*, however, cited "serious weaknesses" (OAG, 2005b) and a "lack of central guidance" (OAG, 2005a, p. 25) in federal efforts to meet its drinking water responsibilities. For example, even though the *Canada Labour Code* and its regulations require that the federal government provide its employees with water that meets the *Guidelines for Canadian Drinking Water Quality* (R.S. 1985, c. L-2, s. 125(1)(j), and regulations, e.g., *Canada Occupational Health and Safety Regulations*, SOR/86-304, s. 9.24), a review of six departments revealed that water testing "was not always done at federal sites"; "when testing was done, it ranged from complete to insufficient, and compliance varied" (OAG, 2005b). A more alarming example is the lack of enforcement of the *Potable Water Regulations for Common Carriers*, which require that Health Canada inspect water quality on passenger trains, aircraft, and cruise ships (C.R.C., c. 1105, s. 10, 11, and 13). The Commissioner of the Environment and Sustainable Development reported in 2005 that such inspections had not occurred since direct funding for this program was removed in the mid 1990s (OAG, 2005a, 4.51), endangering the health of millions of travellers. Equally shocking but more well known in Canada is the state of drinking water in First Nations communities, for which the federal government is responsible but for which it has not legislated any standards. In a confidential 2001 report, Indian Affairs and Northern Canada (INAC) found that three-quarters of First Nations drinking water systems posed a significant health risk (Wallace, 2005b), despite \$560 million in federal spending between 1995 and 2001 to address "urgent" upgrades (INAC, 2003). A recent high-profile outbreak highlights part of the problem. In October 2005, more than 1000 residents of the Kashechewan First Nation in northern Ontario were airlifted for medical treatment for skin conditions aggravated by the high levels of chlorine used to disinfect the community's water supply, which contained high levels of *E. coli*. Ontario premier Dalton McGuinty ordered the evacuations, saying that the federal government had been "missing in action" (Sokoloff, 2005) since problems with Kashechewan's water treatment plant were documented in a federally-commissioned report in 2003 (Wallace, 2005a). "If they're not prepared to assume responsibility," said McGuinty, "then they should talk to not just this premier, but premiers throughout the country and come to a new arrangement" (Sokoloff, 2005). In short, according to the Commissioner of the Environment and Sustainable Development, "When federal responsibility is involved, not all Canadians can assume that their drinking water is always safe" (OAG, 2005a, p. 5). In addition, the federal government's inadequate performance in areas of its own jurisdiction weakens any moral authority it may have gained from leading the creation of national drinking water standards.

Conclusions

The federal government's failure to meet some of its basic responsibilities in safeguarding drinking water is largely the result of poor coordination. The dispersal of responsibilities for drinking water among so many government departments and agencies makes strong interdepartmental coordination essential. At present, however, an interdepartmental coordinating committee with no clear mandate and an outdated *Federal Water Policy* without a government-wide implementation strategy have restricted meaningful communication and integration between the government's drinking water activities and have only perpetuated the culture of "organizational silos and turf protection." If meaningful change is to take place, a new federal policy on drinking water must be created – one that includes a clear delineation of the federal government's responsibilities and the departments in charge of each; clear objectives for each area of responsibility; and clearly defined, allocated, and funded actions that will be taken to achieve each objective. The Interdepartmental Committee on Water, or some surrogate body, should be strengthened and be placed in charge of implementing the new drinking water policy, becoming again the body "to consider and approve all federal government water programs." The federal government, as the leader of the process of creating drinking water guidelines, must also accelerate this process and must ensure that, even if it does not bring forth legislation to make the *Guidelines* binding across the country, it at least makes sure that the standards are met in areas of its own jurisdiction. Unless better coordination and follow-through are achieved, future outbreaks of waterborne disease will be virtually inevitable.

Acknowledgements

This project is funded by a grant from the Canadian Institutes of Health Research.

References

Environment Canada (1987). *Federal Water Policy*. Environment Canada, Ottawa, Canada.

Federal-Provincial-Territorial Subcommittee on Drinking Water (1996). *Guidelines for Canadian Drinking Water Quality*, 6th edn. Health Canada, Ottawa, Canada.

Grover B. and Zussman D. (1985). *Safeguarding Canadian Drinking Waters*. Inquiry on Federal Water Policy Research Paper 4, Environment Canada, Ottawa, Canada.

Indian and Northern Affairs Canada (INAC) (2003). Backgrounder: Water Quality and First Nations Communities. http://www.ainc-inac.gc.ca/nr/prs/m-a2003/02304bk_e.html (accessed 3 September 2005).

Inquiry on Federal Water Policy (1985). *Hearing About Water: A Synthesis of Public Hearings of Inquiry on Federal Water Policy*. Environment Canada, Ottawa, Canada.

Interdepartmental Committee on Water (1994). *Federal Water Policy: Progress Report No. 2*. Environment Canada, Ottawa, Canada.

Mactavish J. S. (1985). *The Federal Role in Water Management*. Inquiry on Federal Water Policy Research Paper 15, Environment Canada, Ottawa, Canada.

O'Connor, D. R. (2002). *Report of the Walkerton Inquiry, Part Two: A Strategy for Safe Drinking Water*. Walkerton Inquiry, Toronto, Canada.

Office of the Auditor General of Canada (OAG) (2001). *2001 Report of the Commissioner of the Environment and Sustainable Development*, Office of the Auditor General of Canada, Ottawa, Canada.

Office of the Auditor General of Canada (2005a). *2005 Report of the Commissioner of the Environment and Sustainable Development*, Office of the Auditor General of Canada, Ottawa, Canada.

Office of the Auditor General of Canada (2005b). News release: Serious weaknesses found in federal efforts to ensure safety of drinking water, 29 September, Office of the Auditor General of Canada, Ottawa, Canada.

Pearse P. H., Bertrand F. and MacLaren J.W. (1985). *Currents of Change: Final Report, Inquiry on Federal Water Policy*. Environment Canada, Ottawa, Canada.

Perkel, C. (2001). Canada said to be lagging in assessing threats to drinking water. St. John's *Telegram*, 12 September 2001, A13.

R. v. Hydro-Québec (1997). 3 S.C.R. 213.

Saunders J. O. (1988). *Interjurisdictional Issues in Canadian Water Management*. Canadian Institute of Resources Law, University of Calgary, Calgary, Canada.

Sokoloff H. (2005). Province vows to probe water on Ontario reserves. Kingston *Whig-Standard*, Kinston, Canada, 27 October 2005, 11.

Wallace J. (2005a). Province, feds knew about bad water. Sudbury *Star*, Sudbury, Canada, 4 November 2005, p. A1.

Wallace J. (2005b). Water woes no secret: Crisis on reserves known 4 years ago. Kingston *Whig-Standard*, Kingston, Canada, 4 November 2005, 1.

Estimation of irrigation water productivity within a rigid rotational delivery system

Noor ul Hassan Zardari and Ian Cordery

School of Civil and Environmental Engineering, The University of New South Wales, Sydney 2052, Australia (E-mail: *Zardari@civeng.unsw.edu.au*; *i.cordery@unsw.edu.au*)

Abstract The rotational irrigation water delivery (the *warabandi*) system practiced in the Indus River Basin of Pakistan has continued for 100 years. The allocated water share under *warabandi* system is attached to the land and is fixed. This rigid allocation system has been partly justified by its contribution to social equity though this may have been at the cost of economic gains. Farmers' survey results presented here indicate a lack of equity within this allegedly "fair" system and the magnitude of those economic losses. Eighty-six farmers on two different watercourses were interviewed. The results suggest that the existence or non-existence of fresh groundwater resources along with other critical variables should be taken into consideration when making water allocation decisions. A framework to allow real equity and efficiency in water allocations – such as considering the gross area of a tertiary canal, sensitivity of crop growth stage to water shortage, crop value, bias of allocation towards most water use efficient areas and water use efficient crops, potential losses from water deficiency, etc. – should be developed as a tool to improve water productivity for Pakistan and for individual farmers.

Keywords Indus River Basin; irrigation allocation system; water right; water value

Introduction to Pakistan's irrigation system

The irrigation system in Pakistan (14.6 M ha) is the largest contiguous irrigation system in the world. It comprises the Indus and its major tributaries, 3 major storage reservoirs, 19 barrages/headworks, 44 canal commands and some 100,000 watercourses. The total length of canals is about 56,073 km, with watercourses, farm channels and ditches running another 1.6 million km in length (Murray-Rust *et al.*, 2000; Khan *et al.*, In Press). The irrigation system uses barrages to divert water from rivers into canals. The main canals are designed for continuous operation at or near full capacity. Distributaries are the lowest level channels directly controlled by the Irrigation Department. The turnouts (*mogha*) are ungated modular outlets designed to deliver fixed amounts of water into watercourses. Each watercourse irrigates 60–250 ha (225 ha average) of land, usually cultivated by 10–150 farmers (average about 50 farmers) (Chaudhary and Robert, 1989; Nakashima, 2000). Each farmer's water share is supposed to be proportional to the size of his land holding.

Problems with the existing system

Since the introduction of the *warabandi* system, little effort has been made to see what has happened to the historical water share granted on the basis of land holdings, despite major political and socio-economic changes (Bandaragoda and Rehman, 1995; Bandaragoda,

1998). Much has been said about the shortage of water resources in relation to crop water requirements, but little has been done to evaluate the effectiveness of water allocation or to find ways and means to improve distribution procedures. The thorough examination of basic rules of the *warabandi* system has shown many drawbacks in the existing delivery system, for example, the size of water share under the *warabandi* system is supposed to be exclusively decided on the basis of the size of the farmland. Water allocation based totally on a single criterion is unlikely to improve productivity from the scarce resource, as the water productivity depends on a variety of critical factors. Bandaragoda and Rehman (1995) indicate that the volume of water available to a farmer is independent of the stage of crop growth and the farmer is forced to either take his/her turn to receive water (whether needed or not), or to forego that water. This means that the system not only leads to wastage of water but also sometimes results in low crop yields due to over-irrigation. Likewise, the rigidity implicit in the *warabandi* system prevents farmers from maximizing economic and social net benefits from the scarce water (Bhatti and Jacob, 1990; Cai *et al.*, 2001).

Objectives of the study

The aim is to evaluate the fundamental rules of the *warabandi* system in order to indicate the benefits that could be gained from utilization of a scarce resource. To do this, the economic value of irrigation water has been calculated against two criteria, i.e. values with respect to land and with respect to water productivity. It is shown that the basic rules of the *warabandi* system control the value of water. Therefore, the study objectives were: (1) to assess the water and farm productivity under that irrigation delivery system; and (2) to recommend some measures that can increase water and farm productivity.

Characteristics of the study area

Salient features of the study area

The study was conducted in two watercourse commands, namely 1-R and 3-R. Both watercourses off take from *Lundo* Distributary. 1-R watercourse is situated at the head of the distributary and 3-R watercourse at the middle of the distributary. *Lundo* Distributary off takes from *Ali Bahar* Branch Canal of *Sukkur* Barrage Circle. The *Lundo* Distributary is 32 km long. The design discharge of the Distributary is 4.6 m^3/s with 13360 ha of cultural command area (CCA), which is equivalent to 224 m^3/ha/week for the whole of the CCA. However, the average water allocation in the survey areas was between 86 and 134 m^3/ha/week. It shows about 50% of water delivered to the Distributary was lost or unfairly allocated elsewhere. Three reasons have been initially found for such huge water losses: (1) Farmers on other watercourses were legally allocated more water; (2) Evaporation and seepage from the channels; or (and) (3) there is significant theft from the Distributary. Determining the magnitude of impact by each factor was beyond the scope of this study.

Farm and water management data was collected for both crop seasons, i.e. *Rabi* (dry) 2004–2005 and *Kharif* (wet) 2004. Crop survey data for both regions shows that cotton was the main crop in *Kharif* and wheat in *Rabi* with some area under sugarcane, banana, corn (maize), sorghum, millet and fodder crops. The groundwater quality was fresh in 1-R region and marginal in 3-R region. A total of 86 farmers were interviewed and a questionnaire was completed for each. Information about the size of 'weekly water share' ('water share' hereafter), proportion of allocated water actually received, the types of crops grown, and other farm and water management practices was collected. The farmers' fields were grouped into three categories namely head, middle and tail according to the distance from the watercourse outlet. Some details of watercourses are given in Table 1.

Table 1 Features of selected watercourses

Number of water-course	Design discharge (m^3/s)	Length (km)	CCA (ha)	No. of farmers interviewed	Average size of farm (ha)	Av. size of weekly water share (m^3/ha)
1-R	0.067	3.1	330	46	4.8	134
3-R	0.025	2.4	166	40	4.1	86

Analysis of weekly water share data

For this study, the allocated water share was calculated on the assumption that both watercourses were flowing with the design discharge. The average water allocation was very low (630 mm depth for 1-R and 470 mm depth for 3-R farms in each year) in both survey areas. The actual amount of water received by tail end farmers was probably smaller, as 20 to 25% of allocated water share was likely to have been lost in both unlined watercourses. Studies on water losses in unlined watercourses by Latif and Sarwar (1994) and Kirpich *et al* (1999) suggest that 20 to 25% of water delivered to the watercourse is lost in the system before reaching the farm-gate. This clearly shows that the structural system ensures that the main principle of *warabandi* system (i.e. equity in water distribution) fails in practice (Note it was suggested above that at least 50% of the water entering the Distributary never reaches the crops).

Water share variation from watercourse to watercourse

Weekly water share data was analyzed to determine equity in the water allocation. The average size of water share with maximum and minimum sizes of water share for both watercourses is shown in Table 2. The average size of water share of all 1-R farms was 134 m^3/ha (13.4 mm depth of water per week). The average water allocation to 3-R farms was 86 m^3/ha (8.6 mm depth of water per week). The analysis shows that 50% of 3-R farms have water share of less than 86 m^3/ha and all 1-R farms have water share above 86 m^3/ha. This uneven water allocation, in the watercourse commands, is shown in Figure 1.

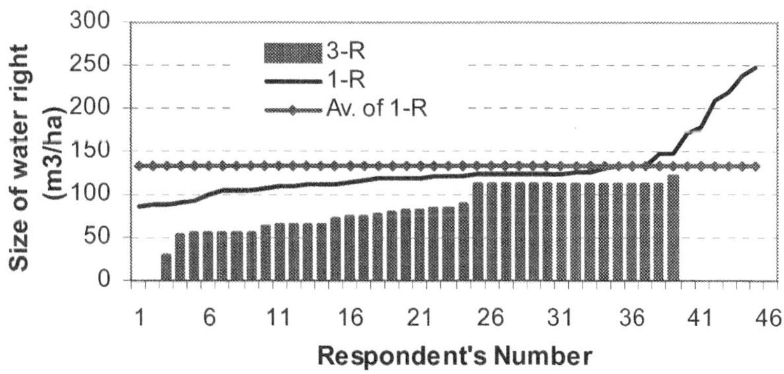

Figure 1 Size of water share (plotted in increasing order) in 1-R and 3-R command areas.

Water share variation within watercourse

Unfair and inequitable allocation of scarce water does not only occur between but also within watercourses. Eighty-five per cent of 1-R tail end farms have water share of less than 134 m^3/ha (i.e. average size of water share of all 1-R farms). The middle farms have greater share in canal water allocation compared to the tail end farms. Head farms were not considered, as the sample size was very small. In case of 3-R farms, none of the head farms had water share above the average size of water share for all 3-R farms. However, 75% of tail end farmers have 30% higher water share than the average. What was the philosophy behind such unequal allocations? The data in hand only gives a single reason for such philosophy. The only resemblance between head and middle farms of 1-R and tail farms of 3-R was the equal size of farms. Medium to large farms were dominant in those locations. So, one can say that the larger farms tend to attract a proportionally higher share of water.

Table 2 Some farmers' survey results

Parameters		1-R watercourse	3-R watercourse
Proportion of farmers with less than High School education		22%	65%
Proportion of farmers with farming experience of 15 years or less		61%	43%
Farm location on watercourse	*Head*	7%	28%
	Middle	28%	37%
	Tail	65%	35%
Average water share (m^3/ha)		134	86
Groundwater quality		Fresh	Marginal
Average size of farm (ha)		4.8	4.1
Proportion of farmers having resources of income other than agriculture	*None*	56%	65%
	Own business	9%	28%
	Employed	35%	7%
Percent of farm area under cultivation		86%	65%
Number of Govt. tubewells		2	None
Number of Private tubewells		36	None
Net crop income (RS/ha)		17150	11365
Value of a unit of water (RS/m^3)		3.9	6.9
Value of a unit of water for each location of farms (RS/m^3)	*Head*	NA	6.9
	Middle	4.31	8.3
	Tail	3.68	5.65
Water shortage faced (proportion of allocated water share)		36%	48%
Water shortage faced by each location of farms	*Head*	NA	34%
	Middle	41%	41%
	Tail	33%	57%

Analysis of water shortage data

Variation of water shortage between watercourses

The second objective of the *warabandi* system, i.e. equitable distribution of any water shortage between all farms, was investigated in the survey areas. It was assumed that the parent channel (*Lundo* distributary) remains closed for six weeks in each year for maintenance reasons. The farmers were entitled to receive only 46 weekly water turns instead of 52. In this paper the term 'water shortage' is the proportion of 'actual number of water turns received' to the 'allocated number of water turns'.

The collected data indicate that the 1-R farms missed 36% of allocated water turns. This figure for 3-R farms was 48%. Hence 3-R farmers faced more water shortage than 1-R farmers (Table 2). This uneven distribution of water scarcity affects the farm area under cultivation and must have a huge effect on farm incomes. Table 2 shows that 65% of 3-R command area was under cultivation, while for 1-R command it was 86%. The farm income of 3-R farmers suffered more than for 1-R farmers. It was probably because the farm income was approximately proportional to area under cultivation. The pattern of actual amount of water delivered to 1-R and 3-R is shown in Figure 2.

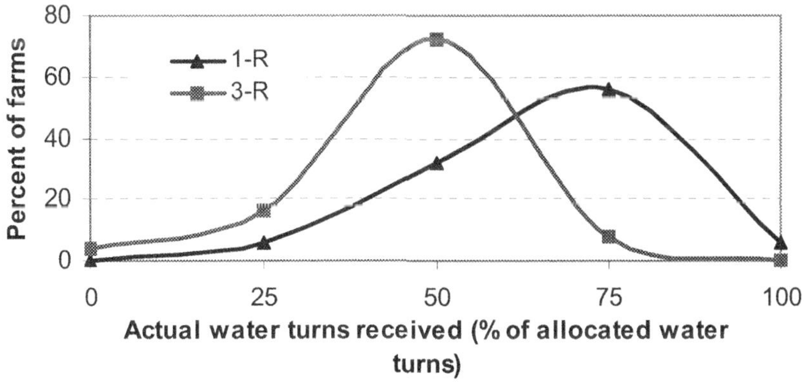

Figure 2 Pattern of allocated water turns received by 1-R and 3-R farmers.

Water shortage at different locations on same watercourse

Analysis of the distribution of water shortage along each watercourse indicates that inequity in water shortage distribution does not stop at the outlet of the watercourse but also continues within watercourses (similarly to uneven water allocations). For example, the farms located on 3-R watercourse shared average water shortage of 48% in this manner: tail end farms did not receive 57% of allocated water turns, middle farms missed 41% and head farms missed 34%. This analysis demonstrates that the tail farmers missed a large portion of their water allocation compared to the head and middle farmers.

Economic analysis from survey data

Farm benefits from water use

In the earlier sections two fundamental rules of the *warabandi* were analyzed and it was shown that those rules have been inadequately applied in the surveyed areas. In this section the impacts of such irregularities on the farm income are estimated. The average annual crop income considering all 1-R farms was 17150 RS/ha. This figure for 3-R farms was 11365 RS/ha (Table 2). It shows that the average farm income of 1-R farmers was 35% more than the average farm income of 3-R farmers. The reasons for higher income by 1-R farmers would be numerous, but the more likely causes were their larger shares in water allocation and lesser share in water scarcity. Another reason for their high crop income was the availability of fresh groundwater, as 36 private and two government tubewells were augmenting canal water supplies where it was insufficient to meet crop water requirements. On the other hand, 3-R farmers had no access to groundwater supplies.

The impacts of the water shortage intensity on the crop income were not always linear. A linear relation between the crop income and water shortage only occurs where the proportion of supplied water to the allocated water was above 75%. The average crop income of head and tail farms of 3-R watercourse was 12798 RS/ha and 9203 RS/ha respectively. The value for the middle farms was 12595 RS/ha. The data regarding the water allocation and the water shortage distribution shows that the tail end farmers were enjoying more water allocation, but were facing more frequent water shortages. The high water share allocated to the tail end farmers does not appear to provide any significant advantage to them, as their income remains low compared to the income of head and middle farms. Farm incomes fell progressively from head to tail of the watercourse. The only apparent reason for this trend was the distribution of water shortage between locations, as the intensity of water shortage was increasing from head to tail of the watercourse. Figure 3 shows annual net crop income of each watercourse.

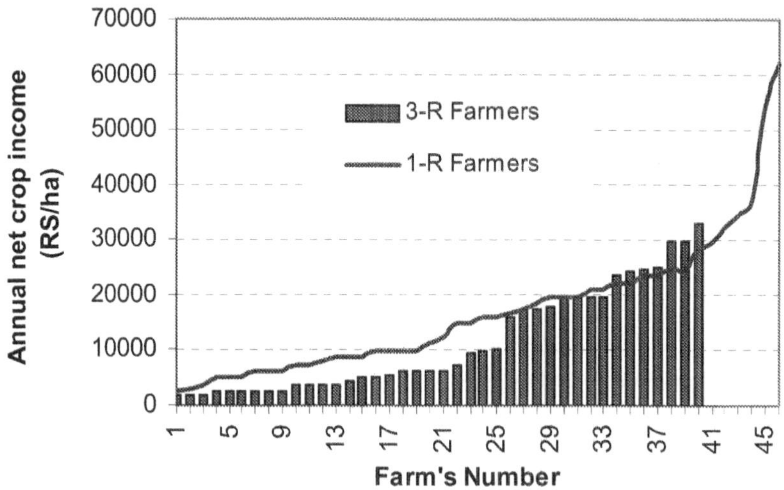

Figure 3 Annual net crop income on 1-R and 3-R (plotted in increasing order).

Economic value of unit of water

The economic value of a unit of water was calculated in order to find the efficient user(s) of irrigation water. Thirty-six private and two government-controlled tubewells were in the 1-R command area. On the other hand, not a single tubewell was found in 3-R command because of non-existence of good quality groundwater. The amount of water delivered to the farms (only 1-R farms) by the private and the government tubewells was calculated on the basis of rough data available from the farmers and the tubewell operators about the tubewells' annual operational hours. It was calculated that each tubewell was annually operating for 170 hours. The average operational cost of a single tubewell was determined as 61 rupees per hour. Each year, on average, each tubewell was adding 10300 m^3 of groundwater into the canal water supplies. This means, a total amount of 370,770 m^3 of groundwater was put into use in the data collection period, which was almost equal to the total amount of water delivered by 3-R watercourse or 30% of total amount of water delivered by 1-R watercourse. The value of a unit of water calculated for each watercourse and location are shown in Table 2. The average net value of 1 m^3 of water consumed by 1-R and 3-R farmers was respectively 3.90 rupees and 6.90 rupees. Similarly, the value of a unit of water for farms situated at different locations on the same watercourse varies significantly. The farms located at the middle of 1-

R and 3-R were more efficient in water than other farms. Therefore, if the data about water share and value of a unit of water could be obtained from a large sample of watercourses the optimal size of water share could possibly be determined.

Factors behind lower and higher farm incomes

It was calculated that the income of 3-R farmers was 35% lower than the income of 1-R farmers. Many factors are thought to be affecting 3-R farmers' income. The most likely factors appear to be: the majority of 3-R farmers fall in the 'low education' category, they have small share in canal water allocation, most of the farmers have no other source of income (less resources for optimal crop inputs), they have no other resource available for irrigation and they missed more water turns. The inverse of this may be true for 1-R farmers. Those variables were compared in order to determine the most likely variables that have affected 3-R farmers' income. The high level of farmer's education did not contribute to farm income, as the majority of farmers at the tail of 3-R watercourse were highly educated but were less efficient in water use. The majority of farmers at the tail of 1-R watercourse had other sources of income (own business, employed, etc) but were not more efficient in generating farm income as compared to those on the head and the middle of the watercourse. So, it does not appear that this factor has any large impact on farm income. However, the analysis of water shortage and water share data produces significant conclusions. Twenty-five per cent shortfall in allocated water share does not have significant impacts on the farm income. When the intensity of water shortage increases beyond 25%, farm income falls rapidly compared to the proportional increase in the water shortage. It was also found that having a larger share in canal water allocation produces higher farm income. Therefore, two critical factors, which most likely have affected 3-R farmers' income, were their smaller share of canal water supplies and the higher degree of water shortage they suffered. The impact on farm income by other factors was minor.

Concluding remarks

Analysis of the survey data showed that the fundamental principles of the *warabandi* were not fully operational in the surveyed regions. Varying rates of water allowances between watercourses and within each watercourse is one example of its ineffectiveness in the real world. Moreover, the variation in the distribution of water shortage between stakeholders particularly contributes to *warabandi's* inefficiency. The inequities in the *warabandi* affect farmers' economic and social outcomes. Farmers having lands on 3-R watercourse were double-handicapped, as they have a lower share in canal water supplies compared to 1-R farmers and have marginal quality groundwater in their locality, which is unfit for raising crops. The theoretical principles of *warabandi* do not provide equal opportunities to all farmers in generating farm benefits. The farms located on 1-R watercourse benefited from the available fresh groundwater resource. The amount of water extracted from the fresh aquifer was equal to 30% of the watercourse supplies. This extra water contributed to the farms' income but produced lower net return from a unit of water.

If improving land use efficiency is the aim of the delivery system then the *warabandi* could produce significant results with higher overall farm income at the cost of lower return per unit of water. If improving water use efficiency is the target, then the *warabandi* system will not produce satisfactory outcomes. As fresh water is becoming scarce day by day, its allocation under the existing type of irrigation system appears to cause waste of this scarce resource. Changing to a new method of water allocation, which might revolutionize the water industry in Pakistan would inevitably face some challenges in its implementation. Therefore, as a first move, it is recommended that the *warabandi* should be made operational in its true spirit before going ahead with other reforms. In addition, it is recommended that a framework that could encompass and rank different possible settings of water allocations, i.e. allocation based on gross area on tertiary canal, allocation based on sensitivity of growth stage to water shortage, allocation based on crop value, allocation to most water use efficient areas, allocation based on most water use efficient crops, allocation based on potential for loss if water is not available, allocation to the most water deficient crops, etc, should be developed as a tool for Pakistan's future water allocation strategy. The *warabandi* system, based on limited criteria, provides no incentive for conscientious efforts and improved efficiency of farmers and is the cause of loss of very large amounts of Pakistan's scarce resources of land, water and human initiative.

References

Bandaragoda, D. J. (1998). *Design and Practice of Water allocation Rules: Lessons from warabandi in Pakistan's Punjab.* Research Report No. 17. International Water Management Institute, Colombo, Sri Lanka.

Bandaragoda, D. J. and Rehman, S. (1995). *Warabandi in Pakistan's Canal Irrigation Systems: Widening gap between theory and practice.* IWMI Country Paper-Pakistan No. 7, International Water Management Institute, Colombo, Sri Lanka.

Bhatti, M. A. and Jacob, W. K. (1990). *Irrigation Allocation Problems at Tertiary level in Pakistan.* ODI-IWMI, Irrigation management Network Paper 90/3C, London: Overseas Development Institute.

Cai, X., Ringler, C. and Rosegrant, M. W. (2001). *Does Efficient Water Management Matter? Physical and Economic Efficiency of Water Use in the River Basin.* Discussion Paper No. 72. Environment and Production Technology Division. International Food Policy Research Institute, Washington DC, USA.

Chaudhary, M. A. and Robert A. Y. (1989). Irrigation water allocation institutions: Pakistan's warabandi system. In *Social, Economic and Institutional Issues in Third World Irrigation Management* (ed. R. K. Sampath and Robert, A. Y.) (*Papers in Water Policy and Management*, No. 15). London, Westview Press.

Khan, S., Tariq, R., Yuanlai, C. and Blackwell, J. (2005). Can Irrigation be Sustainable? *Agricultural Water Management.* Available online at www.sciencedirect.com since 18 August 2005 (with subscription). (In press.)

Kirpich, P. Z., Dorota Z. H. and Stuart W. S. (1999). Problems of irrigation in developing countries. *Journal of Irrigation and Drainage Engineering,* **125**(1), 1–6.

Latif, M. and Sarwar, S. (1994). Proposal for equitable water allocation for rotational irrigation in Pakistan. *Irrigation and Drainage Systems,* **8**, 35-48.

Murray-Rust, D. H., Lashari, B. and Memon, Y. (2000). *Water Distribution Equity in Sindh Province, Pakistan.* Extended Project on Farmer Managed Irrigated Agriculture under the National Drainage Program (NDP). Working Paper No. 9. International Water Management Institute, Colombo, Sri Lanka.

Nakashima, M. (2000). Water users' organization for institutional reform in Pakistan's irrigation sector. *Journal of International Development Studies,* **9**(1), 79–93.

Benchmarking in the Austrian water supply sector: objectives, development, structure and first results of the 2004 project

J. Kölbl*, H. Theuretzbacher-Fritz*, R. Neunteufel, R. Perfler** and R. Krendelsberger*****

* Graz University of Technology, Institute of Urban Water Management and Water Landscape Engineering, Stremayrgasse 10/I, A-8010 Graz, Austria
(Email: *koelbl@sww.tugraz.at*; *theuretzbacher@sww.tugraz.at*)
** University of Natural Resources and Applied Life Sciences Vienna, Institute of Sanitary Engineering and Water Pollution Control, Muthgasse 18, A-1190 Vienna, Austria
(Email: *roman.neunteufel@boku.ac.at*; *reinhard.perfler@boku.ac.at*)
*** University of Applied Sciences Wiener Neustadt, Department of Accountancy and Revision, Johannes Gutenberg-Straße 3, A-2700 Wiener Neustadt, Austria
(Email: *rafael.krendelsberger@fhwn.ac.at*)

Abstract Over the past few years a large number of benchmarking projects has been carried out in the sector of water supply all over the world. They have arisen from different motivations and pursue various objectives. They reflect specific regional structures and framework conditions. The International Water Association (IWA) has developed an overall indicator system (Alegre et al. 2000) and carried out several national field tests in order to adapt the system to practical applications. This set of performance indicators (PIs) has been used as the general basis and has further been developed within the Austrian OVGW project, which started in form of a pilot project with data of 2002. Within the next weeks, the 2004 project ("stage B") will be finished and first results can be presented now.

Keywords Austria; benchmarking; performance; water supply

Basics

What is benchmarking? One definition for benchmarking is the following. Benchmarking is a continuous und systematic measuring process which compares the performance of an enterprise with the "best in class" in order to derive measures for improvements.

Performance indicator systems and benchmarking are management instruments for internal purposes as well as for company comparisons on a regional, national and an international level (Merkel 2001). The aim is to arrange a quasi-competition that is based in Austria upon the principles of voluntary and anonymous participation.

Figure 1 shows the workflow of a benchmarking process. The first step is a comparison of performance indicators of different water undertakings. The variation of the own value from the optimum value represents a theoretical potential for improvements. However, a best possible homogeneous data collection and largest possible samples of similar participants are crucial criteria for gaining good comparability.

Causes, which are related to local characteristics, stay unchangeable. Some causes are not explainable at the moment and thus have to be analysed. For such causes, which are known concrete measures can be defined. In a next step these measures are realised in order to get a future value in the amount of the optimum level plus unchangeable parts. To complete the benchmarking process, an effectivity check-up in form of a new comparison of performance indicators is necessary.

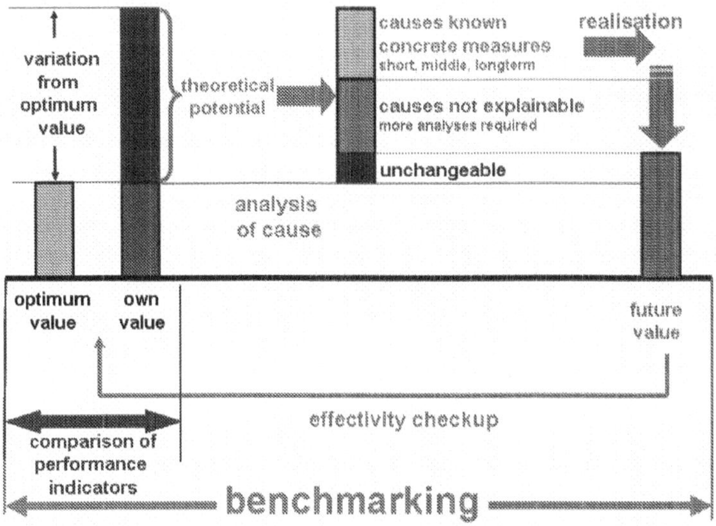

Figure 1 Benchmarking process (Schulz in Hirner & Merkel 2002, amended).

Development and project structure

Similar to many mid-European countries, the Austrian water supply sector is small structured; around 3,000 water undertakings centrally supply 8 million inhabitants in rural, urban and metropolitan areas. Based upon the international and national debates on requirements concerning the improvement of efficiency and the assurance of quality of drinking water services, the Austrian Association for Gas and Water (OVGW) has developed a mid-term strategy for setting up and carrying out benchmarking activities (Figure 2). The pilot study in the year 2002 was followed by the pilot project ("stage A") which was completed in summer 2004. Kick-off for stage B (2004 project), which will be completed in June 2006, was in autumn 2004 (OVGW 2004). Future projects will be organised in time intervals of three years.

The Austrian pilot project "Benchmarking and Best Practices of Austrian Water Supply Enterprises–Stage A" (Neunteufel et al. 2004), in which 23 water supply enterprises (from 40,000 m^3 up to 140 million m^3 supplied water per year) have participated, was launched and conducted by OVGW and largely funded by the Austrian Ministry of Agriculture, Forestry, Environment and Water Management (BMLFUW). OVGW commissioned three academic institutes to operationally run the pilot project as well as the 2004 project from an external, objective and confidential point of view: Graz University of Technology, University of Natural Resources and Applied Life Sciences Vienna and University of Applied Sciences Wiener Neustadt (Theuretzbacher-Fritz et al. 2005).

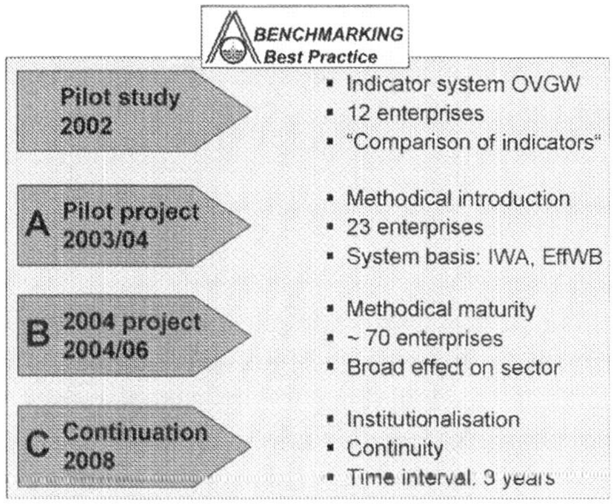

Figure 2 OVGW strategy for benchmarking water supply enterprises (Theuretzbacher-Fritz & Kölbl 2003, amended).

Some of the participating enterprises joined the OVGW benchmarking working group for project supervision and to incorporate practice needs into the indicator system. Thus, the OVGW system can be called a system from water suppliers for water suppliers.

Important for the development of the OVGW system has been the co-operation with the Bavarian (Germany) EffWB project (Rödl & Partner, 2002; Rödl & Partner, 2005; EffWB, 2003), which is also based on the IWA performance indicator system. Due to the similar water supply structure of Austria and Bavaria, a cross border co-operation was defined to develop a compatible system and to conduct trans-national comparisons. The Bavarian-Austrian co operation can be seen as an innovative application and further development of the IWA system generating the chance of large participation (200 participants and more) in these benchmarking activities (Theuretzbacher-Fritz et al. 2005).

In the 2004 project, a presentable amount of 69 water supply enterprises of different sizes (from 20,000 m^3 up to 140 million m^3 supplied water per year) and various legal structures (most of them publicly owned) has been participating. With about 330 million m^3 supplied water per year, these participants represent 45 % of the total amount of supplied water in Austria and about 75 % of supplied water by OVGW members (Figure 3a,b).

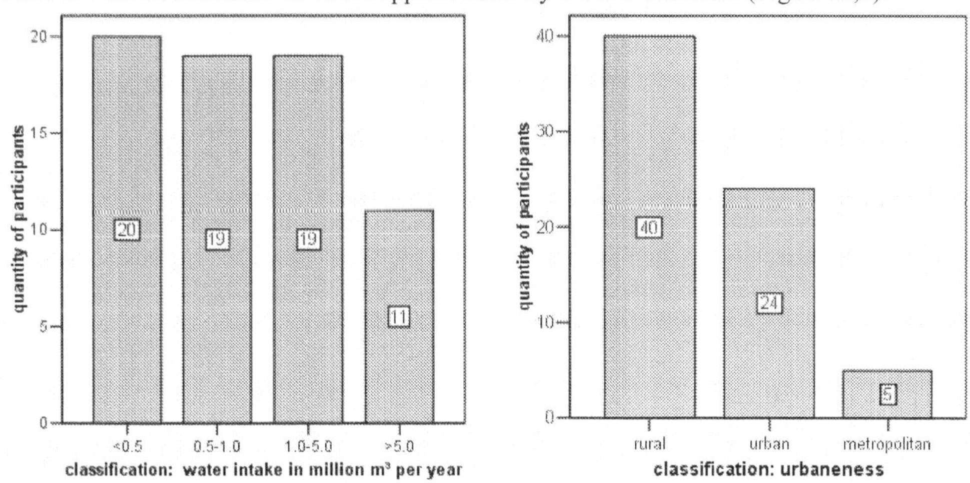

Figure 3a,b Structure of participants in the OVGW 2004 project.

The OVGW system of the 2004 project consists of 75 performance indicators calculated from 190 variables. In addition to these variables, 90 questions about task fulfilment and outsourcing, 75 questions about organisation, 30 questions about customer service and 90 facts as background information for a high comparability complete the system.

Objectives

The main aim of the pilot project was to implement a benchmarking system in Austria's drinking water management in line with the following sub-objectives. The project strongly emphasises the utilisation of benchmarking for internal purposes, serving as a controlling instrument for continual improvement ("learning from the best"). It also provides public transparency of the sector's performance and shows the sector's freewill to arrange a quasi competition, based upon the principles of voluntary and anonymous participation. Strong focus is laid on aspects of comparability (homogeneous data collection, data verification including company visits, classification of similar enterprises etc.), on a holistic approach (according to the IWA performance indicator system, Alegre et al. 2000) and on data security and confidentiality. The aim of the 2004 project is the further improvement and the use of the system to achieve a broad effect on the water sector. Due to the higher amount of participating water undertakings a high representativeness is given.

Results

Pilot project

The results of the pilot project were published in two reports, one individually for each company and a public report (Neunteufel et al. 2004). Methodical aspects had priority, whereas factual results were rather insignificant because of the low amount of participants. The results of the pilot project can be resumed as following.

- The project structure stood the test.
- The developed system of performance indicators is working.
- Successful co-operation with the Bavarian project EffWB.
- Positive experience in data acquisition.
- Company visitations are necessary for project success.
- Increase in the amount of participants is necessary for a better comparability.

2004 Project

The acquisition of data for the year 2004 and their verification was started in summer 2005 and finished in December 2005. On the following pages some examples of first results concerning supply safety, supply quality, sustainability, task fulfilment, outsourcing and efficiency are presented. In the figures reduced box plots are used. The gray boxes represent the medium 50 % of the valid data and the black bar shows the median. The numbers inside the boxes indicate the amount of used values for each box plot.

Supply safety

In comparison to many other countries in the world, Austria is very rich in water resources. Only about 2 % of the yearly regenerated groundwater is used for human consumption. Hardly any surface water need not be abstracted for drinking water purposes. A central water supply for 24 hours a day, 365 days a year in perfect quality and with enough pressure is standard.

Figure 4 presents the resources availability ratios (total water intake per year / total yearly abstraction capacity) for three different groups: water abstraction predominantly from wells (> 80 %), from natural springs (> 80 %) or mixed systems.

In the figure, the median of all three groups is below 50 %. Most of the water undertakings have a high amount of unused reserves and guarantee high supply safety. Even on days with maximum demand (peak days) the medians of used water are between 60 and 70 %. There is no significant difference between small and large enterprises.

Supply quality

Indicators of this category describe water quality, water losses, mains failures, cleaning of storage tanks, inspections etc. Water quality is generally very good, only sporadically single microbiological parameters or chemical parameters like iron or manganese did not keep the conditions of the very strict Austrian drinking water directive (BGBL 2001/304).

The median quantity of mains failures is between 4 per 100 km and year in younger networks and about 9 per 100 km and year in older networks.

Figure 5 shows the real losses per connection and day (l / connection / d) for rural, urban and metropolitan water suppliers. The increase of the median from rural to metropolitan networks goes back to more complex influences in cities (buildings, traffic and other infrastructure networks). It should be mentioned that water losses, which are in general higher in older than in younger networks, can be reduced through active leakage controls.

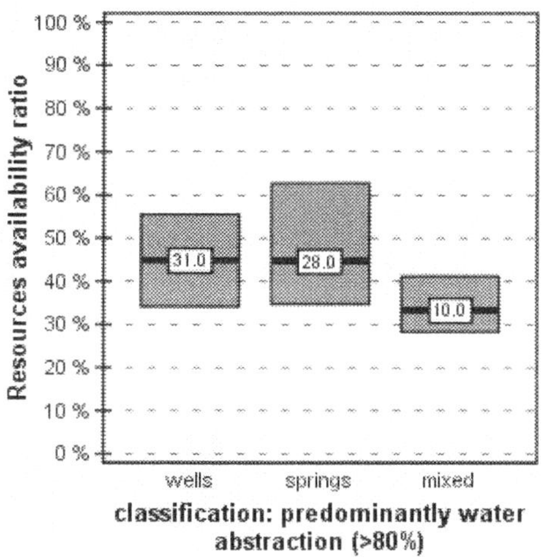

Figure 4 Resources availability ratio.

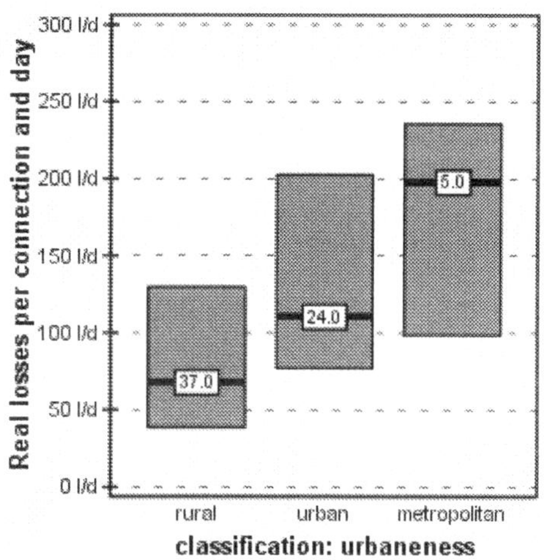

Figure 5 Real losses per connection and day.

Sustainability

This category describes long-term effective measures like water resource protection, technical and economical figures on asset management and cost recovery, different social indicators like personnel qualification and personnel training.

Most of the personnel are well qualified and many of them attend special trainings offered by OVGW (with certifications). Larger water undertakings have a higher rate of personnel with university degree. The average total cost coverage ratio (annual revenues / annual costs) is about 110 % depending on the type of bookkeeping (double bookkeeping or simple entry bookkeeping).

Figure 6 demonstrates average mains rehabilitation ratios of the last three years (% per year) for networks of different age. The higher median in the middle group may has already been a result of rehabilitation actions of the last years. Assuming an average durability of 80 years for mains, an average rehabilitation rate of 1.25 % will be necessary. A reason for the lower rates is the fact that many networks haven't reached this age yet. Nevertheless, it is important to consider money reserves for periods with more than 2 or 3 % rehabilitation per year.

Task fulfilment and outsourcing

For the interpretation of efficiency indicators it is necessary to analyse the practices of task fulfilment and outsourcing. A special point of view should be turned to in-house services, which are common in multi-utility companies as well as in municipal water supply undertakings. Especially administrative tasks like billing or personnel administration are typical in-house services. The participating companies had to state for 90 single tasks, whether these tasks were actually fulfilled in the year 2004 and if so, whether it was done by themselves, by in-house service or by an external company.

The ratio of total task fulfilment (administration and technical tasks) goes up with the increase of water intake. The reason for the higher ratio is that larger companies are on the one hand faced with a broader spectrum of duties and on the other hand they have to fulfil more tasks every year than smaller ones (for example renew or build a new storage tank). Especially some smaller companies have a lack of personnel and thus only fulfil the basic duties.

Figure 7 shows personnel costs per m³ water for enterprises with less (< 15 %), medium (15–30 %) and high (> 30 %) outsourcing ratios. The indirect correlation between personnel costs and outsourcing ratios shows the necessity of outsourcing data when interpreting personnel efficiency. Some outliers (not shown in diagram) with medium or high outsourcing ratios and also relatively high personnel costs need further analyses and will probably find efficiency potentials.

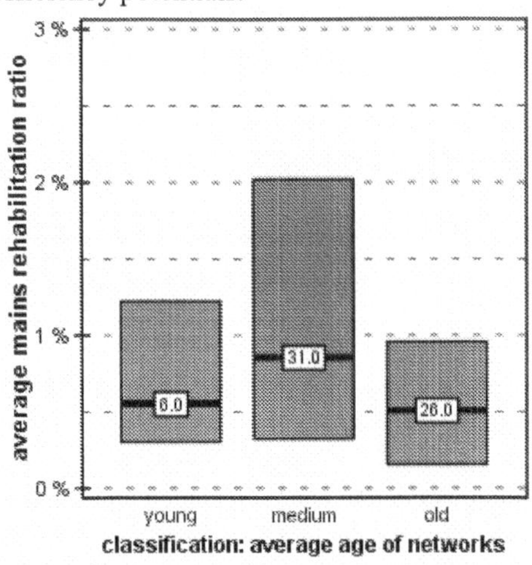

Figure 6 Average mains rehabilitation ratio.

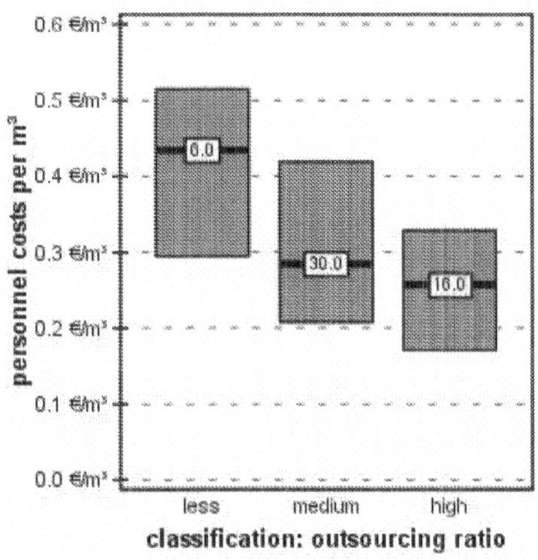

Figure 7 Personnel costs per m³.

Efficiency

There is no doubt that efficiency is one of the crucial chapters of the project. Rising efficiency shall be realised by reducing costs and / or increasing output (for example a higher task fulfilment in future), both under consideration of task fulfilment and outsourcing grades. One part of this chapter is financial analysis and a second one is personnel analysis.

Figure 8 shows the unit total costs (annual running costs + annual capital costs / authorised consumption). Because the mains length is the main cost driver, the chart is classified into three groups of different network delivery rates and into a fourth group of bulk suppliers. There is no significant difference between small and large enterprises, due to criteria like differing task fulfilment and urbaneness compensating scaling effects. Because of the almost perfect raw water quality, treatment plants are only necessary in exceptional cases. Therefore production costs and thus the total costs can be kept low in comparison to many other developed countries.

Figure 9 shows the number of full-time personnel per 100 km mains length for rural, urban and metropolitan water undertakings. Water networks become more complex with increasing urbaneness, which considers network delivery rate and service connection density. This requires more personnel. Other reasons, why metropolitan enterprises have a higher demand of personnel are higher ratios of task fulfilment and lower outsourcing ratios.

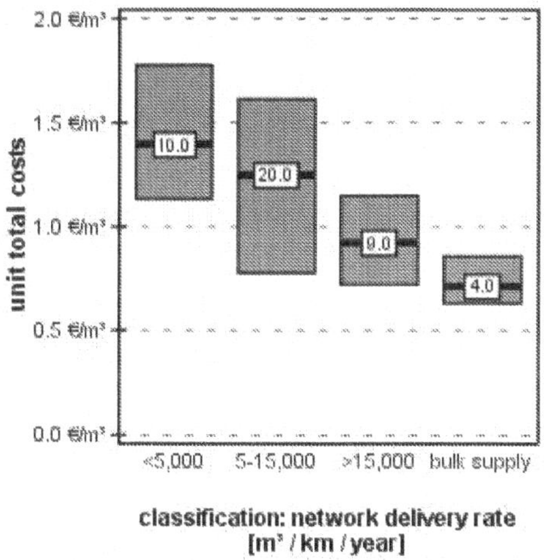

Figure 8 Total unit cost.

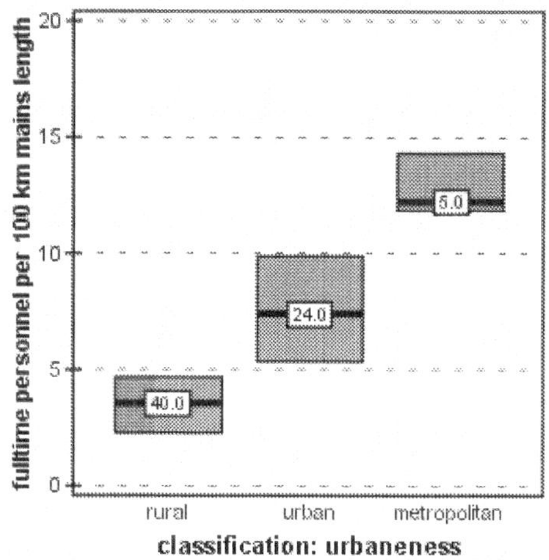

Figure 9 Personnel costs per m³.

Conclusions

With a presentable amount of 69 water supply enterprises participating in OVGW 2004 project, a high representativeness is given (about 330 million m³ water intake per year or 45 % of supplied water in Austria) and the sector's freewill to arrange a quasi competition is clearly documented.

In our experience, a voluntary and anonymous participation is an essential precondition for highly motivated participants and therefore good data quality. Company visitations by the project team proved to be necessary as well for project success.

Generally speaking, the results confirm high quality and efficiency of the Austrian water supply industry even if there are potentials for improvements. First feedback from participants was very positive, and we expect a favourable development for further OVGW benchmarking projects. After final presentation of the 2004 project in June 2006 workshops to several topics will be organised and a follow-up project on process benchmarking is planned to be started in autumn 2006.

References

Alegre, H., Hirner, W., Baptista, J.M. & Parena, R. (2000). *Performance Indicators for Water Supply Services*, Manual of Best Practice, IWA Publishing, London, UK.

BGBL 2001/304 from 21st of August 2001, *Verordnung über die Qualität von Wasser für den menschlichen Gebrauch (Trinkwasserverordnung–TWV)*, Austrian Federal Ministry of Social Security, Generations and Consumer Protection, Vienna, Austria.

Hirner, W. & Merkel, W. (2002). *Benchmarking als Mittel des organisierten Wettbewerbs*, 26. Wassertechnisches Seminar, Berichte aus Wassergüte- und Abfallwirtschaft, **173**/2002, TU München, Germany, pp. 1–30.

Merkel, W. (2001). *Performance assessment in the water industry*, IWA International Report "Kennzahlen und Benchmarking", 2. IWA World Water Congress, Berlin, Germany.

Neunteufel, R., Theuretzbacher-Fritz, H., Teix, P., Kölbl, J. & Perfler, R. (2004). *Benchmarking und Best Practices in der österreichischen Wasserversorgung–Stufe A*, Final Report of OVGW pilot project 2003/04, OVGW, Vienna, Austria.

Rödl & Partner (2002). *Effizienz- und Qualitätsuntersuchung der kommunalen Wasserversorgung in Bayern (EffWB)*, Final Report for year 2000, Rödl & Partner, Nürnberg, Germany.

Rödl & Partner (2005). *Effizienz- und Qualitätsuntersuchung der kommunalen Wasserversorgung in Bayern (EffWB) 2004*, Final Report for year 2004, Rödl & Partner, Nürnberg, Germany.

Theuretzbacher-Fritz, H. & Kölbl, J. (2003). *Benchmarking und Best Practices in der österreichischen Wasserversorgung–Methodische Überlegungen zum laufenden Forschungsprojekt*, Schriftenreihe zur Wasserwirtschaft, vol. **41**, Graz University of Technology, Graz, Austria.

Theuretzbacher-Fritz, H.,Schielein, J., Kiesl, H., Kölbl, J., Neunteufel, R. & Perfler, R. (2005). *Trans-National Water Supply Benchmarking: The Cross-Border Co-operation of The Bavarian EffWB Project and The Austrian OVGW Project*, IWA International Conference on Water Economics, Statistics and Finance, Rethymno, Greece, pp. 161–168.

Project web sites

OVGW (2004). Project homepage with detailed information and link to project report (in German): http://www.boku.ac.at/wv-bench/.

EffWB (2003). Project homepage with detailed information and link to project report (in German): http://www.effwb.de/.

Cost analysis of a night-soil treatment system as a complementary sanitation system to sewerage

H. Harada*, S. Matsui*, Y. Shimizu, T. Matsuda* and H. Utsumi***

* Graduate School of Global Environmental Studies, Kyoto University, Yoshida Honmachi Sakyo, Kyoto 606-8501, Japan (E-mail: *harada@eden.env.kyoto-u.ac.jp*; *matsui@eden.env.kyoto-u.ac.jp*; *matsuda@eden.env.kyoto-u.ac.jp*; *utsumi@eden.env.kyoto-u.ac.jp*)
** RCEQM, Graduate School of Engineering, Kyoto University, 1-2 Yumihama, Otsu 520-0811, Japan (E-mail: *shimizu@biwa.eqc.Kyoto-u.ac.jp*)

Abstract Complementary sanitation systems to sewerage are considered necessary in developing countries since the installation of sewerage requires the decades-long construction and the huge expense. The night-soil treatment system (*i.e.* the NST system) that is a non-pipe sanitation system and uses vacuum truck collection of excreta was analysed as a complimentary sanitation system to sewerage from the aspect of cost. The 33-year costs of NST systems and sewerage systems for 50,000 people were calculated in general conditions of Japan according to a household density range of 10-10,000 household/km^2. Also, the cost efficiencies to remove BOD, T-N and T-P loads were calculated. The calculated total-cost of NST systems was lower than sewerage systems at the density of less than of 1,972 household/km^2. The total-cost ratios of NST systems to sewerage systems were 1.12, 0.88, 0.37 and 0.11 at 10,000, 1,000, 100 and 10 household/ km^2, respectively. The calculated cost efficiency showed that NST systems required less costs to remove pollution loads than sewerage systems at less than 90 household/km^2 for BOD, at less than1,028 for T-N, and at less than 403 for T-P.

Keywords Cost analysis; environmental pollution; household density; night-soil treatment system; sewerage

Introduction

As of 2002, 2.6 billion people had no access to adequate sanitation (WHO and UNICEF, 2005). Treatment of excreta containing pathogens in domestic wastewater is crucial for sanitary improvement. several developing countries have planed to develop sewerage as a sanitation system. Sewerage installation, however, requires the decades-long construction and the huge expense. Complementary sanitation systems to sewerage are considered necessary.

Matsui *et al.* (2004) reported that construction of sewerage piping systems requires a large amount of money and time, a non-pipe sanitation system as is used in Japan, the night-soil treatment system (*i.e.*, the NST system), could be applied to complement sewerage systems in developing countries. The NST system employs vacuum truck collection from pit-toilets and the development of the system has dramatically contributed to sanitary improvement in Japan (Inoue, 2004; Magara, 2003). The system is used in sub-urban and rural areas of Japan, and its treatment performance is at least equal to modern sewerage at present. It has recently been utilized to recover resources in not only excreta but also other organic wastes (Morita, 2004).

One of the crucial issues to apply a sanitation system in developing countries is the cost of the system because several sewerage-development plans in various countries have already been suspended due to huge expense. Some types of sanitation systems are analysed in the relation to sewerage from the aspect of the cost (Engin and Demir, in press; Peter-Frohlich *et al.*, 2004; Berktay and Ellis, 1997). Although studies on the NST system has been done in terms of technology and environmental impact (Matsui *et al.*, 2002), little attention has been given to the cost analysis of the NST system related to the sewerage system.

This study analyses the costs of the NST system and the sewerage system in the conditions of Japan. The costs will vary considerably according to each area condition, and we focus on household density as one of the most influential area conditions. The costs of each system and cost efficiency to remove environmental pollution are calculated depending on household density.

Methodology

Analysed conditions and parameters

The components of the NST system and the sewerage system are various. The analysed conditions of NST systems and sewerage systems are shown in Table 1. One of the major differences between NST systems and sewerage systems involves the treated objectives: human excreta in NST systems, and whole domestic wastewater in sewerage systems. Excreta in the NST system are treated by the high-loading denitrification process, which is the most major process for NST systems in Japan. Rainwater dose not flow into sewerage systems in this analysis. Performance to remove pollution loads of each system is noted in a following section. Sludge in NST and sewerage systems is incinerated. Land costs are not considered in this analysis because they would greatly depend on the local conditions and be difficult to calculate the general costs.

The parameters and their values in this analysis are shown in Table 2. The values are based on general conditions in Japan. In this research, the 33-year cost is calculated for 50,000 people depending on a household density range of 10–10,000 households per km^2. The average, maximum and minimum of household density of sewerage-covered areas in Japan are 1,467, 6,892 and 41 households per km^2.

Table 1 Analysed conditions of NST systems and sewerage systems

Item	Conditions for the analysis
NST system	
Treated objectives	Human excreta
Treatment process	High-loading denitrification process
Sludge treatment	Incineration
Cost in the system	Construction cost of the plant, O&M cost of the plant, Collection cost of human excreta. No land cost.
Sewerage system	
Treatment objectives	Human excreta and greywater, not rain water
Treatment process	Secondary sewerage using activated carbon method
Sludge treatment	Incineration
Pipe system	Separated sewer and no pump station
Cost in the system	Const. cost of the plant and pipe, O&M cost of the plant and pipe. No cost of land and pump station.

Table 2 Parameters for the analysis

	Parameter	Unit	Value in this research
P	Covered population	person	50,000
Hs	Average household size	person	2.67[a]
D	Household density	household/km^2	10–10,000
$Ycal$	Time period for the cost calculation	year	33
Qe	Daily volume of collected excreta	litre/person/day	2.24[b]
Qm	Daily maximum volume of treated sewage	litre/person/day	332.2[c]
Qa	Daily average volume of treated sewage	litre/person/day	439.9[c]
Ypl	Plant duration	year	33[d]
Ypi	Pipe-infrastructure duration	year	72[d]

[a] The average value in Japan (SRTI, 2004); [b] The average value in Japan (MOE, 2004); [c] The values calculated using data from statistics of sewerage in Japan (JSWA, 2002); and [d] The values from a manual of wastewater treatment-facility development in Japan (JSWA, 2001) and the duration of NST plants is assumed to be equal to that of sewerage treatment plants.

Analysis procedure

Cost of NST systems

The total cost of an NST system is composed of construction cost and O&M cost as follows:

$$TCN = C_{NC} + C_{NOM} \quad (1), \qquad C_{NC} = IC_{NCPL} \times \frac{Ycal}{Ypl} \tag{1}$$

$$C_{NOM} = VC_{NOMPL} \times Qe \times 365 \times Ycal + VC_{NOMCL} \times Qe \times 365 \times Ycal \tag{2}$$

where TCN is the cost of a total NST system for $Ycal$ years (yen), C_{NC} is the construction cost of an NST system for $Ycal$ years (yen), C_{NOM} is the O&M cost of an NST system for $Ycal$ years (yen), IC_{NCPL} is the initial construction cost of an NST plant (yen), VC_{NOMPL} is the volumetric cost of NST-plant O&M (yen/litre), and VC_{NOMCL} is the volumetric cost of excreta collection in an NST system (yen/litre). Inflation is ignored in this analysis.

We assumed that IC_{NCPL} is related to plant scale, and VC_{NOMCL} is related to the household density in the area of an NST system. Examining these two relations, two functions for IC_{NCPL} and VC_{NOMCL} were developed. Figure 1 (left) shows the relation between the construction cost of NST plants and the plant scale based on data of six NST plants from a survey, used to satisfy the conditions needed to analyse NST plants. Figure 1 (right) shows the relation between the collection cost of excreta and the household density in an NST system based on data of 11 NST plants from a second survey. Finally, IC_{NCPL} and VC_{NOMCL} are calculated using two functions shown in Table 3. The plant scale of an NST system {S_N (kilo litre/day)} for this analysis is calculated as follows:

$$S_N = 1.2 \times Qe \times P \tag{3}$$

VC_{NOMPL} was not be related to any parameters in this analysis and it was calculated as a constant expression in Table 3 based on statistic data of the total O&M cost and the total volume collected in NST systems in Japan (MOE, 2004). Instead of land cost, the land area for an NST plant is calculated using a function from the literature as shown in Table 3.

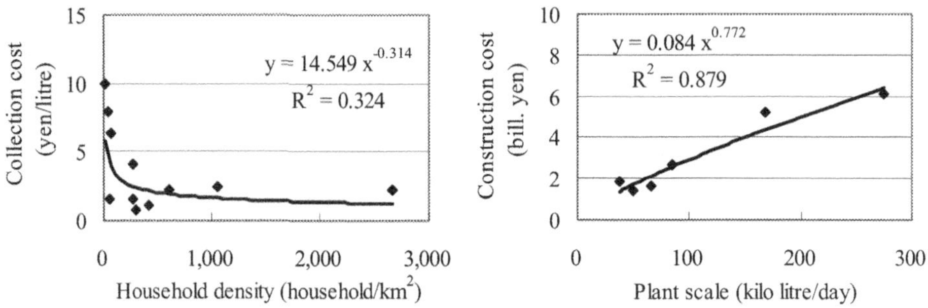

Figure 1 The relations of the NST system between initial construction cost and plant scale (left), and between excreta-collection cost and household density (right).

Table 3 Functions for the analysis of NST systems

	Calculated item	Unit	Function	Ref.
IC_{NCPL}	Initial construction cost of a NST plant	yen	$8.3851 \times 10^7 \times S_N^{0.7719}$	-
VC_{NOMPL}	Volumetric O&M cost of a NST plant	yen/litre	9.275	-
VC_{NOMCL}	Volumetric collection cost of excreta	yen/litre	$14.5489 \times D^{-0.3143}$	-
A_N	Land area of a NST plant	m^2	$1300 \times S_N^{0.477}$	a

[a] A function from JSWA (1998)

Table 4 Functions for the analysis of sewerage systems

	Calculated item	Unit	Function	Ref.
IC_{SCPL}	Initial const. cost of a sewage plant	yen	$7.7847 \times 10^6 \times Qm^{0.7206}$	a
AC_{SOMPL}	Annual O&M cost of a sewage plant	yen/year	$1.1134 \times 10^5 \times Qa^{0.7984}$	a
IC_{SCPI}	Initial const. cost of pipe infrastructure	yen	$75000 \times TL$ (m)	b
AC_{SOMPI}	Annual O&M cost of pipe infrastructure	yen/year	$80 \times TL$	b
A_S	Land area of a sewage plant	m^2	$212.0 \times Qm^{0.4864}$	a
La	Allocated pipe length to a household	m/household	$5454.4 \times D^{-0.8006}$	-

[a] Functions from JSWA (1999); and [b] Functions from JSWA (1998)

Cost of sewerage systems

The total cost of a sewerage system is composed of construction cost and O&M cost, calculated as follows:

$$TCS = C_{SC} + C_{SOM} \tag{4}$$

$$C_{SC} = IC_{SCPL} \times \frac{Ycal}{Ypl} + IC_{SCPI} \times \frac{Ycal}{Ypi} \tag{5}$$

$$C_{SOM} = AC_{SOMPL} \times Ycal + AC_{SOMPI} \times Ycal \tag{6}$$

where TCS is the total cost of a sewerage system for $Ycal$ years (yen), C_{SC} is the construction cost of a sewerage system for $Ycal$ years (yen), C_{SOM} is the O&M cost of a sewerage system for $Ycal$ years (yen), IC_{SCPL} is the initial construction cost of a sewage plant (yen), IC_{SCPI} is the initial construction cost of pipe infrastructure in a sewerage system (yen), AC_{SOMPL} is the annual O&M cost of a sewage plant (yen/year), and AC_{SOMPI} is the annual O&M cost of pipe infrastructure in a sewerage system (yen/year). Inflation is ignored in this analysis.

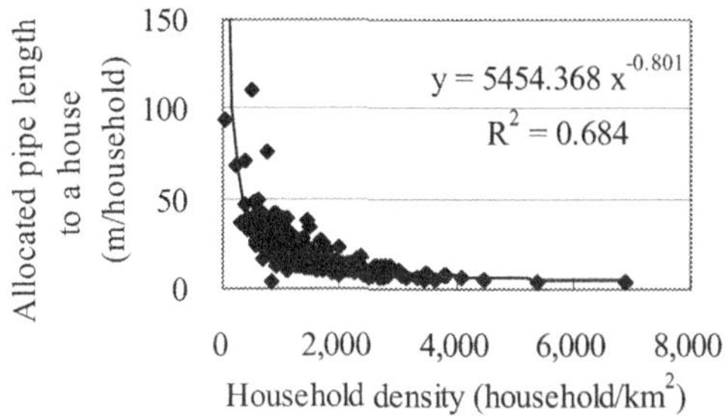

Figure 2 The relation between household density and allocated pipe length in a sewerage area.

Table 5 Pollutions removed by NST systems and Sewerage systems

	High-loading night-soil treatment systems			Secondary sewerage systems		
	Influent conc. (mg/litre)[a]	Effluent conc. (mg/litre)[a]	Pollution removed (mg/litre-influent)[b]	Influent conc. (mg/litre)[c]	Effluent conc. (mg/litre)[c]	Pollution removed (mg/litre-influent)
BOD	9500	1.9	9496.6	187.6	5.5	182.1
T-N	2700	19	2665.8	35.5	15.2	20.3
T-P	360	0.6	358.9	4.72	1.13	3.59

[a] Data from Okazaki *et al.* (2001); [b] Calculated using the dilution factor of 1.9 in the treatment process based on Okazaki *et al.* (2001); and [c] Calculated using sewerage statistics in Japan (JSEA, 2002)

The values of IC_{SCPL}, IC_{SCPI}, AC_{SOMPL} and AC_{SOMPI} are calculated using cost functions from the literatures as shown in Table 4. The functions of IC_{SCPI} and AC_{SOMPI} require the value of TL: the total length of pipe infrastructure in a sewerage system (m). Here, we assumed that pipe length allocated to a household in a sewerage area is related to the household density in the area. Using the allocated-pipe length, TL is calculated by Equation (7):

$$TL = La \times P \tag{7}$$

where La (m/household) is the average-pipe length allocated to a household in a sewerage area. Examining the relation between household density and the allocated-pipe length, a function between household density and the allocated-pipe length was developed and then TL is calculated for IC_{SCPI} and AC_{SOMPI}. Figure 2 shows the relation of household density and allocated length in 415 sewerage areas based on a statistics in Japan (JSWA, 2000). Finally, La is calculated using the function shown in Table 4. Instead of land cost, land area for a plant is calculated using a function from the literature as shown in Table 4.

Cost efficiency for the removal of pollution loads

The cost efficiency of NST systems and sewerage systems, in terms of removal of BOD, T-N and T-P loads, is calculated in Equations (8) and (9):

$$CEN_i = \frac{TCN}{PRN_i \times Qe \times P} \tag{8}$$

$$CES_i = \frac{TCS}{PRS_i \times Qa \times P \times 365 \times Ycal} \tag{9}$$

where i is BOD, T-N or T-P, CEN_i is the cost efficiency to remove the pollution load of i in an NST system (yen/mg), PRN_i is the pollution of i removed from the influent of an NST system (mg/litre-influent), CES_i is the cost efficiency to remove the pollution load of i in a sewerage system (yen/mg), PRS_i is the pollution of i removed from the influent of a sewerage system (mg/litre-influent). The values of PRN_i and PRS_i used in this analysis are shown in Table 5.

Results and discussion

Calculated costs of NST systems and sewerage systems

The calculated total costs and construction costs of NST systems and sewerage systems depending on household density are shown in Figure 3. The total cost of NST systems slightly decreased as household density increased compared to that of sewerage systems, and the total cost of NST systems were lower than sewerage systems at less than of 1,972 households per km^2. The total-cost ratios of NST systems to sewerage systems were 1.12 at 10,000 households per km^2; 1.09 at 5,000 households per km^2; 0.88 at 1,000 households per km^2; 0.37 at 100 households per km^2; and 0.11 at 10 households per km^2.

The construction cost of sewerage systems dramatically decreased in comparison with NST systems as shown in Figure 3. Itemized costs of NST systems and sewerage systems are shown in Table 6. Decreases in the total costs of the NST systems and sewerage systems were due to the costs of collection in NST systems; and the costs of pipe construction and pipe O&M.

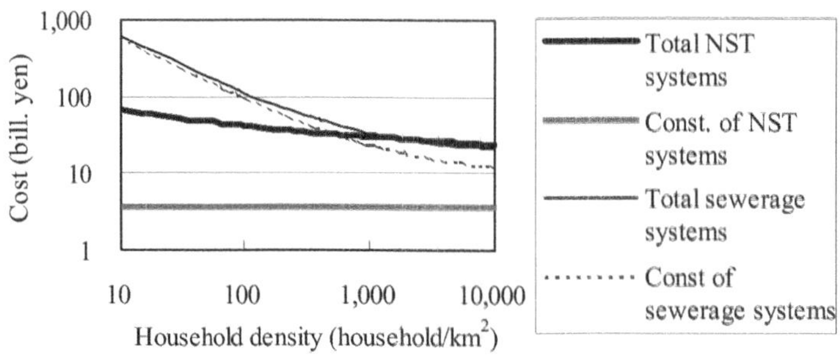

Figure 3 Calculated total costs and construction costs of NST systems and sewerage systems.

Table 6 Itemized costs of NST systems and sewerage systems

	Household density (household/km^2)							
	10		100		1,000		10,000	
	bill. yen	(%)	bill. yen	(%)	bill. yen	(%)	bill. yen	(%)
NST system								
Plant const.	3.69	(5.5)	3.69	(8.7)	3.69	(12.2)	3.69	(15.2)
Collection O&M	48.60	(72.2)	23.57	(55.7)	11.43	(37.9)	5.54	(22.9)
Plant O&M	15.02	(22.3)	15.02	(35.5)	15.02	(49.8)	15.02	(61.9)
Total	67.31	(100.0)	42.28	(100.0)	30.14	(100.0)	24.25	(100.0)
Sewerage system								
Plant const.	10.48	(1.7)	10.48	(9.2)	10.48	(30.6)	10.48	(48.5)
Pipe const.	555.70	(90.0)	87.95	(77.2)	13.92	(40.7)	2.20	(10.2)
Plant O&M	8.76	(1.4)	8.76	(7.7)	8.76	(25.6)	8.76	(40.5)
Pipe O&M	42.68	(6.9)	6.75	(5.9)	1.07	(3.1)	0.17	(0.8)
Total	617.62	(100.0)	113.95	(100.0)	34.23	(100.0)	21.61	(100.0)

In comparison with the collection cost and the pipe-O&M cost, an enormous expansion in the construction cost of the pipe infrastructure is indicated according to the decrease of household density as shown in Table 6. The total cost of the pipe construction at 10 households per km^2 was higher than at 10,000 households per km^2 by 252.6 times, and the NST-collection cost at 10 households per km^2 was higher than at 10,000 households per km^2 by 8.8 times. Land areas for a NST plant and a sewage plant were calculated as 13,500 and 27,400 m^2, respectively. NST systems will require less cost for land than sewerage systems.

It should be clear that sewerage-system installation burdens low household-density areas with huge cost derived from the development of pipe infrastructure, and NST systems sanitize excreta at lower cost than sewerage systems in such areas. The NST system should be a complement to the sewerage system in such areas for sanitary improvement from the perspective of cost.

Caluculated cost effiency for the removal of pollution loads

The calculated costs to remove one kilogram of BOD, T-N and T-P are shown in Figure 4. NST sytems required less cost to remove these three pollution loads at low household denisity, compared to sewerage systems. This means that NST sytems have higher efficiency than sewerage systems at less than 90 households per km^2 for BOD removal; at less than 1,028 for T-N removal; and at less than 403 for T-P removal.

Even though NST systems treat only human excreta and sewerage systems treat whole domestic wastewater, the removal efficiency of NST systems for three pollution loads was still higher than sewerage systems at low household density. The high efficiency of NST systems is considered due to the high concentration of these three pollutions in collected excreta in NST systems; the BOD, T-N and T-P concentrations of NST-system influent were, respectively, 50.6-, 177.6- and 76.3-fold of each of sewerage-system influent as shown in Table 5. It is likely that the NST system may become a complimentary system to sewerage systems in low household-density areas for pollution control from the perspective of cost.

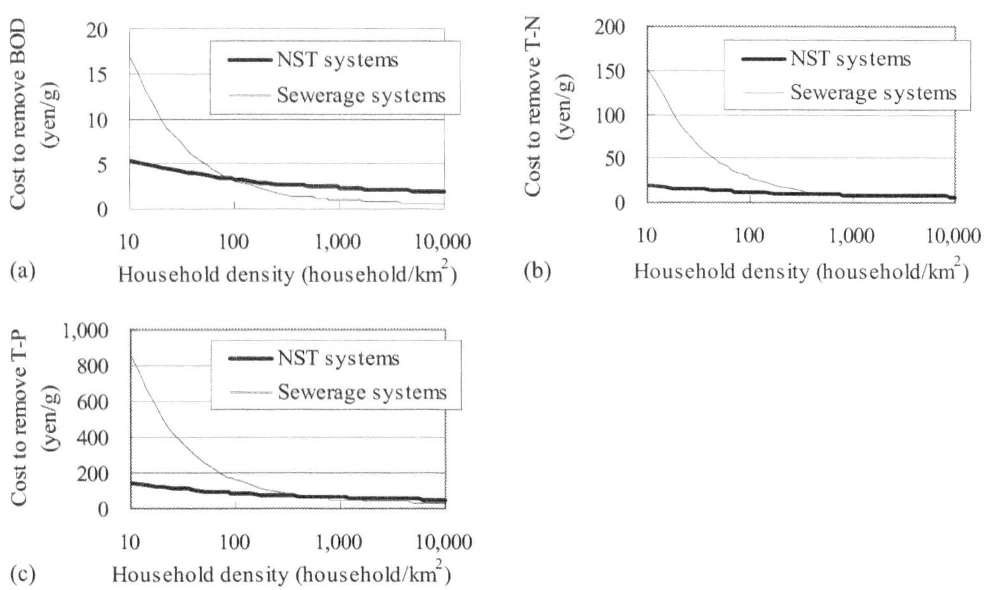

Figure 4 Cost efficiency for the removal of pollution loads depending on household density: (a) for BOD, (b) for T-N, and (c) for T-P.

Conclusions

The costs of NST systems and sewerage systems were analysed depending on household density. It is concluded that the NST system has a great possibility to become a complement to the sewerage system in low household-density areas not only for sanitary improvement but also for pollution control from the aspect of cost.

The NST system will complement the sewerage system not only in low household-density areas but also in areas where sewerage is not an affordable sanitation system. The sanitary situation and water pollution in many developing countries are urgent, and NST systems may contribute to sanitary improvement and pollution reduction in such countries by complementing sewerage systems. Although this study was based on data in Japan, NST-system costs need to be re-analysed based on data in areas where NST systems are applied. Technological feasibility and social acceptance, in addition, should be examined for the application of NST systems.

References

Berktay, A. and Ellis, K.V. (1997) Comparison of the cost of the pressurized wastewater treatment process with other established treatment processes. *Water Science and Technology*, **31** (12), 2973–2978.

Engin, G.O. and Demir, I. (in press) Cost analysis of alternative method for wastewater handling in small communities. *Journal of Environmental Management.*

Inoue, Y. (2004) History of night soil treatment plant and sludge recycling center, *Journal of Japan Waste Management Association*, **57** (261), 431–437.

Japan Sewerage Works Association (1998) *Gesuidoujigyouniokeru hiyoukouka bunseki manual* (draft). JSWA, Tokyo.

Japan Sewage Works Association (1999) *Ryuikibetsugesuidouseibisougoukeikakuchousashishin to kaisetsu*. JSWA, Tokyo.

Japan Sewage Works Association (2002) *Gesuidou toukei* (Statistics of sewerage), **57**(1), JSWA, Tokyo.

Japan Sewerage Works Association (2001) *Kouritsutekina osuisyorisisetsu seibinotameno todouhuken kousou sakutei manual* (draft). JSWA, Tokyo.

Magara, Y. (2003) Status of onsite treatment of domestic wastewater management in Japan. In *Proc. of the Johkasou session, the 3rd World Water Forum*, Kyoto, 17–30 March.

Matsui, S., Harada, H., Utsumi, H., Matsuda, T., and Shimizu, Y. (in press) Advanced sanitation with the vacuum truck collection system of human excreta. *Water and Environmental Management Series* **10**.

Matsui, Y., Yamada, M., Kawamura, K. and Tanaka, M. (2002) Life cycle inventory analysis of treatment facilities for liquid waste. *JSCE Journal of Environmental systems and engineering*, **23** (5), 19–29.

Ministry of the Environment (2004) *Ippanhaikibutusyorijittaichousakekka*. MOE, Tokyo.

Morita, A. (2004) Technology and challenges of sludge recycling center. *Journal of Japan Waste Management Association*, **57** (261), 447–450.

Okazaki, T., Shimizu, T., Morita, A. (2001) Some notes on results of functional inspection of night soil treatment facilities (4). *Bulletin of Japan Environmental Sanitation Center*, **28**, 81–97.

Peter-Frohlich, A., Kraume, I., Luck, F., Lesouef, A. and Oldenburg, M. (2003) Separate discharge and treatment of urine, faeces and greywater–demonstration project in Berlin, Germany. In *Proc. of 2nd IWA Leading-Edge Conference on Sustainability*, 8–10 November 2004, Sydney.

Statistical Research and Training Institute (2004) *Statistical handbook of Japan*. Statistics Bureau, Tokyo.

WHO and UNICEF (2005) *Water for life: making it happen*. WHO and UNICEF, Geneva.

Recent developments in pipeline inspections in Western Australia

Garth Walter*, Barry Holloway*, Gary Munns*, Shaun Brennan*, Jim Armstrong* and Cameron Waters**

* Water Corporation, PO Box 100, Leederville, Western Australia 6902, Australia
(E-mail: *garth.walter@watercorporation.com.au*; *barry.holloway@watercorporation.com.au*; *gary.munns@watercorporation.com.au*; *shaun.brennan@watercorporation.com.au*; *jim.armstrong@watercorporation.com.au*)
** PNDT, PO Box 1089, West Leederville, Western Australia 6901, Australia
(E-mail: *pndt-per@pndt.com.au*)

Abstract Water and wastewater pipelines through out the world continue to age and decline in condition. And although many of these pipelines are nearing, or have exceeded, their economic life; they still remain in service in many locations. Until recently, few cost effective Mild Steel Cement Lined (MSCL) pressure pipeline condition assessment tools have existed. Over the past seven years, the Water Corporation's research and development (R&D) program focussed on the development of a long range in-pipe CCTV inspection system that was capable of inspecting up to 3,000 metres in each direction from the point of entry within a pressurised pipeline. The ability to add other diagnostic and pipeline inspection sensors was considered, resulting in undertaking further R&D. This R&D has resulted in the integration of a new cost effective MSCL pipeline condition assessment tool. Based on a pulsed eddy current, this new tool enables the asset manager to gain an understanding of the condition of their pipeline asset through the insulating cement lining. This tool has been recently used to successfully and reliably assess the condition of wastewater pump station MSCL pressure pipelines that pass under waterways in Western Australia. The condition assessments have resulted in additional replacement and maintenance programs for the assets inspected.

Keywords: Asset; pipeline; inspection; condition assessment; refurbishment; rehabilitation

Introduction

Water and wastewater pipelines through out the world continue to age and decline in condition. And whilst many of these pipelines are nearing, or have exceeded, their economic life; they still remain in service in many locations. Until recently few cost effective Mild Steel Cement Lined (MSCL) pressure pipeline condition assessment tools have existed.

Over the past seven years, the Water Corporation's research and development (R&D) program focussed on the development of a long range in-pipe CCTV inspection system that was capable of inspecting up to 3,000 metres in each direction from the point of entry within a pressurised pipeline. The ability to add other diagnostic and pipeline inspection sensors was considered, resulting in undertaking further R&D.

The result of this R&D program is a "sophisticated remotely operated vehicle (Figures 1 and 2) (ROV) has been developed for closed circuit television (CCTV) inspection and condition assessment of pipelines for a distance up to 5 km from the point of entry" (Hodgkinson, 2002). The vehicle has been designed to enable operation within a pressurised main enabling a survey to be conducted without draining a pipeline of the water contained in the pipeline. This is an important consideration in reducing potable water wastage and reducing the inspection time by not requiring the pipeline to be drained and re-filled. The vehicle has been designed as "a platform for further sophisticated diagnostic capabilities" (Hodgkinson, 2002). The diagnostic capability of the vehicle complements the digital CCTV system to provide a comprehensive asset inspection capability.

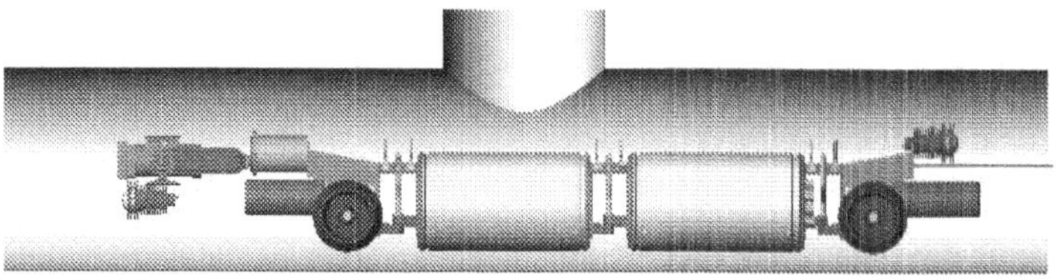

Figure 1 (left) Challenger pipeline inspection vehicle **Figure 2** (right) Challenger vehicle

A 600mm wastewater pressure pipeline near Maxwell Road in Caversham, was reported by a member of the public on Saturday, 16 October 2004. An asset condition assessment was required by the Water Corporation on the remaining sections of the pipeline.

Part of the R&D program for the Challenger vehicle has been to look at what probes were available for integration. For wall thickness there were two modalities were identified. One was with a continuous ultrasonic thickness (UT) probe. However, this modality cannot penetrate a lining, or work through uncleaned surface. The other modality was a pulsed eddy current. Preliminary testing identified this option as promising, however, the concept was not further investigated until the Caversham pressure main failure.

The failed western end of the pipeline river crossing was replaced and the condition of the eastern end needed to be determined. An investigation of available inspection techniques with the ability to assess pipe wall thickness from the inside of the pipeline which became the only access to the river crossing situated under the river and in the river bank without excavation.

A pulsed eddy current probe "monitors the decay of an eddy current pulse within the steel wall. It computes the average thickness of the metal by comparing the transient time of certain signal features with similar calibration tests" (Pilbara NDT, 2005). Figure 3 demonstrates how this average thickness is determined using the INCOTEST® pulsed eddy current probe.

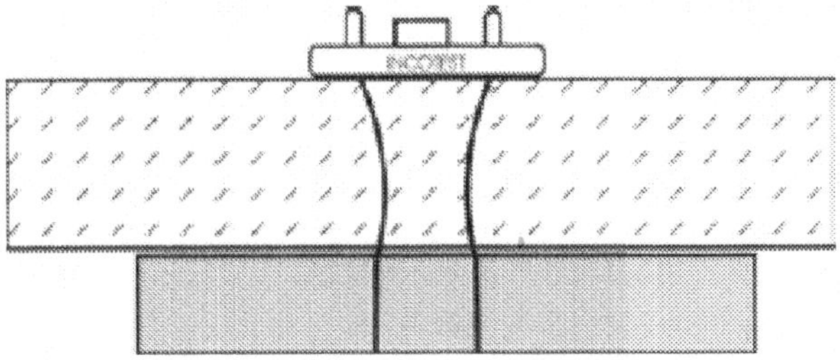

Figure 3 Area of averaging (Waters, 2005).

The pulsed eddy current probe has the potential to become an essential tool for the condition assessment of lined steel pipelines that are inaccessible externally. Where a pipeline is accessible externally an ultrasonic means of measuring steel thickness is available, reliable and cost effective. However, where the mode of access is only internally, a pulsed eddy current probe has the potential to determine a pipelines steel wall thickness and indicate the level of corrosion of that pipeline.

The resulting aims of these investigations were to:

1. To detect the reduction in steel thickness on a section of corroded pipeline using an eddy current probe and compare these measurements with those obtained using an Ultrasonic tool;
2. Determine the reliability and repeatability of a pulsed eddy current probe as a steel pipeline condition assessment technique;
3. Determine the ability of a pulsed eddy current probe to detect a failure in a Mild steel cement lined pipeline;
4. Determine the in field steel thickness of the Caversham pipeline at the eastern side of the River crossing.

Method

Aim 1, 2 and 3

- Relocate western failed pipe to workshop for assessment.
- Mark out pipe in grid pattern with approximately 100mm separation in the direction of the flow marked in eight locations circumferentially (45 degree intervals). Each ring featured 8 locations marked from 1 to 8 starting at the 12 o'clock position, rotating clockwise and stepping at 45 degree intervals.
- Selection of locations for detailed analysis.
- Test each of the locations with ultrasonic steel thickness tester. To obtain the actual steel thickness. (Krautkramer Branson USN 52 with a DA312 Probe).
- Test each of the locations with the cement lining thickness tester (Elcometer). To determine the pulsed eddy current minimum stand-off from the pipe.
- Test locations, inside the pipe with various pulsed eddy current probes, P0.25 with 0 mm stand-off and V1.0 with 25 mm and 50 mm standoff (RTD Incotest ®).

This process was performed for pipeline sections A (curved), B (straight) and C (curved).

Aim 4

- The field deployment of an inspection vehicle is required to gain field measurement of the steel thickness of pipelines. The use of special pipeline access procedures including the bypassing of flow whilst the repairs and inspections is required;
- Select appropriate pulsed eddy current probe;
- Selection of locations for detailed analysis;
- Test each of the locations with pulsed eddy current probe(RTD Incotest ®);
- Report on observed pipe wall thickness.

Results and discussion

The failed pipeline from the Caversham incident was relocated to a workshop to enable a more detailed investigation of the failed pipeline. The pipeline was separated into three sections Pipe A, B and C. These sections joined together in a form of a "gooseneck" where Pipe A was a bend from horizontal to vertical, Pipe B was the vertical riser pipe, and Pipe C was the vertical to horizontal bend.

Each of these pipes was marked out in a grid and the cement lining thickness (Elcometer), actual steel thickness (Ultrasonic Thickness tester), and the pulsed eddy current thickness (Incotest ®) was measured across this grid. A sample of these results for Pipe C is presented in Figures 4–7.

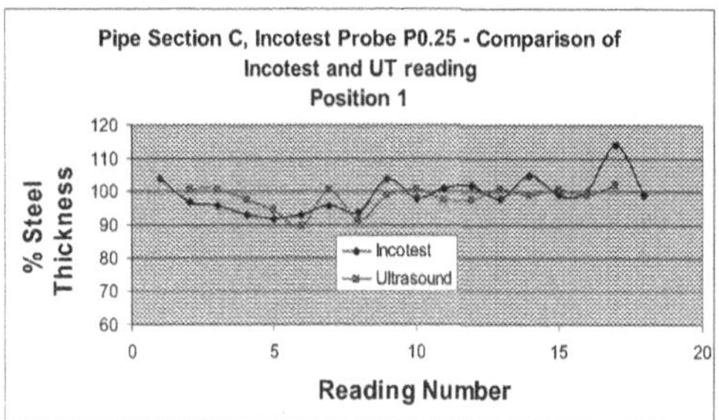

Figure 4.

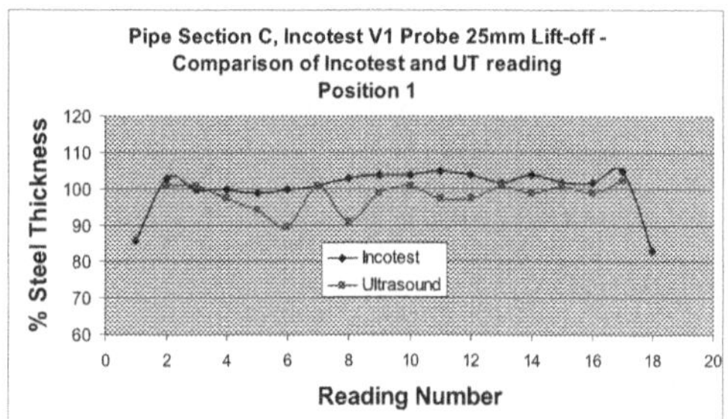

Figure 5.

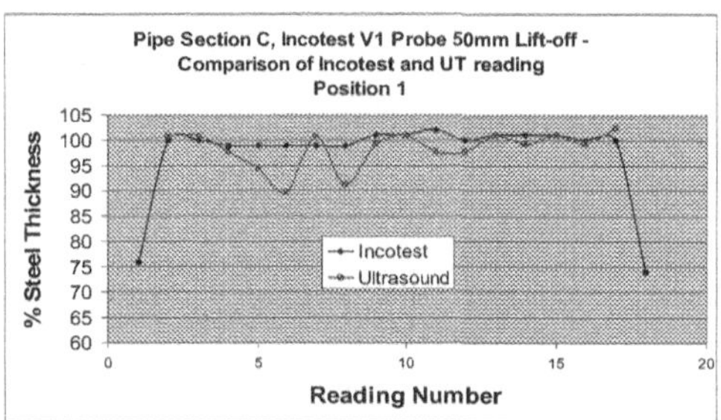

Figure 6.

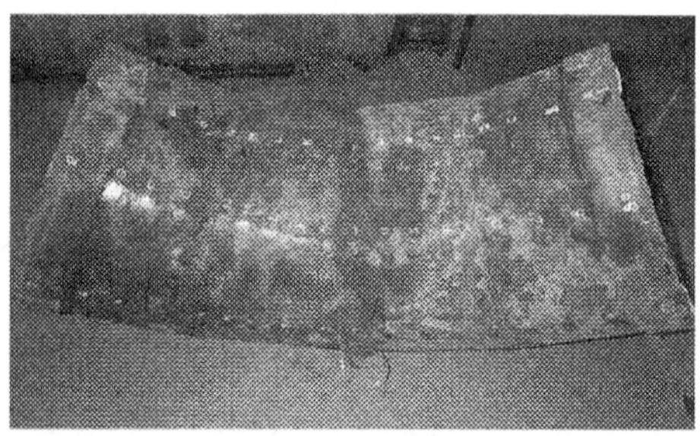

Figure 7 Photograph of Pipe C

Based on the data obtained the P0.25 Probe provided the best correlation to the steel thickness obtained from the Ultrasonic Probe. The following was also found to exist. The Pulsed Eddy Current probe averaged the steel volume within the field of excitation of the pulsed eddy currents (Figure 3). This has the effect of limiting the tools ability to find localised corrosion or "pin holes" in the pipeline. This is clearly demonstrated in Figure 6 where the V1 probe did not indicate the actual pipe wall thickness at Reading 6. Also observed was an "end-effect" of the sensor. This is clearly observed in both Figures 5 and 6 to the left and right of the graphs. This is where the probe is influenced by the end of the pipeline. This effect occurred even though the probe was well within the pipe when the reading was observed, and therefore the area of averaging was within the actual pipe shell. This end effect resulted in an abnormally low relative steel wall thickness readings near to pipe ends.

This review of the results and the selection of the P0.25 Probe gave a good correlation of the actual steel thickness, as measured by the Ultrasonic Probe, and a good repeatability of the pulsed eddy current probe readings. Figure 8 is a combined view of the surveys of the failed pipeline were then combined and the condition of the Western Caversham Pipeline Bend was than able to be determined.

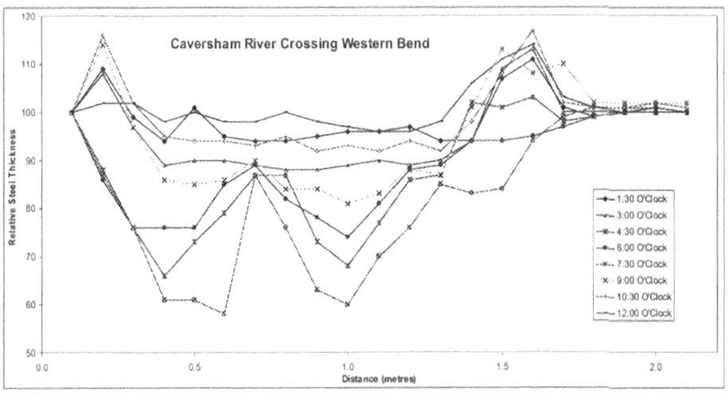

Figure 8 Pulsed Eeddy Current relative steel volume Caversham River Crossing – Western Bend

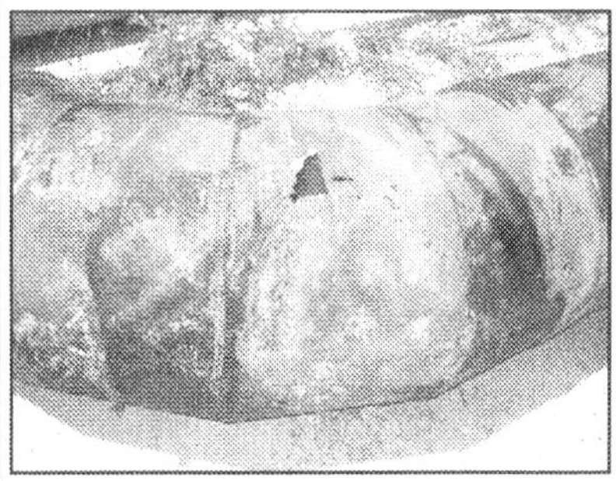

Figure 9 Western Bend failure.

A review of Figure 8 data and correlating this with actual observed failures in the pipeline, Figure 9, enabled the development of a job specific decision matrix for a survey of the remaining Caversham Pipeline Eastern Bend (Table 1). The pulsed eddy current probe was then used in field to determine the steel volume in the Eastern Bend (Figure 10).

Table 1 Caversham Pressure Pipeline – Pulsed Eddy Current Decision Matrix

Minimum average remaining wall thickness	Implication	Response	Comments
>90%	Nil or minimal wall loss	Pipe in "as new" condition: no further inspection or repairs necessary	
80%–90%	Minor wall loss	No urgency: scheduled more detailed assessment	Pulsed Eddy Current at more frequent intervals or focus points.
60%–80%	Moderate wall loss	Planned bypass, more intensive inspection of targeted areas to confirm extent of damage	Likely to require visual inspection and potentially replacement of affected pipe sections
<60%	Extensive wall loss. (Potential perforation)	Urgent: Schedule emergency bypass and repairs	

The decision Matrix, Table 1, was utilised for the assessment of the condition of the eastern pipeline bend. Based on the data obtained this matrix was found to be suitable for the assessment of the Caversham bend, but that it is sensitive to the Probe utilised, the probe's size and stand off from the pipeline, the required design pipeline steel thickness, the location of welds in the pipeline, the location of any welded band joints, other in-ground ferrous sources and any different steel permeability from different pipeline manufacturing processes.

Figure 10 demonstrates some of these variables. The initial graph peak to 128% correlates with a weld in the pipeline. The weld has shown up as a greater volume of steel in the graph. This correlated with the type of weld in the pipe.

At the 6.5 metre mark two peaks are also seen. These peaks correspond to a welded steel band joining a straight pipe and a curved pipe. As the measuring vehicle is entering the curve it is detecting a greater steel volume at the top of the pipe, and when exiting detecting a greater volume as the bottom of the pipe. Therefore the readings are also impacted by the linear distance around a bend in the pipe.

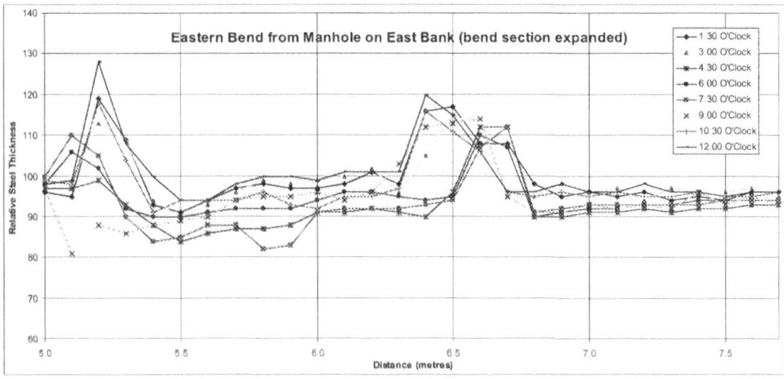

Figure 10 Pulsed Eddy Current relative steel volume Caversham River Crossing – Eastern Bend.

Conclusion

The Water Corporation's ROV system developed for in-pipe CCTV inspections has been able to be successfully modified to include a pulsed eddy current probe. This probe has demonstrated a repeatable test of volume of steel over the probe footprint. This diagnostic tool has been able to operate through the insulating cement based lining inside Mild Steel Cement Lined pipelines, which had not been performed previously in the Water Corporation.

This system is now an essential component of the Water Corporation's condition assessment armoury. When used in conjunction with other decision making tools the pulsed eddy current probe can add real value to the decision making process. Any investigation results obtained can be influenced by probe size, weld effect, welded bands, heat affected zones from welds, other in-ground steel sources, and different steel permeability from different manufacturing processes. The assessment of results from a pulsed eddy current probe will need to be interpreted on a case-by-case basis.

As a result of these investigations, the pulsed eddy current probe was able to provide an indication of the pipeline condition and extent of corrosion, and when used in conjunction with the Decision Matrix, could predict a potential for pipeline failure. The Caversham pipeline bends were replaced, rewrapped and partly concrete encased between 25 October and 7 November 2004. The concrete encased length of pipe under the river and the pipeline bends were then re-lined with a cured-in-place liner to further extend the life of the asset.

References

Hodgkinson, A. (2002). Low Cost Solutions For Condition Assessment, Maintenance And Renovation Of Large Diameter Pipelines. IWA 2002 Conference.

Hodgkinson, A. (2000). Low Cost Solutions for Managing Large Diameter Pipelines - High-tech Systems for Assessment, Maintenance and Refurbishment. 18th International Conference & Exhibition on Trenchless Technology.

Pilbara NDT (2005). http://www.pndt.com.au/rtd_incotest.htm

Waters, C. (2005). , RTD – Incotest – For the Detection of Corrosion Under Insulation. http://www.pndt.com.au/files/brochure/15_INCOTESTTheory.pdf.

A framework for enhancing the effectiveness of water management strategies

D. Marsh* and D. Sharma**

* Institute for Water and Environmental Resources Management, Faculty of Engineering, University of Technology, Sydney, Broadway NSW 2007, Australia
(E-mail: *Debborah.Marsh@uts.edu.au*)
** Energy Planning and Policy Program, Faculty of Engineering, University of Technology, Sydney, Broadway NSW 2007, Australia (E-mail: *Deepak.Sharma@uts.edu.au*)

Abstract Water management strategies are changing in Australia, as the industry is being reformed. The design of these strategies appears to be taking place without consideration of the link that may exist between water and other infrastructure industries, such as electricity. This paper argues that consideration of such link is important because of the mutual interdependencies between the water and electricity industries. This paper presents a framework for accommodating such link and demonstrates its usefulness in terms of designing effective water (and electricity) management strategies. This framework is based on input–output analysis for the New South Wales (NSW) economy. Preliminary results indicate that demand for water by the various electricity sectors vary significantly and should be taken into consideration in decisions over investment in future electricity generation. The results also quantify the electricity demand for the water sectors. Whilst the focus is NSW, the framework and messages are equally relevant for other regions where reforms are underway or are being contemplated.

Keywords Input–output analysis; policy reform; water–energy nexus

Introduction

Water management strategies the world over are changing significantly, in response to industry reforms and increasing uncertainty over future availability of water resources. In Australia, internal reforms in the 1980s and wider microeconomic reform programs initiated by the Federal Government in the mid-1990s have resulted in significant modifications to the structure, ownership and regulatory arrangements of the industry. In terms of structure, vertically-integrated utilities were functionally unbundled, such as the separation of bulk and retail water supply services. Ownership shifted towards the private sector as many utilities were corporatised and private contracts were awarded. Independent control was established over price, service delivery, water quality and environmental regulation. Policies to allocate water for environmental flows, to separate land and water entitlements, and to establish water markets to facilitate trading of entitlements were also introduced (Marsh and Sharma 2003). In 2004, the reform programs were re-invigorated by the Council of Australian Governments (COAG) via the Intergovernmental Agreement on a National Water Initiative (NWI). It was believed that additional benefits could be realised by adopting a "...nationally-compatible, market, regulatory and planning based system of managing surface and groundwater resources…" (COAG 2004).

Marsh and Sharma (2005) contend that the links between water and other infrastructure industries, such as energy, have largely been ignored in the design of water industry reforms. A possible explanation is the 'compartmentalisation' of expertise on water and energy issues that have resulted in reform of the industries occurring in isolation of each other. There is, however, growing evidence that the links between water and energy - termed the water–energy nexus in this paper - has real implications for the efficacy of water (and energy) management strategies, as reported elsewhere: "…water planners at the federal, state, and local levels have largely failed to consider the energy implications of their decisions…the State [of California] appears to not be consciously managing its rapidly evolving water and energy policies in a coherent manner (Cohen *et al.* 2004)…there is acknowledgement of the potential synergies in water and energy management that have largely been neglected, as well as recognition that existing national policies to not to any substantial degree link the infrastructure systems…" (Hagebro and Cederwell 2003). What appears to be lacking is an informed understanding of the nature of the nexus and policy tools to assist decision makers develop more integrated water and energy strategies.

This paper provides a discussion on the nature of this nexus and reviews existing water–energy (electricity) models. An integrated water–energy framework, based on input–output analysis, is also presented. While the focus of this study is New South Wales, the framework and messages are equally relevant for other regions where reforms are underway or are being contemplated.

Nature of the nexus

Increasing pressure on energy and water resources around the world has culminated in an emergence of water–energy nexus issues, providing evidence that strong physical links exist between these two resources. A previous publication by the authors (Marsh and Sharma 2005) classified these links into upstream, transportation and downstream categories, as illustrated in Figure 1 below.

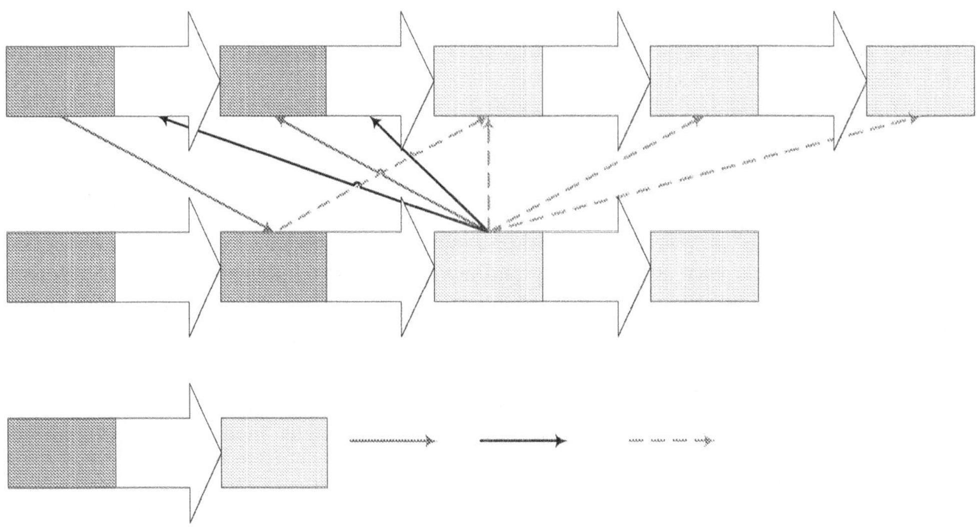

Figure 1 Links between the electricity and water industries.

In the upstream category, two links are apparent. Firstly, raw water is required for cooling in thermal power stations and for hydroelectricity production. This water may also be required by other users, such as irrigation and river ecosystems, that may need to compete for the water, particularly during dry periods. Secondly, bulk water supply requires electricity. This link becomes more apparent when supply options such as desalination are considered by water utilities.

In the transportation category, electricity is required to extract groundwater, and secondly, to convey surface water. Other energy sources, such as diesel, are also used for this purpose, particularly in developing countries. Inadequate provision of water and electricity - which is not uncommon in developing countries - raises further issues. For example, inadequate mains water may be supplemented by tank water, requiring fuel for transportation. If this tank water is stored in underground wells on site, additional electricity (or fuel if electricity supply is also inadequate) would be required for pumping.

Four links are identified in the downstream category. Firstly, retail electricity is required for retail water treatment and supply. Electricity disruptions would have serious public health ramifications. To safeguard against this some water utilities have considered installing microhydro turbines and other electricity generation technologies (James *et al.* 2002), representing the second link. Thirdly, consumption patterns of water end users influence demand for retail electricity, a common example being domestic hot water. Lastly, electricity is required for wastewater treatment and increasingly, energy is being recovered from wastewater processes.

The forgoing examples highlight the strong physical links between water and electricity. A review of recent literature on the subject, however, has indicated that the nature of the link extends beyond the physical dimension, to include economic, regulatory and social issues. In terms of economics, several models have been developed that demonstrate the link between the price of one resource and the consumption of the other resource. For example, electricity prices were shown to improve groundwater productivity in rural India (Kumar 2003) and reduce residential water consumption in Copenhagen, Denmark (Hansen 1996). In the USA, water prices were used as a tool to encourage surface or groundwater consumption, which in return influenced energy consumption (Schuck and Green 2002). In terms of regulation, a lack of regulatory control or even tightening of environmental regulation may induce consumption of one or both resources. For example, lack of regulation over groundwater withdrawals in India has resulted in excessive use of electricity and exploitation of groundwater sources (Kumar 2003). In contrast, tightening of European environmental policies has meant water utilities are required to adhere to more stringent environmental standards, met by more energy-intensive treatment options (Zakkour *et al.* 2002). In terms of social issues, reform of the industries may challenge social norms surrounding water ownership and the right to free or heavily subsidised water or electricity (Kumar 2003).

Status of existing research

Understanding of the nature of the nexus is being facilitated by the development of several water–energy models, a few of which have been mentioned in the preceding section. A closer examination of these models, summarised in Table 1, indicated that many are limited to one industry (e.g. agriculture in models 2, 3, 4 and 5) or one function within an industry (eg wastewater treatment in model 8) and would provide useful tools to assess the interaction between water and electricity at localized levels (Marsh and Sharma 2006). Such an approach, however, may limit the usefulness of the models for policy analysis, because the interactions between these industries and the wider economy are not satisfactorily captured.

An integrated water–energy model, such as one based on input–output (I–O) analysis, would offer such an opportunity and provide policy makers in NSW with a tool to develop more integrated water and energy policies in the future.

Table 1 Summary of existing water–energy models

No, Region	Category	Main objective	Reference
Upstream			
1 USA	Bulk water supply	Estimate energy use for various water sources in urban water cycle	(Cohen *et al.* 2004)
2 USA	Competing uses	Estimate energy impact of irrigation diversions upstream of a hydropower plant	(Cohen *et al.* 2004)
Transportation			
3 India	Water extraction	Analyse impact of electricity pricing on groundwater productivity	(Kumar 2003)
4 USA	Water transfers	Evaluate energy required for three scenarios for water formerly used in retired agricultural lands	(Cohen *et al.* 2004)
5 USA	Water transfers	Examine water price as a tool to encourage surface or groundwater use depending on availability, and the energy implications	(Schuck *et al.* 2002)
Downstream			
6 USA	End users	Estimate energy use for different water conservation options in a commercial or residential building.	(deMonsabert and Liner 1998)
7 Denmark	End users	Estimate impacts of water and energy prices on residential water demand.	(Hansen 1996)
8 Austria	Wastewater treatment	Evaluate energy demand of activated sludge plants that also recover energy.	(Nowak 2003)

Proposed framework

An I–O model is useful for analysing water–energy nexus issues at an economy-wide level, because it recognises the interdependence between electricity and water industries and other sectors in an economy. For example, electricity and water are direct inputs to production sectors, which in turn produce goods and services for end users. End users, therefore, not only consume electricity and water directly themselves, but indirectly through electricity and water embedded in the output of production sectors. These direct and indirect interactions are effectively captured by the input output framework, which is described in more detail below.

Basic input output framework

The I–O table and the coefficients derived from it form the basis of the technique. Essentially, the table depicts three elements of an economic system (see Figure 2): the inter-industry table shows the flow of goods and services between production sectors (outputs form inputs for other sectors and as such is termed intermediate demand); value added refers to earnings of factors of production, such as capital and land and may include imports; and final demand refers to consumption by end users, such as households and government. The following equation describes an economy with n sectors:

$$X_i = \sum_{j=1}^{n} x_{ij} + Y_i = \sum_{j=1}^{n} a_{ij} X_j + Y_i \qquad (1)$$

Outputs to sector j / Inputs from sector i	**Intermediate demand**	**Final demand**	**Total output**
Production sectors	Inter-industry table (x_{ij})	(Y_i)	(X_i)
Value added	(V_j)	GNP	
Total outlay	(X_j)		

Figure 2 Basic structure of an input–output table.

where X_i is the total output from sectors $i = 1$ to n; X_j is the total outlay from sectors $j = 1$ to n; x_{ij} is the interindustry flow from sector i to j; Y_i is final demand for sector i outputs; V_j is value added by sector j; and $a_{ij} = x_{ij}/X_j$ is the direct demand of sector i per unit output of sector j, otherwise referred to as the technical coefficient. In matrix form Equation (1) becomes:

$$X = AX + Y \ldots \Rightarrow \ldots X = (I - A)^{-1} Y \qquad (2)$$

where I is an n x n identity matrix and $(I - A)^{-1}$ (commonly referred to as the Leontief inverse) is the direct and indirect demand for sector i outputs required to meet a unit of final demand for sector j outputs. Direct and indirect demand is also referred to as total demand.

Equation (2) is considered demand-driven in that total output (X) is determined by final demand (Y). The basic assumption of this model is that technical coefficients, a_{ij}, in the A matrix are fixed, ie there is a fixed relationship between the inputs and outputs of the inter-industry table. The economy is assumed to operate under constant returns to scale. Further information about assumptions may be found in Miller and Blair (1985).

Monetary values are commonly used in the input–output table, obscuring the fact that it is actually a 'quantity' model. That is, flows of goods and services may be represented in mixed physical quantities, such as MLs of water or PJs of energy. Where both physical and monetary values are used, the model is considered a 'hybrid'. In this case, X_j comprises of mixed units and cannot be summed. Technical coefficients are therefore calculated using X_i, ie $a_{ij} = x_{ij}/X_i$. In addition to the quantity model, two other equations are required to fully represent an economic system:

$$P = (I - A')^{-1} V \qquad (3)$$

$$P'Y = V'X \qquad (4)$$

Equation (3) represents the price model, where P is a n x 1 vector of unit prices, such as the unit price of a ML of water or a PJ of electricity. If the quantity model comprises of monetary values then the corresponding unit price is equal to 1. The price model may be used to asses the impact of a change in value added costs or a change in technical coefficients on unit prices (P). Equation (4) represents the income model and is derived from Equations (2) and (3). The income model ensures that the value of final demand is equal to total value added.

Natural resources as 'value added' inputs

Value-added, as described previously, refers to the earnings of factors of production such as capital and land that are required by the production sectors to produce their output. Duchin (2004) suggests including natural resources, such as minerals and raw water in the I–O table, in order to acknowledge the contribution of the environment to the economy, rather than considering them 'free gifts of nature'. Adopting this approach, Equations (2)–(4) may be extended to include k natural resource factors of production and their factor prices:

$$X = (I - A)^{-1} Y \tag{2}$$

$$FX = f \tag{2a}$$

$$P = (I - A')^{-1} F'\pi \tag{3a}$$

$$P'Y = \pi' FX \tag{4a}$$

where F is a $k \times n$ matrix of natural resource factor inputs per unit of output in physical units, f is a k vector of total factor use in physical units and π is a k vector of factor prices. Equation (2a) allows one to calculate the natural resources required to satisfy final demand while Equation (3a) allows one to disaggregate unit prices P into portions paid directly and indirectly to each natural resource factor of production.

Water–energy model: a case study of NSW

The water–energy model presented in this paper is based on a 106-sector input output table for the NSW economy for 2000–01. The table was aggregated into thirty-two production sectors, in order to match the level of detail available in water and energy data. The model comprises of three water sectors (irrigation and drainage water providers, bulk and retail water suppliers, and wastewater service providers), six electricity generation sectors (coal fired, combined cycle, cogeneration, gas turbine, hydro and other renewables) and four other energy sectors. There are eleven 'value added' inputs, including surface water, groundwater, instream use, raw coal, six energy import categories, non-energy imports and an 'other value added' category. Surface and groundwater refer to water that has been directly extracted from the environment by production sectors, whilst instream use refers to water that is used *in situ* (e.g. hydro). Water and energy are in physical units, with the remaining in dollar terms.

Preliminary results

Preliminary results presented in this paper focus on the quantity model, in particular the demand for electricity by water sectors and the demand for water by electricity sectors.

Demand for energy by water sectors

Table 2 shows aggregated direct and total demand for electricity by the water sectors. Demand is highest for bulk and retail water suppliers, followed closely by wastewater service providers. As can be expected, it is much lower for irrigation and drainage water providers, whose demand would be confined to extracting groundwater or delivering water to customers in non-gravity systems. Demand for the energy value added inputs of raw coal and energy imports are insignificant; therefore the results have been omitted.

Table 2 Aggregate electricity demand by the water sectors (MWh/ML)

Water sector	Direct demand	Total demand
Irrigation and drainage water providers	0.0074	0.0096
Bulk and retail water suppliers	0.3167	0.3817
Wastewater service providers	0.2581	0.3458

Demand for water by electricity sectors

Table 3 below shows aggregated total demand for water sector outputs and water 'value added' inputs. In terms of water sector outputs, reliance is much higher for combined cycle, compared with the other electricity sectors. Coal fired is second highest, followed by hydro, other renewables, gas turbine and cogeneration. A possible explanation is the high use of water sector outputs for process water by combined cycle, whereas use by other generation sectors is predominantly for potable water consumption. Results for water value added inputs are significantly higher. Hydro, as can be expected is the largest consumer, particularly for insteam use. Coal fired and combined cycles are also high, compared with the remaining sectors. When instream use is excluded, the figures drop significantly, indicating that although electricity sectors other than hydro do not directly use water instream (not apparent in the results below) there is an indirect demand for such inputs, owing to inter-industry transactions. Overall, apart from the very high demand for instream use by hydro, combined cycle and coal fired are the largest consumers of water. The most efficient technologies are gas turbines, followed by cogeneration. At a time of increasing uncertainty over availability of future water resources in NSW, these results would assist energy planners with decisions over future base load capacity.

Table 3 Total water demand by electricity sectors (ML/PJ)

Electricity sector	Water sector	Water value added (including instream)	Water value added (excluding instream)
Coal fired	21.5	190.2	138.3
Combined cycle	141.2	155.0	120.6
Cogeneration	8.9	39.8	12.7
Gas turbine	9.2	35.3	11.0
Hydro	13.3	315,591.3	18.4
Other renewables	13.1	56.2	18.3

Conclusions and further work

The water (and electricity) industry has undergone significant reform over the last two decades. These reforms typify the view that the industry is separate from the economy at large, resulting in limited consideration of the links between water and other infrastructure industries, such as energy. Implications of the nexus, however, are already being felt the world over, and models to quantify the links between water and energy are being developed. These models focus on one industry or a sector within an industry; there are no models that link these industries to the economy at large. In order to address this shortcoming, the model presented in this paper (based on input output analysis) offers a framework that considers water and energy in the wider economy. Preliminary results indicate that the bulk and retail water sectors are the highest electricity users, followed closely by wastewater service providers. Water demand by the electricity sectors varies significantly and should be considered in decisions over investment in future electricity generation. As part of further development, the price model may be used to determine the impact of changes to the cost of water and energy value added inputs to the price of sector outputs. In addition, development of a 'supply side' quantity model may assist in examining the relationship between changes to the supply of value added inputs (e.g. a shortage of surface water) and economic output. Such understanding would benefit the development of future water-management strategies.

References

COAG (2004). *Intergovernmental Agreement on a National Water Initiative*, accessed 27 July 2005, http://www.nwc.gov.au/NWI/index.cfm.

Cohen, R., Nelson, B. and Wolff, G. (2004). *Energy down the drain, the hidden costs of California's water supply*. NRDC and Pacific Institute.

deMonsabert, S. and Liner, B.L. (1998). Integrated energy and water conservation modeling. *Journal of Energy Engineering-ASCE*, **124** (1), 1–19.

Duchin, F. 2004, *Input–output economics and material flows*. Report Number 0424, Rensselaer Polytechnic Institute, New York.

Hagebro, C. and Cederwell, K. (2003). Workshop 7 (synthesis): integrating the water and energy sectors. *Water Science and Technology*, **47** (6), 189–191.

Hansen, L.G. (1996). Water and energy price impacts on residential water demand in Copenhagen. *Land Economics*, **72** (1), 66–79.

James, K., Campbell, S.L. and Godlove, C.E. (2002). *Watergy: Taking Advantage of Untapped Energy and Water Efficiency Opportunities in Municipal Water Systems*. Alliance to Save Energy.

Kumar, M.D. (2003). Impact of electricity prices and volumetric water allocation on energy and groundwater demand management: analysis from Western India. *Energy Policy*. (In press.)

Marsh, D. and Sharma, D. (2003). Water industry reform: some performance issues. *Global Developments in Water Industry Performance Benchmarking*. Office of Water Regulation, Perth, Western Australia.

Marsh, D. and Sharma, D. (2005). Water–energy nexus: identifying the issues. *Environmental Research Event*, Hobart, Tasmania.

Marsh, D. and Sharma, D. (2006). Water–enery nexus: a review of existing models. *1st Australian Young Water Professionals Conference*, University of New South Wales, Sydney.

Miller, R.E. and Blair, P.D. (1985). *Input–output Analysis: Foundations and Extensions*. Prentice-Hall, New Jersey.

Nowak, O. (2003). Benchmarks for the energy demand of nutrient removal plants. *Water Science and Technology*, **47** (12), 125–132.

Schuck, E.C. and Green, G.P. (2002). Supply-based water pricing in a conjunctive use system: implications for resource and energy use. *Resource and Energy Economics*, **24** (3), 175–192.

Zakkour, P.D. *et al.* (2002). Developing a sustainable energy strategy for a water utility. Part I: A review of the UK legislative framework. *Journal of Environmental Management*, 66 (2), 105–114.

Sludge management and waste treatment

Ultrasound sludge disintegration for enhanced biogas production and sludge solids destruction

R.J. Xie*, Y.J. Xing*, Z. Yang, Yahya A Ghani**, K.E. Ooi** and S.W. Ng****

*Department of Water Resources Management, Centre for Advanced Water Technology, Innovation Centre, Blk 2, Unit 241, 18 Nanyang Drive, Singapore 637723
(E-mail: *rjxie@cawt.sui.com.sg*)
**Water Reclamation (Plants) Department, the Public Utilities Board, 40 Scotts Road, #15-00 Environment Building, Singapore 228231

Abstract Ultrasound has been demonstrated to disintegrate sludge solids and enhance anaerobic digestion. A full-scale ultrasonic sludge disintegration system was built and tested at the Ulu Pandan Water Reclamation Plant, Singapore. Two 5,000 m^3 egg-shaped digesters were involved and each was fed up to 200 m^3 d^{-1} mixed primary and thickened secondary activated sludge. Operation conditions were kept the same except one system (test) with and the other without (control) the ultrasonic apparatus to pre-treat the sludge feed. In comparison with the control, the nine-month full-scale study proved that ultrasound pretreatment of the sludge increased the daily biogas production up to 45%. There were no significant differences in compositions of the biogas from the two digesters while gases of minor components increased with time at late stage. Echoing the theoretical calculation, sludge analytical results showed increases in sludge solids removal between 25 and 35%.

Keywords Anaerobic sludge digestion; hydraulic retention time; sludge solids; egg-shaped digester; methane

Introduction

Efforts have been made from various aspects to reduce sludge production in wastewater treatment plants, of which improving anaerobic sludge digestion while recovering biogas is encouraging. A variety of technologies (Müller *et al.*, 2004; Tilche *et al.*, 2006) have been reported to improve sludge properties (e.g., digestibility) and reduce sludge solids for disposal. These include mechanic-mills (Müller, 2000; Lehne *et al* 2001), physical-ultrasound irradiation (Clark and Nujjoo, 2000; STOWA, 2006; Neczaj and Lach, 2003) and high-pressure colliding (Nah *et al.*, 2000), electricity shock or electrolysis (Jung *et al.*, 2001; Joyce *et al.*, 2003) and chemical-ozone dosing (Weemaes *et al.*, 2000). Of these technologies, ultrasound sludge disintegration is the promising and consumes the least energy (ATV-DVWK, 2006; Müller *et al.*, 2004).

Numerous ultrasound disintegration plants have been constructed around the world aiming at reducing sludge production. Reports from these plants have shown that sludge digestibility and methane yields could be enhanced up to 55% while total sludge solids destruction increased by as much as 40% as a result of ultrasonic pre-treatment of the feed to anaerobic digesters. While such effect is remarkable, the practicability of this technology for specific applications in the tropical context has yet to be demonstrated and its economic viability assessed. With several factors controlling the effectiveness of ultrasound-induced disintegration, knowledge of optimal operation conditions is lacking in cost-effectiveness and maximum efficacy of the ultrasound technology.

This study was undertaken to investigate the technical viability of a selected ultrasonic system to pre-treat sludge solids for anaerobic digestion enhancement and thereby to increase biogas production and decrease sludge biomass volume for disposal.

Materials and methods

The ultrasonic sludge disintegration facility at the capacity of 200 m^3 d^{-1} was installed at the Ulu Pandan Water Reclamation Plant (UPWRP), Singapore. At the plant, mixed sludge (1/3 primary sludge and 2/3 thickened activated sludge) is anaerobically digested by a bank of egg-shaped digesters (ESD). Each ESD has a volume of 5,000 m^3 and is designed to feed approximately 200 m^3 d^{-1} of the mixed sludge. The automated mixer in the digester was programmed to mingle the contents upward or downward continuously and periodically.

The sludge flow

The test facility has two process lines: the control and the ultrasound line. The two lines are identical in terms of sludge feed and process except that the control does not involve an ultrasound reactor. The main components of the plant include a 15 m^3 sludge equalization tank made of stainless steel, three sludge pumps (one for the control, one for the ultrasound line, and one stand-by), each having a capacity > 8 m^3 h^{-1}, one ultrasonic reactor for sludge disintegration, and two egg shaped digesters (one for the control and one for the ultrasonic line) and two identical gas flow meters (Flex Masster® ST98, FCI) (one for the control and the other for the ultrasonic line) which were calibrated by the manufacturer and numerous other instruments for data acquisition and process control. The mixed sludge feed was delivered to the equalization tank from which the pumps draw the sludge at the same time and flow rate for the two lines.

The ultrasound facility and the SCADA

The ultrasound system consists of a control panel with programmable logical control, five sets of 20 kHz ultrasound generator, converter, booster and sonotrode and one V shaped reactor chamber in which the five donut shaped horns are stacked one on top of the other with a clearance of 5 cm. The total empty space inside the reactor is approximately 3.5 l. The feed is pumped into the reactor where the sludge is subjected to the high power sonication and sludge disintegration takes place. The power rating of the sonotrode is 6 kW. Under normal operation conditions, the power level of ultrasonic device is controlled at ≤ 50% with an amplitude >70%. Assuming a dead reactor volume of 50% (which was occupied by debris of the sludge), the average hydraulic retention time would be 1.5 seconds at the sludge flow rate of 8.3 m^3 h^{-1}. The Rockwell Automation software was used as part of the supervised control and data acquisition (SCADA) using programmable logic control. The data acquisition system records a range of process operation conditions and performance parameters every five minutes for the two lines, such as sludge flow rate, feed pressure before and after the ultrasound reactor, sludge pH, Redox and temperature, biogas flow rate, power level and electricity frequency for each of the five sonotrodes, etc.

Table 1 Ranges of selected characteristics of the sludge samples taken during operation of the ultrasound sludge disintegration plant (September 2004 – February 2005)

Parameter		Unit	Feed sludge values	Analytical methods
TSS	mg^{-1}	Range	11800–21200	USEPA160.3
		Average	16300	
VSS	mg^{-1}	Range	9380–17100	APHA2540E
		Average	12900	
COD	mg^{-1}	Range	1400–26800	USEPA410.4
		Average	19800	
Total-N	mg^{-1}	Range	900–1800	APHA 4500 norg
		Average	1360	
HPC	cfu $(100\ ml)^{-1}$	Range	$75.5\times10^6 - 1.82\times10^9$	APHA9215Cc
		Average	392×10^6	

Note: the figures were obtained from the analytic results of the samples taken and analyzed weekly during the period.

Sludge analysis

The plant was commissioned and started operation in late September 2004 until late June 2005. Sludge samples were taken once a week at four sampling points: immediately before and after sonication and overflow pipes of the digesters for both the control and the ultrasound line. The sludge samples were taken in the morning every time and sent to the laboratory immediately for determination of various characteristics following standard methods. Table 1 presents the selected properties of the sludge before anaerobic digestion. These are total suspended solids (TSS), volatile suspended solids (VSS), chemical oxygen demand (COD), total nitrogen and heterotrophic count (HPC). The biogas was analysed periodically for compositions of CH_4, CO_2, CO, H_2 and H_2S using a portable gas analyzer (LMSxi G2.18, Gas Data Ltd, Coventry UK).

Results and discussion

Sludge feed

At any given time during the study period, the sludge flow rates were the same for both the control and the ultrasound (test) line (Figure 1). For the entire period, the flow rates were between 90 and 200 $m^3\ d^{-1}$, which were below the designed value of 200 $m^3\ d^{-1}$ for most of the time. At this flow rate, the average hydraulic retention time (about 30 days) of the anaerobic digester might be too long to show the full benefits of the ultrasound sludge pre-treatment for the egg-shaped digesters operated under the tropical conditions. The shortage of sludge feed to the ultrasonic plant was attributed to the irregular pattern of sludge supply at the upstream of the anaerobic digesters.

Biogas production

The sludge fed into the two digesters was identical in terms of volume (Figure 1) and quality as it came from the same sludge holding tank. The test digester, however, constantly produced more (> 200 $m^3\ d^{-1}$) biogas than the control during the study period (Figure 2) even though there were several prolonged or short downtimes of the ultrasound reactor, indicating that sludge ultrasound pre-treatment might have a residual effect in increasing biogas production in the anaerobic digester. The dips in biogas production during the period 20–22 February (Figure 2) were associated with the downtime of the sludge pumps (Figure 1).

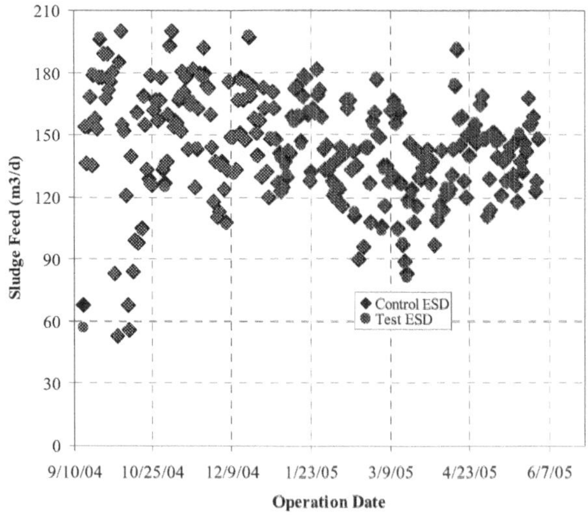

Figure 1 Daily feed to the two egg-shaped digesters (ESD) during the operation of the ultrasonic sludge disintegration facility (values <100 m^3 d^{-1} indicate either a partial record of the daily feed or a stoppage of the sludge pump on that day).

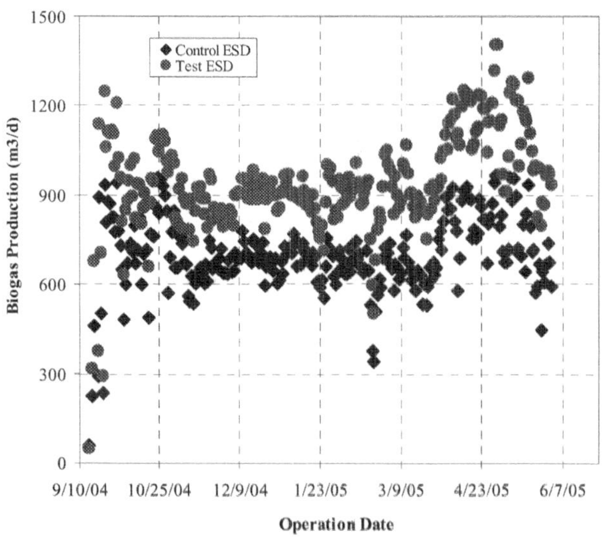

Figure 2 Daily biogas production in both the control and the test egg-shaped digester (biogas production < 400 m^3 before 5 October indicates a partial integration of the daily record).

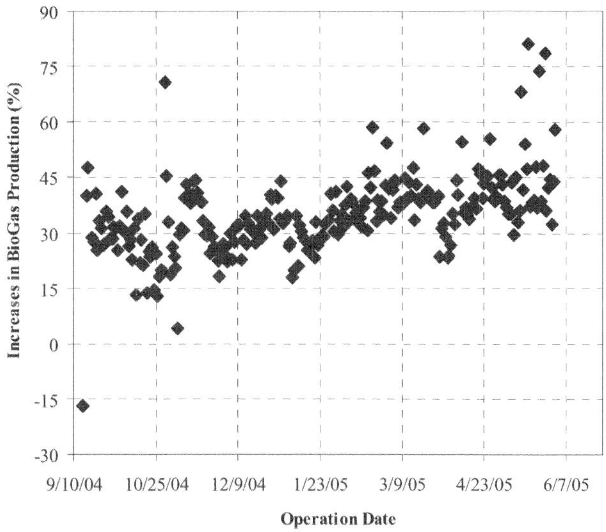

Figure 3 Increases in daily biogas production in the test egg-shaped digester (ESD) as compared to the control ESD.

Assuming that the two ESDs had the same configuration and the sludge matrix in the two digesters was identical, numerous other factors could have influenced the biogas production: (1) the operation history of the two digesters; (2) the quality and quantity of the sludge fed into the two anaerobic digesters; (3) operation conditions; (4) the inaccuracy of the biogas flow meters; and (5) the ultrasound power input. The operation history of the two digesters was the same since they were fed with the same sludge, operated in the same manner and had similar performance before the study. The sludge feed should have not contributed to the variations of biogas production since the volume and contents were the same (Figure 1). Operation conditions should have not made the difference of the biogas production since the sludge temperature, hydraulic retention time, mixing pattern and redox potential were the same for both digesters (data not shown). The gas flow meters were the same and their calibration and installation were done following the identical protocols. Ultrasound power input was the only element that have increased biogas production in the digester fed with the sonicated sludge in comparison with the control.

The range of the increases in daily biogas production was between 15% and 60% and the mode between 20% and 45% during the study period (Figure 3). The increases were greater at the late stage of the study than that at the early stage. The magnitude of the increases in biogas production in this study was comparable with those reported previously (Hogan *et al.*, 2004; Nah *et al.*, 2000; Neis *et al.*, 2000). The credibility of the results from this study relies on the strict field trial facility set-up, i.e., control against ultrasound disintegration line. The main substance of the sludge solids was VSS. For the thickened activated sludge, the VSS is composed of primarily bacterial cells. In the ultrasound reactor, these cells were subject to the high-powered ultrasound irradiation. As a result of such process, the cell walls would be ruptured and cell contents released due to the merging or colliding of the high temperature and high pressure cavities (Harrison, 1991; Joyce *et al.*, 2003; Susklick and Matula, 1999) in the sludge media. The cell contents became the substrate of microbes once being delivered to the anaerobic digester (King and Forster, 1990). In the control digester, however, the cell wall hydrolysis was usually a rate-limiting process under anaerobic conditions (Lafitte-Trouque and Forster, 2002).

Besides the amount of sludge feed (Figure 1), ultrasound power input determined the magnitude of the increases in biogas production. In general, the higher the ultrasound power input, the greater the increase was in biogas production with a short time-lag since sludge digestion is a kinetic process. The variations in the daily power consumption were attributed to the running status and/or increasing power levels of the sonotrodes during the operation (Figure 4).

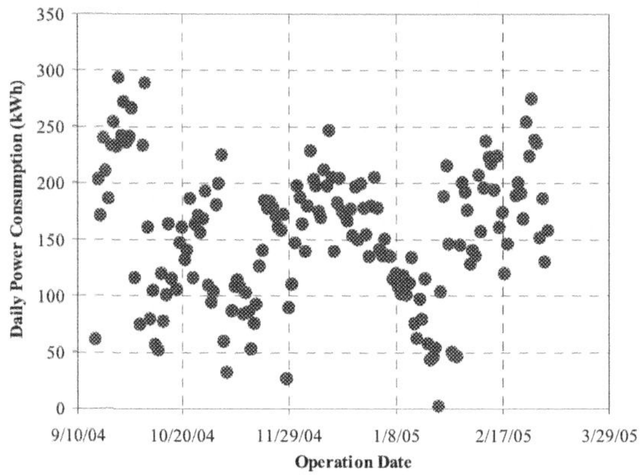

Figure 4 Daily power consumption of the ultrasonic reactor during the operation.

Biogas composition analysis

After removal of water vapor from the biogas, the main components were methane (the average being 67%, v/v) and carbon dioxide (the average being 32.5%, v/v), minor components included carbon mono-oxide (35–145 μl l^{-1}), hydrogen (35–95 μl l^{-1}) and hydrogen sulphide (350–820 μl l^{-1}). The compositions of the biogas from the test and the control digesters did not differ significantly (Figure 5). There were increases in minor gas compositions such as H_2S, CO and H_2 as the operation continued (Figure 5). The reasons behind the increases were not clear.

Sludge solids destruction

The feed sludge was low in total suspended solids concentration (TSS <2.94%) and volatile suspended solids concentration (VSS <1.92%). The results of the sludge analyses near the end of the study period (8.30 a.m., 21 June 2005) showed significant sonication effect on reduction in concentrations of TS (≥28%), TVS (25%), TSS (21%) and TVSS (16%) (Table 2), which were calculated using Equation (1):

$$\Delta S_i\% = (S_{c.i} - S_{t.i})/S_{t.i} \times 100 \quad (1)$$

where

$\Delta S_i\%$ = percentage increase in sludge solids destruction (%);

$S_{t.i}$ = sludge solids concentration in the test digester at *i*th day (mg l^{-1}); and

$S_{c.i}$ = sludge solids concentration in the control digester at *i*th day (mg l^{-1}).

Results of sludge analyses at the previous sampling periods (i.e. before June 2005) were not as significant as that observed here.

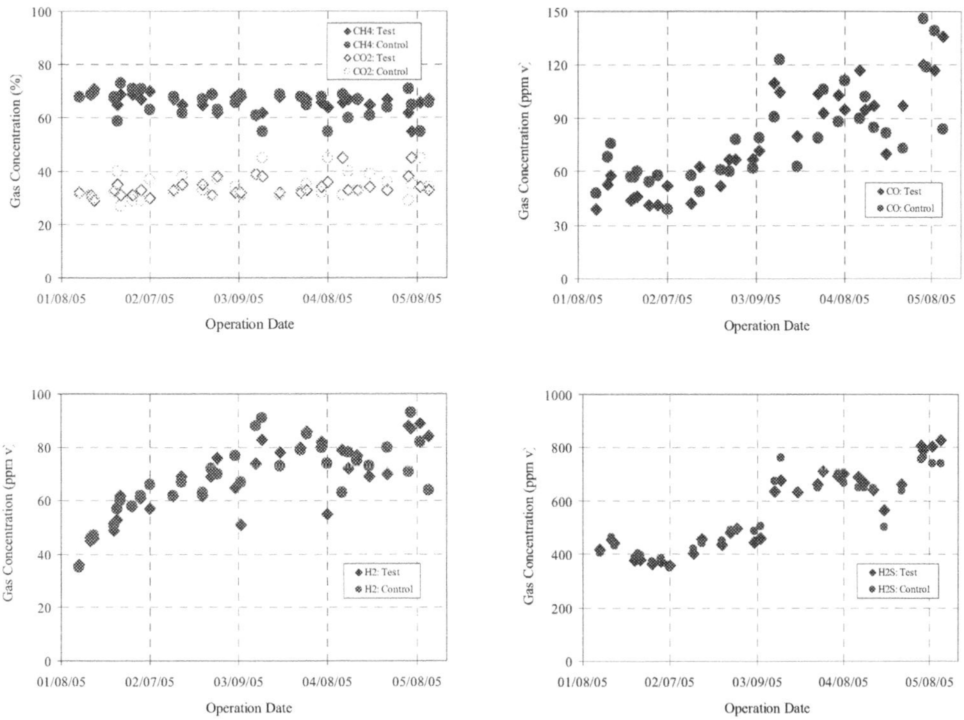

Figure 5 Biogas compositions in both the control and the test ESDs during the operation period: (a) methane and carbon dioxide; (b) hydrogen sulphite; (c) carbon monoxide; and (d) hydrogen gas.

Table 2 Characteristics of sludge solids: samples taken and analysed near the end of the study period[1]

Sampling date	Sludge characteristics	Test ESD	Control ESD	Percentage of difference
		mg^{-1}		%
1 June 2005	TS	17,000	31,000	82.35[2]
21 June 2005[3]	TS	22,900	29,400	28.38
	TVS	15,400	19,200	24.68
	TSS	19,000	23,000	21.05
	TVSS	14,100	16,200	14.89
	COD	20,100	22,500	11.94

[1]Sludge samples taken from the bottom section of the anaerobic digesters, in which the blenders were operated continuously; [2]Based on the figures provided by UPWRP/PUB; [3]Sludge samples were taken at 8:30 a.m., 21 June 2005.

The sampling and the analytical errors could have played a significant role to reduce the sonication effect on concentrations of the solids. Two other factors might have restricted the increases in sludge solids destruction by sonication: (1) the sludge flow rate (Figure 1) being lower than the designed value, resulting in higher hydraulic retention time (HRT). The longer the HRT, the less significant the sonication effect would be. Long HRT (>25 days) could have reduced or masked the impact of the pre-treatment on reductions in sludge solids in both the test and the control digesters (Nah *et al.*, 2000; Chua *et al.*, 2002; Lafitte-Trouque and Forster, 2002); (2) the low concentrations of the feed sludge VSS (1.06–2.08%). This could have increased the sampling error. In addition, it may have limited the source of food for the microbes and resulted in less significant build-up of the microbes in the digesters (Yasui *et al.*, 1996; Yoon *et al.*, 2004).

Theoretically, the increases in biogas production (Figures 2 and 3) had to come from the sludge VSS since the concentrations of dissolved organic carbon were very low: 170–840 mg l^{-1}. During the anaerobic digestion process, the VSS was slowly and continuously hydrolysed - converted into soluble organic substances which were subsequently consumed by the microbes in the digesters. The VSS accounted for 70–81% of the TSS, implying that a minimum of 25–30% increase in total sludge solids destruction should be expected (i.e., $35\% \times 70\% \approx 25\%$; $45\% \times 80\% \approx 36\%$). The sludge analytical results (Table 2) were consistent with the calculations.

Conclusions

Ultrasound pre-treatment of mixed primary and secondary sludge resulted in 15–60% increases in biogas production under controlled operation conditions with an average close to 45% at the late stage of the study. Sonication of the sludge feed had no significant effect on biogas compositions. Sludge analysis showed a 25–35% increase in sludge solids removal in the digester.

Acknowledgements

We thank the staff at the Ulu Pandan Water Reclamation Plant for their cooperation and support in construction and operation of the ultrasonic plant, and the staff at the Research Analytics Department, Centre for Advanced Water Technology, for much analytic work done for this project.

References

Chua, C.P., Lee, D.J., Chang, B.V., You, C.S. and Tay, J.H. (2002). ''Weak'' ultrasonic pre-treatment on anaerobic digestion of flocculated activated biosolids. *Wat. Res.*, **36**(11), 2681–2688.

Clark, P.B. and Nujjoo, I. (2000). Ultrasonic sludge pre-treatment for enhanced sludge digestion. *J. CIWEM*, **14** (February), 66–71.

Harrison, S.T.L. (1991). Bacterial cell disruption: a key unit operation in the recovery of intracellular products. *Biotechnology*, **9**, 217–240.

Hogan, F., Mormede, S., Clark, P.B. and Crane, M. (2004). Ultrasonic sludge treatment for enhanced anaerobic digestion. *Wat. Sci. Technol.*, **50**(9), 25–32.

Joyce, E., Mason, T.J., Phull, S.S. and Lorimer, J.P. (2003). The development and evaluation of electrolysis in conjunction with power ultrasound for the disinfection of bacterial suspensions. *Ultrason. Sonochem.*, **10**(4–5), 231–234.

Jung, M., Weise, Th.H.G.G., Wittenberg, M., Muller, J.A. and Dichtl, N. (2001). Sludge disintegration by electrical shock waves. In: *Proc. International Conference on Pulsed Power Applications, Gelsenkirchen. 27–29 March 2001*, pp B.O1/1–6.

King, R.O. and Forster, C.F. (1990). Effects of sonication on activated sludge. *Enzyme Microb. Tech.*, **12**, 109–115.

Lafitte-Trouque, S. and Forster, C.F. (2002). The use of ultrasound and γ-irradiation as pre-treatments for the anaerobic digestion of waste activated sludge at mesophilic and thermophilic temperatures. *Bioresource Technol.*, **84**, 113–118.

Lehne, G., Muller, A. and Schwedes, J. (2001). Mechanical disintegration of sewage sludge. *Wat. Sci. Technol.*, **43**(1), 19–26.

Müller, J.A. (2000). Disintegration as a key-step in sewage sludge treatment. *Wat. Sci. Tech.*, **41**(8), 123–130.

Müller, J.A., Winter, A. and Strünkmann, G. (2004). Investigation and assessment of sludge pre-treatment processes. *Wat. Sci. Technol.*, **49**(10), 97–104.

ATV-DVWK Arbeitsgruppe AK–1.6 'Klärschlamm Desintegration' (www.klaerschlammdesintegration.de, accessed 7 February 2006).

Nah, I.W., Kang, Y.W., Hwang K.Y. and Song, W.K. (2000). Mechanical pretreatment of waste activated sludge for anaerobic digestion process. *Wat. Res.*, **34**(8), 2362–2368.

Neczaj, E. and Lach, J. (2003). The use of ultrasound to accelerate the anaerobic digestion of waste activated sludge. (http://www.progue2003.fsu.edu/content/pdf/394.pdf, accessed 7 February 2006).

Neis, U., Nickel K. and Tiehm A. (2000). Enhancement of anaerobic sludge digestion by ultrasonic disintegration. *Wat. Sci. Tech.*, **42**, 73–80.

STOWA (2006). (http://www.stowa-selectedtechnologies.nl/Sheets/Sheets/Sludge.Disintegration.Techniques.html, accessed 7 February 2006).

Suslick, K.S. and Matula, T. J. (1999). Acoustic cavitation, sonochemistry, and sonoluminescence. In *Wiley Encyclopedia of Electrical and Electronics Engineering* (ed. Webster, J.G.). Wiley-Interscience, New York, 22, pp. 646–657.

Tilche, A., Bortone, G. and Dohányos, M. (2006). Alternative waste water treatment processes to reduce sewage sludge production (http://europa.eu.int/comm/environment/waste/sludge/workshoppart4.pdf, accessed 7 February 2006).

Weemaes, M., Grootaerd, H., Simoens, F. and Verstraete, W. (2000). Anaerobic digestion of ozonized biosolids. *Wat. Res.*, **34**(8), 2330–2336.

Yasui, H., Nakamura, K., Sakuma., S. and Sakai, Y. (1996). A full-scale operation of a novel activated sludge process without excess sludge production. *Wat. Sci. Tech.* **34**(3–4), 395–404.

Yoon, S.H., Kim, H.S. and Lee, S. (2004). Incorporation of ultrasonic cell disintegration into membrane bioreactor for zero sludge production. *Process Biochem.*, **39**, 1923–1929.

Enhancement of dewaterability with specific enzyme pretreatment: role of proteins and carbohydrates

D. Dursun and S.K. Dentel

Department of Civil and Environmental Engineering, University of Delaware, Newark, DE 19716, USA (E-mail: *derya@udel.edu*; *dentel@udel.edu*)

Abstract The extracellular polymeric substances (EPS) excreted by the microbial content of sludge form a network structure, or *gel*, in which the particles are embedded. These substances are negatively charged macromolecules mainly composed of proteins and polysaccharides. To identify the role of each component on dewaterability of sludge, specific enzymes (proteases and a mixture of carbohydrases) were applied to anaerobically digested sludge samples at various doses to degrade each component selectively. Protein and polysaccharide amounts were then measured both in the sludge and the extracted EPS. Sample dewaterability was quantified before and after enzyme pretreatment by measuring the capillary suction time and the cake solid content after using a bench-scale belt filter press simulator. Incubation time with enzymes was varied from 0 to 16 hours. With 16 hours of protease application at a 25 mg/l dose, the filterability of the sludge was enhanced by 46% and the solid content of dewatered cake increased by 23%. These significant improvements were not observed after pretreatment with carbohydrate degrading enzymes, suggesting that proteins in EPS govern dewaterability. Cryogenic field emission scanning electron microscopy images taken before and after enzyme pretreatment indicated that both enzyme types altered the network structure of the sludge, but the protease required a lower dose for more dramatic changes. Enzyme pretreatment is thus considered as an efficient and environmentally friendly way to treat sludges and enhance dewaterability. The use of specific enzymes also provides insight on the fundamental nature of EPS structure and chemistry.

Keywords: Carbohydrase; dewaterability; protease; sludge; specific enzyme treatment

Introduction

The increase in the number of water and wastewater treatment plants worldwide has led to another waste management issue: the production of excessive amounts of sludge. In order to decrease the final volume of sludge produced, dewatering has become a crucial step in treatment operations, however, recent developments in treatment processes have increased the resistance of sludge to dewaterability. New technologies have been sought to enhance dewaterability and increase the final solid content of the sludge as well as decrease the required amount of chemical conditioner.

Modified mechanical dewatering processes such as the use of higher torque centrifuges or the addition of heat and vacuum to pressure filtration have been developed, though these energy-intensive methods are typically found to be more expensive. They also involve drawbacks such as increased odor or microbial regrowth (Murthy *et al.*, 2003; Qi *et al.*, 2004). Therefore, rather than changing the mechanical dewatering process, it may be worthwhile to modify the structure of the sludge itself in order to increase the efficiency of dewatering.

Sludge flocs are made up of particles, bacteria, and the polymeric substances that are excreted from bacteria (Frølund *et al.*, 1996). These extracellular polymeric substances (EPS) create a network structure, or *gel*, in which the particles are embedded. Greater insight into the gel-like properties of sludge will indicate how to improve its behavior during conditioning and dewatering processes. One such approach is through enzymatic hydrolysis of the EPS that comprises the major organic fraction of sludge flocs.

Researchers (Thomas *et al.*, 1993; Ayol, 2005) have suggested that the application of enzyme mixtures including protease, amylase, lipidase, and anaerobic bacteria might enhance dewaterability by disrupting the network structure of sludge. There is also interest in determining which individual components of EPS impair dewaterability (Higgins and Novak, 1997). Thus, the objective of our work was to determine whether the favorable effects of enzyme treatment might be correlated with the type of enzyme used. Specifically, the effects of proteins and carbohydrates on dewaterability of digested sludge were investigated. It was hypothesized that selective enzymatic degradation of EPS might indicate the chemical constituents primarily responsible for poor dewaterability of sludge.

Methods

Sludge type and collection

Anaerobically digested sludge samples were taken from the Wilmington Wastewater Treatment Facility (Wilmington, DE, USA), which is a conventional activated sludge plant treating domestic and industrial wastewaters. The primary and activated sludges are combined for mesophilic anaerobic digestion at a typical retention time of 15–20 days. Samples were collected directly from the digester and transported to the laboratory, where all experiments were completed within three days of collection. Samples were kept at 4°C before analysis.

Enzyme pretreatment

A protease and a carbohydrase enzyme mixture were used for pretreatment. The chosen protease (Sigma P 6110-EC 3.4.11.1) is a leucyl aminopeptidase from *Aspergillus oryzae* that can hydrolyze peptide bonds under the environmental conditions of anaerobic digesters. Its specific protease activity is optimum around 40°C and pH 5–8. It was used at doses of 5, 15, and 25 mg/L.

An enzyme mixture was essential in order to degrade the various carbohydrates in sludge. The composition and doses of the glycosidic enzyme mixture were based on previous results of Bruno (1994), Dey Estera and Olof (2004), and Sarkar *et al.* (2004), given in Table 1. The two major components were ∀-amylase and cellulase, to cleave ∀- and ∃-glucoside polysaccharide links, respectively. Amylase is a digestive enzyme classified as a saccharidase. It is needed for the breakdown of long-chain carbohydrates (such as starch) into smaller units. Cellulase is an enzyme complex which breaks down cellulose to beta-glucose. Other enzymes were included to degrade other polysaccharides and monosaccharides. This mixture was added to digested sludge in ratios of 1/100, 1/150, and 1/200 corresponding to total enzyme mass additions of 410 mg/l, 273 mg/l and 205 mg/l respectively.

Table 1 Components of enzyme mixture to degrade carbohydrates

Enzyme	Cleavage	%
∀-Amylase (Sigma A6211-EC 3.2.1.1)	1, 4-α-D-glucosidic linkage	2.0
Cellulase (Sigma-EC 3.2.1.4)	1, 4-(1, 3;1,4)-β-D-glucan-4-glucanohydrolase	2.0
Hyalurinase (Sigma H2126-EC 3.2.1.35)	β-N-acetylhexosamine-[1→4] glycosidic bonds in hyaluronic acid	0.012
Dextranase (Sigma D4668-EC3.2.1.11)	1, 6-α-D-glucan 6-glucanohydrolase	0.004
∀-Glucosidase (Sigma G5003-EC 3.2.1.20)	non-reducing 1,4-linked D-glucose residues with release of D-glucose	0.004
Water		95.92

The protease and carbohydrase enzyme mixtures were prepared as stock solutions, and then added to digested sludge samples to give three different dosage levels. All samples were incubated at 35°C under anaerobic conditions, using incubation times of 4, 8, 12, and 16 hours. All analyses were conducted as soon as possible following incubation (generally within two hours) since the enzymes were still active at room temperature.

Filterability and dewaterability

The filterability of samples was evaluated using a capillary suction time (CST) device (Venture Innovations, Lafayette, LA) according to Section 2710G of *Standard Methods* (1995). A Crown Press ®(Neogen Corporation, Lansing, MI) was used in order to simulate the belt filter presses at bench-scale and also assess behaviour during the pressure and shear conditions occurring in dewatering. For these simulations, a 200 ml sludge sample was poured onto a gravity drainage simulator assembly, developed by Severin and colleagues (1999) as modified by Ayol and Dentel (2005). This step simulates the gravity drainage step of a belt press, including the plowing process. The resulting cake was transferred to a Crown Press® and controlled periods of mechanical pressure were applied to the resulting cake between filter cloths. Details about use of the gravity drainage kit and Crown Press to simulate belt filter press dewatering can be found elsewhere (Galla, 1996; Galla *et al.* 1996).

EPS extraction

The EPS were extracted from the activated sludge by mixing with a cation exchange resin (Dowex, Aldrich-Fluka 44445) under turbulent shear. This extraction procedure was based on the method developed by Frølund *et al.* (1996) and has been widely used for EPS extraction from sludge, since it gives high extraction efficiency with little or no cell lysis or polymer disruption. A 0.3 l sample of digested sludge was mixed with Dowex (75 g/g VSS), previously washed in a buffer solution, and transferred to an extraction vessel. The suspension was stirred at 900 rpm for 16 h. The extracted EPS was harvested by centrifugation at 12,000*g* for 15 min and filtration through a 0.45 mm cellulose acetate membrane. The extracted EPS was stored at –20°C. Carbohydrate measurements were determined as glucose equivalent using Dubois' method with phenol–sulfuric acid (Dubois *et al.*, 1956). Protein was measured as bovine albumin equivalent using a protein assay kit Sigma P 5656 following the method of Lowry *et al.* (1951).

Cryo-FESEM

To observe the structural differences between treated samples, a Hitachi 4700 field emission scanning electron microscope (FESEM) was used. The freeze-dry method was selected based on the results of Poxon (1996), indicating that this method conserves structural characteristics of sludge samples. Samples were filtered and mounted dropwise onto an aluminum stub and plunged into liquid nitrogen slush at –130°C. The sample was then warmed to -95°C in a Gatan specimen preparation chamber over a period of 15 min. The prepared sample was then transferred to the cryostage and viewed at –120°C at 0.7 and 1.0 kV.

Analytical methods

The dried solid content of the final cake and filtrate solid content after applying the Crown Press were measured according to Method 2540C of *Standard Methods* (1995).

Results and discussion

Protease pretreatment

Figure 1 gives percent protein degradation in both the sludge and the extracted EPS after various protease doses and incubation times. After 4 h incubation, protein concentrations increased in both the sludge and the EPS, corresponding to negative degradation percents, but this trend reversed at longer incubation times. There is no linear relation observed between protease dose and the protein content, since the highest degradation was observed at 15 mg/l protease addition. This indicates that there is an optimum amount of protease that controls the enzymatic reactions. Increasing the enzyme dose beyond this concentration only led to a decrease in the percent degradation.

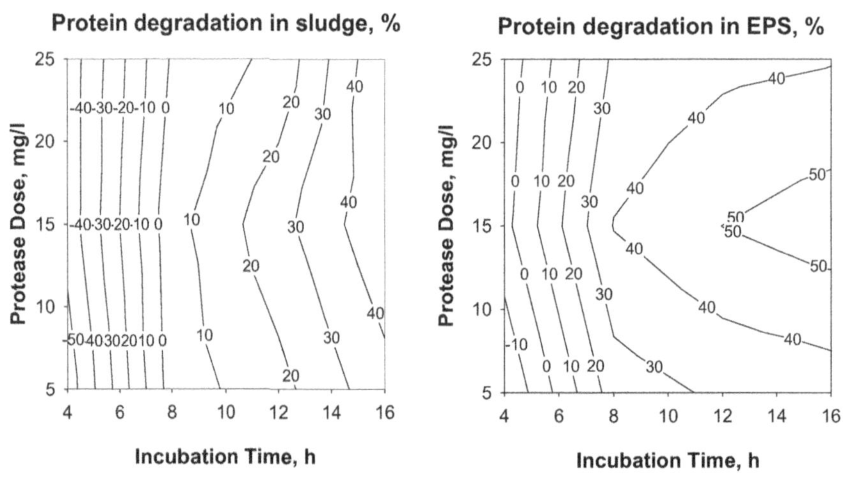

Figure 1 Protein degradation percentages in sludge and EPS after protease pretreatment

Capillary Suction Time is a widely accepted method for determining a sample's water release rate, with lower times indicative of better dewaterability. The CST data was acquired for the highest protease doses (25 mg/l) and four incubation times (4, 8, 12, and 16 hours). Increasing incubation time and protease dose improved the filterability up to 45% by lowering the CST values. The effect of incubation time was significant between 0 and 8 hours, but after 8 hours of incubation the enhancement rate stayed relatively constant even with increasing protease dose (Figure 2). These finding also correlated with the results of protein degradation (Table 1).

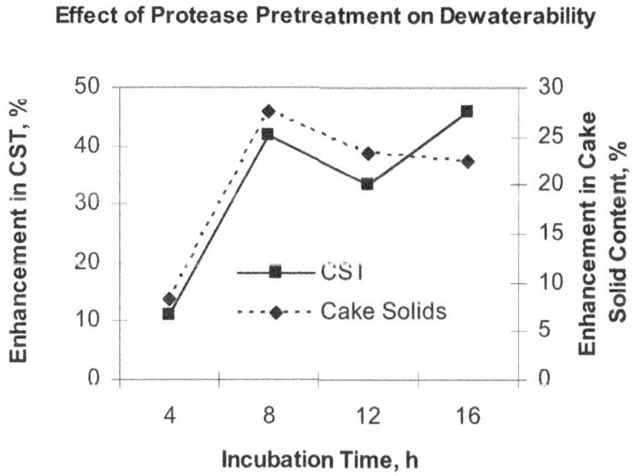

Figure 2 Per cent improvement in filterability (CST) and dewaterability (Crown Press) using protease pretreatment.

For the same sample, the cake solid content after Crown Press® use was determined before and after protease pretreatment, and the change in cake solid content converted to the relative percent increase. After 8 h incubation, the cake solid content of samples was enhanced by over 20%, equivalent to a 20% reduction in the final sludge mass. Figure 2 combines the results of filterability and dewaterability enhancement by protease treatment quantified in terms of CST and cake solid content.

Another concern was the variation of filtrate suspended solids concentrations after enzyme pretreatment since effluent quality is vital to fulfil the regulations. Although the suspended solid concentration again decreased (data not shown) with increasing protease dose and incubation time significantly, the correlation of EPS and sludge protein content with suspended solid concentration were in fact low (Table 1).

Pretreatment with carbohydrase enzyme mixture

The polysaccharide degradation percentages both in sludge and EPS are summarized in Figure 3. Surprisingly, carbohydrase pretreatment did not increase the degradation of polysaccharide in sludge, whereas significant increases in polysaccharide degradation were found in EPS (Figure 3). Polysaccharide concentrations were higher than the initial concentration for both sludge and EPS at 4 hours of incubation, probably due to increased solubility as proposed by Whiteley *et al.* (2003).

Longer incubation times and higher enzyme doses increased the rate of filtration around 30% (for a 410 mg/l dose). However, this positive effect on CST was not observed when dewaterability was assessed with the Crown Press. There was no enhancement shown in the cake solids levels after Crown Press application. Figure 4 depicts the enhancement in filterability and dewaterability of samples with carbohydrase treatment. Furthermore, increasing doses of this carbohydrase mixture tended to increase the suspended solid content of filtrate (data not shown), perhaps because of floc fragmentation. Increasing the incubation time from 0 to 16 hours decreased the value to some extent. We have related this decrease to particle size reductions beyond the ranges characterized as suspended solids.

Effect of enzyme pretreatment on sludge structure

Figure 5a illustrates the network structure observed in cryo-FESEM images of untreated, digested sludge. Figure 5b,c is evidence for the typical eroded structure observed after treatment with 25 mg/l of protease and 410 mg/l carbohydrase mixture. Enzyme pretreatment changed the structure of samples completely, leading to a less consistent and more permeable behaviour. The network structure of raw digested sludge was entirely disrupted. With protease treatment (Figure 5b), structural effects are more visible compared with the carbohydrase treatment, probably due to the greater degree of degradation.

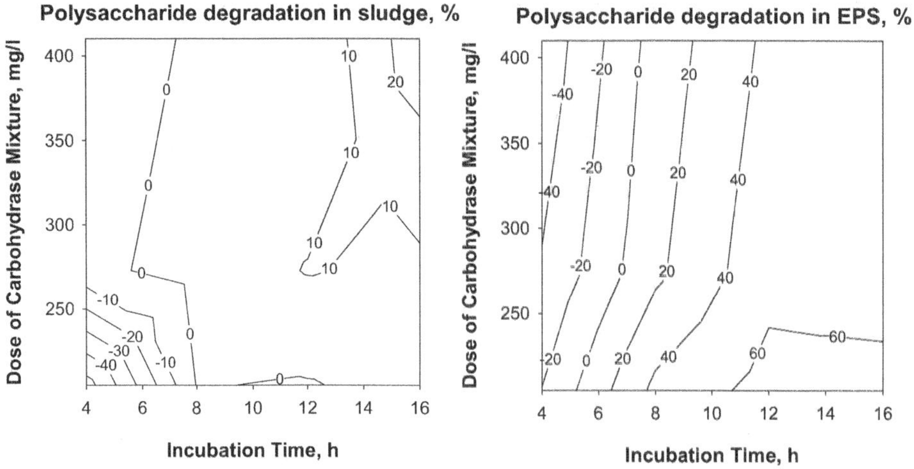

Figure 3 Polysaccharide degradation percentages in sludge and EPS after carbohydrase pretreatment.

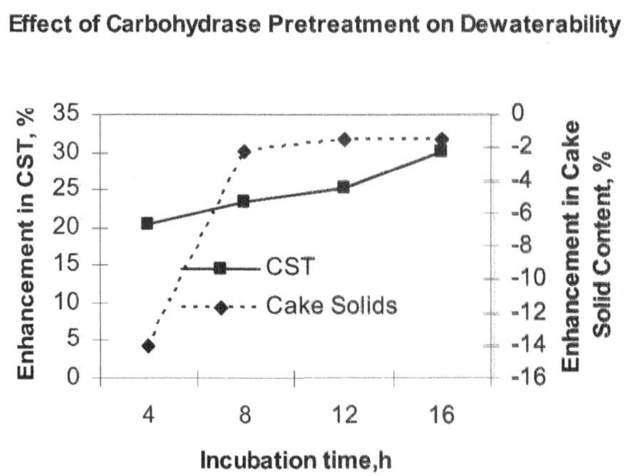

Figure 4 Per cent improvement in filterability (CST) and dewaterability (Crown Press) using carbohydrase pretreatment.

Table 2 Relation between dewaterability parameters and protein and polysaccharide amount both in EPS and sludge, given as square of correlation coefficient (R^2) and the level of significance (p) (n=16 for all R^2 determinations)

		Protein	p	Polysaccharide	p
EPS	**CST**	0.8004	2.92×10^{-6}	0.2218	0.0655
	Cake Solid	0.7244	2.90×10^{-6}	0.1231	0.1827
	Suspended Solid	0.3954	0.0091	0.2353	0.0568
Sludge	**CST**	0.5237	0.0015	0.2931	0.0303
	Cake Solid	0.5279	0.0014	0.2913	0.0309
	Suspended Solid	0.0814	0.2841	0.0016	0.8833

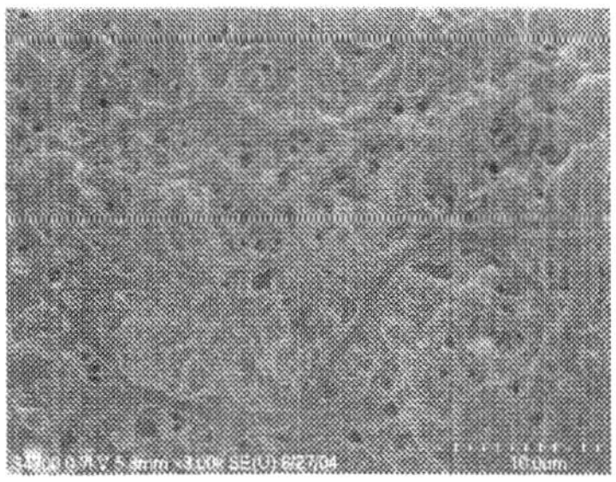

(a) Digested Sludge

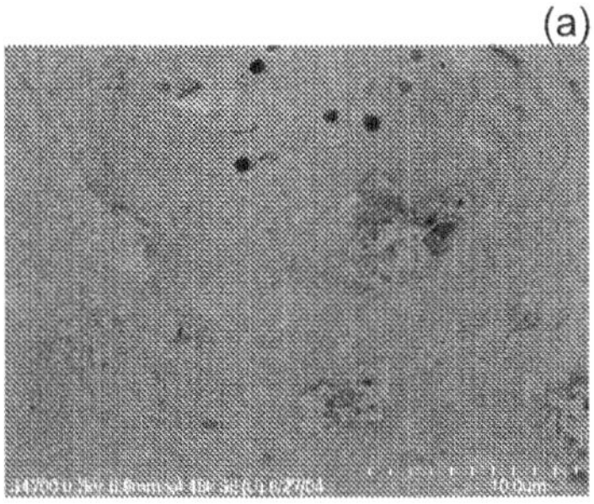

(b) After 16 hours protease treatment

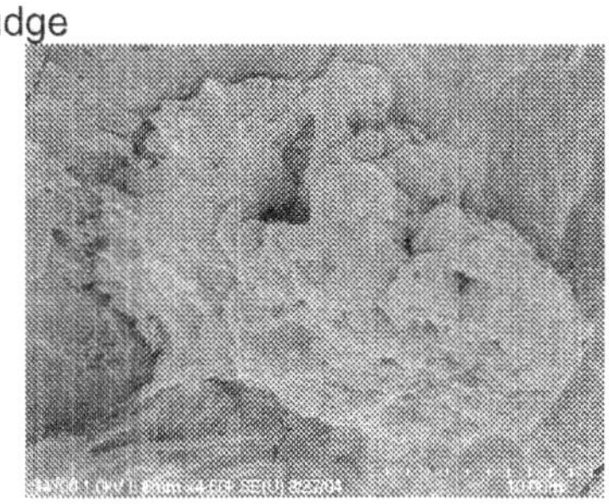

(c) After 16 hours carbohydrase treatment

Figure 5 Structural changes of digested sludge with enzyme pretreatment (all images have the same magnification).

Table 1 gives the correlation coefficients (R^2) between dewaterability parameters and protein or polysaccharide amount in the sludge and extracted EPS. The values for protein were found higher than polysaccharide for the entire samples, indicating the effect of proteins on dewaterability. The protein amount in EPS was more highly correlated with dewaterability than was the protein amount in the overall sludge.

Conclusions

This study investigated the role of two major fractions in EPS, proteins and carbohydrates, on dewaterability of anaerobically digested sludge by selectively degrading them with specific enzyme treatment. The following conclusions can be drawn from our results:

- Both the protease and carbohydrase enzymes degraded their target substrates, with significant reductions effected in both the sludge and the EPS fraction of the sludge. The carbohydrase mixture required much higher dosages.
- Protease application provided the greatest improvements in sludge properties at a dose of 25 mg/l and incubation time of 16 h, increasing filterability by 46% and the solid content of dewatered cake by 23%.

- Carbohydrase, even at much higher doses, improved filterability but not dewaterability, suggesting that proteins in EPS govern dewaterability.
- Reductions in residual protein – especially in the sludge EPS content – are correlated to positive effects on sludge filterability and dewaterability.

As a result of increasing number of wastewater treatment plants and strict environmental regulations, sludge production has increased drastically in recent years. Greater efforts are needed to decrease the final solid mass and reduce energy requirements and chemical conditioner costs in managing these materials. Combining the advantages of biotechnology with the enzyme pretreatment is an efficient and environmentally friendly way to treat biosolids and enhance dewaterability. The use of specific enzymes will also shed light on the fundamental nature of EPS structure and chemistry.

References

Ayol, A. (2005). Enzymatic treatment effects on dewaterability of anaerobically digested biosolids – I: performance evaluations. *Process Biochemistry*, **40**, 2427–2434.

Ayol, A. and Dentel, S.K. (2005). Enzymatic treatment effects on dewaterability of anaerobically digested biosolids – II: Laboratory characterizations of drainability and filterability. *Process Biochemistry*, **40**, 2435–2442.

Bruno, M. (1994). Multienzyme Powdered Composition Containing Bacteria for Treatment of Waste. U.S. Patent 5,464,766.

Dey Estera, S and Olof, N. (2004). A Method for Digestion of Sludge in Water Purification. European Patent WO2004024640.

Dubois, M, Gilles, KA, Hamilton, JK, Rebers, PA, and Smith, F. (1956). Calorimetric method for determination of sugars and related substances. *Anal Chem.* **28**(3): 350–356.

Frølund B., Palmgren R., Keiding K. and Nielsen P.H. (1996). Extraction of extracellular polymers from activated sludge using a cation exchange resin. *Water Research.* **30**, 1749–1758.

Galla, C.A. (1996). *Laboratory Prediction of Belt Filter Press Dewatering Dynamics.* M.S. thesis, University of Illinois at Urbana-Champaign, USA.

Galla, C.A., Freedman D.L., Severin B.F. and Kim B.J. (1996). Laboratory prediction of belt filter press dewatering dynamics. *In Proceedings of the Water Environment Federation 69th Annual Conference, Residuals and Biosolids Management Symposium*, Dallas, TX, USA.

Higgins, M.J. and Novak, J. (1997). Characterization of exocellular protein and its role in bioflocculation. *Journal of Environmental Engineering*, **123**(5), 479–485.

Lowry, O.H, Resebrough, N.J, Farr, A.L and Randall, R.J. (1951). Protein measurement with the folin phenol reagent. *J Biol Chem.*, **193**(2): 265–275.

Murthy, S., Higgins, M., Chin, Y., Toffey, B., and Golembeski, J. (2003). Influence of solids characteristics and dewatering process on volatile sulfur compounds production from anaerobically digested biosolids. *Proceedings of WEF Biosolids Specialty Conference,* Baltimore, MD, USA.

Poxon, T.L. (1996). Structure and dewaterability in anaerobically digested sludge. *In Proceedings of the Water Environment Federation 69th Annual Conference*, Dallas, TX, vol. 2, pp. 167–178.

Qi, Y.N., Herson, D. and Dentel, S.K. (2004) Reactivation and/or growth of fecal coliform bacteria during centrifugal dewatering of anaerobically digested biosolids. *Water Sci. Technol*, **50** (9), 115–120.

Sarkar, J., Shah, J. and Ramesh, M. (2004). Method of Dewatering Sludge Using Enzymes. U.S. Patent 6,733,674.

Severin, B.F., Jeffrey, V.N. and Kim, B.J. (1999). Model and analysis of belt drainage thickening. *Journal of Environmental Engineering*, **125**, 807–815.

Standard Methods for the Examination of Water and Wastewater (1995). 19th ed. American Public Health Association/American Water Works Association/Water Environment Federation, Washington, DC.

Thomas, L., Jungschaffer G. and Sprossler B. (1993). Improved sludge dewatering by enzymatic treatment. *Wat. Sci. Technol.* **28** (1), 189–192.

Whiteley, C.G., Burgess, J.E., Melamane, X., Pletschke, B. and Rose, P.D. (2003). The enzymology of sludge solubilisation utilizing sulphate-reducing systems: the properties of lipases. *Water Research,* **37**, 289–296.

Luxury uptake of phosphorus by microalgae in waste stabilisation ponds

N. Powell, A. Shilton, S. Pratt and Y. Chisti

Centre for Environmental Technology and Engineering, Massey University, Private Bag 11-222, Palmerston North, New Zealand (E-mail: *N.Powell@massey.ac.nz*; *A.N.Shilton@massey.ac.nz*; *S.Pratt@massey.ac.nz*; *Y.Chisti@massey.ac.nz*)

Abstract Waste stabilisation ponds (WSP) effectively remove organic carbon and pathogens but their capacity to remove phosphorus is generally limited. Some phosphorus can be removed by biological means but there has been very little research into the biological mechanisms involved. Organisms remove phosphorus for growth but there are also some microalgae that are capable of storing phosphorus as polyphosphate under certain conditions; a process known as luxury uptake. This paper investigates if luxury uptake can occur in WSP microalgae. Reactors were used to simulate the conditions that occur in WSP and the solids and phosphate in the effluent were measured. Initially 100% phosphorus removal was achieved but a change in dominating species from *Scenedesmus* to filamentous microalgae resulted in a decrease in removal to approximately 60%. Three techniques which have not previously been used for studying phosphorus removal in WSP were applied. These techniques were staining cells, measurement of the percentage phosphorus in the cells, and phosphorus fractionation. Staining microalgal cells with a stain specific for polyphosphate granules showed a large number of granules were present in the cells. Normally microalgal cells contain 1% phosphorus per dry weight but the cells in the reactors contained 1.3% and 2.0% for filamentous and *Scenedesmus* respectively. The polyphosphate fractions were also extracted from the cells and showed a large amount of polyphosphate present. All these techniques used to detect luxury uptake clearly implied that luxury uptake was occurring in these reactors. With further study of this mechanism it is possible that in the future WSP could be better designed for phosphorus removal.

Keywords Luxury uptake; microalgae; phosphorus; waste stabilisation ponds

Introduction

Waste stabilisation ponds (WSP) are used extensively for wastewater treatment in small communities around the world due to their simplicity and low cost. WSP provide excellent organic carbon and pathogen removal but phosphorus (P) removal is generally low, varying between 15 and 51% (Garcia *et al.*, 2000; Picot *et al.*, 1992; Racault *et al.*, 1995).

Phosphorus is often the limiting nutrient for growth in fresh waterways so phosphorus removal is becoming increasingly important. Effluents high in phosphorus can lead to algal blooms and eventually eutrophication of the receiving water bodies.

Phosphorus removal mechanisms

Phosphorus is removed by both chemical and biological mechanisms in WSP. As microalgae grow they consume inorganic carbon and so shift the carbonate/bicarbonate equilibrium causing the pH to increase. At elevated pH chemical precipitation of phosphate ions with calcium, magnesium or iron can occur. Also, phosphorus is a part of many biological compounds so a small amount of phosphorus will be removed from the wastewater for organism growth. Many researchers studying WSP and high rate algal ponds (HRAP) assume that the contribution of biological assimilation to overall phosphorus removal is negligible and, instead, that phosphorus removal in WSP is due to precipitation (Moutin *et al.*, 1992; Picot *et al.*, 1992; Shelef *et al.*, 1982; Sukenik *et al.*, 1985). However, other researchers have found that phosphorus removal due to organism growth is more significant (Mesple *et al.*, 1995).

Another biological phosphorus removal mechanism that could occur in ponds is luxury uptake. This is the accumulation of phosphorus in the cells in the form of polyphosphate which is stored in granules (voluin granules). This mechanism is fundamental for PAOs (polyphosphate accumulating organisms) which are bacteria used in the wastewater treatment process known as enhanced biological phosphorus removal (EBPR) (Mino *et al.*, 1998). While PAOs are unlikely to function in the conditions found in WSP, it is possible that other phosphate accumulating organisms could. It has been found that some green microalgae store phosphorus that can then be used as an internal source of phosphorus when the external phosphorus becomes limiting (Kuhl, 1974). With regard to overall phosphorus removal in WSP, the significance of this mechanism is unknown.

In this work, phosphorus removal by microalgae, which are subjected to conditions that occur in WSP, is assessed. Emphasis is placed on the mechanisms of biological phosphorus uptake with the view of quantifying the contribution of luxury uptake to phosphorus removal in WSP.

Methods

Two continuous reactors were set up to simulate the algal band of a WSP in the laboratory with a retention time of ten days. The reactor had a volume of twelve litres and was fed with synthetic wastewater according to Davis and Wilcomb (1967). This synthetic wastewater contained chelating chemicals to prevent precipitation so that the biological mechanisms can be studied.

Daylight fluorescent lights were used to simulate day and night (15 hours light and 9 hours dark) and the light intensity at the surface was 150 $\mu E/m^2.s$. The light intensity was monitored using an Irradiance Sensor (Biospherical Instruments QSL-2101). The reactor was in a constant temperature room at 25°C and slowly stirred with a mechanical stirrer. The sides of the reactor were covered to prevent growth on the walls. Wastewater from a domestic facultative WSP in Ashhurst, New Zealand was used as the inoculum. A diagram of a reactor is shown in Figure 1.

The solids, phosphate in the feed, and phosphate in the effluent were monitored. The phosphate was measured using ion chromatography (Dionex ICS-2000, using an IonPac AS11-HC column) in duplicate. The solids were measured according to Lee and Shen (2004). The microbial species present in the reactors were also regularly examined under the microscope.

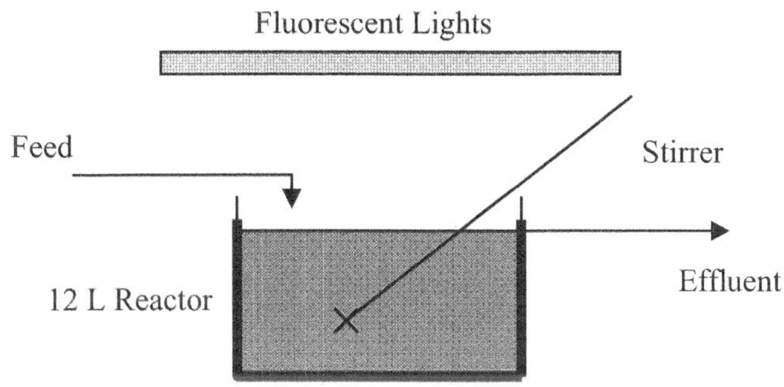

Figure 1 Diagram of reactor.

Techniques for investigating luxury uptake

Luxury uptake can be investigated in a number of ways. The cells can be stained for polyphosphate granules and then examined under the microscope, the percentage phosphorus in the cells can be measured, or the polyphosphate can be extracted from the cells and measured.

All three techniques were applied in this study. These methods have not previously been applied to studying phosphorus removal in WSP.

Polyphosphate granule stain

Luxury uptake can be confirmed by using a stain specific for polyphosphate granules in microalgae. Bolier *et al.*, (1992) developed a staining technique which stains the polyphosphate granules dark brown. The cells are then examined under the microscope and the presence of polyphosphate can be quickly confirmed.

Percentage phosphorus in the cells

Microalgae require phosphorus as a nutrient for growth and the amount of phosphorus removed for growth can be calculated as microalgal cells contain approximately 1% phosphorus by dry weight (Borchardt and Azad, 1968; Kaplan *et al.*, 1986). The amount of phosphorus in the cells can be measured using the total phosphorus test according to standard methods (APHA *et al.*, 1995) and measuring the dry weight. The value can then be compared to the 1% phosphorus per dry weight for normal growth and any value higher that this indicates that luxury uptake is occurring.

Polyphosphate extraction

To confirm the presence and quantify the amount of polyphosphate in the microalgae a method of fractionating the different types of phosphorus in the algal cells was used. This method was developed by Aitchison and Butt (1973) and Kanai *et al.,* (1965) to study phosphorus metabolism in microalgae. The different forms of phosphorus are extracted using trichloroacetic acid (TCA), ethanol, ethanol: ether (3:1), and potassium hydroxide (KOH). The fractions that are determined using this method are acid soluble polyphosphate, acid insoluble polyphosphate, other cell phosphorus along with the phosphate (PO_4) and the total phosphorus in the liquid.

Results and discussion

Phosphate in effluent

When the laboratory reactor reached the first steady state there was no detectable phosphate in the effluent as shown in Figure 2. This period of 100% removal was sustained for the first 4–5 months after which phosphate began to appear in the effluent. Towards the end of the experiment the effluent phosphate concentration fluctuated but was approximately 2 mgP/l (57% removal efficiency). Phosphorus removal in WSP systems is usually between 15-50% (Garcia *et al.*, 2000; Picot *et al.*, 1992; Racault *et al.*, 1995) so this level of removal would not be untypical; however, the initial 100% removal is exceptional.

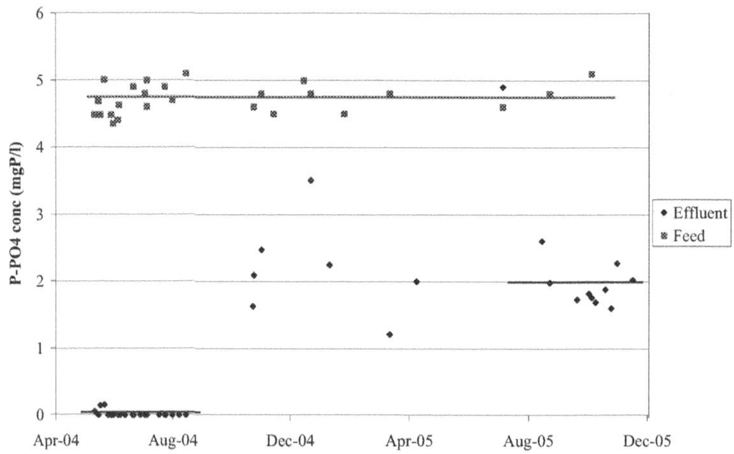

Figure 2 Phosphate in the feed and effluent of Reactor 1.

The phosphorus removal achieved is not due to precipitation as chelating chemicals were added to prevent precipitation occurring at the high pH levels achieved, and high phosphorus removal could still be obtained. This means that the phosphorus has been removed due to growth alone or due to growth with luxury uptake of phosphorus occurring.

Solids concentration

Initially the reactor reached a steady-state solids concentration of approximately 260 mg/l. At that stage the dominating species was *Scenedesmus*. There was then a slow change in species from *Scenedesmus* to filamentous green algae. This caused an increase in solids concentration and the steady state solids concentration became approximately 670 mg/l. The cause for the species change is unknown but it may be due to the filamentous algae reaching a significant population by growing on the walls of the reactor. The solids concentrations are shown in Figure 3.

When the species began to change a second continuous reactor (seeded from the original reactor) was started. The operating conditions were the same as the original reactor. It was observed that the second reactor was dominated with *Scenedesmus* with a small population of the filamentous species. This second reactor was then used for measurements to compare the performance of the two dominating species.

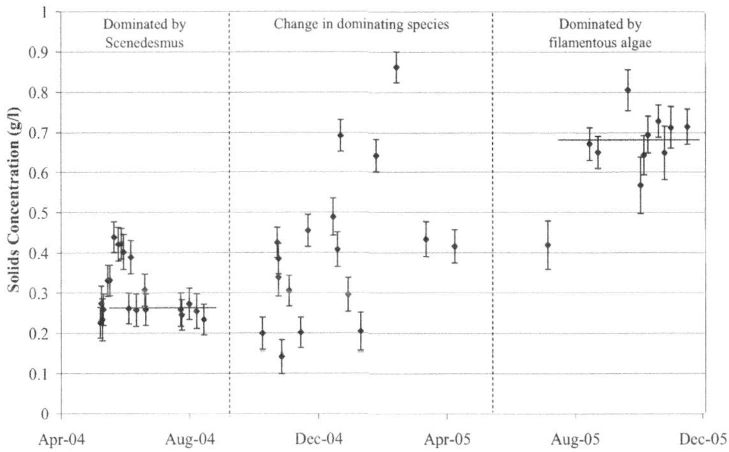

Figure 3 Solids concentration and dominating microalgae in Reactor 1.

The solids concentration in the laboratory reactor is much higher than the solids found in a WSP. WSP generally contain an algal biomass of 10-100 mg/l and high rate algal ponds (HRAP) can get up to 1500 mg/l (Abeliovich, 1986). In a WSP only the top 20–30 cm of the pond contains enough light for microalgal growth to occur but the ponds are generally 1–1.5 m deep so the algae that grow in the top algal band are then diluted by the rest of the pond. The laboratory reactors were only 20 cm deep so there were no dilution effects. If the same dilution occurred for the laboratory experiments the dry weights would have been 52 mg/l and 134 mg/l for *Scenedesmus* and the filamentous microalgae respectively.

Confirmation of luxury uptake using staining

Figure 4 shows the large number of polyphosphate granules stained dark brown in the microalgal cells. This observation has been reported in a pervious paper (Powell *et al.*, 2005). The figure indicates that luxury uptake is occurring but does not offer a means to quantify the amount of polyphosphate stored in the cells.

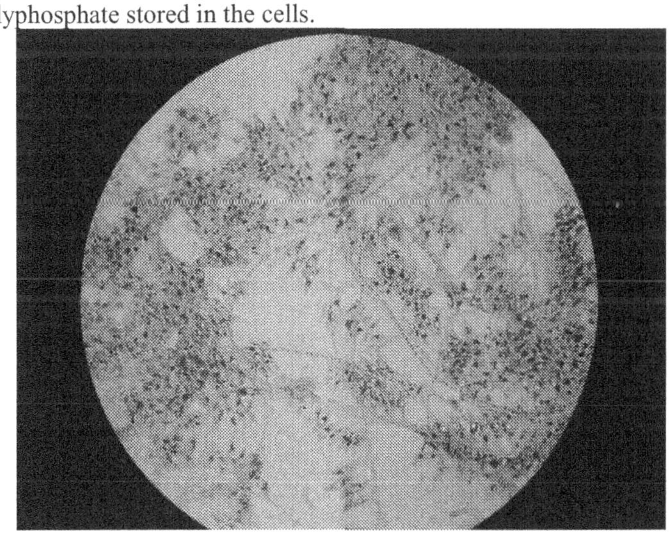

Figure 4 Microalgal cells stained for polyphosphate granules (Powell *et al.*, 2005).

Measurement of percentage P in cells

Algae contain approximately 1% phosphorus per dry weight for normal metabolism (Kaplan *et al.*, 1986). Measurement of the total phosphorus of the cells and the dry weight show that the filamentous algae contained an average of 1.29% phosphorus and *Scenedesmus* contained 2.02% phosphorus. The result indicates that phosphorus removal beyond that required for biomass growth occurred.

Confirmation of luxury uptake using P fractionation

The amount of polyphosphate can be determined by chemical fractionation of the phosphorus within the microalgal cells. This method analyses the fractions as acid soluble polyphosphate, acid insoluble polyphosphate, other cell phosphorus along with the phosphate (PO_4) and the total phosphorus in the liquid. Figure 5 shows the phosphorus fractions in two different samples taken for each dominating species. Even at this latter, less effective stage when filamentous microalgae dominated, the cells contain a large amount of polyphosphate, which is indicative of luxury uptake.

Figure 5 shows that both species are capable of storing phosphorus as polyphosphate. The phosphorus fractionation method splits the polyphosphate into acid soluble and acid insoluble. Polyphosphate or volutin granules consist of mainly acid soluble polyphosphate (Cembella *et al.*, 1984) but may also contain RNA, proteins, lipids and magnesium ions (Kuhl, 1962). Polyphosphates are not only present as granules but also in non-granular deposits in the cytoplasm and as dissolved inclusions in vacuoles or chloroplasts (Cembella *et al.*, 1984). The acid insoluble polyphosphates are more actively involved in metabolism (Kuhl, 1962).

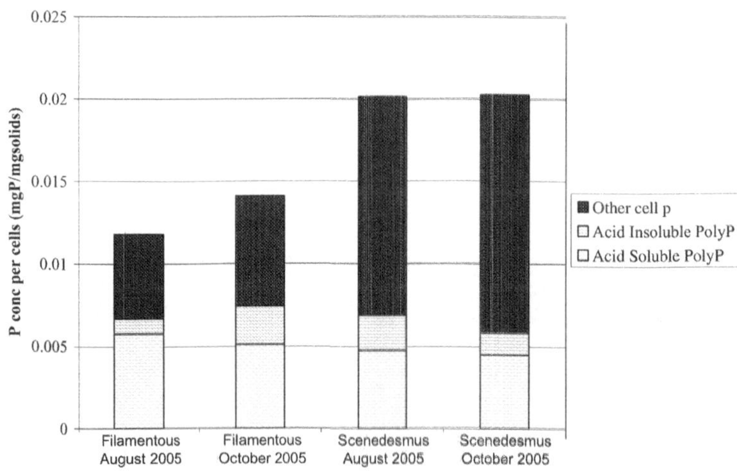

Figure 5 Phosphorus fractions in two microalgal species.

Figure 5 also shows that *Scenedesmus* cells contain more phosphorus than the filamentous species, which may explain why there was better phosphorus removal when *Scenedesmus* dominated. The amount of polyphosphate in the two species is similar but there seems to be some other form of phosphorus storage which is identified as the other cell phosphorus fraction. The large amount of other cell phosphorus shown in Figure 5 may be due to incomplete extraction of the polyphosphate or accumulation of some other cellular component that contains phosphorus. Other cellular compounds that contain phosphorus include lipids, nucleotides, nucleic acids, DNA, RNA and proteins. The amounts of the other cellular phosphates can change according to the environmental conditions. Rhee (1973) found that the concentration of RNA P, lipid P, nucleotide P, and DNA P were effected by growth rate.

Future implications

WSP can be upgraded for phosphorus removal which is generally done using chemical dosing. Chemicals such as alum are added to the effluent to chemically precipitate out the phosphorus. The disadvantages of using chemicals are that a chemical sludge is produced which is difficult to dispose of and the process becomes more complex and no longer simple to operate. This makes chemical dosing inappropriate for long term use (Crites and Reed, 2002). The other option is to install EBPR but this also significantly increases cost and complexity. Ideally a new solution is required which maintains the low cost and simplicity of the WSP system. Optimising luxury uptake, which in this work has been demonstrated to be possible in conditions found in WSP, offers potential for such a solution.

Conclusions

The experimental results show that WSP microalgae can achieve 100% phosphorus removal over a 4–5 month period. It is possible that the change in phosphorus removal performance is due to the change in algal species from *Scenedesmus* to filamentous microalgae. This is because the drop in performance occurred at the same time as the change in species and that the detailed phosphorus fractions in the two species show that *Scenedesmus* contains more phosphorus per amount of cells than the filamentous species.

All three techniques for detecting luxury uptake indicate that polyphosphate can be stored by WSP microalgae under the conditions that occur in the WSP. The staining technique shows a large number of granules, the percentage phosphorus in the cells is above 1%, and the phosphorus fractions show that there is a significant amount of polyphosphate in the cells.

These methods have not previously been applied to studying phosphorus removal in WSP and the results show that luxury uptake of phosphorus by microalgae may be a significant phosphorus removal mechanism in WSP.

References

Abeliovich, A. (1986). Algae in wastewater oxidation ponds. In A. E. Richmond (Ed.), *CRC Handbook of microalgal mass culture* (pp. 331–338). Florida: CRC Press.

Aitchison, P. A., and Butt, V. S. (1973). The relation between the synthesis of inorganic polyphosphate and phosphate uptake by Chlorella vulgaris. *Journal of Experimental Botany*, **24**(80), 497–510.

APHA, AWWA, and WEF. (1995). *Standard methods for the examination of water and wastewater*, 19th edn. Washington, D.C.: Water Environment Federation.

Bolier, G., de Koningh, M. C. J., Schmale, J. C., and Donze, M. (1992). Differential luxury phosphate response of planktonic algae to phosphorus removal. *Hydrobiologia*, **243/244**, 113–118.

Borchardt, J. A., and Azad, H. S. (1968). Biological extraction of nutrients. *Journal of the Water Pollution Control Federation*, **40**(10), 1739–1754.

Cembella, A. D., Antia, N. J., and Harrison, P. J. (1984). The utilisation of inorganic and organic phosphorous compounds as nutrients by eukaryotic microalgae: A multidisciplinary perspective: Part 2. *CRC Critical Reviews in Microbiology*, **11**(1), 13–81.

Crites, R. W., and Reed, S. C. (2002). Land treatment systems for municipal and industrial wastes.

Davis, E. M., and Wilcomb, M. J. (1967). Enzymatic degradation and assimilation of condensed phosphates by green algae. *Water Research*, **1**, 335–350.

Garcia, J., Mujeriego, R., Bourrouet, A., Penuelas, G., and Freixes, A. (2000). Wastewater treatment by pond systems: experiences in Catalonia, Spain. *Water Science and Technology*, **42**(10–11), 35–42.

Kanai, R., Aoki, S., and Miyachi, S. (1965). Quantitative separation of inorganic polyphosphates in *Chlorella* cells. *Plant and Cell Physiology*, **6**, 467–473.

Kaplan, D., Richmond, A. E., Dubinsky, Z., and Aaronson, S. (1986). Algal Nutrition. In A. E. Richmond (Ed.), *CRC handbook of microalgal mass culture*. Florida: CRC Press.

Kuhl, A. (1962). Inorganic phosphorus uptake and metabolism. In R. A. Lewin (Ed.), *Physiology and biochemistry of algae* (pp. 211–230). New York: Academic Press.

Kuhl, A. (1974). Phosphorus. In W. D. P. Stewart (Ed.), *Algal physiology and biochemistry*. Oxford: Blackwell Scientific.

Lee, Y. K., and Shen, H. (2004). Basic culturing techniques. In A. E. Richmond (Ed.), *Handbook of microalgal culture: biotechnology and applied phycology* (pp. 40–56). Oxford: Blackwell Science Limited.

Mesple, F., Troussellier, M., Casellas, C., and Bontoux, J. (1995). Difficulties in modelling phosphate evolution in a high rate algal pond. *Water Science and Technology*, **31**(12), 45–54.

Mino, T., van Loosdrecht, M. C. M., and Heijnen, J. J. (1998). Microbiology and biochemistry of the enhanced biological phosphate removal process. *Water Research*, **32**(11), 3193–3207.

Moutin, T., Gal, J. Y., El Halouani, H., Picot, B., and Bontoux, J. (1992). Decrease of phosphate concentration in a high rate pond by precipitation of calcium phosphate: theoretical and experimental results. *Water Research*, **11**, 1445–1450.

Picot, B., Bahlaoui, A., Moersidik, B., Baleux, B., and J., B. (1992). Comparison of the purifying efficiency of high rate algal pond with stabilization pond. *Water Science and Technology*, **25**(12), 197–206.

Powell, N., Shilton, A., Pratt, S., Chisti, Y., Bilby, S., and Pepper, C. (2005). Upgrading waste stabilisation ponds for phosphorus removal. Paper presented at the New Zealand Water and Wastes 47th Annual Conference, Auckland, New Zealand.

Racault, Y., Boutin, C., and Seguin, A. (1995). Waste stabilisation ponds in France: a report on fifteen years experience. *Water Science and Technology*, **31**(12), 91–101.

Rhee, G. Y. (1973). A continuous culture study of phosphate uptake, growth rate and polyphosphate in *Scenedesmus* sp. *Journal of Phycology*, **9**, 495–506.

Shelef, G., Azov, Y., and Moraine, R. (1982). Nutrients removal and recovery in a two stage high rate algal wastewater treatment system. *Water Science and Technology*, **14**, 87–100.

Sukenik, A., Schroder, W., Lauer, J., Shelef, G., and Soeder, C. J. (1985). Coprecipitation of microalgal biomass with calcium and phosphate ions. *Water Research*, **19**(1), 127–129.

Nitrifying and denitrifying biomass characterization of an organic biofilter treating pig manure

Y. Gilbert*, Y. Le Bihan, G. Aubry*, C. Duchaine***, and P. Lessard***

* Département de génie civil, Pavillon Pouliot, Université Laval, Québec G1K 7P4, Canada (E-mail: *paul.lessard@gci.ulaval.ca*)
** Centre de Recherche Industrielle du Québec, 333 Franquet, Québec G1P 4C7, Canada (E-mail: *yann.lebihan@criq.qc.ca*)
*** Centre de Recherche de l'Hôpital Laval, 2725 chemin Ste-Foy, Québec G1V 4G5, Canada (E-mail: *Caroline.Duchaine@crhl.ulaval.ca*)

Abstract Organic biofiltration of pig manure shows a complete removal of organic matter but partial nitrogen elimination. In order to enhance nitrification and denitrification, it is necessary to characterize the biomass responsible for its transformation inside the process. Respirometry, acetylene blockage technique and real-time PCR have been used to respectively measure nitrifying and denitrifying activities and to enumerate nitrogenous microorganisms inside a pilot biofilter. Nitrogen in gas and liquid phases has also been measured at different depths to monitor nitrogen transformations in the system. Results showed that the nitrifying and denitrifying biomasses mainly colonize the layer near the inlet of the biofilter and that they can be found together, having colonized the same ecological niche. Potential activities measured are much greater than that can be observed inside the process, suggesting the possibility to enhance nitrogen removal. It can be seen that denitrifiers containing *nirS* gene and AOB are a part of the biomass respectively responsible for the denitrifying and nitrifying activity, as their number evolution statistically follows that of their respective potential activity. Data obtained will allow the optimization of nitrogen removal by showing what the biomass is capable of and where it is inside the process.

Keywords Activity; biofilter; denitrification; nitrification; real-time PCR

Introduction

Nitrogen in industrial and agri-food wastewaters, especially in pig manure, poses serious environmental problems when not treated. Accumulation of this element in ecosystems is in part responsible for eutrophication of lakes and rivers and underground water contamination, often used as potable water sources.

It is possible to use pig manure as a fertilizer by spreading it on agricultural soil; it constitutes an excellent nutrient source for many plants. However, because canadian laws limit the volume that can be spread on a given land, some regions produce manure in excess compared with the available land. Spreading this surplus increases the risks of water and environmental contamination. It thus must be treated to reduce the risks associated with its presence.

Organic biofilters, developed under the label BIOSOR™-Manure, constitutes an interesting alternative by treating *in situ* the pig manure. This process uses indigenous microorganisms of the filtering media and its physico-chemical properties to retain and transform the pollutants in less damaging compounds, such as carbon dioxide (CO_2) and gaseous nitrogen (N_2) (Rock *et al.*, 1984; Gilbert *et al.*, 2005; Garzón-Zúñiga *et al.*, 2005). Already operating BIOSOR^{TM} biofilters have shown an organic matter removal of 95 to 99%, evaluated as BOD_5 (biological oxygen demand after 5 days) and allowed an incomplete elimination of TKN (Total Kjeldahl nitrogen) as N_2.

For nitrogen removal, a sequence of nitrification and denitrification is necessary. The first is the aerobic oxidation, via two steps, of ammonia (NH_4^+) to nitrate (NO_3^-), mostly performed by autotrophic bacteria. The primary step, realized by the *Nitroso-* group, *Nitrosomonas* being a well known member, allows the conversion of NH_4^+ to nitrite (NO_2^-), while the second step, NO_2^- to NO_3^-, is performed by the *Nitro-* group, *Nitrobacter* and *Nitrospira* being the most studied microorganisms of this group. Denitrification is mostly accomplished by heterotrophic microorganisms that are capable of using NO_3^- and NO_2^- as electron acceptors in lack of oxygen. Many reduction steps are necessary to transform these nitrogen oxides to gaseous compounds, each one being realized by a specific reductase (Zumft, 1997).

Garzón-Zúñiga *et al.* (2005) studied the behavior of nitrogen compounds in an aerated organic biofilter treating pig manure. After the start-up period, the process was found to totally oxidize the ammonia in NO_3^- and to partly reduce the latter in N_2. They hypothesized that a simultaneous nitrification and denitrification (SND) occurred inside the system under aerobic conditions. In order to optimize nitrogen transformations by the biofilter and for a better understanding of its behavior inside the system, characterization of the nitrifying and denitrifying biomass was needed.

However, except for compost (Kowalchuk *et al.*, 1999; LaMontagne *et al.*, 2002), few studies relate to the monitoring of the biomass inside this type of media, particularly when treating pig manure. This is due to a lack of monitoring techniques really adapted to this particular heterogeneous media.

The principal objective of this work was thus to realize a microbiological monitoring of nitrification and denitrification inside an organic biofilter treating pig manure. It was however essential to adapt at first existing techniques to realize a characterization of the biomass present on the media particles and involved in nitrogen transformations. This paper is a summary of the results obtained for the monitoring of a laboratory biofilter. A brief description of the adapted techniques is made in the Methods section.

Methods

Biofilter operation

A pilot biofilter was constructed and operated during 180 days (Figure 1). It consisted of two principal layers, for a total height of 1.9 m. At the bottom, there was a mix of peat, wood chips, and calcite for 1.2 m called the "organic layer". A 0.7 m coarse layer, mostly consisting of pozzolana, surmounted it in order to prevent an early clogging of the organic media. The biofilter was fed by percolation at a hydraulic rate of 0.017 m^3 m^2 d^{-1} with pig manure. It was a physico-chemically pretreated manure having a BOD_5 of approximately 14,000 mgO_2 L^{-1}, a BOD_5:TKN ratio of 6 and a chemical oxygen demand (COD) of about 20,000 mgO_2 L^{-1}. It was kept at 4°C for its preservation. A countercurrent aeration of 1.67×10^{-3} m^3 m^{-2} s^{-1} brought the oxygen to the biomass. To get a better comprehension of the nitrogen transformations inside the system, the monitoring has been done at different depths and at various durations of operation; the organic layer has thus been divided in three sub-layers.

Fiberglass wicks were placed at regular intervals inside the filter to sample, by capillarity, the liquid phase at different heights as described by Zhu *et al.* (2003). Holes were also located at regular intervals between the wicks to get biomass samples at the depth studied. For each depth, three sample holes around the biofilter allowed an alternate sampling of the studied zone. Hermetic lids were installed at each end of the biofilter to collect gas and treated water.

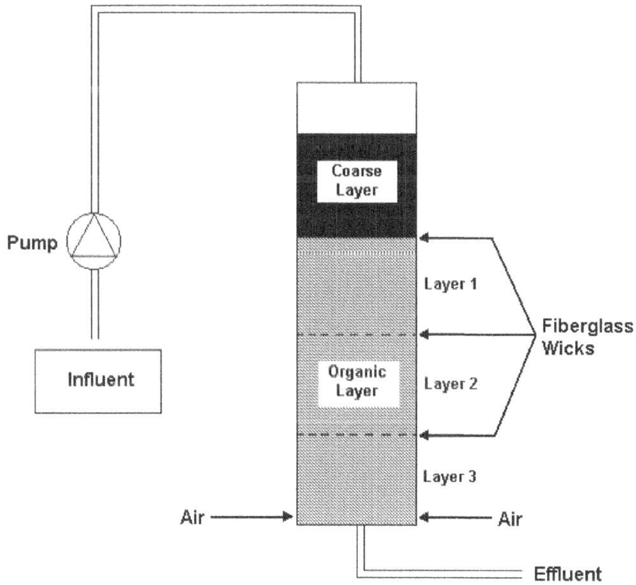

Figure 1 Schematic view of the biofilter.

Ammonia (NH_3), nitrous oxide (N_2O) and carbon dioxide (CO_2) were analyzed weekly with a gas analyzer using photo-acoustic spectroscopy (Brüel and Kjœr model 1301) at each depth and in the inlet and outlet air flow. Nitrite (NO_2^--N), nitrate (NO_3^--N) and aqueous ammonia (NH_4^+-N) concentrations were analyzed twice a month for each wick and in the liquid influent and effluent. Samples were kept at –20°C for no more than two months before being analyzed. NO_2^--N and NO_3^--N were measured by HPLC as described by Fernández-Boy *et al.* (1998) while NH_4^+-N was evaluated according to Standard Methods (APHA 1992). These data have been used to determine the nitrifying and denitrifying performances of a given layer by calculating a mass balance for nitrogen.

Biomass samples were collected every three weeks for a microbiological characterization, i.e. evaluation of the potential nitrifying and denitrifying activities and molecular quantification. Activities were evaluated the moment the sample was taken out of the process. Approximately 2 g of filtering media were freezed at –20°C for the quantification, up to 6 months before DNA extraction. New media was added to the biofilter in sample holes in order to maintain biofilter structure and to reduce preferred channeling.

Biomass activity

Denitrifying activity was measured *ex-situ* by an adapted method implying the acetylene inhibition technique (Gilbert *et al.*, submitted). The acetylene inhibits the last step of denitrification, N_2O to N_2, giving an easy way to evaluate denitrification by measuring N_2O instead of N_2 (Yoshinari and Knowles, 1976). Samples were put inside a cylindrical cell

(Fig. 2) and saturated with a synthetic nutritive solution containing sodium acetate, potassium nitrate and salts. A flow of nitrogen (N_2) was applied to the cell, at the same rate than the aeration of the biofilter, to create optimal conditions for denitrifiers. N_2O concentration was monitored during 1 hour, at the gaseous inlet and outlet of the cell, to evaluate N_2O produced by the biomass under anoxic conditions, associated with an incomplete denitrification. N_2O was analyzed with a gas analyzer using photo-acoustic spectroscopy (Brüel and Kjœr model 1301). An acetylene flux, with a partial pressure of 10 kPa (acetylene 10% / argon 90%) was then applied to the cell to inhibit the last reduction step. Again, N_2O concentration was monitored, for two to three hours. This represents the N_2O produced by complete and incomplete denitrification and thus, the potential denitrifying activity. The cumulative N_2O-N produced by the biomass was calculated using Equation 1 and expressed on a dry weight of filtering media basis and then used to calculate the potential denitrification rate of a layer.

$$N - N_2O_{cumul} = \frac{28}{44} Q \sum_{i=1}^{n} T_i (N_2O_i - N_2O_{inlet}) \tag{1}$$

N-N_2O_{cumul} is the cumulative mass of nitrogen produced from t = 0 to t = i (µg), N_2O_i is the concentration of N_2O at t = i (µg m^{-3}), N_2O_{inlet} is the concentration of N_2O of the gaseous inlet (µg m^{-3}), Q is the gaseous flow (m^3 s^{-1}), 28/44 is the factor to obtain the mass of N instead of N_2O and T_i is the incremental time at t = i (s). The maximum activity is reached when a linear accumulation can be detected.

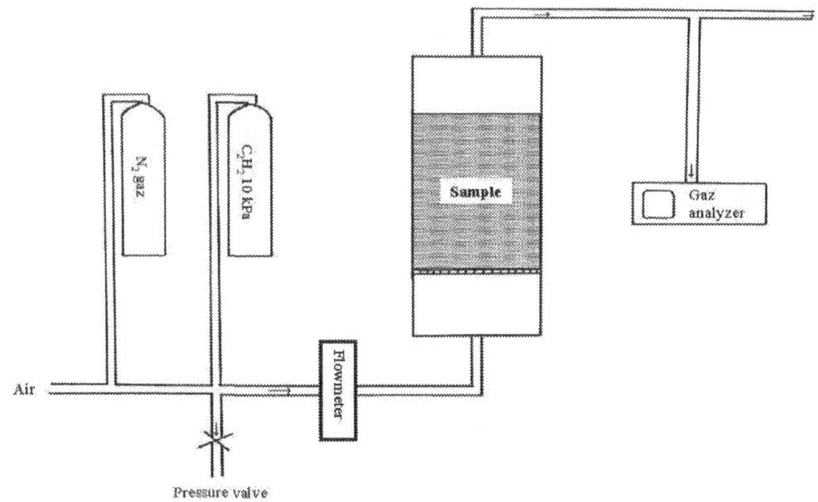

Figure 2 Schematic view of the cell used to evaluate potential denitrification.

Nitrifying activity was also evaluated *ex situ*, by LSS respirometry, as described by Gilbert *et al.* (2005). Ammonia was injected to the reactor and dissolved oxygen was followed for 30 minutes or until a drop of 2 mg O_2 L^{-1}, to evaluate the respiration of the biomass. Allylthiourea (ATU) was then injected to specifically inhibit nitrifiers (McCarty, 1999). The difference of the oxygen consumption before and after the addition of ATU gives the potential nitrifying activity. Specific activity was expressed on a volatile suspended solid (VSS) basis and then used to calculate the activity of a layer. A PCM 800 Orion Sensorlink data acquisition system and an Orion 083005D dissolved oxygen probe (Thermo Electron Corporation) were used during the assays.

Biomass quantification

Real-time polymerase chain reaction (PCR) was used to quantify microorganisms fixed on the filtering media. Ammonia oxidizing bacteria (AOB), *Nitrospira* and *Nitrobacter* were chosen as reference microorganisms for nitrifiers quantification while organisms containing *nirS* gene were used as such for denitrifiers.

AOB 16S rDNA was amplified using primers CTO189f A/B, CTO189f C and RT1r as described by Hermansson and Lindgren (2001) and Harms *et al.* (2003). The PCR program was 2 min at 50°C, 2 min at 95°C, 45 cycles at 95°C for 30 s, 60°C for 30 s, and 72°C for 30 s.

Nitrobacter 16S rDNA was amplified using primers FGPS872 and FGPS1269' as described by Féray *et al.* (1995). The PCR program was 2 min at 50°C, 2 min at 95°C, 45 cycles at 95°C for 1 min s, 50°C for 1 min, 72 °C for 1 min and 80°C for 15 s for florescence reading.

Nitrospira 16S rDNA was amplified using primers NSR1113f and NSR1264r and quantified with Taqman probe NSR1143Taq as described by Harms *et al.* (2003). The PCR program was 2 min at 50°C, 2 min at 95°C, 45 cycles at 95°C for 30 s and 63°C for 30 s.

nirS was amplified using primers cd3aF and R3cd as described by Throbäck *et al.* (2004). The PCR program was 2 min at 50°C, 2 min at 95°C, 45 cycles at 95°C for 30 s, 63°C for 30 s and 80°C for 15 s for florescence reading.

For each reaction, the PCR mix, with a total volume of 25 µL, contained Platinum SYBR Green qPCR SuperMix UDG or Platinum qPCR SuperMix UDG (Invitrogen Life Technologies, Carlsbad, CA), adequate concentration of primers and 1 to 100 ng of sample DNA or dilutions of pDrive cloning vector (Qiagen) carrying an insert generated using the same primers, as standard (from 10^1 to 10^8 copies of the vector). Real-time PCR assays were run using a DNA Engine Opticon 2 Real-time PCR Detection System (MJ Research, Waltham, MA). Because some of the vectors have been constructed using environmental DNA (AOB, *Nitrobacter* and *Nitrobacter*), they all have been sequenced to be assured that the amplicon cloned really came from the desired microorganism. Vectors for *nirS* gene have been constructed with amplicons from the *nirS* gene amplification of *Pseudomonas aeruginosa* ATCC 10145. Sequences are available on Genbank under accession numbers DQ386157-DQ386160. DNA was extracted using a commercial kit (FastDNA SPIN kit for soil, Qbiogene Inc, CA) following supplier instructions.

Results and discussion

The process showed satisfying purification performances all along its operation, with a BOD_5 under 6 mgO_2 L^{-1} and a removal of more than 95% of the COD. Ammonia stripping was observed for the first month of operation but it gradually diminished to a stable concentration (~15 ppm) at the outlet air flow during the next thirty days, corresponding to the beginning of the nitrification. Mass balances on nitrogen showed that from the moment NO_3 could be observed, denitrifiers transformed a part of it into gaseous nitrogen.

Data obtained after 180 days of operation shows that the denitrifying and nitrifying biomass mainly colonized the coarse layer (Figure 3), evaluated by the potential activity of the layer. The potential nitrification and denitrification activities were found to be much higher in this layer than in the organic biofilter itself, up to 400 times higher for nitrification and more than 20 times higher for denitrification. COD analysis showed that 97% of the organic matter was removed by the coarse layer's biomass. Because most of the denitrifying bacteria is heterotrophic, it seems reasonable to imagine that there is not enough organic matter to sustain colonization in the organic media as important as that found in the coarse layer. That could explain why the potential denitrifying activity is less in the organic layers. It was however surprising to found that nitrifiers could grow at the top of the biofilter, knowing that their growth rate is much less than that of heterotrophic microorganisms and that they have to compete for oxygen (Metcalf and Eddy, 2003). We hypothesized that the biofilm retained them from being washed out of the top layer and that the carbon dioxide produced in large quantity by the biodegradation of organic matter promotes their growth.

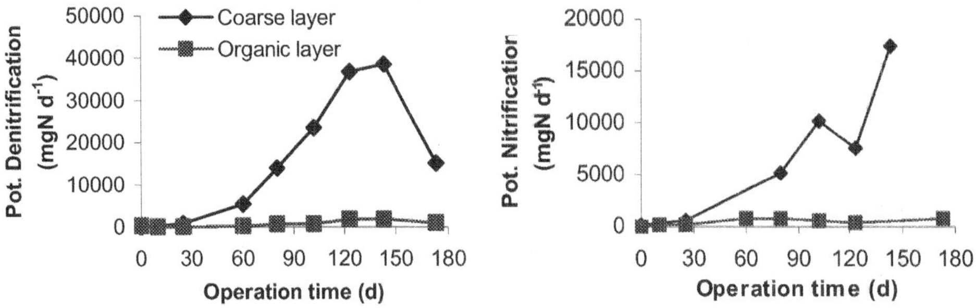

Figure 3 Potential denitrifying and nitrifying activities of each section of the biofilter.

We also found that a microbiological gradient was found all along the height of the process, being mostly colonized near the inlet (Figure 3) and diminishing from top to bottom (Figure 4), sub-layer 1 being more colonized than sub-layer 2. This trend was observed for both nitrifiers and denitrifiers. Data for sub-layer 3 is not described on the figures since its activity was negligible.

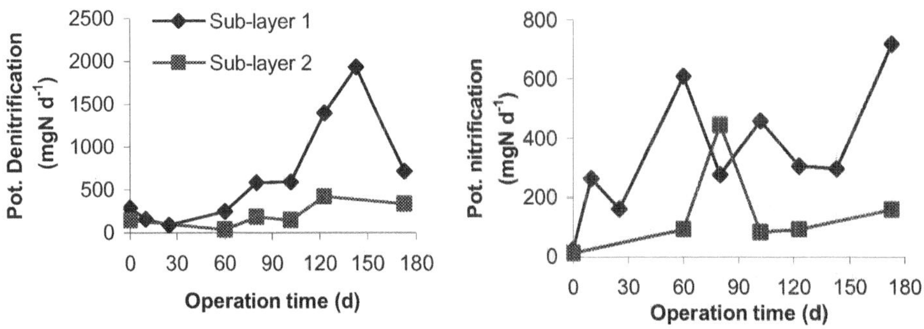

Figure 4 Potential denitrifying and nitrifying activities of the first two layers of the organic filtering media.

The nitrogen mass balances calculated for each layer of the biofilters demonstrated that a part of the nitrogen was lost in the gaseous phase. Taking into account the microbial nitrogen assimilation estimated (Garzón-Zúñiga *et al.*, 2005), it is supposed that this lost is due to the production of N_2O and N_2 by denitrification, showing the simultaneous nitrification and denitrification taking place in the biofilter under aerobic conditions. This SND is probably possible in thick biofilm and in anoxic micropores of the filtering media as proposed by many authors (Pochana and Keller, 1999; Garzón-Zúñiga *et al.*, 2005). The potential nitrifying and denitrifying activities obtained corroborate these results, even showing that the biomass was able to transform much more nitrogen than what was being used and that, for all the layers being studied. Actually, the coarse layer had the capability to nitrify up to five times what it was really doing and denitrify up to 16 times more. The biomass of the organic layer was able to nitrify and denitrify five times more nitrogen than that was being utilized. These data suggest that there is place to optimize nitrogen removal by the organic biofiltration process.

The molecular quantification also supports the hypothesis of SND inside the process, nitrifiers and denitrifiers being found in large quantity in the same samples. Moreover, the variation of the log-number of gene copies, quantified by real-time PCR, statistically followed the same trend than that can be observed for its associated activity ($p > 0.05$), particularly with *nirS* for denitrifiers and with AOB (Figure 6) for nitrifiers, for all layers studied. These data strongly suggest that those microorganisms are largely responsible for the activities measured. *Nitrospira* was in insufficient number to be quantified by this method, its concentration being less than the minimum detectable threshold, 4.0×10^5 copies g^{-1} (dry weight basis). The number of copies of 16S rDNA for *Nitrobacter* did not statistically follow the trend of the potential nitrifying activity ($p < 0.05$), suggesting that this microorganism is not the one limiting nitrification.

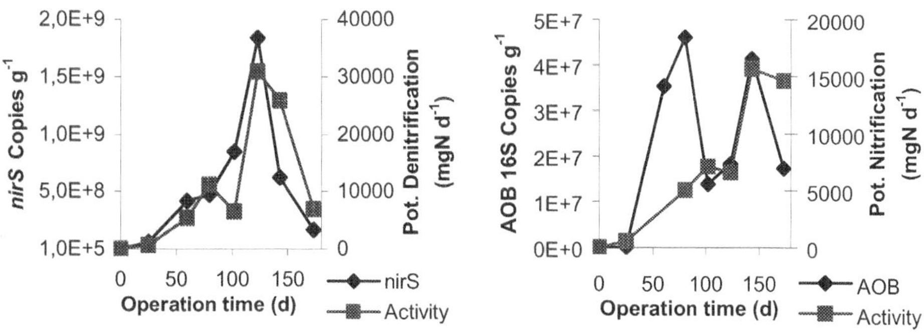

Figure 6 Evolution of the potential activities and genes quantity in the coarse layer for nitrifiers and denitrifiers.

Conclusion

The characterization made on the filtering media described how the nitrifying and denitrifying biomass colonized the process. It has been showed that microorganisms implicated in nitrogen transformations mainly colonized the top of the biofilter during the first 180 days of operation. Nitrification began approximately after 50 days of operation and denitrification appeared as soon as nitrate was being produced.

Nitrifiers and denitrifiers have demonstrated high potential activities in the coarse layer compared to deeper ones, and they have been found in large numbers in it. Data also demonstrated that the biomass colonizes the biofilter with a gradient from top to bottom.

Relationship between observed nitrifying and denitrifying activities strongly supports the hypothesis that SND is a possible mechanism inside this kind of process and that it occurs inside the biofilm and anoxic microzones. Moreover, the fact that nitrifiers and denitrifiers can be found together in important number in the same samples also supports it.

The potential activities obtained, for nitrification and denitrification, were much higher than that could be observed. It suggests that it is possible to enhance nitrogen removal. Because nitrification of the ammonia is already fully completed by the process, this optimization could be done particularly by promoting denitrifiers activity. For example, diminishing aeration in some layers, thus creating more anoxic zones in the process, would help denitrifiers to use more nitrate as electron acceptor. However, according to our findings, that all the biodegradation processes occur at proximity from one another, precaution has to be taken to increase denitrification without influencing nitrification and biodegradation of organic matter.

References

APHA (1992) *Standard Methods for the Examination of Water and Wastewater*, 18th edition, American Public Health Association/American Water Works Association/Water Environment Federation, Washington DC, USA.

Féray, C., Volat, B., Degrange, V., Clays-Josserand, A. and Montuelle, B. (1999) Assessment of three methods for detection and quantification of nitrite-oxidizing bacteria and Nitrobacter in freshwater sediments (MPN-PCR, MPN-Griess, Immunofluorescence). *Microbial Ecology*, **37**, 208–217.

Fernández-Boy, M. E., Cabrera, F. and Moreno, F. (1998) Analysis of inorganic anions in drainage water and soil solution by single-column ion chromatography. *Journal of Chromatography A*, **823**, 285–290.

Garzón-Zúñiga, M, Lessard, P, Aubry, G and Buelna, G. (2005) Nitrogen elimination mechanisms in an aerated peat biofilter treating pig manure. *Environmental Technology*, **26**(4), 361–371.

Gilbert, Y., Le Bihan, Y., Buelna, G. and Lessard, P. (2005) Application of respirometry to Monitor the Biomass Activity of a Biofilter on an Organic Bed Used to Treat Pig Manure. *Water Quality Research Journal of Canada*, **40**(2), 155–163.

Gilbert, Y., Le Bihan, Y. and Lessard, P. (Submitted) Acetylene blockage technique as a tool to determine denitrification potential of a biomass fixed on an organic media treating wastewater. *Journal of Environmental and Engineering Science.*

Harms, G., Layton, A. C., Dionisi, H. M., Gregory, I. R., Garrett, V. M., Hawkins, S. A., Robinson, K. G. and Sayler, G. S. (2003) Real-time PCR quantification of nitrifying bacteria in a municipal wastewater treatment plant. *Environmental Science and Technology*, **37**(2), 343–351.

Hermansson, A. and Lindgren, P.-E. (2001) Quantification of ammonia-oxidizing bacteria in arable soil by real time PCR. *Applied and Environmental Microbiology*, **67**(2), 972–976.

Kowalchuk, G. A., Naoumenko, Z. S., Derikx, P. J. L., Felske, A., Stephen, J. R. and Arkhipchenko, I. A. (1999) Molecular analysis of ammonia-oxidizing bacteria of the β subdivision of the class Proteobacteria in compost and composted materials. *Applied and Environmental Microbiology*, **65**(2), 396–403.

LaMontagne, M. G., Michel, F. C., Holden, P. A. and Reddy, C. A. (2002) Evaluation of extraction and purification methods for obtaining PCR-amplifiable DNA from compost for microbial community analysis. *Journal of Microbiologial Methods*, **49**, 255–264.

McCarty, G. W. (1999) Modes of action of nitrification inhibitors. *Biol. Fertil. Soils*, **29**, 1–9.

Metcalf and Eddy (2003) *Wastewater engineering, treatment and reuse*, 4th edition. McGraw Hill, New York.

Pochana, K. and Keller,J. (1999) Study of factors affecting simultaneous nitrification and denitrification (SND) *Water Science and Technology*, **39**(6), 61–68.

Rock, C. A., Brooks, J. L., Bradeen, S. A. and Struchtemeyer, R. A. (1984) Use of peat for on-site wastewater treatment: I. Laboratory evaluation. *Journal of Environmental Quality*, **13**(4), 518–523.

Throbäck, I. N., Enwall, K., Jarvis, A. and Hallin, S. (2004) Reassessing PCR primers targeting nirS, nirK and nosZ genes for community surveys of denitrifying bacteria with DGGE. *FEMS Microbiology Ecology*, **49**(3), 401–417.

Yoshinari, T. and Knowles, R. (1976) Acetylene inhibition of nitrous oxide reduction by denitrifying bacteria. *Biochemical and Biophysical Research Communication*, **69**(3), 705–710.

Zhu, Y., Fox, R. H. and Toth, J. D. (2003) Tillage effects on nitrate leaching measured by pan and wick lysimeters. *Soil Science Society of America Journal*, **67**, 1517–1523.

Zumft, W. G. (1997) Cell biology and molecular basis of denitrification. *Microbiology and Molecular Biology Reviews*, **61**(4), 533–616.

Enhancement of food waste anaerobic digestion in a hybrid anaerobic solid–liquid (HASL) system

Xue-Yan Liu, Olena Stabnikova and Jing-Yuan Wang

Division of Environmental and Water Resources Engineering, School of Civil and Environmental Engineering, Nanyang Technological University, 50 Nanyang Avenue, Singapore 639798 (E-mail: *Liux0012@ntu.edu.sg*; *Costab@ntu.edu.sg*; *Jywang@ntu.edu.sg*)

Abstract The hybrid anaerobic solid–liquid (HASL) system was developed to reclaim energy and minimize waste amount for disposal in Singapore. Thermal pre-treatment of food waste at 70 °C for 2 h (Experiment E1) and at 150 °C for 1 h (Experiment E2) enhanced the hydrolytic and acidogenic processes in the acidogenic reactor (Ra) and methanogenesis in the methanogenic reactor (Rm) in the HASL system. The highest dissolved COD concentrations in the leachate from Ra were 17575, 19980 and 24235 mg/l in control with food waste without thermal pre-treatment, E1 and E2, respectively. The highest VFA concentrations in the leachate from Ra were 12470, 17818 and 19795 mg/l in control, E1 and E2, respectively. The maximum concentrations of methanogens in Rm were 2.3×10^7, 3.8×10^7 and 4.3×10^7 cells/ml in control, E1 and E2, respectively. Use of thermally pre-treated food waste as feedstock halved the time to produce the same quantity of methane in comparison with anaerobic digestion of fresh food waste in the HASL system. The direct counting of methanogens and the fluorescence measurement of coenzyme F_{420} were used to monitor the changes of methanogens in Rm during the experiments. It was a reliable statistical linear correlation (coefficient of linear correlation was 0.829) between fluorescence of F_{420} and methanogens concentration in Rm.

Keywords Co-enzyme F_{420}; food waste; HASL; methanogens; thermal pre-treatment

Introduction

The total food wastes output in Singapore amounted 1,092,500 tonnes in 2000 (http://www.nea.gov.sg). Currently, almost all food wastes are disposed by incineration. The hybrid anaerobic solid–liquid (HASL) system was developed to be used in industrial scale operation to minimize the amount of food waste for disposal in Singapore (Wang *et al.*, 2005a,b). The HASL system includes an acidogenic reactor (Ra) to treat solid food waste and an upflow anaerobic sludge blanket (UASB) reactor as a methanogenic reactor (Rm) to treat the leachate from Ra. Part of the effluent from Rm is used for the dilution of the acid leachate from Ra to maintain optimal pH for methanogenesis in Rm. The rest of the effluent from Rm is recycled into Ra to avoid addition of water for food waste hydrolysis. The schematic diagram of the hybrid anaerobic solid–liquid system is shown in Figure 1.

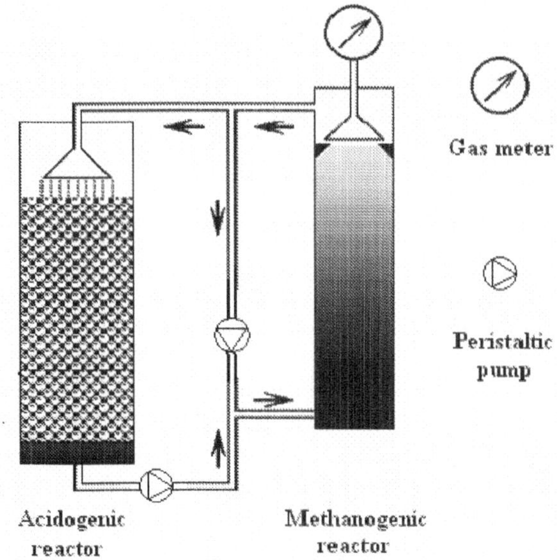

Figure 1 Schematic diagram of the hybrid anaerobic solid–liquid system.

Liquefaction/acidification and acetogenesis/methanogenesis phases are separated in the HASL system. Hydrolysed and acidified leachate is obtained in the first acidogenic reactor. The leachate from the acidogenic reactor will be degraded in the second methanogenic reactor, and biogas with high content of methane will be produced there. Hydrolysis, the first step in anaerobic biodegradation of organic matter, ensures the breakdown of biopolymers of food waste to water-soluble products, which could be easily consumed by microorganisms. However, hydrolysis and liquefaction of organic waste often are the rate-limiting steps in anaerobic process (Verstraete *et al.*, 1996; Shin *et al.*, 2001). To enhance anaerobic digestion of organic waste, mechanical, biological and physico-chemical pre-treatment methods could be used (Mata-Alvarez *et al.*, 2000). Heating of food waste appears to be the simplest pre-treatment method (Tsukahara *et al.*, 1999; Schieder *et al.*, 2000). Main limitation for thermal pre-treatment of food waste is energy cost. To cover these expenses, biogas produced from the HASL system, could be used partially for the combustion and heating of food waste.

The aim of present research was to evaluate the effect of thermal pre-treatment of food waste on its anaerobic digestion in the HASL system.

Materials and methods

Food waste used in this study was collected from a university canteen and shredded into particles with average size of 6.0 mm in a Robot-Coupe Shredder (CL50 Ultra, Hobart, France). The food waste mixture was consisted of 50% vegetable roots, 20% orange peel, 15% rice and 15% noodles. The total solid (TS) was 16.3% and the volatile solid (VS) was 88.0% of TS in the mixed waste. Fresh food wastes were used in control (C), while pre-heated at 70 °C for 2 h and at 150 °C for 1 h were used in E1 and E2, respectively.

The seed sludge used as inoculum in Ra was taken from an anaerobic digester in local wastewater treatment plant. The suspended solids (SS) and volatile suspended solids (VSS) of the seed sludge were 4600 mg/l and 3100 mg/l, respectively. Anaerobic granules in Rm were collected from an UASB reactor, which had been operated for more than 6 months and adapted to high concentration of volatile fatty acids (VFA) (Wang *et al.*, 2003).

800g of food waste and 1 litre of sludge were placed in each acidogenic reactor of three identical HASL systems operated at 35±1 °C in a temperature room. The leachate from Ra and the effluent from Rm were sampled for analyses daily. Chemical oxygen demand (COD) was determined by standard methods (*Standard Methods*, 1998). Volatile fatty acids were analyzed using HPLC (Perkin Elmer, Series 200, Norwalk, CT, USA). Biogas production was monitored by a wet gas meter (Ritter TG05, Germany), while gas composition was analysed by a GC (HP5890A HACH, USA). Enumeration of methanogens was performed by epifluorescence microscope BX-FLA-3 (Olympus, Tokyo, Japan) at 900-fold magnification under UV wavelength from 330 to 385 nm (Solera *et al.*, 2001; Stabnikova *et al.*, 2005). The fluorescence measurement of co-enzyme F_{420} was made using a fluorescence/luminescence spectrometer LS-50B (Perkin Elmer, Wellesley, MA, USA) (Ivanov *et al.*, 2005). The excitation wavelength was changed from 360 to 480 nm and the emission wavelength was synchronously changed from 380 to 500 nm.

The specific growth rate of methanogens was estimated by different ways:

(1) μ was calculated using data of the direct count of methanogens by the equation:

$$\mu = (\ln N - \ln N_0)/(t - t_0) \qquad (1)$$

where N_0 and N are the initial and final concentrations of cells at time t_0 and t, respectively, and

(2) μ_F was calculated using data of autofluorescence of coenzyme F_{420} by the equation:

$$\mu_F = (\ln F - \ln F_0)/(t - t_0) \qquad (2)$$

where F_0 and F are the initial and final fluorescences of F_{420} at time t_0 and t, respectively.

Results and discussion

The changes of COD and VFA concentrations in the leachate from the acidogenic reactors and in the effluent from the methanogenic reactors in control and experiments were shown in Figure 2. The initial dissolved COD concentrations in Ra were 7221, 8544 and 13325 mg/l in control, E1 and E2, respectively (Figure 2a). The initial VFA concentrations in Ra were 1382, 2578 and 3008 mg/l in control, E1 and E2, respectively (Figure 2b). The higher initial COD and VFA concentrations in experiments indicated the intensification of hydrolysis process due to thermal pre-treatment of food waste. The highest dissolved COD concentrations in the leachate from Ra were 17575, 19980 and 24235mg/l on the fourth, third and second day in control, E1 and E2, respectively (Figure 2a). The highest VFA concentrations in the leachate from Ra were 12470, 17818 and 19795 mg/l on the fourth, fourth and third day in control, E1 and E2, respectively (Figure 2b). The peaks of dissolved COD and VFA concentrations were higher and reached faster in E1 and E2 in comparison with control (Figure 2a,b), which meant the thermal pre-treatment enhanced food waste digestion in the HASL system. The dissolved COD and VFA concentrations in the effluent from Rm were lower in E1 and E2 than that in control (Figure 2a,b). The overall COD removal was 78.2%, 80.3% and 82.2% in control, E1 and E2, respectively.

The lowest pH in the leachate from Ra observed was 4.8, 5.7 and 6.2 on the fifth, third and second day in control, E1 and E2, respectively, which matched with the VFA changes shown in Figure 2b. Usually, pH in the leachate from Ra in batch HASL system dropped rapidly to about 5.0 probably owing to cells and vacuoles lysis, and after that slowly increased to neutral value, probably due to hydrolysis of proteins and buffering of pH by released aminoacids (Stabnikova *et al.*, 2005). Therefore, lower pH values in control can be hypothetically explained by deeper hydrolysis of proteins after thermal pre-treatment of food waste.

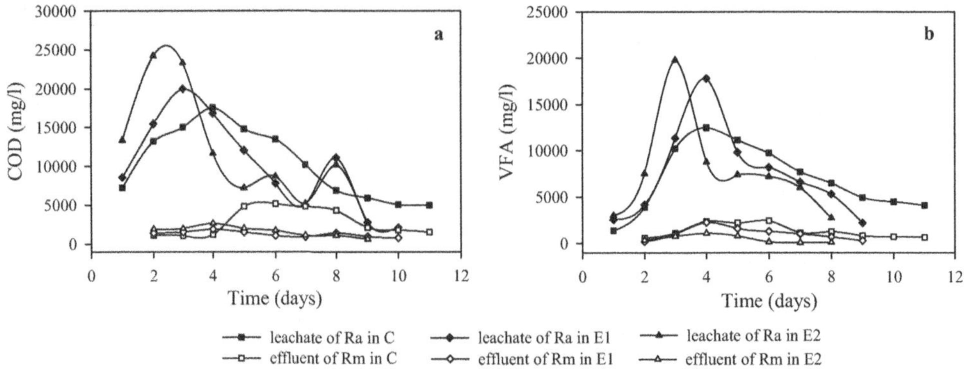

Figure 2 Changes of COD (a) and VFA (b) in the leachate from Ra and in the effluent from Rm in control and experiments.

The lowest pH in the leachate from Ra observed was 4.8, 5.7 and 6.2 on the fifth, third and second day in control, E1 and E2, respectively, which matched with the VFA changes shown in Figure 2b. Usually, pH in the leachate from Ra in batch HASL system dropped rapidly to about 5.0 probably due to cells and vacuoles lysis, and after that slowly increased to neutral value, probably because of hydrolysis of proteins and buffering of pH by released aminoacids (Stabnikova *et al.*, 2005). Therefore, lower pH values in control can be hypothetically explained by deeper hydrolysis of proteins after thermal pre-treatment of food waste.

Total methane production increased only by 3.3% and 11.5% in E1 and E2 in comparison with control. However, use of thermally pre-treated food waste as feedstock halved the time to produce the same quantity of methane in comparison with anaerobic digestion of fresh food waste in the HASL system (Figure 3a). According to the concentrations of methanogens estimated by direct cell count which was shown in Figure 3b, thermal pre-treatment of food waste increased the concentrations of methanogens in Rm. The maximum concentrations of methanogens in Rm were 2.3×10^7, 3.8×10^7 and 4.3×10^7 cells/ml on the eleventh, ninth and seventh day of waste processing in control, E1 and E2, respectively. Thus, pre-treatment of food waste increased yield of methanogens in the methanogenic reactor, which ensured higher methane production in E1 and E2 than in control (Figure 3a).

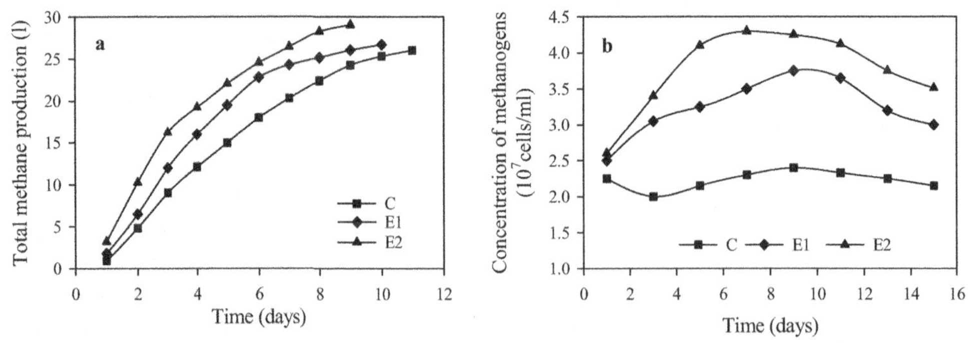

Figure 3 Accumulative methane production (a) and methanogens concentrations (b) in Rm in control and experiments.

To evaluate the changes of methanogenic activity in the methanogenic reactor, independently from direct cell count, measurement of autofluorescence of coenzyme F_{420} was used, because F_{420}, a 7,8-didemethyl-8-hydroxy-5-deazaflavin electron transfer coenzyme, is an essential component of methanogenic biomass. The autofluorescence measurement of coenzyme F_{420} can be used to monitor methanogens during anaerobic digestion of food waste (Heine-Dobbernack *et al.*, 1988; Ivanov *et al.*, 2005). According to the intensity of autofluorescence of F_{420} shown in Figure 4a, the concentration of methanogens was highest in E2, medium in E1, and lowest in control, which had the same trend as the changes of methanogens in Rm estimated by direct count (Figure 3b).

The maximum growth rates, μ_{max}, calculated by equation (1) were 0.07, 0.10 and 0.13 d^{-1} for control, E1 and E2, respectively. The maximum growth rates, μ_{Fmax}, calculated by equation (2) were 0.18, 0.29 and 0.33 d^{-1} for control, E1 and E2, respectively. The differences in the growth rates determined by direct count of methanogens and by fluorescence of F_{420} can be explained that intracellular content of coenzyme F_{420} varies widely among methanogens (Heine-Dobbernack *et al.*, 1988; DiMarco *et al.* 1990), meanwhile direct count shows the presence of F_{420} in microbial cells. Higher growth rates determined by fluorescence F_{420} demonstrate increasing content of F_{420} in biomass during phase of exponential growth.

However, notwithstanding hypothetical changes of F_{420} content in biomass during exponential growth, correlation between autofluorescence of F_{420} and methanogens concentration was statistically reliable (Figure 4b). The coefficient of linear correlation was 0.829. Therefore, the fluorescence measurement of F_{420} can be used for the comparative estimation of methanogens during anaerobic digestion of food waste.

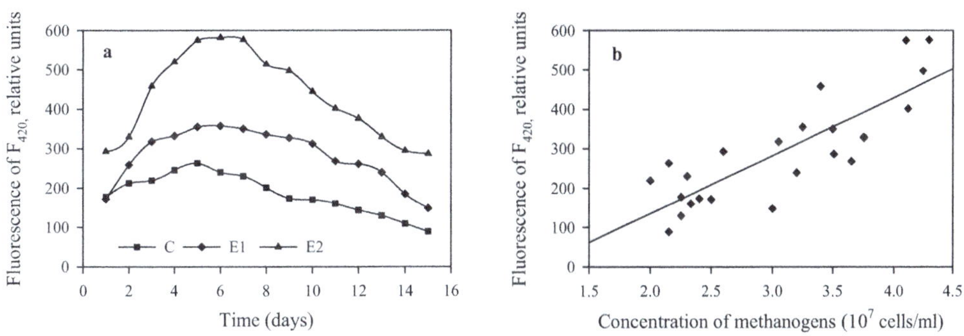

Figure 4 Fluorescence of coenzyme F_{420} in the samples from Rm in control and experiments (a) and correlation between fluorescence of coenzyme F_{420} with the concentration of methanogens (b).

Conclusions

Thermal pre-treatment of food waste enhanced the hydrolytic and acidogenic processes in the acidogenic reactor and increased the methanogens concentrations in the methanogenic reactor, which led to faster anaerobic digestion of food waste in the HASL system. Use of thermal pre-treated food waste halved the time needed to produce the same quantity of methane in comparison with the anaerobic digestion of fresh food waste. The fluorescent methods of F_{420} can be used for the comparative estimation of methanogens during anaerobic digestion of food waste.

References

DiMarco, A.A., Bobik, T.A. and Wolfe, R.S. (1990). Unusual coenzymes of methanogenesis. *Annual Review of Biochemistry*, **59**, 355–394.

Heine-Dobbernack, E., Schoberth, S.M. and Sahm, H. (1988). Relationship of intracellular coenzyme F420 content to growth and metabolic activity of *Methanobacterium bryantii* and *Methanosarcina barkeri*. *Applied and Environmental Microbiology*, **54**(2), 454–459.

Ivanov, V.N., Wang, J.Y., Stabnikova, O., Tay, S.T.L. and Tay, J.H. (2004). Microbiological monitoring in the biodegradation of sewage sludge and food waste. *Journal of Applied Microbiology*, **96**(4), 641–647.

Mata-Alvarez, J., Macé, S. and Llabrés, P. (2000). Anaerobic digestion of organic solid wastes. An overview of research achievements and perspectives. *Bioresource Technology*, **74**(1), 3–16.

Ministry of the Environment, Singapore. The Annual Report. (2000). http://www.nea.gov.sg (accessed 24 Nov 2005).

Schieder, D., Schneider, R. and Bischof, F. (2000). Thermal hydrolysis (TDH) as a pretreatment method for the digestion of organic waste. *Water Science and Technology*, **41**(3), 181–187.

Shin, H.S., Han, S.K., Song, Y.C. and Lee, C.Y. (2001). Multi-step sequential batch two-phase anaerobic composting of food waste. *Environmental Technology*, **22**(3), 271–279.

Solera, R., Romero, L.I. and Sales, D. (2001). Determination of the microbial population in thermophilic anaerobic reactor: Comparative analysis by different counting methods. *Anaerobe*, **7**(2), 79–86.

Stabnikova, O., Ang, S.S., Liu, X.Y., Ivanov, V., Tay, J.H. and Wang, J.Y. (2005). The use of hybrid anaerobic solid–liquid (HASL) system for the treatment of lipid-containing food waste. *Journal of Chemical Technology and Biotechnology*, **80**(4), 455–461.

Standard Methods for the examination of Water and Wastewater. (1998). 20th edition, American Public Health Association/American Water Works Association/Water Environmental Federation, Washington DC, USA.

Tsukahara, K., Yagishita, T., Ogi, T. and Sawayama, S. (1999). Treatment of liquid fraction separated from liquidized food waste in an upflow anaerobic sludge blanket reactor. *Journal of Bioscience and Bioengineering*, **87**(4), 554–556.

Verstraete, W., de Beer, D., Pena, M., Lettinga, G. and Lens, P. (1996). Anaerobic bioprocessing of organic wastes. *World Journal of Microbiology and Biotechnology*, **12**(3), 221–238.

Wang, J.Y., Zhang, H., Stabnikova, O. and Tay, J.H. (2003). Removal of ammonia in a modified two-phase food waste anaerobic digestion system coupled with an aerated submerged biofilters. *Waste Management and Research*, **21**(6), 527–534.

Wang, J.Y., Zhang, H., Stabnikova, O., Ang, S.S. and Tay, J.H. (2005a). A hybrid anaerobic solid–liquid (HASL) system for food waste digestion. *Water Science and Technology*, **52**(1–2), 223–228.

Wang, J.Y., Zhang, H., Stabnikova, O. and Tay, J.H. (2005b). Comparison of lab-scale and pilot-scale hybrid anaerobic solid–liquid systems operated in batch and semi-continuous modes. *Process Biochemistry*, **40**(11), 3580–3586.

Monitoring and data analysis

Deletion analysis of the microcystin synthetase bi-directional promoter from the cyanobacterium *Microcystis aeruginosa* PCC 7806 suggests a role for iron and nitrogen in transcription regulation

Hannah Root and Brett Neilan

Centre for Cyanobacterial and Astrobiology Research, School of Biotechnology and Biomolecular Science, University of New South Wales, Sydney, Australia
(E-mail: *patriciahannah@yahoo.co.uk*)

Abstract Promoter deletion studies and real-time PCR were utilised to analyse the relative expression conferred by the *mcyA/D* bi-directional promoter within the microcystin synthetase gene cluster of *Microcystis aeruginosa* PCC 7806. Results yielded an essential minimal region of promoter for efficient transcription. In addition, putative transcription binding sites were identified which may suggest a direct role for nitrogen and iron availability in the regulation of microcystin production in *M. aeruginosa* PCC 7806. This may aid the control of toxic cyanobacterial blooms and further the understanding of the function of microcystin.

Introduction

Nutrient overloading of freshwater systems is considered a major problem. This eutrophication promotes excessive growth of cyanobacteria. Cyanobacterial proliferation results in visible blooms, which are exacerbated by high levels of sunlight due to deforestation surrounding water bodies. Of particular concern are toxins (cyanotoxins) produced as secondary metabolites by harmful cyanobacteria (1). One such cyanotoxin is microcystin, which may be produced, by some but not all species, of the genera *Microcystis*, *Anabaena* and *Planktothrix* (2).

Cyanotoxins are classified according to their chemical structures and include cyclic peptides, alkaloids, and lipopolysaccharides (3). These toxins are highly toxic to humans and animals, and exhibit bioactivity against algae, bacteria, fungi and mammalian cell lines (1). Microcystin contamination of water has been linked to cancer in China (4) and hepatic poisoning has resulted in deaths in Brazil (5). Microcystin acts by specifically inhibiting protein phosphatases 1 and 2A, which may cause degradation of the cytoskeleton of liver cells, resulting in liver haemorrhage and death (1).

There are approximately 65 isoforms of microcystin, the non-ribosomal synthesis of which proceeds via a large modular multi-enzyme complex, encoded within a 55 kb gene cluster (6). This gene cluster is regulated by a 750 bp bi-directional promoter, known as *mcyA/D*.

At present, the *in vivo* role and function of microcystin is under debate. A key step in the understanding of microcystin is the elucidation of transcriptional regulation in the synthesis of the product. A growing amount of information is available detailing cyanobacterial promoters. The –10 bp position from the transcriptional start point shows adherence to the *E. coli* paradigm (7). An extended –10 bp region, with the consensus TGnTATAAT, may enable recognition by the sigma factor of RNA polymerase alone, without the –35 hexamer contact (8), which is a motif shown to be commonly absent in cyanobacterial promoters. A growing array of transcription factors have also been identified and are suggested to regulate transcription in cyanobacteria (9-11), specifically enabling environmental conditions to dictate the transcription of relevant genes.

The minimal region required for basal transcription can be studied in promoter deletion studies. In this method, a series of reduced promoter lengths are tested with a promoterless reporter gene. Such experiments have been carried out on cyanobacterial promoter regions (12). In addition to resolving the important basal motifs in the promoter, deletion studies can be used to detect changes in expression of the reporter gene which can then be correlated with the absence or presence of DNA segments upstream. Further promoter analysis and identification of regulatory binding domains may allow the possible environmental control mechanisms to be identified as key factors in microcystin production, which may in turn provide further evidence for a role and function for microcystin *in vivo*.

Understanding microcystin production in *M. aeruginosa* PCC 7806 will also aid the control and management of toxic blooms that consistently cause water supply problems and risk human and animal health.

Materials and methods

To analyse the expression conferred by the *mcyA/D* promoter from the microcystin synthetase gene cluster from *M. aerugionosa* PCC 7806, the promoter region was cloned into a vector containing a *gfp* reporter gene. In this method, RNA produced during transcription of *gfp* may be measured by real-time PCR. Utilising SYBR Green chemistry, the level of transcript can be related to the level of fluorescence of SYBR Green detected during the PCR amplification process. This enables varying promoters, with differing transcription efficiencies, to be compared to a standard. In this study, transcription levels conferred by deleted lengths of the *mcyA/D* promoter was compared to that of the entire promoter, to elucidate the essential areas of the promoter that enable transcription.

The arrangement of deletions constructed for this study can be seen in Figure 1. Results from real-time PCR were analysed using the Delta-delta method for relative quantitation of transcription levels (13), incorporating the PCR efficiencies for both genes. Detailed Materials and Methods are available on request.

Results and discussion

Promoter deletion analysis utilising the deletion series pPd1-6, based upon the entire *mcyA/D* promoter from *M. aeruginosa* PCC 7806, indicated the minimal region of the *mcyA/D* promoter necessary for the highest transcription efficiency, in a heterologous *E. coli* host. Figure 2 is the relative transcription of *gfp* conferred by promoters cloned into deletion constructs pPd1-6. Deletion fragment Pd5 conferred the highest transcription level relative to the entire promoter present in pPGF, and showed a 4.28 fold increase.

Fragment Pd6 conferred the next highest transcription level, with a fold increase of 1.11, with deletion constructs Pd1-4 showing a fold decrease of 0.16, 0.75, 0.57 and 0.40 respectively. The minimal region of the *mcyA/D* promoter necessary for efficient transcription in an *E. coli* host was identified. Figure 2 indicates the area designated Pd4 appears to be essential for efficient transcription, owing to the reduced transcription conferred by promoters Pd1-3. However, the promoter must include the region upstream of Pd4, hence Pd5 showed the highest level of transcription.

Characterisation of sequence data of the *mcyA/D* promoter showed that several regions were identified as putative binding sites for the global nitrogen regulator transcription factor NtcA (Figure 3, in bold). NtcA regulates the transcription of nitrogen responsive genes and is generally seen as an activator (14). Possible binding and regulation by NtcA was suggested by identification of areas with high similarity to the tightly conserved sequence $GTAN_8TAC$ (15,16). Contained within deletion fragment Pd5 were three putative NtcA binding sites, one of which lay in the region designated Pd4 (Figure 1).

Deletion fragment Pd6 conferred less transcription than Pd5, with a 1.11 fold increase. Two putative binding sites of the Ferric uptake regulator transcription factor Fur were also identified in the promoter sequence (Figure 3, underlined in large font). Fur regulates iron responsive genes and acts generally as a repressor of transcription (17,18). These regions were homologous to the consensus sequence GATAATGATAATCATTATC found to be the binding site of Fur in prokaryotic promoter regions (19). Two putative Fur binding regions were identified in deletion fragment Pd6 with the third only present in the entire promoter construct PGF (Figure 1).

The function and mechanism of control of microcystin have yet to be elucidated. It was previously hypothesised that environmental factors such as nitrogen or iron may play a role in the transcription of the toxin. The results from this study suggest a role of the regulatory proteins Fur and NtcA in the transcription control of microcystin from *M. aeruginosa* PCC 7806. Homologous cyanobacterial regulatory proteins are present and shown to be active in the *E. coli* host (18), and may have affected the relative transcription of the marker gene, *gfp*. Further understanding may be obtained with the repetition of the experiment utilising a cyanobacterial host, to clarify if transcriptional regulation of the microcystin synthetase gene cluster is conferred by NtcA and Fur activity.

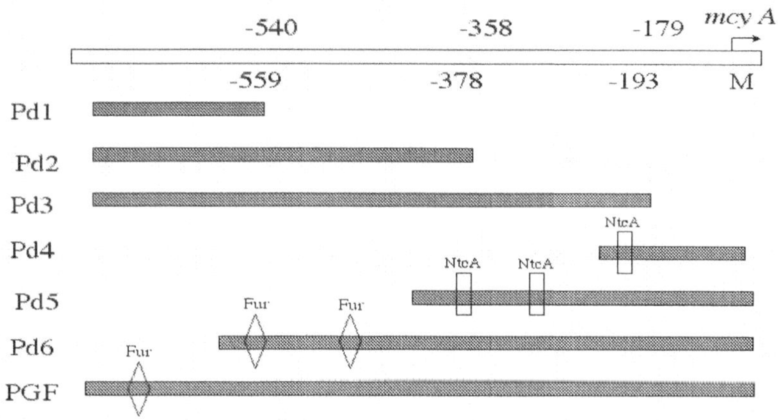

Figure 1 Arrangement of deleted promoter lengths generated from entire 750 bp *mcyA/D* promoter, designated PGF, from *M. aeruginosa* PCC 7806. Also indicated are putative NtcA and Fur binding sites identified in this study.

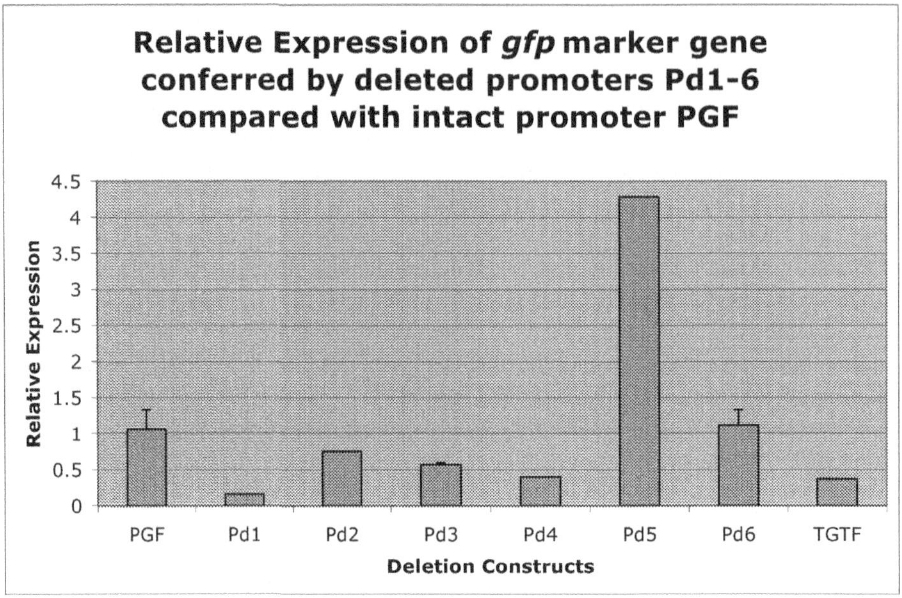

Figure 2 Expression of marker gene *gfp,* shown by relative increase or decrease of transcription, conferred by promoter deletion constructs, compared with the entire promoter construct contained within pPGF.

catagtgttagaatcgacttggaaaagaataattattgcgactgacggggtgacaagcagatggaaagtgaaacagg
gtgtagagtgtcgggtttagggaaaaagcttgagactttcgccaaaagataacgagggaatttggttttgtctag
taagtcgattaatttgatggatcacagtgaggaaatttttcccccacctcacttaaacttcaacctcgttgtcaccc
cttcagctattacgaccagacagctaatcgtacctgatcaaggtagtaattgtcaatagacatctgcaataaacgt
ttatggggtgtggcatcctaagctctgctctcttggtctcgcgcaagcttatctttaaatgtcacacttctgcacttcttaat
atttaattaatgattttactaatttattgg**gttcagtggtttcta**cagtgaagatttttgtcaaaacatactagggaatgta
aaaatatgtaaaagtatatggagatgtgcagaatgtcg**gttagtatgctac**aatgtcgaggctcagaacaattttgga
gaagcgacagaaaccctgaccttagccgtagtcgggtttcctgtagttcaaatagcaataattccactcgtcagagac
cggaattatcgctttaagggaactgggaacggggaaaaaagcattgtacccc**atgactctgaatac**cgccatcaac
cactatttagggaaaaagttagaacagcaatg

Figure 3 Sequence of entire *mcyA/D* promoter from *M. aeruginosa* PCC 7806. Large type underlined sequence indicates putative Fur binding sites, identified by similarity to the consensus GATAATGATAATCATTATC. Bold type sequence indicates putative NtcA binding sites, identified by similarity to consensus $GTAN_8TAC$.

References

1. Carmichael, W. W. (1992) *J. Appl. Bacteriol* **72**, 445–459.
2. Rouhiainen, L., Vakkilainen, T., Siemer, B. L., Buikema, W., Haselkorn, R., and Sivonen, K. (2004) *Appl. Environ. Microbiol.* **70**, 686–692.
3. Kaebernick, M., and Neilan, B. A. (2001) *FEMS Microbiol. Ecol.* **35**, 1–9.
4. Ueno, Y., Nagata, S., Tsutsumi, T., Hasegawa, A., Watanabe, M. F., Park, H. D., Chen, G. C., Chen, G., and Yu, S. Z. (1996) *Carcinogenesis* **17**, 1317–1321.
5. Pouria, S., de Andrade, A., Barbosa, J., Cavalcanti, R. L., Barreto, V. T., Ward, C. J., Preiser, W., Poon, G. K., Neild, G. H., and Codd, G. A. (1998) *Lancet* **352**, 21–26.
6. Tillett, D., Dittmann, E., Erhard, M., von Dohren, H., Borner, T., and Neilan, B. A. (2000) *Chem. Biol.* **7**, 753–764.
7. Harley, C. B., and Reynolds, R. P. (1987) *Nucl. Acids Res.* **15**, 2343–2361.
8. Mazouni, K., Bulteau, S., Cassier-Chauvat, C., and Chauvat, F. (1998) *Mol. Microbiol.* **30**, 1113–1122.
9. Hernandez, J. A., Lopez-Gomollon, S., Bes, M. T., Fillat, M. F., and Peleato, M. L. (2004) *FEMS Microbiol. Lett.* **236**, 275–282.
10. Luque, I., Vazquez-Bermudez, M. F., Paz-Yepes, J., Flores, E., and Herrero, A. (2004) *FEMS Microbiol. Lett.* **236**, 47–52.
11. Nakahira, Y., Katayama, M., Miyashita, H., Kutsuna, S., Iwasaki, H., Oyama, T., and Kondo, T. (2004) *Proc. Natl Acad. Sci. USA* **101**, 881–885.
12. Shibato, J., Agrawal, G. K., Kato, H., Asayama, M., and Shirai, M. (2002) *Mol. Genet. Genomics* **267**, 684–694.
13. Livak, K. J., and Schmittgen, T. D. (2001) *Methods* **25**, 402–408.
14. Vega-Palas, M. A., Flores, E., and Herrero, A. (1992) *Mol. Microbiol.* **6**, 1853–1859.
15. Luque, I., Flores, E., Herrero, A. (1994) *EMBO J* **13**, 2862–2859.
16. Herrero, A., Muro-Pastor, A. M., and Flores, E. (2001) *J. Bacteriol.* **183**, 411–425.
17. De Lorenzo, V., Herrero, M., Giovannini, F., Neilands, JB. (1988) *Eur. J. Biochem.* **173**, 537–546.
18. Bagg, A., and Neilands, J. B. (1987) *Biochemistry* **26**, 5471–5477.
19. Calderwood, S. B., Mekalanos, J.J. (1987) *J. Bacteriol.* **169**, 4759–4764.

Continuous monitoring of trace metals in wastewater effluents using voltammetric system

K. Strašunskienė, Ø. Mikkelsen, K. H. Schrøder and M. Rydningen

Department of Chemistry, Norwegian University of Science and Technology, Trondheim, Norway (E-mail: *kristina.strasunskiene@chem.ntnu.no*; *oyvind.mikkelsen@nt.ntnu.no*; *knut.schroder@chem.ntnu.no*; *marionr@stud.ntnu.no*)

Abstract An automated voltammetric system has been applied for continuous remote monitoring of trace metals in wastewater. The concentration of trace metals has been continuously monitored every 30 minutes for more than 7 months in wastewater effluents after primary treatment at Høvringen wastewater treatment plant (Trondheim, Norway). The system operates unattended for 5 days and is capable of transmitting data to computer networks. Concentration of trace metals has been measured by electrochemical method, using voltammetric system with three electrodes. The working electrode is environmental safe solid dental amalgam electrode. The system is simple to use and requires minimal maintenance. A speciation study has been performed by comparing data from manually collected samples, analyzed by ICP-MS and data obtained from continuous voltammetric measurements performed at the monitoring station.

Keywords Monitoring; trace metals; speciation; voltammetry; wastewater

Introduction

The common techniques for trace metal analyses in water samples are atomic absorption spectroscopy (AAS) and inductively coupled plasma–mass spectroscopy (ICP–MS). These techniques are sensitive, but very expensive, require manual sampling, sample preparation, is time consuming and results available long after sampling. Therefore these methods are not suitable for continuous field monitoring, which is very important due to possibilities of long time monitoring and especially to monitor short time pollutions or accidents. However, electrochemical methods can potentially be used in field, because they are simple to use, easy to automate, inexpensive, require no sample preparation, and results are immediately available after sampling and short analysis time.

Therefore electrochemical techniques are suitable for the determination of trace metals in natural and process waters whereas voltammetry is the most common method. Stripping analysis is defined as an extremely sensitive electrochemical technique for measuring trace metals (Wang, 2000). Four to six metals can be measured simultaneously in various matrices at concentration levels down to 10^{-10} M, using relatively inexpensive instrumentation.

Electrochemical techniques are based on measuring of the current response to an applied potential. These techniques can measure any chemical species that is electroactive, e.g., that can be made to reduce or oxidize. The current is proportional to metal concentration in the sample.

Stripping analysis is a two-step technique. The first step involves the electrolytic deposition of a small portion of the metal ions on the electrode surface. This is followed by the stripping step, which involves the dissolution of the deposit. Anodic stripping voltammetry (ASV) is one of the most widely used forms of stripping analysis. In this method preconcentration is done by cathodic deposition at a controlled time and potential (Wang, 2000).

Since the development of polarography, liquid mercury amalgams or mercury films has been used as electrode material for voltammetry. This is due to high hydrogen overvoltage, extended cathodic potential window, highly reproducible, readily renewable and smooth surface. Disadvantages of the use of mercury are its limited anodic range and its toxicity (Wang, 2000). The use of such compounds is restricted and cannot be included in voltammetric systems for field applications. Introduction of solid dental amalgams and alloy electrodes (Mikkelsen and Schrøder, 2000) has created new possibilities for continuous trace metals monitoring in field. The main components of amalgam are silver and mercury, but due to the special properties of dental amalgam compared to liquid mercury itself it is not toxic. It has been found that dental amalgam electrodes have very high hydrogen overpotential and allows to carry out trace analysis at sufficiently low potentials to determine trace concentrations of metals. These electrodes have been used in automatic voltammetric monitoring stations in river and sea (Mikkelsen *et al.* 2005). The contribution of this paper is an analysis and overview of an ongoing research project of voltammetry application for continuous trace metal monitoring in wastewater.

Considerable interest has been shown to metal speciation in natural waters. Many reports have been published about heavy and trace-metal speciation in natural waters (Florence, 1981, Nurnberg, 1983, Achterberg and Braungardt, 1999). However, there is a more challenging analytical problem to assess speciation in wastewater, where formation of organometallic complexes must be taken into account. Natural waters contain less than 20 mg/L organic material, while wastewater may contain soluble organic material ten times higher. The concentration of suspended and colloidal particles in wastewater is usually significantly higher than in natural water. These differences in matrix composition may have important implications for the approach to heavy metal speciation in complex wastewater matrices (Sterritt and Lester, 1984).

Relatively little attention has been paid to metal speciation in sewage effluents due to the incidence of matrix interference effects and uncertainty about the validity of sensitive and specific analytical techniques applied to these complex samples (Sterritt and Lester, 1984). The reasons for studying heavy metal speciation in wastewater are different from those of studying speciation in natural waters. In natural waters, obtaining data immediately amenable to biological interpretation is probably the most common reason for considering speciation (Neubecker and Allen, 1983). In wastewater, approaches to speciation have been concerned with understanding the partition of heavy metals into various forms or fractions that will ultimately determine their dispersion and impact on the environment. (Sterritt and Lester, 1984).

Experimental

The experiments have been performed at Høvringen wastewater treatment plant (Trondheim, Norway) to measure the concentration of heavy and trace metals in the effluents (outlet to the sea). The wastewater treatment plant collects discharge from two-thirds of the city, including private household, organizations and industry, and serves about 110,000 persons from an agglomeration of 138,500. The plant is designed for primary wastewater treatment and sludge processing. During the mechanical treatment process suspended particles are removed by 50% and organic material measured as BOD_5 is removed by 20% (Trondheim municipality, 2005). The treated wastewater is discharged through a diffuser system into a large fjord basin at 48–65 m depth that exchanges water effectively with Norwegian Sea. The receiving waters for treated discharges from Trondheim are identified by the Norwegian authorities as a less sensitive area (Tangen and Arff, 2003, Trondheim municipality 2004). The receiving fjords are characterized by their large water depths, large water volumes, and very good water exchange with a deep ocean area outside the fjord mouths.

The primary source for most metals is through the sewage system. It receives domestic wastewater, urban runoff and industrial discharge in varying proportions. Industrial sources have been important for all metals, but increasing control has lowered the concentrations of most metals discharged to the sewer system. However, for some of the metals, for example Pb, Cd, Hg, Cr, Zn, Ni, Cu that enter the sewage system today, industrial sources are still significant. Domestic wastewater is also a significant source of certain metals, for example Cu and Zn. Pb also enters the sewage system through urban runoff.

The voltammetric system used for the continuous online monitoring has been placed at the end of mechanical treatment process, right after sedimentation tanks to monitor trace metals concentration in the effluents, which are discharged to Trondheim fjord.

The system is constructed and developed by the group of Analytical Chemistry at NTNU. The system consists of external pump, control box, electrochemical cell, dosage pump and computer (Figure 1). The main part is a 50 ml electrochemical cell with three electrodes. The working electrode is solid dental amalgam electrode, made of 51% mercury and 49% silver, the reference electrode is silver/silver chloride/saturated silver chloride/potassium chloride Ag/AgCl/KCl (3M), the counter electrode is made of a platinum wire. The external pump is immersed in wastewater and supplies the system with wastewater samples. It is connected to the control box and controlled automatically. Water samples is filtered through a coarse filter (Schott Duran) in a cross-flow system (made in NTNU laboratories) to avoid fouling in the cell, valves and tubes. 375 μL of supporting electrolyte ammonium acetate NH_4Ac (2M) is added by use of a dosage pump (ProMinent Beta). Analysis is performed immediately after sampling and adding of the electrolyte. The system is automatically washed with 0.01 M citric acid ($C_6H_8O_7$) 4 times a day to clean the electrodes, the tubes and the cell.

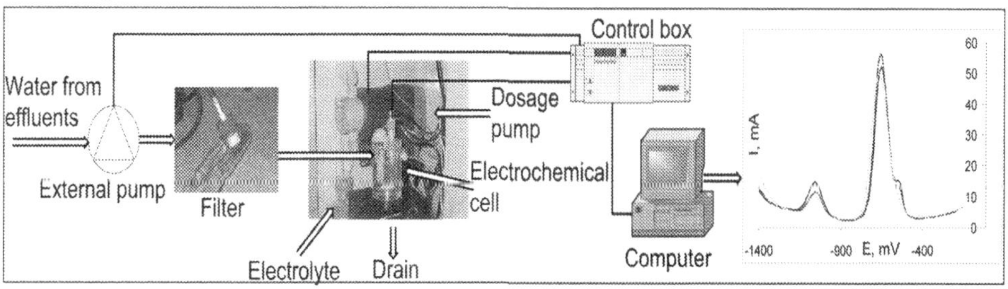

Figure 1 The automated voltammetric monitoring system.

The voltammetric monitoring system operates unattended for 5 days and is configured to transmit data to computer networks. Once in 5 days electrodes are polished with a fine polishing paper (Struers, P4000), the cell is rinsed with hydrochloric acid HCl (1M) and distilled water, filter is changed. Calibration is performed once a week, supporting electrolyte is refilled once a month.

Standard solutions for calibration have been prepared from analytical grade AAS standard solutions (Merck) by diluting 5 mL of 1000 mg/L Zn or Pb standard solution with distilled water till 100 ml. Standard solution for iron (II) has been made by dissolving 250 mg of iron (II) sulfate $FeSO_4$ in distilled water, purged for 5 minutes with nitrogen and added 250 mg of hydroxylamine hydrochloride NH_4OHCl.

Continuous voltammetric measurements have been performed every 30 minutes, using different pulse anodic stripping voltammetry (DPASV) with 600s deposition time at –1400 mV, scan rate 20 mV/s, modulation pulse 75 mV, equilibrate time 10 s. Supporting electrolyte NH_4Ac (0.02 M). The software program is Voltammetry Analyser (Oceanor, Norway).

Results

The voltammetric system was fully automated by the end of June and it has successfully been working for more than 7 months. To obtain full information about complex sample matrix it is intended to continue the experiment for a few years.

From the continuous voltammetric measurements of the monitoring system has been obtained unique data, demonstrating varying concentration of trace metals during the day and during all period of measurements (Figures 2 and 3).

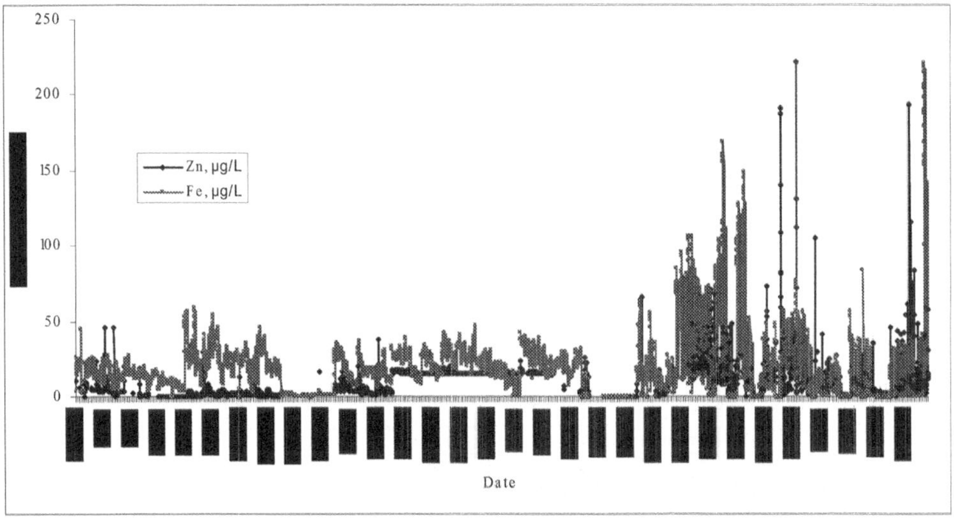

Figure 2. Continuous voltammetric measurements of zinc and iron at the Høvringen wastewater treatment plant (Trondheim, Norway). The measurements have been performed every 30 minutes, DPASV with 600s deposition time, -1400 mV, scan rate 20 mV/s, modulation pulse 75 mV, equilibrate time 10 s. Supporting electrolyte NH_4Ac (0.02 M).

The variations during the day and during all period of measurements are assumed to relate to the water treatment cycle in the plant and other factors e.g. year-seasons, precipitation, activities in industry. The observed concentration range for iron is from less than 0.032 µg/L to 220.17 µg/L, average 21.65 µg/L, zinc – from less than 0.038 µg/L to 221.00 µg/L, average 8.93 µg/L.

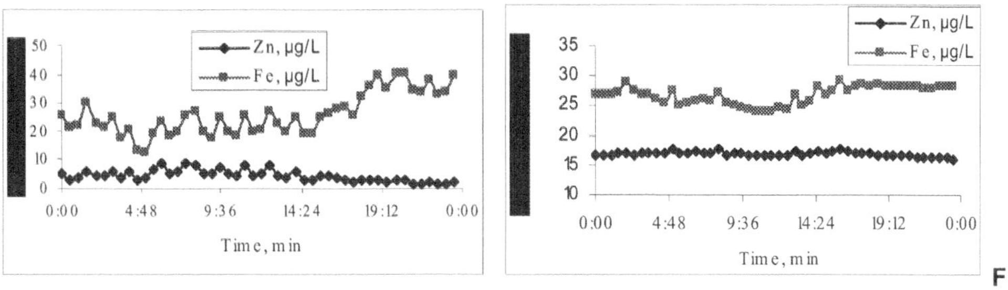

Figure 3 Variations in concentration of zinc and iron during day. Measurements performed 8 August and 17 September.

The decrease in concentration of both zinc and iron in the period from 23 August to 3 September can be explained by the changed operating parameters in the electrochemical system, when by accident deposition time was set 10 times less than it was normally used. The current response is dependant of operating parameters. Too short deposition time gives low current response and too long deposition time would increase possibility of not related sample matrix reactions. For measuring low trace metals concentrations 600 seconds deposition time was experimentally selected to be optimal to obtain maximum current response and to avoid disturbing sample matrix reactions.

The absence of measurements from 28 October to 12 November is explained by technical problems of the system. The significant increase in both iron and zinc concentrations after 12 November is most likely influenced by introducing citric acid to system. The citric acid has been introduced to automatically rinse voltammetric cell every 6 hours in order to minimize maintenance. The additional function is that electrode area had become active for longer because the acid prevents electrode fouling. It is also possible that some acid remains in the cell and releases more free ions from the weak complexes.

Since the automatization of the system, it works without any maintenance for 5 days. Filtration is the critical factor. Because of the high concentration of suspended and colloidal particles filter must be changed after 5 days. Calibration is performed once a week in order to ensure stability of the measuring system. Calibration curves show good linearity (Figure 4). Calibration of iron was sometimes complicated because of the marked tendency of Fe(II) to oxidise to Fe(III). It is also possible, that free iron ions are bound rapidly to organic complexes, colloidal and suspended particles, concentration of which in the sample is high.

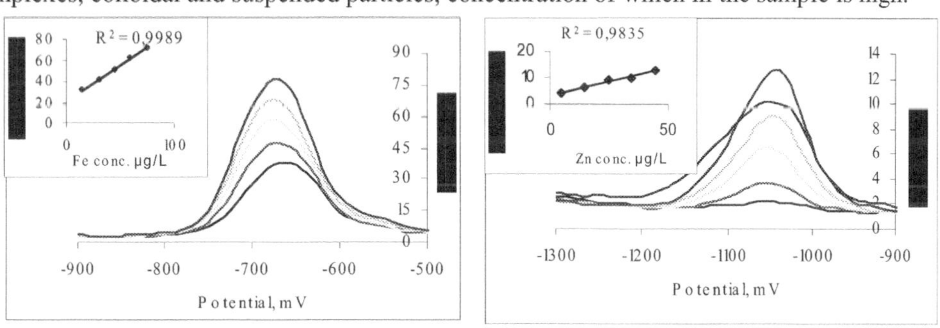

Figure 4 Calibration curve and calibration plot for Zn and Fe. All operating parameters are the same as in Figure 2. Increasing peaks represent standard additions of 0.1–0.5 mL 5µg/L Zn (II) solution or 0.1-0.5 mL 2µg/L Fe (II) solution.

The major type of interferences in analyses has been overlapping stripping peaks caused by a similarity in the oxidation potentials, the presence of surface-active organic compounds that adsorb on the electrode surface and inhibit the metal deposition and the formation of intermetallic compounds which affects peak size and position. Similarity in oxidation potentials of lead and iron and the fact that concentration of iron significantly higher than concentration of lead has made difficult to measure the concentration of lead. It was observed that response of zinc peak is very dependent on clean electrode surface. The fine Zn peaks were obtained right after cleaning and polishing electrodes and response was observed to decrease after few days because of absorption and fouling of electrode surface and therefore reduced the active electrode area. The iron peak has been observed to be not so sensitive to electrode surface and fine iron peaks were usually obtained independently of electrode surface conditions. The feature of copper rapidly bound to other complexes made it impossible to measure its concentration, although it was obvious some electroactive copper presence in the samples.

Speciation analysis

Speciation in water sample is necessary for understanding of toxicity, bioavailability, bioaccumulation and transport of a particulate element (Florence, 1982). Speciation is very important in evaluating ecological effects. The chemical behaviour of any element in the environment depends of the nature of its compounds and species. Physiological, ecological and toxicological effects of a metal are usually strongly structure specific; that is, they depend on the species. Variables such as total concentrations of the metal, pH, alkalinity, concentration of ligands, concentration of competing metals, organic complexes and presence of adsorptive compounds all can affect the concentration of free metal ions (Stumm, and Morgan, 1996)

Control tests have been carried out by comparing data received from the continuous electrochemical system with the data obtained by ICP-MS analyses of manual samples. From this it is possible to get unique information about the metal speciation in the complex sample matrix. For 7 months continuous voltammetric measurements were performed every 30 minutes. For the same period samples to ICP-MS analysis have been collected manually once a week. Samples to ICP-MS analyses were collected into disposable containers and acidified with nitric acid (HNO_3) to pH 2. Four samples were collected every week. Two samples were filtered through membrane filter (Supor® Membrane Disc) with 0,45μm pore size and analyzed for total dissolved metal concentration. Other two samples were acidified and decomposed for 40 minutes in a microwave oven to decompose organic material and then analyzed for total metal concentration.

The results (Figures 7 and 8) show distribution of iron and zinc species in the complex wastewater matrix. Results are obtained by comparison of voltammetric measurements from monitoring station and ICP-MS measurements.

It should be noted, that ICP-MS reports the total element concentration, but the voltammetric method measures the labile amount. Combination of these two analytical methods gives a possibility to speciation studies. The labile metal ion concentration includes free aquo ions, inorganic as well as weak organic complexes (Stumm and Morgan, 1996).

Measurements only of total concentration of trace element provide no information about its bioavailability or its interaction with sediments and suspended particles (Florence, 1982).

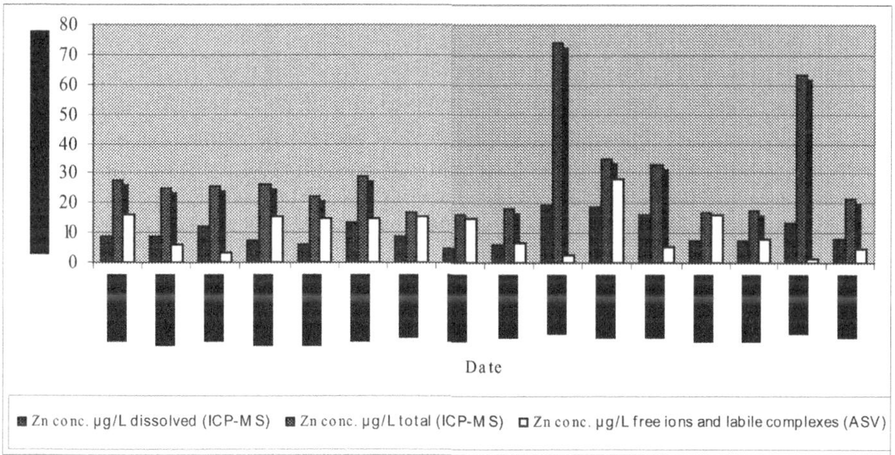

Figure7 Distribution of zinc species in wastewater sample.

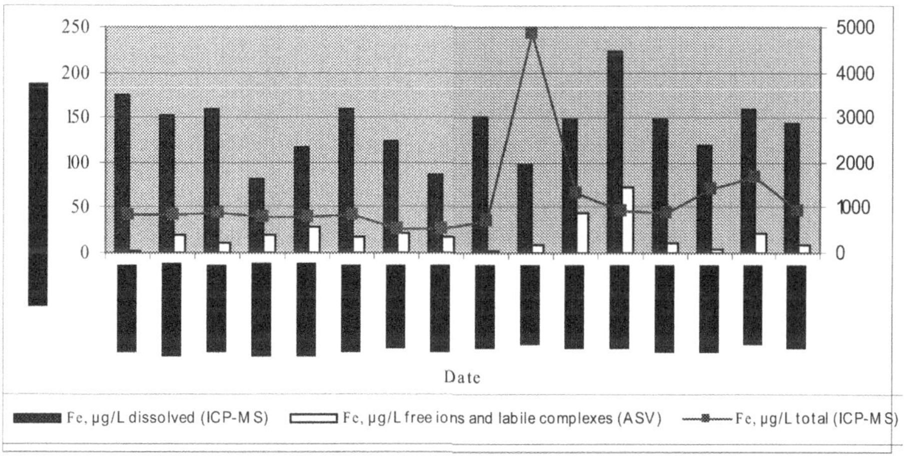

Figure 8 Distribution of iron species in wastewater sample. Secondary Y axis indicates concentration of total iron.

Anodic stripping voltammetry (ASV) is useful speciation method because of its sensitivity and ability to distinguish labile metal species from strong (non labile) organic complexes. Furthermore, this technique is very useful to distinguish dissolved species from species bound to colloidal or particulate phases (Goncalves *et al.* 1985). Comparison of results obtained by these two different methods shows an obvious difference in the labile, total dissolved and total amount for all measured metals. Results indicate a tendency of most metals to be bound to complexes and suspended particles (Figures 7 and 8).

Future work and conclusions

It is expected that in the near future water-quality legislation for trace metals will include statements relating to their speciation and that it may be possible to tolerate higher concentrations of some metals as long as the labile or bioavailable fraction is below a certain level (Florence, 1982). Choice of suitable analytical technique is necessary to monitor the free metal concentration. Voltammetry is very useful for this task and application in field opens new possibilities of monitoring long-term variations and warning about possible environmental accidents. However, the complex wastewater matrix makes voltammetry application more challenging so in the future research will be focused on improvement of both the analytical measurement system and technical equipment. It includes improvement of the filtering system, which would reduce the concentration of colloidal and suspended particles in the sample hence making easier to perform analytical measurements and reducing maintenance of the monitoring station. Further work will also focus on inclusion of monitoring more elements. Metals like Cd, Hg and Pb is of utmost importance and interest both by Trondheim authorities and present authors.

The huge data collection of long time monitoring, obtained by voltammetric measurements and by ICP-MS analyses gives possibilities to study relations between different species, different elements and correlate them against other parameters, like BOD_5, concentration of suspended particles, precipitation and others. This requires more investigation and more thorough statistical analysis.

References

Achterberg, E.P., van den Berg, C.M.G. (1992) Automated voltammetric system for shipboard determination of metal speciation in sea water. *Analytica Chimica Acta.*

Achterberg, E.P., Braungardt, C. (1999) Stripping voltammetry for the determination of trace metal speciation and in-situ measurements of trace metal distributions in marine waters. *Analytica Chimica Acta.*

Florence, T.M. (1982) The speciation of trace metals in waters. *Talanta*, **92**, 345.

Goncalves, M.S., Sigg, L., Stumm, W. (1985) Voltammetric methods for distinguishing between dissolved and particulate metal ion concentrations in the presence of hydrous oxides. A case study on lead (II). *Environmental Science and Technology* **19**, 141–146.

Lester, J.N. (1987a) *Heavy metals in wastewater and sludge treatment processes. Volume I. Sources, analysis and legislation.* CRC Press, Inc.

Lester, J.N. (1987b) *Heavy metals in wastewater and sludge treatment processes. Volume II. Treatment and disposal.* CRC Press, Inc.

Mikkelsen, Ø., Schrøder, K. H. (2000) Dental amalgam in voltammetry – some preliminary results. *Analytical letters*, **33** (15), 3253–3269.

Mikkelsen, Ø., Skogvold, S. M., Schrøder, K. H. (2005). Continuous heavy metal monitoring system for application in river and seawater. *Electroanalysis*, **17**, 431–439.

Neubecker, T.A. and Allen, H.E. (1983) The measurement of complexation capacity and conditional stability constants for ligands in natural waters. *Water Research*, **17**, 1–14.

Nurnberg, H.W. (1983) Investigations on heavy metal speciation in natural waters by voltammetric procedures. *Frezenius Z. Anal. Chem.*

Sterritt, R.M. and Lester, J.N. (1984) *Science of the Total Environment*, **34**, 117–141.

Stumm, W., Morgan, J.J. (1996) *Aquatic chemistry, Environmental Science and Technology.* 1022 pp.

Tangen, K. and Arff, J. (2003) Høvringen wastewater plant and the environmental quality of the Trondheimfjord.*Oceanor Rep. OCN R-23027*, 52 pp.

Trondheim municipality. (2004) Wastewater from the agglomerations of Trondheim city. *Directive 91/271/EEC (2004).*

Trondheim municipality (2005) Høvringen wastewater treatment plant http://www.trondheim.kommune.no/content.ap?thisId=1096576536 (accessed 26 October 2005).

Wang, J. (2000). *Analytical Electrochemistry*, Wiley-VCH, 203 pp.

Development of novel faecal source tracking techniques and their application to a temperate river catchment in southern England

J.E. Ebdon and H.D. Taylor

School of the Environment, Cockcroft Building, University of Brighton, Lewes Rd, Brighton BN2 4GJ, UK (E-mail: *je3@brighton.ac.uk*)

Abstract Bacterial water quality across much of Europe is currently monitored using a number of parameters prescribed by the European Union. While the parameters indicate the presence or absence of faecal contamination, they provide little information as to its origin. Determining the origin of faecal contamination would help to assess the risk posed to public health and would provide a more effective approach to the remediation of waters contaminated with faecal material. Two recently developed faecal source tracking (FST) techniques, antibiotic resistance analysis (ARA) and the enumeration of bacteriophages infecting a newly isolated *Bacteroides* host strain, were applied in order to identify the origin of faecal pollution in river water samples. ARA appeared to be particularly good at classifying bacteria originating from pigs (83% correctly classified). According to ARA, the most common sources of faecal pollution present in river samples were livestock (cattle and sheep) and wild birds. Counts for bacteriophages infecting *Bacteroides* strain GB124 were highest at two sites both downstream from municipal wastewater treatment works. More recent (unpublished) data suggests that host strain GB124 may be human-specific. The ARA system could be used in combination with phage detection to offer a low-cost and rapid means of identifying human and non-human sources of faecal pollution within the study catchment.

Keywords Antibiotic resistance; bacteriophage; catchment; faecal; source

Introduction

The contamination of surface waters with untreated or partially treated faecal material from point and non-point sources can lead to deterioration in water quality. Contamination may originate from the faeces of humans, domestic and wild animals, and may impact on drinking water supply, fishing, aquaculture, and recreational water use. While bacterial water quality parameters prescribed by the European Union Bathing Water Directive (76/160/EEC) are used to detect the presence or absence of faecal contamination, monitoring to comply with the Directive provides little information as to its source. Determining the origin of faecal contamination would help to assess the risk posed to public health. It would also further understanding of the transport of faecal pollution in river catchments, and would ensure a more targeted approach to the remediation of impaired waters.

EU legislation has led to significant improvements in the sanitary quality of surface waters during the past 30 years, mainly through improvement in the treatment of municipal wastewaters. However, the reduction of wastewater inputs has highlighted the presence of non-point inputs such as those arising from animal husbandry and wild animals. Consequently, numerous faecal source tracking (FST) techniques have emerged in recent decades, largely in response to stricter environmental legislation and changes in predominant pollution sources. However, to date no single technique has emerged, that is capable of determining the origin of faecal material in all aquatic environments. Although several techniques appear to be promising, their application to European river catchments has been limited. This study assesses the effectiveness of two recently developed FST techniques to determine the origin of faecal pollution in water samples from a temperate river catchment in southern England. The first technique, antibiotic resistance profiling was developed as part of a Ph.D. study (J. Ebdon) at the University of Brighton and funded by a European Commission Fifth Framework Project, 'Tracking the Origin of Faecal Pollution in Surface Waters' (TOFPSW) (EVK1-CT-2000-00080). The project involved testing numerous potential approaches, the results of which suggested that accurate faecal source tracking may best be achieved using a combination of techniques. Among the most promising potential faecal source indicators to emerge from the 'TOFPSW' project was a technique involving the detection of bacteriophages infecting bacteria of the genus *Bacteroides*. However, problems with geographical stability meant that it was not until Payan *et al.* (2005) isolated a more suitable strain, that the technique could be applied to the catchment featured in this study.

Methods

Thirteen water quality monitoring sites were selected at various points along the course of the River Ouse and its tributaries in East Sussex, UK (Figure 1). On each sampling occasion 1 litre of river water was collected from approximately 30 cm below the water surface in sterile polyethylene sampling bottles. All samples were stored in the dark at approximately 4 °C, and analysed for the presence of enterococci or phages infecting *Bacteroides* strain GB124 within four hours of sampling. All media and reagents were prepared in accordance with the manufacturers' instructions, or as described in Ebdon *et al.* (2004), Jofre *et al.* (1986) and Payan *et al.* (2005).

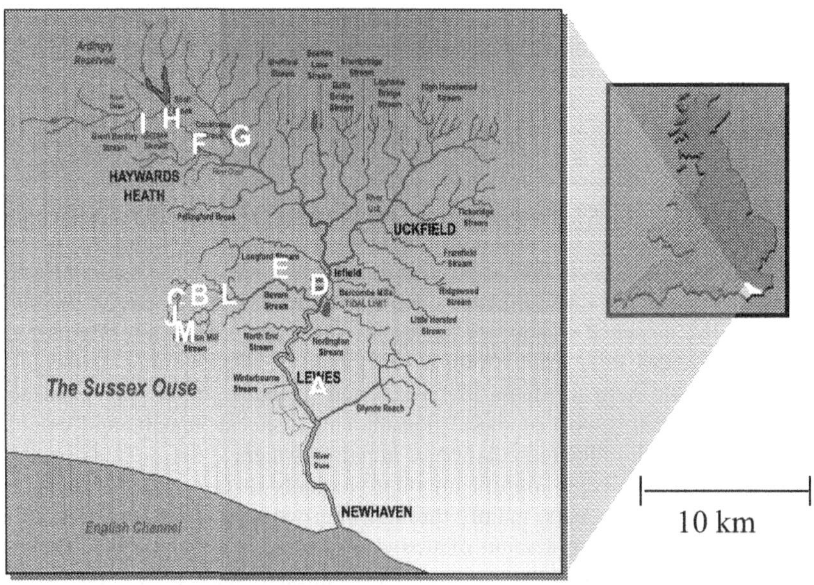

Figure 1 River Ouse catchment and sampling sites. A, Lewes; B, Wales Farm; C, Streat Lane; D, Hamsey; E, Barcombe; F, Sloop Inn; G, Freshfields Bridge; H, East Mascals 1; I, East Mascals 2; J, Spatham Lane; K, Swansyard Farm; L, Plumpton; M, Ditchling Spring.

The River Ouse drains an area of 396 km^2 to its tidal limit at Barcombe and is approximately 67 km in length (14 km of which are tidal). The catchment was chosen because it is located close to the laboratory facilities and is in a region that encompasses a range of potential pollution sources and land management practices. While agriculture is the dominant land use in the catchment, there are also a number of municipal wastewater treatment works discharging partially treated wastewater into the river and its tributaries.

Antibiotic resistance analysis (ARA) of *Enterococcus*

Antibiotic resistance analysis is based on the observation that a bacterial population that has had prolonged exposure to one or more antibiotics may exhibit a unique antibiotic resistance 'pattern', when tested against a battery of antibiotic agents. The use of different classes and concentrations of antibiotic agents in animal husbandry (either for growth promotion, prophylaxis, and/or therapeutic treatment) and in human medicine makes it possible to determine the most likely origin of a bacterial population.

While ARA techniques have been applied successfully in the U.S.A. to determine sources of faecal contamination in river catchments (Wiggins 1996; Hagedorn *et al.*, 1999; and Harwood *et al.*, 2000), its use outside of the U.S.A. has been limited. Phenotypic approaches such as ARA are generally cheaper than genotypic approaches but previous protocols were labour-intensive and the manual handling of data was susceptible to human error. A novel low-cost ARA technique was therefore developed that combined semi-automated reading of data with simple statistical analyses (Ebdon *et al.*, 2004).

Antibiotic resistance analysis was carried out on presumptive enterococcal isolates following isolation on 0.45 μm membrane filters (Gelman GN-6), incubation on M-Enterococcus agar (Difco) and growth on bile aesculin agar in accordance with ISO 7899/2. Individual aesculin positive colonies were transferred using toothpicks into a series of 96-well microplates, each containing a 0.01% solution of bromocresol purple (pH 6.8), suspended in tryptone soy broth (Oxoid, Hampshire, UK). A single antibiotic concentration was then added to each microplate, allowing up to 96 enterococcal isolates to be tested per plate. Twenty-one antibiotic agents (at up to four concentrations each) were selected for inclusion in the initial ARA database, as they represented classes of both broad and narrow-spectrum agents used in either human medicine and/or animal husbandry in the UK. However, it was shown (Ebdon *et al.*, 2004) that this number could be significantly reduced once the most discrimant antibiotic concentrations had been determined.

Therefore, for the analysis of river water samples, a battery of thirty-one antibiotic concentrations was selected as this appeared to offer the best compromise between accuracy of classification and cost. These were (in μg/ml): ampicillin (sodium salt) (16), apramycin (sulphate salt) (16), cephalothin (sodium salt) (8, 16), erythromycin (16), kanamycin monosulphate (16, 32), nitrofurantoin (4, 16, 32), nitrofurazone (128), ofloxacin (2, 32), oxytetracycline hydrochloride (64), penicillin G (sodium salt) (2, 16), rifampicin (2, 32), streptomycin sulphate (16, 64), sulfathiazole sodium salt (16, 64), tetracycline hydrochloride (2, 32, 64), tylosin tartrate (16, 32), vancomycin hydrochloride (4, 8, 32), and virginiamycin (2). All antibiotics were obtained from Sigma-Aldrich.

Once the antibiotic plates had been incubated at 37 (± 2) °C for 44 (± 4) h colour change was read using a transmissive flatbed scanner (Umax Astra 3450) and stored as bitmap images. Specific software (PhPWIN ver. 4.2, Microplate Techniques AB, Stockholm) enabled the plate images to be assigned absorbency values, which could then be converted into binary data for statistical analysis. Statistical analysis of the antibiotic resistance patterns was carried out in Minitab (version13.31) using discriminant function analysis (DFA). In total, 2739 enterococcal isolates of known origin and 480 of unknown origin (river water samples) were successfully analysed for their sensitivity to a combination of antibiotic concentrations, creating a database of 3219 isolates and approximately ¼ million test reactions. Statistical analysis revealed how many of the 'known' and 'unknown' isolates were correctly grouped into each source category, on the basis of their antibiotic resistance patterns.

Bacteriophages infecting *Bacteroides* strain (GB124)

The second faecal source tracking method involved the enumeration of bacteriophages (viruses) capable of infecting and lysing a specific *Bacteroides* host strain. *Bacteroides* is a genus of Gram-negative, anaerobic bacterium, commonly found in the intestines of humans and animals and can therefore be regarded as an indicator of faecal contamination. While the survival of *Bacteroides* spp. within aquatic environments is poor and their host range not limited to human sources, it has been recognised that bacteriophages capable of infecting certain *Bacteroides* spp. may be host specific (Jofre *et al.* 1986; Tartera and Jofre, 1987). Bacteriophages are also more persistent within the environment than the bacterial hosts they tend to infect, but the bacterial strains appear to demonstrate a high level of geographical specificity. For instance, bacteriophages infecting *B. fragilis* strain HSP40 have been successfully detected in the Mediterranean region (Cornax *et al.*, 1990, Duran *et al.*, 2002, Tartera *et al.*, 1989, Tartera and Jofre, 1987) and in South Africa (Grabow *et al.*, 1995), but have not been detected in significant numbers in Northern Europe (Puig *et al.*, 1999) and the U.S.A. (Kator and Rhodes, 1992).

Recent research (Payan *et al.*, 2005) identified a potentially suitable *Bacteroides* host strain (GB124) from a municipal wastewater sample obtained from the UK catchment featured in this study. The next step was to determine whether phages infecting strain GB124 were present in the catchment in significant numbers and whether they were host specific. If this were the case then the technique would be useful for identifying the origin of river water pollution. Phage enumeration was carried out using a soft agar plaque count procedure, as described by Jofre *et al.*, (1986). On each occasion the samples were passed through a 0.22μm syringe filter unit (Millex GP) to remove background bacterial contamination. The samples were then mixed with strain GB124 in soft agar, before being plated on *Bacteroides* phage recovery media (BPRM) agar (Scharlau, Spain). Samples were incubated in anaerobic jars in accordance with ISO/FDIS 10705-4. The presence of phages within a river water sample resulted in the development of plaques (zones of lysis) on a confluent lawn of host strain. The results were reported as the mean number of plaque forming units (PFU) per 100 ml of sample. All samples were analysed in duplicate.

Results and discussion

Antibiotic resistance analysis

When classifying isolates from both known and unknown sources using this library based approach, there will always be a number of isolates that are misclassified. It is therefore important to establish a rate of misclassification, above which it is possible to be confident that classification into each of the source groups is not occurring as a result of misclassification. Misclassification rates were established by averaging the percentage of misclassified isolates in faecal samples from each source category (Table 1). Isolates classified below this misclassification threshold were not considered as genuine sources within the river water samples. When the ARA database was split three-ways into municipal wastewater (MW), livestock and wild bird, 73% of the 2739 isolates (from known sources) were correctly classified into one of the three source categories on the basis of their antibiotic resistance patterns. Wild bird isolates were correctly classified at the highest rate followed by livestock and municipal wastewater isolates. Municipal wastewater and wild bird isolates were most frequently misclassified as having originated from livestock (20% and 13%, respectively), while livestock isolates were misclassified equally into both municipal wastewater and wild bird categories. However, the percentage of isolates correctly classified into each source category was well above the level expected had classification been based on chance alone (33.33% for three source categories) and demonstrated that source identification was possible.

Table 1 Classification of 2739 database isolates into three source categories (31 concentrations)

		Percentage of isolates classified as:		
Source	*n*	**MW**	**Livestock**	**Wild bird**
MW	590	**64**	20	16
Livestock	1479	13	**74**	13
Wild bird	670	10	13	**77**
Misclassification		12%	17%	14%

MW, municipal wastewater; ARCC (average rate of correct classification) = 73%;
n, number of isolates, figures in **bold** denote correctly classified isolates.

When the UK database was broken down further into six source categories, the ARCC fell slightly from 73% to 64% (Table 2), however, this value is still well above that expected had classification been based upon chance alone (16.67% for six source categories). Individual sources were correctly classified at rates ranging from 54% for municipal wastewater, to 83% for pig. The results of six-way classification revealed that municipal wastewater isolates were most commonly misclassified as wild bird and sheep isolates, while poultry isolates were most similar to municipal wastewater. Cattle isolates on the other hand were most similar to sheep isolates and *vice versa*. The lower rate of classification observed amongst municipal wastewater isolates can be attributed to misclassification into poultry, sheep and wild bird categories. Pig isolates were the most successfully classified source category, with the lowest rate of misclassification. The highest rate of misclassification occurred among cattle isolates (mainly because of similarities to sheep isolates).

Table 2 Classification of 2739 database isolates into six source categories (31 concentrations)

		Percentage of isolates classified as:					
Source	*n*	**MW**	**Cattle**	**Pig**	**Poultry**	**Sheep**	**Wild bird**
MW	590	**54**	9	2	10	12	13
Cattle	541	6	**58**	1	7	17	11
Pig	434	5	4	**83**	5	2	1
Poultry	187	10	8	8	**63**	3	8
Sheep	317	2	15	0	4	**74**	5
Wild bird	622	6	6	6	6	14	**62**
Misclassification		6%	8%	3%	6%	10%	**8%**

MW, municipal wastewater; ARCC (average rate of correct classification) = 64%;
n, number of isolates, figures in **bold** denote correctly classified isolates.

Four hundred and eighty enterococcal isolates were obtained from river samples on four separate occasions over a six-month period from November 2003 to April 2004. The geometric mean of enterococcal counts for river samples ranged from 2.54 to 4.22 $\log_{10}$ colony forming units (CFU) per 100ml, at sites F (Sloop Inn) and B (Wales Farm), respectively. The classification of river water samples into municipal wastewater, livestock and wild bird revealed that livestock appeared to be the most common source at most sites on most occasions (Table 3).

Table 3 Classification of 480 river water samples into three source categories (31 concentrations)

Source	Date	Percentage of isolates classified as:		
		MW	**Livestock**	**Wild bird**
A. Lewes	23/11/03	**25**	**42**	**33**
B. Wales Farm	23/11/03	**13**	**30**	**57**
C. Streat Lane	23/11/03	**32**	**17**	**51**
D. Hamsey	12/01/04	11	**64**	**25**
E. Barcombe	12/01/04	**23**	**60**	**17**
F. Sloop Inn	12/01/04	**38**	**49**	**13**
F. Sloop Inn	19/04/04	8	**38**	**54**
G. Freshfields	12/01/04	**20**	**72**	8
G. Freshfields	19/04/04	4	**63**	**33**
H. East Mascals 1	19/04/04	8	**42**	**50**
I. East Mascals 2	19/04/04	8	**67**	**25**

MW, municipal wastewater. Figures in **bold** are above misclassification rate.

For several of the samples, the percentage of municipal wastewater (MW) and wild bird isolates present fell below the misclassification threshold and so were not considered to be significant sources of faecal contamination on these occasions. However, the highest proportion of isolates classified as having originated from municipal wastewaters (38%) was encountered at site F (Sloop Inn) on 12/01/04 in a sample containing 4.00 $\log_{10}$ CFU/100ml. This site is located downstream from the discharge from the largest municipal wastewater treatment works in the catchment (p.e. 37,327), at Scaynes Hill. Although these works deal primarily with human and industrial waste, the combined sewerage system may also contain a significant component of animal waste from urban run-off, particularly after heavy rainfall. High municipal wastewater classifications were also observed in samples from site C (Streat Lane) on 23/11/03 and at site A (Lewes) on 23/11/03. Again both sites receive inputs of treated wastewater from a number of municipal sewage treatment works. The highest percentages of isolates classed as livestock were found at site G (Freshfields Bridge on 12/01/04) and at site I (East Mascals on 19/04/04). Wild bird appeared to be the dominant source at sites B and C (Wales Farm and Streat Lane) on 23/11/03, and sites F and H (Sloop Inn and East Mascals 1) on 19/04/04.

The further break-down of classification into cattle and sheep, pig, poultry, wild bird, MW and "other" yielded additional important information regarding source attribution (Table 4). Cattle and sheep emerged as the dominant sources present in most river samples. Isolates classified as MW were observed in samples from site G (Freshfields Bridge), site A (Lewes), site F (Sloop Inn) and site E (Barcombe). The highest percentage of isolates classified as poultry was found at site C (Streat Lane), which is situated in a rural location downstream of a poultry farm. For all classifications (Tables 3 and 4), site D (Hamsey) demonstrated no appreciable municipal wastewater signal (i.e., one that was above the rate of misclassification). River samples from site D identified 'livestock' as the dominant source of faecal pollution. Site D is largely rural with arable farming and livestock rearing being the dominant land uses.

Table 4 Classification of 480 river water samples into six source categories (31 concentrations)

Source	Date	Percentage of isolates classified as:					
		MW	Cattle and sheep	Pig	Poultry	Wild bird	Other*
A. Ouse	23/11/03	**13**	**27**	**9**	**14**	**9**	**28**
B. Wales Farm	23/11/03	4	**42**	2	**7**	**25**	**20**
C. Streat Lane	23/11/03	6	**42**	**0**	**18**	**9**	**25**
D. Hamsey	12/01/04	0	**73**	2	7	**18**	0
E. Barcombe	12/01/04	**8**	**58**	**17**	0	**15**	2
F. Sloop	12/01/04	**11**	**70**	2	2	**13**	2
G. Freshfields	12/01/04	**21**	**57**	**6**	0	13	2

MW, municipal wastewater. Figures in **bold** are above misclassification rate. *Other includes dog and donkey.

ARA appeared to be particularly good at classifying enterococci originating from pigs (83% correct classification). According to ARA the most common sources of faecal pollution present in the river samples tested were livestock and wild birds. Enterococcal isolates originating from municipal wastewaters were present in many of the river samples but were not found to be the dominant source on these particular sampling occasions. The analysis of isolates contained within the UK database revealed that the ARCC values achieved (Tables 1–4) were comparable with those obtained by previous investigators in the U.S.A. (Wiggins, 1996; Wiggins *et al.*, 1999; Hagedorn *et al.*, 1999; Harwood *et al.*, 2000).

Bacteriophage analysis

In a subsequent stage of our research, ninety-two river water samples from seven sites were tested for the presence of bacteriophages on 15 separate occasions over a six month period between April and September 2005. Phages infecting *Bacteroides* strain GB124 were detected in 55% of all river water samples tested (51 out of 92). The highest levels of bacteriophage infecting strain GB124 were found at sites (J, C and E), which were all situated less than 1.6 km downstream from municipal wastewater treatment works (Table 5).

Table 5 Enumeration of bacteriophage infecting strain GB124 in river water samples (n = 92)

Source	*n*	Mean log10 PFU per 100ML	Range (max–min) Log10 PFU per 100ML	Percentage positive samples
B. Wales Farm	15	0.17	2.48–0	7
C. Streat Lane	15	2.71	3.90–0	93
E. Barcombe	15	2.20	2.60–2.00	100
J. Spatham Lane	15	3.33	4.30–2.48	100
K. Swansyard Farm	15	0.93	2.30–0	47
L. Plumpton Mill	15	0.73	2.95–0	29
M. Ditchling Spring	2	0	0	0

Bacteriophages were detected in 98% of samples (44 out of 45) from these sites. No bacteriophages were found in samples from site M (Ditchling Spring), which is less than 100 m from where the stream emerges from the South Downs. Interestingly, counts were also very low at site B (Wales Farm), which was downstream of a bridge over which dairy cattle passed four times a day for milking. This was despite the presence of high counts of enterococci (3.70 - 4.71 $\log_{10}$ CFU/100ml), at this site especially after heavy rainfall, when cattle faeces were observed to be running into the stream.

Conclusions

These initial results suggest that the host strain GB124 is specific to human faeces and is therefore a potential indicator of human faecal contamination within the river catchment. This is supported by recent studies on municipal wastewaters and animal faeces (in preparation). Therefore, the strain may be a useful tool for identifying the contribution of wastewater discharges, leaking sewers and septic tanks to pollution problems in rural catchments. Further work is currently underway to determine whether bacteriophages infecting strain GB124 can be isolated from the final effluent of municipal wastewater treatment works from across Southern England and beyond. If so then the method may also prove useful for determining anthropogenic sources of faecal contamination beyond the catchment featured in this investigation. The phage technique is low-cost, rapid (requiring no sample concentration) and unlike ARA is database independent.

However, the ARA system could be used in combination with phage detection to offer a low-cost and rapid means of identifying animal (livestock and wild bird) sources of faecal pollution. As such, the techniques may be of benefit to developing countries where water quality has a significant impact on health and where cost is the decisive factor in the choice of environmental management technology. To date, both techniques have only been applied to surface waters, but they could prove useful in determining sources of faecal contamination in groundwater, food-stuffs (e.g. shellfish), and soils.

This work is currently being further developed in a more comprehensive catchment study (EU Interreg IIIA project 'Advanced Monitoring and Control of Microbial Water Quality') that combines extensive monitoring with chemical, biological and physical parameters. This Franco-British project aims to improve the prediction and management of faecal contamination through the development of a predictive model, using a range of microbiological and environmental test parameters linked together spatially using geographical information systems (GIS). The model will combine digital elevation data, soil drainage data, rainfall data, river flow data, sunshine data, and agricultural land use data with a 'toolbox' of faecal detection methods. This 'toolbox' approach, in which a number of methods are used in combination, will provide important information regarding the most likely sources of faecal contamination present within water samples from different sites within the study catchment.

Acknowledgements

We acknowledge the support of the European Commission Fifth Framework Programme, 'Tracking the Origin of Faecal Pollution in Surface Waters' (TOFPSW) (EVK1-CT-2000-00080) and EU Regional Development Fund (Interreg IIIA project AMACOM 'Advanced Monitoring and Control of Microbial Water Quality'). We also thank the Sussex Ouse Conservation Society for their support and permission to use the map shown in Figure 1.

References

Cornax, R., Morinigo, M.A., Paez, I.G., Munoz, M.A., and Borrego. (1990). Application of direct plaque assay for detection and enumeration of bacteriophages of *Bacteroides fragilis* from contaminated water samples. *Applied and Environmental Microbiology*. **56**, 3170–317316.

Duran, A.E., Muniesa, M., Mendez, X., Valero, F., Lucerna, F, and Jofre, J. (2002). Removal and inactivation of indicator bacteriophages in fresh waters. *Journal of Applied Microbiology*. **92**, 338–347.

Ebdon, J.E., Wallis, J.L., and Taylor, H.D. (2004). A simplified low-cost approach to antibiotic resistance profiling for faecal source tracking. *Water Science and Technology*, **50** (1), 185–191.

EEC. 1976. Council Directive of 8 December 1975 concerning the quality of bathing water 76/160/EEC. *Official Journal of the European Communities*, L31/1 (5 February 1976).

Grabow, W.O.K., Neubrech, T.E., Holtzhausen, C.S., and Jofre, J. (1995). *Bacteroides fragilis* and *Escherichia coli* bacteriophages: Excretion by humans and animals. *Water, Science and Technology*, **31**, 223–230.

Hagedorn, C., Robinson, S.L., Filtz, J.R., Grubbs, S.M. Angier, T.A., and Reneau, Jr., R.B. (1999). Using antibiotic resistance patterns in faecal streptococci to determine sources of faecal pollution in a rural Virginia watershed. *Applied and Environmental Microbiology*, 65, 5522–5531.

Harwood, V.J., Whitlock, J., and Withington, V. (2000). Classification of antibiotic resistance patterns of indicator bacteria by discriminant analysis: Use in predicting the source of fecal contamination in subtropical waters. *Applied and Environmental Microbiology*, **66**, 3698–3704.

ISO 7899/2: Water quality. Enumeration of faecal enterococci by membrane filtration. International Organisation for Standardisation, Geneva, Switzerland.

ISO/FDIS 10705-4. Water quality – detection and enumeration of bacteriophages – Part 4: enumeration of bacteriophages infecting *Bacteroides fragilis*. International Organisation for Standardisation, Geneva, Switzerland.

Jofre, J., Bosch, A., Lucerna, F., Girones, R., and Tartera, C. (1986). Evaluation of *Bacteroides fragilis* bacteriophages as indicators of the virological quality of water. *Water Science and Technology*, **18**, 167–173.

Kator, H., and Rhodes, M. (1992). Evaluation of *Bacteroides fragilis* bacteriophage, a candidate human-specific indicator of fecal contamination for shellfish-growing waters. Final Report. Virginia Institute of Marine Sciences, College of William and Mary, Gloucester Point, Vancouver, Canada.

Payan, A., Ebdon, J.E, Taylor, H., Gantzer, C., Ottoson, J., Papageorgiou, G., Blanch, A.R., Lucena, F., Jofre, J., and Muniesa, M. (2005). A method for the isolation of suitable bacteriophage host strains of *Bacteroides* for tracking sources of fecal pollution in water *Journal of Applied and Environmental Microbiology*, **71** (9), 5659–5662.

Puig, A., Queralt, N., Jofre, J., Araujo, R. (1999). Diversity of *Bacteroides fragilis* strains in their capacity to recover phages from human and animal wastes and from fecally polluted wastewater. *Applied and Environmental Microbiology*, **65**, 1172–1176.

Tartera, C., and Jofre, J. (1987). Bacteriophages active against *Bacteroides fragilis* in sewage polluted waters. *Applied and Environmental Microbiology*, **53**, 1632–1637.

Tartera, C., Lucerna, F., and Jofre, J. (1989). Human origin of *Bacteroides fragilis* bacteriophages present in the environment. *Applied and Environmental Microbiology*, **55**, 2696–2701.

Wiggins, B.A. (1996). Discriminant analysis of antibiotic resistance patterns in fecal streptococci, a method to differentiate human and animal sources of fecal pollution in natural waters. *Applied and Environmental Microbiology*, **62**, 3997–4002.

Wiggins B.A., Andrews R.W., Conway R.A., Corr C.L., Dobratz E.J., Dougherty D.P., Eppard J.R., Knupp S.R., Limjoco M.C., Mettenburg J.M., Rinehardt J.M., Sonsino J., Torrijos R.L., and Zimmerman M.E. (1999). Use of Antibiotic Resistance Analysis to Identify Nonpoint Sources of Faecal Pollution. *Applied and Environmental Microbiology*, Aug. 1999, 3483–3486.

Ground water potential zone identification in Lower Palar Basin, Tamilnadu, using remote sensing and GIS

S. Saravanan[*] and P. Suresh Babu[**]

* Research Scholar, Civil Engineering Department, IIT, Roorkee, Uttranchal, India (E-mail: *ssaravananirs@hotmail.com*)
** CSIR-SRF, Institute of Remote Sensing, Anna University, Chennai-600025, Tamilnadu, India (E-mail: *suba_babu@yahoo.com*)

Abstract Water is one of the important natural resource, which controls the human development activities, and the development of the country. In India, the surface water resource is inadequate to fulfil the water demand and productivity through ground water is high compared with surface water. Over the years, increasing population, urbanization and expansion in agriculture has accentuated the usage of Ground Water. Unscientific way of exploring groundwater leads to large-scale decline of groundwater levels, which ultimately cause a water stress condition. It is essential to maintain a proper balance between the groundwater quantity and its exportation. So it is important to identify the potential zones for ground water. The study focuses on development of remote sensing and GIS techniques to identify the potential groundwater zones in hard rock areas of lower Palar basin in Tamilnadu, India. Utilizing the remote sensing data and GIS techniques, thematic maps, such as geomorphology, geology, land use, slope, drainage density, lineament density and soil type were prepared on 1:50,000 scale. Each unit in the thematic maps were assigned to a suitable rank and weightages depending on the groundwater occurrence. These maps were overlaid two at a time and integrated coverage was classified into four categories based on the ground water potential; namely very good, good, moderate and poor. Later on, the final results were compared with the Groundwater level data from the bore wells collected by Ground water Board, Tamilnadu, India. It was found that the integrated assessment of thematic maps using a model developed based on GIS techniques is the most suitable method for groundwater potential zone prediction for hard rock terrain, but was less effective in the alluvium environment.

Keywords GIS; groundwater; remote sensing

Introduction

Precipitation in India is confined to only about three of four months in the year and varies from 10 cm in the western parts of Rajasthan to over 1000 cm in northeastern part. Due to the topographic and climatic effect the availability of water is highly uneven in both space and time. The assessment of potential zone for ground water is essential for effective management of water supplies, irrigation and industries. Remote sensing and Geographical Information System (GIS) are playing an important role in the field of hydrology and water resources development. The technique of integration of remote sensing and GIS has proved to be extremely useful for groundwater studies (Ravindran *et al.*, 1993; Krishnamurthy *et al.*, 1996; Saraf and Chaudhury, 1998; Saraf *et al.*, 2004). Ground water exploration is essentially a hydrologic inference operation and is dependent on the correct interpretation of the hydrological indicators and evidence. The occurrence of groundwater in hard rock regions is erratic and its analysis needs information of geomorphology and other associated parameters (Rao, 1991; Pradeep, 1998; Kumar *et al.*, 1999). Assessment of ground water availability in hard rock regions primarily a function of fracture-controlled permeability and zone of weathering (Dhokariker, 1991). The depth of fracture zone and weathering determine the availability of ground water.

Keeping this in view, the present study attempts to identify the favourable sites for groundwater exploration using an integrated approach of GIS and remote sensing for a hard rock terrain in Lower Palar Basin, Kanchipuram district of Tamilnadu, India.

Methods

Study area

The study area is bounded by longitudes E 79°57′10″ and E 80°12′10″ and latitudes N 12°28′30″ and N 12°41′05″, Kanchipuram district in Tamilnadu, India. The total geographical area of watershed is 360.5 km^2 and shown in Figure 1. The area is fed by Northeast monsoon rainfall, which starts in last week of September and extends until the end of December. The average annual rainfall is 957.7 mm and temperature ranges from 20.9°C to 34.5°C. The climate of the region varies between arid and semi-arid region. The topography of the area is characterized by uniform and very gentle slope and the eastern part shows undulating rolling. The general elevation of the area ranges from 60 m to 240 m above mean sea level with a gentle gradient from west to east. The area is mainly drained by Palar River flowing towards the eastern parts and finally effluences in to the Bay of Bengal. There are few small drainages in the central part of the block contributing to the numerous tanks. Buckingham canal flows from North East to South East part. All the tanks in this block are mainly fed by rainfall only.

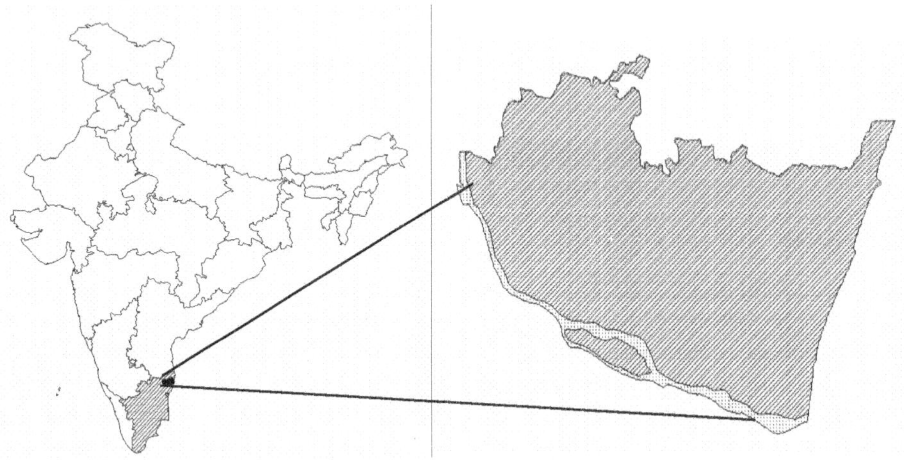

Figure 1 Study area map of Lower Palar River basin, Tamilnadu, India.

Data used and methodology

The delineation of features from the satellite image is based on the colour, tone, texture, pattern and association. The geomorphology and lineaments maps have been prepared by visual interpretation on 1:50,000 scale using IRS 1C LISS-III data. The Survey of India toposheet no. 57P/14, 66D/2, and 66D/3 have been used for preparation of drainage and base map. The geological map of the study area has been prepared from the geological map prepared by Geological Survey of India (1978). All primary input maps such as geomorphology, lineament, geology soil, landuse, and drainage, were digitized in Arc/Info, GIS software from ESRI. Polygons in each of the thematic layers were labelled and categorised separately. Suitable weightages were assigned depending on the geospatial characteristics to delineate groundwater prospective zones (Krishnamurthy *et al.*, 1996, 1997). To group the final weights of the final thematic map into four classes of ground water potential zone, such as, (i) very good (ii) good (iii) moderate and (iv) poor, statistical analysis was made by assuming normal distribution. The overall methodology adopted for the present study has been depicted in Figure 2 below.

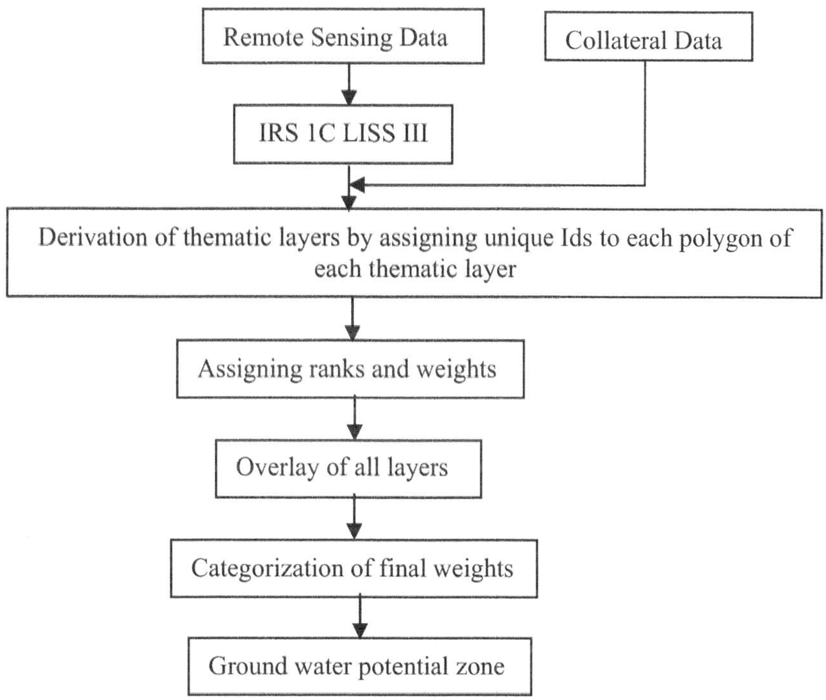

Figure 2 Overall methodology of the present study.

Results and discussion

Geological setup

Geologically, the area consists of charnockites of Achaean age and Alluvium deposit of recent age, found along the western part and southern part (Figure 3). Marine formations of Quaternary age are found along the southeastern part. Charconite shows medium grey tone and medium to coarse texture. River alluvium exists along the Palar River course consists unconsolidated materials composed of various proportions of sand, silt and clay, which forms fertile soil strata that support agriculture. This deposit forms very good water bearing horizon.

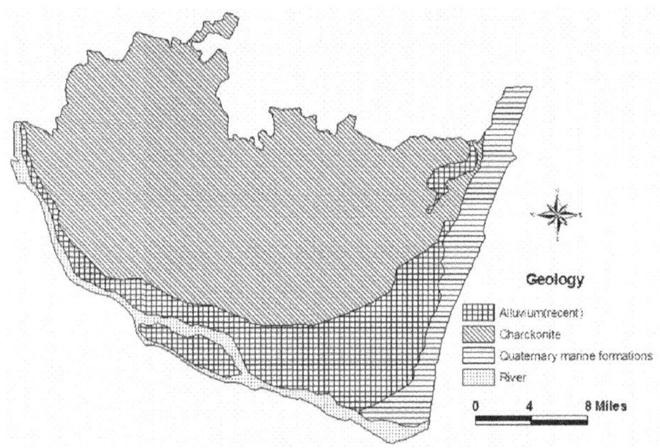

Figure 3 Geology of Lower Palar River basin, Tamilnadu.

Geomorphology

Geomorphology is the highly influencing parameter of ground water occurrence. The study area predominantly consists of buried pediments deep, followed by alluvial plains, pediments, buried pediments shallow and residual hills. The buried pediment is formed due to weathering of the hornblende, biotite gneisses under semi-arid climatic condition (Sankar, 2002). In this, infiltration is moderately good. The thickness of weathered zone varies from 10-15 m and favours a good amount of water to circulate with in this zone before reaching the deeper fractured zone. Ground water potential of this zone is very good. Alluvial plains are along the river course of the Palar River; it is the youngest geological unit and including various landforms formed by fluvial action. The flood plains have deposition of loose unconsolidated materials such as sand and silt, which is having more pore spaces and highly permeable zone. It is very good in ground water prospect. Pediments are isolated residual hillocks that consists very low weathered zone with thickness of varying up to 4 m. These areas have high runoff potential, but poor ground water recharge. Residual hills are the end products of the process of pediplanation, which reduces the original mountain masses into series of scattered knolls standing on the pediplains (Thornbury, 1990). Groundwater prospect is very poor in this area. The mud flat, beach and coastal plains are found in the eastern part of the study area. Figure 4 depicts the geomorphologic features of the study area.

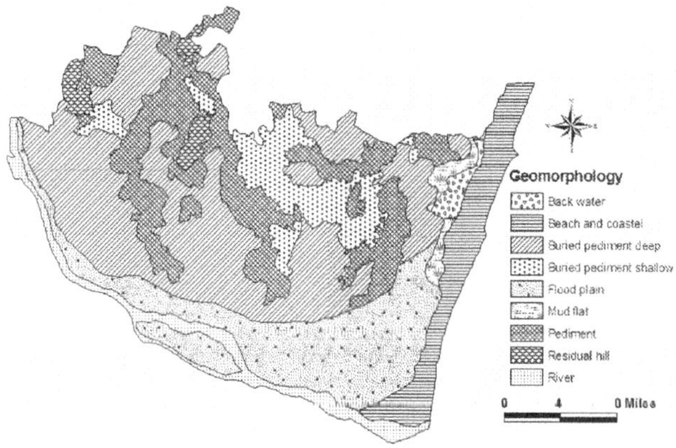

Figure 4 Geomorphology of Lower Palar River basin, Tamilnadu.

Lineament

Lineament is defined as topography of the underlying linear structural features. Lineaments provide the pathways for ground water movements and are hydro geologically very important (Sankar *et al.*, 1996). The lineaments intersections are considered as good ground water potential zones. Numerous lineaments have been mapped and shown in Figure 5. The lineaments are trending in the direction of northwest to southeast; also a few of them are oriented in northeast to southwest direction. The lineaments are more in north, south central part and the length of the lineaments varying from few km to several km. The lineament present in the hard rock region is very significant for ground water occurrence. In the present study area, it is a function of a fracture - induced secondary permeability and hydraulic conductivity.

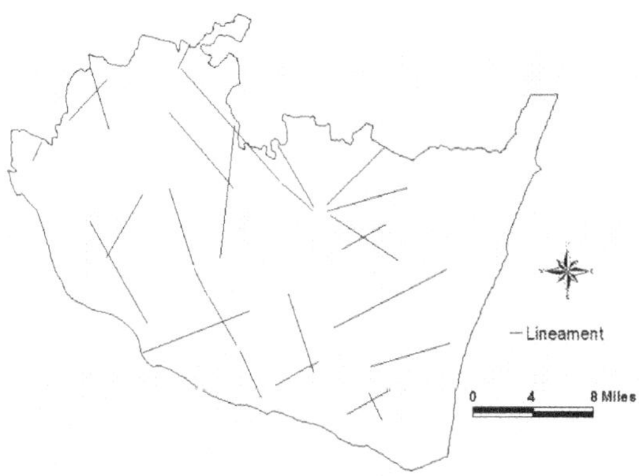

Figure 5 Lineament of Lower Palar River basin, Tamilnadu.

Drainage density

Drainage density provides a direct measure of surface runoff and permeability. Increasing drainage density reduces the recharge capabilities of ground water regime. It was seen that the least drainage density occurs in the alluvial plains of the study area, which have the maximum ground water prospect.

Slope

Infiltration and runoff capacities are controlled by slope of the terrain, soil and land-use practices. Slope with lesser gradient tend to spread the overland flows thus favouring the infiltration and ground water prospects. The areas with 0–1% and 1–2% of slope are excellent from the point of occurrence of ground water, while the areas with slope greater than 30% have poor ground water prospect.

Landuse and soil type

Figure 6 shows the land-use map of the study area. The study area predominantly has agricultural land use, other than, settlement, degraded forest, Fallow harvested land, salt affected land, sand-inland coastal, upland with or without scrub, water logged area, and few tanks. To note that agricultural land, and forest allows more water to infiltrate than barren land.

The study area predominantly contains hydrological soil group B, followed by A and C. The porous soil (such as A and B) allows more water to infiltrate than clay soil (like D).

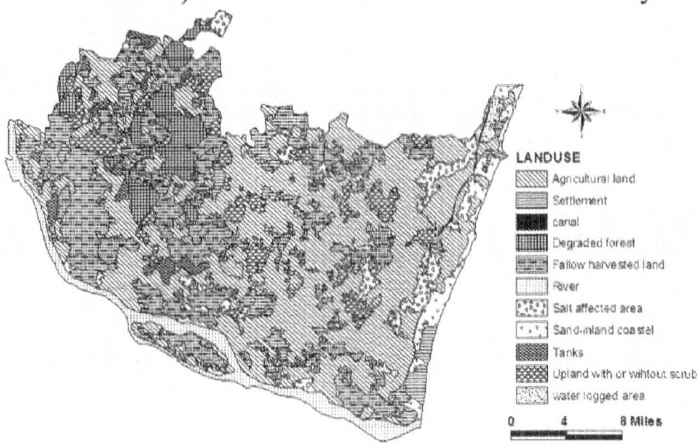

Figure 6 Landuse of Lower Palar River basin, Tamilnadu.

In order to produce the final map for ground water potential zone identification, a GIS model has been used, to integrate the above thematic maps such as geomorphology, geology, land use, slope, drainage density and soil type. Each thematic layer consists of a number of polygons, which correspond to different features. The polygons in each of the thematic layers have been categorized, depending on the suitability/relevance to the ground water potential, and suitable weights were assigned based on the reference of Krishnamurthy *et al.* (1996; 1997). Subsequently, Table 1 shows the ranking and weightages assigned to the study area features. Finally, all the thematic layers were integrated and the ground water potential zone map was arrived, based on the statistical analysis of final weights. Figure 7 presents the final map of ground water potential zone of the Lower Palar basin with four classes, namely very good, good, moderately good, and poor.

The summary of the results shows that almost all alluvial plains and valley fills associated with high density of lineaments and lineament intersections have been classified as relatively high groundwater prospective zone. The low groundwater potential zone matches with steeply mountainous areas underlain by granite with low lineament density. Furthermore, it shows that in hard rock areas, the groundwater potential is high with high lineament density and low drainage density. Finally to verify the ground water potential zone map, it is compared with the ground water level fluctuation map (Figure 8) generated from the water level measured from the bore well of the study area. Visual comparison of both maps proves that the analysis yield reliable ground water potential zone map for the study area.

Table 1 Ranking and weightages of various themes

Theme	Rank 1 4 * W	Rank 2 3 * W	Rank 3 2 * W	Rank 4 1 * W	Final Weight
Geomorphology	Buried Pediment (Deep), FloodPlain, Bazada, Valleyfills, Buried Channel, Palaeo Channel,	Buried Pediment (Shallow), Medium Dissected Plateau	Pediment, Pediment Inselberg, Highly Dissected Plateau	Structural Hills, Denudational Hills, Deflection Slope, Ridges	28
Hydrological Soil Group	A	B	C	D	7
Slope	0-3%	3-5%	5-10%	>10%	14
Lineament Density	High	Moderate	Less Moderate	Low	21
Geology	Alluvium, Schisto Gneiss	Charnockite, Kankar, Limestone	Pegmatite, Laterite	Dolerite, Granite, Quartzite	10
Landuse	Wet crop, Plantation	Dry crop, Fallow, Harvested land	Scrub, Barren	Rock outcrops, Forest & Others	10
Drainage Density	Low	Less Moderate	Moderate	High	10

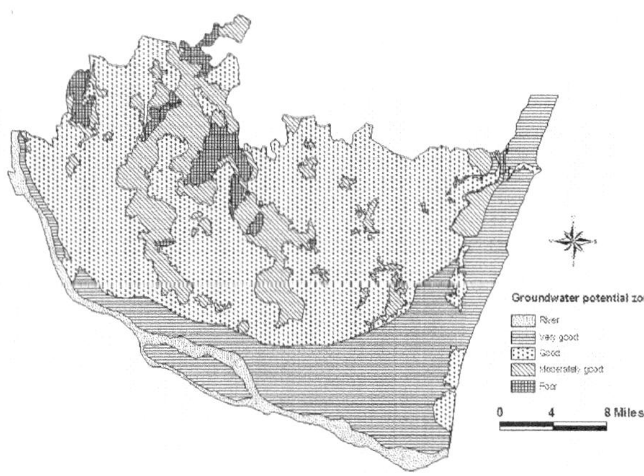

Figure 7 Ground water potential zone of Lower Palar River basin, Tamilnadu.

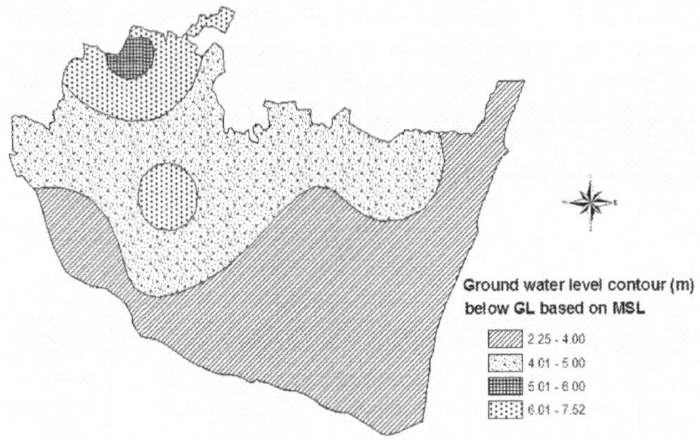

Figure 8 Ground water level fluctuation map of Lower Palar River basin, Tamilnadu.

Conclusion

The study has concluded that the remote sensing satellite imageries are found to be very useful in studies related to targeting potential zones for groundwater. The GIS techniques has been used effectively to generate thematic maps on geology, geomorphology, and hydro geomorphologic zones which, offers the basic concept for delineating groundwater potential zone in a complex terrain of hard rock region such as, lower Palar basin of Tamil Nadu, India. Further, the present study concludes the following:

(1) The present result shows that the deep buried pediments, and alluvial flood plains are the most prospect zone for ground water.
(2) The analysis of the surface water bodies, lineaments, and intersection of lineaments are directly influencing the ground water potential.
(3) Satellite data has been proven to be very informative and useful for surface study, especially in detecting a surface features and characteristics such as lineaments and land use.
(4) Integrated assessment of thematic maps using a model developed based on GIS techniques is the most suitable method for groundwater potential prediction zoning.
(5) The methods and results of this study were effective only for groundwater zone prediction in hard rock terrain, but were less effective in the alluvium environment.

Furthermore, the comparative analysis with the ground water level fluctuation map generated from water level data of bore well in the study area, shows that the analysis gives satisfactory results for delineation of potential ground water zone.

References

Dhokarikar, B.G. (1991). *Groundwater resource development in basaltic rock terrain of Mahastra.* Water Industry Publication, Pune, 275pp.

Geological Survey of India. (1978). Know your district (published), Tamilnadu and Pondicherry, Geological Survey of India.

Krishnamurthy, J., Arul Mani, M., Jayaraman, V., and Manivel, M. (1997). Selection of sites for artificial recharge towards groundwater development of water resource in India. In *Proceedings of the 18th Asian Conference on Remote Sensing, Kuala Lumpur, 20–24 October.*

Krishnamurthy, J., Kumar, V.N., Jayaraman, V., and Manivel, M. (1996). An approach to demarcate ground water potential zones through remote sensing and GIS. *International Journal of Remote Sensing*, **17**(10), 1867–1884.

Kumar, A., Tomar, S., and Prasad, L.B. (1999). Analysis of fractures inferred from DBTM and remotely sensed data for ground water development in Godavari sub-watershed, Giridith, Bihar. Journal of Indian Society Remote Sensing, **26**(2), 105–114.

Pradeep, K.J. (1998). Remote sensing techniques to locate ground water potential zones in Upper Urmil river basin, District Chatarpur, Central India. *Journal of the Indian Society of Remote Sensing*, 26(3), 135-147.

Rao, U.R. (1991). Remote sensing for national development. Curr. Sci., **61**(3 & 4), 121–128.

Ravinderan, K.V. and Jayaram, A. (1997). Ground water prospects of Shahbad Teshil, Baran District, Eastern Rajasthan: A remote sensing approach. *Journal of the Indian Society of Remote Sensing*, **25**(4), 239–246.

Sankar, K. (2002). Evaluation of Ground water potential zones using Remote sensing data in upper Vaigai river basin,Tamil nadu, India. *Journal of the Indian Society of Remote Sensing*, **30**(3), 119–129.

Sankar, K., Jagatheesan, M.S., and Balasubramanian, A. (1996). Geoelectrical Resistivity studies in the Kaniyakumari District, Tamilnadu. *Journal of Applied Hydrology*, **IX** (1 & 2) 83–90.

Saraf, A.K., and Chaudhury, P.R. (1998). Integrated Remote Sensing and GIS for Groundwater Exploration and Identification of artificial recharge sites. *International Journal of Remote Sensing*, **19**(10), 1825–1841.

Saraf, A.K., Chaudhury, P.R., Roy, B., Sharmab,Vijay, S., and, Choudhury. (2004). GIS based surface hydrological modeling for Groundwater recharge zone. *International Journal of Remote Sensing*, **25**(24), 5669–5770.

Thronbury, W.D. (1990). *Principles of Geomorphology*. Wiley Eastern Limited, New Delhi, 594 pp.

Multivariate data analysis: a tool for assessing sludge dewatering properties

A. Nikolic* **and B. Hultman****

* Department of Land and Water Resources Engineering, Royal Institute of Technology, Brinellvägen 32, S-100 44 Stockholm, Sweden (E-mail: *nikolic@kth.se*)
** Department of Land and Water Resources Engineering, Royal Institute of Technology, Brinellvägen 32, S-100 44 Stockholm, Sweden (E-mail: *bgh@kth.se*)

Abstract In wastewater treatment plants (WWTP) many parameters can influence sludge properties. To analyse large collected process datasets an appropriate method should be used. Multivariate data analysis (MVDA) with its PCA and PLS methods was used to analyse and identify the most influential parameters to sludge properties, as well as to evaluate two different sludge dewatering techniques, traditional with flocculation and newly developed Kemicond with a combination of dissolution, precipitation and flocculation. A Kemicond dataset consisted of parameters from pilot scale trials (June 2003-March 2004). Key parameters from Käppala WWTP were collected as weekly values in the same period. MVDA was found to be an appropriate statistical tool for analysing large matrixes of data. It was shown that influent sludge quality to dewatering had a great influence on both conventional and Kemicond dewatering. Sludge quality and quantity was fluctuating during the year due to seasonal changes but also due to three different operational modes of digesters (that had a great influence on parameters from digesters). The most influential variables on sludge properties were fixed solids fraction (inorganic part) of excess sludge, sludge volume indexes, temperature of influent wastewater, sludge age, and the mass proportion of excess sludge.

Keywords dewatering; digested sludge; multivariate analysis; PLS; PCA; sludge properties

Introduction

Käppala wastewater treatment plant (WWTP) near Stockholm in Sweden produces around 38 000 tons of dewatered, digested sludge per year with a dry solid content (DS) around 20%. Since it is forbidden in Sweden from 2005 to deposite organic waste on landfills the need for better dewatering technology is increasing. Today sludge from the Käppala plant is used mainly for covering of old landfills, for soil production where soil is used on construction sites and golf courses, and on farmland. Sludge handling costs at Käppala depend much on the obtained sludge volume due to a long transport distance before sludge disposal and are about €1.4 million each year. Thus better dewatering will decrease the operational costs.

From 2006 the Käppala WWTP will use a new technology for sludge conditioning and dewatering called Kemicond that has been developed by the company Kemira. In the period from June 2003 to March 2004 Kemira performed several pilot-scale tests at Käppala WWTP investigating how different operational conditions influence the Kemicond process.

Multivariate data analysis (MVDA) has been used previously as a statistical tool for monitoring and control, design and operation of biological wastewater treatment processes (Tomita *et al.*, 2002; Miettinen *et al.*, 2004; MacGregor *et al.*, 2005; Rosen, 2005; Gut, 2006). It has also been used to model the dynamic nature of the system by use of multivariate time series analysis (Van Dongen and Geuens, 1998). The aim of the project was to use MVDA to find which parameters from the plant have the largest influence on sludge properties and to investigate how variations in sludge properties influence conventional dewatering on belt presses as well as the Kemicond process efficiency, both owing to yearly fluctuations and owing to different operation of the plant.

Background

Käppala wastewater treatment plant

The Käppala WWTP treats domestic wastewater from approximately 500,000 people equivalents together with industrial wastewater mechanically, biologically and chemically. The average flow to the plant during 2005 was 5,400 m^3/h. The plant consists of the old part with six biological basins, BB1-BB6, and the new part and its five new basins, BB7-BB11. Two-thirds of the total flow passes through the new part and 1/3 through the old part. Primary (mechanical) treatment consists of coarse screens, grit removal and primary sedimentation. The biological treatment is an activated sludge process with pre-denitrification and biological phosphorous removal (only in the old part of the plant, i.e. BB1-6), UCT-process (University of Cape Town), followed by secondary sedimentation. Phosphorous is removed by use of chemical precipitation with iron sulphate in the new part of the plant. The sludge treatment consists of sludge stabilisation conducted in two digesters R1 and R2, and sludge-dewatering conducted on three belt presses. Primary sludge is collected from the primary sedimentation unit and fed into a digester. Since the summer of 2002 excess sludge together with chemical sludge are centrifuged before being digested, to increase the dry solids content. The operation of digesters is described in Figure 2.

Kemicond process

The Kemicond process consists of a chemical part and a dewatering part (see Figure 1) (Gillberg *et al.*, 2003; Manhem and Palmgren; 2004). In the chemical part first sulphuric acid is added (to reach pH 5) in order to dissolve water-binding organic and inorganic substances like iron compounds. Ferrous phosphate, a jelly-like precipitate, which originates from chemical phosphorus removal is dissolved. Then hydrogen peroxide is added to oxidise Fe^{2+} to Fe^{3+}, which then precipitates as ferric phosphate, a compound that holds less water. By that way dewatering properties are improved which has been also reported by (Jansson and Palonen, 2002). Furthermore, in this strong oxidative environment further release of water from oxidised organic substances is accomplished, and sludge is disinfected and odour reduced. In the dewatering part conditioned sludge is mixed with a polymer and led to a roller drum (a pre-dewatering step) and then to a screw press.

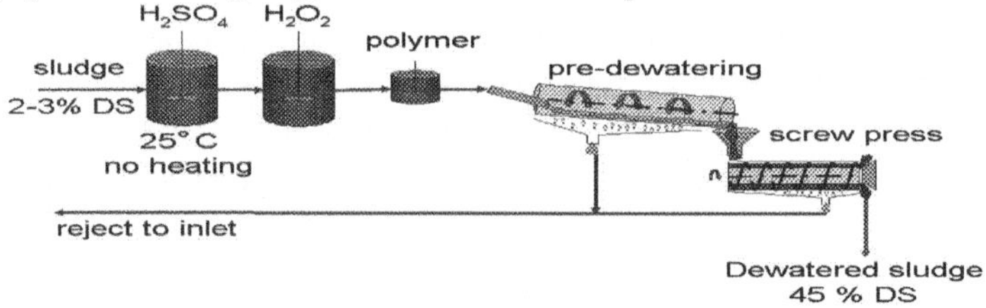

Figure 1 The Kemicond process scheme.

Multivariate data analysis (MVDA)

MVDA is a method for extracting information from small or large tables of data. It can be presented as points (observations) in multidimensional spaces (variables) which are approximated by models and projected onto planes as a two dimensional window like Principal Component Analysis (PCA) or Partial Least Squares projections to latent structures (PLS) (Kourti and MacGregor, 1995; Eriksson *et al.*, 2001). Each PCA and PLS is presented with score and loading plots. Scores (t) are co-ordinates in the plane and loadings (p) define the orientation of the plane and are used to interpret the scores. Using PCA it is possible to get an overview of a process dataset: if observations are grouped with trends in behaviour (represented with score plots) or if outliers exist (observations with deviating variable patterns). Further, are variables correlated (represented with loading plots) and well explained (represented with values R^2X and R^2Y) (Wold *et al.*, 1987). By use of PLS it is possible to find how the X-variables influence the Y-variables (the response of the process), and to find interrelations as all variables can be correlated against each other (Wold *et al.*, 2001). The predicting power of PLS is presented as Q^2.

Material and methods

The Simca-P+ package was used as a MVDA tool with its PCA and PLS plots. The X-matrix, analysed with PCA, consisted of all parameters that might exert an influence on the dewatering process, whereas Y-matrix presented the responses of the process i.e. dry solids reached after dewatering. The covariance between these two matrixes was modelled with PLS plots that were obtained after plotting one 'Y' variable 'DS after dewatering' against all X parameters. In all cases only one 'Y' variable was chosen in order to avoid hidden interactions of parameters with an individual Y variable.

Two different X-matrixes were analysed: I) The Käppala dataset consisted of key parameters from wastewater treatment (mechanical and activated sludge steps) and sludge treatment processes (digestion and dewatering) that were collected as weekly values (an average value during a week) from the Käppala database called 'Waste' in the period from February 2003 to April 2004. In total 54 variables and 57 observations were analysed. During the chosen observation period the two digesters R1 and R2 were run in three different operational modes, as shown in Figure 2, where Mode 1 was operated during the longest period of time until January 2004, Mode 2 for only one month from January to February 2004, and Mode 3 from February until now.

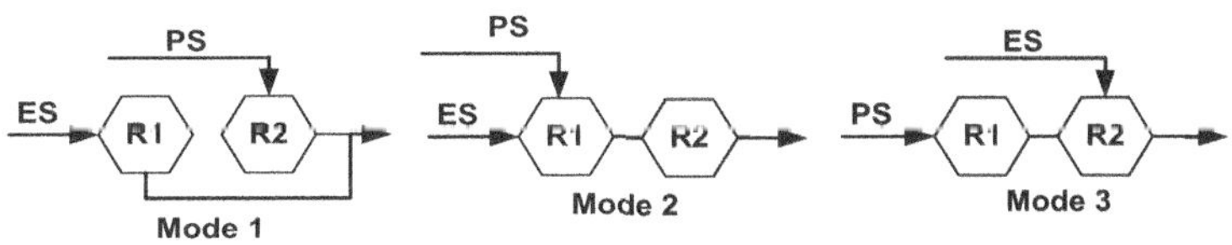

Figure 2 Digesters operational modes (PS = primary sludge, ES = excess sludge).

II) The second X-matrix was the Kemicond dataset that consisted of 50 different variables that were followed during pilot-scale trials conducted at Käppala WWTP. Variables were analysed like concentrations of P, Fe, Al, S, COD_{Cr}, TOC, DS, and loss on ignition (the organic content or volatile solids) in filtrate after the chemical conditioning step, in supernatant after the roller drum and in the cake after the screw press. Reference values were supernatant and cake after belt presses in Käppala's conventional dewatering unit. 112 observations were analysed i.e. measuring dates when trails were performed.

Pre-processing of data involved lagging of variables of digesters and dewatering with calculated average retention time in digesters. Since only weekly values were used, retention time in the biological basins that is around 1.5 days was neglected. Furthermore, data were transformed in order to obtain approximately normal distribution of the data, autoscaled i.e. mean centred and scaled to unit variance (UV-scaling) so that all variables measured in different scales gain equal importance. The output model was validated by methods of cross-validation and response permutation testing for checking statistical significance of prediction (in case of PLS modelling). The improvement of models was done by excluding unimportant variables. For PLS models observed vs predicted values as well as the model residuals were also checked. For all models before the interaction of all parameters and a response variable Y has been analysed with PLS plots, each X block was evaluated with a PCA method.

Results and discussion

Analyses of data only for WWTPs

The variations of all parameters from Käppala WWTP were first analysed using a PCA method where score and loading plots are investigated (Figures 3 and 4). The first two principal components (PC) explained most of variations of data. In Figure 3 can be seen that observations are grouped and correlated according to different operational modes of digesters. It can be also seen that there were no strong outliers. This spreading of observations is influenced by the position of variables presented with PCA loading plots for two PC as shown in Figure 4. Observations from the first operational mode are grouped and divided into three subgroups. The first subgroup consists of observations from the end of March until the beginning of June 2003, which are separated from other observations mostly with high values of primary sludge flows, and flows of polymer added to the centrifuges. The second subgroup consists of observations from November and December 2003 where the lowest flows of polymer added to the centrifuges, and the highest alkalinity in R1 have been noticed. The third subgroup (June–November 2003) is characterized by high values of retention time in biological basins (BB), temperature, pH in the first digester R1, and fixed solids in excess sludge in the new part of the plant BB7-11. Furthermore, the highest DS after dewatering are observed in this period, which can be due to the highest retention time in the first digester i.e. excess sludge that is difficult to dewater was digested for the longest period.

Observations from the second and third operational modes are grouped and are separated from other observations with high values of alkalinity, fixed solids (FS), DS and pH in the second digester R2 (the influence of added excess sludge), high values of DS in excess sludge, DS load in ES and mass and flow proportions of ES, as well as SV and SVI in BB 7-11. Furthermore, flow to BB, DS before presses and SS in reject after dewatering were high.

The influence of parameters from Käppala WWTP on dewatering has been analysed using PLS plots, as shown in Figure 5, where parameters with a positive influence are listed above the zero line and with a negative influence under the zero line. A model with good prediction power (Q^2 67%) was obtained with two PC and high R^2X 58% and R^2Y 76% (well explained). Sludge quality was fluctuating during the year due to seasonal changes but also due to three different operational modes of digesters. Conventional dewatering was better during the warmer months (June–November 2003) when the temperature of influent wastewater was the highest (17 °C) occurring together with the lowest flows to the biological step and highest retention times. The sludge properties were better during that period due to lower sludge volume indexes (SVI) in the plant. Generally, sludge sedimentation properties are better in the new part (BB7-11) than in the old part of the plant (BB1-6) where biological phosphorus removal is achieved. In the new part 2/3 of the influent wastewater is directed, where due to chemical phosphorus precipitation inorganic content i.e. fixed solids fraction (FS) of excess sludge is higher, and sludge has better sedimentation properties. Furthermore, during the summer period concentrations of phosphorus and nitrogen in influent wastewater are higher due to lower flows. Temperature is a very influential variable since it co-varies with several other variables.

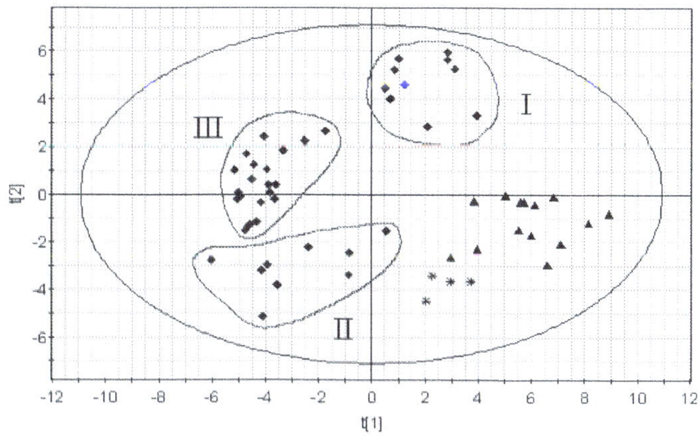

Figure 3 PCA score-observations plot (including confidence intervals based on Hotelling's T^2 distribution) grouped according to the variable "operational modes of digesters", where rhomboids represents observations during Mode 1 with three subgroups, stars = Mode 2, and triangles = Mode 3.

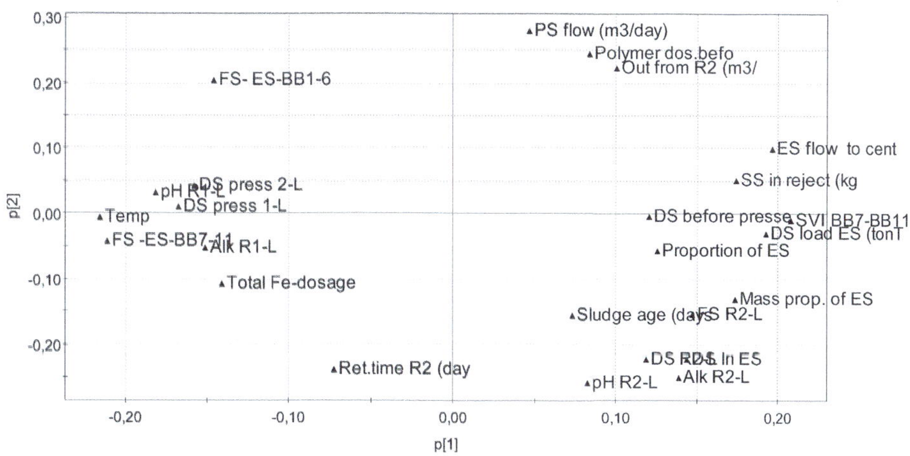

Figure 4 PCA loading plot of all parameters from a WWTP.

Other negatively correlated variables with dewatering are suspended solids in supernatant and sludge age that is lower during the summer i.e. it is more difficult to dewater sludge during winter when sludge age is higher. It was also shown that it was much easier to dewater primary sludge and that an increase of a mass and flow proportion of excess sludge made dewatering worse.

The influence of parameters of digesters like pH, DS, FS and alkalinity on dewatering were mostly governed by the operational modes of the digesters, and the most influential was the first operational mode that was operated in the longest period of time until January 2004. This is probably due to the largest number of observations from the summer period that had a great influence on sludge quality. Therefore, it cannot be concluded how different operational parameters of digestion influence dewatering properties but only which of the operational modes had the largest influence.

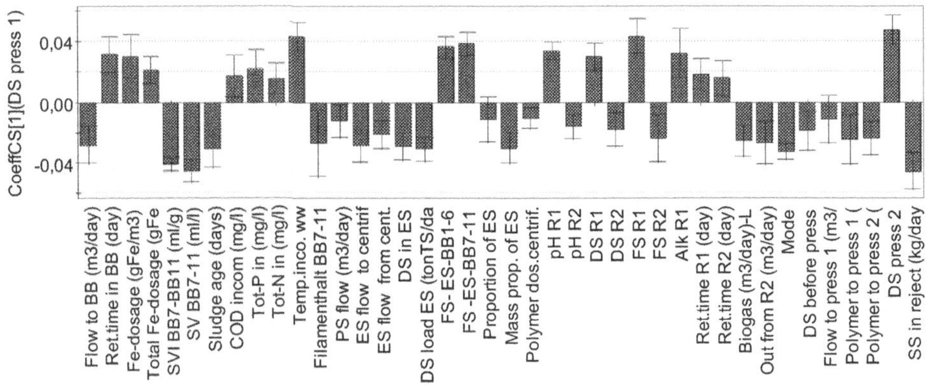

Figure 5. Regression coefficients of scaled and centred variables of Käppala WWTP for the first principal component for the response Y 'DS after belt press 1'. Bars indicate 95% confidence intervals.

Analyses of Kemicond data

Here, all Kemicond data from the trials in the period from June 2003 until March 2004 were plotted together. From a PLS score plot for a 'Y' variable DS after a screw press (DS-Cake) it can be seen that there exist several strong outliers due to either the highest or the lowest DS. In Figure 6 can be seen variables that are strongly positively correlated with the 'Y' variable (listed above the zero line) like Al concentration (in%) in cake (Al-Cake). Furthermore, with increasing amounts of cake from the roller drum ($V_{cake.drum}$) in kg/h DS after the screw press decreases. DS after the roller drum (DS-Drum) and DS in cake after the belt press in Käppala's dewatering unit (DS-BeltpCake) are correlated well with DS after the screw press, i.e. DS after Kemicond are positively correlated with DS after Käppala's present dewatering practice.

Furthermore, DS after the screw press strongly increased with decreasing concentration of suspended solids in supernatant (TSS-ScrewRej), which can be a good way to follow up efficiency of dewatering. High dosages of chemicals and high loss on ignition (organic content) in cake (LOI-cake) decrease dry solids after presses, i.e. with an increase of the inorganic content in the cake due to precipitation of iron, phosphorus, aluminium and other ions, better dewatering is achieved. High total nitrogen concentration (tot-N-cake) and ammonium concentration in the cake after the screw press (NH_4-cake) had a negative effect on dewatering, and this can be due to a presence of some organic matter but also due to the operation of the plant. High retention time in digesters can for instance decrease the nitrogen content in sludge. It was also shown that concentrations of dissolved and precipitated Fe, P and Al after additions of acid and peroxide, respectively, are important and positively correlated with dewatering, i.e. higher concentration of dissolved Fe, Al and P facilitate dewatering.

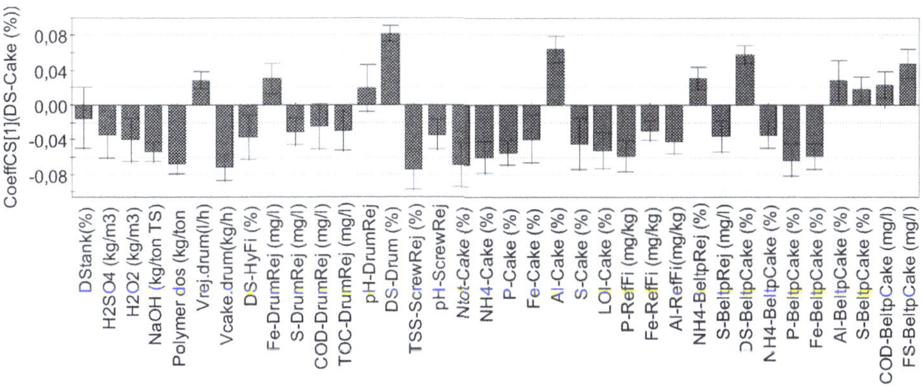

Figure 6 PLS loading plot: Regression coefficients of scaled and centred variables of Kemicond for the first principal component for the response Y 'DS after a screw press'. Bars indicate 95% confidence intervals. (R^2Y = 89%, Q^2 = 83%)

Analyses of Kemicond data together with WWTP data

The comparison of the two datasets showed that due to a lot of missing data from the Kemicond set on a weekly basis a model with a low predicting power was obtained. Therefore another analysis was preformed with only observations when Kemicond trials were preformed. A new model was obtained with two PC and high R^2X 60%, R^2Y 83%, Q^2 63%. It was shown a good correlation between Käppala conventional dewatering and Kemicond dewatering. The influence of variable sludge properties on Kemicond pilot-plant dewatering data can be based on Käppala's full-scale dewatering data.

Conclusions

Multivariate data analysis (MVDA) with PCA and PLS methods can be a useful modelling tool for statistically analysing large process data matrixes, finding correlations between different sets of variables, and obtaining the most influential parameters on the process. In that way a number of analysed parameters can be reduced.

Analyses of the Käppala dataset with PCA showed that sludge quality and quantity were fluctuating during the year due to seasonal changes but also due to three different operational modes of digesters. The most influential was the first operational mode, probably due to the largest number of observations from a summer period that had a great influence on sludge quality. Therefore general conclusions about the influences of different parameters of digesters on dewatering could be misleading. Since digesting of sludge is very important in determining the sludge quality, future investigation should include analyses of the key parameters from the plant for the years 2004 and 2005 when digesters were operated in the same mode.

Multivariate analysis with both PCA and PLS of key parameters of the Kemicond process gave the most influential variables on dewatering like a positive influence of Al content in the cake, which could be due to the precipitation of $AlPO_4$ and therefore excess Fe in the solution, or some other factor that influenced Al content in the sludge. Furthermore, suspended solids in supernatant had very strong negatively correlation with dewatering which can be used as a controlling variable. In addition the organic matter in the cake (ammonium, total nitrogen, loss on ignition) had a negative effect on dewatering, as well as chemicals addition (acid and to some extent peroxide) that were overdosed.

Analyses of all data together with a PLS method revealed that the influent sludge quality to the dewatering step had a great influence even though the conditioning and dewatering techniques were different, either only flocculation with traditional dewatering or a combination of dissolution, precipitation and flocculation with Kemicond. The most influential parameters on sludge properties were fixed solids fraction (inorganic part) of excess sludge, sludge volume indexes, temperature of influent wastewater, sludge age, and the mass proportion of excess sludge. It was also shown that the efficiency of dewatering could be followed by analyses of the suspended solids in the supernatant after dewatering.

Acknowledgements

We thank Göran Karlsson and Ingemar Karlsson at Kemira for providing all material needed, for valuable comments and discussions. Thanks are given to Per Manhem, Torsten Palmgren and Anna Maria Borglund at Käppala WWTP for their contribution and co-operation. The authors are grateful for the financial support from Kemira.

References

Eriksson L., Johansson E., Kettaneh-Wold N. and Wold S. (2001) *Multi- and Megavariate Data Analysis. Principles and Applications.* Umetrics AB, Umeå, Sweden. ISBN 91-973730-1-X.

Gillberg L., Hansen B., Karlsson I., Enkel A.N. and Pålson A. (2003) *About water treatment.* Kemira Kemwater, ISBN 91-631-4344-5.

Gut L. (2006) *Assessment of a partial nitritation/Anammox system for nitrogen removal.* Licentiate Thesis, Royal Institute of Technology, Stockholm, TRITA-LWR Licentiate thesis 2034.

Jansson K. and Palonen H. (2002) Effect of PH and Oxidation on Sludge Dewatering. In: *Chemical Water and Wastewater Treatment VII,* H.H. Hahn, E. Hoffmann and H. Ødegaard (Eds.), IWA Publishing, London, ISBN: 1 84339 009 4, pp. 309–330.

Kourti T. and MacGregor J.F. (1995) Process analysis, monitoring and diagnosis, using multivariate projection methods. *Chemometrics and Intelligent Laboratory Systems*, **28**, 3–21.

MacGregor J. F., Yu H., Muñoz S. G. and Flores-Cerrillo J. (2005) Data-based latent variable methods for process analysis, monitoring and control. *Computers and Chemical Eng.*, **29**, 1217–1223.

Manhem P. and Palmgren T. (2004) Kemicond process at the Käppala wastewater treatment plant, Lidingö, Sweden. In: *Chemical Water and Wastewater Treatment VIII,* H.H. Hahn, E. Hoffmann and H. Ødegaard (Eds.), IWA Publishing, London. ISBN: 1 84339 068 X, pp. 337–344.

Miettinen T., Hurse T.J., Connor M.A., Reinikainen S.-P. and Minkkinen P. (2004) Multivariate monitoring of a biological wastewater treatment process: a case study at Melbourne Water's Western Treatment Plant. *Chemometrics and Intelligent Laboratory Systems*, **73**, 131–138.

Rosen C. (2005) *A Chemometric Approach to Process Monitoring and Control–With Applications to Wastewater Treatment Operation.* PhD thesis, Department of Industrial Electrical Engineering and Automation, Lund University, Sweden, ISBN: 91-88934-20-9.

Tomita R.K., Park S.W. and Sotomayor O.A.Z. (2002) Analysis of activated sludge process using multivariate statistical tools–a PCA approach. *Chemical Eng. J.* **90**, 283–290.

Van Dongen G. and Geuens L. (1998) Multivariate time series analysis for design and operation of a biological wastewater treatment plant. *Wat. Res.* **32** (3), 691–700.

Wold S., Esbensen K. and Geladi P. (1987) Principal Component Analysis. *Chemometrics and Intelligent Laboratory Systems*, **2**, 37–52.

Wold S., Sjöström M. and Eriksson L. (2001) PLS-regression: a basic tool of chemometrics. *Chemometrics and Intelligent Laboratory Systems*, **58**, 109–130.

Modelling

Development of land-use export coefficients for the modelling of nonpoint source pollutant loads in Fuji River Basin, Japan

S. Shrestha and F. Kazama

Department of Ecosocial System Engineering, Interdisciplinary Graduate School of Medicine and Engineering, University of Yamanashi, Yamanashi, 400-8511, Japan
(E-mail: *sangam@ccn.yamanashi.ac.jp*; *ktutaba@ccn.yamanashi.ac.jp*)

Abstract We developed land-use export coefficients of biochemical oxygen demand, chemical oxygen demand, nitrate nitrogen and inorganic dissolved phosphorus specific to Fuji River basin using available hydrological, statistical, and spatial tools. The BTOPMC, a distributed hydrological model, was used to estimate discharge for those stations where continuous discharge data was not available. A seven-parameter log linear model was used to estimate the loadings using historical water quality concentration, observed and estimated discharge for each sub-basins.The annualized and normalized loadings as dependent variable and land-use proportion in each sub-basin as independent variables were used to derive the land-use export coefficients through multiple regression technique. Most of the land-use export coefficients are significant at α equal to 0.05 and the land-use categories used in the multiple regression models explained more than 85% variability in loadings. Therefore, the derived coefficients can be used to estimate the nonpoint source pollution loadings from major land uses particularly given the short time schedule and limited data for the management practices to improve the water quality within the Fuji River basin.

Keywords BOD; COD_{Mn}; NO_3-N; PO_4-P; land-use export coefficients

Introduction

Land-use export coefficients representing the rate of pollutant loadings by land area are often recommended as a way to estimate loadings from nonpoint sources (Loehr *et al.*, 1989; Reckhow *et al.*, 1990). Several empirical, semi-distributed and distributed water quality/pollution loading models rely heavily on land-use export coefficients that are not region-specific or use limited data from literatures in order to estimate pollution loadings. However, use of land-use export coefficients for estimating pollution loads is based on the knowledge that, for a given climatic regime, specific land-use types will yield or export characteristics quantities of specified organic matters or nutrients to a downstream water body over an annual cycle (Rast and Lee, 1983). Therefore, it is necessary to develop regionally specific land-use export coefficients to estimate the pollutant loadings.

The waters of Fuji River basin represent a valuable water resource contributing to the prefectural economic base through industrial, agricultural, recreational activities and hydropower operation. Nevertheless, the long-term consequences of resulting pressure and demands on the prefectural water resource remain unknown. Of particular concern is the response of waters to increasing nonpoint source pollutant loadings due to watershed development and land-use activities. While the watershed nutrient budget measurement has been continued, the recent cut in funding has limited the resources for current and future watershed diagnostic studies. So far, no attempt has been made to review the existing data provided from long term water quality monitoring and investigate whether basin-wide land-use export coefficients can be developed. Therefore, the major objective of this study is to develop land-use export coefficients of organic matters (biochemical oxygen demand and chemical oxygen demand) and nutrients (nitrate nitrogen and inorganic dissolved phosphorus) specific to Fuji River basin using readily available statistical, hydrological and spatial tools.

Study area description

Fuji River basin study area, drained by the Fuji River, is located in the central part of Japan (Figure 1). The basin area is 3,570 km^2 and the mainstream length is 128 km.The river originates, as the Kamanashi River, from Mount Komagatake in the north of the Southern Alps, and as the Fuefuki River from the north of Yamanashi Prefecture. These two rivers flow together in the south of Kofu Basin as the Fuji River and subsequently, flows to Pacific Ocean at Suruga Bay. The summers are hot and humid and winters are cold, with average temperatures of 26°C and 3°C, respectively. The basin receives a mean annual precipitation of ~2100mm.

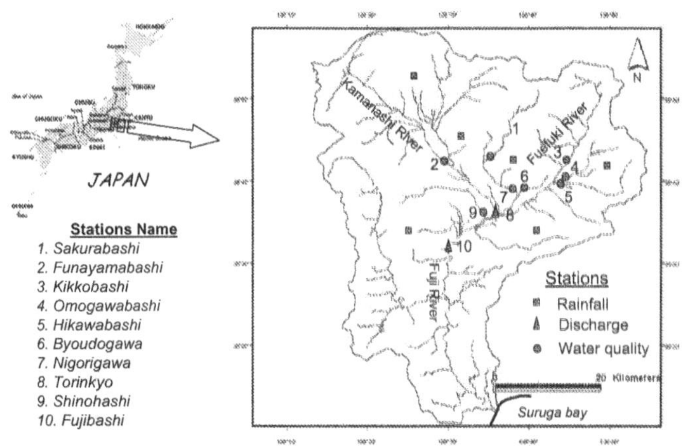

Figure 1 Location map, rainfall, discharge and water quality monitoring stations of the study area.

Methods

The semi-monthly/monthly water quality data of biochemical oxygen demand (BOD), chemical oxygen demand (COD_{Mn}), nitrate nitrogen (NO_3-N) and inorganic dissolved phosphorus (PO_4-P) from eight stations and daily discharge of three stations during 6 years (1995–2000) were obtained from Environment Divison of Yamanashi Prefecture (EDYP) and Ministry of Land Infrastructure and Transportation (MLIT), Japan respectively. Rainfall and other meterological data were obtained from MLIT, Japan. Similarly, the landuse data were obtained from Ministry of Environment, Japan. The other relevent information were obtained from literature survey.

The methodology outlined below was followed to estimate the land-use export coefficients for nonpoint source of BOD, COD_{Mn}, NO_3-N and PO_4-P.

(a) Determine the major land uses and their proportion in the draingae area above each monitoring site within the river basin. A sub basin map defining drainage area for each sampling site was generated from a digital elevation map (DEM) (250 m) using ArcGIS 8.3 and Arc Hydro tools according to water quality monitoring stations. Land cover map was overlayed and proportion of major landuses in each subbasins were calculated.

(b) Estimate discharge at ungauged sites within the basin using distributed hydrological model in those stations where continuous discharge is not available. In order to estimate the discharge at six ungauged sites, a distributed hydrological model, Blockwise use of TOPMODEL with Muskingum-Cunge routing (BTOPMC) (Takeuchi *et al.*, 1999) was used. Rainfall and meteorological data of six stations, DEM, land use map, soil map were used as input to the model.

(c) Estimate loadings by regression modelling using the measured, estimated discharge and water quality concentration. A seven-parameter log-linear model (Cohn *et al.*, 1992) was used to estimate loadings from each sub basin using water quality concentrations, observed and estimated discharges. The model calculates loads of nutrients (and other constituents) carried by rivers employing a statistical regression model, where the constituent concentrations are estimated based on stream flow and time/season.

(d) Apply multiple regression model to determine the optimal estimates of export coefficients for major land uses using estimated loadings and land-use information. Typically, land-use export coefficients are determined by monitoring land uses, such as forest, row crop or urban, using field plots to isolate individual land uses (Reckhow *et al.*, 1980). However, monitoring single land uses watershed may be ideal, most watersheds even small ones, are generally comprised of variety of different land uses (McFarland and Hauck, 2001). Therefore to isolate the loading contribution from the heterogeneous drainage areas, multiple regression techniques were used to develop the nutrient export coefficients for the major land uses in the watershed based on procedures described by Hodge and Armstrong (1993). The dependent variable was the annualized nutrient loading and independent variables were land-use proportion in each sub-basin. All multiple regression models were developed using a forced zero intercept, thus, giving a loading of zero when all independent variables equal zero.

(e) Compare estimated land-use export coefficients with available literature values to evaluate the reasonableness of derived coefficients.

Results and discussion

Determination of land uses above stream sampling sites

The size of the drainage area and percentage of major land uses above each sampling site is presented in Table 1. The land-use categories and their area included in the classification are forest, agriculture, grassland and urban.

Table 1 Land uses associated with the drainage area above sampling sites in Fuji River basin

Site	Forest (%)	Agriculture (%)	Grassland (%)	Urban (%)	Other (%)	Total Area ($\times 10^3$ ha)
Kikkobashi	83.91	8.70	4.89	1.57	0.93	24.34
Hikawabashi	78.27	12.72	4.59	2.86	1.55	11.11
Omogawabashi	59.05	32.41	1.56	6.28	0.70	10.88
Sakurabashi	82.75	9.58	2.78	0.85	4.04	15.50
Funayamabashi	77.23	15.56	2.06	2.91	2.24	48.05
Shinohasi	84.79	6.32	3.81	2.37	2.70	5.95
Nigorigawa	4.62	66.07	1.82	27.08	0.41	1.57
Byoudogawa	65.25	17.55	3.45	12.58	1.17	23.15

While choosing the sampling sites, a careful consideration was given to exclude the major point sources influence from the basins. Although some variations in land use are expected to occur over time, but for this study, land use was assumed to be stable.

Estimation of discharge

A model calibration was carried out to optimize the model inputs so that the differences between estimated and observed data could be minimized and better accuracy of model predictions accomplished. A detail description on this can be found in Shrestha *et al.* (2005). The whole basin was divided into two sub basins and calibration of the model for 1995-2000 was carried out by comparing the observed and estimated discharge at Fujibashi and Torinkyo stations. The performance of the model was evaluated by Nash and Sutcliffe (NE) efficiency. The model is good in simulating both the low flows and high flows (Figure 2). The Nash efficiencies are 70% and 75% for Fujibashi and Torinkyo stations respectively. Similarly, the ratio of simulated volume of discharge to the observed volume of discharge (Qsim/Qob) is more than 80% in both stations. In other words, model has slightly underestimated (<20%) the volume of discharge. Therefore, discharge at ungauged sites located at the upstream of the calibrated stations was taken for the further study.

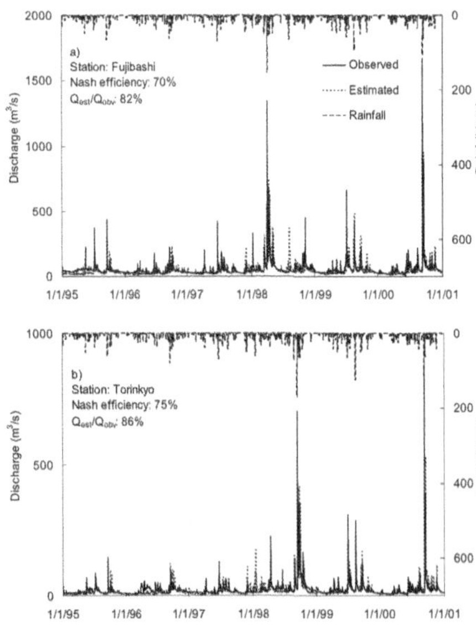

Figure 2 Comparison between observed and estimated discharge at a) Fujibashi and b) Torinkyo.

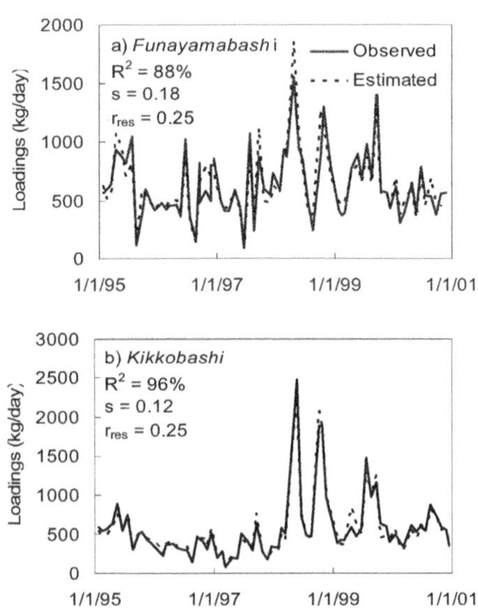

Figure 3 Comparison between observed and estimated NO_3-N loadings in a) Kikkobashi and b) Funayamabashi.

The model performs very well in estimating loadings. The coefficients of determinations (R^2) for loadings in all basins are >90%. Figure 3 shows a typical case from two sub basins for NO_3-N loading estimation. The lack of fit in estimation was also tested by plotting probability plot correlation coefficient (PPCC) (Vogel, 1986) of the residuals, and found to be normally distributed. Similarly, residuals are plotted against the explanatory variables and against the predicted values; they appeared to be reasonably homoscedastic.

Loadings from each sub basin were annualized and normalized by the basin area (Table 2) in order to set it as dependent variables in the multiple regression modeling to derive the land-use export coefficients.

Table 2 Annualized and normalized nutrient masses (kg/ha/yr) associated with stream sampling sites in the Fuji river basin for 1995–2000

Site	BOD	COD_{Mn}	NO_3-N	PO_4-P
Kikkobashi	8.79	19.01	9.48	0.11
Hikawabashi	9.35	18.00	na	1.01
Omogawabashi	22.56	36.97	14.80	0.45
Sakurabashi	2.57	12.47	1.91	0.05
Funayamabashi	6.23	7.43	5.47	0.21
Shinohasi	8.59	13.90	4.60	0.29
Nigorigawa	38.52	53.30	13.98	0.91
Byoudogawa	15.38	26.19	12.67	0.64

Determination of export coefficients for different land uses

The results of multiple regression analysis are summarized in Table 3. Most of the coefficients for the land-use variables are significant at α equal to 0.05 and has coefficient of determination (R^2) more than 85% for all regression models. This indicates that the land-use categories used as independent variables explain a large proportion of the variability in loadings. For forest landuse, all the coefficients could not be statistically defined as significantly different from zero. However, the positive values still represented the optimized coefficients for the current data set.

Table 3 Export coefficient (kg/ha/yr) estimates for land-use variables versus loadings using a zero-intercept multiple regression model

Landuse	BOD	COD_{Mn}	NO_3-N	PO_4-P
Forest	1.00	0.75	2.12	0.71
Agriculture	0.62*	0.64*	1.06*	0.42*
Grassland	0.54*	0.52*	0.46*	0.59*
Urban	2.63*	1.48*	3.80*	1.73*
R^2	0.96	0.97	0.89	0.85

*Significant at α equal to 0.05; R^2 is the variability explained by landuse categories on loadings in multiple regression models.

Land-use export coefficients from different studies within Japan has been reviewed and compared with the present study (Table 4). The wide variability in literature values within categories reflects site-specific variations in management and environmental conditions. Due to unavailability of published nitrate and phosphate export coefficients, the derived coefficients were compared with published total nitrogen and total phosphorus export coefficients. The proportion of nitrate nitrogen to the total nitrogen and inorganic dissolved phosphorus to the total phosphorus are calculated from some least impacted sites in the Fuji River basin and it is found that total nitrogen contains more than 85% of nitrate nitrogen and total phosphorus contains more than 80% inorganic dissolved phosphorus. Therefore, it is assumed reasonable to compare the coefficients. The export coefficients of NO_3-N for forest and urban land use and PO_4-P for forest and agriculture land use are within the literature values. Similarly, the export coefficients of COD_{Mn} for forest and urban land uses are comparable with literature values.

Table 4 Range of available literature values for coefficients (kg/ha/yr) compared with calculated values for land uses in the Fuji river basin

Land use	COD_{Mn}	TN	NO_3-N	TP	PO_4-P	Source
Forest	0.91	0.9–38.0	na	0.01–0.90	na	Ukita and Nakanishi, 1999
	0.75		2.12		0.71	Present study
Agriculture	3.80	5.5–52.5	na	0.36	na	Ukita and Nakanishi, 1999
	0.64		1.06		0.42	Present study
Urban	3.65	5.3–6.9	na	0.18	na	Ukita and Nakanishi, 1999
	1.48		3.80		1.73	Present study

The derived export coefficients of NO_3-N and COD_{Mn} for agriculture are lesser and PO_4-P for urban land uses is higher when compared with literature values. These wider differences in comparison can be attributed to the specific basin characteristics for loading of these water quality constituents and the methodology used for deriving export coefficients. For example, a recent study conducted by Nakamura *et al.* (2005) shows that the NO_3-N concentration, applied as fertilizers in agriculture area, in groundwater is higher than the river water in Fuji River basin. Therefore, a lower rate of nitrate loading is expected as compared to the actual field studies in other basins. Similarly, in this study the concentration of PO_4-P was influenced by the domestic wastewater discharge but the literature values do not account the influence of domestic wastewater. Therefore, a higher rate of PO_4-P loading is expected as compared to the actual field studies in other basins.

Conclusion

The export coefficients of biochemical oxygen demand, chemical oxygen demand, nitrate nitrogen and inorganic dissolved phosphorus for major land uses specific to Fuji River basin are developed using available statistical, hydrological and spatial tools. Most of the land-use export coefficients are significant at $\alpha = 0.05$ and the land-use categories used in the multiple regression models explained more than 85% variability in loadings.These export coefficients represents the average of conditions and practices (e.g., soils, slopes, tillage practices, fertilization timing and amounts, etc.). These land-use export coefficients are also compared with available literature values to evaluate its resonableness. Although the methodology in this study does not indicate the specific mechanisms causing surface runoff and pollutant transport, the derived coefficients can be used to estimate the nonpoint source pollution loadings from the major land uses particularly given the short time schedule and limited data for the management practices to improve the water quality within the Fuji River basin.

Acknowledgements

We extend sincere thanks to Yuki Hiraga for her help in the database development; Tim Cohn, Hazard Theme Coordinator of United Geological Survey (USGS), for providing a seven parameter log linear model. We are also thankful to The Fuji Xerox Setsutaro Kobayashi Memorial Fund for providing fund and 21st Century Center of Excellence (COE), University of Yamanashi for giving opportunity to conduct this study.

References

Cohn, T., Caulder, D.L., Gilroy, E.J., Zynjuk, L.D., Summers, R.M. (1992). The validity of a simple statistical model for estimating fluvial constituent loads: an empirical study involving nutrient loads entering Chesapeake Bay. *Water Resource Research*, **28** (9), 2353–2364.

Hodge, T.A., and, Armstrong, L.J. (1993). *Use of a Multiple Linear Regression Model to Estimate Stormwater Pollutant Loading*. In New Techniques for Modelling the Management of Stormwater Quality Impacts, W. James (Editor). Lewis Publishers, Boca Raton, Florida, pp.201–214.

Loehr, R.C., Ryding, S.O. Sonzogni, W.C. (1989). *Estimating the Nutrient Load to a Water body. In: The Control of Eutrophication of Lakes and Reservoirs*, S.O. Ryding and W. Rast (Editors). Volume I, Man and the Biosphere Series, United Nations Educational Scientific and Cultural Organization, Paris, France and The Parthenon Publishing Group, Park Ridge, New Jersey, pp. 115–146.

McFarland, A.M.S., and, Hauck, L.M. (2001). Determining nutrient export coefficients and source loading uncertainty using in-stream monitoring data. *Journal of the American Water Resource Association*, **37**(1), 223–236.

Nakamura, T., Kazama, F., Satake, H., Shrestha, S. (2005). Tracing Nitrate Transport in the Ground-water Aquifers using Isotope Techniques– a case study in Western Kofu Basin, Japan. *In Proc* The Third International Symposium on Southeast Asian Water Environment, 6–8 December 2005, Asian Institute of Technology, Bangkok, Thailand

Rast, W., and, Lee, G.F. (1983). Nutrient loading estimates for lakes. *Journal of Environmental Engineering*, **109**: 502–517.

Rekchow, K.H., Beaulac, M.N., Simpson, J.T. (1980). *Modeling phosphorus loading and lake response under uncertainty: A manual and compilation of export coefficients*. U.S. Environmental Protection Agency, Clean Lake Section, Washington, DC, EPA 440/5-80-011, June 1980, pp.214.

Shrestha, S., and, Kazama, F. (2005). Development of land use export coefficients for the estimation of nutrient loads in Fuji river basin, Japan. In *Proceedings of the 14th Korea/Japan Symposium on Water Environment*, pp. 52–69. Masan, South Korea, 25–27 September.

Takeuchi, K., Ao, T.Q., Ishidaira, H. (1999). Introduction of block-wise use of TOPMODEL and Muskingum-Cunge method for the hydro-environmental simulation of a large ungauged basin. *Hydrological Sciences Journal*, **44**: 633–646.

Ukita, M., and, Nakanishi, H. (1999). *Pollutant load analysis for the environmental management of enclosed sea in Japan*. In Proc. The Fourth International Conference on the Management of Enclosed Coastal Seas.

Vogel, R.M. (1986). The probability plots correlation coefficient tests for the normal, lognormal, and Gumbel distributional hypotheses. *Water Resources Research*, **22** (4), 587–590.

Statistical modelling of nutrient removal and correlations to water quality parameters in a small-scale constructed wetlands system

Rupali Mohansingh*, Keith E. Dennett* and George C.J. Fernandez**

* Department of Civil and Environmental Engineering / MS 258, University of Nevada Reno, Reno, Nevada 89557, USA (E-mail: *rmohansingh@hotmail.com*; *kdennett@unr.edu*)
** College of Agriculture, Biotechnology and Natural Resources / MS 204, University of Nevada Reno, Reno, Nevada 89557, USA (E-mail: *gcjf@unr.edu*)

Abstract Steamboat Creek (Reno, Nevada, USA) is contaminated by agricultural and stormwater runoff. Exploratory, descriptive, and statistical modelling was performed using SAS® software (ver. 9.13) in order to predict the performance of a small-scale constructed wetlands. The performance of 6 constructed wetland designs with different combinations of sediment and water were considered. Water sampling was conducted twice each month over the period from September 2000 through August 2004. The removal of total nitrogen (TN), total phosphorus (TP), and total suspended solids (TSS) within the wetlands were monitored. Data exploration using box plots were examined to check for trends, outliers, and distribution patterns. The removal of TN, TP, and TSS, associated with each wetland design, seasonal changes, and interactions between design and season were analyzed using mulitvariate analysis of variance (MANOVA). Canonical correlations between operational conditions such as flow, pH, and temperature and the removal of TN, TP, and TSS were analyzed. Scatter plots were used to explore the linear trends, significance, and the range of the majority values in a correlation. The multivariate test statistic was used to check for the overall significance in canonical correlations. Following canonical correlations, multiple regression models were developed for TN, TP, and TSS with respect to each of the predictor variables. The results from the statistical analyses performed on the data from the small-scale constructed wetlands demonstrated that the construction of a large-scale wetlands system would be an effective method to improve water quality in Steamboat Creek by reducing the loadings of TN, TP, and TSS into the lower Truckee River.

Keywords Constructed wetlands; nutrient removal; statistical modelling

Introduction

Steamboat Creek is a major tributary of the Truckee River in northern Nevada, USA. It is the most polluted tributary of the Truckee River and has been affected by land development and mining activities, and artificial lowering of the water surface in the Truckee River. The purpose of this project was to evaluate the overall impact of incorporating constructed wetland systems into flood control and watershed restoration projects. The performance of different wetland designs during the four seasons of the year were evaluated for the removal of TN, TP, and TSS.

Methods

Data analysis

The overall objective of this project was to quantify the performance of the small-scale constructed wetlands system in terms of the removal of TN, TP, and TSS. Statistical analyses were conducted on the data for the removal of TN, TP, and TSS obtained from monitoring the performance of the wetlands system over a period of four years from September 2000 through August 2004. The performance of six different wetland designs with different combinations of sediment and source water is discussed. The efficiency of each wetland design with respect to the removal of TN, TP, and TSS, are reported. All of the data for the removal of TN, TP, and TSS was analyzed in terms of loading rates.

Four seasons were considered in the analyses. Fall season consisted of data collected during the months of late September, October, and early November. Winter season consisted of data collected during late November, December, January, and early February. Spring season consisted of data collected from late February, March, April, and May. Summer season consisted of data collected during June, July, August, and early September.

The effects of operating conditions such as flow, pH, and temperature were also analyzed with respect to the removal of TN, TP, and TSS. Correlations and regression models were developed between the removal of TN, TP, and TSS and the operating conditions.

Before performing MANOVA, the data set was checked for outliers and distribution patterns of the multiple responses. The results are summarized in box plots. All of the statistical analyses were performed using the SAS® (version 9.13) macro applications, 'MANOVA' (Fernandez, 2005b) and 'CANON' (Fernandez, 2005a).

Results and discussion

Data exploration and descriptive statistics

Graphical analyses were performed to identify extreme outliers that could influence the overall results. The box plots were very useful to explore univariate data, check for distribution patterns, trends, and outliers (Fernandez, 2002). A sample box plot is shown in Figure 1. Unusually large extreme observations were removed from the dataset before performing the MANOVA.

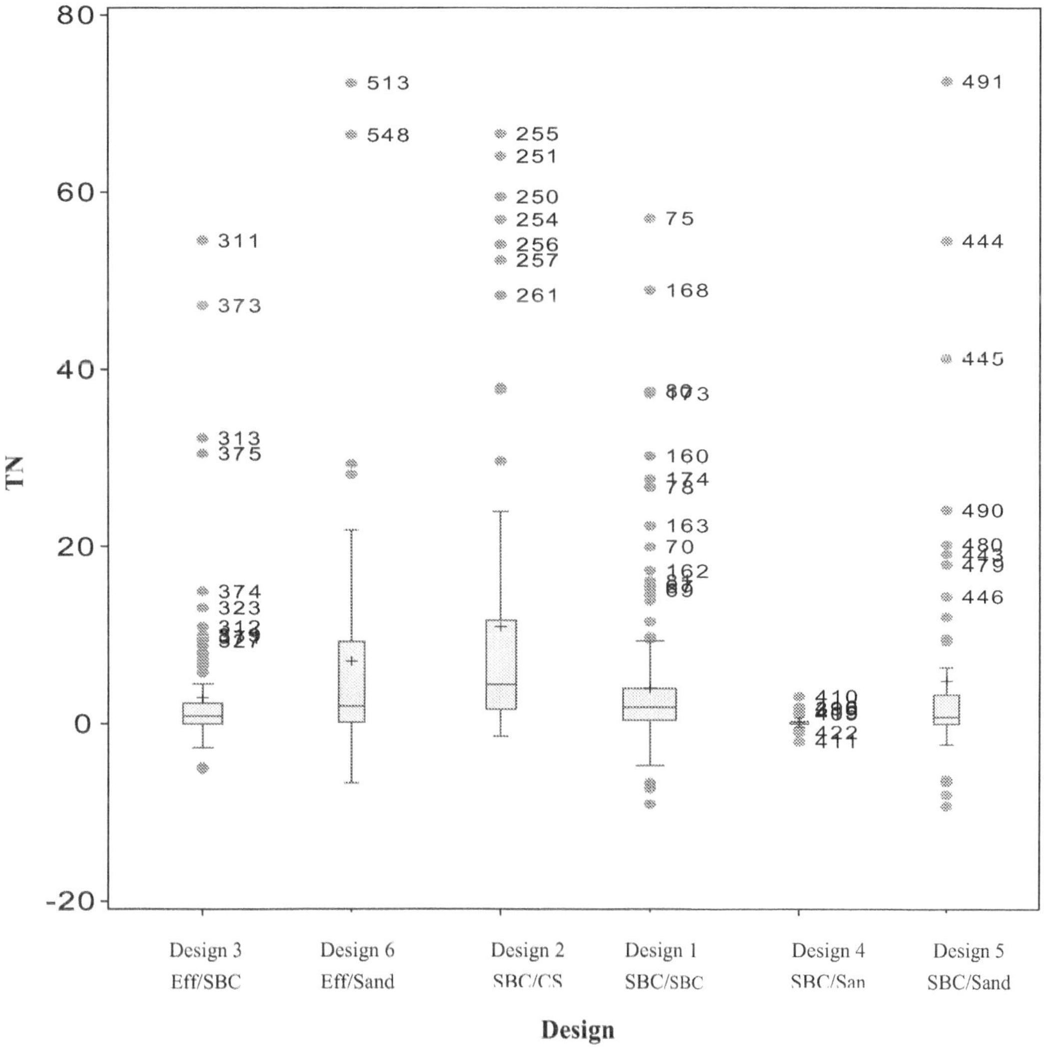

Data exploration and descriptive statistics with wetland design and seasons

The statistical analyses were performed based on: 1) the wetland design; and 2) the source of the water. Design 1 (SBC/SBC) included Trains 1 and 2 and received water from Steamboat Creek (SBC) and were constructed using sediments obtained from SBC. Design 2 (SBC/CS) included Train 3 and received water from SBC and was constructed using clean sand and gravel (CS). Design 3 (Eff/SBC) included Trains 4 and 5 received TMWRF effluent (Eff) and were constructed using sediments from SBC. Design 4 (SBC/San) included Train 6 and received water from SBC and was constructed using clean sand and also had no vegetation (San). Design 5 (SBC/Sand) included Trains 7 and 8 and received water from Steamboat Creek and was constructed using clean sand (Sand). Design 6 (Eff/Sand) included Trains 9 and 10 and received TMWRF effluent and was constructed using clean sand.

Removals of TN, TP, TSS and TOC for various wetland designs

The performance of the various wetland designs is summarized in Table 1. All of the least square means were estimated at the mean values of the covariates. Design 2 achieved the highest removal of total nitrogen (TN) at 10.00 g/d. Design 2 was filled almost completely with sand and aggregate. It is believed that since both aerobic and anaerobic conditions existed through the permeable soil, this design was very effective in the removal of TN (Crites and Tchobanoglous, 1998). Design 6 achieved the highest removal of TP at 0.31 g/d. Phosphorus is stored within the roots and rhizomes and is mobilized into the shoots of plants during the growing season (Mitsch and Gosselink, 1993). Design 1 achieved the highest removal of TSS at 35.22 g/d. The density of the vegetation increases resistance to flow, increases retention time, and facilitates the settling of suspended particulates (Gopal, 1999).

Table 1 Removals of TN, TP, and TSS based on wetland design

Design	TN LSMEAN and standard error	TP LSMEAN and standard error	TSS LSMEAN and standard error
D1	4.32 ± 0.66	0.20 ± 0.04	35.22 ± 2.94
D2	10.00 ± 0.90	0.20 ± 0.05	32.90 ± 4.00
D3	1.92 ± 0.74	0.20 ± 0.04	21.15 ± 3.30
D4	−2.80 ± 1.51	0.01 ± 0.09	17.38 ± 6.75
D5	6.17 ± 1.10	0.16 ± 0.06	21.48 ± 4.91
D6	9.17 ± 1.11	0.31 ± 0.06	24.86 ± 4.96

Removals of TN, TP, and TSS during different seasons

The effect of season on the performance of the wetlands system is summarized in Table 2. When the performance of all of the wetland designs was considered together, the highest removal of TN was achieved during the fall season with an average removal of 8.02 g/d. The highest removal of TP was achieved during the fall season with an average removal of 0.30 g/d. The highest removal of TSS was achieved during the summer season with an average removal of 29.91 g/d.

Table 2 Average Removal of TN, TP, TSS, and TOC with seasonal changes

Season	TN LSMEAN and standard error	TP LSMEAN and standard error	TSS LSMEAN and standard error
Fall	8.02 ± 1.12	0.30 ± 0.06	24.53 ± 5.00
Spring	5.28 ± 0.62	0.11 ± 0.03	28.37 ± 2.76
Summer	2.85 ± 0.63	0.26 ± 0.03	29.91 ± 2.80
Winter	3.03 ± 0.97	0.06 ± 0.05	19.18 ± 4.33

Removal of TN, TP, and TSS based on design and season interactions

The least squares means of the removals of TN, TP, and TSS within the wetland designs during different seasons are summarized in Table 3. The highest removal of TN for wetland designs receiving TMWRF effluent was achieved in Design 6 during the fall season with an average removal of 21.62 g/d. The highest removal of TN for wetland designs receiving water from Steamboat Creek was achieved in Design 2 during the spring season with an average removal of 13.08 g/d. The highest removal of TP for wetland designs receiving TMWRF effluent was achieved in Design 6 during the summer season with an average removal of 0.53 g/d. The highest removal of TP for wetland designs receiving water from Steamboat Creek was achieved in Design 2 during the fall season with an average removal of 0.31 g/d. The highest removal of TSS for wetland designs receiving TMWRF effluent was achieved in Design 6 during the fall season with an average removal of 31.62 g/d. The highest removal of TSS for wetland designs receiving water from Steamboat Creek was achieved in Design 1 during spring season with an average removal of 44.36 g/d.

The removal of TN was significant with respect to Designs 4 and 5 (p from <0.0001 to 0.0025). The removal of TP was not significant (p from 0.1090 to 0.9688) with respect to the different wetland designs during the different seasons. The removal of TSS was not significant (p from 0.3360 to 0.9929) with respect to the different wetland designs during the different seasons.

Table 3 Removal of TN, TP, and TSS based on wetland design during different seasons

Design	Season	TN LSMEAN and Standard Error	TP LSMEAN and Standard Error	TSS LSMEAN and Standard Error
D1	Fall	6.54 ± 1.30	0.27 ± 0.07	29.93 ± 5.77
D1	Spring	5.96 ± 1.10	0.14 ± 0.06	44.36 ± 4.88
D1	Summer	1.98 ± 1.01	0.30 ± 0.06	36.93 ± 4.53
D1	Winter	2.80 ± 1.63	0.12 ± 0.09	29.64 ± 7.27
D2	Fall	10.22 ± 2.14	0.31 ± 0.12	28.40 ± 9.55
D2	Spring	13.08 ± 1.48	0.07 ± 0.08	42.00 ± 6.61
D2	Summer	8.17 ± 1.43	0.25 ± 0.08	35.01 ± 6.40
D2	Winter	8.54 ± 2.00	0.15 ± 0.11	26.18 ± 8.88
D3	Fall	1.34 ± 1.53	0.18 ± 0.09	16.91 ± 6.82
D3	Spring	2.35 ± 1.18	0.13 ± 0.07	15.96 ± 5.25
D3	Summer	2.36 ± 1.27	0.39 ± 0.07	22.45 ± 5.68
D3	Winter	1.65 ± 1.72	0.11 ± 0.10	29.29 ± 7.70
D4	Fall	1.62 ± 4.27	0.35 ± 0.25	15.66 ± 19.04
D4	Spring	−0.78 ± 2.13	−0.001 ± 0.12	27.88 ± 9.52
D4	Summer	−1.82 ± 2.01	−0.04 ± 0.12	30.77 ± 8.98
D4	Winter	−10.23 ± 3.22	−0.23 ± 0.19	−4.79 ± 14.38
D5	Fall	6.80 ± 3.02	0.43 ± 0.17	24.64 ± 13.47
D5	Spring	4.41 ± 1.53	0.20 ± 0.09	21.32 ± 6.83
D5	Summer	1.76 ± 1.55	0.18 ± 0.09	31.76 ± 6.91
D5	Winter	11.70 ± 2.41	−0.16 ± 0.14	8.20 ± 10.77
D6	Fall	2.16 ± 3.10	0.22 ± 0.18	31.62 ± 13.83
D6	Spring	6.65 ± 1.53	0.12 ± 0.09	18.69 ± 6.82
D6	Summer	4.67 ± 1.53	0.53 ± 0.09	22.53 ± 6.84
D6	Winter	3.75 ± 2.24	0.38 ± 0.13	26.60 ± 9.98

Canonical correlations for entire small-scale wetland system

Correlations between the effects of flow, pH, and temperature and the removals of TN, TP, and TSS were examined using canonical correlations. Over the period from January 2001 through August 2004, the flow through the wetlands varied from 136 L/d to 34,560 L/d. Some of this variation was due to systematic manipulation of the operation of the wetland system in order to identify the effects of variation in flow. The pH ranged from 6.6 to 9.0. The average air temperature ranged from 32.7 to 78.2°F (0.55–25.7°C).

Data exploration

Effect of Flow on the Removals of TN, TP, and TSS

Scatter plots were prepared in order to examine the range of the variables, the simple linear correlation coefficient, the direction, and the significance of the regression slopes. A scatter plot fits a linear regression and 95% confidence band for the regression line. If the mean line falls within the 95% confidence interval, then the correlation between 'x' and 'y' is not considered significant, indicating that there is no correlation between 'x' and 'y'. The variations in flow significantly affected the removal of TN, TP, and TSS. Within the range of flows that was observed (136 and 34,560 L/d), the removal of TN ranged between –10 and 20 g/d. The correlation between the removal of TN and flow was positive and was significant with an estimated sample coefficient of 0.40. Within the range of flows that was observed, the removal of TP ranged between –0.5 and 0.75 g/d. The correlation between the removal of TP and flow was positive and was significant with an estimated sample coefficient of 0.18. Within the range of flows that was observed, the removal of TSS ranged between –25 and 100 g/d. The correlation between the removal of TSS and flow was positive and was significant with an estimated sample coefficient of 0.11.

Effect of pH on the removals of TN, TP, and TSS

The removal of TN had no correlation with pH within the pH range 6.6–9.0 since the mean was within the 95% confidence interval. The removal of TP ranged from –1 to 1.5 g/d and was influenced by changes in pH within the range 6.6–9.0. The correlation between the removal of TP and pH was slightly significant with a negative estimated sample coefficient of 0.16, indicating that for an increase in pH within the range 6.6– 9.0, the removal of TP decreased. The removal of TSS was influenced by changes in pH within the pH range 6.6–9.0. The removal of TSS ranged between –25 and 100 g/d within the observed pH range. The correlation between TSS and pH was significant with an estimated sample coefficient of 0.40, indicating that for an increase in pH within the range 6.6–9.0, the removal of TSS increased.

Effect of temperature on the removals of TN, TP, and TSS

The removal of TN had no correlation with respect to temperature since the mean was within the 95% confidence interval. Similarly, the removal of TP had no correlation with respect to temperature since the mean was within the 95% confidence interval. The removal of TSS had a significant negative correlation with respect within the temperature range 32.7–78.2°F (0.55–25.66°C) with a negative value of 0.23. A negative correlation indicates that for an increase in temperature, the removal of TSS decreased.

Canonical correlations between the removals of TN, TP, and TSS and flow, pH, and temperature

The data were tested for overall correlations between the predictor variables (i.e., flow, pH, and temperature) and the overall effects on the response variables (i.e., removal of TN, TP, and TSS). The first canonical correlation between the removal of TN, TP, and TSS and flow, pH, and temperature was 0.476. This canonical correlation is highly significant based on the MANOVA statistics summarized in Table 4. Wilk's Lambda is a commonly used statistic. The *p value* observed in Table 4 for the overall canonical correlation for overall removal of TN, TP, and TSS was significant with respect to the overall predictor variables.

Table 4 MANOVA test statistic for testing overall significance of canonical correlation between predictor and response variables for the entire wetlands system

Multivariate statistics	*p*
Statistic	**Pr > F**
Wilks' Lambda	<0.0001

Multiple linear regression

The canonical correlations between the response and predictor variables were highly significant. This correlation was verified individually with respect to each response variable (i.e., removal of TN, TP, and TSS) and each predictor variable (i.e., flow, pH, and temperature). Table 5 summarizes the Raw Regression Models based on the multiple linear regression models.

Table 5 Regression models using raw regression coefficients

Response variable	Regression model	R^2	p
Total nitrogen	$36.71 + 0.76\ \text{flow}^* - 3.47\ \text{pH}^* - 0.18\ \text{Temp}^*$	0.197	<0.0001
Total phosphorus	$0.93 + 0.006\ \text{flow}^{NS} - 0.11\ \text{pH}^{S} + 0.001\ \text{Temp}^{NS}$	0.025	0.005
Total suspended solids	$-177.28 + 1.49\ \text{flow}^* + 30.34\ \text{pH}^* - 0.59\ \text{Temp}^{S}$	0.122	<0.0001

*Highly significant: $p < 0.0001$; ^{S}significant: $p < 0.05$; ^{NS}not significant: $p > 0.05$.

Conclusions

Comparing the overall wetland performance of the different wetland designs with respect to water quality improvement in Steamboat Creek, Design 2 performed the best in terms of the overall removals of TN, TP, TSS, and TOC. Designs 3 and 6, which received TMWRF effluent, performed similarly with respect to the removal of TN, TP, and TSS.

Seasonal changes affected the performance of the wetlands system. The removal of TN was most effective during the spring season. The removal of TP was most effective during the summer season. The removal of TSS was the most efficient during summer and spring season. The spring season was the most effective season in the overall removal of TN, TP, and TSS in the wetlands. Growth of the wetland vegetation along with increases in temperature and microbial activity make the spring season the most effective season in terms of the removal of TN, TP, and TSS within the wetlands system.

Comparing the performance of the two wetland designs which received TMWRF effluent, the performance of both the designs was statistically similar during all the seasons with respect to the removal of TN, TP, and TSS.

Comparing the performance of the wetland designs which received water from Steamboat Creek, the removal of TN was the highest within Design 2 (Train 3) for most seasons. The removal of TP was the highest within Design 2 (Train 3). The removal of TSS was the highest within Design 1 (Trains 1 and 2), but not significantly different from Design 2, except during the winter season.

The overall correlations between the removal of TN, TP, and TSS and operational parameters such as flow, pH, and temperature were highly significant. An increase in flow rate (within the range 136–34,560 L/d) increased the removal of TN, TP, and TSS. The effect of pH did not have any significant influence on the removal of TN. An increase in the pH (within the range pH 6.6–9.0) decreased the removal of TP. The effect was small but was significant. The effect of pH within the range of pH 6.6–9.0 influenced the removal of TSS and was significant. With an increase in pH, the removal of TSS increased

The effect of temperature within the range 32.7–78.2°F (0.55–25.7°C) did not have any correlation with respect to the removal of TN. The removal of TP increased as the temperature increased within the range of 32.7 and 78.2°F (0.55 and 25.7°C). The correlation was small indicating the effect of temperature had a small influence on the removal of TP. The removal of TSS decreased with increasing temperature with a significant correlation

The performance of Designs 1 and 2 on the removal of TN, TP, and TSS was better than the other wetland designs. The results from the statistical analyses performed on the data from the small-scale constructed wetlands demonstrated that the construction of a large-scale wetlands system would be an effective method to improve water quality in Steamboat Creek by reducing loadings of TN, TP, and TSS into the lower Truckee River.

References

Crites, R. and Tchobanoglous, G. (1998) *Small Decentralized Wastewater Management Systems.* McGraw Hill, New York.

Fernandez, G. (2005a) *CANCORR.SAS* http://www.ag.unr.edu/gf/.

Fernandez, G. (2005b) *MANOVA.SAS* http://www.ag.unr.edu/gf/.

Gopal, B. (1999) *Water Science Technology,* **40**, 27–35.

Mitsch, W. J. and Gosselink, J. G. (1993) *Wetlands.* International Thomson Publishing, New York.

Mathematical modelling of rotating biological contactor performance

Sanjay Dutta, Erhard Hoffmann and Hermann H. Hahn

Department of Aquatic Environmental Engineering, Institute for Water and River Basin Management , University of Karlsruhe (Technical University), Adenauerring 20, 76128 Karlsruhe, Germany
(E-mail: *dutta@iwg.uni-karlsruhe.de*; *Erhard.Hoffmann@iwg.uni-karlsruhe.de*)

Abstract The high biomass packing density in biofilms result in a compact system which can sustain the fluctuations arising in wastewater characteristics and substrate concentration. Process optimisation and adaptability under different environmental conditions and influent characteristics remain challenging tasks. Mathematical modelling helps to predict the system performance under various external conditions. Modelling of the physical oxygen transfer through the liquid film (pure water) developed on a rotating disc revealed that the oxygen transfer coefficient K_L increased with the rotational speed. Similarly, ambient temperature held a large influence on oxygen transfer coefficient. At lower rotational speed, K_L value decreased with increasing temperature since boundary layer thickness decreased. The biofilm model was implemented for the first stage of a three-stage rotating biological contactor. The results from the multi-species biofilm model confirmed that under high organic loading conditions, heterotrophs offered strong competition to autotrophs and dominated in the surface layers while autotrophic population could grow only underneath. The particulate fraction of inerts was high in the deeper layers due to higher decay and inactivation of heterotrophs. The penetration depth of oxygen remained at 250–300µm after which anoxic conditions set in and resulted in some denitrification. Dissolved oxygen concentration was high in the bulk (above 3.0 $gO_2\ m^{-3}$) while it was slightly higher in the liquid film. Results showed 85% removal of organic substrate at an influent loading rate of 12g sol.COD $m^{-2}\ d^{-1}$, while nitrification was 55% at a loading rate of 2.60g NH_4-N $m^{-2}\ d^{-1}$ (Ist stage RBC). The results are comparable with the experimental findings. Further removal can be easily achieved by considering multiple stages of the RBC.

Keywords: Biofilm modelling; denitrification; multi-culture biofilm (MCB); nitrification; oxygen transfer coefficient (K_L); rotating biological contactors (RBC)

Introduction

Aerobic treatment of wastewater using fixed biofilm technology has been established to be an efficient and proven technology with relatively stable end-products. Biofilm mechanisms have inherent advantages because of the high biomass packing density and compactness of the system. They offer an ideal alternative as a secondary or tertiary biological treatment unit for simultaneous removal of organic substances, nitrogen and other nutrients in municipal wastewater (Masuda *et al.* 1990, Müller *et al.* 1980, Boller *et al.* 1990). Rotating biological contactors are a unique adaptation of the fixed film process wherein a series of discs mounted on a common shaft undergo continuous rotations. These discs provide the surface for biofilm growth. A part of the discs is immersed in bulk liquid inside the reactor and the rotations generate a thin liquid film acting as a boundary layer over the biofilm surface (Figure1). The large specific surface area of biofilm (150–250 m^2/m^3) facilitate high removal efficiency even at short contact periods. The economic advantage and compactness of the system compared to conventional treatment processes such as activated sludge plants make it a favourable choice especially suited to the concept of decentralized wastewater treatment systems (Patwardhan, 2003). Apart from the inherent advantages of a biofilm mechanism, rotating biological contactors present an added advantage in aerobic wastewater treatment by facilitating easy and effective oxygen transfer. Aeration is achieved by oxygen transfer through diffusion at the air–liquid boundary facilitated Process optimisation and adaptability under different environmental conditions and feed constitution remain challenging tasks as ever. Previous research shows that oxygen concentration remains one of the most limiting substrates in aerobic biofilm treatment processes. Therefore, it is important to understand the process of physical oxygen diffusion into the boundary layer through the air-liquid interface. By way of mathematical modelling, the phenomenon of oxygen diffusion has been studied. The oxygen transfer coefficient and the resulting mass transfer characteristics affect the system performance to a great extent. An attempt has been made to study the physical oxygen transfer at different temperatures and speed of rotation. Ambient temperature is one of the important physical factors affecting the diffusion process and biofilm kinetics.

The transient and dynamic nature of oxygen transfer in RBC makes the boundary conditions for the mixed culture biofilm model extremely difficult to specify. Although there are a number of models available for predicting biofilm behaviour, RBC modelling is unique because of this dynamic nature of the gas transfer through the boundary layer. It became necessary to suitably modify the boundary considerations in a biofilm model to make it specifically suited for RBCs.

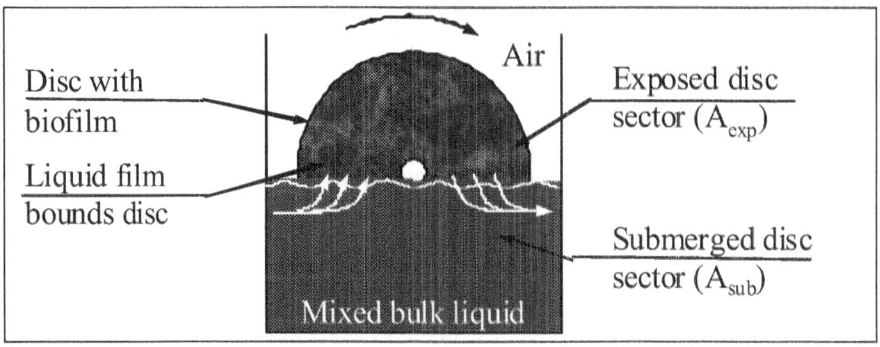

Figure 1 Sketch of RBC disc with liquid film and biofilm in operation.

Method

Physical oxygen transfer into the liquid film boundary layer

In fixed-film processes like RBCs, the mass transfer effects govern the functioning of the system to a significant extent when compared to biological reaction kinetics. These transfer aspects associated with both the liquid phase as well as biofilm result in concentration gradients from bulk liquid to reaction sites, and normally control the system performance. In the first place, the characteristics of mass transfer over the liquid film have been investigated. Several studies since 1950s have concerned gas–liquid diffusion processes. The elementary mathematical equation for diffusion of oxygen into the thin liquid film developed over a flat rotating plate of RBC is dependent on time and space coordinates. The diffusion equation for concentration distribution of oxygen in a liquid film (Crank, 1955) is as follows:

$$\frac{\partial C}{\partial t} = D\frac{\partial^2 C}{\partial x^2} \tag{1}$$

with initial condition: at t = 0 and $0 < x < \delta$, $C = C_0$; and boundary conditions: at t > 0 and x = 0 (surface), $C = C_S$; and at t > 0 and $x = \delta$ (media support), $\left(\frac{\partial C}{\partial x}\right) = 0$

where:
δ is the thickness of the liquid film boundary layer [L];
D is the diffusivity coefficient of oxygen in liquid film i.e. water $[L^2T^{-1}]$;
C is the concentration of dissolved oxygen in the liquid film $[ML^{-3}]$;
C_0 is the Initial concentration of oxygen in liquid film;
C_S is the Saturation concentration of oxygen in air;
x is the Space coordinate from surface (x = 0) [L].

The time-dependent flux of the solute (oxygen) at the boundary layer surface (x = 0) is given by Equation (2) below.

$$N_t = -D\left(\frac{\partial C}{\partial x}\right)_{x=0} \tag{2}$$

The oxygen transfer coefficient may be calculated using Equation 3.

$$K_L = \frac{\left|\overline{N_t}\right|}{(C_S - C_0)} \tag{3}$$

where:

$$\overline{N_t} = \frac{\int_0^{t_R} N_t\,dt}{\int_0^{t_R} dt} \tag{4}$$

N_t is the Oxygen flux through the boundary layer $[ML^{-2}T^{-1}]$;
K_L is the Oxygen transfer coefficient from air to liquid film $[LT^{-1}]$;
$\overline{N_t}$ is the time averaged flux of oxygen into liquid film over time interval t_R.

Yamane and Yoshida (1972) first obtained the approximate solution of the above equations based on some assumptions. The basic assumptions were that the velocity of rotation of the liquid film remained same as that of the rotating disc and there was complete mixing of the liquid film into the bulk liquid once the disc re-enters the trough. Another important assumption was that the thickness of the liquid film entrained over the disc surface is uniform throughout the air-liquid mass transfer phase. In the current physical oxygen transfer model, the solution to the diffusion Equation (1) is obtained numerically by dividing the liquid film into a number of layers and applying the boundary conditions. However, to obtain a solution, it became necessary to estimate the thickness of the liquid film layer, as well as average time of exposure t_R of the liquid film in air. The average contact time was calculated using Equation 5, based upon the dimensions of the RBC disc as shown in Figure2 below. Above the water surface, when the disc rotates in direct contact with air, the average liquid film thickness can be obtained from the correlation given by Zeevalkink *et al.* (1978) as represented by Equation 6. It was first obtained by Bintanja *et al.* (1970) using the flat plate withdrawal theory (Groenveld, 1970), where the film thickness was calculated to be a function of the effects of gravity, centrifugal force (circumferential velocity) and physical properties of the fluid. Zeevalkink *et al.* (1979) modified it to be a function of the rotational velocity and depth of immersion in addition to the forces of gravity and viscosity at a given temperature. K_L can be easily obtained using Equation (3) provided C_S is known at a given temperature and $\overline{N_t}$ is obtained from Equations (2) and (4) for a given value of t_R and δ.

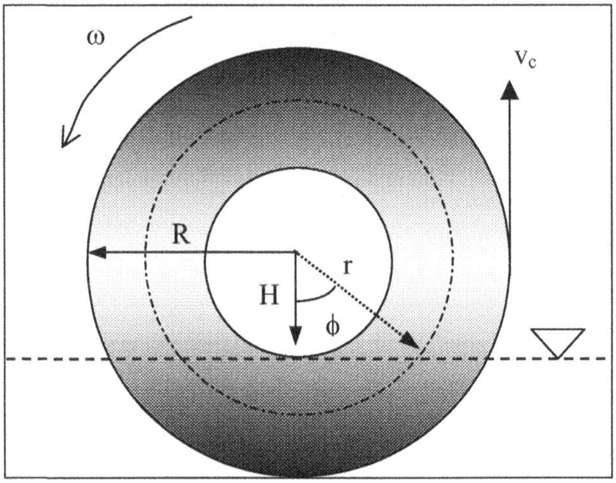

Figure 2 Schematic diagram of RBC disc in water.

$$\overline{t_R} = \frac{1}{\omega}\left(1 - \frac{1}{\pi}\cos^{-1}\left(\frac{H}{r}\right)\right) \tag{5}$$

$$\delta = \frac{4\sqrt{2}}{15}\left(\frac{\eta . v_C}{\rho . g}\right)^{1/2} \tag{6}$$

where

$$v_C = \omega\sqrt{\left(R^2 - H^2\right)} \tag{7}$$

where

η is the dynamic viscosity of water $[ML^{-1}T]$;
v_c is the vertical component of the peripheral velocity at the point of emergence of disc from the liquid surface $[LT^{-1}]$;
ρ is the density of water $[ML^{-3}]$;
ω is the rotational speed in rpm $[T^{-1}]$.

Theoretical determination of molecular diffusivity coefficient D of a gas in water is based on the physical properties of the gas and water. Bird *et al.* (1960) cited the empirical correlation developed by Wilke and Chang (1955). on the basis of the Stokes–Einstein equation as represented by Equation (8).

$$D^0{}_{O_2-H_2O} = 7.4\times10^{-8}\frac{\left(\phi_{H_2O}.m_{H_2O}\right)^{0.5}T}{\eta_{H_2O}.v_{O_2}{}^{0.6}} \tag{8}$$

where

T is the absolute temperature (°K);

ϕ_{H_2O} is the association factor for the solvent (water);

m_{H_2O} is the molecular weight of solvent (18 g/mole);

$\eta._{H_2O}$ is the dynamic viscosity of solvent at T (c.poises);

v_{O_2} is the molar volume of oxygen at its normal boiling temperature(−183°C) 25.6 cm^3/g.mole.

This has been used for calculation of D for oxygen at different temperatures. Essentially, the value of $\frac{D\eta_{H_2O}}{T}$ remains constant for a given temperature T (K). Wilke and Chang recommended that ϕ_{H_2O} be taken as 2.6 when the solvent is water. However, Perry *et al.* (1984) state that the Wilke–Chang equation is most accurate when ϕ_{H_2O} is 2.26 for a solute diffusing into water. The results of the physical model are based upon an RBC disc of 25 cm diameter and 42% disc submergence in water at a temperature of 19°C.

Biofilm model

The biofilm model developed in the current work is essentially based on mathematics of the one-dimensional mixed-culture biofilm model formulated by Wanner and Gujer (1986) and RBC model of Gujer and Boller (1990). Suitable modifications and process customisations have been incorporated based on the activated sludge model no. 3 (IAWQ, 1999) for reaction terms i.e. kinetics and stoichiometry and the definition of the boundary conditions for the dynamic RBC biofilm. The liquid film boundary has been assumed to be static and completely mixed over the interfacial area, with a part submerged in bulk and another part exposed to air. The modelling of the dissolved components through the boundary layer into the biofilm is essentially based on diffusion Equation (1) because diffusion is the dominant process here. An additional process for reaction is added to model the conversion processes occurring due to the biomass in each layer. However, for particulate solids, the transport is assumed to occur as a combination of diffusive and advective forms of movement. Four particulate species have been considered, namely heterotrophs, ammonium oxidising autotrophs, nitrite oxidising autotrophs and particulate inerts. The solids transport through the biofilm is represented mathematically using Equation (9) below.

$$\frac{\partial X_j^{Bf}}{\partial t} = u\big|_{x=0}\cdot\frac{\partial X_j^{Bf}}{\partial x} + D_{X_j}\frac{\partial^2 X_j^{Bf}}{\partial x^2} + R_{X_j} \tag{9}$$

where

X_j^{Bf} is the concentration of particulate species j within biofilm [ML^{-3}];

u is the advective velocity of displacement of particulate species j [LT^{-1}];

D_{X_j} is the effective diffusive coefficient of particulate species j [L^2T^{-1}];

R_{X_j} is the rate of production of particulate species j within biofilm [$ML^{-3}T^{-1}$].

The biofilm thickness is essentially modelled as a function of attachment and detachment velocities occurring at the biofilm surface as shown in 10.

$$\frac{d\delta_{Bf}}{dt} = u\big|_{x=0} + u_{att} - u_{\det} \tag{10}$$

where

u_{att} is the attachment velocity at biofilm surface [LT^{-1}];

$u_{\det}$ is the detachment velocity at biofilm surface [LT^{-1}];

δ_{Bf} is the thickness of biofilm at time t [L].

The detachment of solids is modelled as a constant shear depending upon the biofilm thickness and the density of the particulate component at the surface. Attachment depends upon the concentration of solids in the tank. The attachment and detachment velocities are based upon the surface fluxes as described in the RBC model of Gujer *et al.* (1990). Suitable considerations for changes in the concentration of the solids in the surface biofilm layer due to attachment and detachment have been incorporated. The model has been developed in Matlab using finite difference solution methodology. For the discretization of the second order spatial derivatives, the method of lines has been employed. Owing to the stiff nature of the ode system, a variable step ode solver based on stiff algorithm has been used. It is essentially based upon numerical differentiation formulas as well as backward differentiation formulas (Gear's method) for stability generation.

Results and discussion

Physical oxygen transfer simulations

In Figure 3, the variation of the K_L value with liquid film thickness has been simulated at different rotational velocities. At a given rotational velocity, (i.e. constant time of exposure t_R), the oxygen transfer coefficient increases rapidly with increased film thickness within a certain domain of film thickness. After that, it stabilises and remains almost constant. It can be inferred that increased film thickness beyond a certain significant point does not really enhance the oxygen transfer coefficient. This significant film depth value (100–120μm) decreases as the rotational speed increases.

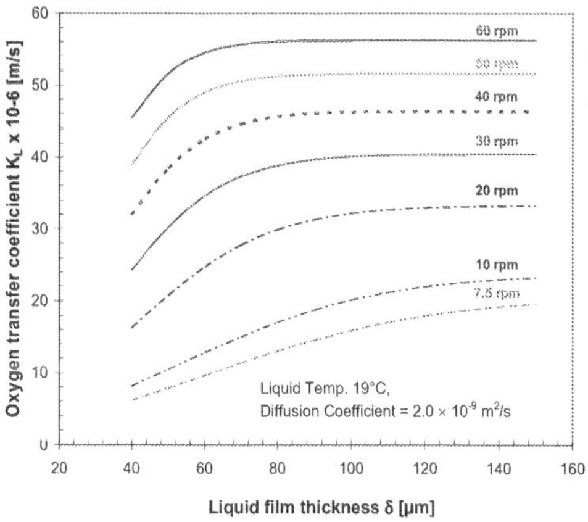

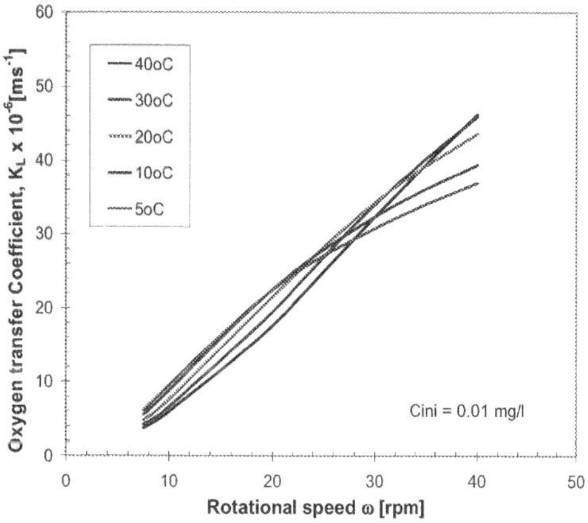

Figure 3 (left) Variation of oxygen transfer coefficient with liquid film thickness at constant temperature.
Figure 4 (right) Variation of average K_L on disc surface with rotational speed at different temperatures.

At high rpm (ω), the oxygen is not able to penetrate into the innermost layers of the film as the film depth increases. Whereas at low rotational velocities, there is more exposure time to enable deeper penetration. Moreover, increase in rotational speed uplifts the transfer rate since the exposure time of the disc to air gets reduced. The oxygen transfer rate is affected by changes in temperature as well. Diffusivity coefficient D, the liquid film thickness δ and the saturation oxygen concentration C_0 are three important physical parameters which control the oxygen transfer coefficient value K_L (Equations 2 and 3). Any temperature change will affect the viscosity of the fluid which will in turn affect the δ and D values (Equations 6 and 8). Figure 4 displays the variation of average K_L on the disc surface with ω and temperature. The liquid film thickness has been calculated for a given temperature and rotational speed using Equations 6 and 7 at different radial positions over the disc surface and then averaged to obtain a mean value at a given rpm. It follows from the curves that at low rotational speeds, oxygen transfer coefficient is lesser at higher temperatures. This may be substantiated by the fact that at high temperatures, the film entrained on the disc has a lesser thickness due to reduced viscosity. Additionally, the diffusion is faster due to higher diffusivity coefficient. Therefore, the film is quickly saturated and this lowers the average oxygen transfer speed K_L. However, as the rotational speed increases, film thickness increases. Although film thickness should decrease with increasing temperature, the rise of rotational speed compensates this effect and eventually reduced exposure time causes partial saturation of the film. The liquid film is no longer fully saturated in spite of reduced exposure time. Hence the average K_L value shows a rise with temperature reversing its earlier trend. The decrease in the driving force (i.e. the concentration gradient, C_S–C_0) is over-ruled by the increase in the diffusivity coefficient D with increasing temperature. The behaviour of the oxygen transfer coefficient at 5°C, 20°C and 30°C as studied through experimental measurements by Rittman *et al.* (1983) show a similar trend although the measurements were done in the trough and not measured and averaged over the disc surface.

Biofilm model results

Simulations were conducted with the biofilm model for a 1-stage RBC set-up fed with an influent organic substrate loading of 12 g sol.COD m^{-2} d^{-1} and an ammonium loading rate of 2.60 gNH4-N m^{-2} d^{-1}. The organic substrate is assumed to be soluble and readily biodegradable. The hydraulic loading rate on the first stage was kept at 0.058 m^3 m^{-2} d^{-1}. The total dry density of solids in the biofilm was taken as 45 kg COD m^{-3} based on experimental measurements. The kinetic and stoichiometric parameters were based upon different literature values including ASM no.3 and suitably corrected for temperature. The liquid film thickness was assumed as 100 μm on average for a rotational speed of 2 rpm. Experimental determination of δ over the RBC disc surface was difficult to obtain due to surface irregularity. The volume of the RBC stage was 5.8 l. Figure 5 displays the spatial oxygen profiles for a complete RBC cycle of 60 s. There is higher oxygen concentration during the exposed cycle when the disc comes in contact with air which gradually diminishes after it enters the reactor. The plots generated after 12 days simulation show that penetration depth for oxygen is of the order of 250–300 μm and is nearly insignificant thereafter. As the disc emerges out from the bulk, oxygen diffusion from air through the liquid boundary is faster than the uptake by the bacteria, and this results in higher initial concentration (top curves). But as the disc moves on, high uptake rate by the aerobic bacteria lower the surface concentration, so that it decreases towards the end when the disc re-submerges into bulk. There is no further aeration inside bulk and so oxygen concentration keeps decreasing thereafter. Figure 6 shows the distribution of particulate species and the removal rate of substrate in the biofilm. The effective diffusion coefficient for all the particulate species D_X was taken as 6×10^{-9} m^2/d, based upon a similar value (i.e. 5×10^{-9} m^2/d) in the extended MCB model of Wanner and Reichert (1996). It allowed the species to get mixed inside the biofilm from the starting. The effect of attachment and detachment of the microbial cells and particles at the biofilm surface could be easily carried forward from the surface into the deeper layers. The deeper biofilm layers have mostly inerts and the heterotrophic species are concentrated in the surface layers of the biofilm. It is evident that heterotrophs offer strong competition at high nutrient loadings and autotrophic organisms are able to grow only in the deeper layers of the biofilm and possibly in the latter stages of the RBC when the organic load is substantially reduced. Nitrite oxidising micro-organisms (NOM) appear lesser in fraction compared to ammonia oxidising micro-organisms (AOM) in their relative distribution. Siegrist *et al.* (2000) showed similar distribution in their RBC model results except that they had a completely nitrifying RBC model. The active aerobic zone for all the species is about 300μm after which anoxic activities take prominence. Owing to high population of heterotrophs and availability of dissolved oxygen in the upper layers, most of soluble substrate is degraded towards the surface, and the biofilm gets substrate limited in the deeper layers, as revealed in Figure 7. Wanner and Reichert (1996) showed a similar distribution for bacteria and substrate in their model results. The penetration depth of substrate does not change much with biofilm thickness unless the biofilm gets very thin or the biofilm density varies by a large amount. This may occur at low influent loading rates of nutrient or at high rotational speeds. The above simulation scenario is obtained after 300 days when the biofilm thickness is relatively stable and compares well with experimental thickness of 1500 μm in the discussed RBC stage. The overall removal efficiency of organic substrate in the bulk is about 85% with a removal flux of 10.35 g sol.COD m^{-2} d^{-1} and compares well with the experimental findings obtained. The nitrification in the system is only up to 55% with an ammonia removal flux of 1.44 gNH_4-N m^{-2} d^{-1}. The nitrification is only partial although both heterotrophs and autotrophs are responsible for the removal as per the ASM no. 3 process kinetics. It has been observed that significant nitrification will not occur in the RBC process unless the soluble BOD_5 concentration is reduced below 15 mg/l (Metcalf and Eddy, 1995). The denitrification rate in the first stage is predicted by the model to be 0.176 gN m^{-2} d^{-1}. Heterotrophs are mainly responsible for denitrification under anoxic conditions in the deeper biofilm layers when they function as facultative bacteria allowing nitrate as the electron acceptors in place of oxygen. The model showed a DO content in the bulk at 3.90 gO_2 m^{-3} which was found close to experimental result. Figures 8 and 9 show the performance of the RBC under varying influent loading conditions of organic substrate and ammonia-nitrogen as a sensitivity analysis. The plots are comparable with the results of Gujer *et al.* (1990). It is found that for a given hydraulic loading rate on the first-stage RBC, the removal fluxes increase with influent loading rate up to a limit. For organic carbon substrate, the maximum removal flux comes to 24 gCOD m^{-2} d^{-1} while for ammonium, the maximum nitrification rate is about 1.5 gNH_4-N m^{-2} d^{-1}. Any further increase in influent loading does not increase the removal efficiency at the given temperature and loading rate. However, further removal can be achieved in the subsequent stages of a multi-stage RBC. Temperature and boundary layer thickness pronounce a big effect on the removal efficiency.

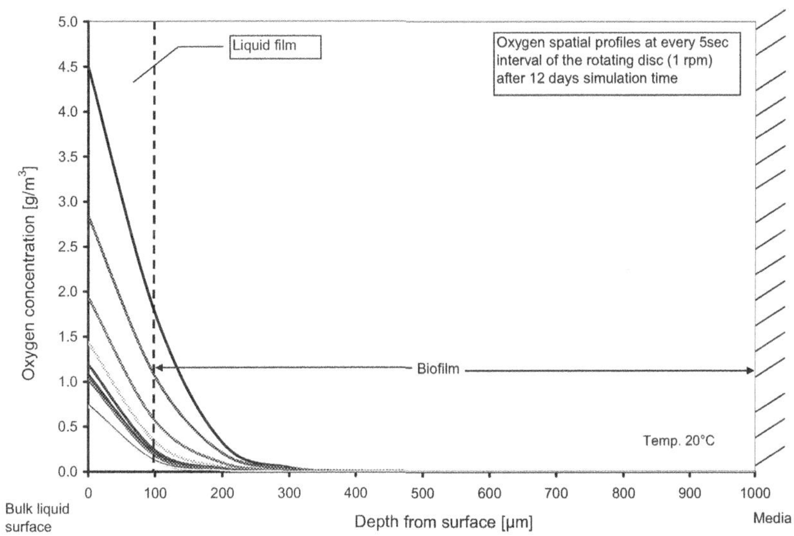

Figure 5 Diffusion of oxygen through the liquid film into the RBC biofilm at 20°C.

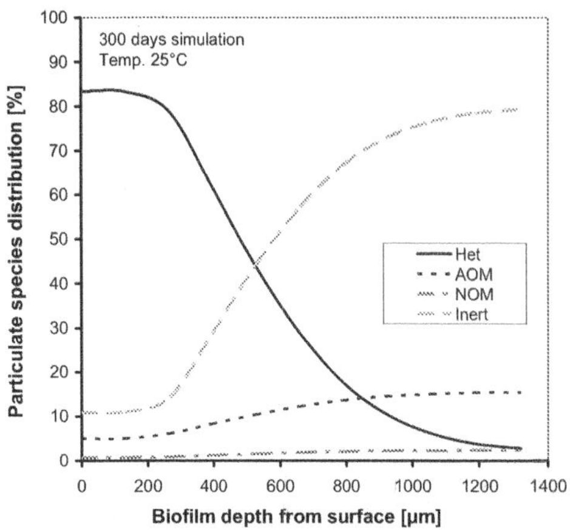

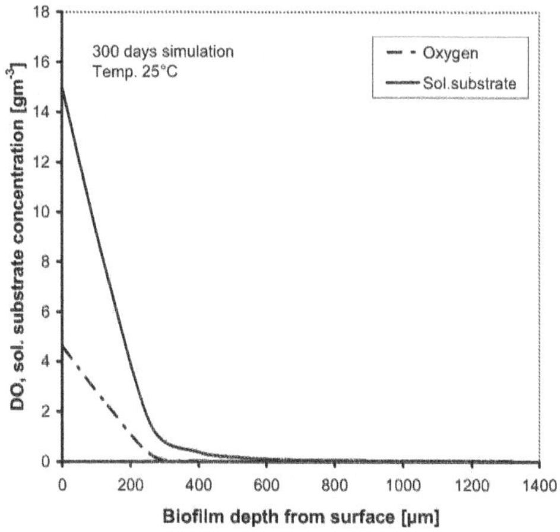

Figure 6 (left) Relative abundance of particulate species inside biofilm matrix.
Figure 7 (right) Dissolved components concentration inside biofilm matrix.

Conclusions

Higher thickness of boundary layer resulted in increased K_L value up to a limiting point. At lower rotational speed, K_L value decreased with increasing temperature since boundary layer thickness decreased. This trend was reversed at higher rotational speed when the liquid film thickness increased. However, with biofilm on the disc, the rotational speed for RBCs is generally limited to 2–3 rpm to prevent complete shearing of the biomass from the disc surface. High uptake rate of the diffused dissolved oxygen by the microorganisms resulted in its limitation beyond the surface layers. High DO content in the liquid film and the bulk reveal that RBCs provide good aeration and requirement of additional aeration in the reactor may be avoided. Like oxygen, organic substrate removal occurred only in the surface layers and got limited in the deeper layers. Results showed 85% removal of organic substrate at an influent loading rate of 12 g sol.COD m^{-2} d^{-1}, while nitrification was 55% at a loading rate of 2.60 g NH_4-N m^{-2} d^{-1} in a single-stage RBC. Heterotrophs are the dominant species in the biofilm matrix at high nutrient loadings and autotrophs can flourish once the organic load is substantially reduced. At a given hydraulic loading rate and temperature, removal efficiency increases with increasing F/M ratio, but up to a certain limiting value. Further improvement in nitrification and organic substrate degradation may be achieved by considering a number of stages of the RBC. In the end, it can be concluded that RBCs offer an ideal alternative as a secondary or tertiary biological treatment unit for simultaneous removal of organic substances and nitrogen in domestic sewage. The economic advantage and compactness of the system compared to conventional treatment processes such as activated sludge plants make it a favourable choice especially suited to the concept of decentralized wastewater treatment systems gaining popularity today.

Acknowledgements

The model is part of the PhD research sponsored under International Postgraduate Studies in Water Technology (*IPSWaT*) programme of German Federal Ministry of Education and Research (BMBF), Germany.

References

Bintanja, H.H.J., van der Erve, J.J.V.M. and Boelhouwer, C. (1975) Oxygen transfer in a rotating disc treatment plant. *Water Research* **9**, 1147–1153.

Bird, R.B., Stewart, W.E. and Lightfoot, E.N. (1960) *Transport phenomena.* John Wiley, New York.

Boller, M., Gujer, W. and Nyhuis, G. (1990) Tertiary rotating biological contactors for nitrification. *Water Science and Technology* **22** (1/2), 89–100.

Crank, J. (1955): *The mathematics of diffusion*, 45, Oxford University Press.

Famularo, Mueller, J.A. and Mulligan, T. (1978) Application of mass transfer to rotating biological contactors. *Journal of Water Pollution Control Fed.* **50**, 653–671.

Groenveld, P. (1970) Dipcoating by withdrawal of liquid films. Thesis, Delft.

Gujer, W., Boller, M. (1990) A mathematical model for rotating biological contactors. *Water Science and Technology* **22** (1/2), 53–73.

Müller, J.A., Paquin, P. and Famularo, J. (1980) Nitrification in rotating biological contactors. *Journal of Water Pollution Control Fed.* **52**, 688–710.

Patwardhan, A.W. (2003) Rotating biological contactors: a review. *Indian Engng Chem. Research* **42**, 2035–2051.

Perry, R H., Green, D.W. and Maloney, J.O. (1984) *Chemical Engineers' Handbook*, 6th edition, McGraw-Hill Book Co., New York, 3–273 and 3–286.

Rittmann, B.E., Suozzo, R. and Romero, B.R. (1983) Temperature effects on oxygen transfer to rotating biological contactors. *Journal of Water Pollution Control Fed.* **55**, 270–277.

Wanner, O. and Gujer, W. (1986) A multispecies biofilm model. *Biotechnology and Bioengineering* **XXVIII**, 314–328.

Wanner, O. and Reichert, P. (1996) Mathematical modeling of mixed-culture biofilms. *Biotechnology and Bioengineering* **49**, 172–184.

Yamane, T. and Yoshida, F. (1972) Absorption in a rotating disc gas-liquid contactor. *Journal of Chem. Engng Japan* **5**, 55–59.

Zeevalkink, J.A., Kelderman, P. and Boelhouwer, C. (1978) Liquid film thickness in a rotating disc gas-liquid contactor. *Water Research* 12, 577–581.

Zeevalkink, J.A., Kelderman, P., Visser, D.C. and Boelhouwer, C. (1979) Physical mass transfer in a rotating disc gas-liquid contactor. *Water Research* **13**, 913–919.

Toward modelling of secondary settling tank performance

Miklós Patziger*, Harald Kainz*, János Józsa and Michaela Hunze*****

* Institute of Urban Water Management and Landscape Water Engineering, Graz University of Technology, Stremayrgasse 10/I, A-8010 Graz, Austria
(E-mail: *patziger@sww.tugraz.at*; *kainz@sww.tugraz.at*)
** Department of Hydraulic and Water Resources Engineering, Budapest University of Technology and Economics, Műegyetem rkp. 3. Kmf. 4, H-1111 Budapest, Hungary
(E-mail: *jozsa@vit.bme.hu*)
*** FlowConcept GmbH., Vahrenwalder Strasse 7, 30165 Hannover, Germany
(E-mail: *hunze@flow-concept.de*)

Abstract In the design of secondary settling tank (SSTs) the detailed inner hydrodynamic behaviour is still often poorly addressed. Coupled fine-scale flow and concentration measurement in operating tanks can give more insight into the interconnected flow and concentration pattern, furthermore, provide validation data for advanced numerical models. The latter can serve then as design tool to optimize the performance by finding proper tank geometry, sludge recycling ratio etc. This paper presents the first results of such an investigation of one of the SSTs at the wastewater treatment plant (WWTP) of Graz. Acoustic Doppler velocity and optical turbidity measurements gave information on the flow and concentration pattern in the SST. FLUENT-based axisymmetric flow and coupled activated sludge transport modelling could then reasonably reproduce measured distributions and patterns. The calibrated and validated numerical SST model was then used to investigate the impact of various inlet geometry features on the SST performance in peak flow conditions.

Keywords ADV; biological wastewater treatment; CFD; secondary settling tank

Introduction

As is known, in the activated sludge process it is necessary to separate the treated wastewater from the activated sludge mass, thereby producing a clear effluent. This solid/liquid separation phase is traditionally achieved by gravity sedimentation in secondary settling tanks (SST). The SST is therefore a vital component of an activated sludge system.

A SST usually combines the function of (1) a thickener to produce a continuous underflow, (2) a clarifier to produce clarified final effluent and a (3) storage tank to store sludge during peak flows. Should the settling tank fail in one of this three functions, suspended solids (SS) will be carried over the effluent weirs and escape with the effluent. Besides delivering an effluent of poor quality excessive loss of SS (several hundred mg/l) could affect the behaviour of the biological process by uncontrolled decrease in mixed liquor suspended solids (MLSS), and hence the sludge age to values below that required for a proper plant performance. The behaviour of the settling tank in its clarification, thickening and storage functions is influenced by the settling tank design features and its operation. The factors affecting these are: the hydraulic features such as flow rate, (2) physical features such as inlet and sludge collections arrangements, (3) site conditions such as wind and temperature and (4) sludge characteristics (Ekama *et al*., 1997).

Figure 1 shows some most relevant physical processes and characteristic zones in a SST. On the flow pattern of a SST the density difference between the inflow and the medium in the tank exerts a great influence (Larsen, 1977; Krebs, 1991; Hunze and Schumacher, 2004). Due to this density difference the inlet jet plunges in the near field of the inlet (1). A density current (3) is induced in the direction of the outer zone. The velocity of this jet is higher than the average velocities in the SST. Due to the mass conservation a backflow (5) will be induced in the upper zone of the SST. The still settled sludge flows at the bottom of the tank (7) in the direction of the sludge hopper (9) and will be returned into the aeration tank.

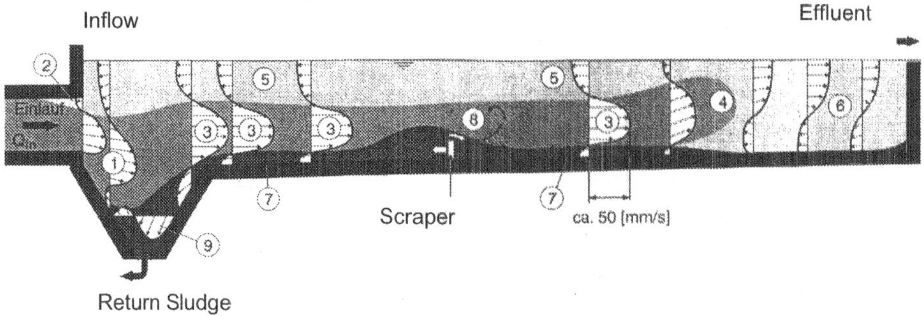

Figure 1 Typical physical processes in a secondary settling tank (after Winkler, 2001).

Study tank and measurements

The investigations were carried out at the WWTP Graz. The measurements were performed in one of the 4 circular SSTs of the WWTP Graz (Patziger *et al.*, 2005) (Figure 2).

During the flow and concentration measurements in the tanks the hydraulic load, inflow and the return sludge concentration as well as the effluent quality were controlled. All the investigations including flow and SS concentration measurements were carried out in a period by a nearly steady hydraulic load. The inflow concentrations ranged between 4.2 and 4.5 g/l in the measuring period.

Figure 2 Top view on the secondary settling tanks in the WWTP of Graz (Vermessungsamt Graz, 2004).

Circular secondary settling tanks can be reasonably investigated in axisymmetric approach. The tangential velocities in the tank are generated by the scraper velocity. These tangential velocities usually range about 3 cm/s in the outer part, whereas the inner part is dominated entirely by the inlet velocities. 3D velocity measurements were carried out by a Nortek-Vector ADV. 90 sec long velocity vector time series with 16 Hz sampling rate were obtained in the nodes. SS concentrations were measured in the measuring profiles of the measuring raster by an Endress & Hauser optical turbidity meter, calibrated for the SS features of the tank. Concentration profiles were obtained at every 2 cm in the measuring verticals by winching the sensor at constant speed. The sludge volume index (SVI) in the investigated Plant is usually about 70 l/g. From the numerical modelling aspect the most important sludge characteristic in addition to SVI is the settling velocity V_s of the sludge. As is known the settling velocity is the function of the local concentration (Krebs, 1991). The settling velocity of the SS of the Plant at different concentrations was tested in a 2000 mm high 300 mm diameter cylinder.

Hydrodynamic modelling

The axisymmetric turbulent model equations are solved by the CFD code FLUENT 5, in which Hunze and Schumacher (2004) have implemented special modules for settling, density and rheology features. The implicit unsteady segregated solver solves the Navier–Stokes equations for the mean flow with a k–ε turbulence closure. For modelling of transport processes the advection–diffusion equation is be applied with a term describing settling and thickening. This term includes a modified version of the settling function of Patry and Takács (1992).

Results

Model calibration and validation

The SSTs at the WWTP of Graz receive daily average flow Q =800 l/s as dry weather flow, with an average inflow SS concentration of 4.3 g/l. The return sludge flow is kept constant at Qr = 1000 l/s. In the SST clear water was defined as initial condition (t = 0).

In the filling-up phase of the tank the simulation showed some well-known hydraulic phenomena (Armbruster *et al.*, 2001). Owing to momentum and density differences, the inlet jet initially turns downwards and impinges on the bottom of the tank. It is split then into two density currents, towards the effluent and towards the sludge hopper. The mass of the settled sludge detains the bottom current. The congregated sludge mass on the bottom begins to flow slowly towards the sludge hopper. Thus the concentration in the sludge hopper becomes higher and higher. The stabilised velocity and concentration patterns in the model preserve the wavy pattern as seen in Figure 3, respectively, in a way similar to the measured ones. Figure 4 shows the comparison of measured and calculated concentration in some representative measuring profiles.

Figure 5 shows the calculated and measured flow pattern in the near field of the inlet. The inlet jet plunges with considerable downward momentum component initially facilitated also by the lower density of the medium at the inlet height in the SST. A wavy pattern is induced by the inlet jet, first impinging into the settled sludge, then being forced to move upwards by buoyancy. These phenomena lead to the large-scale erosion the settled sludge layer resulting in its considerable dilution at least in the near field zone of the inlet and lead to a lower returned sludge concentration X_{RS}. The consequences of all that is increased sludge mass M_{SST} [kg] in the SST and decreased SST performance.

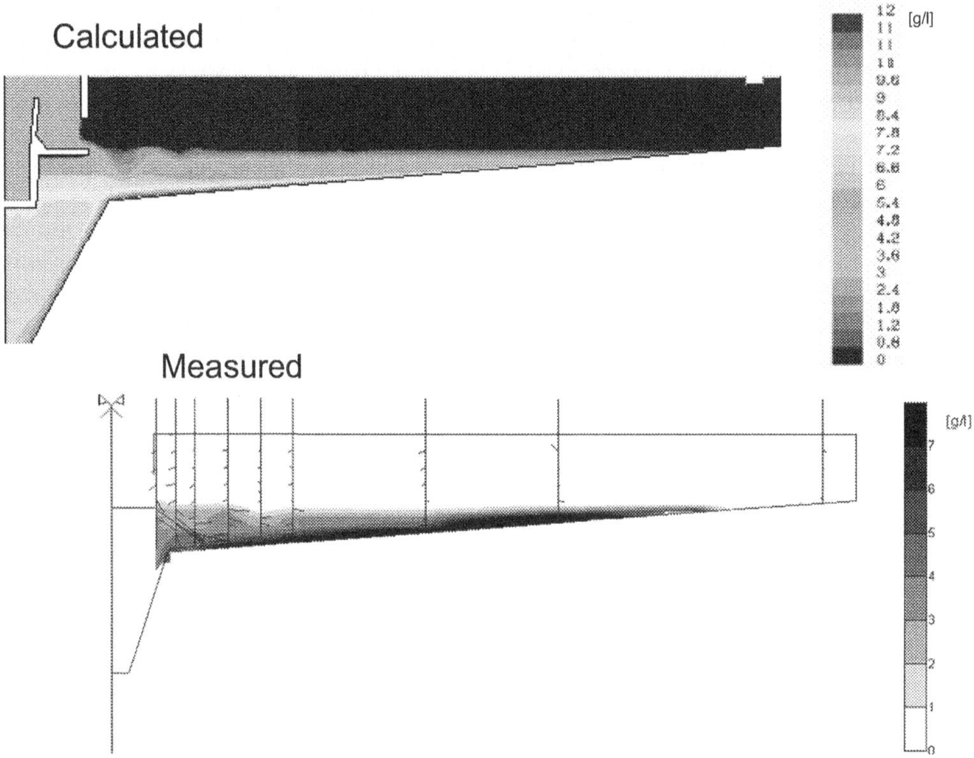

Figure 3 Calculated concentration pattern.

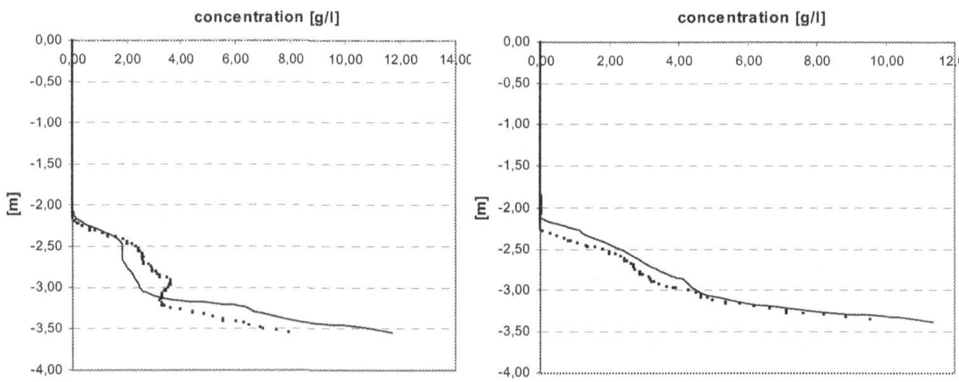

Figure 4 Comparison of the measured (dots) and calculated (continuous line) concentrations.

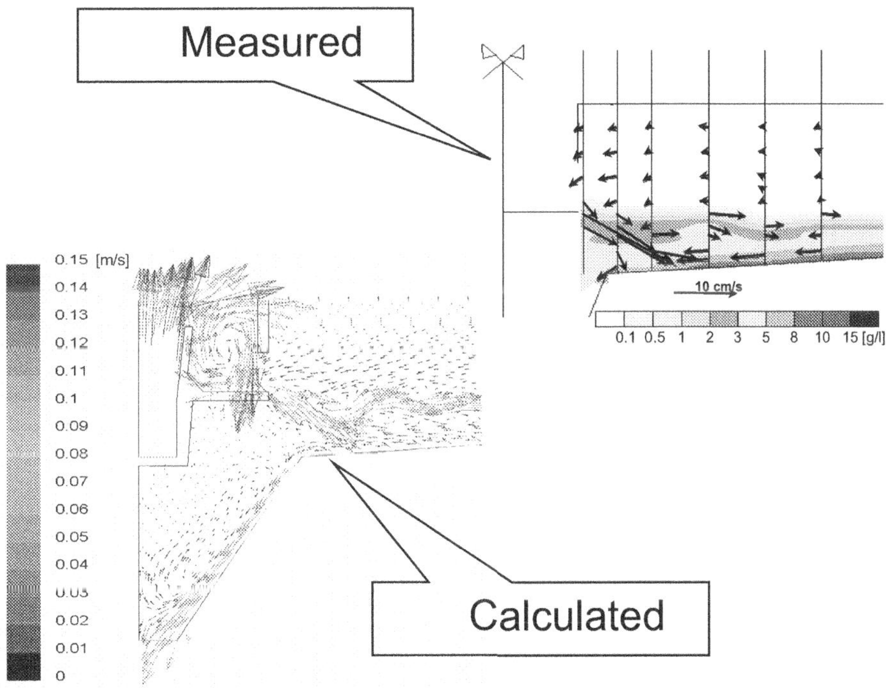

Figure 5 Comparison of the measured (dots) and calculated (continuous line) concentrations.

Model investigations

In case of peak flow conditions due usually to storm events, the SST's performance becomes a very important issue in the effectiveness of the treatment plant. Peak flow events lead to a high hydraulic load of WWTPs inducing high inlet velocities and bringing increased sludge mass in the SST, all requiring operation at full capacity.

Figure 6 shows the modelled flow and concentration pattern (1) in the investigated SST in case of an increased hydraulic load (Q=2500 l/s, Q_{RS}=1400 l/s; SS_o=4,3g/l). Because of the increased sludge level the medium straight at the inlet section in the SST has a higher density then the one of the inlet jet itself. The jet turns thus upward and induces a wavy sludge water interface. The high energy content of the inlet jet effects settling and thickening. This results in a low returned sludge concentration (Figure 7) and high activated sludge mass in the SST, M_{SST} = 18 tonnes in our particular case (Figure 8).

As a first step toward optimising the shape of the inlet jet a baffle was installed in the inlet facility (2) to ensure better energy dissipation. Although the inlet velocities and the sludge mass decrease whereas the returned sludge concentration increases due to the baffle, the large return flow region induced in such a way at the inlet section causes an instable sludge water interface. By the extension of the scum baffle (3) the instabilities of the sludge water interface could be reduced. By a lower position of the baffle and a discharge well designed by the use of the so-called densimetric Froude number (Krebs, 1991), (4) the layering of the inlet jet could be reasonably optimized. The optimal layering of the inlet jet and its low energy assure better settling and thickening conditions and consequently a higher returned sludge concentration. These lead to a decreased sludge mass in the SST and a higher SST performance by about 20 percent (Figure 8).

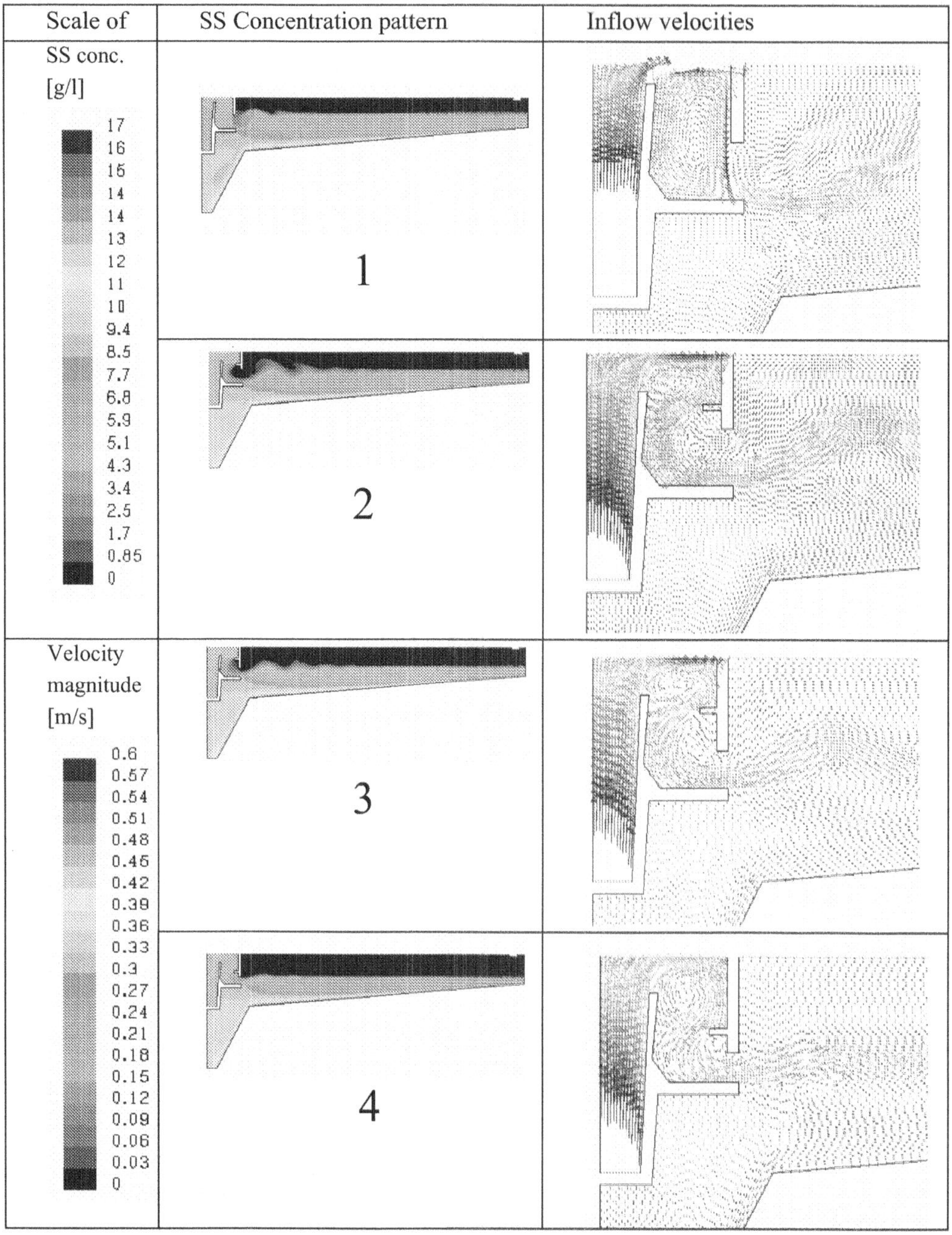

Figure 6 Influence of different variants of inlet geometry on the flow and concentration pattern in the investigated SST in peak flow conditions.

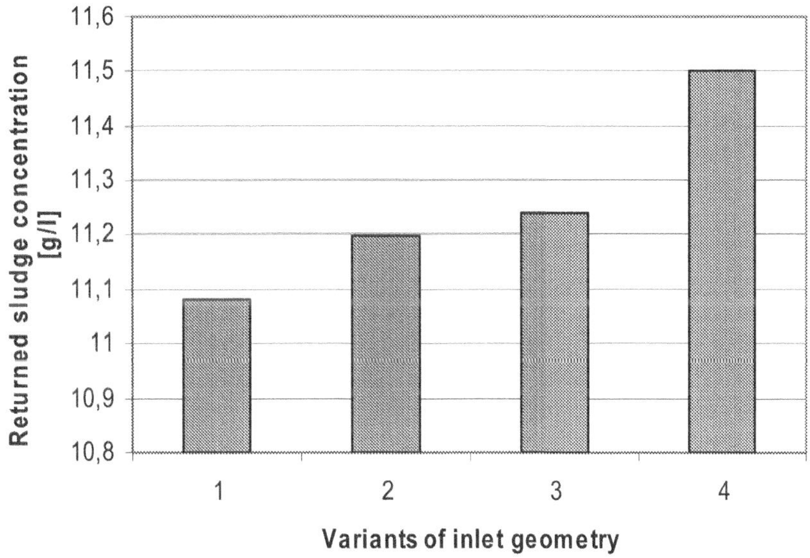

Figure 7 Influence of different variants of inlet geometry on the returned sludge concentration X [g/l].

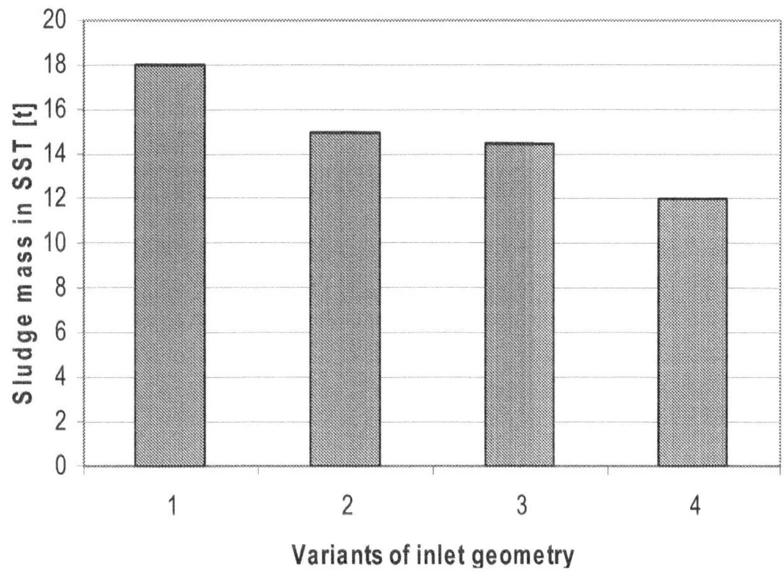

Figure 8 Influence of different variants of inlet geometry on the sludge mass M_{SST} [g/l].

Conclusions

It has been presented that coupled fine scale flow and concentration measurement in operating SSTs can give valuable insight into the interconnected flow and concentration pattern, furthermore, provide validation data for advanced CFD modelling. The first results of such an investigation of one of the SSTs at the WWTP of Graz have brought both overall information and detailed evidences on the wavy character of the sludge-water interface, supporting the applicability of axisymmetric numerical flow and coupled activated sludge transport modelling. The FLUENT 5 based numerical model proved a useful numerical tool for investigating secondary settling tank performance by varying the inlet geometry. As an important step toward optimising SST performance, a reasonable modification of the inlet geometry could result in the suppression of the large scale structures of the sludge water interface thus providing an SST performance enhanced by about 20%.

References

Armbruster, M., Krebs, P. and Rodi, W. (2001). Numerical modelling of dynamic sludge blanket behaviour in secondary clarifiers. *Water Science and Technology*, **43** (11), 173–180.

Ekama, G. A. *et al.* (1997). *Secondary Settling Tanks: Theory, Modelling, Design and Operation. IAWQ Scientific and Technical Report No. 6.* London: IAWQ.

Hunze, M. and Schumacher, S. (2004). „CFD Modellierung in der Abwasserreinigung, *Vorlesungsscriptum*, FlowConcept GMBH, Hannover.

Krebs, P. (1991). The hydraulics of final settling tanks, *Water Science and Technology*, **23** (4–6), 1037–1046.

Larsen, P. (1977). On the hydraulics of rectangular settling basins. *Report No. 1001*, Dep. Water Research Engineering, Lund Institute for Technology, University of Lund, Sweden.

Patry, G.G. and Takács, I. (1992). Settling of flocculent suspensions in secondary clarifiers, *Water Research* **26** (4), 473–479.

Patziger, M., Kainz H., Józsa J. (2005) Analysing secondary settling tank behaviour by in situ measurements and numerical modeling. In *Proceedings XXXI IAHR Congress, September 11–16, Seoul, Korea.*

Water management optimization by a probabilistic forecast model based on Bayesian networks

L. Mediero and L. Garrote

Escuela de Caminos, Ciudad Universitaria, s/n. 28040, Madrid, Spain
(E-mail: *ljmediero@yahoo.es*; *garrote@caminos.upm.es*)

Abstract. Optimization of water management requires reservoir operation during floods that minimizes downstream damages and maximizes the water volume stored at the end of the flood event. Reservoir operation can be improved by a probabilistic forecast model that predicts future inflows and simulates the flood control process. Decisions about floodgate operation can be made in terms of risk, incorporating the probability of occurrence of future discharges and the different operation strategies considered. A probabilistic forecast model based on Bayesian networks has been developed to obtain the best operation strategy that achieves the flood control objectives, depending on the inflow discharge, measured in the past and forecasted in the near future, and the initial water level in the reservoir. Bayesian networks allow the simulation of reservoir flood control in a probabilistic way that enables the consideration of the uncertainty inherent to every forecast. The methodology was applied to the real case of the *Conde de Guadalhorce* reservoir located in the south of Spain. The forecast model was validated with real data, quantifying its application as a decision-making tool.

Keywords: Bayesian networks; probabilistic forecast; reservoir operation

Introduction

An essential task of flood management is the determination of an effective reservoir operation strategy that minimizes downstream damages and maximizes the water volume stored at the end of the flood event. The lack of knowledge about future inflows into the reservoir makes difficult this issue. Actual hydrologic forecasts in Spain are produced through supervised simulation with deterministic rainfall-runoff models, spending much time in calibration and slowing down the decision making process. Currently, strong efforts are being made in developing flood forecast models.

Water resources systems are characterized by nonlinearity and stochastic properties. A flood control system based on artificial intelligence approaches seems to be a promising technique for reservoir control (Karaboga *et al.*, 2004). Neural networks and fuzzy logic have been applied widely to forecast and optimization of reservoir operation (Jain *et al.*, 1999; Bagis and Karaboga, 2004; Jeong *et al.*, 2004; Chandramouli and Deka., 2005).

Bayesian networks are another computational structure taken from the field of artificial intelligence. Bayesian networks are directed acyclic graphs representing probabilistic dependencies among values of the variables. Variables are represented by nodes and relationships between them by links, which are evaluated as conditional probability distributions (Pearl, 1988; Charniak, 1991). The probabilistic nature of Bayesian networks allows the consideration of forecast uncertainties. Bayesian networks have been applied successfully on different scientific areas: Medical and pharmaceutical sciences, educational area, military planning, risk analysis, biological science and water quality forecast.

Models based on Bayesian networks have been applied scarcely as a decision making tool on flood forecasting (Garrote and Molina, 2004; Molina *et al.*, 2005), despite the fact that they provide some advantages over other forms of artificial intelligence like neural networks: Bayesian networks allow the value of a variable to be entered as a known input or to calculate its likelihood as an output of the system without redesigning the network. Furthermore, Bayesian networks accept evidence at any point in the system and provide output at each other (McCabe *et al.*, 1998).

Optimization concept implies two definitions: One is related to the real existing surrounding conditions, among which stands out the type of information really available. The other may be established *a posteriori*, once the event is fully known and can be processed by appropriate simulation models (Giron *et al.*, 2000). In the case of the present paper, the optimization concept implies the two definitions, flood routing in the reservoir has been simulated with a reservoir operation module applied over a large data set of synthetic inflow hydrographs, obtaining a probabilistic forecast that optimizes the reservoir operation strategy from the real data available in the basin.

A selection methodology for finding the best reservoir operation strategy during a flood event is presented, coupling a rainfall-runoff forecast model with a reservoir operation model, both of them in a probabilistic framework based on Bayesian networks.

Uncertainty consideration

Estimation of inflow hydrographs by a rainfall-runoff model can have great errors because of the uncertainty in the parameter estimation (Jeong *et al.*, 2004). Complex systems involve many uncertainties but they can be classified into two types of uncertainty: aleatory and epistemic (Ferson and Ginzburg, 1996; Apel *et al.*, 2005). Aleatory uncertainty or variability represents the natural variability of the system and epistemic uncertainty or ignorance represents the incomplete knowledge of the system. Relevant uncertainties of the system studied in the present paper are presented in Table 1.

Table 1 Sources of uncertainty in the system separated into the different modules considered

Modules	Aleatory uncertainty	Epistemic uncertainty
Rainfall-Runoff module	Rainfall variability	Measurement errors in the observed rainfall Estimation of the initial moisture Estimation of parameters that describe the basin behaviour Model errors because of no perfect simulation of basin hydrological processes Different basin response of discharge to the same rainfall event
Reservoir operation module	Variability of the input discharge hydrograph	Measurement errors in the observed water level data at the reservoir Errors in the stage-outflow discharge table Errors in the stage-volume table

Rainfall variability, estimation of parameters that describe basin behaviour and identification of basin state are the most important uncertainties derived from the rainfall-runoff model. The former has been taken into account by a stochastic model of rainfall generation, creating a set of possible temporal series of rain events with similar parameters to the observed rain data. The second uncertainty has been considered by the statistical analysis of the parameter values obtained in the calibration process and the determination of the probability density function that characterizes the variability of every rainfall-runoff model parameter. The latter has been quantified by a probabilistic intelligence approach method based on Batesian networks, estimating the basin response variability as conditional probabilities that depend on the initial basin conditions. Model errors are inherent to the deterministic model and can only be reduced by a better representation of the hydrological physic processes, like the use of a spatially distributed hydrological model.

In the case of the reservoir operation module the most important uncertainty is the variability of the input hydrograph and it has been taken into account by the results of the rainfall-runoff process given as probabilistic inflow hydrographs. Errors in the stage tables have not been considered because of their minor influence in the simulation.

Other uncertainties, as measurement errors and initial conditions estimation, can be easily considered by a probabilistic density function of the observed or estimated values, but it has not been the aim of this paper for the sake of simplicity.

Rainfall-runoff forecast module

Probabilistic future inflow discharges in the reservoir have been forecasted by a rainfall-runoff forecast module generated in a Bayesian framework. The module has been implemented by a deterministic rainfall-runoff model run in a repetitive mode, taking a set of 4,000 temporal series of rain as input data and giving a set of 4,000 series of inflow hydrographs as output data. Rain temporal series have been obtained by a stochastic rainfall generator that creates rain cells by a Poisson-cluster process, which belongs to a type of stochastic rainfall models called as pulse-based (Wójcik and Buishand, 2003).

Model parameter values in every run of the repetitive simulation have been generated by a Monte Carlo simulation applied over the probability density functions that characterize every parameter as a randomized process. Probability density functions can be selected from the knowledge about the parameter variability obtained after the calibration process. Curve Number (CN) and time to peak have been represented by Gaussian probabilistic density functions with mean and standard deviation values presented in Table 2. The initial abstraction coefficient (K) is represented by a uniform probability density function in the range 0.0–0.26 (Ponce, 1989) because of the lack of knowledge about this parameter in the observed data (Soong, 2004, p. 192).

Table 2 Parameter values of the probability density functions selected

Parameters	Curve Number	Time of peak (min)
Mean (m)	70	600
Standard deviation (s)	10	50

The Batesian network structure that achieves a better representation of the basin hydrological behaviour is presented in Figure 1. A dependence analysis has shown that CN, time to peak and initial abstraction have no significant influence on the inflow discharge forecast, but initial moisture in the basin has a high influence. This Bayesian network structure was learned from the discretization of the data series generated by the repetitive run of the rainfall-runoff model.

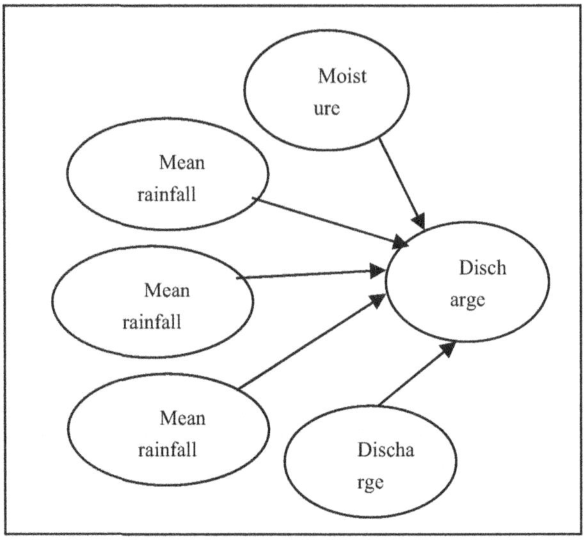

Figure 1 Batesian network structure for the rainfall-runoff forecast module.

Reservoir operation module

Flood control operations were simulated by the Volumetric Evaluation Method (MEV), based in three principles: a) Releases must be less than the inflow discharge when inflow discharge is increasing; b) If inflow discharge increases, outflow discharge must increase accordingly; c) The higher the reservoir level, the greater should be the increment of outflow discharge. This method can be applied to every reservoir and type of flood, making the best use of the surcharge pool from the available data (Giron, 1988).

MEV empties the surcharge pool in a progressive way, increasing the releases while the available surcharge pool volume is decreasing. The method will try to keep the reservoir level until every spillway gate has been fully opened when the surcharge pool has been filled. If we consider a linear increment of the outflow discharge, it must increase the following amount every time step:

$$\Delta Q_{out}^{t} = \frac{\Delta V_t^2}{V_t^d \Delta t}$$

ΔQ_{out}^{t} is the outflow discharge increment at t time step, ΔV_t is the reservoir volume increment at t time step and V_t^d is the available surcharge pool volume at t time step.

The method is based in the determination of the reservoir level at which the releases begin, called as Activation level (ACT). ACT must be located under the top level of the flood control pool and avoid that little floods cause high releases. A maximum gradient of discharge (MGD) has been added to limit the outflow discharge increment between two consecutive time steps, avoiding sudden releases that could be dangerous for people staying near the downstream river channel. Operation reservoir strategies will be defined by different combinations of ACT and MGD.

Application to the Conde de Guadalhorce reservoir

Conde de Guadalhorce reservoir is located on the Turón River in the south of Spain with a total storage capacity of 86 hm^3. The reservoir is operated and maintained by *'Confederación del Sur'* Basin Authority. The dam has a curve concrete gravity section with a height of 74 m and length of 160 m. The top of the flood control pool is located at 341.30 m, the top of the surcharge pool at 342.90 m and the top of dam at 344.10 m. Water from the reservoir is released through three outlet facilities: one outlet work located at 304.0 m with two 1,000-mm-diameter conduits and capacity of 24 m^3/s, another outlet work for hydroelectric power generation with one 3,000 mm diameter conduit located at 319.90 m and capacity of 13 m^3/s and, finally, one controlled spillway with a length of 32 m and two gates with a height of 15 m, the spillway crest elevation is located at 338.40 m and it has a capacity of 600 m^3/s when the water level coincides with the top of the surcharge pool.

The MEV implemented for the Conde de Guadalhorce reservoir has been applied over the set of 4,000 series of inflow hydrographs to simulate flood control operations. Six reservoir operation strategies have been considered by combinations of ACT and MGD (Table 3).

Table 3 Determination of the variables values for every reservoir operation strategy

Variables	Strategy 1	Strategy 2	Strategy 3	Strategy 4	Strategy 5	Strategy 6
ACT	340.3	340.3	340.3	341.3	341.3	341.3
MGD	50	100	200	50	100	200

The initial volume of the reservoir and the reservoir operation strategy used in every event of the set have been determined by a Monte-Carlo method, randomizing these variables from their probability density functions. The former has been represented by a gamma probability density function with parameters m=1 and l=1 and upper limit equal to the top level of the flood control pool, so the higher volumes have more probability of occurrence than the lower ones, taking into account that flood control operations are more problematic as the volume is nearer to the top level of the flood control pool. The latter has been represented by a uniform distribution because every strategy has the same probability of being used. The MEV output is a set of 4,000 temporal series of outflow discharges and water levels at the reservoir. The Batesian network structure (Figure 2) was learned from the discretization of these data sets.

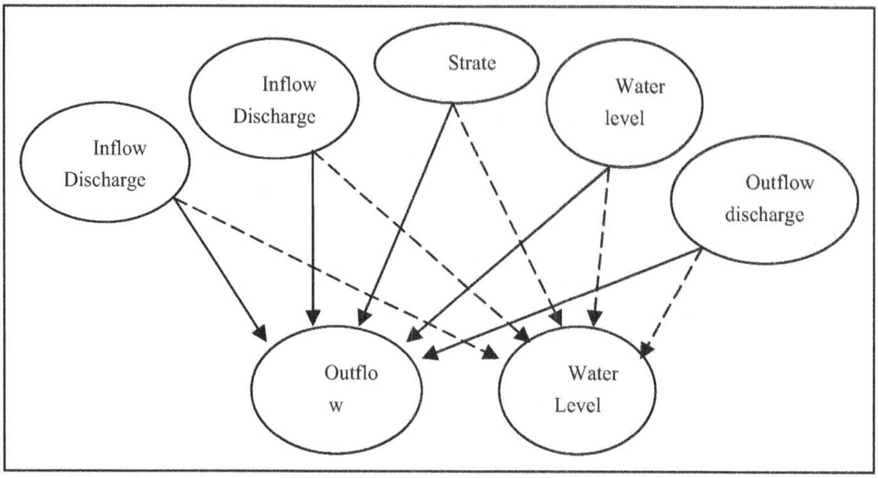

Figure 2 Batesian network structure for the reservoir operation module.

Results

The methodology proposed was applied to a real flood event given by the observed data as average rainfall values in the basin and the input hydrograph to the Conde de Guadalhorce reservoir. An hourly time step was used. The forecast can be made from the rainfall observed in the past 10 hours, because it is the lag time of the basin calculated in the calibration process.

The example described below shows a simulation of flash-flood forecasting and decision making process in real-time, realised at the twentieth time step from the flood begin. Before this time step the inflow hydrograph was measured, but after it the inflow hydrograph is represented by a probabilistic forecast of discharges given as output data of the rainfall-runoff model (Figure 3).

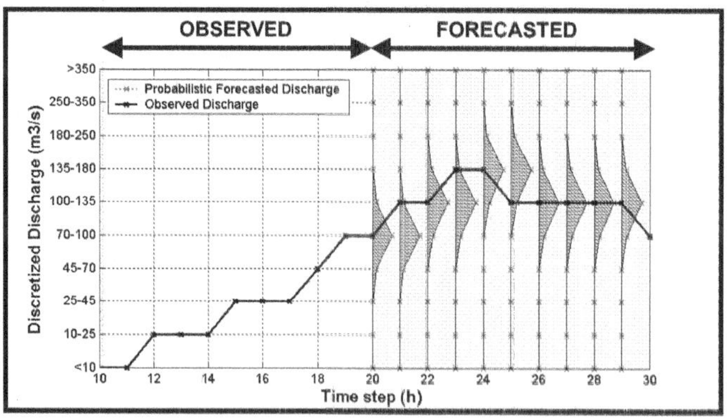

Figure 3. Probabilistic forecast of inflow hydrograph in the Conde de Guadalhorce reservoir. Vertical filled lines drawn in future time steps represent the forecast probability of occurrence of every bin of discharge. Horizontal solid line represents measured discharges.

The probabilistic inflow hydrograph in the future and the initial conditions defined by inflow discharge, water level and outflow discharge in the current time step were introduced as input data in the reservoir operation module, giving the probabilistic outflow discharges and water levels for every strategy considered as output data for the next nine time steps in the future (Table 4).

Table 4 Cumulative probabilities of occurrence of outflow discharge for the reservoir operation strategy 4

Forecast	Forecast time step (hours)								
probabilities	1	2	3	4	5	6	7	8	9
$Q_{out} > 0$	1.000	1.000	1.000	1.000	1.000	1.000	1.000	1.000	1.000
$Q_{out} > 25$	0.009	0.022	0.032	0.042	0.053	0.121	0.111	0.087	0.085
$Q_{out} > 50$	0.009	0.022	0.032	0.042	0.053	0.121	0.111	0.087	0.085
$Q_{out} > 75$	0.009	0.022	0.032	0.042	0.053	0.121	0.111	0.087	0.085
$Q_{out} > 100$	0.009	0.022	0.032	0.042	0.053	0.121	0.110	0.086	0.084
$Q_{out} > 150$	0.009	0.022	0.032	0.041	0.052	0.121	0.110	0.085	0.083
$Q_{out} > 200$	0.000	0.001	0.002	0.005	0.008	0.011	0.027	0.029	0.019
$Q_{out} > 300$	0.000	0.001	0.002	0.005	0.007	0.011	0.027	0.028	0.018
$Q_{out} > 400$	0	0	0	0	0	0	0	0	0

Flood control objective is defined in terms of the risk of dam and downstream damages that can be assumed. The objective is fixed by the maximum probability of occurrence of an outflow discharge or a water level or both at the same time, which can be assumed by their associated damages. In the present example, the flood control objective has been defined as a maximum probability of 25% of occurrence of an outflow discharge higher than 150 m^3/s calculated over the cumulative probabilities in all time steps of the forecast (Table 5). Reservoir operation strategy 3 is the only that does not exceed the flood control objective defined, so this is the best reservoir operation strategy for the present flood event in terms of the assumed risk of downstream damages associated to a peak discharge higher than 150 m^3/s.

Table 5 Output cumulative probabilities of occurrence over all time steps forecasted of outflow discharge for the different reservoir operation strategies

Forecast	Reservoir operation strategy					
probabilities	1	2	3	4	5	6
$Q_{out} > 0$	10.00	10.00	10.00	10.00	10.00	10.00
$Q_{out} > 25$	5.99	4.98	5.05	0.64	0.69	0.90
$Q_{out} > 50$	5.99	4.98	5.05	0.64	0.69	0.90
$Q_{out} > 75$	5.77	4.97	4.87	0.64	0.69	0.90
$Q_{out} > 100$	5.39	4.59	3.93	0.64	0.69	0.90
$Q_{out} > 150$	0.45	0.48	0.21	0.63	0.69	0.90
$Q_{out} > 200$	0.04	0.02	0.02	0.63	0.68	0.89
$Q_{out} > 300$	0.00	0.00	0.00	0.12	0.14	0.11
$Q_{out} > 400$	0.00	0.00	0.00	0.11	0.13	0.10

Conclusions

A methodology to select the best reservoir operation strategy during a flood event is presented. The methodology is based in coupling a rainfall-runoff forecast model with a reservoir operation module, both of them implemented in a probabilistic Bayesian framework. The rainfall-runoff forecast module gives the probabilistic inflow discharges in the future time steps from the average rainfall measured at the basin. If it is available, a rainfall forecast given as deterministic or probabilistic rainfall values can be introduced as input data of the rainfall-runoff module. The reservoir operation module takes the probabilistic inflow hydrograph and the initial conditions at the reservoir as input data and gives the probabilities of occurrence of discharges and water levels for the different strategies considered.

The best reservoir operation strategy is selected in terms of risk by the determination of probability thresholds of maximum outflow discharge and maximum reservoir level that must not be exceeded during the flood control process. Thresholds are defined depending on the dam and downstream damages that can be assumed.

The methodology proposed can be introduced as a decision-making tool in real-time during a flash-flood event, forecasting the future inflow discharges at any reservoir and selecting the best reservoir operation strategy depending on the risk associated to damages that can be assumed.

Acknowledgments

This research work was carried out thanks to the financial support of the Spanish Ministry of Science and Technology (project REN 2003-09021-C03-01). Authors also wish to thank the Organizing Committee of the IWA Young Water Professionals Programme for the financial assistance provided.

References

Apel, H., A.H. Thieken, B. Merz and G. Blöschl (2005). Flood risk assessment and associated uncertainty. *Natural Hazards and Earth System Sciences* **4**, 295–308.

Bagis A. and D. Karaboga (2004). Artificial neural networks and fuzzy logic based control of spillway gates of dams. Hydrological Processes 18 (13), 2485–2501.

Chandramouli V., P. Deka (2005). Neural network based decision support model for optimal reservoir operation. *Water Resources Management* **19** (4), 447–464.

Charniak E. (1991). Batesian networks without tears. *AI Magazine* **12** (4), 50–63.

Ferson, S. and L.R. Ginzburg (1996). Different methods are needed to propagate ignorance and variability. *Reliability Engineering and System Safety* **54**, 133–144.

Garrote L. and M. Molina (2004). A framework for making probabilistic forecasts using deterministic rainfall-runoff models. Hydrological risk: Recent advances in peak river flow modelling, prediction and real-time forecasting. Ed. Bios, Bolonia.

Giron, F. (1988). The evacuation of floods during the operation of reservoirs. Transactions of the Sixteenth International Congress on Large Dams, Volume IV, Report 75, pp. 1261–1283. San Francisco.

Giron, F., Yagüe, J. and Martínez R. (2000). Flood routing in reservoirs based on hydrological forecasting. Transactions of the Twentieth International Congress on Large Dams, Volume IV, Report 25, pp. 403–417. Beijing.

Jain S.K., A. Das and D.K. Drivastava (1999). Application of ANN for reservoir inflow prediction and operation. *Journal of Water Resources Planning and Management* **125** (5), 263–271.

Jeong C.S., W.J. Koh and J.H. Heo (2000). A study on real-time forecasting of reservoir inflow based on artificial neural network. Watershed Management and Operations Management 2000. Fort Collins, Colorado, USA

Karaboga, D., A. Bagis and T. Haktanir (2004). Fuzzy logic based operation of spillway gates of reservoirs during floods. *Journal of Hydrologic Engineering* **9** (6), 544–549.

McCabe, B., S.M. AbouRizk and R. Goebel (1998). Belief networks for construction performance diagnostics. *Journal of Computing in Civil Engineering* **12** (2), 93–100.

Molina M., R. Fuentetaja and L. Garrote (2005). Hydrologic models for emergency decision support using Bayesian networks. Symbolic and quantitative approaches to reasoning with uncertainty. *Proceedings Lecture Notes in Computer Sciences* no. 3571, 88–99.

Pearl, J. (1988). Probabilistic reasoning in intelligent systems: Networks of Plausible Inference. Morgan Kauffman, San Francisco.

Ponce, V.M. (1989). *Engineering hydrology. Principles and practices*. Ed. Prentice Hall. New Jersey, USA.

Soong, T.T. (2004). *Fundamentals of probability and statistics for engineers*. Wiley & Sons.

Wójcik R., T.A., Buishand (2003). Simulation of 6-hourly rainfall and temperature by two resampling schemes. *Journal of Hydrology* **273**, 69–80.

Reducing uncertainties in water distribution network models using fuzzy sets and genetic algorithms

Nemanja Branisavljević and Marko Ivetić

Faculty of Civil Engineering, University of Belgrade, Bulevar kralja Aleksandra 73, 11000 Belgrade, Serbia and Montenegro
(E-mail: *nbranisavljevic@yahoo.com*; *marko.ivetic@hikom.grf.bg.ac.yu*)

Abstract One of the most important steps in the development of water distribution systems simulation models is providing the dependence of results of simulation on the uncertainty of input values. Information about propagation of uncertainty through the simulation model provides a basis for general model improvement in an efficient and economic (low cost) way. Determination of model output uncertainties provides the possibility of reducing it. Predictions of uncertainties in water distribution network models procedure make the modeling easier and the models more reliable. In this paper uncertainties are represented by fuzzy set parameters, which do not have direct correspondence with statistical background, though, indications of uncertainty can be easily recognized. In the analysis presented in the paper, applied to the model water distribution system of the Becej community, roughness of old pipes has been selected as the uncertain parameter, since the correct values and their statistical distribution are unknown. Since fuzzy uncertainty analyses for nonlinear models require efficient optimization methods, a method based on evolutionary genetic algorithm has been used.

Keywords Water distribution systems, uncertainty analysis, fuzzy sets, genetic algorithms

Introduction

Uncertainty modelling contributes to better water supply management by quantifying the uncertainties and determining the sources of significant errors. Also it has great effect on quality of predicting water distribution network behaviour and network conditions, which affect reliability and safety of water distribution. There are several reported methods for uncertainty estimation in water distribution network models (Kottegoda and Rosso, 1998); the most common are the Monte Carlo method and the First Order Second Moment (FOSM) method. But, applications of these methods have significant limitations.

Since the models are mostly developed for precise, crisp, input values, and uncertain values are usually represented by a minimum of two variables (often more than two), this problem requires rather more parameters to yield definition. In the Monte Carlo method, uncertain parameters are presented with distribution function (mean value and standard deviation at least) and in FOSM only standard deviation is required. In the fuzzy set theory, depending on the shape of fuzzy number, number of data can be much larger, so the uncertain input value can be presented in detail, including the asymmetry as well. However, representing uncertainty using fuzzy set theory, estimating uncertainty of model output parameters may, under certain conditions, be significantly simplified, and cover a wider range of input parameters. Although, it cannot give us statistical information about uncertainty appearance (phenomenon), fuzzy set information about uncertainty quantity is more significant. After all, the fuzzy set approach can be extremely useful for characterizing the uncertainties in a linguistic way, which may be more adequate for uncertain variables.

Water distribution network model uncertainties and its reduction

Owing to the complexity of water distribution networks, most of the simulation models are developed from much simpler versions with precise crisp input values. Models with uncertain input parameters are rare and not popular in engineering practice due to complexity and duration of calculations. However, all input parameters do have a certain amount of uncertainty, starting with pipe lengths, diameters, connections etc. Uncertainties of input parameters may be divided into two groups: (1) those that depend on the accuracy of measuring instrument; and (2) those in which uncertainty is caused by a lack of information. Besides pipe lengths, the first group also contains pump characteristics, reservoir heads and pipe diameters (Kapelan, 2004) (Ivetic, 1996). In the second group, pipe roughness and nodal demands are the most significant. Although measurements can be erroneous, measured data are usually considered satisfactory, indicating that the majority of uncertainties come from the second group.

Significant numbers of uncertainties may be incorporated with model simplification and conceptualization. For example, individual user demands are aggregated and greatly simplified by using a concept of nodal demand, rather than entering each user connection and collect all data in a model. There is always an open question on how to define the value of nodal demand, and it has to be admitted that this may be a major source of model uncertainties.

Estimation of model parameter uncertainties is therefore followed by reducing it. Since the resources (money and time) are usually limited, the most significant achievement would be to determine the hierarchical order of the most significant parameters that will reduce uncertainties of a model output.

Fuzzy numbers as uncertainty presentations

Beside exact, crisp values, some parameters may be presented as sets that represent ranges of possible values. Since not all of the members are distributed evenly, an additional membership function has to be incorporated. So, the difference between a regular and a fuzzy set is that regular sets have binary representation of 'member ness' ('is a member' or 'is not a member'), and fuzzy sets have a function to represent 'member ness' of a value to a set.

The membership function only gives information about how much a variable belongs to the set. So, the fuzzy set representation is convenient to present uncertain values with no data available for calculating statistical distribution.

If the fuzzy set is standardized, the membership values will be between 0 and 1. The interval of values (ie the one with the largest range), for the value of $\mu=0$ is called the support of the fuzzy number. The interval (for the triangular membership function this will be just one value) where $\mu=1$ is called the most likely value/s of a fuzzy set. So, the membership function can give us information how close the variable is to be a member of a fuzzy set. Any interval for μ between 0 and 1 is called the α cut interval.

As parameters are presented in fuzzy way, the most convenient is to assume that the uncertainty of a fuzzy set is its support, and that the most probable value of a precisely determined parameter would be its fuzzy representation of the most likely value.

α Cut method

The challenge is to solve complicated mathematical problems that incorporate numerical analysis, with fuzzy input values. One of the approaches is schematically described in Figure 1:

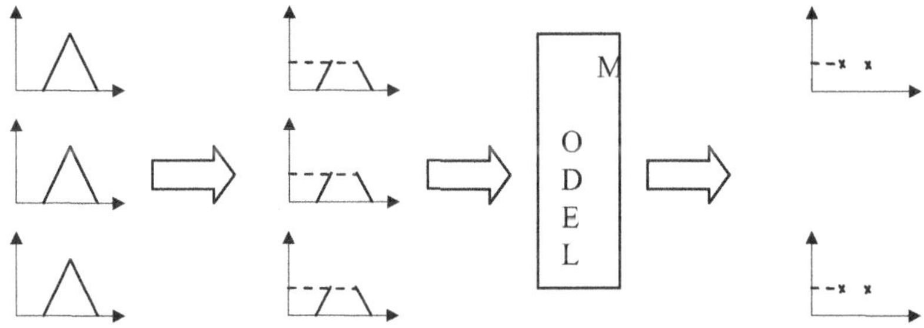

Figure 1 α Cut methodology.

Figure 1 presents basic steps in calculating fuzzy output parameters by the α cut methodology (Revelli and Ridolfi, 2002). Calculating extremes of output values of interest (when the model is designed for crisp input values and actual input values are intervals), requires interval mathematics and optimization methods. There are two possible solutions (Abebe *et al.*, 2000; Maskey *et al.*, 2004) (Figure 2):

- if the model output value is monotonic on the field of input parameter intervals then extremes are calculated only with extreme values of input parameter intervals as input data;
- if not, optimization method should be used for finding output extremes.

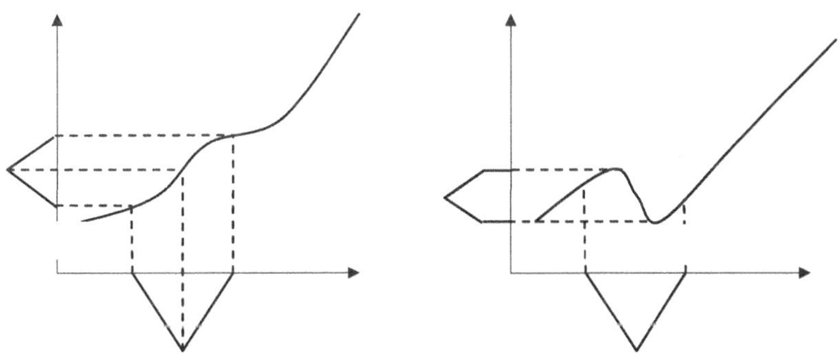

Figure 2 Monotonic and non monotonic function.

This procedure should be repeated for every α cut that input fuzzy numbers are divided in (Figure 3).

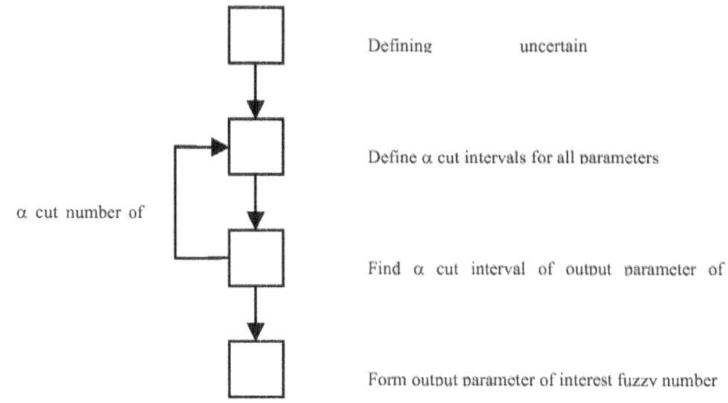

Figure 3 α cut method procedure.

The greater the number of α cut intervals, the more precise will be the output fuzzy set. The slopes of fuzzy number membership are rarely straight, except when the function is monotonic with the same slope on input parameter intervals. In general, this is not the case, so the slopes of the fuzzy output number represent the sensitivity of a model to α cut intervals.

For better performance and for the case of simplicity, fuzzy numbers were represented in triangular form, with three coefficients, the most likely value, and coefficients c_1 and c_2. The support of the fuzzy set was formed with the most likely value multiplied by coefficients $[a \cdot c_1, a \cdot c_2]$ (Figure 4).

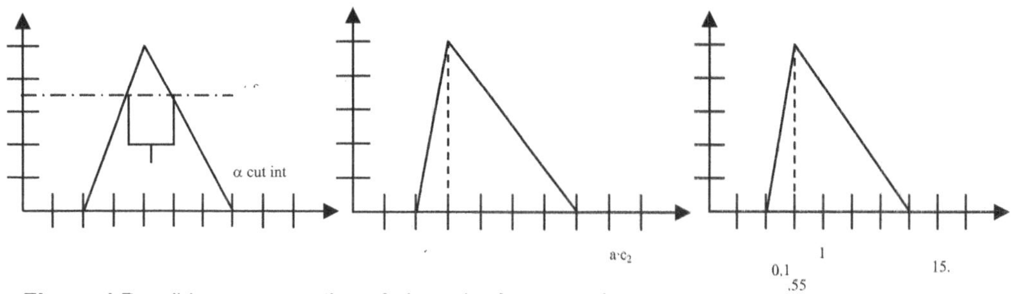

Figure 4 Possible representation of triangular fuzzy number.

Membership functions are standardized with the maximum at value 1, the minimum at value 0, and if it is triangular or trapezoidal, α cut intervals can be determined by the following expression (1):

$$[a(\alpha + \frac{c_1}{1-c_1})(1-c_1), a(\frac{c_2}{c_2-1} - \alpha)(c_2 - 1)] \tag{1}$$

Case study

The water distribution network of Becej is represented by an EPANet model that has 453 nodes and 616 pipes. There is a pump station to maintain pressure in the network, but it was more convenient to model it with a fixed head reservoir. The average system flow is 90 l/s.

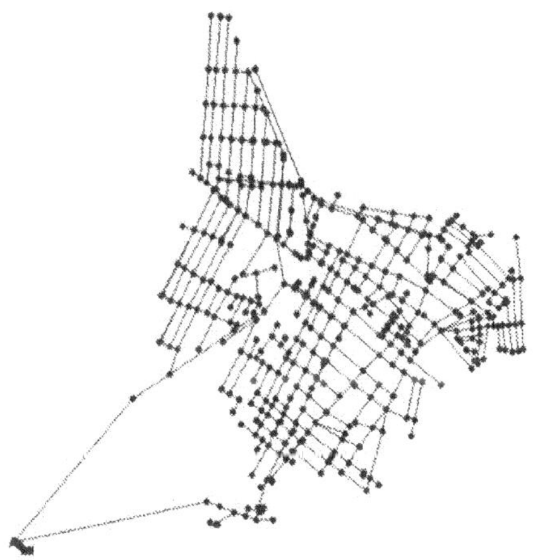

Figure 5 Model of water distribution network of town Becej.

The model was developed in EPANet format (Rossman), and EPANet was used as the engine in a form of dynamic link library in C++ computer program. Input data were held in memory during program execution, which resulted in better and faster program work. A library of genetic algorithm components, also written in C++ (Wall 1996), was used for optimization.

Pipe roughness (k [mm]) was chosen as the uncertain input parameter. After pipe diameter, pipe roughness has the strongest impact on friction head loss, so small mistakes in roughness assumption can make great differences between model result and actual measured real data. After a number of years in the ground, it is not easy to determine the condition of buried pipes. It is clear that material of the pipe cannot become smoother. In this case coefficient c_1 has value that is less than 1 and coefficient c_2 much greater than 1. The most likely value is determined using manufacturers' data on pipe roughness. Fuzzy numbers that represent uncertainty of pipe roughness are not symmetric, as shown in Figure 4(b).

Fuzzy numbers are assigned to every single pipe in the network, as a first step in pre-processing. Pipes in Becej are on average more then 10 years old, so coefficients for both boundaries of support interval are hard to determine. The left boundary of the support interval is equal to the most likely value multiplied by the coefficient 0.1 and the right boundary is equal to the most likely value multiplied by 10. The example of fuzzy number that represents uncertain pipe roughness value is given in Figure 4(c).

Calculation was proceeded for 6 α cuts (6 different membership values) $\mu=\{0,0.2,0.4,0.6,0.8,1.0\}$. So, the input intervals were determined with formula 1 where $\alpha=\mu$. Output values of interest were node pressure and link discharge. After simulation for node 366, and after fuzzy number reconstruction from 5 intervals and one crisp value, results are presented in Figure 6 (Branisavljevic and Ivetic, 2005):

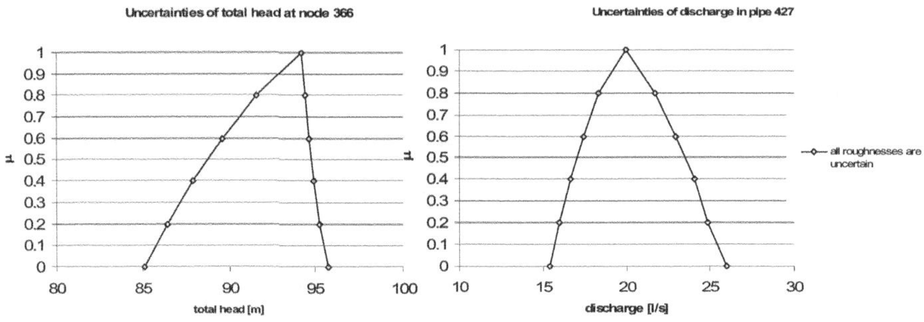

Figure 6 Uncertain total head at node 366 and uncertain discharge in pipe 427.

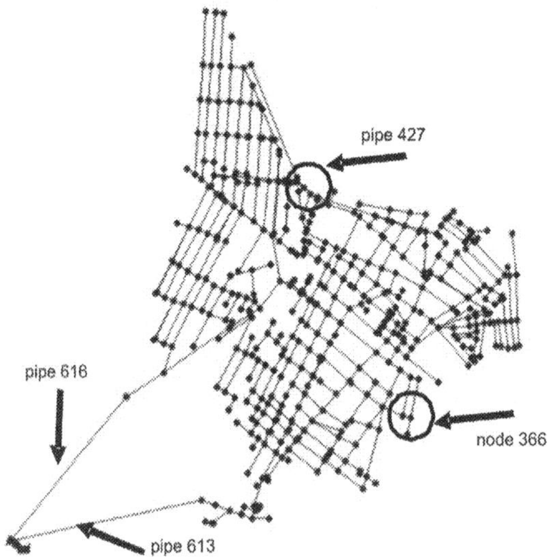

Figure 7 Node and pipes of interest.

Two output values (node 366 total head and pipe 427 discharge) are chosen by chance, but to be on the boundaries of a network, for uncertainty to be large enough for presentation.

After determination of uncertain output model values it was tested what is the impact of determining more accurate pipe roughness values of two inlet pipes (representing it as exact/crisp value of a most likely value of its fuzzy representation), on pressure\head in node 366 and on discharge in pipe 427. As the measure of uncertainty of a fuzzy number the most convenient is the support of a fuzzy number, the α cut for $\mu=0$. And the most likely value might be the most probable value of a model parameter ($\mu=1$) when it happen to be estimated more carefully. The results are presented in Figure 8:

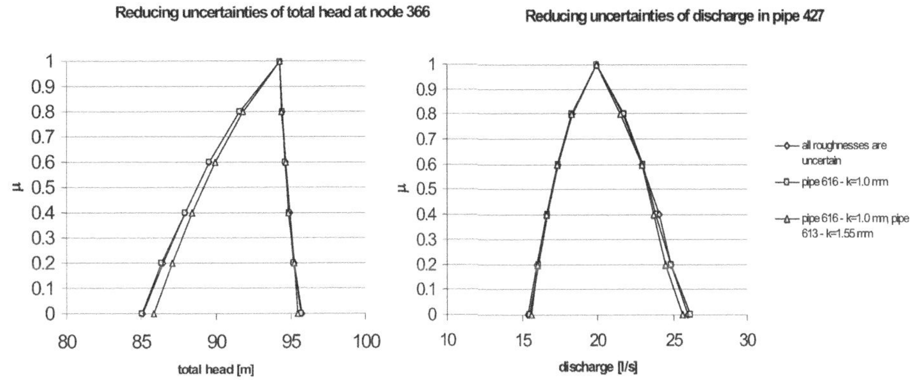

Figure 8 Reduced uncertainty of model output values.

It can be seen that reducing uncertainty in inlet pipes roughness has significant effect on reducing uncertainty of nodal head, but have minor effect on discharge in certain pipes.

Next step would be to investigate the hierarchical order of needed interventions on network parameter estimation in order to reduce the head uncertainties in certain node.

Since the head uncertainty lesser than 0.05 m (5 mbar) has only the theoretical significance (the most field measuring devices has greater uncertainties in measuring results), the procedure terminates when the head uncertainty comes below 5 cm.

Pipe	Node 366 head uncertainty (m)
614	11.54
483	10.09
369	9.47
613	8.88
482	8.33
573	7.81
598	7.30
218	6.92
485	6.54
489	6.18
64	5.88
...	...
438	0.08
513	0.07
149	0.05
434	0.04

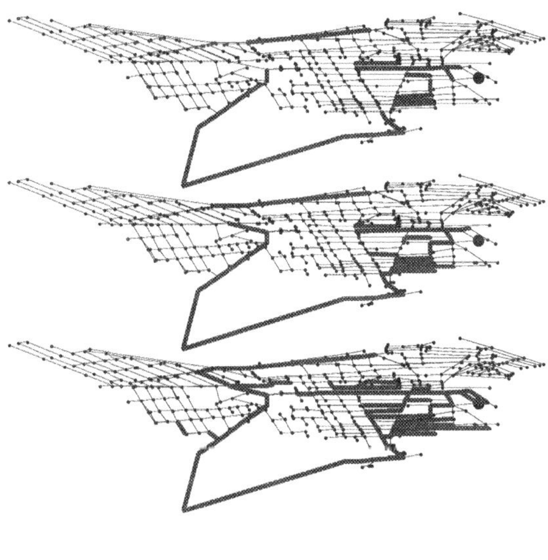

Table 1 and Figure 9 Hierarchical order of pipe roughness significances for total head uncertainty reduction in node 366.

Results are presented in a tabular (Table 1) and in graphic form (Figure 9). In table 1 first and last several pipes, from 84 significant pipes, whose roughness should be estimated more carefully are presented along with the reduced head uncertainty in the observed node.

In the right column of Table 1 it can be seen how the uncertainty decreases as the roughnesses of pipes become crisp values.

In the right column of Table 1, uncertainty reduction is presented with a wideness of the output fuzzy set support. As the wideness of a fuzzy set support decreases the output value becomes the crisp value.

In Figure 9 the hierarchical order of pipes that should be investigated with more care is presented graphically. In Figure 9, the significant pipes are divided into three groups in order of significance. The first group (upper scheme) are the most significant pipes (presented darker than others) that have the greatest impact in reduction of uncertainties. The pipes from that group should be investigated first. It can be noticed that the most significant are the mains.

Conclusions

In this paper it is shown how this simple method can be applied on a real network, and it can be successfully used for better network model input parameters determination. There are two types of output variables in water distribution network modelling. First, there are node pressures (total head) that depend on input variables in a monotonic way, and there are pipe discharges/velocities that depend on input values in a non monotonic way. The proposed method addresses these two types of variables in a different way. For the first type, the α cut method is more convenient since it requires only two model executions for a single α cut, and for the second type, an additional step with optimisation method is required, so the application of the presented method is a bit more complicated, but still, very efficient. Ordering of pipes according to their significance in reducing the uncertainty of the model is very helpful in planning an efficient field measurements campaign.

References

Abebe A., Guinot V., Solomatine D. (2000). Fuzzy alpha-cut vs. Monte Carlo techniques in assessing uncertainty in model parameters. In *Proc. 4th International Conference on Hydroinformatics, Iowa City, USA, July*.

Branisavljevic N., Ivetic M. (2005). Fuzzy Approach In The Uncertainty Analysis Of The Water Distribution Network Of Becej. CCWI2005 Computing & Control in the Water Industry Conference 5–7 September 2005, Exeter UK.

Ivetic M. (1996). *Computational Hydraulics, Pipe Flow* (in Serbian). Faculty of Civil Engineering University of Belgrade, Belgrade.

Kapelan Z. (2004). Calibration of water distribution system hydraulic models. Phd thesis.

Kottegoda N., Rosso R. (1998). *Statistics, Probability and Reliability for Civil and Environmental Engineers*. McGraw-Hill , International Editions.

Maskey S., Guinot V., Price R. (2004). Treatment of precipitation uncertainty in rainfall-runoff modeling: a fuzzy set approach. *Advances in Water Resources* **27**, 889–898.

Revelli R., Ridolfi L. (2002). Fuzzy approach for analysis of pipe networks. *Journal of Hydraulic Engineering*, January, 93–101.

Rossman L. EPANET 2, Water Supply and Water Resources Division National Risk Management Research Laboratory, Cincinnati, OH 45268.

Wall M. (1996). GAlib: A C++ Library of Genetic Algorithm Components, Documentation, August.

Water quality and wastewater treatment

Kinetics and mechanisms of oxidation of lead(II) solid phases by chlorine in drinking water

H. Liu*, G.V. Korshin, J.F. Ferguson** and W. Jiang***

* Department of Architecture and Environmental Engineering, Sichuan University, Chengdu 610065, China (E-mail: *capriaty001@163.com*; *jwj@263.net*)
** More Hall 305, Department of Civil and Environmental Engineering, BOX 352700, University of Washington, Seattle 98115-2700, USA
(E-mail: *korshin@u.washington.edu*; *jferg@u.washington.edu*)

Abstract Lead contamination in drinking water distribution system is typically thought to be caused by the dissolution of Pb(II) solid phases, notably hydrocerussite $Pb_3(CO_3)_2(OH)_2$ which is formed via the corrosion of Pb(0) present in lead pipe and other lead-bearing materials. In contrast with that, Pb(IV) oxides have very low solubility and appear to be beneficial for lead release control. Residual chlorine in drinking water plays a crucial role in the formation of Pb(IV) oxides. It is therefore important to gain a better understanding of the chemical processes that involve Pb(II) and Pb(IV) solid phases and their transformations in the presence of chlorine. In this study, studies of the oxidation of hydrocerussite were conducted in chlorinated water with varying alkalinities and pHs. X-ray Diffraction (XRD) and Scanning Electron Microscope (SEM) analyses of the solids formed in these reactions demonstrated that α-PbO_2 (scrutinyite) was formed when Pb(II) phases were oxidized by chlorine. Chlorine consumption profiles showed a complex reaction kinetics that included a lag phase, a rapid transition phase and a final steady-state phase. The lag phase was primarily associated with the transformation of hydrocerussite to cerussite $PbCO_3$. In the rapid transition phase, the oxidation of $PbCO_3$ and formation of α-PbO_2 occurred. Dissolved inorganic carbon (DIC) was found to greatly impact the oxidation rate of hydrocerussite and increase the duration of the lag phase. The rate of chlorine consumption rate was decreased at high pHs. High resolution SEM images indicated that the formation of PbO_2 was localized at specific Pb(II) surface sites possibly active in the generation of reactive Pb(II) hydroxo complexes. Based on that, a hypothesis was developed that lead (II) hydroxo species were more reactive with chlorine than lead carbonate species.

Keywords Chlorine; corrosion scales; crystalline structure; drinking water; kinetics; lead

Introduction

Health effects caused by lead released into drinking water from lead-bearing materials (LBM), notably lead pipes, have been well documented (Raloff, 1995; Needleman *et al.*, 1996). Despite the general success in controlling lead in drinking water, the phenomenon of a several-fold increase of lead in Washington D.C. drinking water was recently observed. It was apparently caused by subtle changes of its chemistry, specifically the introduction of chloramines instead of chlorine as the residual disinfectant (Edwards and Dudi, 2004; Rener, 2004).

In most situations, the level of lead in tap water depends on the time during which the water has been in contact with lead-containing materials, their age, surface area, metallurgical and electrochemical prosperities, and overall drinking water chemistry (Kuch and Wagner, 1983; Van Der Leer *et al.*, 2002, Lytle and Schock, 2000). The effects of stagnation time, pH, alkalinity and concentration of phosphate and sulphate on lead release, as well as the roles of chlorine and chloramines, have been studied (Lin *et al.*, 1997; Cantor *et al.*, 2003; Edwards and Dudi, 2004). However, the sudden occurrence of the lead "crisis" in Washington D.C. associated with chlorine/chloramine change indicates that the current theory of lead release in drinking water is incomplete.

It is typically assumed that lead contamination in drinking water distribution systems with lead pipes occurs via the dissolution of Pb(II) solid phases such as litharge PbO, cerussite $PbCO_3$ and hydrocerussite $Pb_3(CO_3)_2(OH)_2$ (Lytle and Schock, 2000, Schock and Harmon, 2001), with little impact of Pb(IV). However, structural analyses using specimens of lead service lines removed from drinking water distribution systems has shown the coexistence of Pb(II) and Pb(IV) species in the corrosion scales (Schock and Harmon, 2001; Davidson *et al.*, 2004). In contrast with Pb(II) solids, Pb(IV) oxides have very low solubility and are likely to be beneficial to for lead control. Thermodynamic data suggest that Pb(IV) oxides can form and be stable only in the presence of strong oxidants such as free chlorine. Therefore, chemical processes that involve Pb(II) solids and chlorine to from Pb(IV) oxides are crucially important to lead solubility control and are likely to play a central role in lead species transformation in drinking water system.

Prior research indicates that two Pb(IV) dioxides polymorphs formed in presence of chlorine (Lytle and Schock, 2000), but how the mechanism and kinetics of this process depend on the nature of the involved Pb(II) solids and water properties such as pH, concentration of chlorine and dissolved inorganic carbon has not been determined. The goal of this paper is to present experimental data that help quantify the oxidation of Pb(II) solids by chlorine and ascertain the effects of key water properties on this process.

Materials and methods

Pb(II) oxidation experiments were carried out using powders of hydrocerussite and cerussite. A desired amount of these solids was suspended in a chlorine-containing solution and placed in 200 ml glass headspace-free vials. All solutions were prepared using reagent-grade chemicals and high-purity water (resistivity 18.2 MΩ/cm). The concentration of chloride in the synthetic water was 105 mg/L. The concentration of dissolved inorganic carbon (DIC) was varied from 3.6 mg/L and 120 mg/L as carbon and controlled by additions of sodium bicarbonate. The pH was varied from 7 and 9 and adjusted with 0.1mol/L perchloric acid or sodium hydroxide. The temperature was 20 °C.

Before the injection of chlorine, Pb(II) solids were equilibrated in each solution for 10 hours. Concentrations of free chlorine were varied from 4 to 55 mg/L. The pH of the stock solution of chlorine was adjusted to the same pH as that the synthetic water. Solutions containing suspended Pb(II) solids and chlorine were transferred into 10 ml headspace-free glass test tubes that were kept in a shaker. At desired reaction times, they were filtered through a 0.45 μm filter and analysed for pH, free chlorine concentrations, particle size distributions. Scanning electron microscopy (SEM) and X-ray diffraction (XRD) examination of the solids retained by the filters was also carried out. In selected experiments, lead particles were first treated with 0.2 mol/L hydrochloric to remove Pb(II) solids and retain Pb(IV) phases only.

Free chlorine was measured with a Hach DR/4000U spectrophotometer using the standard DPD method. SEM images of the lead solids were obtained using J JSM 5200 and Jeol JSM-7400F SEM microscopes. XRD data to identify crystalline lead phases were obtained with a Philip 1820 XRD. XRD pattern analysis was performed with Jade^{+} V.6 software. Reference XRD patterns were from the 1995 version of International Center for Diffraction Data database.

Results and discussion

Formation of new Pb(IV) solid phase

Since hydrocerussite (HC) is a typical lead (II) solid formed on the surface of lead service lines in Washington D.C. and many other drinking water utilities, it was used to quantify changes of lead (II) solids caused by chlorine. In the initial experiments, 0.5 g/L HC suspended in water with DIC concentration 12 mg C/L and pH 7.0 was exposed to 55 mg Cl_2/L. The reaction was accompanied by visually discernible changes. The colour of lead solids changed from initially white to light yellow and, at reaction time 8 and 24 hours, to orange-red and black, respectively.

To identify all lead phases formed in this reaction, they were examined using SEM. It showed that before induction of chlorine, HC particles had a hexagonal symmetry as thin plates (Figure 1). The platy hexagonal crystals had typical dimensions of 3–10 μm and were very thin.

(1) (2) (3)

Figure 1 SEM images of lead solid phases at different reaction times: (1) beginning of reaction; (2) 4 hours of reaction; (3) 16 hours of reaction.

In the sample obtained from reaction in 4 hours, both the prismatic structures and hexagonal HC crystallites were present (Figure 2). The prisms had smooth surfaces and distinct orthorhombic arrangement of the facets. Their typical length and width were 5 μm and 2 μm. XRD analysis showed that the prisms were cerussite $PbCO_3$. When reaction lasted for 16 hours, more prisms formed, while hexagonal crystallites gradually disappeared (Figure 3). Simultaneously, numerous small crystalline structures began to form on the surface of the prisms. These small crystallites were less than 1 μm in dimension.

To determine the nature of these particles, solid samples obtained for a 16 hour reaction time were treated with hydrochloride acid to remove Pb(II) compounds. SEM showed that the small crystals remained unchanged in the acid-treated samples (Figure 2). Their XRD spectrum (Figure 3) indicated that they were solely scrutinyite α-PbO_2. When the reaction time was 24 hours, only scrutinyite and cerussite were found (Figure 4) High resolution SEM image (Figure 5) showed the existence of distinct attack sites on the crystalline surfaces of cerussite and formation of scrutinyite particles at them. SEM data were in an agreement with the results of XRD that indicated the existence of transformation of hydrocerussite to cerussite followed by the formation of scrutinyite.

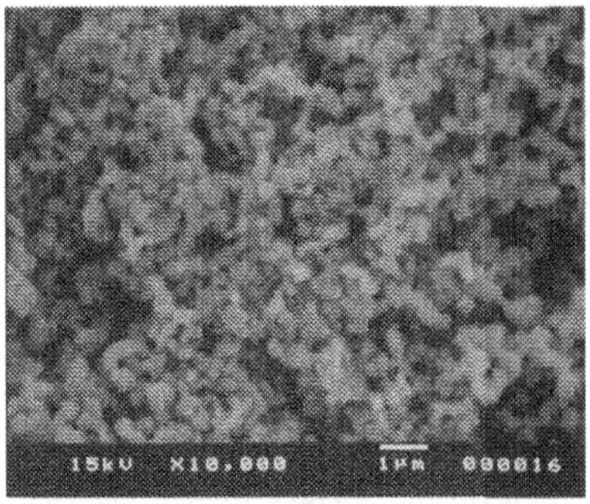

Figure 2 SEM image of the small crystals at reaction hour 16 during hydrocerussite oxidation reaction. Acid-treated sample.

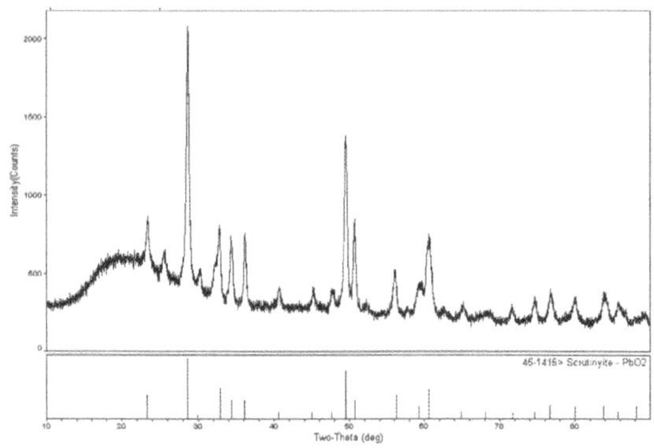

Figure 3 XRD pattern of PbO_2 collected in acid selected sample at hour 16 during oxidation of

Figure 4 SEM image of cerussite crystals covered by massive lead dioxide small crystals at reaction hour 24 during hydrocerussite oxidation. reaction.

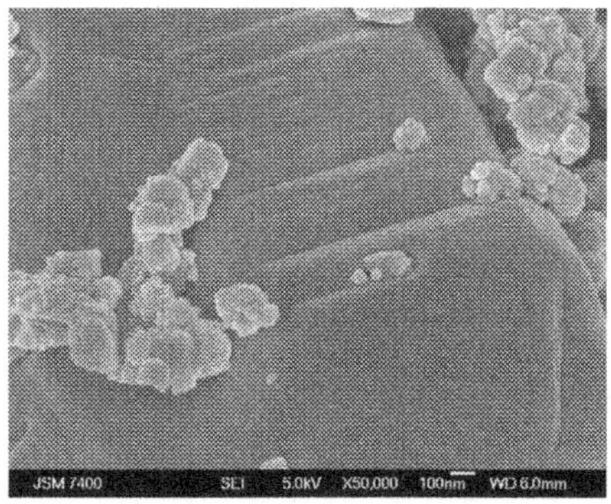

Figure 5 Higher magnification SEM image of micro site cerussite crystals covered by massive lead dioxide small crystals at reaction hour 24 during hydrocerussite oxidation reaction.

Chlorine consumption rate

To explore the kinetics of hydrocerussite oxidation by chlorine and to ascertain the effects of dissolved inorganic carbon (DIC) and pH, 0.5 g/L hydrocerussite suspensions with initial pH 7.0 and DIC values of up to 120 mg C/L were exposed to 55 mg Cl_2/L. Chlorine consumption profiles generated in these experiments demonstrated the existence of three stages (Figure 6). These included a first lag phase when chlorine concentration was largely constant; a second phase in which the chlorine was rapidly consumed, and a final reaction phase during which the residual chlorine concentration stabilized at a low steady-state level. The final phase was determined to last days to weeks during which time the residual chlorine disappeared. Examination of the chlorine consumption profiles together with the SEM and XRD data discussed above showed that the lag phase was associated with the transformation of hydrocerussite to cerussite. In the second reaction phase, the concentrations of chlorine decreased a result of its reactions with Pb(II) to form α-PbO_2. In the final phase, residual cerussite was present together with large amounts of α-PbO_2.

The chlorine consumption profiles also demonstrated that DIC concentrations greatly affected the apparent oxidation rates. As the DIC values increased, chlorine was consumed more slowly. The lag phase was highly sensitive towards DIC and pH variations. For instance, it increased from 6 hours to >70 hours when the DIC concentration was increased from 12 mg /L to 36 mg/L as carbon (Figure 6). The duration of the lag phase tended to increase with the pH, especially at pH>8 (Figure 7) and for pH 9 the chlorine consumption rate in and morphological changes of lead solid phases, were slow.

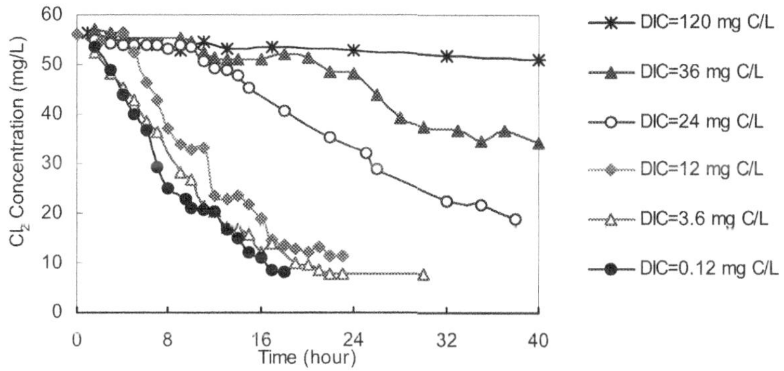

Figure 6 Chlorine consumption profile at varying DIC concentrations for hydrocerussite oxidation with initial pH 7.0 and hydrocerussite concentration 0.5 g/L.

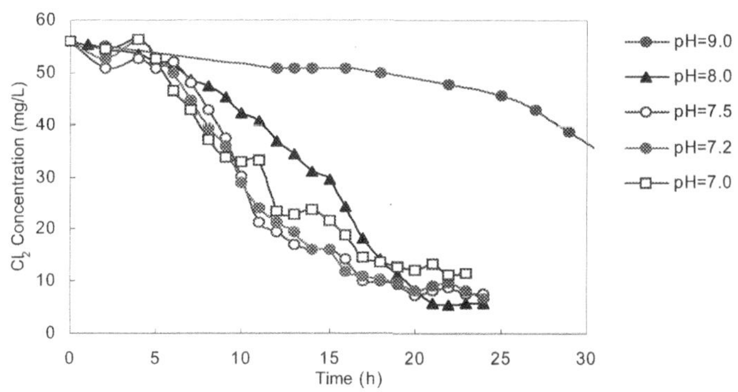

Figure 7 Chlorine consumption profile at varying pH values for hydrocerussite oxidation with initial DIC concentration 12 mg C/L and hydrocerussite concentration 0.5 g/L.

Another series of experiments was carried out using a smaller amount of hydrocerussite (0.1 g/L), which was done to stabilize the pH of the system during the oxidation and compare effects of the initial chlorine dose. The chlorine consumption profiles generated for initial chlorine concentrations from 4 to 55 mg/L showed the presence of the three distinct stages for all conditions, but the duration of the lag phase was approximately inversely proportional to the concentration of chlorine at the inception of the reaction.

All major features of the chlorine consumption profiles were in agreement with the results of SEM and XRD measurements. They appeared to indicate that hydrocerussite was largely non-reactive in chlorine oxidation, and chlorine was consumed only after the phase transformation of hydrocerussite to cerussite had taken place. However, experiments with cerussite exposed to chlorine showed that the oxidant was consumed by $PbCO_3$ very slowly. Correspondingly, cerussite *per se* can not necessarily be assumed to be a reactive phase in oxidations by chlorine, and the formation of reactive intermediate of yet unidentified nature is likely to affect the kinetics of oxidation of Pb(II) solids.

We currently hypothesize that soluble or adsorbed lead hydroxo complexes are the reactive species that initiate the entire reaction sequence. This hypothesis is based on independent observations that these complexes were highly reactive in homogeneous chlorine oxidations. Further modelling of the impact of the soluble or possibly adsorbed lead hydroxo complexes will provide more information to ascertain the nature of reactions that govern the Pb(II)/Pb(IV)/chlorine system.

Conclusion

The results presented above indicate that lead (II) solid species can be oxidized by chlorine to form PbO_2. In the case of hydrocerussite $Pb_3(CO_2)(OH)_2$, the oxidation proceeds through a lag phase, a transition phase and a final quasi-steady state phase. The lag phase corresponds to the transformation of hydrocerussite to cerussite $PbCO_3$; the transition phase is accompanied by a rapid formation of PbO_2; the final phase corresponds to a steady state. During the lag phase, chlorine concentration is largely stable, while it decreases rapidly during the transition phase to achieve a low steady-state level characteristic for the final phase. A nearly linear relationship was determined to exist between DIC concentrations and the duration of the lag phase. The duration of the lag phase also tended to increase at pH>8 and was approximately inversely proportional to the initial chlorine dose.

The transition from the predominance of hydrocerussite $Pb_3(CO_3)_2(OH)_2$ to that of cerussite $PbCO_3$ during the lag phase is of great interest for the mechanistic description of processes associated with changes of disinfectants in drinking water distribution system. Since the hydrocerussite/cerussite transition seems to contradict to the stability of the lead (II) phases predicted based on the thermodynamic data, it is hypothesized that lead hydroxo complexes, which were proven in independent experiments to be reactive in chlorine oxidations, can initiate the sequence of the oxidation reactions.

References

Cantor A.F., Park J.K., and Vaiyavatjamai, P. (2003). Effect of chlorine on corrosion in drinking water systems. *J. American Water Works Association*, **95** (5), 112–123.

Davidson C.M., Peters N.J., Britton A., Brady L., Gardiner P.H., and Lewis B.D. (2004). Surface analysis and depth profiling of corrosion products formed in lead pipes used to supply low alkalinity drinking water. *Water Science and Technology*, **49** (2), 49–54.

Edwards M., and Dudi A. (2004). Role of chlorine and chloramines in corrosion of lead-bearing materials. *J. American Water Works Association*, **96** (10), 69–81.

Kuch A., and Wagner I. (1983). A mass transfer model to describe lead concentration in drinking water. *Water Research*, **17** (10), 1303–1307.

Lin N.H., Torrents A., Davis A.P., Zeinali M. and Taylor F.A. (1997). Lead corrosion control from lead, copper-lead solder, and brass coupons in drinking water employing free and combined chlorine. *J. Environmental Science. Health A*, **32** (4), 865–884.

Lytle D.A., and Schock M.R. (2000). Impact of stagnation time on metal dissolution from plumbing materials in drinking water. *J. Water Supply: Research and Technology—AQUA*, **49** (6), 243–257.

Neddleman H.L., Riess J.A., Tobin M.J., Biesecker G.E., and Greenhouse J.B. (1996). Bone lead levels and delinquent behaviour. *J. Medicine Association*, **275** (5), 363–370.

Raloff J. (1995). Lead may foster immune attack on brain. *Science News*, **147** (2), 23–24.

Renner R. (2004). Plumbing the depth of D.C.'s drinking water crisis. *Environmental Science and Technology*, **38** (12), 225A–227A.

Schock M.R., and Harmon S.M. (2001). Tetravalent Lead: A Hitherto Unrecognized Control of Tap Water Lead Contamination. Water Quality Technology Conference, American Water Works Association, Nashville, TN.

Van Der Leer D., Weatherill N. P., Sharp R.J., and Hayes C.R. (2002). Modelling the diffusion of lead into drinking water. *Applied Mathematical Modelling*, **26**, 681–699.

Health consequences of inadequate water service delivery: the case of the cholera epidemic in KwaZulu-Natal, South Africa

M.D. Said*, S.N. Venter*, I. Booysen and T.E. Cloete***

* Department of Microbiology and Plant Pathology, University of Pretoria, Pretoria 0001, South Africa (E-mail: *maryam_dsaid@yahoo.com*)
* Department of Geography, Geoinformatics and Meteorology, University of Pretoria, Pretoria 0001, South Africa

Abstract The aim of this study was to identify the importance of water supply and the related sanitation services, or lack thereof, as a factor that exposed the KwaZulu-Natal (KZN) communities in South Africa to the risk of cholera. An ecological study was conducted to test the potential of health and socio-economic databases as management tools to control the future spread of cholera in rural communities. Statistical and spatial exploration of the databases revealed the distribution of cholera cases in KZN corresponded with areas showing inadequacies in water services, as well as those having inadequate sanitation. Households using water from dams, rivers, streams and boreholes were positively correlated to the incidence of cholera. Positive correlations to the incidence of cholera were also found with households using the bucket toilet sanitation or had no specified form of sanitation. The results demonstrate that cholera was often associated with rural areas and the poor communities, which lacked basic water supply and sanitation services.

Keywords Cholera, GIS, management tool, poverty, river water

Introduction

Cholera has long being acknowledged as a disease that reflects a complicated transmission pattern, in that multiple factors including poor living conditions may play a role in its spread. Factors implicated in one area as being important to the spread of cholera may thus not be significant in another. It has been recognised that in a given region, *Vibrio cholerae* may be established as an endemic or a recurrent pathogen, suggesting the possible existence of at least a transient environmental reservoir for cholera in the area affected (Shapiro *et al.*, 1999). Owing to this constant risk of reoccurring outbreaks, the spread of cholera within such a region can only be limited by identifying and addressing socio-economic factors that are conducive to the spread of the disease in that region.

Since 1970, when the seventh cholera pandemic first crossed over to the African continent, South Africa had experienced several imported cholera cases from neighbouring countries of Angola, Malawi, Mozambique, Zambia and Zimbabwe where the disease was considered to be endemic (Barua, 1992; Isaäcson, 1986; Küstner *et al.*, 1981; WHO, 2000). Most of these cases are thought to be a consequence of migrant labourers coming to seek employment on South African mines or plantations (Isaäcson *et al.*, 1974). During this period several small-scale outbreaks were also reported in parts of the country. It was established that cholera mainly affected socio-economically deprived communities in densely populated, rural, tropical and subtropical areas, with the peak incidence occurring after the onset of the main summer rains (Isaäcson, 1986; Küstner and du Plessis 1991).

Recently South Africa experienced a major cholera epidemic in the eastern coastal province of KwaZulu-Natal (KZN) (Figure 1). The epidemic started in August 2000 with cases continuing to be reported well into 2004. A cholera outbreak started from the more urban centres of KZN, and rapidly proceeded to the interior of the province where no community was spared from the scourge. The start of the outbreak was attributed by some sources to an incident where a local authority stopped supplying free water and introduced a charge to the very poor people living in an informal settlement near the town of Empangeni (Hemson, 2000; Sidley, 2001, Cottle and Deedat 2002). Several reports linked cholera to water privatisation, and that the communities affected were mostly those who could not afford to pay for water services thus resorted to use natural water sources (Nhlapo-Hlope, 2001; Ka-Min, 2000). This implied that a link exists between the use of natural water sources, and the spread of cholera. Despite prompt medical intervention, health education and media awareness campaigns, cholera continued to spread throughout the province of KZN. By March 2004, the official statistics of cholera cases in KZN registered 158 895 cases (Dept-KZN Health, 2000). The death toll stood at 575 persons translating to a per cent case fatality rate (%CFR) of 0.36%, the lowest compared to the previous epidemics recorded in South Africa (Küstner *et al.*, 1981; Küstner and du Plessis, 1991).

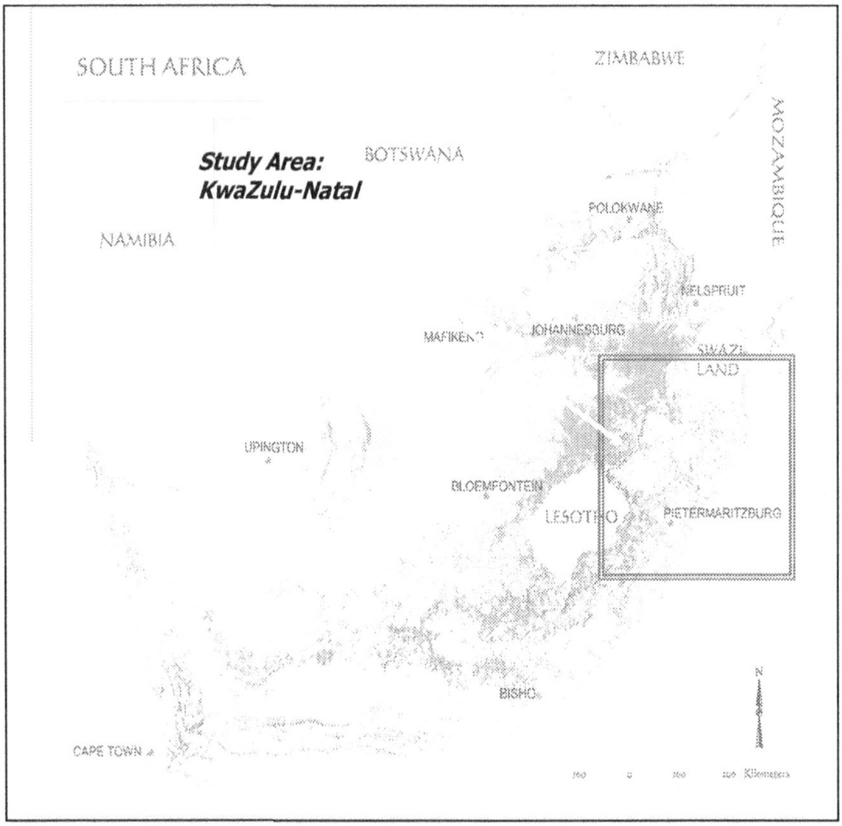

Figure 1 Orientation map showing the study area of KwaZulu-Natal in South Africa.

The latest population statistics show KZN accommodating over 9.6 million residents, the largest (20.6%) provincial population in South Africa (STATSSA, 2005). The majority of the people in KZN live in crowded, rural and underdeveloped conditions that favour the spread of the disease (Laskow, 2001). In addition, many people in KwaZulu-Natal still rely on river water for cooking, drinking and washing practices which have been demonstrated to be positively associated with an increased risk for transmission of cholera (Küstner *et al.*, 1981; Shapiro *et al.*, 1999).

Objectives

The purpose of the investigation was to test the suitability of using health and socio-economic data as a management tool to control the future spread of cholera in rural communities. The main aim of this study was to assess the importance of water supply and sanitation services, or lack thereof, as factors that expose the KZN communities to the risk of cholera.

Methodology

An ecological or correlational study was conducted based upon disease incidence using GIS technology. It was realised that the major limitation of an ecological study using health surveillance data is the potential for substantial bias in the end result. As such knowledge of disease and human behaviour was used to explain any associations observed so as to avoid the drawing of inappropriate conclusions. The study used the Department of Health of KwaZulu-Natal's (South Africa) official Cholera Database for the period August 2000 to February 2004 in which 158,895 cholera cases were recorded. The analyses presented in this paper though, is based on the records of only 136,793 cholera cases reported from all the nine District Councils (DCs) of KZN over the same period of time. The cases omitted did not have complete data entries in all the required fields. The Cholera database was consolidated with the Population Census database of South Africa of 1996 (STATSSA, 1998). The choice of the 1996 Census database was based on the correspondence of geographical information (names of cities, towns and villages) with the information used in the creation of the Cholera Database. The names and number of places for which cholera was reported/recorded were the same as those represented in the 1996 Population Census. The Population Census data for 2001 had a different list of place names due to the demarcation changes that were instituted after the 1996 Population Census, Differences in the socio-economic data between 1996 and 2001sensus were investigated and were found not to be significant enough to influence the analyses (see Table 1).

Table 1 The socio-economic variables of Census 1996 and Census 2001 of KZN

	Census 1996	Census 2001	Diff. 96–01	% Change
Water Supply	**Households**	**Households**	**Households**	**Households**
Piped Water to the Dwelling	650,677	618,267	32,410	–5.2
Piped Water inside yard	145,237	413,535	268,298	64.9
Piped Water to Community Stand (public tap)*	304,502	494,966	190,464	38.5
Borehole (Well)*	110,755	88,065	22,690	–25.8
Rain-water Tank	N/A	15,787	N/A	N/A
Dam/ river/ stream/ spring	402,822	384,448	18,374	–4.8
Water vendor	20,059	18,126	1,933	–10.7
Other	26,882	53,056	26,174	49.3
Toilet Facility	**Households**	**Households**	**Households**	**Households**
Flush toilet or chemical toilet	693,130	978,016	284,886	29.1
Pit latrine with/without ventilation	690,560	747,091	56,531	7.6
Bucket latrine	15,713	24,025	8,312	34.6
None	250,956	337,119	86,163	25.6
Unspecified/Other	10,575	N/A	N/A	N/A

Source: STATSSA, 1998, STATSSA, 2003

Statistical and spatial analyses

Per cent cumulative incidence rates (%CIR) were calculated for each Magisterial District (MD) of KZN to standardise the infection rates between the areas. This ensured that comparisons and correlations were also standardised. The relationship between the study variables and the incidence of cholera were determined using Partial Spearman's correlations using SAS® (version 8.2). The statistical correlations were performed at the MD level. The MD data was used to avoid the introduction of a bias considering the fact that some place-names existed where not a single cholera case was reported. Outputs of the statistical analyses formed the basis for the creation of spatial maps to illustrate the epidemic using ArcView GIS 3.3. A water and sanitation predictive model based on the 52 MDs was developed to find potential cholera areas in the rest of KZN where conditions conducive to cholera existed. Based on the statistical correlations of the water and sanitation situation in relation to the incidence of cholera, the spatial queries tool in ArcView 3.3 was used to select the MDS that had a high proportion of its population dependent on the natural water resources and had inadequate sanitation. The spatial queries are described below as part of the results.

Results and discussion

The distribution of water services in KZN was such that only 15–47% of households in the non urban MDs had piped water in their dwellings, 6–14% depended on boreholes while the majority at 23–55% sourced their water from rivers, streams and springs (Figures 2–4). The most affected areas had between 1–33% of households with piped water in their dwelling.

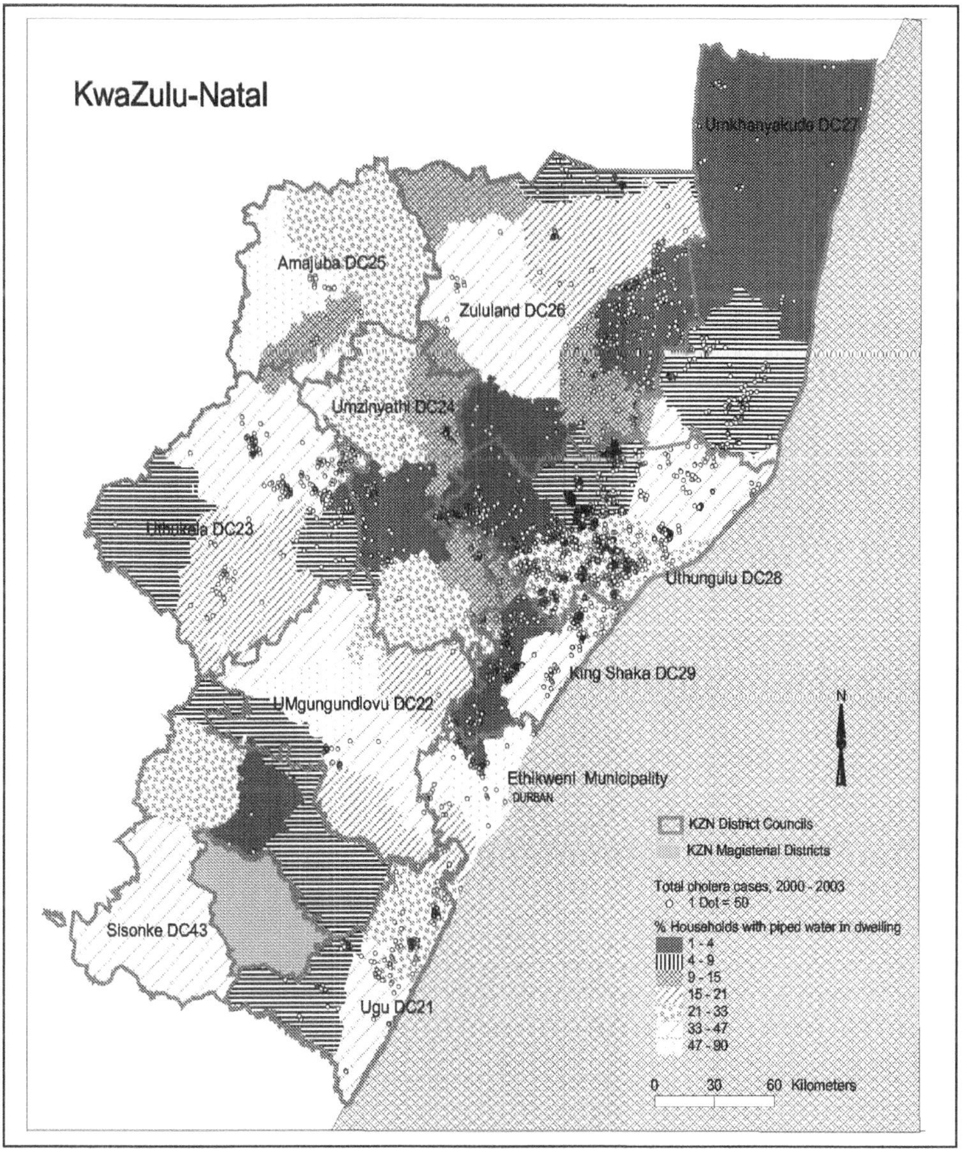

Figure 2 Per cent households in KZN with piped water in dwelling.

There was a total of six water services/options considered in the study, of which three were positively correlated with the %CIR of cholera, i.e. having piped water inside the dwelling, using boreholes and using water from rivers, dams, streams and springs (Table 2). The positive correlation of river water to the incidence of cholera ($r = 0.507$, $p = 0.0016$), stresses the risk polluted natural water sources pose to communities' dependant on them (Tshibangu, 1987; Sitas, 1986; Bradley *et al.*, 1996; Shapiro *et al.*, 1999; Patel and Isaäcson, 1989). This information is in agreement with the reports of people getting cholera because they could not afford to pay for water and had thus to resort to natural water sources which were most probably polluted with pathogens including the cholera organisms (Ka-Min, 2000; Nhlapo-Hlope, 2001; Sidley, 2001; Cottle and Deedat 2002).

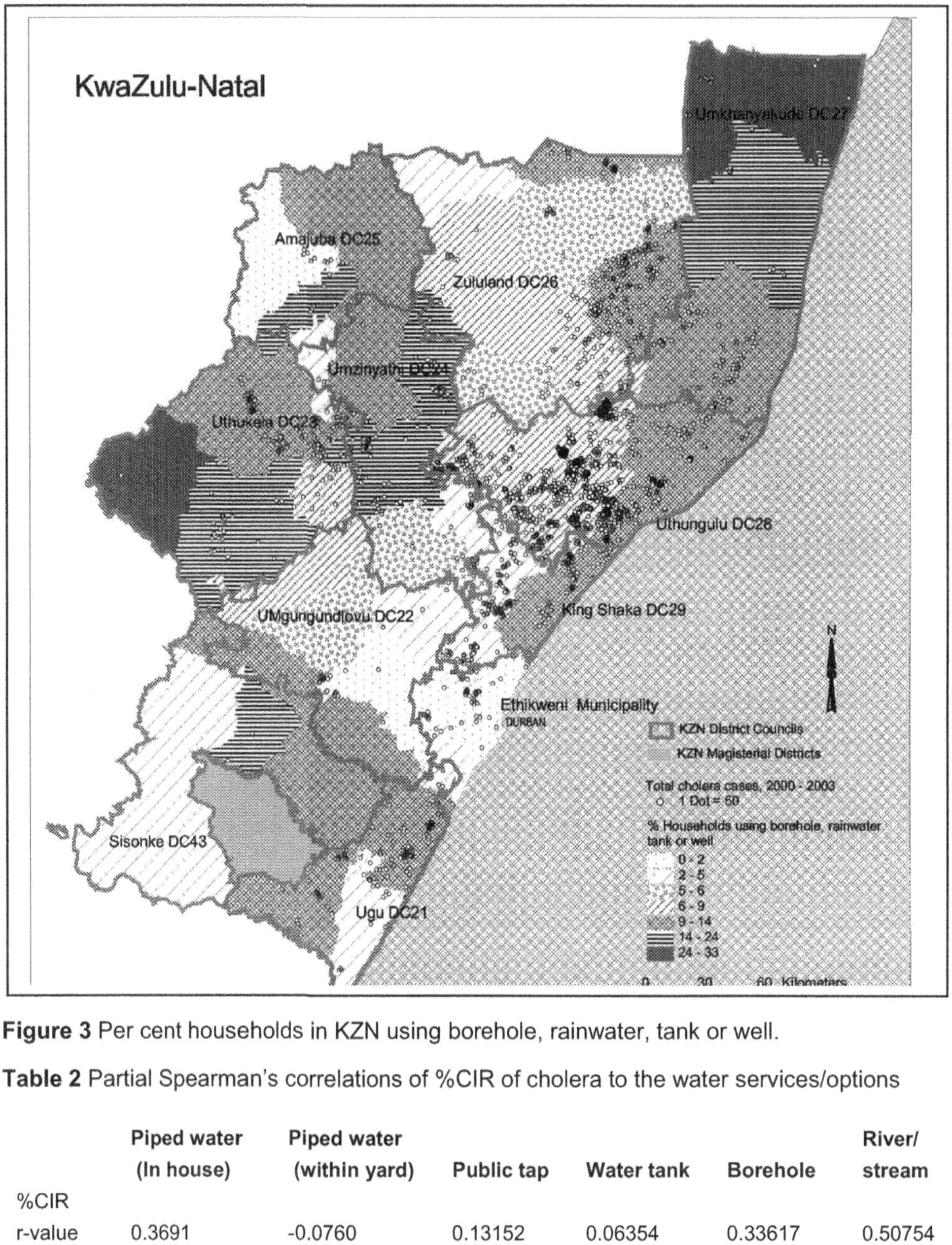

Figure 3 Per cent households in KZN using borehole, rainwater, tank or well.

Table 2 Partial Spearman's correlations of %CIR of cholera to the water services/options

	Piped water (In house)	**Piped water (within yard)**	**Public tap**	**Water tank**	**Borehole**	**River/ stream**
%CIR r-value	0.3691	-0.0760	0.13152	0.06354	0.33617	0.50754
%CIR p-value	0.0267	0.6594	0.4445	0.7128	0.045	0.0016

The positive correlation of piped water (in house) and the incidence of cholera ($r = 0.369$, $p = 0.026$) has also been documented by Acosta *et al* (2001) from a case-control study in Southern Tanzania. The risk of contracting an infectious disease like cholera in households with piped water (in house) may highlight an indirect association with the lack of education in matters of basic hygiene especially when handling water. In rural areas a piped water supply within the house or within a close proximity to the house may still require that water be stored and used from containers such as buckets and drums (Tshibangu, 1987; Patel and Isaäcson, 1989). As such, the cleanliness of the containers where the water is stored is important to reduce the risk of contracting infectious diseases like diarrhoea and cholera (Sánchez and Taylor, 1997; Quick *et al.*, 1999; Sobsey *et al.*, 2003). Another aspect to consider with the piped water (in house) service is that, although the database records the presence of piped water in a dwelling, it may just be reflecting the fact that the infrastructure for piped water exists. In reality, the data does not guarantee that there is a reliable supply of water nor does it guarantee that the quality of water is of the standard acceptable for human consumption. The various reports at the start of the cholera epidemic claiming that households had their water supplies terminated because of non-payment also adds credence to the notion that having piped water inside a dwelling does not necessarily guarantee a supply of clean water (Hemson, 2000; Morris, 2001; Cottle *et al.*, 2002; Pauw, 2003). Water sourced from borehole also had a positive correlation to the incidence of cholera ($r - 0.336$, $p = 0.045$), implying that the water sourced from the boreholes might have been of questionable quality, possibly contaminated with cholera. This could be due to the heavy rains that were recorded in the KZN region just before the start of the cholera epidemic (Kriner, 2001; Sidley, 2001). The surface run off that resulted from the heavy rains has the potential to contaminate the water table, which is usually the source of the water from boreholes. A similar situation was also being experienced in neighbouring Mozambique during 2001when heavy rains had flooded rural areas including their pit latrines and sewage systems (Bateman, 2002). Likewise, the risk of cholera was expected to increase after the heavy rains contaminated environmental water sources.

The spatial distribution of cholera cases in KZN corresponded well with areas that have households using boreholes (Figure 3), concurring with the positive correlation described above between the use of boreholes and %CIR of cholera (0.336 $p = 0.045$). It is quite probable then that if the natural water sources around the areas where cholera was rampant were polluted, this could have affected the quality of water sourced from such boreholes. There was also the possibility of wells and boreholes being polluted because of the heavy rains and the subsequent surface run off that occurred around the same time period as the cholera epidemic (Kriner, 2000; Bateman, 2002).

Issues of sanitation go hand in hand with those of water supplies, in that poor sanitation is more often than not a source of water supplies contamination. Five types of sanitation services/options were considered and correlated with the %CIR of cholera (Table 3). It was reasonable to assume that sanitation options like the flush or chemical toilets were mostly available in the urban areas because of the high maintenance costs associated with them. MDs that reported high cholera case clusters had up to 4% of households using the bucket toilet (Figure 5). This is a point of concern considering that the bucket toilet has been documented to have a high potential to expose households to unsanitary conditions particularly in areas where the bucket maintenance (emptying) service is not regularly provided by the local authorities concerned (DWAF, 2002). On the negative extreme of the sanitation options is the 1–6.3% of households in the most affected MDs having had no specified form of sanitation (Figure 6). An important aspect to consider especially in an epidemic environment, is that people with no specified sanitation options will probably also include disease carriers who may continuously seed the aquatic and terrestrial environmental with pathogens like *V. cholerae*. The potential of such environmental pollution could exacerbate the contamination of surface and groundwater that recharges wells and boreholes. Inevitably then, people using untreated water from such sources would be at risk of contracting waterborne diseases such as cholera. In support of the type and distribution of sanitation options or lack thereof and the associated risks, the use of bucket toilets was positively correlated to %CIR of cholera ($r = 0.30$ p 0.063, Table 3). Similarly, households with unspecified sanitation were also positively correlated to %CIR of cholera (r = 0.411 p = 0.0094). The areas most affected by cholera were also the ones with a high proportion of households that require serious attention as far as their sanitation needs are concerned (Figures 5 and 6).

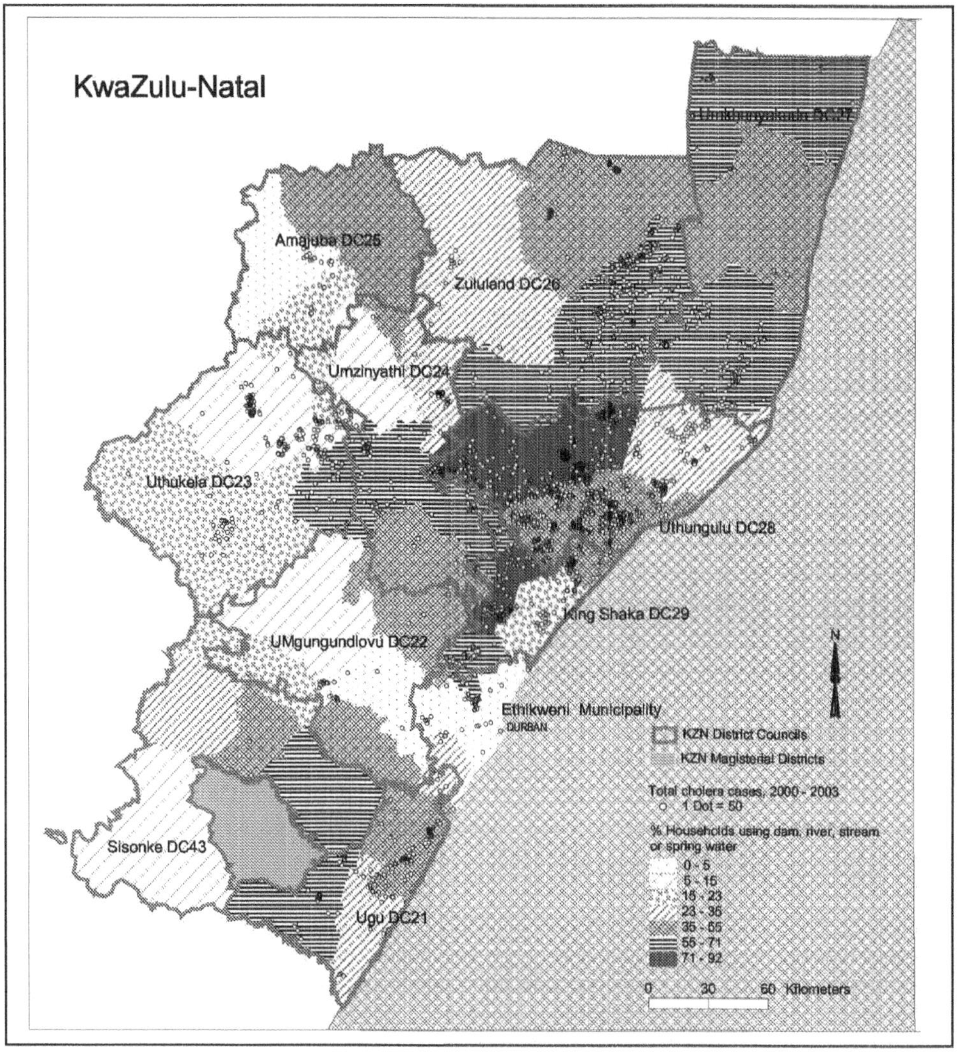

Figure 4 Per cent households in KZN using dam, river, stream or spring.

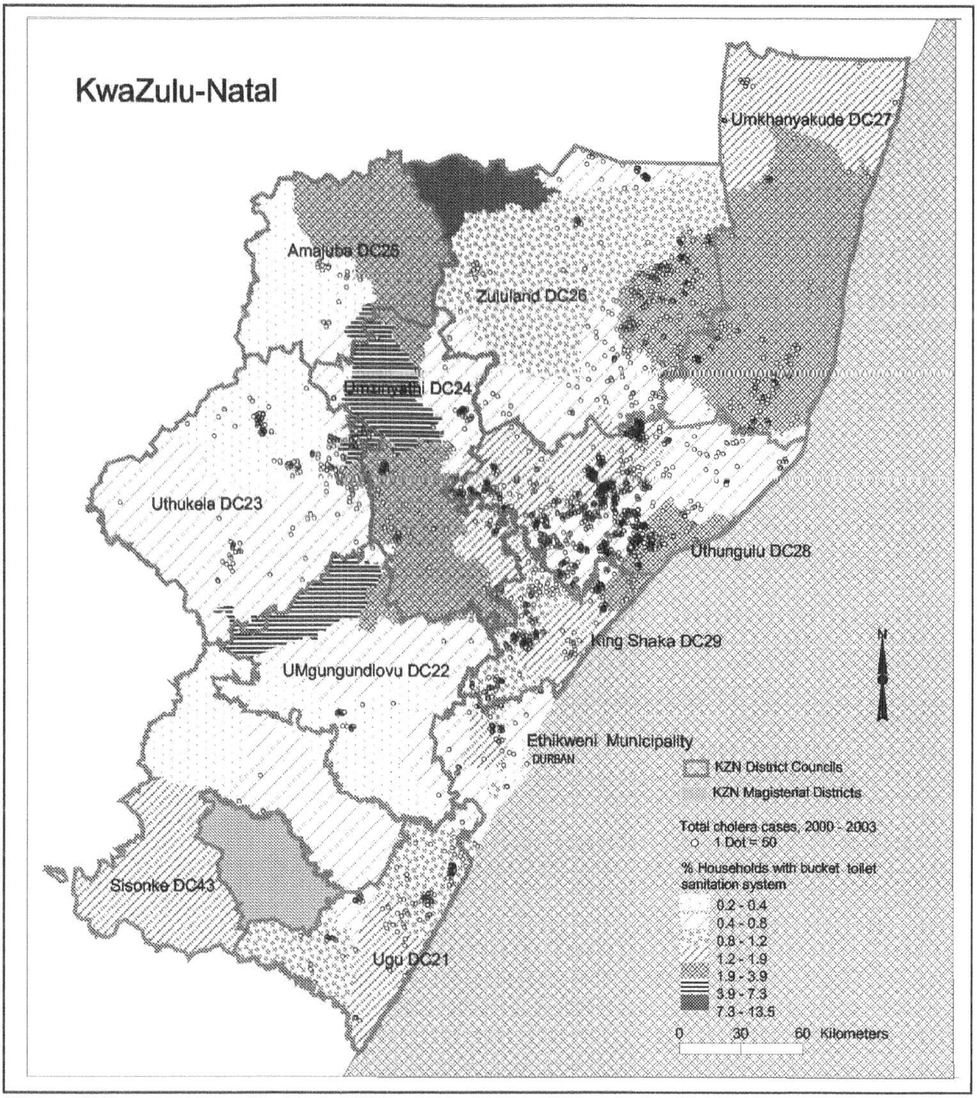

Figure 5 Per cent households in KZN with bucket toilet sanitation system.

Table 3 Partial Spearman's correlations of %CIR of cholera to the sanitation options

	Flush toilet	**Pit latrine**	**Bucket system**	**Other**	**None**
%CIR r-value	0.01655	0.01785	0.30015	NOT	0.41089
%CIR p-value	0.9204	0.9141	0.0634	CONSIDERED	0.0094

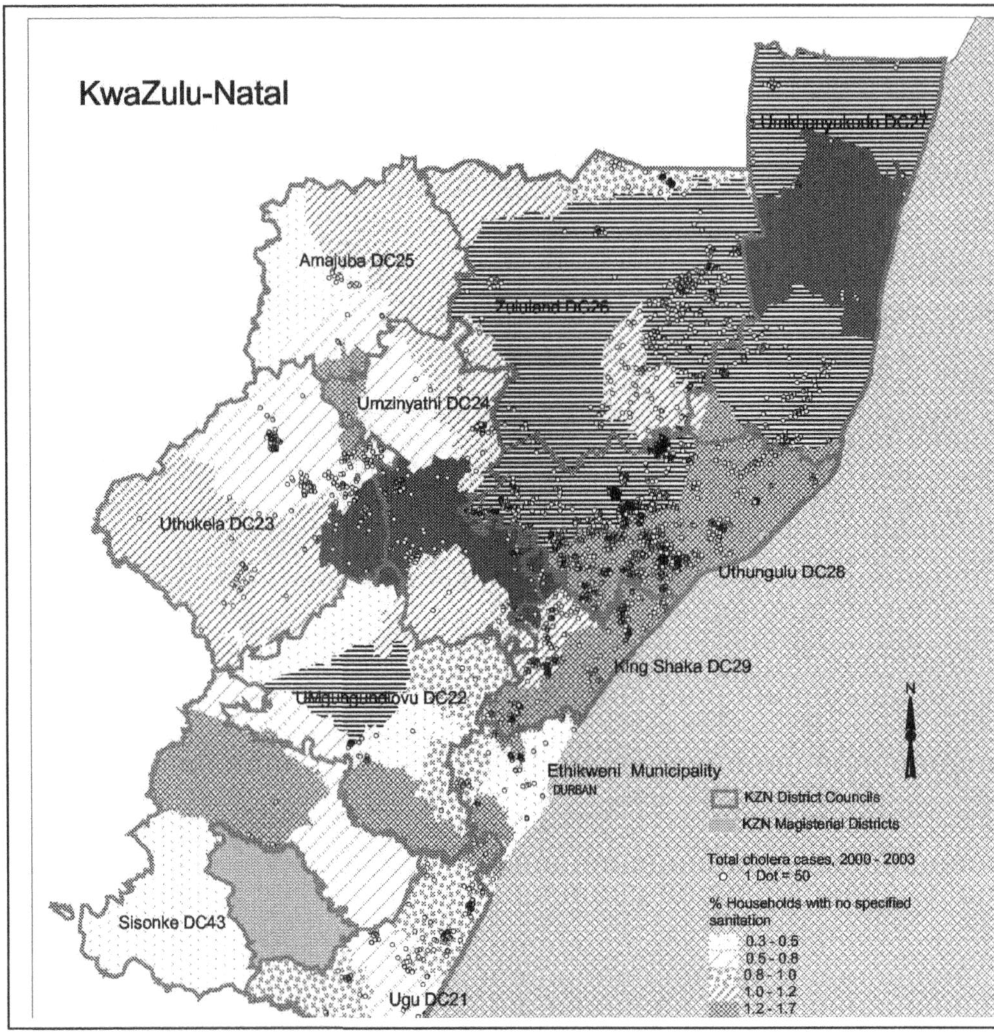

Figure 6 Per cent households in KZN with no specified sanitation.

GIS maps of the water and sanitation options that demonstrated to have a positive correlation to the incidence of cholera were reconciled and merged with the reported cholera cases to show the relative spatial distribution of the epidemic. The spatial queries tool in ArcView GIS 3.3 was used to select the statistically verified parameters of water and sanitation, so as to show poverty stricken areas, where sanitation was inadequate, and communities were mostly dependent on their environment for their water requirements. A spatial model was thus created based on the highest classification of water (natural resources) and sanitation services inadequacies as demonstrated by the spatial maps using the following formula. Based on data at the MDs level, the spatial query tool was to select an area; if $\geq 24\%$ of households used borehole water or $\geq 71\%$ of households used dams, river, stream or spring water and $\geq 7\%$ of households used the bucket toilet or 3% of households had no specified sanitation (Figure7).

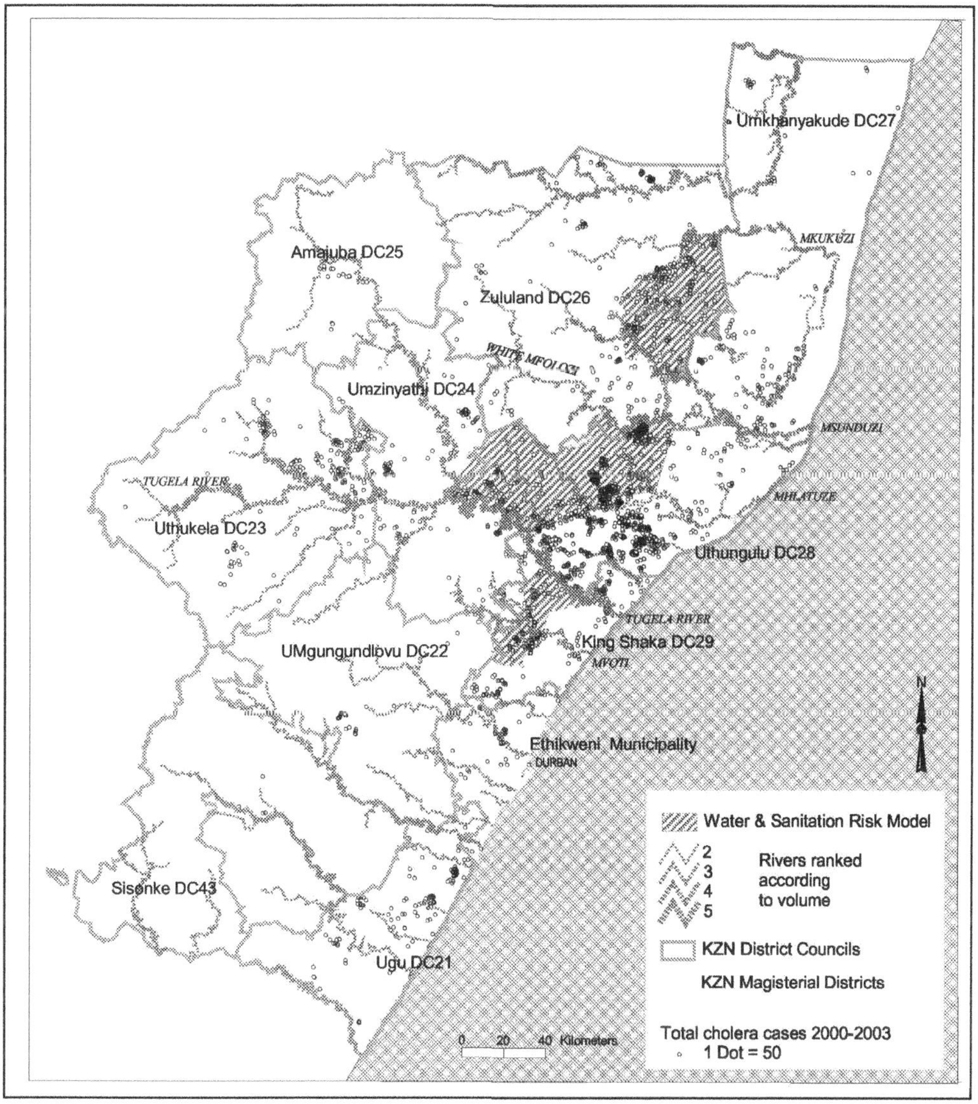

Figure 7 Spatial model based on inadequate water and sanitation variables showing high-risk cholera areas.

The results demonstrate that cholera was more severe in the rural areas and the poor communities of KZN. Majority of the high-risk cholera areas were characterised by a high level of poverty. In addition, the social-economic factors supporting the outbreaks clearly point to the low level of delivery of water and sanitation services in the affected communities possibly due to the inability to afford such services (Cottle and Deedat 2002). The development of man-environment transmission cycle that stems from unsatisfactory disposal of sewage contributes to contamination of the environment with *Vibrio*-bearing faeces (Crowcroft, 1994). Thus, the establishment of a transmission cycle through pollution of water sources appears to have been the most common form, though soil and food may also have been contaminated.

Conclusions

To date, there is no conclusive evidence as to the exact source, be it aquatic or terrestrial, of the 2000–2004 cholera epidemic. It is widely accepted that the environmental reservoir of cholera is primarily the aquatic environment (Colwell, 1996; Colwell *et* al., 1981; Colwell and Huq, 2001; Hood *et al.*, 1981; Huq *et al.*, 1983, 2005). Marine studies on the occurrence of *V. cholerae* and its link to outbreaks of cholera in South Africa are yet to be established. If the primary source was not environmental, it may have well have been a traveller from one of the neighbouring countries like Mozambique where cholera is known to be endemic. Alternatively, cholera may have already been endemic in KZN communities, lying in wait for optimum conditions that supported the proliferation of *V. cholerae* to infective doses; sufficient for an outbreak.

Whether cholera originated from an undiscovered endemic site or from a fresh importation is unknown. The cholera epidemic persisted for four years, so permanent endemicity in KZN must be considered a definite possibility. The continued survival of *Vibrio cholerae* makes new epidemic outbreaks probable in the future. The issues that were proven to be significant in this study related to the inefficiencies in the provision of water and sanitation, which go hand in hand with poverty. The issue of poverty was therefore indirectly reflected in the data and statistically supported as an issue that compounded the cholera epidemic. In South Africa, the lack of a social security and high levels of unemployment mean that poor households and communities often slip further and further into poverty and deprivation.

Reliable data, in addition to being used for monitoring and evaluation purposes can also be used in disease management issues. As such, ecological or correlation analyses can also be used to evaluate population interventions. These findings can be compared with findings of other observational studies designed to test the same etiologic hypothesis. If the estimated effect is consistent across studies involving different designs and different populations, a causal interpretation is enhanced (Morgenstern, 1982). This study made efficient use of available data resources to develop a disease prediction and management tool based on the understanding of the relationships between the incidence of cholera and the different factors previously implicated in its spread to epidemic proportions. This study has also shown that the development of databases that combine information about communicable diseases, demography, population movements and climate may be used as future disease management tools that will improve response capability. Such a system can also identify communities at high risk to epidemics and will help with the prioritization of interventions to reduce the impact of the disease.

The scarcity of human resources, and time, to combat cholera necessitates the selection of particular high-risk target populations as the primary concern of control programmes. Timeous warning systems based on the health monitoring programme will lead the way to environmentally sound public health policies and interventions (Epstein, 2001). It is hoped that the findings of this study will contribute in the use of health and socio-economic data as a management tool in the prediction and management of diseases like cholera at the community level.

References

Acosta, C. J., Galindo, C.M., Kimario, J., Senkoro, K., Urassa, H., Casals, C., Corachán, M., Eseko N., Tanner, M., Mshinda, H., Lwilla, F., Vila, J. and Alonso, P.L. (2001). Cholera outbreak in Southern Tanzania: Risk factors and patterns of transmission. *Emerg. Infect.Dis.* **7**(3), 583–587.

Barua, D. (1992). History of cholera. In: *Cholera.* Edited by D. Barua and WB Greenough III. Plenum Medical Book Company, New York. pp. 1–36.

Bateman, C. (2002). Mozambique cholera will affect region. *S. Afr. Med. J.* **92**(2), 104–106.

Bradley, M., Shakespeare, R., Ruwende, A., Woolhouse, M.E.J., Mason, E., and Munatsi, A. (1996). Epidemiologic features of epidemic cholera (El Tor) in Zimbabwe. *Trans. R. Soc. Trop. Med. Hyg.* **90**, 378–382.

Colwell, R.R. (1996). Global climate and infectious disease: the cholera paradigm. *Science* **274**, 2025–2031.

Colwell, R.R., Seidler, R.J., Kaper, J., Joseph, S.W., Garges, S., Lockman, H., Maneval, D., Bradford, H., Roberts. N., Remmers, E., Huq, I. and Huq, A. (1981). Occurrence of *Vibrio cholerae* serotype O1 in Maryland and Louisiana estuaries. *Appl. Environ. Microbiol.* **41**, 555–558.

Colwell, R.R and Huq, A. (2001). Marine ecosystems and cholera. *Hydrobiologia.* **460**(1–3), 141–145.

Dept-KZN Health. (2000). KwaZulu-Natal Cholera Data August 2000–Present: GIS Unit, KwaZulu-Natal Health Department, Pietermaritzburg, South Africa.

Isaäcson, M. (1986). Plague and cholera surveillance in southern Africa. *S. Afr. Med. J.* (Supplement), 43–46.

Isaäcson, M., Clarke, K.R., Ellacombe, G.H., Smit, W.A., Smit, P., Koornhof, H.J., Smith, L.S. and L.J. Kriel. (1974). The recent cholera outbreak in the South African gold mining industry: A preliminary report. *S. Afr. Med. J.* **48**, 2557–2560.

Hood, M.A., Ness, G.E. and Rodrick, G.E. (1981). Isolation of *Vibrio cholerae* serotype O1 from the Eastern Oyster, *Crassostrea virginica. Appl. Environ. Microbiol.* **41**(2), 559–560.

Huq, A., West, P.A., Small, E.B., Huq, M.I. and Colwell R.R. (1983). Ecological relationships between *Vibrio cholerae* and planktonic crustacean copepods. *Appl. Environ. Microbiol.* **45**, 275–283.

Huq, A., Sack, R.B., Nizam, A., Longini, I.M., Nair, G. B., Ali, A., Morris Jr., J.G Khan, M.N.H., Siddique, A.K., Yunus, M., Albert, M.J., Sack, D.A., and Colwell R.R. (2005). Critical factors influencing the occurrence of *Vibrio cholerae* in the environment of Bangladesh. *Appl. Environ. Microbiol.* **71**(8), 4645–4654.

Cottle, E and Deedat, H. (2002). The cholera outbreak: a 2000–2002 case study of the source of the outbreak in the Madlebe Tribal Authority areas, uThungulu Region, KwaZulu-Natal. Health Systems Trust. Durban, South Africa.

Crowcroft, N.S. (1994). Cholera: current epidemiology. *Communicable Disease Report* **4**(13), 157–164.

DWAF. (2002). Sanitation technology options. *Sanitation.* Department of Water Affairs and Forestry Pretoria, South Africa.

Epstein, P.R. (2001). Climate change and Emerg Infect Dis. *Microbes. Infect.* **3**, 747–754.

Hemson, D. (2000). Policy and practice in water and sanitation. *Indicator SA* **17**(4), 38–43.

Ka-Min, L. (2000). Water charges blamed for cholera outbreak in South Africa. Third World Network. Penang. Malaysia.

Nhlapo-Hlope, K.J. (2001). Cholera and water privatisation. *The Shopsteward* **10**(2).

Kriner S. (2001). Cholera sweeps through South Africa townships. *Disaster Relief.* Internet: http://www.disasterrelief.org/Disasters/010309choleraSA/

Küstner, H.G.V., Gibson, I.H.N., Carmichael, T.R. van Zyl, L. Chouler, C.A., Hyde, J.P. and du Plessis, J.N. (1981). The spread of cholera in South Africa. *S. Afr. Med. J.* **60**(3), 87–90.

Küstner, H.G.V. and du Plessis, G. (1991). The cholera epidemic in South Africa, 1980–1987. *S. Afr. Med. J.* **79**, 539–544.

Laskow, S. (2001). Cholera sweeps across South Africa. *Global Perspectives.* **121**(9).

Morgenstern, H. (1982). Uses of ecological analysis in epidemiologic research. *Am J Public Health* 72(12), 1336–1344.

Morris, K. (2001). Prevention fails to halt South Africa's well-treated cholera epidemic. *Lancet* **357**, 290.

Nhlapo-Hlope, K.J. (2001). Cholera and water privatisation. *The Shopsteward* **10**(2).

Patel, M. and Isaäcson, M. (1989). Survival of *Vibrio cholerae* in African domestic water storage containers. *S. Afr.Med. J.* **76**, 365–367.

Pauw, J. (2003). The politics of underdevelopment: metered to death–how a water experiment caused riots and a cholera epidemic. *Int. J. Health. Serv.* **33**(4), 819–830.

Quick, R.E., Venczel, L.V., Mintz, E.D., Soleto, L., Paricio, J., Gironaz, M., Hutwagner, L., Greene, K., Bopp, C., Maloney, K., Chavez, D., Sobsey, M., Tauxe, R.V. (1999). Diarrhoea prevention in Bolivia through point-of-use water treatment and safe storage: a promising new strategy. *Epidemiol. Infect.* **122**(1), 83–90.

Sánchez, J.L. and Taylor, D.N. (1997). Cholera. *Lancet* **349**, 1825–1830.

Shapiro, R.L., Otieno, M.R., Adcock, P.M., Phillips–Howard, P.A., Hawley, W.A. Kumar, L., Waiyaki, P., Nahlen, B.L.and Slutsker, L. (1999). Transmission of epidemic *Vibrio cholerae* in rural western Kenya associated with drinking water from Lake Victoria: an environmental reservoir for cholera? *Am. J. Trop. Med. Hyg.* **60** (2), 271–276.

Sidley, P. (2001). Cholera sweeps through South African province. *Br. Med. J.* **322**, 71.

Sitas, F. (1986). Some observations on a cholera outbreak at the Umvoti Mission Reserve, Natal. *S. Afr. Med. J.* **70**, 215–218.

Sobsey, M.D., Handzel, T and Venezel L. (2003). Chlorination and safe storage of household drinking water in developing countries to reduce waterborne disease. *Wat. Sci. Technol.* **47**(3), 221–228.

STATSSA (1998). The people of South Africa population census 1996. Statistics South Africa, Pretoria, South Africa.

STATSSA. (2003). Population census (2001). Statistics South Africa, Pretoria, South Africa.

STATSSA (2005). Mid-year population estimates P0302. Statistics South Africa, Pretoria, South Africa.

Tshibangu, N.N. (1987). Water supplies and sanitary facilities in rural Transkei. *S. Afr. Med. J.* **71** (6), 368–369.

WHO (2000). Cholera. Fact Sheet N107. World Health Organisation, Geneva.

The effect of fouling on concentration polarization in reverse osmosis at constant flux operation: NaCl tracer response technique

T.H. Chong*, J.B. Liew, F.S. Wong***, and A.G. Fane***

* School of Civil and Environmental Engineering, Nanyang Technological University, 50 Nanyang Avenue, 639798, Singapore (E-mail: *Q020001@ntu.edu.sg*; *AGFane@ntu.edu.sg*)
** School of Engineering, Ngee Ann Polytechnic, 535 Clementi Road, 599489, Singapore (E-mail: *s10006558@np.edu.sg*)
*** Institute of Environmental Science and Engineering, Innovation Centre Block 2 Unit 237, 18 Nanyang Drive, 637723, Singapore (E-mail: *wongfs@ntu.edu.sg*)

Abstract This paper presents a simple technique, the NaCl tracer response test, to determine the effect of fouling on concentration polarization level in a reverse osmosis system operated in the constant flux mode. Alginate was used as a model bio-foulant in this study. It was found that the formation of the 'unstirred' fouling layer greatly enhanced the concentration polarisation level in RO, with a more pronounced effect at high flux operation. The results suggested that operation at high flux can significantly reduce apparent permeability, but that this is through a dramatic increase in CP and consequent loss in driving force rather than through an additional hydraulic resistance. The study also shows that the 'salt' spiking response is a promising technique that can be applied as a non-invasive tool to detect and monitor fouling development in reverse osmosis.

Keyword Concentration polarisation; constant flux; fouling; NaCl tracer test; reverse osmosis

Introduction

Reverse osmosis (RO) is widely used in desalination and water reclamation to convert seawater or wastewater to potable quality water. However, membrane fouling is still a bottleneck in RO technology. Fouling is linked to concentration polarization, which in turn is controlled by flux and mass transfer. In an RO system, high cross flow or flow channel spacers are used to increase the mass transfer, thus reducing the polarization effect. However, in the event of formation of a fouling layer, which can be considered as an unstirred layer, the back diffusion of salt ions from the membrane surface to the bulk solution is hindered. Consequently, the concentration at the membrane surface is significantly increased and so is the osmotic pressure. The increase in osmotic pressure means a decrease in the effective transmembrane pressure. This is known as 'cake-enhanced osmotic pressure', which was introduced by Hoek *et al.* (2003) and was further studied by Ng and co-workers (2004). In their study, colloidal silica of uniform size was used to form a cake layer on the membrane in the presence of sodium chloride under constant pressure operation. The cake-enhanced osmotic pressure was analysed by a lengthy mathematical iteration process only applicable for particles with known size and a uniform cake layer. In our study, a rather simple NaCl tracer response technique is used to study the effect of fouling on concentration polarization in an RO system at constant flux operation. This technique makes use of the unique property of a RO membrane to reject NaCl salts. Tracer testing is a well established method which was commonly used to study the residence time distribution of a reactor (Levenspiel, 1972); and recently applied to detect fouling in a membrane module by comparing the arrival time and magnitude of the peak from a pulse-spike of a used membrane against a new membrane (Roth *et al.* 1999). Alginate, which has a broad species size distribution, was used as a model bio-foulant in our study. Alginate has been identified as an extracellular polymeric substance produced by *Pseudomonas* and it can play an important role in the formation of microbial biofilms (Nivens *et al.*, 2001).

Theory

Before the NaCl spike, flux can be expressed as

$$J_v = \frac{\left(TMP - CP \cdot \Delta\Pi_b\right)}{\eta\left(R_m + R_f\right)} \tag{1}$$

where CP is the concentration polarisation modulus, $(C_w - C_p)/(C_b - C_p)$. When NaCl is injected into the system, this causes an increase in the osmotic pressure. In order to maintain constant flux operation, the transmembrane pressure of the system has to be raised. Therefore,

$$J_v = \frac{\left(TMP_s - CP \cdot \Delta\Pi_{bs}\right)}{\eta\left(R_m + R_f\right)} \tag{2}$$

where $\Delta\Pi_{bs}$ is the osmotic pressure and TMP_s is the transmembrane pressure of the system during the spike. Combining the above two equations, yields

$$CP = \frac{\left(TMP_s - TMP\right)}{\left(\Delta\Pi_{bs} - \Delta\Pi_b\right)} \tag{3}$$

Therefore, by knowing the transmembrane pressure of the system as well as the conductivity of feed and permeate (conductivity can be easily converted to osmotic pressure) before and during the NaCl spike, CP can be determined.

Materials and methods

A commercial RO membrane (FilmTec, model BW30) was used in all the tests. The membrane came in 1 m × 1 m size and was cut to fit the RO cell. Membranes were stored dried in a refrigerator at 4 °C and soaked in Milli-Q water (Millipore) for 24 hours before use. Alginate, as sodium alginate (Hayashi Pure Chemical Industries) was used as the model bio-foulant and sodium chloride (Sigma Aldrich) was used as background ionic solution and for the tracer test. Milli-Q water was used to prepare the concentrated alginate and NaCl solutions and the solutions were filtered with a 0.45 μm filter (Millipore, model HAWP04700) before use.

The set up of the reverse osmosis system is shown in Figure 1. The RO cell has parallel plate geometry and a flow channel of 310 mm × 60 mm × 0.8 mm. The effective membrane area is 0.0186 m^2. Test solution was delivered from the 10 L feed tank using a high pressure pump (Cats Pump, model 277) to the RO cell. The temperature of the feed water was kept constant at 25 ± 1°C with cooling water from a chiller (AR Engineering, custom made). An overhead stirrer (IKA, model Eurostar) was installed at the feed tank to ensure the solution was well mixed. System pressure and crossflow rate was controlled by a back-pressure regulator (Veriflo Corporation, model ABP 1ST52BP4) and a flow control valve (Cole Parmer, model CP-32505-40) respectively. Pressures of the feed and permeate stream were monitored with pressure transducers (Bourdon Haenni, model E913); meanwhile flow rate of the stream was monitored with a turbine flow meter (McMillan, model 114). The amount of permeate withdrawn, which fixes the flux, was controlled with a mass flow controller (Brooks Instrument, model 5882), which consists of a thermal mass flowmeter and a micro control valve, located at the permeate side of the RO cell. To achieve constant flux operation, the RO system was operated at excess feed pressure so that the mass flow controller self adjusted the permeate side pressure to meet the set permeate flow rate; i.e. only a transmembrane pressure of 3 bar was required to obtain a flux of 10 L/m^2 h, but instead the applied pressure was set to 20 bar, so the mass flow controller regulated the permeate side pressure to 17 bar. The quality of feed and permeate was monitored with conductivity meters (Mettler Toledo, model Seven Multi and Rosemount Analytical, model Solu Comp II, respectively). During tests, concentrate and permeate were recycled back to the feed tank. For the sodium chloride tracer test, sodium chloride solution was injected into the high pressure feed line to the RO cell via a metering pump (Pulsafeeder, model Pulsatron). During the tracer test, concentrate and permeate samples were removed and feed solution was topped up in order to maintain a constant volume. Pressures, crossflow rate, flux, and conductivity readings were recorded with a data acquisition system (National Instrument, model PCI 6014 and LabView).

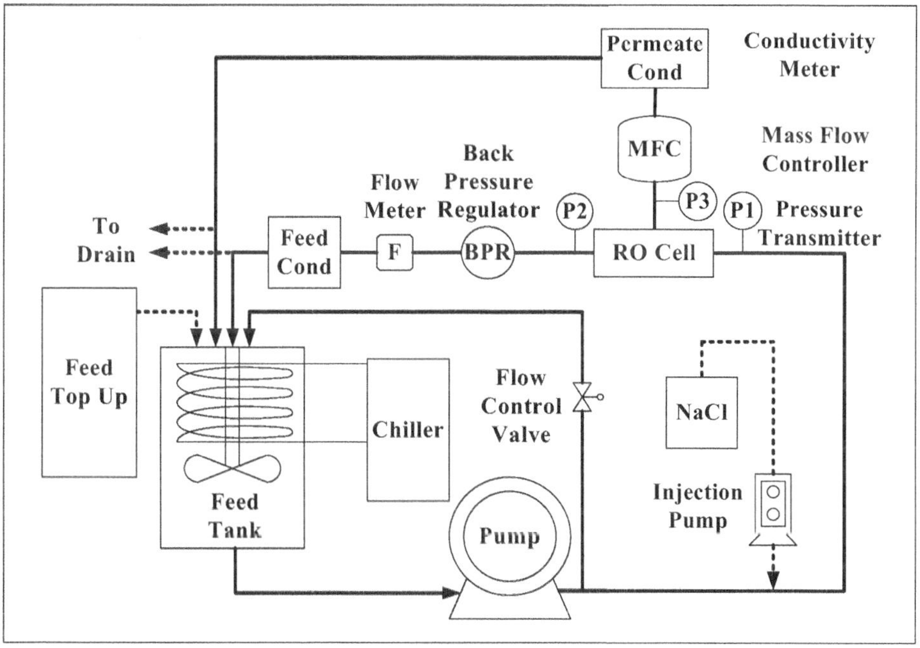

Figure 1 Schematic diagram of RO set up.

The experiment was commenced with the compression of the membrane at 20 bar and a crossflow rate of 0.5 L/min for 2 hours with Milli-Q water. Then, flux was set to the desired value, and the system was left to equilibrate for 12 hours. For test runs with NaCl in the background, this was followed by the addition of concentrated NaCl solution to achieve a final concentration of 2000 ppm NaCl and the system was left to equilibrate for another 2 hours. Subsequently, the tracer test was carried out and this was used as a reference point for the 'clean' membrane i.e. CP value determined here was taken as the reading at t = 0 for the fouling test. In the tracer test, 200 g/L of NaCl was injected into the high pressure feed line via the metering pump, to give an additional of 1000–1500 ppm NaCl to the feed solution entering the RO cell. The spike length lasted for 30 min. When the spiking was stopped, the system TMP and conductivity readings returned to the previous values and the system was left to equilibrate for 1 hour. Then the fouling test was initiated by adding sodium alginate solution to the feed tank at a final concentration of 200 ppm (60 ppm as TOC) and 20 ppm (6 ppm as TOC) for runs without and with background NaCl, respectively. The TOC reading was measured with Total Organic Carbon Analyser (Shimadzu, model TOC VHS-C). Several tracer tests were carried out during each fouling run. Samples were collected from the feed tank and the permeate line and analysed for TOC contents.

Results and discussion

Alginate fouling test was initially conducted without NaCl in the background. An example of the TMP, feed conductivity, and permeate conductivity profiles during one of the NaCl spikes is shown in Figure 2a–c, respectively. The CP values determined based on the TMP and conductivity profiles over a period of steady alginate fouling are shown in Figure 3. Only a slight change in the CP value was observed for a relatively low flux of 15 L/m^{2}h whereas the polarization effect at a higher flux of 25 L/m^{2}hr was greatly enhanced by the presence of the alginate layer on the membrane surface. Interestingly, instead of quickly reaching a steady value as shown in the data at flux of 15 L/m^2 h, the CP values continued to increase during the spike. The slope of the plot, d(CP)/dt, also increased as fouling progressed. A possible reason for this was due to the interaction between the Na^+ ions and the alginate layer on the membrane surface. Two models have been suggested: (i) mannuronate and isolated guluronate may complex with Na^+; (ii) Na^+ induced association between polyguluronate chain sequences was formed, which was described in terms of an 'egg-box' model in which there was regular packing and coordination of the cations in the interstices of associated and, extended segments of the polymer chains (Zhang *et al.*, 1998). The Na^+ ions would be slowly accumulated and 'entrapped' in the vicinity of the alginate layer and therefore cause a steady increase in osmotic pressure.

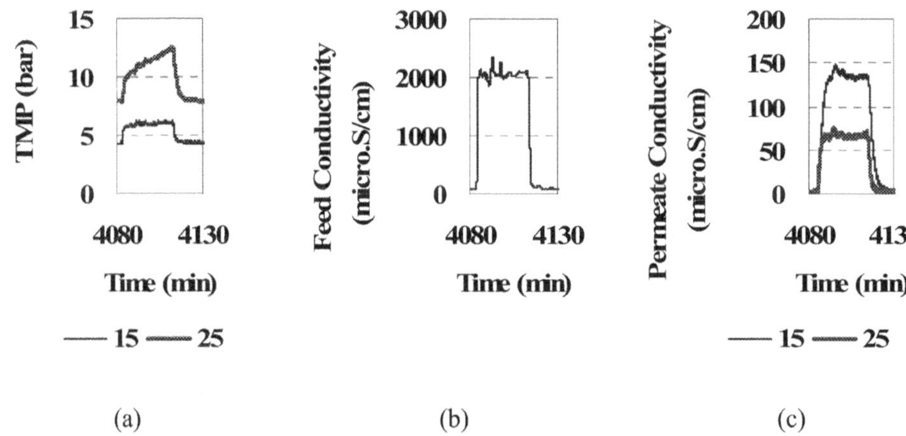

Figure 2 (a) TMP, (b) feed conductivity, and (c) permeate conductivity profile during one of the NaCl spikes in alginate fouling test without background sodium chloride.

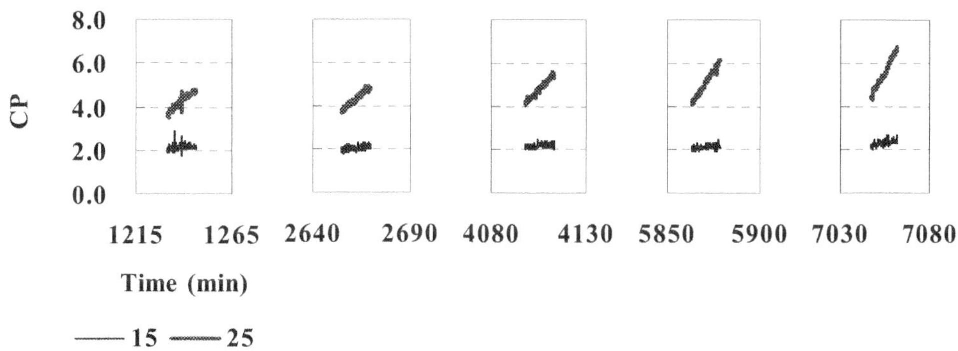

Figure 3 CP values determined from TMP and conductivity profile during one of the NaCl spikes in alginate fouling test without background sodium chloride.

The experiment was repeated with 2000 ppm of NaCl in the background. This would be a more typical condition for brackish RO. An example of the TMP, feed and permeate conductivity profiles during one of the NaCl spikes is shown in Figure 4a,b, respectively. Before the NaCl spike, the TMP was 10.3 bar at flux of 20 L/m^2 h. When NaCl was spiked into the system, the conductivity of the feed increased from 4310 to 7800 μS/cm. Owing to the increase in osmotic pressure, in order to maintain constant flux operation, a higher TMP was required; therefore the transmembrane pressure was increased to 13.1 bar. Meanwhile, the conductivity of the permeate increased due to higher salts passage. The CP values determined based on the TMP and conductivity profiles over a period of steady alginate fouling are shown in Figure 4c. It should be noted that the CP values quickly reached a steady reading during the spike, and did not continue to rise as in the test without NaCl in the background. After knowing the CP values, the fouling resistances of the alginate layer, R_f, can be obtained from Equation 2 and are shown in Figure 4d.

Experiments at different flux levels were conducted to investigate the extent of fouling on concentration polarization. The average CP changes due to alginate fouling at different flux levels are shown in Figure 5. The CP values increased linearly with time at all flux levels, except at a flux of 10 L/m^2 h, where the CP values remained constant throughout the test since insignificant fouling was occurring. The change in CP was more pronounced at high flux levels as shown in the plot of d(CP)/dt vs J_v (Figure 7). When comparing the change in R_f as plotted in Figure 6, the R_f values also increased linearly with time at all flux levels, except at the flux of 10 L/m^2 h. However, unlike the change in CP, there was negligible change in R_f when the flux was increased from 25 to 30 L/m^2 h. The results suggested that operation at high flux can significantly reduce apparent permeability, but that this is through a dramatic increase in CP and consequent loss in driving force rather than through an additional hydraulic resistance.

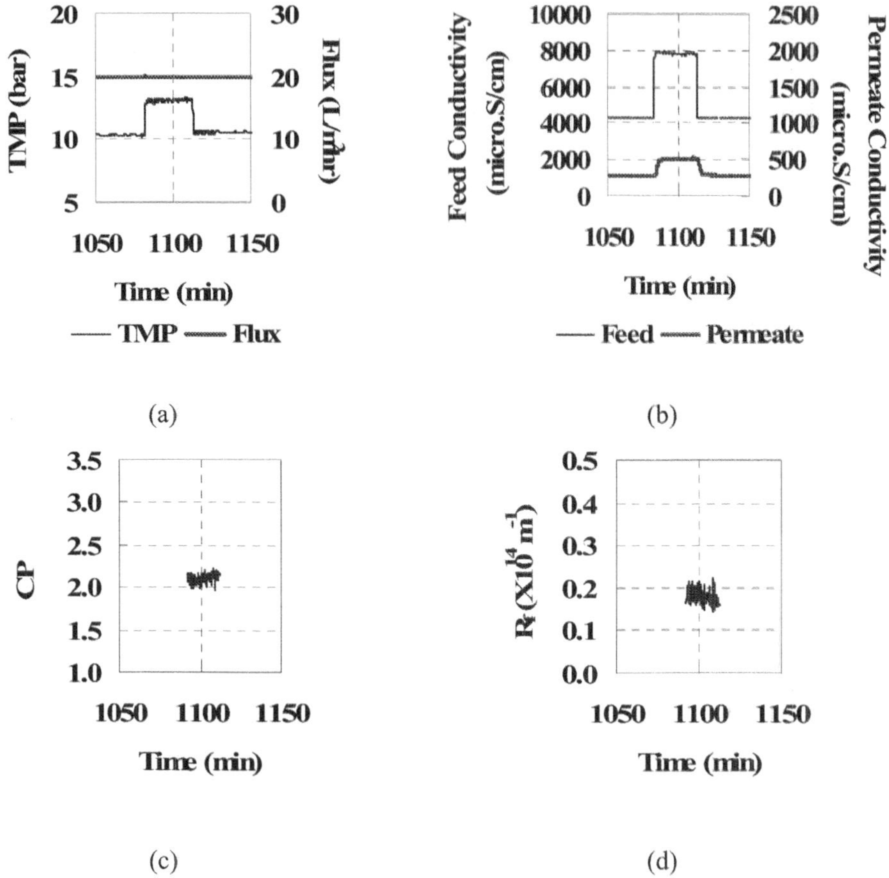

Figure 4 (a) TMP and flux, (b) conductivity of feed and permeate, (c) CP, and (d) R_f profile during one of the NaCl spikes in alginate fouling test with background 2000 ppm NaCl at J_v = 20 L/m^2 h.

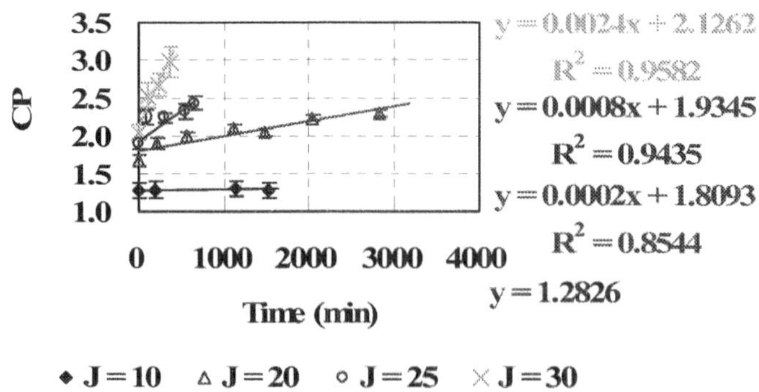

Figure 5 CP changes due to alginate fouling at different flux levels.

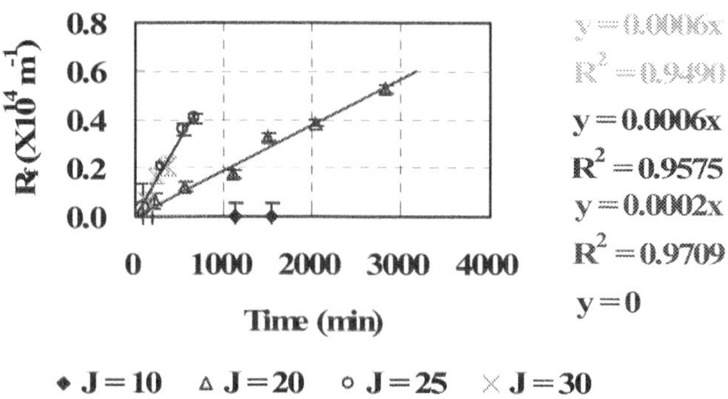

Figure 6 Fouling resistance at different flux levels.

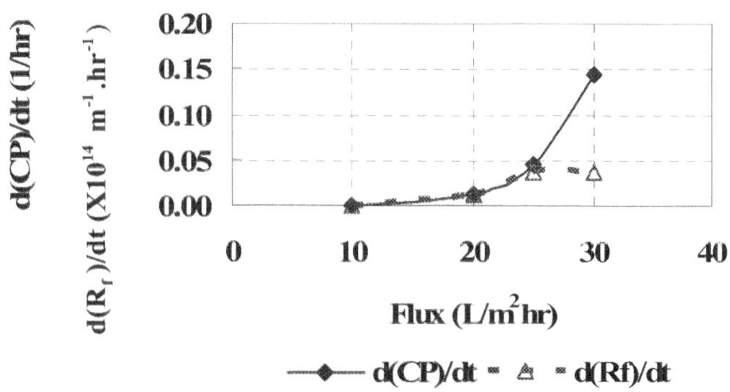

Figure 7 d(CP)/dt and d(R_f)/dt vs J_v.

Conclusions

The effect of fouling on concentration polarization in reverse osmosis at constant flux operation was studied by applying a NaCl tracer response technique. Alginate fouling tests with NaCl as background ionic solution showed that negligible fouling occurred at a low flux of 10 L/m^2 h and therefore no change in CP level was observed. Meanwhile at mid-range fluxes, both CP levels and fouling resistance increased at a similar rate. However, the increase in CP was more obvious at a high flux level. The results confirmed that at high flux, typical 'bio'fouling layers in RO could significantly reduce apparent permeability, but that this is through a dramatic increase in CP and consequent loss in driving force rather than through an additional hydraulic resistance. It is also suggested that the 'salt' spiking response can be used as a diagnostic for fouling phenomena as it can distinguish between hydraulic resistance and cake enhanced osmotic effects.

Acknowledgements

We acknowledge financial assistance from A*STAR (Agency for Science, Technology and Research, Singapore) through support for the Temasek Professor Programme in Membrane Technology for Sustainable Water.

References

Hoek, E. M. V. and Elimelech, M. (2003). Cake enhanced concentration polarization: A new fouling mechanism for salt-rejecting membranes. *Environ. Sci. Technol.*, **37**, 5581–5588.

Levenspiel, O. (1972). *Chemical reaction engineering*. New York, John Wiley & Sons.

Ng, H. Y. and Elimelech, M. (2004). Influence of colloidal fouling on rejection of trace organic contaminants by reverse osmosis. *J. Membr. Sci.*, **244**, 215–226.

Nivens, D. E., Ohman, D. E., Williams, J. and Franklin, M. J. (2001). Role of alginate and its O Acetylation in formation of *Pseudomonas aeruginosa* microcolonies and biofilms. *J. Bacteriology*, **183**(3), 1047–1057.

Roth, E., Kessler, M., Fabre, B. and Accary, A. (1999). Sodium chloride stimulus-response experiments in spiral wound reverse osmosis membranes: A new method to detect fouling. *Desalination*, **121**, 183–193.

Zhang, H., Zheng, H., Zhang, Q., Wang, J. and Konno, M. (1998). The interaction of sodium alginate with univalent cations. *Biopolymers*, **46**, 395–402.

Metal decontamination of soil and sediment by EDTA washing coupled with EDTA electrochemical regeneration/recycling

R. Cesaro* and G. Esposito**

* Department of Hydraulic and Environmental Engineering Girolamo Ippolito, University of Naples Federico II, Via Claudio, 21, 80125 – Naples, Italy (E-mail: *racesaro@unina.it*)
** Department of Mechanics, Structures and Environmental Engineering, University of Cassino, Via Di Biasio 43, 03043 – Cassino (FR), Italy (E-mail: *giovanni.esposito@unicas.it*)

Abstract The present research was aimed at investigating at a laboratory scale the feasibility of a solid washing process operated with electrochemically regenerated EDTA washing solutions for heavy metal extraction from natural contaminated soil (excavated from Bellolampo, Palermo, Italy) and River sediment (dredged from Sarno river, Naples, Italy). The use of fresh EDTA for the washing of Bellolampo soil gave Pb and Cu removal efficiencies of 70.0 and 32.0%, respectively. A decrease of the Pb removal efficiency from 70.0 to 62.5% occurred when using regenerated EDTA, owing to the limited performance of the EDTA regeneration process. A lower decrease of the Cu removal efficiency from 32.0 to 31.5 was observed as a consequence of a higher electrochemical EDTA separation from Cu in the electrolytic regeneration cell. An incomplete electrochemical regeneration of the washing solution resulted in a decrease of the heavy metal extraction efficiency from the Sarno sediment as well. Multi-step washing experiments carried out on Bellolampo solids using fresh and multi-regenerated EDTA gave comparable Pb and Cu removal efficiencies, indicating that the use of electrochemically regenerated EDTA is appropriate for multi-step soil washing.

Keywords Electrochemical; metals; recovery; sediment; soil; washing

Introduction

Heavy metals contamination, originated from industrial activities (i.e. metallurgy, mining, paint factories, military operation fields, battery-recycling facilities, etc.), involves about 50% of all contaminated sites suffering from a mixed organic/metals pollution and about 16% of sites containing only heavy metals (USEPA, 1997).

Soil remediation is becoming an important industrial activity both in Europe and the United States, with potential for expansion within the next 20 years illustrated, for instance, by the start-up of dozens of soil washing installations over the last decade in Europe (Vandevivere *et al.*, 2001). Soil and sediment washing is a physicochemical treatment in which excavated soil or dredged sediment are washed by means of Mass Separating Agent (MSA) in an aqueous solution such as acids or chelating agents (Van Benschoten *et al.*, 1997). The advantages of organic complexing agents over strong mineral acids are comparable removal efficiencies, biodegradability, and lower aggressiveness towards the solid matrix. However, the costs of the agents, which depend almost linearly on the demand of the MSA, are limiting factors for their use in technical applications (Holbein, 1994; Bradbury and Scrivens, 1995, Arevalo *et al.*, 2002).

Therefore the MSA regeneration/recycling is highly attractive for obtaining feasible treatment costs as it can drastically lower the MSA chemical consumption (Arevalo *et al.*, 2002). One effective way to achieve this is the organic complexing agent regeneration by means of an electrochemical process (Scott, 1995; Thöming *et al.*, 2000; Arevalo *et al.*, 2002; Hansen *et al.*, 1997).

However, the efficiency of a soil washing process coupled with MSA electrochemical regeneration is limited by the MSA drag-out due to the treated solid leaving the process, which makes not possible a closed loop operation (Arevalo *et al.*, 2002). Besides this, the process efficiency depends on: (i) the characteristics of the solid; (ii) the MSA applied; (iii) the type and extent of metal contamination; (iv) the operational conditions of the washing process and (v) the efficiency of the electrochemical MSA recovery.

The present research was aimed at investigating at a laboratory scale the feasibility of a solid washing process operated with electrochemically regenerated EDTA washing solutions as MSA for heavy metal extraction from natural contaminated soil (excavated from Bellolampo, Palermo, Italy) and River sediment (dredged from Sarno river, Naples, Italy). The efficiency of a multi-step extraction of heavy metals from Bellolampo soil using multi-regenerated EDTA washing solutions was also assessed.

Methods

Experimental set-up

Sediment washing

Sediment washing tests were carried out at laboratory temperature (22 ± 2 °C) using a Jar Test equipment (model JF/6, ISCO, Italy) with a mixing rate of 150 rpm for the continuous mechanical mixing of 1l samples.

Milli-Q water and EDTA acid were used to prepare the EDTA 0.05 M (unless specified otherwise) washing solution. The latter was added with 25 g of the investigated soil or sediment (all solid samples were dry weight based) to reach a solids concentration of 2.5%. Solids used in the experiments were as follows:

- Soil with a very high Pb (12893.0 mg kg^{-1}) and Cu (59528.0 mg kg^{-1}) contents excavated close to a landfill in Bellolampo, Palermo, Italy;
- Sediment with a discrete Pb (141.8 mg kg^{-1}) and Zn (350.3 mg kg^{-1}) contents dredged from Sarno River in the south of Italy.

During the extraction process 5 ml samples were taken at various times (see the experimental design) and pre-filtered with a 0.45 μm membrane filter before metal analyses. After the extraction process the washing solution was separated from the solid sample by means of sedimentation (1h) and electrochemically regenerated.

Electrochemical regeneration of the washing solution

The electrolytic regeneration of the washing solution was performed batchwise in a laboratory scale divided electrolysis cell as proposed by Arevalo *et al.* (2002). The volume of the cathodic and anodic chambers was 50 ml each; they were divided by a cation exchange membrane NAFION* 450 (DuPont). A three dimensional cathode was used consisting of a 45 mm × 70 mm × 2 mm graphite plate and a fixed bed of 2 mm graphite particles (35 g). The anode was a Ti/RuO2 mesh plate (DeNora Deutschland GmbH).

The EDTA solution was pumped upflow through the cathodic chamber in a flow-by cell configuration (Fig. 1), where heavy metals are reduced and deposited on the surface of the graphite. The solution is collected in a stirred vessel and recirculated through the cell. A 50 ml stagnant sulphuric add solution (0.2 M) was used as anolyte.

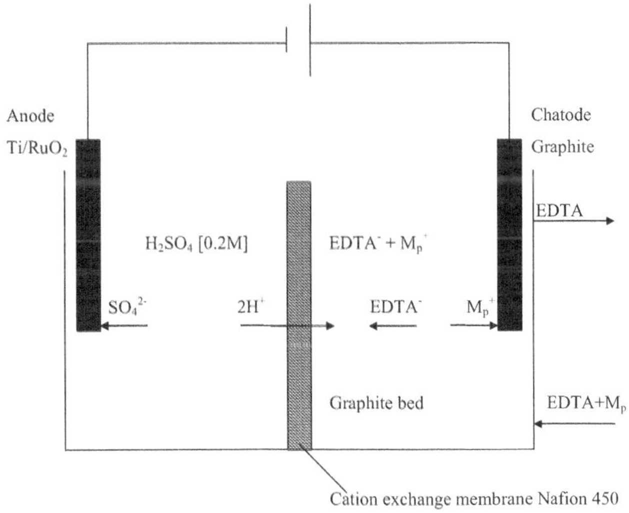

Figure 1 Schematic representation of the electrolytic regeneration process.

Experimental design

Effect of pH and EDTA dosage on the washing process of Bellolampo soil and Sarno sediment

The effect of the initial pH of the EDTA solution on the washing process of Bellolampo soil and Sarno sediment was investigated in the range 2.0–6.3 using 0.01 and 0.05 M EDTA solutions (Table 1). HCl and NaOH were used to reach the pH set-point before solids addition. Washing tests were run for 72 h taking 5 ml samples at 1, 3, 6, 24, 48 and 72 h.

Table 1 EDTA concentration and pH of the washing solution used in preliminary experiments

EDTA concentration (M)	pH	Sarno sediment	Bellolampo soil
0.01	2.0	X	X
0.01	3.0	X	
0.01	6.3*	X	X
0.05	2.0	X	X
0.05	3.0	X	
0.05	5.8*	X	X

*No addition of HCl and NaOH.

Comparison between fresh and regenerated EDTA for the washing of Bellolampo soil and Sarno sediment

Washing tests were carried out in order to assess the capability of regenerated EDTA solutions to treat Bellolampo soil and Sarno sediment. The treatment procedure consisted of three extraction processes and two electrochemical EDTA regenerations as reported in Figure 2A. A fresh 0.05 M EDTA solution, pH 5.8, was used as initial washing solution and new solid samples were used for each extraction process. The washing process was run for 24 h.

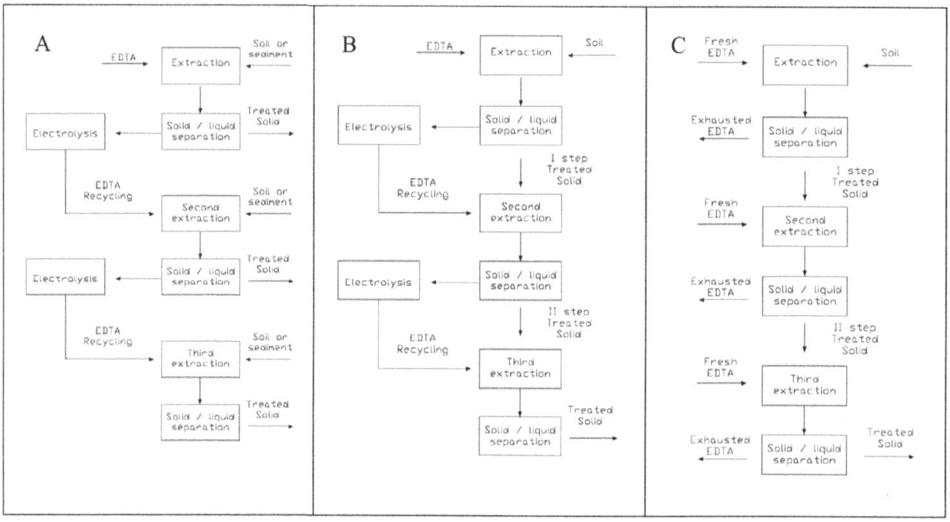

Figure 2 Treatment procedures applied for the washing of fresh Bellolampo soil and Sarno sediment using multi-regenerated EDTA (A) and multi-step washing of Bellolampo Soil using multi-regenerated (B) or fresh (C) EDTA.

Use of multi-regenerated EDTA for multi-step washing of Bellolampo Soil

Further 24 h washing experiments were carried out in order to assess the feasibility of multi-step washing to increase the heavy metal removal from Bellolampo soil. The treatment procedure was still consisting of 3 extraction processes and 2 electrochemical MSA regenerations (Fig. 2B) but the washing processes were applied to the same sample of solids. A fresh 0.05 M EDTA solution, pH 5.8, was used as initial washing solution.

Use of fresh EDTA for multi-step washing of Bellolampo Soil

Control experiments were carried out to compare the use of fresh and multi-regenerated EDTA for multi-step washing of Bellolampo Soil. The treatment procedure was the same as above described but a fresh EDTA washing solution (0.05 M, pH 5.8) was used for all extraction processes (Fig. 2C).

Heavy metal analyses

Liquid samples were filtered with a 0.45 μm membrane filter and analysed by means of a Parkin-Elmer Zeeman/3030 Flame Atomic Absorption Spectrophotometer (FAAS) to measure the concentrations of Cu, Pb and Zn.

Solid samples (500 mg) were dried at 106°C for 24 h and added with 9 ml of 65% HNO_3 and 3 ml of 50% HF in high-temperature, high-pressure reinforced Teflon vessels before digestion with aqua regia in a microwave oven (MARS5-CEM, NC, USA) for 3 h. The liquid samples from the microwave oven were then analysed for metals by FAAS.

Results and discussion

Effect of pH and EDTA dosage on the washing process of Bellolampo soil and Sarno sediment

Preliminarily washing tests were carried out on Bellolampo soil and Sarno sediment to optimise the operational conditions of the extraction process in terms of initial pH and EDTA concentration of the washing solution. Experiments were run for 72 h but 24 h were shown to be sufficient to reach the maximum heavy metal removal efficiencies (data not shown).

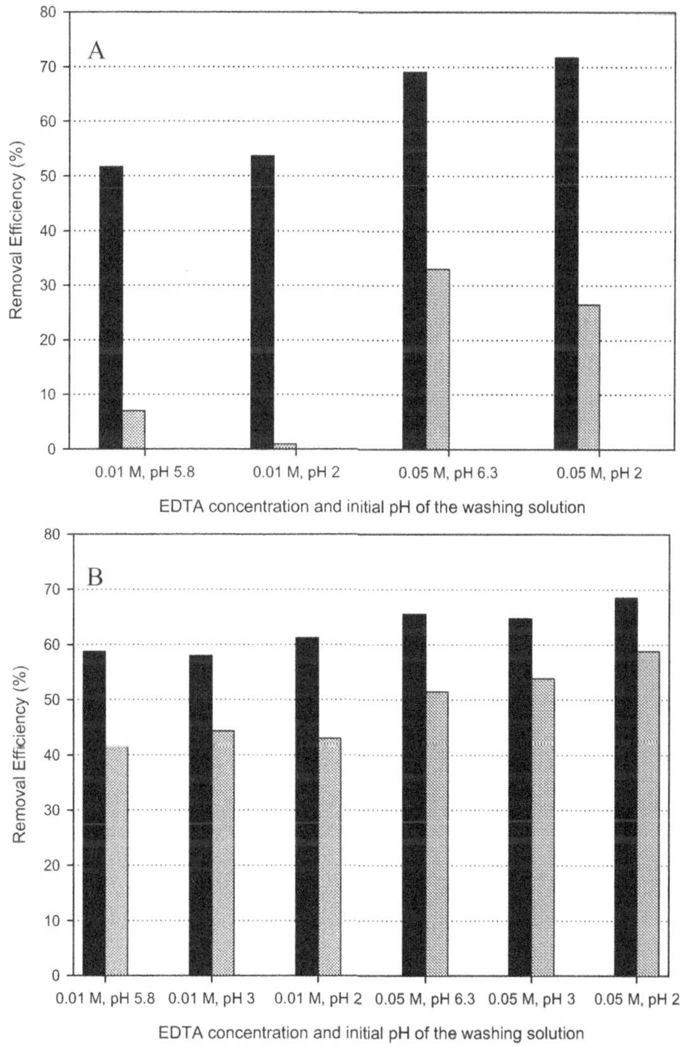

Figure 3 Effect of the operational conditions on the efficiency of Pb (■) and Cu (▒) extraction from Bellolampo soil (A) and Pb (■) and Zn (▒) extraction from Sarno sediment (B).

The results (Fig. 3) show that the Pb removal efficiency ranged between 51.7 and 71.7 and between 58.0 and 68.6%, respectively, for Bellolampo soil (Fig. 3A) and Sarno sediment (Fig. 3B). The removal efficiencies of Cu from Bellolampo soil (Fig. 3A) and Zn from Sarno sediment (Fig. 3B) ranged between 0.9 and 33.0 and between 41.4 and 58.9%, respectively. The initial pH of the washing solution hardly affected the metal removal efficiency (Fig. 3). This is due to the high acid neutralising capacity of both Bellolampo soil and Sarno sediment, as indicated by a fast increase of the pH of the washing solution to neutral levels through the experiments (data not shown).

Increasing the EDTA concentration from 0.01 M to 0.05 M resulted in an important increase of both Pb and Cu removal efficiencies from Bellolampo soil (Fig. 3A). A slighter effect of the EDTA dosage was shown on Pb and Zn removal from Sarno sediment. This is probably due to the much higher heavy metal contamination of Bellolampo soil if compared with Sarno sediment, which implies a higher EDTA demand. In fact EDTA, such as other chelants (e.g. EDDS, NTA), is known to form 1:1 chelates with metals, thus the theoretical minimum chelant dosage is the one that binds the metals stoichiometrically. However, a several-fold excess of chelant is usually needed to maximize metal extraction (Linn and Elliott, 1988). This might explain why a washing solution with an EDTA concentration (0.01 M) in excess of the metal concentration in Bellolampo soil was not sufficient to maximize the metal removal efficiency.

Comparison between fresh and regenerated EDTA for the washing of Bellolampo soil and Sarno sediment

Three sediment washing cycles were applied to fresh solids from Bellolampo soil and Sarno sediment using fresh EDTA for the first cycle and regenerated EDTA for the following cycles (Fig. 2A). Washing experiments were run for 24 h, which was shown to be the optimal treatment length in previous tests.

The use of fresh EDTA for the washing of Bellolampo soil gave Pb and Cu removal efficiencies of 70 and 32%, respectively (Table 2). A decrease of the Pb removal efficiency occurred throughout the washing cycles when using regenerated EDTA (Table 2) due to the limited performance of the EDTA regeneration process which did not reach a 100% efficiency (Fig. 4). The decrease of the Cu removal efficiency throughout the washing cycles was much lower (Table 2) as a consequence of the higher electrochemical EDTA separation from Cu in the electrolytic regeneration cell if compared with separation from Pb (Fig. 4).

Table 2 Heavy metal extraction from fresh Bellolampo solids using fresh EDTA (Cycle I) and regenerated EDTA (Cycles II and III) washing solutions

	Pb		Cu	
	Residual concentration (mg kg^{-1} TS)	Removal efficiency (%)	Residual concentration (mg kg^{-1} TS)	Removal efficiency (%)
Cycle I	3867.9	70.0	40479.0	32.0
Cycle II	4834.9	62.5	40776.7	31.5
Cycle III	7091.2	45.0	41431.5	30.4

An incomplete electrochemical regeneration of the washing solution (Fig. 5) resulted in a decrease of the heavy metal extraction efficiency throughout the washing cycles from the Sarno sediment as well (Table 3).

Furthermore, Figure 5 indicates a higher Pb recovery efficiency in the electrolytic cell if compared with Zn when the electrochemical regeneration process was applied to the Sarno sediment washing solution. Indeed, the feasibility of metal electro-deposition in the cathode is determined by several factors, such as the presence of other metals, the chemical species of the dissolved metals, their concentration, their standard redox potential, etc. (Arevalo *et al.*, 2002), which differently affect the recovery efficiency of different metals.

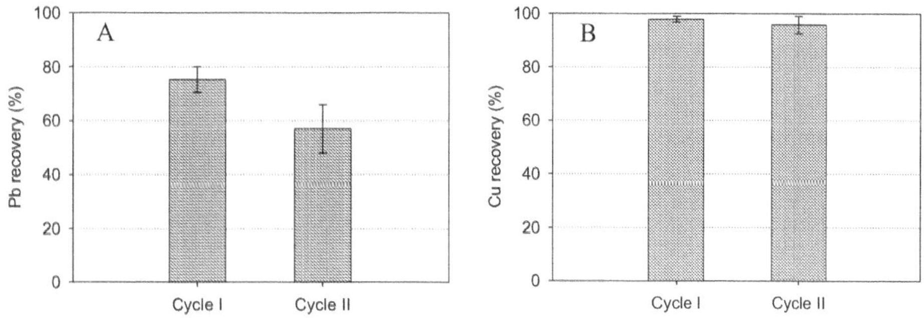

Figure 4 Efficiency of Pb (A) and Cu (B) electrochemical recovery from EDTA washing solutions applied to the treatment of Bellolampo soil.

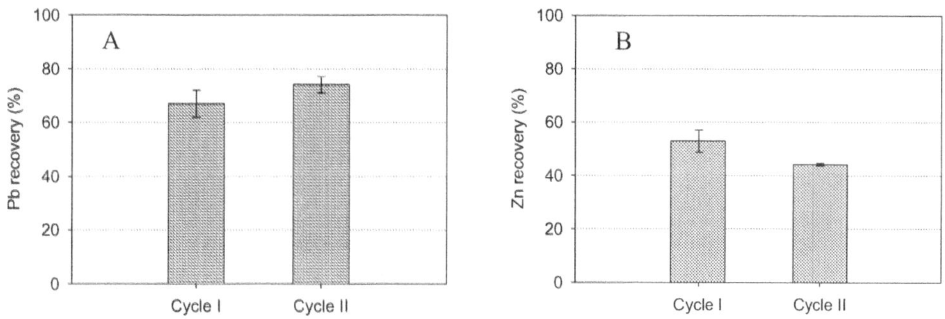

Figure 5 Efficiency of Pb (A) and Zn (B) electrochemical recovery from EDTA washing solutions applied to the treatment of Sarno sediment.

Table 3 Heavy metal extraction from fresh Sarno solids using fresh EDTA (Cycle I) and regenerated EDTA (Cycles II and III) washing solutions

	Pb		Zn	
	Residual concentration (mg kg^{-1} TS)	Removal efficiency (%)	Residual concentration (mg kg^{-1} TS)	Removal efficiency (%)
Cycle I	46.7	67.1	169.9	51.5
Cycle II	52.0	63.3	265.2	24.3
Cycle III	59.0	58.4	279.2	20.3

Use of multi-regenerated EDTA for multi-step washing of Bellolampo Soil

Multi-step washing experiments were carried out on Bellolampo solids using regenerated EDTA as MSA according to Fig. 2B. The results show total Pb and Cu removal efficiencies over three washing cycles (Fig. 2B) of 76 and 60 %, respectively (Table 4). Fresh EDTA used in control experiments (Fig. 2C) gave comparable removal efficiencies throughout the washing cycles (Table 4), indicating that the electrochemical regeneration of EDTA is a suitable recovery process of the extracting agent for multi-step soil washing.

Table 4 Heavy metal removal efficiency in multi-step washing of Bellolampo soil

Treatment procedure	Metal	Removal efficiency (%)			
		Cycle I	Cycle II	Cycle III	Total
Multi-step washing (Fig. 2B)	Pb	62	10	4	76
Multi-step washing (Fig. 2B)	Cu	30	17	13	60
Control (Fig. 2C)	Pb	62	12	5	79
Control (Fig. 2C)	Cu	30	16	15	61

Conclusions

- EDTA washing coupled with EDTA electrochemical regeneration/recycling is a suitable process to reduce the complexing agent consumption in the heavy metal decontamination of soil and sediment.
- The efficiency of the electrochemical EDTA recovery affects the efficiency of metal extraction from soil and sediment.
- The use of multi-regenerated EDTA is appropriate for multi-step soil washing.

References

Arevalo, E., Stichnothe, H., Thöming, J. and Calmano, W. (2002). Evaluation of a leaching process coupled with regeneration/recycling of the extractant for treatment of heavy metal contaminated soil. *Environmental Technology*, **23**, 571–581.

Bradbury, D. and Scrivens, S. (1995). Remediation of radioactive and heavy metal contaminated soils and groundwater using the ACT*DE*CON and MAG*SEPSM processes. Proc., Contaminated Soil '95, W. J. van den Brink, R. Bosnian and F. Arendt (eds.), Kluwer, Dordrecht, The Netherlands.

Hansen, H. K., Ottosen, L. M., Kliem, B. K. and Villumsen, A. (1997). Electrodialytic remediation of soils polluted with Cu, Cr, Hg, Pb, and Zn. *Journal of Chemical Technology and Biotechnology*, **70**, 67–73.

Holbein, B. (1994). Vitrokele™ process successfully removes metals from contaminated soils. *The Hazardous Waste Consultant*, **12**, 1.10–1.13 (May/June).

Linn, J. H. and Elliott, H. A. (1988). Mobilization of Cu and Zn in contaminated soil by NTA. *Water Air Soil Pollution*, **37**, 449–458.

Scott, K. (1995). *Electrochemical Processes far Clean Technology*, 1st edn, The Royal Society of Chemistry, London, United Kingdom.

Thöming, J., Stichnothe, H., Mangold, S. and Calmano, W. (2000). Hydrometallurgical approach to soil remedetion – process optimization applying heavy metal speciation. *Land Contamination and Reclamation*, **8**, 19–31.

U.S. Environmental Protection Agency (1997). *Recent developments for in-situ treatment of metal contaminated soils*, Report Contract 68-W-0055, Office of Solid Waste and Emergency Response and Technology Innovation Office, Washington, D.C., USA.

Van Benschoten, J. E., Matsumoto, M. R. and Young, W. H. (1997). Evaluation and analysis of soil washing for seven lead-contaminated soils. *Journal of Environmental Engineering*, **123**(3), 217–224.

Vandevivere P., Hammes F., Verstraete W., Feijtel T. and Schowanek D. (2001). Metal Decontamination of soil, sediment and sewage sludge by means of transition metal chelant [S, S] – EDDS. *Journal of Environmental Engineering*, **127**(9), 802–811.

Pilot engineering of an Anammox-SBR: from *de novo* enrichment to online process control

T. Gaul, I. Wesoly, N. Weinobst and S. Kunst

University of Hannover, Environmental Biotechnology Group, Institute of Water Quality and Waste Management (ISAH), Welfengarten 1, 30167 Hannover, Germany
(E-mail: *tobias.gaul@environmental-biotechnology.net*; *wesoly@biop.uni-hannover.de*; *nele.weinobst@o2online.de*; *kunst@isah.uni-hannover.de*)

Abstract The microbial process of anaerobic ammonium oxidation (Anammox) allows for autotrophic nitrogen removal from concentrated wastewater streams with unfavourable C/N-ratio at low oxygen demand and without an additional organic electron donor. Since there is no full-scale installation available to serve as a source of biomass for inoculation, strategies for *de novo* enrichment of Anammox organisms are important. Sensitivity of the organisms to inhibition requires a proper process control. In this study, it could be shown that an Anammox sequencing batch reactor can be started up from scratch under realistic conditions with wastewater from anaerobic sludge dewatering within three months. A cycle strategy of feeding the influent at once resulted in the enrichment of biomass showing low sensitvity to nitrite inhibition. The results suggest that selection between inhibition-prone Anammox organisms like *Candidatus* 'Brocadia Anammoxidans' and the less sensitive *Candidatus* 'Kuenenia stuttgartiensis' is governed by the availability of nitrite. It could be demonstrated that the process can be easily monitored by standard online parameters. Conductivity, pH and oxidation reduction potential are well suited for reliable, automated real-time process control.

Keywords Autotrophic nitrogen removal; anaerobic ammonium oxidation; Anammox ecology; SBR; online process control; operation strategy

Introduction

The anaerobic ammonium oxidation (Anammox) is a microbial pathway catalyzing the oxidation of ammonium with nitrite as electron acceptor to dinitrogen gas in the absence of oxygen, with the intermediates hydrazine and hydroxylamine. Initially described and enriched in The Netherlands more than a decade ago, it has gained increasing interest during the last years, as it is very likely to play a key role in the global nitrogen cycle and appears to be a promising option for technological processes for nitrogen removal from wastewater (see Kartal *et al.* 2004 and Jetten *et al.* 2005 for a review).

The stoichiometry of the Anammox reaction including biomass formation as found experimentally (Strous *et al.* 1998) reads:

$$1\ NH_4^+ + 1.32\ NO_2^- + 0.066\ HCO_3^- + 0.13\ H^+ \rightarrow 1.02\ N_2\uparrow + 0.26\ NO_3^- + 0.066\ CH_2O_{0,5}N_{0,15} + 2.03\ H_2O \quad (1)$$

Ammonium and nitrite are removed without any organic electron donor. This offers interesting possibilities for nitrogen removal from wastewater with high NH_4-N concentration and low C/N-ratio, which are hardly suitable to be treated by conventional nitrification/denitrification systems, such as sludge digestor supernatant, landfill leachate or certain industrial wastewaters. If combined with conventional aerobic ammonium oxidation to supply NO_2-N, a completely autotrophic N-removal is possible, which results in reduced oxygen demand without any need of an additional carbon source. Only two of the Anammox organisms detected in wastewater treatment systems have been enriched and physiologically characterized so far: *Candidatus* 'Brocadia Anammoxidans' (Strous *et al.* 1999a,b) and Candidatus 'Kuenenia stuttgartiensis' (Schmid *et al.* 2000, Egli *et al.* 2001).

At present, there are only few reports on successful start-up of Anammox reactors from scratch under realistic conditions (Imajo *et al.* 2004); there is no full scale installation available yet that may serve as a biomass source to inoculate new systems to accelerate start-up. Current Anammox research in wastewater treatment is mainly based on studies with synthetic wastewater or on reactors inoculated with Anammox biomass from other lab-scale systems. Furthermore, nitrite does not only act as electron acceptor for the Anammox reaction, but also exhibited strong inhibitory up to lethal effects on the organisms, if certain threshold concentrations were exceeded, which requires a proper process control.

There were two objectives in this study: (a) Testing a start-up strategy to enrich Anammox biomass from scratch under realistic conditions without synthetic media and 'laboratory tricks' only feasible in small scale; and (b) developing an easy-to-operate control strategy for Anammox reactors to allow for successful technical application.

Methods

SBR lab-scale plant and wastewater composition

The enrichment experiments were carried out in two identical sequencing batch reactors (SBR) made of glass, with a total volume of 12 l (working volume 10 l), equipped with three impellers for gentle mixing (120 rpm) and a lid of stainless steel bearing fittings for introducing probes and tubes for influent and effluent. The vessels were not gas-tight. The simple operation cycle consisted of 1) settling (10 min.), 2) withdrawal (2 min.) 3) filling (3 min.) and 4) reaction (345 min.) resulting in total cycle time of 6 hours. Influent and effluent were conveyed by peristaltic pumps and the cycle time was controlled by a simple clock timer. The whole wastewater load was fed within the 3 minutes filling period at the beginning of the cycle (feed-at-once). The hydraulic retention time was fixed at 2.5 days.

The enrichment SBR was inoculated with 2 kg of UASB granules from an industrial wastewater treatment plant (WWTP) (as suggested by Imajo *et al.* 2004) and with activated sludge of a nitrifying/denitrifying municipal WWTP (sludge age 10–12 days). No initial Anammox activity could be detected in the inocula. The second SBR was inoculated with homogenized biofilm sampled from a rotating disc reactor treating landfill leachate, which definitely contained Anammox organisms (Helmer-Madhok *et al.* 2002). This SBR served as a control reactor, to verify whether the applied enrichment strategy was successful. Influent and effluent were sampled 3–5 times a week; at least once a week the complete reaction phase was sampled every 30–60 min. to calculate the actual conversion rates.

The influent wastewater was typical sludge digestor supernatant (NH_4-N-rich reject water from sludge dewatering after anaerobic digestion) from a municipal WWTP (> 500,000 P.E.). The remaining COD was non-degradable in WWTPs as well as in the Anammox reactors described in this study. To vary the influent concentration to the pilot reactors, the supernatant was diluted with tap water during start-up. The main characteristics of the raw wastewater are summarized in Table 1. Nitrite was added in appropriate concentration from a concentrated $NaNO_2$ stock solution.

Table 1 Average characteristics of undiluted sludge digestor supernatant (average of one year, 60 samples. Mean and standard deviation)

pH	Alkalinity as CO_3^{2-} (mmol/l)	NH_4-N (mg/l)	PO_4-P (mg/l)	Kjeldahl-N total (mg/l)	Kjeldahl-N dissolved (mg/l)	COD soluble (mg/l)
7.7 ± 0.3	67.9 ± 6.0	839 ± 165	31.4 ± 3.5	981 ± 71.4	978 ± 70	385 ± 76

Batch tests, analytical methods and online data

To characterize the biomass with regard to favourable environmental conditions, short-term batch tests (3–4 hours) were carried out to determine the conversion rates of NH_4-N, NO_2-N and NO_3-N. To ensure comparable conditions, biomass was sampled, centrifuged and resuspended in a standard HCO_3^- - based, COD-free medium with trace elements as described by Egli *et al.* 2001, except the pH-tests, where a TRIS/HCl buffer was used. All tests were run in 110 ml glass bottles after repeated evacuation and reflushing with Ar/CO_2, sealed with gas-tight rubber septa, and shaken in a water bath at 30 °C.

Ammonium, nitrite, nitrate, ortho-phosphate and COD were determined spectrophotometrically using commercially available ready-to-use test kits (LCK series, Hach Lange GmbH, Duesseldorf, Germany); the hydrazine test was obtained from Merck (Spectroquant®). Dissolved oxygen concentration (DO) and pH were measured with standard probes (WTW, Weinheim, Germany). Samples were filtered using 0.45 μm nylon membrane filters. All chemicals used were of analytical grade.

Conductivity, pH and oxidation reduction potential (ORP, Ag/AgCl type) data were recorded with standard probes every two minutes and were conveyed via transducers and a bus-system (all Greisinger electronic GmbH, Regenstauf, Germany) to a standard PC for data storage. Probes were flushed daily with distilled water during the settling period of the reactors, and re-calibrated once a week.

Results and discussion

De novo enrichment of Anammox biomass

Both SBRs were operated identically at 30 °C and a hydraulic retention time of 2.5 days. Feeding strategy and influent composition were the same for both reactors (diluted sludge digestor supernatant supplemented with nitrite). Although no sparging with inert gases was applied, the O_2 concentration within both reactors was below the detection limit. So the apparent key selection factors favourable for Anammox organisms were given: Simultaneous availability of ammonium and nitrite and absence of dissolved oxygen.

Anammox activity was detected first in the control reactor, which had been inoculated with already Anammox containing biomass, after 50 days (not shown). This indicated that the operation strategy applied was principally suitable for Anammox enrichment.

In the reactor used for *de novo* enrichment, the first weeks of operation were dominated by endogenous processes. Decaying biomass served as electron donor for nitrite denitrification, which resulted in an additional NH_4-N release: Up to day 60, nitrite consumption and a slight simultaneous ammonium production were observed (Figure 1a,b). After depletion of this internal C-source, heterotrophic denitrification was no longer possible due to a lack of electron donors. An increase in nitrate concentration was measured (Figure 1c), together with simultaneous removal of ammonium and nitrite which is specific for the Anammox reaction as depicted in Equation (1). As there is no other reaction known to produce nitrate under anoxic conditions and a nitrate production rules out significant presence of denitrifiers, increasing nitrate concentration in the effluent of an Anammox reactor during start-up may be a good indicator for successful enrichment.

Further balancing of the SBR cycles revealed a subsequent, approximately exponential, increase of Anammox activity, which is equal to the net growth rate of the organisms (Figure 1d). A true exponential net growth rate could not be expected, as the biomass was subjected to additional losses by regular sampling.

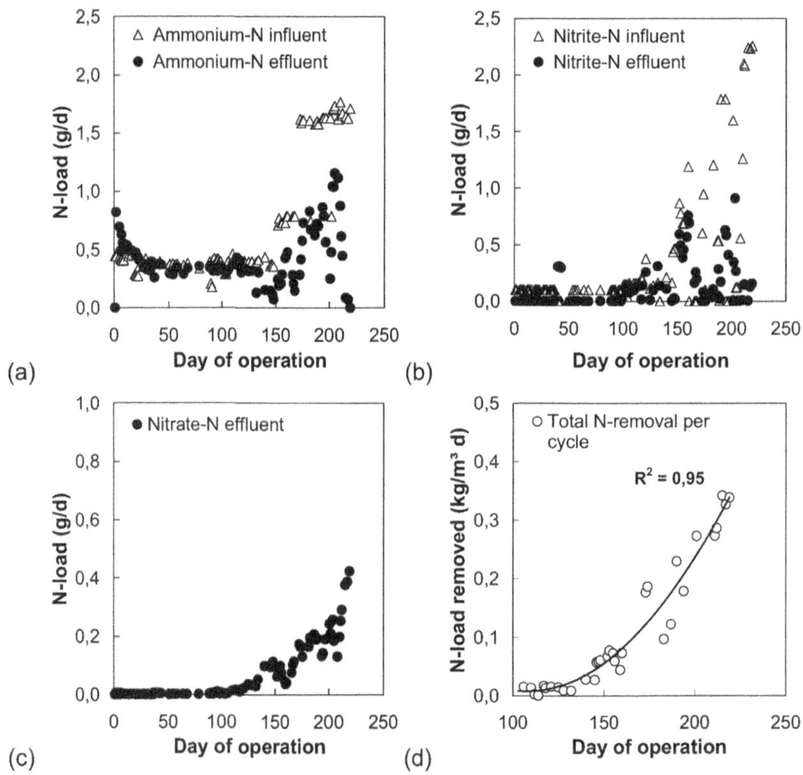

Figure 1 Start-up of the Anammox-SBR: (a)–(c) daily influent/effluent loads of nitrogen compounds; (d) increase of Anammox activity after 100 days of operation.

An orientating population analysis of the biomass by fluorescence *in situ* hybridization (FISH) was carried out on day 230 (Schmider 2004, data not shown). Approx. 25–50% of the cells detected by a probe mix specific for most Eubacteria also hybridized with probe Kst157 (Schmid *et al.* 2001) indicating the presence of 'Kuenenia' like Anammox organisms. No positive signal was observed with probe Ban162 (Schmid *et al.* 2001) specific for *Candidatus* 'Brocadia Anammoxidans'. The same observation was made with biomass from the inoculated control reactor, although the share of Kst157-positive bacteria appeared to be rather 50% than less. This suggests that the identical operation strategy of feeding at once and creating a substrate gradient in both reactors resulted in a selective advantage for these types of Anammox bacteria.

Moreover, the use of fast settling UASB granules as inoculum may improve biomass retention, as they serve as growth carriers for Anammox organisms. Because the granules are retained in the reactor (i.e. within an environment favourable for Anammox), they offer a specific habitat for these organisms. Attachment to the granules may thus be a mechanism to withstand the hydraulic pressure of being washed out during effluent withdrawal.

Physiological characterization

Short-term batch tests (3–4 hours) of small volume were carried out to investigate the influence of different environmental parameters (temperature, pH, substrate concentration, inhibitory effects) on the actual specific Anammox conversion rate. A synthetic medium was used to create identical conditions besides the single parameter under study. An initial concentration of 70 mg/l NH_4-N and 140 mg/l NO_2-N was set as a reference value, as pre-testing had shown that the highest conversion rate could be measured under these conditions. All other rates were compared to this 100% value. No difference could be detected between the SBR for *de novo* enrichment and the control reactor. Therefore, it will be only referred to the first below.

The characteristics obtained were in many respects quite similar to results already published (reversible inhibition by oxygen, stoichiometry of the conversion under non-limiting conditions). However, the biomass showed a shift in reaction stoichiometry under elevated nitrite concentration (Figurc 2).

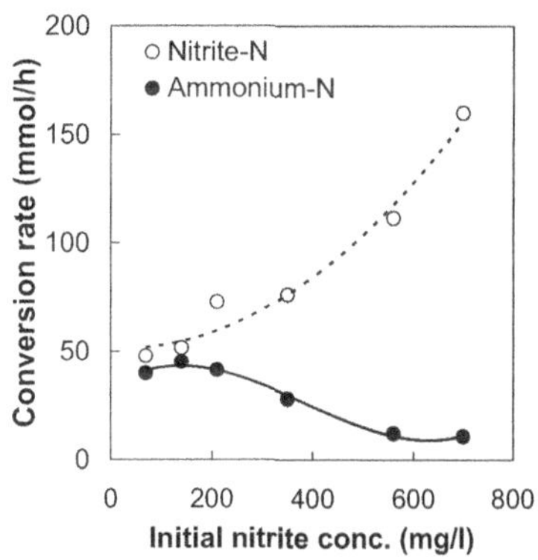

Figure 2 Conversion rates of NH_4-N and NO_2-N under different initial NO_2-N concentrations in anoxic short-term batch tests with the enriched Anammox biomass.

A similar effect has been described by Strous *et al.* 1999b and was hypothesized to be due to an internal electron donor used for nitrite denitrification. Their observation can be confirmed for the enrichment investigated here, although an explanation is still missing. ^{15}N labelling experiments would be an appropriate method to trace back the pathway of NO_2-N removal under these conditions (van de Graaf *et al.* 1997, Gaul *et al.* 2002).

Remarkably, it took an NO_2-N concentration of 350 mg/l N to significantly inhibit the ammonium conversion rate, which is much higher than reported for the sensitive *Candidatus* 'Brocadia' (Strous *et al.* 1999b) or the less sensitive *Candidatus* 'Kuenenia' (Egli *et al.* 2001). However, mass transfer limitation has to be taken into account compared to previous

'Kuenenia' results, as those were obtained with non-aggregated biomass. Some key data are summarized in Table 2.

Table 2 Main physiological characteristics of the enriched Anammox biomass compared to *Candidatus* 'Brocadia Anammoxidans'

Parameter	Unit	*Cand.* „Brocadia' (Strous *et al.* 1999b)	*Cand.* 'Kuenenia' (this study)
Temperature range	°C	20–43	15–45
– optimum	°C	40	40
pH range	–	6,7–8,3	< 7,0–9.0
- optimum	–	8.0	7.5
Apparent maximum growth rate	1/h	0.0027	0.0016–0.0028
Threshold concentration for complete inhibition by			
NO_2^-	mmol	7	> 30
PO_4^{3-}	mmol	2	> 50

These data confirm the results published by Egli *et al.* 2001 obtained from laboratory enrichments using synthetic media, showing a significant higher tolerance of 'Kuenenia'-like Anammox organisms to elevated concentrations of nitrite and phosphate than 'Brocadia'. This gives rise to the question how the different ecological niches of both organisms could be described. The apparent disadvantage of 'Brocadia' of being very sensitive to inhibition needs to be counter-balanced by some other competitive advantage. The present data do not contain enough information on the maximum specific growth rate of 'Kuenenia', as this parameter is strongly biased by temporary substrate limitation which was unavoidable due to the mode of SBR operation employing feed-at-once. Taking this into account, for 'Kuenenia' a higher real growth rate has to be assumed than it appeared during these experiments.

On the other hand, Strous *et al.* 1999b reported a high substrate affinity of their 'Brocadia' enrichment (half saturation constant < 0.1 mg/l for NH_4-N and NO_2-N each). To avoid mass transfer limitation, cell aggregates were disrupted by magnetic stirring. Either N_2H_4 or NH_2OH as Anammox intermediates were added in trace amounts to kick-start the Anammox reaction again which had been regularly lost after aggregate disruption. However, similar experiments with the 'Kuenenia' biomass enriched in this study failed completely. Although large efforts were made to find a gentle method of homogenization (stirring, pestle homogenizer, thurrax and sonication) without destroying the cells, Anammox activity always disappeared as soon as the biomass aggregates lost their integrity. Neither NH_2OH nor N_2H_4 addition showed any positive effect (data not shown). It was thus impossible to even estimate the substrate affinity constants for the 'Kuenenia' biomass.

In other experiments with reactors containing mixed aerobic/anaerobic ammonium oxidizing biofilms, 'Brocadia' like organisms were enriched if they had to compete for nitrite with aerobic nitrite oxidizers (Gaul *et al.* 2005). In a similar system, even a population shift from 'Kuenenia' to 'Brocadia' like organisms could be observed depending on nitrite availability (Gaul *et al.*, in preparation). It is therefore hypothesized, that substrate affinity may be at least one important factor for differentiating the ecological strategies of both types of Anammox organisms, in addition to differing sensitivity to inhibition. This hypothesis is supported by the environmental conditions prevailing in the reactors where the respective organisms had been initially observed: To avoid nitrite inhibition of their (then still unidentified) 'Brocadia' enrichment, Strous *et al.* (1998) operated an SBR under strict nitrite limitation, which favours organisms with high substrate affinity. Contrary to that, *Candidatus* 'Kuenenia stuttgartiensis' had been found at first in an anoxic fixed-bed trickling filter fed with a concentrated NH_4-N/NO_2-N mixture (Schmid *et al.* 2000).

There is thus some reasonable indication, that a strategy of shock-loading an Anammox-SBR at once at the start of the cycle, as applied in the present study, will select for 'Kuenenia' like Anammox organisms showing a high tolerance to inhibition. In terms of operational stability under practical conditions, such a robust biomass may be considered advantageous.

Options for real-time process control using standard probes

Although high substrate availability at the beginning of an SBR cycle is likely to favour non-sensitive organisms, a major disadvantage of this type of feeding strategy is that the cycle time becomes a critical factor. If the reaction has not been completely finished at the end of a time-fixed cycle, the ongoing Anammox reaction will continue to produce N_2 leading to biomass flotation and eventually to biomass run-off (Dapena-Mora *et al.* 2004). The effluent concentration will also be negatively affected. On the other hand, if the cycle time is too long, the capacity of the reactor cannot be completely utilized. Furthermore, idle periods without substrate do not allow for biomass growth, limit the actual growth rate and thus lead to more extended start-up periods than essentially necessary. Consequently, a flexible cycle duration adapted to the actual state of the process would increase the reactor's efficiency and minimize operational problems. To avoid the off-time always linked to chemical analyses even if automated analyzers are used, a real-time control by sensors is a promising concept.

So-called biosensors have been developed recently, which make use of biological mechanisms (immobilized biomass or enzymes) which are transduced to electrical signals (e.g. Nielsen *et al.* 2004). However, these sophisticated devices are not easily available yet.

An interesting alternative are well-known electrochemical sensors like probes for pH, conductivity and ORP. The latter principle has been employed in controlling simultaneous nitrification/denitrification processes for a long time (e.g. Lie and Welander 1994, Plisson Saune *et al.* 1996), while pH is an important parameter for the control of nitrification processes (e.g. Kishida *et al.* 2003). Conductivity also belongs to the parameters that are routinely measured in the influent of many WWTPs.

A reactor fed at the beginning of the cycle within a short period will show specific changes over time as the reaction proceeds. From the stoichiometry in Eq. (1) it is clear that the Anammox reaction a) *eliminates* reduced and oxidized ions (NH_4^+ and NO_2^-), b) *produces* oxidized ions (NO_3^-) and c) *consumes* a small amount of protons. There should be a change in conductivity κ (ion elimination), pH (H^+ consumption) and ORP, accordingly. These parameters were measured using commercial probes in the SBR studied here for more than one year.

It could be shown that pH and conductivity are very reliable parameters to indicate the end of the reaction phase, as they show a significant inflection point after the Anammox reaction stopped due to substrate limitation (Figure 3).

Neither variation in operating conditions showed any influence on this correlation. The inflection point could always be observed and was reliably correlated to the Anammox reaction as theoretically predictable. By differentiation of this function, a process control system is able to automatically detect the end of the cycle and to initiate the sedimentation step and subsequent withdrawal and re-filling.

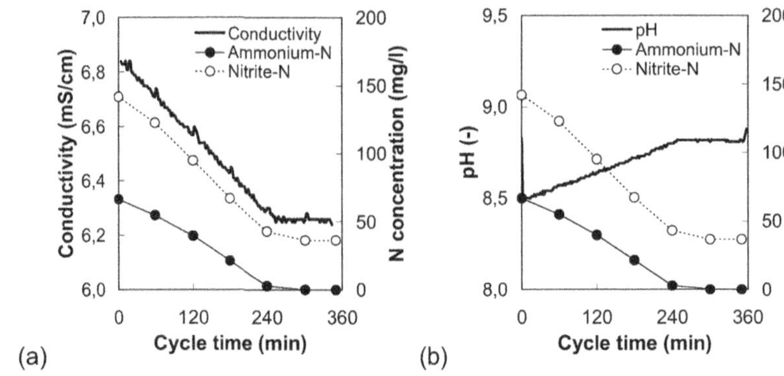

Figure 3 Course of (a) conductivity and (b) pH during a typical, ammonium-limited SBR cycle.

Furthermore, there is a strong correlation between Δ κ and the N-load removed (Figure 4). As this correlation does not depend on the loading rate (only the difference between cycle start and end is important, but no absolute values. There is no correction for hydrogen standard potential necessary), it is possible to automatically calculate the N-load actually removed and to make it available to the plant operator and the plant's process control system.

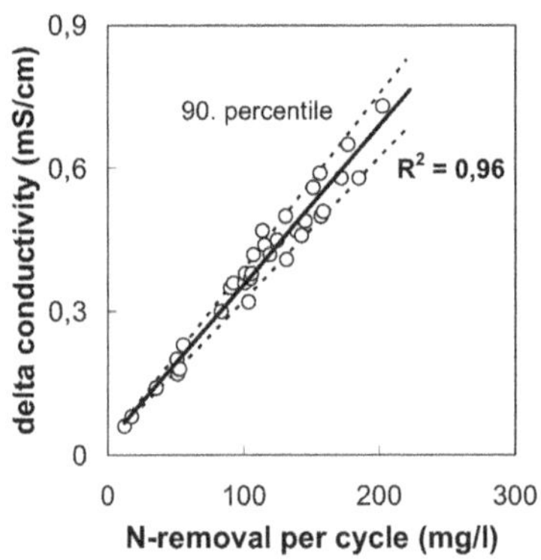

Figure 4 Correlation between change in conductivity (Δ κ) and N-load removed within the SBR cycles.

Consequently, from the slope of conductivity decrease the actual reaction rate can also be derived (not shown). This allows predicting the time of cycle duration and the time when the reactor could be re-filled, which represents valuable information for upstream processes (preceding nitrification step or even sludge dewatering).

Although ORP has also some significance and characteristics, it does not seem suitable for being automatically evaluated. There are too many factors influencing the ORP course during a cycle, with dissolved oxygen introduced during the filling period as the most important. ORP can, however, be interpreted by an experienced plant operator, as differing shapes of ORP curves at least indicate severe changes in reactor operation. In order to keep a control strategy as simple as possible and error-proof, ORP should be displayed, but not be used as a parameter for automated control of the process.

Conclusions

Anammox biomass can be enriched in SBRs within a reasonable span of time (a few months) without prior inoculation with Anammox organisms, under realistic conditions, feeding real wastewater from sludge digestion, and without using 'laboratory tricks'. Adding fast settling, easily available UASB granules as biological carrier material for adhesion and biomass retention may be beneficial. There is some evidence that feeding the reactor at once at the start of the cycle will select for less inhibition-prone organisms of the *Candidatus* 'Kuenenia' type instead of more sensitive *Candidatus* 'Brocadia' like organisms, which might be an advantage for operational stability.

pH and conductivity turned out to be parameters ideally suited for automated process control, as they deliver reliable information on the end of the reaction, the N-load actually removed, and the process rate. The probes are cheap, reliable, well-known to plant operators in operation and maintenance, and they provide real-time information.

Acknowledgements

This work was supported by Deutsche Forschungsgemeinschaft (DFG, KU 602/6-1 and KU 602/7-4) and Deutsche Bundesstiftung Umwelt (DBU, Az. 18531). I.W. is also supported by a DBU grant. The authors wish to thank Nadja Schmider and Dr. Natuschka Lee from Technical University of Munich for the FISH analyses. Cooperation and support by Prof. Dr. Angelika Anders, Dr. Ralf Pätzold and Heidi Bliedung (Institute of Biophysics, University of Hannover) is also greatfully acknowledged.

References

Dapena-Mora, A., Arrojo, B., Campos, J.L., Mosquera-Corral, A. and Mendez, R. (2004). Improvement of the settling properties of Anammox sludge in an SBR. *Journal of Chemical Technology and Biotechnology* **79** (12), 1417–1420.

Egli, K., Fanger, U., Alvarez, P.J., Siegrist, H., van der Meer, J.R. and Zehnder, A.J. (2001). Enrichment and characterization of an Anammox bacterium from a rotating biological contactor treating ammonium-rich leachate. *Archives of Microbiology* **175** (3), 198–207.

Gaul, T., Filipov, E., Schlosser, N., Kunst, S. and Helmer-Madhok, C. (2002). Balancing of nitrogen conversion in deammonifying biofilms through batch tests and GC/MS. *Water Science and Technology* **46** (4–5), 157–162.

Gaul, T., Märker, S. and Kunst, S. (2005). Start-up of moving bed biofilm reactors for deammonification–the role of hydraulic retention time, alkalinity and oxygen supply. *Water Science and Technology* **52** (7), 127–133.

Helmer-Madhok, C., Schmid, M., Filipov, E., Gaul, T., Hippen, A., Rosenwinkel, K.H., Seyfried, C.F., Wagner, M. and Kunst, S. (2002). Deammonification in biofilm systems: population structure and function. *Water Science and Technology* **46** (1–2), 223–231.

Imajo, U., Tokutomi, T. and Furukawa, K. (2004). Granulation of Anammox microorganisms in up-flow reactors. *Water Science and Technology* **49** (5–6), 155–163.

Jetten, M.S., Cirpus, I., Kartal, B., van Niftrik, L., van de Pas-Schoonen, K.T., Sliekers, O., Haaijer, S., van der Star, W., Schmid, M., van de Vossenberg, J., Schmidt, I., Harhangi, H., van Loosdrecht, M., Kuenen, J.G., Op den Camp, H. and Strous, M. (2005). 1994–2004: 10 years of research on the anaerobic oxidation of ammonium. *Biochemical Society Transactions* **33** (Pt 1), 119–123.

Kartal, B., van, Niftrik, L., Sliekers, O., Schmid, M.C., Schmidt, I., van de Pas-Schoonen, K.T., Cirpus, I., van der Star, W., van Loosdrecht, M.C.M., Abma, W., Kuenen, J.G., Mulder, J.-W., Jetten, M.S.M., Op den Camp, H., Strous, M. and van de Vossenberg, J (2004). Application, eco-physiology and biodiversity of anaerobic ammonium-oxidizing bacteria. *Reviews in Environmental Science & Bio/Technology* **3** 255.

Kishida, N., Kim, J.H., Chen, M.X., Sasaki, H. and Sudo, R. (2003). Effectiveness of oxidation–reduction potential and pH as monitoring and control parameters for nitrogen removal in swine wastewater treatment by sequencing batch reactors. *Journal of Bioscience and Bioengineering* **96** (3), 285–290.

Lie, E. and Welander, T. (1994). Influence of dissolved oxygen and oxidation–reduction potential on the denitrification rate of activated sludge. *Water Science and Technology* **30** (6), 91–100.

Nielsen, M., Larsen, L.H., Jetten, M.S.M. and Revsbech, N.P. (2004). Bacterium-based NO_2-biosensor for environmental applications. *Applied and Environmental Microbiology* **70** (11), 6551–6558.

Plisson Saune, S., Capdeville, B., Mauret, M., Deguin, A. and Baptiste, P. (1996). Real-time control of nitrogen removal using three ORP bending-points: signification, control strategy and results. *Water Science and Technology* **33** (1), 275–280.

Schmid, M., Twachtmann, U., Klein, M., Strous, M., Juretschko, S., Jetten, M., Metzger, J.W., Schleifer, K.H. and Wagner, M. (2000). Molecular evidence for genus level diversity of bacteria capable of catalyzing anaerobic ammonium oxidation. *Systematic and Applied Microbiology* **23** (1), 93–106.

Schmid, M., Schmitz-Esser, S., Jetten, M.S.M. and Wagner, M. (2001). 16S–23S rDNA intergenic spacer and 23S rDNA of anaerobic ammonium-oxidizing bacteria: implications for phylogeny and in situ detection. *Environ. Microbiol.* **3** (7), 450–459.

Schmider, N. (2004): Analysis of the population structure of ammonium oxidizing bacteria in selected wastewater treatment systems by molecular methods (in German). Diploma thesis, Department of Microbiology, Technical University of Munich, Germany.

Strous, M., Heijnen, J.J., Kuenen, J.G. and Jetten, M.S.M. (1998). The sequencing batch reactor as a powerful tool for the study of slowly growing anaerobic ammonium-oxidizing microorganisms. *Applied Microbiology and Biotechnology* **50** (5), 589–596.

Strous, M., Fuerst, J.A., Kramer, E.H., Logemann, S., Muyzer, G., van de Pas-Schoonen, K.T., Webb, R., Kuenen, J.G. and Jetten, M.S. (1999a). Missing lithotroph identified as new planctomycete. *Nature* **400** (6743), 446–449.

Strous M, Kuenen JG and Jetten MS (1999b). Key physiology of anaerobic ammonium oxidation. *Appl Environ Microbiol* **65** (7), 3248–3250.

van de Graaf, A.A., de Bruijn, P., Robertson, L.A., Jetten, M.S.M. and Kuenen, J.G. (1997). Metabolic pathway of anerobic ammonium oxidation on the basis of ^{15}N studies in a fluidized bed reactor. *Microbiology* (143), 2415.

Trace organics removal from surface water by adsorption and photo-catalysis with low-pressure membranes

Sejin Park[*,**], **S. S. Chin**[**], **Y. Jia**[**] **and A.G. Fane**[**]

* Environmental Research Center, In-Ha University, YonghyunDong 253, NamGu, Inchon, 402-751, Republic of Korea (E-mail: *SeJinPark@ntu.edu.sg*)
** School of Civil and Environmental Engineering, Nanyang Technological University, Innovation Centre, Block 2, Unit 237, 18 Nanyang Drive, Singapore – 637723
(E-mail: *agfane@ntu.edu.sg*)

Abstract This study was done as a preliminary investigation prior to the application of continuous hybrid membrane system containing PAC, TiO_2, and submerged hollow fibres for the removal of trace organics from surface water. The combined system with PAC adsorption and TiO_2 photo-oxidation and in-situ regeneration was studied as a batch test. Bisphenol A (BPA), an endocrine disrupting compound present at low ppm levels, was used as a model compound. TiO_2 addition did not decrease the adsorption capacity of PAC for BPA. When UV light and 0.5 L/min of oxygen were introduced to the PAC/TiO_2 batch system, the 20 mg/L of BPA could be removed within 90 minutes. The results were 2.3 to 3.1 times faster than that of neat TiO_2 oxidation in the pH range of 4 to 10. For the regeneration of PAC pre-saturated by BPA, it was found that 89% of the adsorption sites could be recovered by 12 hours of photo-oxidation.

Keywords Endocrine disruptors; hybrid membrane; insitu-regeneration; powdered activated carbon; photocatalysis; titanium dioxide

Introduction

Membrane technology is now widely adopted for water production and reclamation. An import reason for this is the efficiency of removal of species across the spectrum of contaminants. However, trace organics, including persistent organic pollutants (biocides, endocrine disruptors, carcinogens etc) present a major challenge in water and wastewater treatment. This is because the allowable levels are very low and they are not completely removed, even by high-pressure reverse osmosis. For this class of compound hybrid membrane processes that combine membranes with another unit operation may be more effective. Two combinations are particularly interesting, membranes coupled with adsorbents and membranes coupled with photocatalysts. This opens up the possibility of using submerged low pressure membranes located in a tank containing the facilitating agent (adsorbent and catalyst) in suspension. In our previous study, the hybrid TiO_2/UV /submerged membrane (a non-modified hydrophobic PVDF membrane) was able to achieve mineralization of more than 90% of 10ppm of bisphenol-A (BPA) within 100min of UV illumination (Chin *et al.*, 2005). On the other hand, a rapid trace organics removal was achieved by batch dosing of powdered activated carbon (PAC) while breakthrough was gradual due to the saturation of the carbon particles (Jia *et al.*, in press). A third option is the integration of membranes with both adsorption and photocatalysis. It has been reported that PAC improved the TiO_2 oxidation by mass transfer from the PAC to the TiO_2 through the interfacial surface between TiO_2 and PAC, which compensates for the inhibition of UV irradiation by the presence of PAC (Matos *et al.*, 2001). It is anticipated that the coupled processes will be synergistic and out-perform either of the individual processes. In effect the PAC could be regenerated in situ by photo oxidation, and the presence of this PAC could increase the retention time of organics and improve oxidation efficiency. In this study, the combined system with PAC adsorption and TiO_2 photo-oxidation and the possibility of in situ regeneration was evaluated as a preliminary investigation before the application of a continuous hybrid membrane system containing PAC, TiO_2, and submerged hollow fibres for the removal of trace organics from surface water.

Methods

Bisphenol A, a model pollutant of endocrine disrupting compounds, was purchased from Merck-Schucharatt at the highest purity grade and used as received. The photocatalyst was TiO_2 Degussa P-25, mainly anatase (*ca.* 70%) in the shape of nonporous polyhedral particles of ca, 30-nm mean size with a surface area of 50 m^2/g. The commercial activated carbon, HydroDarco C (25 μm mean diameter with a surface area of 600 m^2/g), was purchased from Norit. The PAC used in this study, was pulverized by grinding machine (RockLabs) for 21 min to 10-μm mean diameter with a pore size of 330 nm. Hydrogen peroxide, sodium hydroxide and perchloric acid were used at the highest purity grade for photo-oxidation and pH adjustment, respectively.

The photocatalytic oxidation system (PCOS) was a cylindrical flask made of pyrex of *ca.* 1000 mL capacity equipped with an aeration distributor for mixing the suspension and providing air (0.5 L/min). It was equipped with 0.425 mW/cm^2 of UV irradiation by UV-BLB fluorescent lamps.

The adsorption rate and isotherm tests were conducted by a jar tester at 180 rpm that was equivalent to the same level of stirring as 0.5 L/min of aeration in PCOS. The comparison of photo-oxidation kinetics of TiO_2 neat and mixed PAC/TiO_2 was carried out in the PCOS at 20 mg/L of initial BPA concentration. Regeneration was implemented in three steps. In the first step PAC/TiO_2 pre-saturated by BPA for 3 days. This was followed by the additional injection of BPA into the pre-saturated suspension followed by 30 min of dark adsorption before UV irradiation (second step). Finally, 5 mg/L of BPA was added to evaluate the re-adsorption rate.

The amount of BPA was measured by use of a high-perfomance liquid chromatograph (HPLC, Waters), equipped with a photodiode array detector and a Waters Xterra RP18 column (3.9 mm × 150 mm, 5 μm). The elution was monitored at 275 nm. The eluent used was a mixed solvent of acetonitirile and water (5/5, v/v). The PAC and TiO_2 particles were removed from the solution by filtration, and the resulting solution was analyzed by HPLC.

The BPA in the solid phase (PAC/TiO_2) was extracted by methanol. The solid phase in a 50 mL of sample was recovered by centrifugation at 4000 rpm for 20 min. The 40 ml of supernatant was drained follow by adding 10 mL of methanol. The extraction was conducted by sonication at 30 °C for 15 min. The separation of solid phase (centrifugal, 4000 rpm, 10 min) was done and 10 mL of supernatant was collected, followed by addition of 10 mL of methanol, and these extraction steps were repeated 3 times. The collected BPA extraction was measured by the same HPLC method mentioned above.

Results and discussion

Adsorption of BPA

0.5 g/L of TiO_2 was an optimum concentration that showed maximum removal efficiency of BPA and reduced fouling potential of the submerged membrane system in our previous study (Chin *et al.*, 2005). It was also confirmed that a ratio of PAC/TiO_2 of 0.2 could minimize attenuation of UV light in the PCOS by UV-Vis spectra evaluation (results not shown here). Ohko *et al.* (2001), Matos *et al.* (2001) and Araña *et al.* (2003) mentioned similar levels of TiO_2 and PAC as an optimum concentration and ratio in their studies.

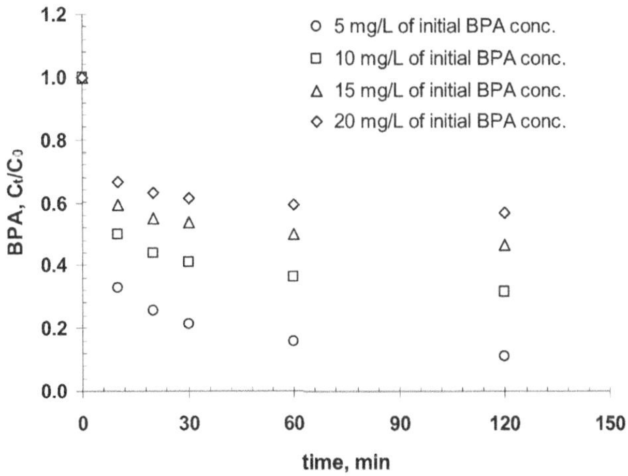

Figure 1 BPA Adsorption rate of PAC at different initial concentration. PAC = 0.1g/L, BPA conc. = 5~20 mg/L, pH = 7.

Adsorption rates of 0.1 g/L of PAC were compared among 5~20 mg/L of initial BPA concentrations as shown in Figure 1. Low levels of initial target concentration could induce a higher adsorption rate. The adsorption capacities in 3 days were shown to be 56.8, 92.3, 105.1, 114.0 mg BPA/g PAC when 5, 10, 15, 20 mg/L of BPA were added respectively. During the initial 30 min of adsorption time, the PAC could reach the 77.6 to 84.2% of its adsorption capacity.

At 5 mg/L of initial BPA concentration, the effects of pH on the adsorption rate of PAC and PAC/TiO_2 are shown in Figure 2. The adsorption rates of PAC and PAC/TiO_2 decreased at the basic condition (pH 10) compared to acidic and neutral condition. TiO_2 and PAC in water have isoelectric points (pH_{IEP}) of about 6.0. At low pH ($pH < pH_{IEP}$), TiO_2 and PAC particles possess positive charge while a negative charge is expected at higher pH ($pH > pH_{IEP}$). At high pH, the charge of BPA is negative because of the ionization of its functional groups (i.e. hydroxyl and methyl groups, pK_a=9.2, 10.2) (Kosky *et al.*, 1991). From Figure 2, it is clear that PAC and TiO_2 can absorb BPA at low and neutral pH more efficiently.

Results contradict with Matos *et al.*'s study (2001), which indicated a competition between the pollutants and titania for approaching the PAC's surface. The mean pore size of PAC used in our study was 1.7 times higher than that of the PAC used in their study and this might be the reason why we do not observed the effect of competition. In addition, the TiO_2 particles have much lower adsorption activity compared to PAC particles as can be seen from the nearly straight line of adsorption by only TiO_2 in this figure, so that the existence of TiO_2 did not apparently affect the overall adsorption rate. In fact, the addition of TiO_2 in our case seems to be slightly advantageous to the overall adsorption rate.

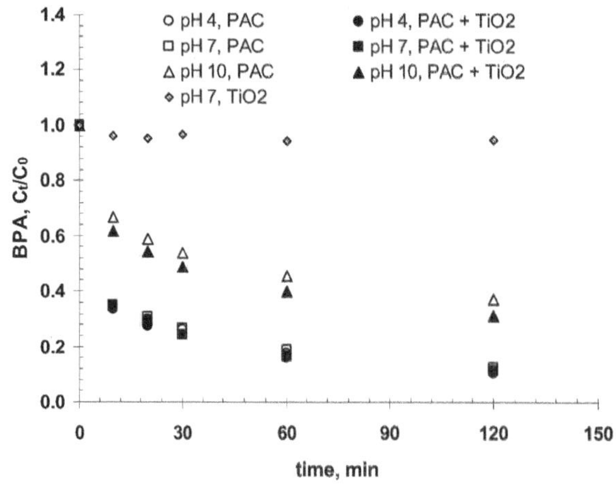

Figure 2 The effects of pH on the adsorption rate of PAC and PAC/TiO_2. Initial BPA cocn. = 5 mg/L, PAC = 0.1 g/L, TiO_2 = 0.5g/L, PAC/TiO_2 ratio = 0.2, pH = 4~10.

The photocatalytic degradation and adsorption of BPA in PCOS

Figure 3 shows the BPA decomposition in the presence of UV-irradiated TiO_2 (without PAC) at different pH and 10 mg/L of initial BPA (C_0) in the PCOS.

The kinetic curves in Figure 3 are of apparent first order as confirmed by a correlation coefficient (close to 0.99), obtained from the linear transformation of $\ln(C_t/C_0) = f(t)$, giving apparent rate constants equal to $k_{app} = 5.58 \times 10^{-2}$, 3.93×10^{-2}, 1.67×10^{-2} min^{-1} at pH 4, 7, 10, respectively.

As mentioned above, the solution pH was expected to affect the degradation process since pH affects BPA adsorption on the TiO_2 due to surface charges that vary with pH of the solution. Acidic pH was found to enhance the photodegradation rate of BPA (Figure 3). This agrees with Watanabe *et al.* (2003) and Chiang *et al.* (2004). However, different types of TiO_2 might result in different surface properties that have a different effect on the pH study (Kaneco *et al.*, 2004). On the other hand, an increase in the concentration of hydroxyl ions at the TiO_2 particle/solution interface with increasing pH can promote the photocatalytic reaction by facilitating the generation of OH radical as follows.

$$TiO_2 \xrightarrow{hv} e^- + h^+ \quad (1)$$

$$OH^- + h^+ \longrightarrow OH^\bullet \quad (2)$$

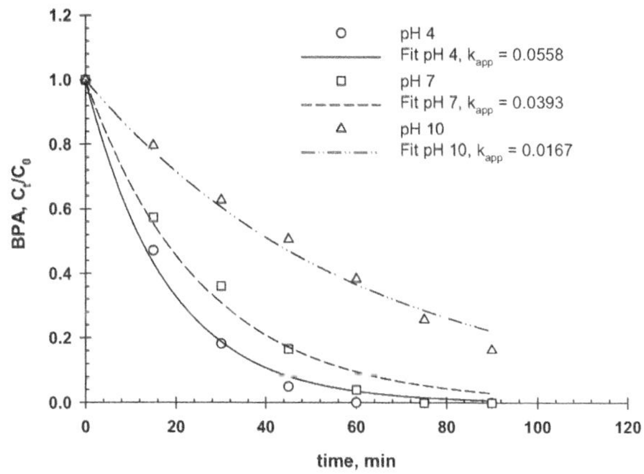

Figure 3 BPA decomposition in PCOS (without PAC) at different pH. Initial BPA conc. = 10 mg/L, TiO_2 = 0.5 g/L, pH = 4–10.

The influence of PAC on the photocatalytic reaction of TiO_2 was investigated at different pH and double the initial BPA concentration (20 mg/L) as shown in Figure 4.

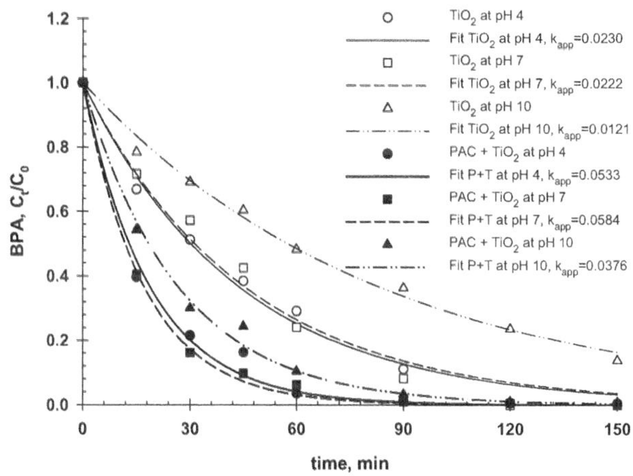

Figure 4 BPA decomposition in PCOS at different pH. Initial BPA conc. = 20 mg/L, TiO_2 = 0.5 g/L, PAC = 0.1 g/L pH = 4~10.

The kinetic rates of neat TiO_2 at 20 mg/L were lower than those of neat TiO_2 at 10 mg/L of initial BPA. However, the removal rate of PAC/TiO_2 was higher than that of neat titania, and the addition of 0.1g/L PAC to 0.5 g/L TiO_2 created a positive synergy effect with an increase in the rate constant by a factor of ca. 2.3, 2.6, and 3.1, respectively, at pH 4, 7, 10. The improvement of rate constant increased with increasing pH. Thc trends are rationalized by considering that the high adsorption rate of PAC can prevent the sharp decreasing of oxidation rate of PAC/TiO_2 although both TiO_2 oxidation and BPA adsorption rate decline at high pH. The synergy effect of PAC in our study is in accordance with the results of Matos *et al.* (2001), who reported that even if active carbon perturbs the transmission of UV light to the surface of titania, it largely compensates this inhibition by strong beneficial effects in the adsorption of phenol followed by a transfer of phenol to titania.

They evaluated that photocatalytic activity was improved by the addition of PAC. However, it is not easy to clearly distinguish between PAC adsorption and TiO_2 oxidation, even though they estimated the apparent kinetic rates after 1hr of PAC/TiO_2 dark adsorption. PAC adsorption rate might be increased according to the alteration of BPA concentration on the PAC particle/solution interface when BPA was decomposed by photo-oxidation. In addition, it is not confirmed that BPA adsorbed by PAC could be completely transferred to the surface of TiO_2.

We therefore implemented the extraction test of PAC/TiO_2 particles after the photocatalytic oxidation tests as shown in Figure 5. After the PAC/TiO_2 was presaturated by 5 mg/L of BPA for 3 day (-30min. in Figure 5), 5 mg/L of BPA was added into the photocatalytic system before UV irradiation. BPA concentration in the liquid phase was not detected after 2 h of UV irradiation as shown. However, about 34% of the total BPA injected remained in the solid phase of PAC/TiO_2 until 6 h UV irradiation. The results indicate that mass transfer of the adsorbed BPA by diffusion from the PAC to oxidation sites on the surface of TiO_2 could be limited because PAC ha a strong affinity to adsorb BPA.

From the viewpoint of in-situ regeneration of PAC by photocatalytic oxidation, the adsorbed BPA in PAC particles was found to decrease from 70% to 34% of the overall BPA added to the system after 6 h of photo-oxidation. This indicates that the UV oxidation facilitated the regeneration of the PAC up to 50 %.

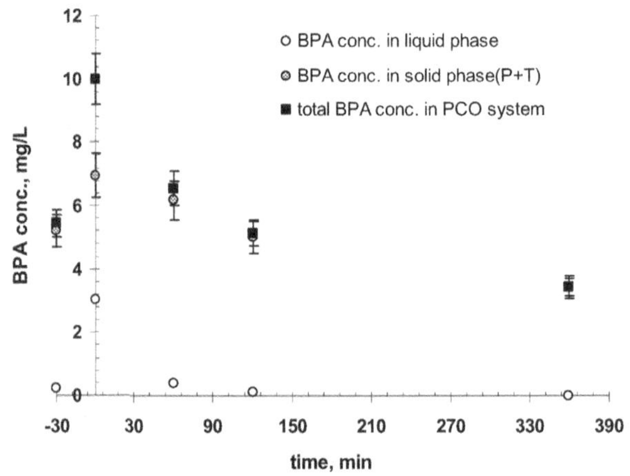

Figure 5 Concentration of BPA in liquid and solid phase. Pre-saturated BPA conc. = 5 mg/L, Additional BPA conc. = 5 mg/L, PAC = 0.1 g/L TiO_2 = 0.5 g/L, pH = 7, UV-irradiation 6 h.

Hydrogen peroxide is decomposed to OH radical and the hydroxyl ion by UV irradiation and TiO_2 catalyst. This species can promote the photocatalytic reaction. Figure 6 shows the effect of concentration of hydrogen peroxide added to the photocatalytic system at pH 4 and 20 mg/L of BPA.

When hydrogen peroxide of 50 mg/L to 150 mg/L was added, the rate constant of TiO_2 neat reaction increased by 1.3 times to 2.8 times, respectively. Also, the PAC/TiO_2 system was slightly improved by 1.2 times to 1.4 times, respectively. The effect of hydrogen peroxide on the PAC/TiO_2 systems was seen to be limited. This may be because the rate of PAC adsorption was very fast compared to the hydroxyl radical oxidation rate and thus when PAC was added the effect of hydrogen peroxide was not prominent when compared to neat TiO_2 system.

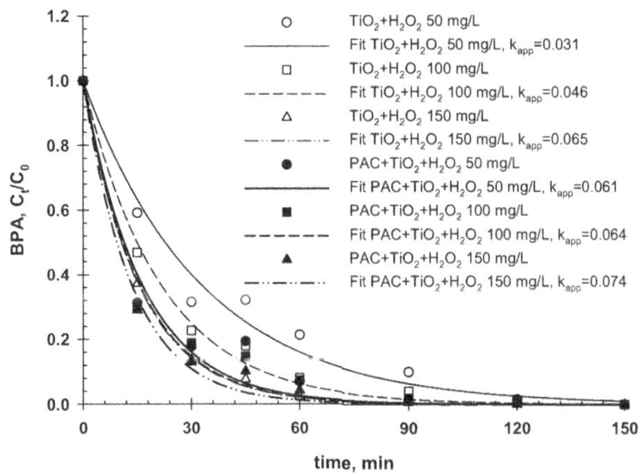

Figure 6 effect of concentration of hydrogen peroxide in PCOS. Initial BPA conc. = 20 mg/L, PAC = 0.1 g/L, TiO_2 = 0.5 g/L, pH = 4.

In situ regeneration of PAC in PCO

After 4–12 h of UV irradiation, 5 mg/L of BPA adsorption rates compared to virgin PAC and PAC/TiO_2 at pH 7 are shown in Figure 7.

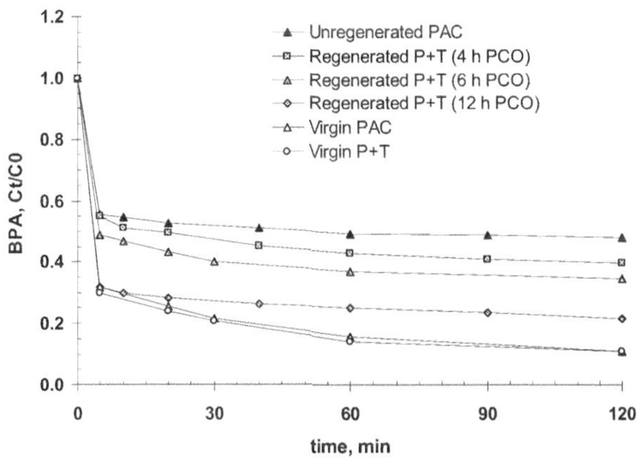

Figure 7 Effect of photo-oxidation time on re-adsorption rate of PAC. Pre-saturated BPA conc. = 5 mg/L, Additional BPA conc. = 5 mg/L, PAC = 0.1 g/L TiO_2 = 0.5 g/L, pH = 7.

Virgin PAC/TiO_2, regenerated PAC/TiO_2 for 4, 6, and 12 h absorbed 89%, 60%, 66% and 79% of added BPA, respectively. This means that 4, 6, and 12 h of regeneration by PCOS can recover about 67%, 74%, and 89% of PAC adsorption site, respectively. This clearly shows time is a crucial factor when using TiO_2 as an *in situ* tool for regeneration of the PAC surface.

The adsorption rate of regenerated PAC/TiO_2 rapidly decreased after 5–10 min of adsorption compared to virgin PAC and PAC/TiO_2. This indicates that the regenerated PAC has a diffusion-limiting step for re-adsorption. It could be explained by three kinds of mechanisms as follows. First, BPA and its intermediates moved deeply into the PAC pores where they cannot transfer to neat TiO_2 surfaces. Secondly, intermediates of BPA produced by photocatalytic oxidation are not completely decomposed but are adsorbed onto PAC adsorption sites. These intermediates such as 4-isopropylphenol, formate, acetate, and acetealdehyde have smaller molecule size than BPA. Thus they move easily into the irreversible adsorption sites deep inside the pore. The other reason is a possibility that PAC pores were physically blocked during the UV-irradiation with TiO_2. However a BET test did not reveal any evidences of the third reason, i.e. that the pore size and surface area were changed after UV-irradiation.

Conclusion

Compared with PAC particles, TiO_2 has nearly no function of adsorption based on both the adsorption capacity and adsorption rate studies. The advantage of the PAC addition to TiO_2 photocatalytic oxidation process is ambiguous in the current stage due to the fact that BPA, the parent compound, was found to remain inside the PAC even after 6 h of UV irradiation. However, the basic study of the interaction between PAC and TiO_2 has led to the possibility of in-situ regeneration of PAC by TiO_2 oxidation. It was found that 67%, 74%, and 89% of the adsorption sites could be recovered by 4, 6 and 12 h of photocatalytic oxidation, respectively. The effect of different parameters on the regeneration of PAC by TiO_2 is under further study in our lab and will be reported in a future publication.

References

Araña, J., Doña-Rodríguez, J. M., Tello Rendón, E., Cabo, G., González-Díaz, O., Herraera-Melián, Pérez-Peña J., Colón, G. and Nació, J. A. (2003). TiO_2 activation by using activated carbon as a support. Part I. Surface characterisation and decantability study. *Appl. Catalysis B Environ.*, **44**, 161–172.

Chiang, K., Lim, T. M., Tsen, L. and Lee, C. C. (2004). Photocatalytic degradation and mineralization of bisphenol A by TiO_2 and platinized TiO_2. *Appl. Catalysis A General*, **261**, 225–237.

Chin, S. S., Lim, T. M., Chiang, K., Fane, A. G. (2005). Hybrid low-pressure submerged membrane-photoreactor for the removal of Bisphenol A. *Proceedings of the 5th International Conference on Wastewater Reclamation and Reuse for Sustainability*, 7–11 Nov., Jeju, Korea.

Jia, Y., Wang, R., Fane, A. G. and Krantz, W. B. (in press). Effect of air bubbling on atrazine adsorption in water by powdered activated carbons – competitive adsorption of impurities. *Seperation and Purification Tech.*

Kaneco, S., Rahman, M. A., Suzuki, T., Katsumata, H. and Ohta, K. (2004). Optimization of solar photocatalytic degradation condition of bisphenol A in water using titanium dioxide. *J. Photochem. Photobiol. A: Chemistry*, **163**, 419–424.

Kosky, P. E. and Guggenhiem, E. A. (1991) The aqueous phase in the interfacial synthesis of polycabonates. 1. Ionic equilibria and experimental solubilities in the BPA–NaOH–H_2O system. *Ind. Eng. Chem. Res.*, **30**(3), 462–467.

Matos, J., Laine, J. and Herrmann, J.-M. (2001). Effect of the type of activated carbons on the photocatalytic degradation of aqueous organic pollutants by UV-irradiated titania. *Journal of Catalysis*, **200**, 10–20.

Ohko, Y., Ando, I., Niwa, C., Tatsuma, T., Yamamura, T., Nakashima, T., Kubota, Y. and Fujishima, A. (2001). Degradation of bisophenol Ain water by TiO_2 photocatalyst. *Environ. Sci. Technol.*, **35**, 2365–2368.

Watanabe, N., Horikoshi, S., Kawabe, H., Sugie, Y., Zhao, J. and Hidaka, H. (2003). Photodegradation mechanism for bisphenol A at the TiO_2/H_2O interfaces. *Chemosphere*, **52**, 851–859.

Pilot, bench-scale and field testing to integrate 740 Ml/d of UF membranes, advanced oxidation and biological treatment

Jesus Garcia-Aleman* and James Lozier**

*CH2M HILL Canada, 255 Consumers Rd., Toronto, ON M2J5B6, Canada (E-mail: *jgarciaa@ch2m.com*)
CH2M HILL, 2625 South Plaza Dr., Suite 300, Tempe, AZ, USA (E-mail: *jlozier@ch2m.com*)

Abstract This paper reviews key findings of the bench and pilot-scale testing and conceptual design that culminated in the design of the two largest North American drinking water facilities that employ and incorporate low pressure membrane filtration, specifically submerged ultrafiltration (SUF) into an integrated treatment scheme with ozone, and biological filtration to produce the highest-quality drinking water, with emphasis on the control of multiple water quality parameters, including pathogen, disinfection-by-products, taste and odor, and emerging contaminants. The first is a 360 ML/d expansion of the Lakeview WTP west of Toronto, Canada and the second is the 380 ML/d green field zero-liquid discharge Twin Oaks Valley WTP in San Diego, California. The process was thoroughly evaluated during a 6-month pilot study in 2003 with the participation of three industry-leading membrane suppliers. The robustness of these treatment approaches was demonstrated during the study, especially during water quality excursions. Key performance, design, cost and operational criteria were developed as a result of the study, and utilized to procure, design, construct and operate these facilities. This treatment process offer the best value to projects where water quality (both for current and future drinking water regulations), performance, environmental footprint, social benefits, schedule and cost efficiency are paramount.

Keywords Advanced oxidation; biological pre-treatment; pilot testing; ultrafiltration

Introduction

An environmental assessment (EA) was conducted in 2003, which compared 22 alternative treatment options as part of the treatment expansion of the Lakeview WTP. These options were narrowed down to two viable integrated treatment solutions to achieve the client's treatment objectives. Each solution was leveraged for its strengths and optimized for performance and cost. The integrated process was selected based on the choice of disinfectant (ozone versus UV) and choice of particulate removal (membrane filtration versus conventional treatment). At the end of the EA, ozone, biologically activated carbon contactors (BACC), and membrane filtration were deemed to offer the best value to the client in terms of water quality, performance, environmental footprint, social benefits, and cost efficiency. This treatment train provides multiple barriers for disinfection (membranes and ozone), taste & odour control (ozone and BACC), and turbidity removal (membranes and BACC), while exceeding current regulations, and providing biologically stable water in the distribution system. Residual management issues are mitigated by not using coagulant and producing innocuous waste from which 95% meets regulatory requirements to be returned directly to the source water without requiring further treatment (Garcia-Aleman *et al.*, 2005).

Although both the Lakeview WTP and the Twin Oaks Valley WTP utilize the same basic treatment processes, the process sequence differs, reflecting differences in source water quality (Figs. 1 & 2) and treatment objectives. The treatment scheme at Lakeview WTP is ozone, biological filtration (BACC) (using granular activated carbon) and SUF, followed by free chlorine disinfection. Lakeview WTP receives raw water from Lake Ontario, a source that has low turbidity, moderate alkalinity and low TOC and DBP precursors. Pilot testing of the selected sequence in parallel with SUF/ozone/BF showed that cost savings from enhanced UF flux following BACC more than compensated for slightly higher ozone dose necessary to achieve desired Cryptosporidium inactivation and taste and odor removal directly with the raw water. In contrast, the Twin Oaks WTP is served by three surface sources: Colorado River and State Project (both regional) and Lake Hodges (local). Levels of turbidity and TOC in these sources are higher than for Lake Ontario and range widely and seasonally. The optimized treatment sequence places UF first, followed by ozone (with hydrogen peroxide for advanced oxidation) and BACC. UF minimizes ozone demand, while providing a means to achieve a high level of pathogen removal without pre-coagulation or clarification. Avoidance of coagulation significantly reduces solids residuals and cost for residuals hauling and disposal. The plant does not generate liquid waste (zero liquid discharge). Ozone/AOP/BF provides a multiple treatment approach to taste and odor and trace contaminant removal and an additional pathogen disinfection capability.

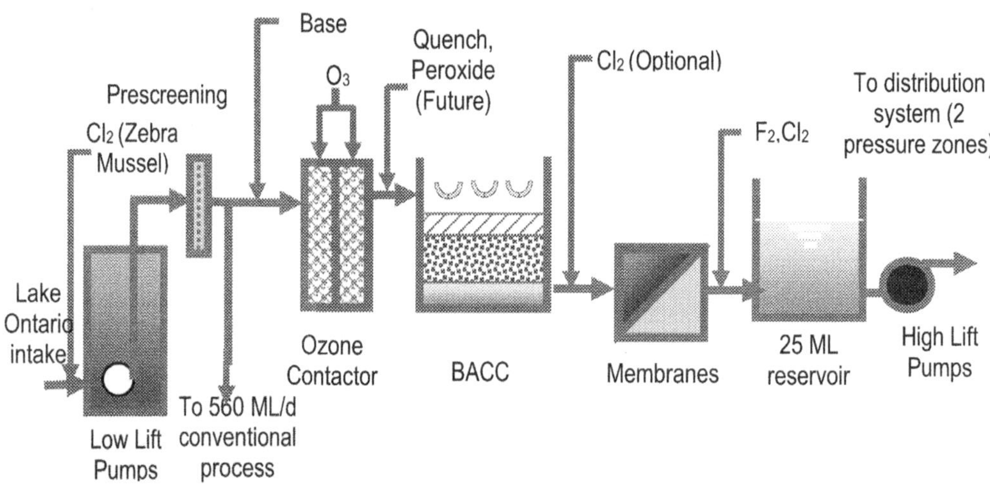

Figure 1 Flow diagram of ozone, BACC, UF membrane treatment process for the 363 ML/d expansion of the Lakeview WTP.

These two projects exemplify the trend in drinking water treatment towards the engineering of large-scale, large-capacity treatment works that address multiple objectives. First and foremost is finished water quality, however, incorporation of processes and operations that (1) optimally leverage plant staff and maximize automated plant control through high level water quality monitoring tied to PLC and SCADA functionality, (2) reduce plant footprint and social impact (e.g., noise and chemical deliveries and truck traffic) and (3) minimize chemical-based residuals, residuals production and energy consumption footprint are equally important to provide a high level of regulatory, public and governmental acceptance. These facilities have set industry-wide benchmarks for technological innovation, environmental footprint (residual management, energy efficiency) and membrane equipment costs. Both facilities are being designed using the state of the art 3D CADD software and are being designed and constructed in under 30 months. The Lakeview WTP is currently under construction with a substantial completion date of June 2007. The Twin Oaks Valley WTP is currently under design and construction with a substantial completion date of April 2008.

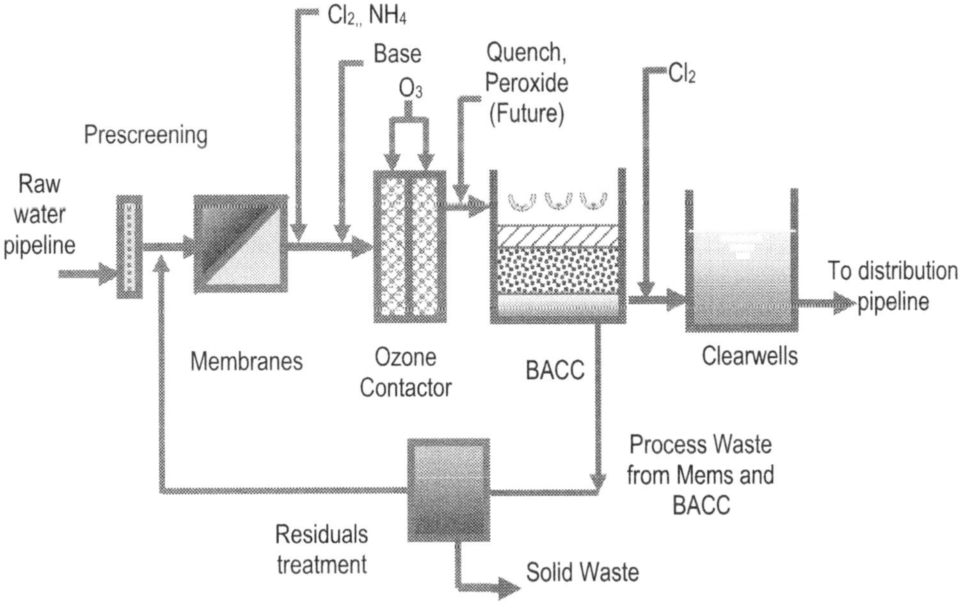

Figure 2 Flow diagram of UF membranes, ozone, and BACC treatment process for the 380 ML/d zero-liquid discharge Twin Oaks Valley WTP.

A handful of full-scale facilities in the World are currently operating with a similar treatment process (< 5 years): Clane Lane WTP in the UK treating GWUDI (Oxtoby, 2002) with ozone, GAC and UF; and, the Sandhurst and Trentham WTPs in Australia treating pre-coagulated reservoir water with UF, ozone and GAC. Pilot and experimental data for the three processes (separately) is widely available. The most relevant long term studies were performed by the National Institute of Public Health of Japan (Kunikane *et al.*, 1995, 1996) and the Amsterdam Water Supply (van der Hoek *et al.*, 2000). Most authors have reported benefits of pre-treated carbon contactors with ozone and the enhanced performance of membranes after pre-treatment with activated carbon (Park *et al.*, 1997; Park 2002). Water quality improvements have also been reported, with improvements in TOC and THM formation as a function of ozone dose and carbon contact time (Takeuchi *et al.*, 1997; Hishijima *et al.*, 1998; Vahala *et al.*, 1999; Rittman *et al.*, 2002).

Methods

Pilot testing was critical to proceeding with the detailed design of the expansion to the Lakeview WTP due to the limited experience documented when using ozone with membranes at full-scale, and no experience specific to the Great Lakes. . The pilot study had the following objectives (1) confirm the performance of the proposed treatment train with respect to the treatment objectives; (2) assess the impact of pre-treating surface water with ozone and BAC contact on membrane performance; (3) assess the impact of membrane filtration on ozone system sizing; and, (4) identify design and operating criteria. Results from the pilot program were used to provide performance information and criteria to support membrane equipment procurement. The Lakeview pilot study extended for six months from the Spring to Fall of 2003 and was divided into three stages. In the first stage the membrane pilots treated direct raw water from Lake Ontario to establish an operational baseline. In subsequent stages, the performance of the membrane pilot on ozone and BAC pre-treated feed water was analyzed during an optimization period and a demonstration period of the long-term performance of this treatment strategy.

The treatment process was evaluated in two trains with two sequences: 1) ozone, BACC, membranes (OBM) and 2) membranes, ozone, BACC (MOB). It is noted that neither of the trains was operated with coagulation. Water from the OBM train was directed to one of three membrane filtration pilots (one from each of the participating membrane equipment suppliers), with all pilots running in parallel (USFilter/Memcor CMFs, Zenon ZW-1000 and Ionics/Norit XIGA). Owing to space and resource limitations, only one membrane equipment supplier (Zenon) was used in the control or MOB train. A more detail description of the pilot facility and design criteria are presented in Garcia-Aleman *et al.* (2004). Extensive sampling and bench-scale testing programs were developed and implemented in the pilot study. An internet-based data management system was utilized to manage the data and information from the different laboratories, suppliers and agencies involved in this project.

Results and discussion

A summary of the performance of the pilot trains is presented in this section. All membrane pilot units consistently produced drinking water with turbidities < 0.1 NTU and with less than 5–10 counts/mL +96% of the time when treating raw or pre-treated water. Feed water quality excursions did not significantly affect permeate quality, as discussed later. Water quality performance for both process sequences are presented in Table 1. Requirements of the Ontario Drinking Water Standards (ODWS) and the United States Environmental Protection Agency (USEPA) are included for comparison.

Table 1 Comparison of US and Ontario regulations vs. pilot plant performance

Parameter	Ontario Drinking Water Standards	USEPA regulations	OBM Pilot Train	MOB Pilot Train
Turbidity	1.0 NTU 95% of monthly CFE samples	0.03 NTU 95% of the time	<0.04 NTU	0.1 NTU
TOC/DOC/ BDOC	None yet defined	None yet defined	2.1/2.0/0.2 mg/L	2.0/1.9/0.7 mg/L
Total THMs	100 ug/L	< 80 ug/L	3 – 9 ug/L	3 – 9 ug/L
HAAs	None yet defined	< 60 ug/L	< 2 ug/L	< 2.5 ug/L
Bromate	10 ug/L	10 ug/L	≤ 5 ug/L	≤ 6 ug/L
Crypto	2.0-log reduction = 2.0-log removal	2-log defined in the IESWTR	> 4.0-log removal	> 4.0-log removal
Giardia	3.0-log reduction = 2.5-log removal + 0.5-log inactivation	3-log reduction as defined in the SWTR	> 4.0-log removal	> 4.0-log removal

Table 1 Comparison of US and Ontario regulations vs. pilot plant performance

Parameter	Ontario Drinking Water Standards	USEPA regulations	OBM Pilot Train	MOB Pilot Train
HPC	≤ 500 cfu/mL		2,400 CFU/100 mL	1,700 CFU/100 mL
Viruses	4.0-log reduction = 2.0-log removal + 2.0-log inactivation	4.0- log reduction as defined in the SWTR.	Not measured	Not measured
Geosmin	None yet defined	None yet defined	2.2 ng/L	1.9 ng/L
MIB	None yet defined	None yet defined	< 1 ng/L	< 1 ng/L
EDC	None yet defined	None yet defined	Not measured	Not measured
Algal Toxins	None yet defined	None yet defined	Not measured	Not measured

Impact of pre-treatment on membrane filtration

All four membrane pilots were initially operated directly on the conditioned source water to establish flux rates and cleaning regimes. Upon start-up of ozonation and BAC contact upstream of the membrane pilots as part of the OBM train, it was observed that in general, membrane fouling rates decreased; furthermore, two of the three membrane equipment suppliers were able to sustain substantially higher fluxes with essentially the same cleaning regimes. Fouling rates continued to decrease for the duration of the pilot program, and this was observed to correlate with improved performance of the BAC contactor as measured by the filter run lengths. The OBM and MOB BAC contactors were typically backwashed every 7 and 30 calendar days, respectively. These lower fouling rates resulted from the removal of particulate matter and the oxidation of organic matter by ozonation. The oxidation step is believed to degrade high molecular weight NOM fractions into hydrophilic fractions, with the latter having a smaller potential of fouling the membranes.

Although fouling rates were similar in both the OBM and MOB trains at low temperatures (< 10°C), higher fouling rates were observed in the membrane pilot of the MOB train (essentially treating raw water without the benefit of ozone and BACC) when temperatures increased beyond 10°C. For a five day period during the pilot program the ozone system was removed from service for maintenance needs, while the BAC contactor and membrane pilots continued operations. During this period, a compromise in performance was observed in all three membrane pilots of the OBM train although none of the pilots reached terminal trans membrane pressure (TMP). During this three day period, membrane fouling was observed to be sensitive to the loss of ozone pre-treatment: fouling rates over several chemically enhanced backwash (CEB) cycles increased for two of the three membrane equipment suppliers (Ionics/Norit and Zenon) and were comparable to the fouling rates observed on the MOB train. Fouling rates decreased to normal level once ozonation was restored. The membrane feed turbidity increased 0.4 NTU during this period (ozone loss). However, BAC contactor run time was not affected by the increased turbidity.

Water quality excursions

Relevant raw water turbidity events are summarized in Table 2, the worst of which is discussed here (event 4). Membrane fouling rates and BAC filter runs did not change significantly during all other water quality events.

The fourth turbidity event was caused by Hurricane Isabel passing through the region. This event was characterized by increased turbidity and a temperature swing with significant amounts of sand, algae and particulate matter in the raw water. At the start of the event, an increased chlorine demand was observed in the Lake water, which in turn resulted in lower chlorine levels measured in the source water (intake well) used for the pilot and the dechlorination agent used for the purpose of the pilot plant was over dosed. The excess agent exerted an ozone demand, which ultimately resulted in lower ozone residuals through the ozone contactor (< 0.1 mg/L). Increased feed turbidity to the BACC (> 18 NTU) limited the filter run time to less than 2 hours necessitating a backwash cycle. During this 2 hour period, despite the compromised performance of both the ozone (where UV transmittance was reduced from >95% to 80%) and BACC (2 hour filter run), the BACC effluent was < 1.3 NTU. During this same period, there was no appreciable change in membrane fouling rates nor permeate quality. The BACC was able to achieve adequate organic removal and perhaps more importantly, served as a "roughing filter" for the membrane pilots. Filter run times in the BACC contactor increased from 2 to 12 hours once the ozone residual was restored. Source water and BACC effluent turbidities peaked at 40 NTU and 1.5 NTU, respectively, during this 12 hour period. Filter run times for the BACC were subsequently increased to more than 7 days for the balance of this turbidity event. No change fouling rates was observed in the OBM membrane pilots. In contrast, the fouling rate of MOB membrane pilot increased in excess of 20% over a CEB cycle during the first 27 hours of this 2-day event.

Table 2 Turbidity excursions

Event	Date	Duration (h)	Raw Water Turbidity		Pre-treated Water Turbidity		Turbidity Reduction (%)
			Peak	Avg.	Peak	Avg.	
1	Aug. 12	14	8.7	2.6	1.3	0.85	85
2	Sept. 8–9	26	2.1	0.9	0.6	0.34	71
3	Sept. 10 -19	9 days	0.5	0.5	0.4	0.3	40
4	**Sept. 19–20**	**48**	**40**	**9.5**	**1.5**	**0.46**	**93**
5	Sept. 20–24	72	–	1.9	–	0.34	82

Role of membrane filtration on ozone demand and decay

Ozone demand and decay testing was conducted at bench-scale with conditioned raw water as used in the OBM train and repeated with MOB membrane permeate. The results from this testing were used to determine the ozone generator capacity needed to provide for 1.0-log *Cryptosporidium* inactivation. Membrane filtration upstream of ozonation served to decrease the ozone generator size by 10% when compared with ozonation of raw water.

Organics

The profile of assimilable organic carbon (AOC) and biodegradable dissolved organic carbon (BDOC) through the OBM pilot train as measured on 22 October 2003 is presented in Figure 3. As expected, the concentration of AOC and BDOC increased following ozonation; BACC reduced the AOC concentration (by half) to levels less than 0.20 mg/L. However, the concentration of BDOC increased following BACC, which was not expected. The BDOC concentration in the contactor effluent was significantly reduced through the membranes to approximately 0.20 mg/L. Similar results were observed for the effectiveness of the BACC at controlling AOC and BDOC in the MOB train.

For both trains, TOC levels through the process were reduced marginally and results were inconclusive owing to the experimental error associated with the method used. Reductions of approximately 10% were observed through the OBM and MOB train, with average source levels of 2.2 mg/L TOC reduced on average to 1.9 mg/L TOC. A similar trend was observed for DOC, as most of the organics present in the source water are dissolved in nature.

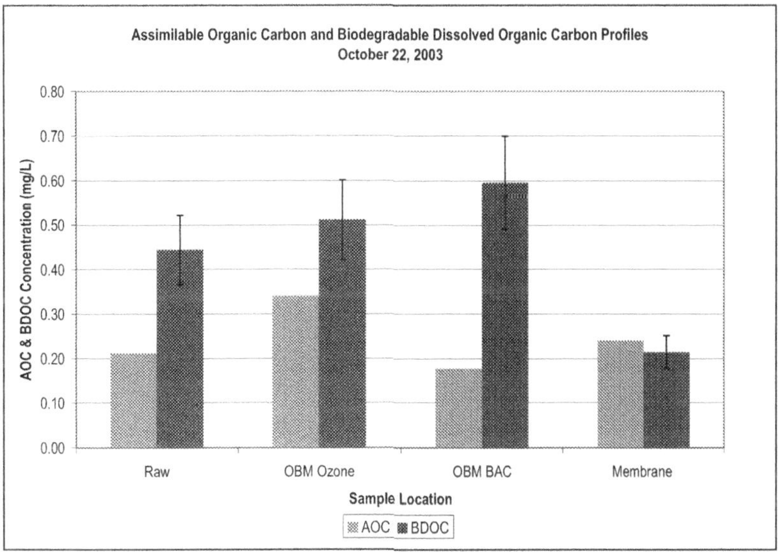

Figure 3 AOC and BDOC (OBM Train).

Disinfection by-products (AOX)

Adsorbable organic halides (AOX) were measured at various locations through both process trains to provide an indication of their formation and decay. The AOX concentration in the effluent from each train is displayed in Figure 4, with levels typically less than 0.1 mg/L. Source water AOX was attributed to the use of previously chlorinated water from the intake well. The raw water AOX concentration was reduced during ozonation through the OBM train (as high as 0.2 mg/L to 0.05 mg/L), with a further reduction through the BACC where concentrations of less than 0.04 mg/L were generally observed in the effluent. As such, no further reduction was observed by membrane filtration in the OBM train. In the MOB train, the raw water concentration of AOX was reduced by membrane filtration (from as high as 0.2 mg/L). Further reductions in AOX levels were observed during ozonation. AOX levels following ozonation were less than 0.04 mg/L.

In general, lower DBPs were observed in samples treated by the pilot plant (both MOB and OBM), demonstrating the impact that this process had on precursor removal. Again, samples treated with ozone and BACC showed a marginal increase in THMs relative to their non-ozonated counterparts. Lower concentrations of DBPs may be experienced at full-scale if the filters are operated longer in a biological mode. Levels of bromate in all three samples were 0.005 mg/L, which is less than the interim maximum acceptable concentration (set in Ontario) for bromate in drinking water of 0.010 mg/L.

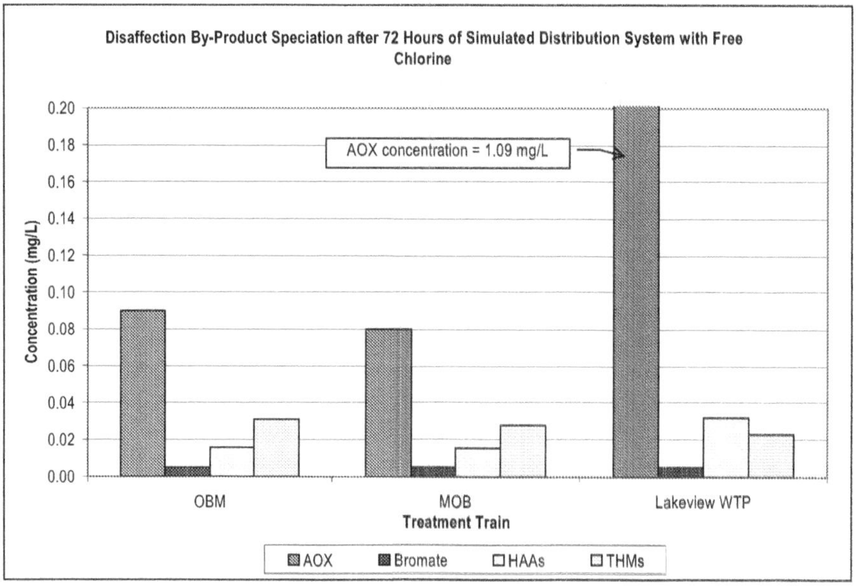

Figure 4 Disinfection by-product formation.

Taste and odour

A series of taste and odor (T&O) samples were collected from various locations within the pilot and the full-scale plant for Geosmin and MIB. The maximum Goesmin and MIB levels measured in the source water during the T&O event were < 20 ng/L and 2 ng/L, respectively. As expected, Gesomin was significantly reduced by ozone and BAC treatment. Geosmin concentrations of < 4 ng/L were generally observed after the pre-treatment train. Membrane filtration did not significantly contribute to the reduction of Geosmin in the finished water. This observation illustrates the importance of incorporating ozonation/BACC pre-treatment in the WTP for taste & odor control. Geosmin levels in the full-scale plant effluent ranged from 1.0 ng/L to 4.7 ng/L, which are 50% higher than the treated water from the pilot plant.

Key findings of pilot study

The key findings of the study are summarized here: The treatment train is robust, as demonstrated during water quality excursions that included, elevated turbidity events (up to 40 NTU), prolonged taste and odour events (3 weeks), rapid temperature fluctuations (±10°C in 24hrs), and elevated pathogenic risk (influence of WWTP). The multiple barrier approach is effective. During elevated turbidity events the pre-treatment train alone removes over 96 percent of the raw water turbidity. The membranes with or without pre-treatment are capable of producing water quality with turbidity less than 0.05 NTU.

Membrane fouling is predominantly organic in nature. Air-enhanced backwashes are effective in controlling short-term particulate fouling between CEBs. Daily chlorine CEBs are effective in controlling cumulative particulate and organic fouling. CIP intervals are greater than the required minimum of 6 weeks for all membrane pilots. Pretreatment can reduce membrane surface area by > 10% based on higher flux rates, longer filtration times and lower fouling rates observed during the pre-treated water demonstration.

Asimilable organic carbon levels following ozonation increased by 38%; however, levels following BAC contact were reduced by more than 50%. Similar observations were observed for biodegradable organic carbon. The OBM and MOB processes achieve lower DBPs (except for THMs) than the existing water treatment process. Reductions in the level of organic compounds, such as TOC and DOC, are generally low, and approximately 10%. Pretreatment with ozone/BAC is very effective at reducing the Geosmin concentration. Typically, the Geosmin levels in the BAC effluent were < 4 ng/L.

Conclusions

Pilot, bench-scale and design data were developed for the ozone, BAC, membranes treatment process. This process has been applied to the Lakeview and Twin Oaks Valley WTP, two large-scale drinking water installations in North America, each with a different treatment sequence. This sequence is driven by raw water quality, treatment objectives, and project specific need. These projects exemplify how fundamental science, technical know-how and sound engineering can be combined throughout the conceptualization, evaluation and design process to create an advanced treatment system that is cost effective while providing the highest-quality drinking water from variable and challenging source waters.

References

Garcia-Aleman, J., K. Mains, and A. Farr (2005) Integrating 360 ML/d of membranes, ozone and biological pretreatment at Lakeview WTP. In *Proceedings of the AWWA Membrane Technology Conference and Exposition 2005*, Phoenix, Arizona.

Garcia-Aleman, J., J. Sladic, and A. Farr. (2004); Pre-treatment of membrane filtration with ozone and biological filtration – the Lakeview WTP piloting experience' *Proceedings of the OWWA Conference*, Niagara Falls, Ontario.

Hishijima, W., *et al.* (1998) The performance of an ozonation-biological activated carbon process under long-term operation. *Water Science and Technology* **38**(6), 163–169.

Kunikane, S., *et al.* (1995) A comparative study on the application of membrane technology to the public water supply. *Journal of Membrane Science* **102**, 149–154.

Kunikane, S., *et al.* (1996) The role of micro- and ultrafiltration in drinking-water treatment. *Water Supply* **14**(3/4), 473–486.

Nishijima, W. and M. Okada (1998) Particle separation as a pretreatment of an advanced drinking water treatment process by ozonation and biological activated carbon. *Water Science and Technology* **37**(10), 117–124.

Oxtoby, S. (2002) Experiences in integrity testing ultrafiltration membranes at a large potable water treatment works. Vivendi Water Partnership.

Park, Y.G. (2002) Effect of ozonation for reducing membrane-fouling in the uf membrane. *Desalination* **147**, 43–48.

Park, Y.K., *et al.* (1997) Purification of polluted river water by ultrafiltration-ozonation-biological activated carbon filtration. *Water Science and Technology* **35**(7), 179–186.

Rittmann, B.E., *et al.* (2002) Treatment of a colored groundwater by ozone-biofiltration: pilot studies and modeling interpretation. *Water Research* **36**, 3387–3397.

Takeuchi, Y., *et al.* (1997) Removal of organic substances from water by ozone treatment followed by biological activated carbon treatment. *Water Science and Technology* **35**(7), 171–178.

van der Hoek, J.P., *et al.* (2000) Benefits of ozone-activated carbon filtration in integrated treatment processes, including membrane systems. *Journal of Water Supply Research and Technology—AQUA* **49**(6), 341–356.

Vahala, R. *et al.* (1999) Controlling adsorbable organic halogens (AOX) and trihalomethanes (thm) formation by ozonation and two-step granule activated carbon (GAC) filtration. *Water Science and Technology* **40**(9), 249–256.

Boron mitigation strategies in seawater desalination

R. Huehmer*, F. Wang, J. Lozier*** and L. Henthorne******

*CH2M HILL, 13921 Park Center Rd, Herndon, VA 20171, USA
(E-mail: *robert.huehmer@ch2m.com*)
**CH2M HILL Singapore Pte Ltd, 150 Beach Road, Gateway West 34th Floor, Singapore 189720 (E-mail: *felix.wang@ch2m.com*)
***CH2M HILL, 2625 S. Plaza Dr., Ste 300, Tempe, AZ 85282, USA
(E-mail: *james.lozier@ch2m.com*)
****CH2M HILL International, Al Thuraya Tower, Ste 606, Dubai Internet City, P.O. Box 500554, Dubai, Unite Arab Emirates (E-mail: *lisa.henthorne@ch2m.com*)

Abstract Boric acid is found in natural seawaters at concentrations typically ranging from 4 to 7 mg/L. Current World Health Organization (WHO) drinking water concentrations for boron are limited to 0.5 mg/L. Commercially available seawater reverse osmosis (SWRO) membranes possess a wide range of rejection characteristics for boron in seawater under ambient temperature and pH, ranging from approximately 50% for low-energy membranes to greater than 90% for the newest high rejection membranes. This level of rejection is typically insufficient to reduce boron concentrations in natural seawater to less than recommended levels. One technique that may be utilized to mitigate boron concentrations is to dissociate boric acid by increasing pH before SWRO. Utilizing this technique, the authors tested commercially available SWRO membranes from three different manufacturers. The authors found that all three SWRO membranes were able to produce permeate complying with WHO regulations, with a cost of up to \$0.05/m^3 of permeate produced. However, with current total water cost (TWC) ranging from \$0.50/m^3 - \$1.60/m^3, this technique represents up to 10% of the TWC and may not be appropriate for large scale desalination facilities.

Keywords boron; desalination; pH adjustment; rejection; reverse osmosis; seawater

Introduction

Recent improvements in seawater reverse osmosis (SWRO) membrane science have yielded commercially available membranes possessing total dissolved solids rejection of greater than 99.7%. This has resulted in the design and development of desalination facilities that achieve adequate permeate quality utilizing a single-pass SWRO system, rather than a second pass or partial second pass commonly in use. Unfortunately, the rejection of boric acid in SWRO membranes is lower than sodium chloride and other ionic constituents typically found in seawater. Commercially available SWRO membranes possess a wide range of rejection characteristics for boron in seawater under ambient temperature and pH, ranging from approximately 50% for low-energy membranes to greater than 90% for the newest high rejection membranes. In many current designs, permeate boron concentrations are the limiting factor when determining the plant design, requiring a partial second pass or other mitigation technique to produce acceptable potable water.

One method of increasing boron rejection in SWRO applications is to promote dissociation of boric acid into ionic species by increasing the pH of the feed water. In order to evaluate this technique, the authors conducted testing at a SWRO pilot facility located in South-East Asia. The objectives of the study were to determine the boron permeate concentration versus feed water pH, evaluate the boron rejection (based upon [$C_{feed\text{-}concentrate}$]) versus pH and to determine the operating-cost impact of increasing the pH to mitigate permeate boron concentrations in excess of 0.5 mg/L, in a single stage system, for each of the manufacturers studied.

Regulatory background of boron

Several regulatory bodies have implemented drinking water standards for boron, which World Health Organization (WHO) reports may inhibit foetal development and/or cause testicular lesions (WHO, 2003). WHO first provided a health based guideline value of 0.3 mg/L in 1993. In 1998, WHO increased the recommended concentration to 0.5 mg/L, while revising it to a provisional value due to the limited ability of conventional water treatment technologies to remove boron. WHO is currently re-evaluating, with future guidance expected, the recommended boron concentration. The European Union first established a boron limit of 1.0 mg/L in 1980. The maximum boron concentration was reiterated in Council Directive 98/83/EC (European Union, 1998). The European Union sets minimum regulatory standards for its member states. Currently, there is no United States Environmental Protection Agency (USEPA) rule regulating boron concentrations in drinking water. Boron has been identified as a candidate contaminant by USEPA; future regulatory action may occur (USEPA, 2005). Although the United States does not possess boron regulations, the state of California has established regulatory language. California Department of Health Services (CDHS) has established an action level of 1 mg/L for boron (CDHS, 2005). Under CDHS's action level regulations, systems producing water with greater than 1 mg/L of boron may be required to implement one or more mitigation strategies. Table 1 summarizes the regulation of boron by these agencies.

Table 1 Allowable boron concentrations in drinking water by regulatory agency

Regulatory Agency	Regulatory Limit
World Health Organization	0.5 mg/L
European Union	1.0 mg/L
United States	None
California, USA	1 mg/L

Boron occurrence and chemistry

Seawater boron concentrations range from non-detect to approximately 7 mg/L, with typical concentrations in the range of 4–6 mg/L. Monitoring data in the greater Los Angeles area determined that boron was present in concentrations ranging from 4 to 4.75 mg/L (Huehmer and Henthorne, 2004). Boron is usually present in water as boric acid, with the level of dissociation into ionic species dependent primarily upon pH.

Equations 1 through 3 describe the dissociation of boric acid (Weast, 1989) at 20°C. Equations 2 and 3 have a marginal impact at pHs utilized in desalination processes. Figure 1 illustrates the concentration of each species in Equations 1 through 3 versus pH in distilled water.

$$H_3BO_3 \Longleftrightarrow H^+ + H_2BO_3^- \qquad pK_a = 9.14 \tag{1}$$

$$H_2BO_3^- \Longleftrightarrow H^+ + HBO_3^{2-} \quad pK_a = 12.74 \qquad (2)$$
$$HBO_3^{2-} \Longleftrightarrow H^+ + BO_3^{3-} \quad pK_a = 13.8 \qquad (3)$$

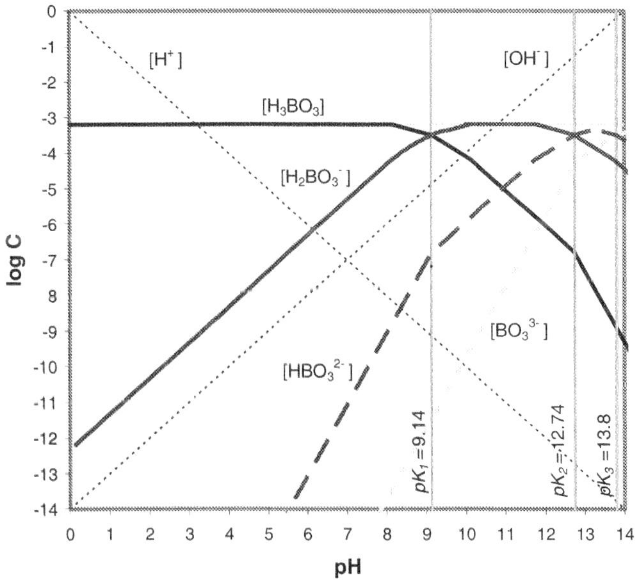

Figure 1 Log Concentration versus pH boron.

Seawater temperature and ionic strength have lesser impacts upon boric acid dissociation than pH (Busch *et al.*, 2003). Figure 2 displays the effects of pH on dissociation of boric acid into dihydrogen bromate at 20°C. At typical seawater pH of 8.0, over 90% of boric acid is non-dissociated.

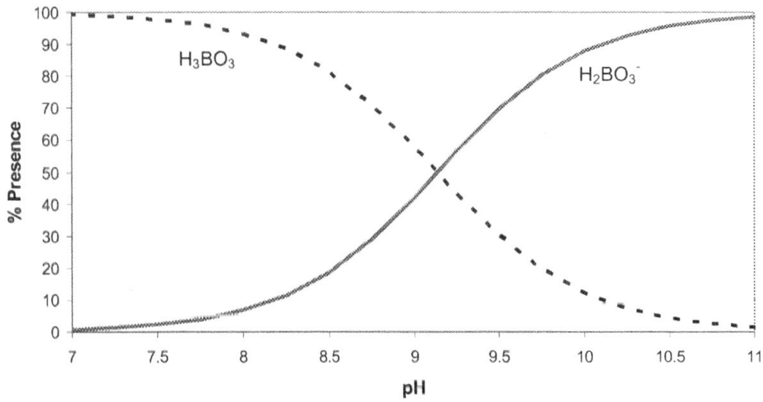

Figure 2 Dissociation of boric acid as a function of pH.

Removal of boric acid using reverse osmosis

Rejection of boron by reverse osmosis membranes is influenced by several factors including feed water pH, TDS, temperature, membrane type, permeate flux and system recovery. Non-dissociated charge neutral boric acid molecules possess poor rejection characteristics using reverse osmosis membranes. Ionization of the boric acid into borate species (using pH adjustment) substantially increases the level of membrane rejection. Busch *et al.* (2003) studied the rejection of boron at standard conditions. They found that boron rejection at typical seawater pH (pH 8) resulted in rejections ranging from 55% using low-pressure, high flux membranes to 90% using high rejection seawater membranes. At a pH of 9.5, the rejection ranged from 84% to 97%) with increasing rejection experienced at higher pH. Busch *et al.* (2003) also found that temperature can have an effect on the rejection of boron. With increasing temperature, the pK_a decreases, increasing ionization of boron and therefore the SWRO rejection. This effect is of lesser impact than the increase in solute mass transfer through the membrane resulting from increased temperature. Research performed by Fusaoka *et al.* (2003) reports upon the development of high rejection SWRO membranes achieving over 95% rejection of boron at 25°C and ambient pH (pH 8) and system recovery of up to 60%. These membranes operate at pressures of approximately 8–10 MPa, resulting in substantial increases in power consumption over conventional membrane desalination. Bench scale testing performed by the Long Beach Water Department (Tseng *et al.*, 2004) examined the mitigation of boron in a two-pass NF seawater desalination system. They found that pH adjustment of the second stage feed water to 9.5 using 15 mg/L of NaOH increased boron rejection sufficiently to maintain permeate concentrations of less than 1.0 mg/L. In limited testing, they found that boron rejection substantially deteriorates with increasing water temperatures.

Several different mitigation strategies have been employed to address boron concentrations in permeate. Mitigation strategies include the use of a second pass, pH adjustment before either the first or second pass, reduction of recovery or the use of ion exchange. Recently, a four-stage membrane process with pH adjustment has been successfully used to achieve boron concentrations of less than 0.3 mg/L (Busch *et al.*, 2003; Redondo *et al.*, 2003).

Methods

The pilot facility utilized for the boron rejection testing consists of an open water intake with bar screens and nominal 100 μm screens. Coagulation using approximately 3 mg/L ferric chloride was utilized to coagulate colloidal particles. The coagulated water was filtered using a pressurized ultrafiltration membrane system. Filtrate from the ultrafiltration system was transferred into a filtrate holding tank. Feed water transfer pumps transferred the water from the tank to the SWRO. Before the high-pressure pump, the pH was adjusted utilizing sodium hydroxide or sulphuric acid, an inline injection point and inline static mixers. The pH-corrected water was filtered through a 5 μm cartridge filter, then pressurized to approximately 50 bar(g) using a positive displacement pump.

The reverse osmosis system consisted of a three single-stage RO systems in parallel, each consisting of two 10 cm (4") diameter pressure vessel configured for six membranes in each vessel (6M design). Each system possessed independent feed throttle valves, concentrate throttle valves, and instrumentation. Sample taps were located at the feed, concentrate and permeate streams of each system. The systems were designed such that all three systems receive identical feed water. Membranes from three different SWRO membrane manufacturers were utilized in the study. Table 2 summarizes the key performance data outlined in the manufacturer's specification sheet. Each SWRO system is loaded with membranes from a single manufacturer, such that the performance of the three membrane products can be tested in parallel.

Table 2 SWRO Membrane Specifications

Parameter	Manufacturer A	Manufacturer B	Manufacturer C
Nominal Salt Rejection	99.4%	99.5%	99.75%
Permeate Production	7.4 m^3/d	4.5 m^3/d	4.5 m^3/d
Membrane Area	7.4 m^2	6.5 m^2	7 m^2
Factory Test Conditions			
- Temperature	25°C	25°C	25°C
- NaCl concentration	32,000 mg/L	32,000 mg/L	32,000 mg/L
- Pressure	5.5 MPa	5.5 MPa	5.5 MPa
- Recovery	8%	10%	8%

At the time of the testing, the three SWRO trains had been in operation for approximately 9 months. Before performing the testing, the membrane systems were cleaned, and the membranes were probed to ensure that the membrane systems were integral. Since seasonal variations in seawater quality were observed during the study, the testing was performed during a period of the year when relatively stable water quality was expected. Before testing, all pH probes, flow meters, conductivity probes and pressure transmitters were cleaned, verified and calibrated (or replaced in the case of the pressure transmitters) utilizing the manufacturers recommended procedures.

On each day of testing, the system SWRO flow and recovery were adjusted so that each system was operating at 45% recovery and 13.5 l m h. The system was operated for 30 minutes to allow process stabilization to occur. The pH was then adjusted to 6.5 using sulphuric acid. Following each pH change, the SWRO systems were operated for 30 minutes to stabilize performance at the adjusted pH. Once the system stabilized, the permeate and concentrate flows, feed, permeate and concentrate pressures, feed and permeate conductivity, and feed pH were recorded. Water samples were collected from the combined feed, combined concentrate and each SWRO permeate stream for laboratory analysis. The chemical feed-rate was determined using drawdown tests to determine the acid or sodium hydroxide dose. Once the samples were collected, the pH was adjusted to the next increment using either acid or base (as required). In all, the test was performed using pHs from 6.5 to 9.5 in half pH unit increments. Two days were required to perform the testing.

To evaluate the rejection data, the feed-concentrate concentration (Equation 4) was first calculated. Equation 5 was utilized to determine the rejection of the system during testing.

$$C_{feed-concentrate} = C_f \cdot \frac{\ln \frac{1}{1-Y}}{Y} \tag{4}$$

where, Y = recovery (%), C_f = feed concentration

$$R = 1 - \frac{C_{permeate}}{C_{feed-concentrate}} \tag{5}$$

where, R = rejection

The resultant data for each pH value was plotted to determine the required pH dose point to achieve a permeate boron concentration of less than 0.5 mg/L.

Results and discussion

After the completion of field testing and analysis, the data was analyzed to determine the ratio of the permeate concentration over the feed water concentration and plotted with respect to pH. Figure 3 presents the specific boron concentration versus pH. The permeate boron concentration of from the system loaded with membranes from Manufacturer A outperformed the other manufacturers' membranes in terms of permeate boron concentration. With an average feed water boron concentration of 4.25 mg/L, Manufacturer A's membranes produced permeate boron concentrations of less than 0.5 mg/L at pH 8.

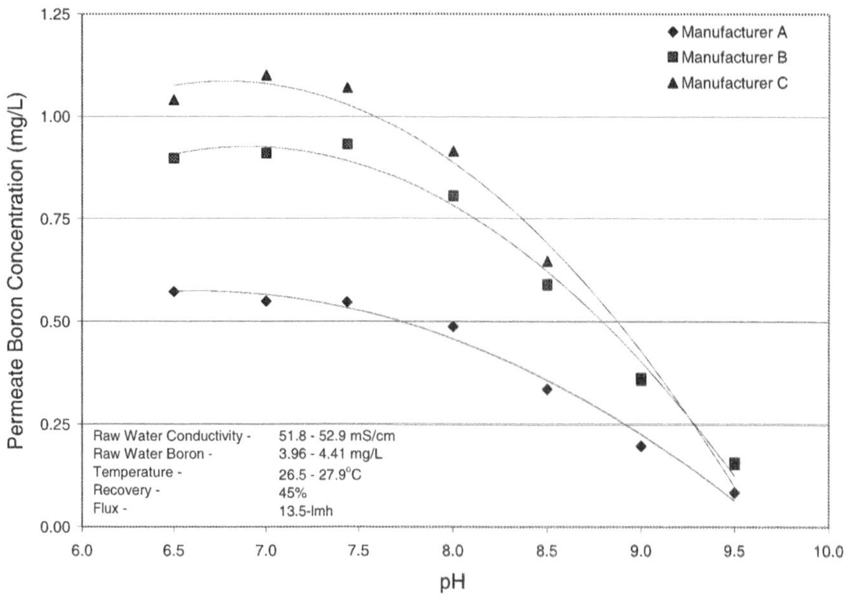

Figure 3 Permeate boron concentrations for SWRO membranes tested.

Figure 4 represents the membrane rejection of both total dissolved solids (TDS) (as estimated using conductivity) and boron. Each manufacturer's membranes achieved TDS rejections of greater than 99.4%. The system loaded with Manufacturer A's membranes outperformed its specifications (Table 2). The system loaded with Manufacturer B's membranes met the published specifications. Unexpectedly, Manufacturer C's membranes performed well below the specifications for the elements. For all three manufacturers, the membranes exhibited a slight increase in rejection with respect to increasing pH.

The systems loaded with membranes from Manufacturer B and Manufacturer C performed poorly at ambient seawater pH (~7.9) and lower with respect to boron rejection. To achieve permeate boron concentrations of less than 0.5 mg/L, pH adjustment to 9 was required for both of these systems. The membrane rejection was then evaluated using Equations 4 and 5. The resultant boron rejection versus pH is presented in Figure 4. Manufacturer A outperformed the other manufacturers in terms of boron rejection, with boron rejection exceeding 90% even at pH 6.5. Membranes from both Manufacturer B and Manufacturer C possessed boron rejections of less than 85% at pH 6.5. As expected from

Figure 2, increasing pH resulted in improved performance of the membranes, with rejection exceeding 97% at pH 9.5 for all three manufacturers.

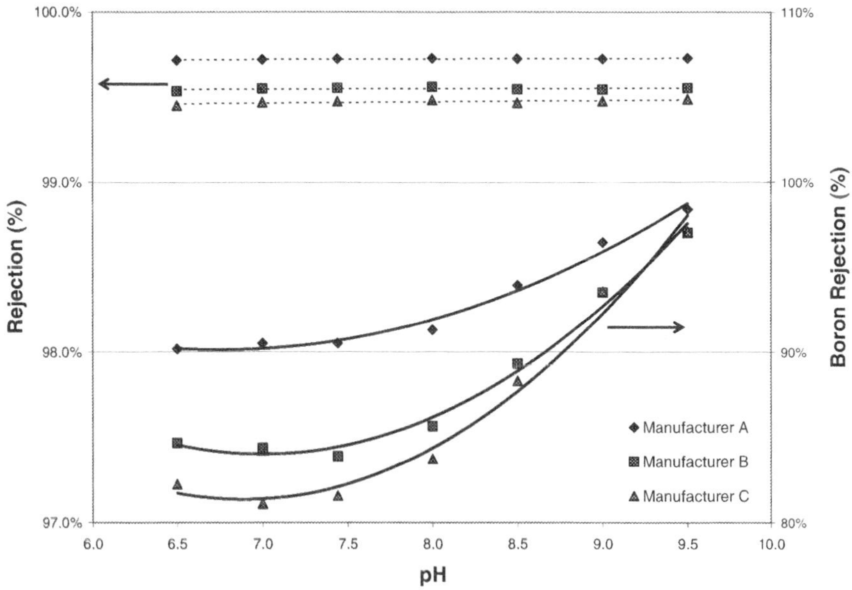

Figure 4 Percentage rejection for SWRO membranes tested.

Titrations performed during testing indicated that sodium hydroxide doses of 21.1 mg/L, 42.6 mg/L and 62.7 mg/L were required to raise the pH from seawater ambient to 8.5, 9.0 and 9.5, respectively. Based upon bulk sodium hydroxide costs of $0.33/kg, this equates to operating cost increases of $0.0155/m^3 (pH 8.5), $0.0312/m^3 (pH 9.0) and $0.046/m^3 (pH 9.5) to implement this boron mitigation technique. With total water costs on the order of $0.50/ m^3 for recently constructed desalination facilities (Gluckenstein and Priel, 2003), this operations component would add an additional 3–9% to the total water cost.

Conclusions

The degree of dissociation of boric acid in seawater is a major factor in the rejection of boron using SWRO. Of the three manufacturers' membranes studied, the highest rejection of boron observed was approximately 90% at ambient seawater pH. With increasing pH, all three manufacturers' membranes were capable of achieving boron concentrations of less than 0.5 mg/L.

The use of pH correction of SWRO feed water to increase ionization and rejection of boron is technically feasible. With estimated operating cost increases of $0.0155/m^3 (pH 8.5), $0.0312/m^3 (pH 9.0) and $0.046/m^3 (pH 9.5) to implement this boron mitigation technique, it is not likely a feasible solution for large desalination systems. With bulk sodium hydroxide prices increasing from $0.06/kg in April 2004 to $0.33/kg in February 2005 (Mitsch, 2005), use of alternative methods of boron mitigation may be more economical than utilizing first-stage feed pH adjustment. Future work will examine alternative boron mitigation techniques.

References

Busch, M., Mickols, W., Jons, S., Redondo, J. and De Witte, J. (2003). Boron Removal in Seawater Desalination. In: *Proceedings of the 2003 International Desalination Association Annual Conference*, Bahamas.

California Department of Health Services (CDHS) (2005). *DHS Drinking Water Notification Levels: An Overview*. California Department of Health Services - Drinking Water Program, September 30, 2005.

European Union (1998). COUNCIL DIRECTIVE 98/83/EC of 3 November 1998 on the quality of water intended for human consumption. In: *L 330/32 EN, Official Journal of the European Communities*, 5 December 1998.

Gluckenstein, P. and Priel, M. (2003). Potential cost reduction of seawater desalination, *Water Science and Technology: Water Supply*. **3** (5–6) 39–47.

Fusaoka, Y., Ooto, K., Sugita, K. and Kurihara, M. (2003). Progress of the High Performance SWRO Elements for Boron Removal. In: *Proceedings of the 2003 International Desalination Association Annual Conference*, Bahamas.

Huehmer, R. and Henthorne, L. (2004). *LADWP Seawater Desalination Facility Feasibility Study for the Scattergood Generating Station in Playa Del Rey*, Report for the Los Angeles Department of Water and Power, Los Angeles, CA, USA.

Mitsch, F. (2005). *Reacting to Chemicals*. www.CNNmoney.com (accessed 13 February 2006).

Redondo, J., Busch, M. and De Witte, J.P. (2003). Boron Removal from Seawater using FILMTEC High Rejection SWRO Membranes. *Desalination*, **156**, 229–238.

Tseng, T.J., Cheng, R.C., Voung, D.X. and Wattier, K. (2004). Bench and Pilot-Scale Investigation of Boron Removal for Seawater Membrane Desalination. In: *Proceedings of the 2004 American Water Works Association Annual Conference and Exhibition*, Orlando, FL, USA June 15, 2004.

United States Environmental Protection Agency (USEPA) (2005). *Drinking Water Contaminant Candidate List 2; Final Notice, Federal Register: February 24, 2005*, Volume 70, Number 36, http://www.epa.gov/fedrgstr/EPA-WATER/2005/February/Day-24/w3527.htm (accessed 13 February 2006).

Weast, R. C. (ed.) (1989). *CRC Handbook of Chemistry and Physics*, 70th edition. CRC Press, Boca Raton, Florida.

World Health Organization (WHO) (2003). *Boron in Drinking-water: background document for development of WHO Guidelines for Drinking-water Quality (2003).*

Fouling behaviours of ultrafiltration membranes during dead-end filtration of coagulated water with different initial mixing conditions

H. C. Kim, J. H. Hong and S. Lee

Water Environment and Remediation Center, Korea Institute of Science and Technology, P.O. Box 131, Cheongryang, Seoul 130-650, Korea
(E mail: *animaplus@hanmail.net*, *jhhong@kist.re.kr*; *seocklee@kist.re.kr*)

Abstract The fouling characteristics in the UF membrane filtration of waters pre-treated by chemical coagulation using different initial mixing conditions were compared and the influence of natural organic matter (NOM) on the fouling of membranes was investigated. It was suggested that organic matter in the molecular weight ranges 300–2,000 Daltons was mainly responsible for the fouling. The fouling was greater for hydrophobic than hydrophilic membranes. ATR-FTIR analysis of the fouled hydrophobic membrane indicated aliphatic amide and alcoholic compounds as well as polysaccharides contributed to significant membrane fouling. These adsorptive foulants are considered as neutral fractions present in hydrophobic and hydrophilic NOM components. In the case of similar hydrophilic fractions, water pre-coagulated with a high hydrophobic content resulted in greater flux decline, which was presumed to be due to the organic matter with neutral properties contained within the hydrophobic fraction. The relative concentrations of each NOM fraction in coagulated water are important. Mechanical mixing for chemical coagulation, with a backmixing-type, rather than pump diffusion mixing, with an in-line type, is likely to be more effective at reducing the fouling caused by NOM.

Keywords Coagulation; fouling; natural organic matter; ultrafiltration

Introduction

Membrane filtration is a promising method because denaturing of organic matter is minimized and it does not need any additions of chemicals during operation. Fouling is one of main disadvantages in membrane filtration processes, which is defined as the loss of membrane permeability due to the accumulation of solutes onto the surface of the membrane and/or into its pores. Many studies have suggested that natural organic matter (NOM) is a most important foulant due to its possible interactions between membrane surfaces. Factors potentially affecting membrane fouling by NOM include properties of the feed constituents (molecular weight distribution, hydrophobicity and charge density) and the membrane (hydrophobicity, surface roughness, porosity), as well as the solution phase (pH, ionic strength, Ca^{2+} concentration) (Cho and Amy, 1999; Lee *et al.*, 2005). Although studies on the role of solution composition have been undertaken, there is little agreement on the role of the each NOM fraction. For instance, Fan *et al.* (2001) found that the major contribution to fouling was from the hydrophilic neutral fraction within bulk NOM; however, others have reported that the hydrophobic fraction, such as humic substances, is a major foulant controlling the rate and extent of fouling (Yuan and Zydney, 1999).

The removal of NOM before membrane filtration has became an important issue for the reduction of membrane fouling. The most common technique for treating water, including NOM, is coagulation/sedimentation with aluminium or ferric salts. According to Carroll *et al.* (2000), coagulation treatment improves the removal of NOM and reduces membrane fouling. In the study performed by Pikkarainen *et al.* (2004), the specific cake resistance decreased with increasing dosage of coagulants used for the removal of NOM. The effectiveness of a coagulation process in removing NOM varies according to the nature of the NOM (its molecular weight, charge density, hydrophobicity and so forth), and the physicochemical characteristics of the water and the operational conditions, such as the initial mixing intensity and duration (Letterman *et al.*, 1999). Above all, the initial mixing conditions are critical in a coagulation process, since rapid and uniform dispersion of metal coagulants has advantages for the formation of charge-neutral products. In a previous study, differences in the removal of NOM have been found between pump diffusion and mechanical mixer of the in-line type and backmixing-types, respectively, for chemical coagulation. It was reported that an in-line pump diffusion mixer is effective for the removal of the non-humic fraction within bulk NOM in comparison to a backmixing-type mixer (Kim *et al.*, 2005).

Membrane fouling might also be influenced by different NOM compositions, since NOM components could be selectively removed by chemical coagulation using different initial mixing conditions, as mentioned above. In recent, related research, Park *et al.* (2002) and Oh and Lee (2005) filtered a large volume of coagulated waters through a membrane filter under different initial mixing conditions. The results of these studies differed, with very little agreement on the consistency of in-line pump diffusion mixing being a good system for the reduction of flux decline compared to back-mixing. However, the water samples used in their studies contained both large particles (i.e., flocs) and dissolved organic matter, and little attention was given to the contribution of NOM toward membrane fouling. Therefore, the main purposes of this paper were to compare the membrane performance in the filtration of the pre-treated waters obtained using different initial mixers for the chemical coagulation, and investigate the influence of NOM characteristics on the fouling of membranes.

Materials and Methods

Membranes and filtration apparatus

The water samples were filtered with a 0.45 μm filter before UF membrane filtration. Two types of UF membrane, hydrophobic (UF-HPO, Polyethersulfone) and hydrophilic (UF-HPI, Regenerated cellulose), supplied by Millipore, were used to compare the flux decline in relation to their properties. The molecular weight cut-off (MWCO) pore size of both these membranes was 100 kDa. Flux decline tests were performed using a dead-end stirred-cell filtration unit (Model 8200, Amicon) connected to a feed reservoir and nitrogen gas bomb. The permeate discharges from the bottom of the cell, under constant pressure (14 psi) and temperature, as well as the flux were checked at 15 second intervals.

Initial mixing conditions for chemical coagulation

A pilot plant, designed to treat 1,000 m^3/day, was built in the a water treatment plant, optimized for energy efficiency and mixing intensity, and operated for the study of the feasibility of introducing pump diffusion flash mixing (PDFM), with in-line mixing, as shown in Figure 1. PDFM is an effective in-line mixing method for the quick dispersion of a hydrolyzing coagulant, which also has the potential to solve problems, such as the noise, high-energy consumption and high maintenance cost, associated with the use of mechanical mixers. The water treatment plant employs conventional water treatment, consisting of coagulation/sedimentation, sand filtration and chlorination, and uses Han River water as a water source. The Han River water was pre-chlorinated to satisfy the CT value for viruses within the aqueduct.

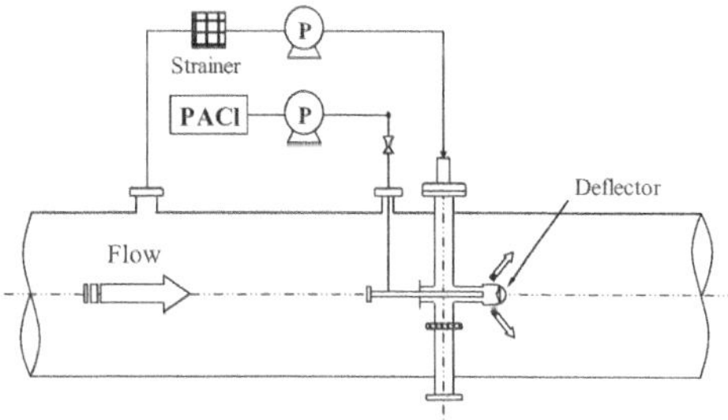

Figure 1 Schematic diagram of the PDFM system.

A scaled-down mechanical mixing basin was used for comparison with PDFM; this mixing type is defined as conventional rapid mixing (CRM) in this study. CRM consists of a square baffled container constructed from polymethylmethacrylate, with a 6 flat-bladed stainless steel turbine impeller, one-third the standard geometry, similar to that of the turbine impeller used in a real plant. The typical residence time (*t*) and velocity gradient value (*G*) for mechanical mixing were applied for adequate CRM operation. The operational conditions for PDFM and CRM for coagulation treatment are shown in Table 1. Commercially available liquid polyaluminium chloride (PACl, 17% as Al_2O_3) solution was used as the coagulant in all experiments, with the optimal coagulant dose for PDFM and CRM operation decided using the jar-test.

Table 1 Operation conditions for CRM and PDFM[a]

Mixing type	PACl dosage	Rapid mixing		Slow mixing	
		G (s^{-1}) [a]	t (sec) [b]	G (s^{-1}) [a]	t (min) [b]
CRM	15 ppm (v/v) [c]	320	30	25	30
PDFM	15 ppm (v/v) [c]	680	1	25	30

[a]Velocity gradient; [b]residence time; [c]3.1 mg/L as Al_2O_3.

NOM characterization and identification of foulants

NOM isolation

The pH of the raw and treated waters was adjusted to 2 by the addition of hydrochloric acid, following filtration through a 0.45 μm membrane filter. The sample solution was then isolated into three fractions by sequentially passing through DAX-8 and XAD-4 resins. The hydrophobic (HPO) and transphilic fractions (TPI) are DAX-8 and XAD-4 absorbable, respectively. The effluent collected is termed as the hydrophilic fraction (HPI).

ATR-FTIR spectroscopy

An infrared spectrometer (Infinity Gold 60AR, Thermo Mattson, USA), fitted with a Single Reflection Horizontal ATR (PIKE technologies, USA), was used to identity the functional groups of the clean and fouled membrane. After the membranes were washed to remove any cake layer on their surfaces filtration, then placed in sterilized Petri dishes and stored in desiccators for 48 hours. Infrared spectra of various membranes were obtained by scanning with an attenuated total reflection-Fourier transform infrared (ATR-FTIR) spectrometer. All spectra were normalized after acquisition to a maximum absorbance of 1.0 for comparative purposes.

Analytical methods

The dissolved organic carbon (DOC) was determined using a total organic carbon (TOC) analyzer (TOC-V CPN, Shimadzu, Japan), after filtration of the samples through a 0.45 μm membrane filter. A UV-visible spectrophotometer (Cary 50 Probe, Varian, USA) was used to determine the absorbance at 254 nm (UV_{254}). Turbidity was measured using a turbidimeter (2100P, Hach, USA). The concentrations of calcium were measured using inductively coupled plasma spectrometry (ICPS-7500 spectrometer, Shimadzu, Japan).

Molecular weight separation was performed by high performance size exclusion chromatography (HPSEC), using a HW-50S column. A high performance liquid chromatograph (LC-20AD, Shimadzu, Japan) was used, with UV detection (SPD-20AD, Shimadzu, Japan) at 254 nm and an on-line DOC detector (Modified Sievers Total Organic Carbon Analyzer 800 Turbo). Polyethylene glycols (PEG) were used to calibrate the relationship between the molecular weight and the retention time.

Results and discussion

NOM removal with different mixing types

The DOC, specific ultra-violet absorbance (SUVA) and zeta potential, as well as the turbidity remaining in solution after coagulation/sedimentation under different mixing conditions, are shown in Table 2. In the case of PDFM, the residual DOC was relatively low compared to that of CRM, while the SUVA, which has been used as an indicator of humic content in water environmental systems, was reduced to a lesser extent than in the case of CRM. The SUVA value was calculated from the UV_{254} divided by the DOC of the water sample, with the UV_{254} mainly caused by electro-rich sites, such as aromatic functional and double-bonded carbon groups in the NOM molecules. In general, the non-humic fraction of bulk NOM has less electron-rich sites than the humic fraction, implying the non-humic fraction in bulk NOM is more effectively removed by PDFM than CRM. This assumption was identified by isolation of NOM components from raw and treated waters, as shown in Figure 2. Figure 2 shows the variation in the distribution of hydrophobic, transphilic and hydrophilic fractions during water treatment, as determined in relation to the DOC concentration. The relative concentration of the hydrophilic fraction increased with the removal of the hydrophobic and transphilic fractions via CRM. In the case of PDFM, with the transphilic fraction mainly removed, the relative concentrations of the hydrophobic and hydrophilic fractions were increased. From the result of NOM isolation, the transphilic fraction was more effectively removed by PDFM, while the hydrophobic fraction was effectively reduced by CRM. However, no significant difference in the removal of the hydrophilic fraction was observed between the different mixing conditions.

Table 2 Variation in water quality of raw and treated waters[a]

Parameter	Raw water	CRM	PDFM
Turbidity	6.43	0.46	0.36
DOC (mg/L)	2.53	2.04	1.88
UV_{254} (cm^{-1})	0.0281	0.0172	0.0186
SUVA (L/mg/m)	1.11	0.84	0.99
pH	7.9	7.8	7.8
Conductivity (μs/cm)	171	179	176
Zeta potential (mV)	−15.1	-4.0	-0.5
Ca^{2+} (mg/L)	0.05	0.03	0.04
Si (mg/L)	2.22	1.85	1.93

[a] Filtered with a 0.45 μm MF filter before analysis

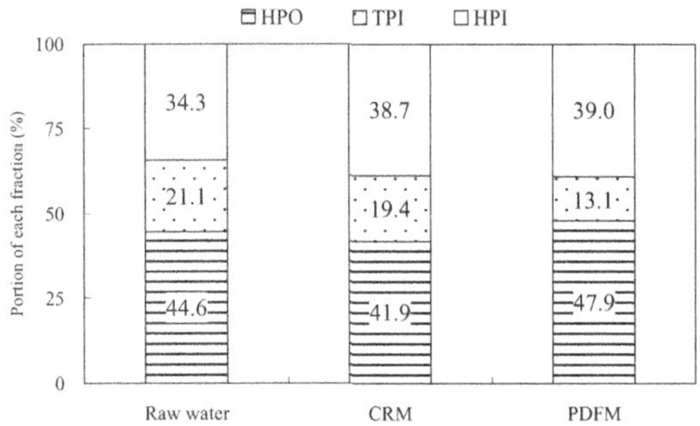

Figure 2 Distribution of NOM components with different mixing conditions.

Amirtharajah (1981) suggested that the smaller eddies formed at higher G values might be more successful at transporting preformed aluminium hydrolysis species to the surface of the particles responsible for turbidity, with the rapid-mix G value being important in charge neutralization. Consequently, the coagulation mechanism in the case of CRM was dominated by conventional entrapment (i.e., sweep coagulation), while PDFM was more likely influenced by charge neutralization due to the quick dispersion of aluminium salts prior to the formation of the aluminium hydroxide precipitate. It was suggested that due to the different reactivities of NOM fractions toward cationic aluminium hydrolysis products, the transphilic fraction was more effectively removed by PDFM than CRM.

Organic fouling of UF membranes

Figure 3 shows the ratio of permeate to the initial flux (J/Jo), as a function of the delivered DOC (cumulative feed-water DOC mass per unit area of membrane surface = feed water DOC (mg/L) × cumulative volume (L)/membrane surface (m^2)) for the filtrations of raw and treated waters using UF-HPO/HPI membranes. The rate of flux decline for the UF-HPO was considerably greater than that for the UF-HPI. In general, NOM components smaller than the membrane pores can be adsorbed to surfaces, which increases the flux decline, while larger components can block the pore entrances, which contributes to cake formation. Therefore, it is important to know the composition and size distribution of NOM components.

Cho *et al.* (1998) reported that negative-charged membranes can be easily fouled by the hydrophilic neutral fraction. Fan and co-workers (2001) also proposed a mechanism for membrane fouling by the hydrophilic neutral NOM fraction, which involved a combination of adsorption of small molecules onto the membrane pore wall, with pore blockage by colloidal organics within the membrane pore structure. They also identified the potential foulants in order: hydrophilic neutrals > hydrophobic acids > transphilic acids. In the case of PDFM, in contrast to CRM, the treated water contained relatively low transphilic and high hydrophobic fractions, as shown in Figure 2.

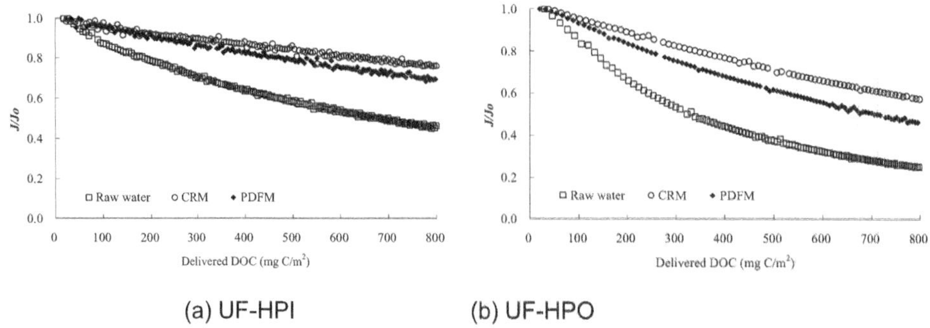

Figure 3 Comparison between flux decline of raw and treated waters with different ultrafiltration membranes.

The fouling effect was higher with PDFM than CRM for similar hydrophilic concentration and molecular weight distribution (Figures 2 and 4). It is reasonable to suppose that significant fouling occurred due to the hydrophobic as well as the hydrophilic fraction, whether the fouling is irreversible or not. Therefore, CRM is more effective in respect to the reduction in fouling by NOM compared to PDFM. Although the relative contributions to fouling by residual NOM and flocs could be influenced by the morphological characteristics of the flocs, the relative concentrations of each fraction in the NOM components are also important in hybrid membrane systems that employ pre-coagulation treatment.

Identification of foulants

The relative molecular weight distribution is shown in Figure 4 as determined by the PEG standards. For further interpretation, the chromatograms were divided into four bands, based on previous research (Huber, 1998). Substances in bands (I) and (II) were assigned as polysaccharides and humic substances, as large molecules (or aggregates), respectively. Substances in band (III) were assigned as the building blocks of refractory organic substances, while the low molecular weight substances (e.g., carbohydrates, aldehydes or alcohols) were assigned to band (IV). Bands (I) and (II) were effectively reduced through coagulation treatment, irrespective of the initial mixing conditions, which caused decreases in the flux decline. It is suggested that these bands included significant organic foulants.

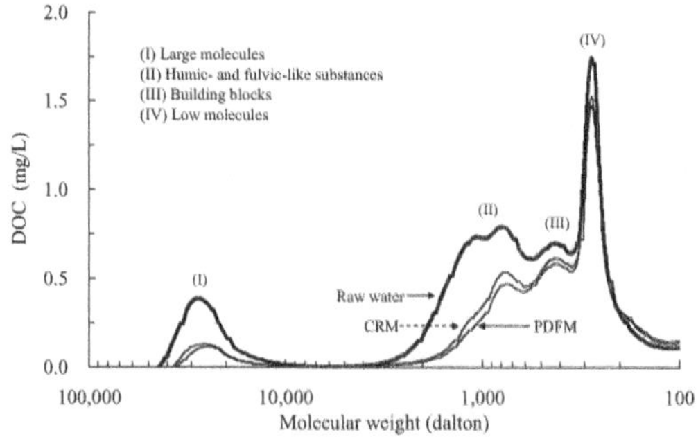

Figure 4 HPSEC-DOC chromatograms for raw and treated waters.

The FT-IR spectra of a clean UF-HPO, as well as one fouled during filtration of the raw water, are shown in Figure 5. Interpretation of the absorption bands of the NOM was performed, as described by Aiken *et al.* (1985) and Stevenson (1994). The peak at 1660 cm^{-1} was attributed to the C=O stretching of amide groups (amide I band), and this stretching vibration was coupled with the N-H (amide II band) bonding vibration peak that appeared at 1580 cm^{-1}. The bands at 1070 and 1010 cm^{-1} were attributed to the C–O stretching of alcoholic compounds. The alcoholic C-O bonds may have originated from polysaccharide-like substances (Her *et al.*, 2004). Also, bands in the 1280–1137 cm^{-1} region were attributed to the C-O stretching of esters, ethers and phenols, and those at 800 and 630 cm^{-1} to OH stretching vibration of carboxylic groups and deformation of COOH, respectively.

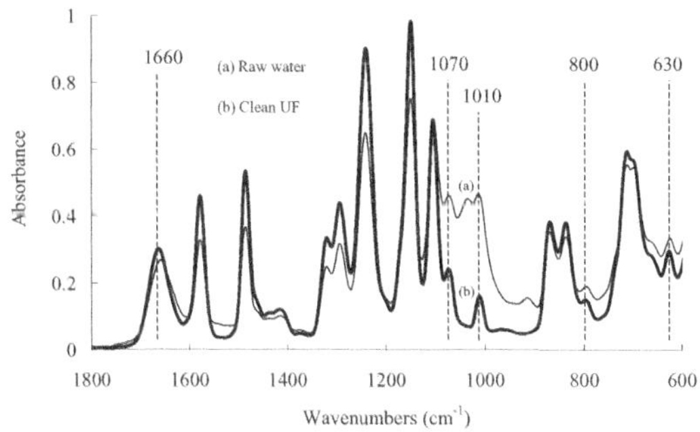

Figure 5 ATR-FTIR spectra for fresh and fouled UF-HPO membranes.

The areas of the peaks at 1660, 1070, 1010, 800 and 630 cm^{-1} in the spectrum of the clean membrane were increased by irreversible fouling, while those at 1240 and 1150 cm^{-1} decreased, as shown in Figure 5. In summary, this indicates that aliphatics, such as short and/or long chain aliphatic amides and alcohols, rather than humic substances, including aromatic groups, caused the irreversible NOM fouling. The foulants are also considered neutral NOM components, based on the fractionation of NOM components (Świetlik *et al.*, 2004). Therefore, it is possible that hydrophobic (e.g., $>C_5$ aliphatic amides and alcohols) as well as hydrophilic neutral fractions (e.g., short chain ($<C_5$) aliphatic amides and alcohols; polysaccharides) in the bulk NOM were adsorbed onto the membrane surface or within the pores.

Conclusions

The fouling effect for a hydrophobic membrane was greater than that for a hydrophilic membrane with a similar pore size, due to fouling caused by adsorption. In the case of a hydrophobic membrane, alcoholic compounds, aliphatic amides and polysaccharides were the main contributors to significant membrane fouling due to adsorption. These adsorptive foulants are considered as the neutral fraction present in hydrophobic and/or hydrophilic NOM components. Although the morphological characteristics of flocs in coagulated water can alter the relative contributions to fouling due to residual NOM and flocs, the relative concentrations of each NOM fraction in the coagulated water are also important. For the case of UF membranes with treated water containing few large particles (i.e., flocs), following the coagulation/sedimentation process, mechanical mixing with the backmixing-type, compared to pump diffusion mixing with the in-line type, is more effective in the reduction of fouling due to residual NOM when the coagulant dosage has been optimized for the removal of turbidity-causing materials.

References

Aiken, G. R., McKnight, D. M., Wershaw, R. L. and MacCarthy, P. (1985). *Humic Substances in Soil, Sediment, and Water: Geochemistry, Isolation and Characterization*. Wiley, New York.

Amirtharajah, A. (1981). Initial mixing: coagulation and filtration: back to the basics. In Proceedings of AWWA, Denver, USA.

Carroll, T., King, S., Gray, S. R., Bolto, B. A. and Booker, N.A. (2000). The fouling of microfiltration membranes by NOM after coagulation treatment. *Water Research*, **34**(11), 2861–2868.

Cho, J., Amy, G., Pellegrino, J. and Yoon, Y. (1998). Characterization of clean and natural organic matter fouled NF and UF membranes, and foulants characterizations. *Desalination*, **118**, 101–108.

Cho, J. and Amy, G. (1999). Interactions between natural organic matter (NOM) and membranes: rejection and fouling. *Water Science and Technology*, **40**(9), 131–139.

Fan, L., Harris, J. L., Roddick, F. A. and Booker, N. A. (2001). Influence of the characteristics of natural organic matter on the fouling of microfiltration membranes. *Water Research*, **35**(18), 4455–4463.

Her, N., Amy, G., Park, H. R. and Song, M. (2004). Characterizing algogenic organic matter (AOM) and evaluating associated NF membrane fouling. *Water Research*, **38**, 1427–1438.

Huber, S. A., (1998). Evidence for membrane fouling by specific TOC constituents. *Desalination*, **119**, 229–234.

Kim, H. C., Lee, S., Byun, S. J. and Yu, M. J. (2005). Application of the improved rapid mixing for enhanced removal of dissolved organic matter and the DBPFPs control. In *Proceedings of the Particle Separation 2005; IWA International Conference on Sustainable and Innovative Technologies to Meet MDG, IWA, Seoul, Korea.*

Lee, N., Amy, G., Croué, J.-P. and Buisson, H. (2005). Morphological analyses of natural organic matter (NOM) fouling of low-pressure membranes (MF/UF). *Journal of Membrane Science*, **261**, 7–16.

Letterman, R. D., Amirtharajah, A. and O'Melia, C. R. (1999). *Coagulation and Flocculation in Water Quality and Treatment*, 5th edn. McGraw-Hill, New York.

Oh, J. I. and Lee, S. (2005). Influence of streaming potential on flux decline of microfiltration with in-line rapid pre-coagulation process for drinking water production. Journal of Membrane Science, **254**, 39–47.

Park, P. K., Lee, C. H, Choi, S. J., Choo, K. H., Kim, S. H. and Yoon, C. H. (2002). Effect of the removal of DOMs on the performance of a coagulation-UF membrane system production for drinking water. *Desalination*, **145**, 237–245.

Pikkarainen, A. T., Judd, S. J., Jokela, J. and Gillberg, L. (2004). Pre-coagulation for microfiltration of an upland surface water. *Water Research*, **38**, 455–465.

Stevenson, F. J. (1994). *Humus Chemistry: Genesis, Composition, Reactions*. John Wiley & Sons Inc., Canada.

Świetlik, J., DąBrowska, A., Raczyk-Stanisławiak, U. and Nawrocki, J. (2004). Reactivity of natural organic matter fractions with chlorine dioxide and ozone. *Water Research*, **38**, 547–558.

Yuan, W. and Zydney, A. (1999). Humic acid fouling during microfiltration. *Journal of Membrane Science*, **157**, 1–12.

Drinking water quality decline due to internal corrosion of copper pipes

A. Panico[*], **M. Fabbricino**[*] **and M. Trifuoggi**[**]

[*] University of Naples *Federico II*, Department of Hydraulic and Environmental Engineering *Girolamo Ippolito* , via Claudio 121, 80125 Naples, Italy
(E-mail: *anpanico@unina.it*; *fabbrici@unina.it*)
[***] University of Naples *Federico II*, Department of Chemistry, Complesso Universitario di Monte Sant'Angelo, via Cinthya, 80125 Naples, Italy (E-mail: *marco.trifuoggi@unina.it*)

Abstract This paper deals with the effects, caused by electrochemical corrosion processes occurring in distribution pipes, on drinking water quality. These studies focus on copper tubes, as this metal is widely used as plumbing material and is considered toxic by the WHO (World Health Organization), that recommends a maximum drinking water concentration of 2 mg l^{-1}. To investigate the dependence of copper release on both water composition and contact time between water and copper, several experiments were carried out and a mathematical model was developed. On metal release, both effects of different concentrations of bicarbonates, sulphates, phosphates and chlorides and different stagnation times, between 15 and 90 minutes, were studied. As experimental apparatus, a special corrosion cell was used. To verify the existence of a direct correlation between the corrosion potential of the metal surface and the total dissolved copper concentration in the stagnating water, measurements of both parameters, respectively, were taken during and at the end of each stagnation episode. The experimental results allow the copper concentration in drinking water to be quantified by measuring the potential corrosion, instead of analyzing water, reducing in this way the monitoring costs and time.

Keyword Copper; corrosion; drinking water; metal release; monitoring

Introduction

The main effects of the electrochemical corrosion on copper pipes are represented by pipe breakings and risks for human health. Contaminatinated by copper water, in fact, is responsible for health problems, such as abdominal pains, nausea, vomiting, diarrhea, headache, and dizziness (Sidhu *et al.*, 1995), and is considered to be responsible for a particularly heavy form of liver cirrhosis, known as non-Indian childhood cirrhosis (Zietz *et al.*, 2003). This toxicity, scientifically recognized, led the WHO (World Health Organization), in 1993, to put the copper in the list of elements dangerous to human health and to fix for it a maximum drinking water concentration of 2 mg l^{-1}.

However, already two years before, in 1991, the Environmental Protection Agency (EPA) promulgated in the USA, the so-called Lead and Copper Rule, fixing 1.3 mg·l^{-1} as the maximum admissible copper concentration in drinking water after 6 hours of stagnation.

Some years later, in 1998, the European Commission fixed a value of 2 mg l^{-1} in drinking water for a period of stagnation established differently by each country, ranging from 4 hours in Germany, to 12 hours in Denmark (Hoekstra *et al.*, 2003).

Parallel to the introduction of regulations limiting the copper presence in drinking water, the international scientific community produced studies about this theme aimed to increase knowledge which, however, is still fragmentary .

The most important research concerns the effect of stagnation time (Merkel *et al.*, 2002; Lytle and Schock , 2002) and water composition (Broo *et al.*, 1999; Edwards *et al.*, 2002; Shin and Kim, 2004; Pehkonen *et al.*, 2002) on copper release, or deal with the effort of predicting and monitoring the exposure to copper from drinking water (Broo *et al.*, 1997; Lagos *et al.*, 1999; Xiao *et al.*, 2004; Kirmeyer *et al.*, 2004).

A direct proportionality between the concentration of copper in stagnating water, and the stagnation length, appears clear at least for short stagnation periods. For stagnation periods longer than 48 h the process seems to reach a plateau (Merkel *et al.*, 2002). Instead, the effects of water composition on copper release appears unclear because they are often conflicting. Edwards *et al.* (2002) found that low phosphate concentration inhibits the corrosion process until the alkalinity is less than 300mg/l as $CaCO_3$, whereas other experimental studies (Merkel *et al.*, 2002) showed that alkalinity reduces the corrosion rate. In contrast with these findings, some experimental results obtained in Florida (Xiao *et al.*, 2004) showed a direct proportionality between water alkalinity and copper concentration in stagnating water, pointing out also the beneficial effect of sulphates in reducing copper pipe corrosion. As far as the concern of presence of chlorides, once more contrasting results have been obtained. Broo *et al.* (1997) found that the presence of chlorides tends to increase the phenomenon of copper release, while Pehkonen *et al.* (2002) showed that the corrosion process is somehow inhibited by chloride concentration.

To better understand the effect of water characteristics on the corrosion process taking place in copper pipes, the present paper illustrates and discusses the results of an experimental study carried out in laboratory controlled conditions. The experiments have highlighted the existence of a relationship between the electrochemical potential of the copper surface and the total dissolved copper concentration in the stagnating water. This correlation offers the possibility of monitoring the copper levels in the drinking water trough the measurement of a surrogate parameter, determining in this way, costs and times lower than water analysis. The experimental study is completed by the development of a mathematical model able to describe both chemical and physical processes responsible for generating different copper levels in drinking water.

Methods

Corrosion potential stagnation flow (CPSF) theory

The CPSF theory is based on the quantification of copper release during stagnation periods, monitoring the electrochemical potential of the corroding surface. The method assumes that the corrosion is caused only by the oxygen dissolved in the water. The corrosion potential of the metal is therefore defined by the following balance condition:

$$i_{Cu} + i_{O2} = 0 \qquad (1)$$

where i_{Cu} is the copper oxidation current, and i_{O2} is the oxygen reduction current.

Coupling this equation with the Butler-Volmer expressions of i_{Cu} and i_{O2}, and considering the relationship between the electrochemical potential of an electrochemical couple M_{OX}/M_{RED} (oxidized form / reduced form) and the concentration of M_{OX} in the water described by the Nernst equation, the following expression of the corrosion potential is obtained:

$$E_{corr} = \frac{1}{\alpha_1 + \beta_2}\left(\log\left(\frac{i_1}{i_2}\right) + \alpha_1 E_1 + \beta_2 E_2 - \beta_2 \Phi pH\right) - \frac{0.5\Phi\alpha_1}{\alpha_1+\beta_2}\log\Theta + \\ + \frac{0.25\Phi}{\alpha_1+\beta_2}\left(2\alpha_1 \log[Cu_{TOT}] + \left(\beta_2 + \frac{4}{\Phi}\right)\log[C_0 - 0.5Cu_{TOT}]\right) \tag{2}$$

which is valid for steady state conditions, neglecting the cathodic branch of the copper's Evans plot and the anodic branch of the oxygen's Evans plot.

In equation (2) E_{corr} is the corrosion potential; α_1 is the anodic Tafel slope for copper oxidation; β_2 is the cathodic Tafel slope for oxygen reduction; i_1 and i_2 are, respectively, the standard exchange current for oxygen reduction and the standard exchange current for copper oxidation; E_1 and E_2 are, respectively, the standard potential for copper and the standard potential for oxygen; Φ is the Nernst potential; Θ is the ratio between the total dissolved copper concentration, and the concentration of free copper, assumed to be completely in the form of Cu^{+2}; Cu_{TOT} is the total dissolved copper concentration and C_0 is the initial oxygen concentration.

The obtained E_{corr} expression is equivalent to the formal relationship:

$$E_{corr} = \varepsilon_0 + \mu \cdot \log[Cu_{TOT}] \tag{3}$$

where ε_0 and μ are numerical coefficients involving electrochemical constants and chemical parameters depending on water characteristics.

Equation (3) is the master equation of the CPSF theory: it shows the existence of a direct correlation between the corrosion potential and the logarithm of the total copper concentration, and therefore allows the quantification of copper concentration in the water, measuring the corrosion potential whenever the two coefficients, ε_0 and μ, are known or estimated.

Experimental

Experiments were performed on a corrosion cell designed by the Civil and Environmental Engineering Department of the University of Washington (Kirmeyer *et al.*, 2004). The cell was made by two rectangular 30.5x7.6x0.64 cm copper plates, separated with a 1.5x0.16 cm thick plastic gasket. In this way ,a thin rectangular channel was created through which the water could flow or stagnate. The internal volume of the cell was about 200 ml, which corresponded also to the volume of water exposed to copper release during stagnation. The area of the metal in contact with water was 245 cm^2.

A double junction standard calomel reference electrode (Beckman Instrument Inc., Model 511105) used to measure the electrochemical potential of the corroding surface was inserted in the middle of the upper copper plate and supported by an acrylic block.

All experiments were performed using synthetic water, prepared adding chemical reagents high purity degree, to deionised water, produced by Millipore Deionizer (model Rios 30). Sodium sulphate (Na_2SO4) and sodium bicarbonate (NaHCO3) were utilised, respectively, as source of sulphates and bicarbonates, while dibasic sodium phosphate (Na_2HPO_4) and sodium chloride (NaCl) were used, respectively, as sources of phosphates and chlorides.

The water was pumped through the cell from a HDPE plastic tank using a peristaltic pump (KNF, model ND 100 KT.18). The flow rate was set to 0.8 $l \cdot minutes^{-1}$ and controlled using a pump speed controller

A stagnation/flow sequence (Figure 1), characterized by varying stagnation periods, was adopted to identify the effect of stagnation time on copper release.

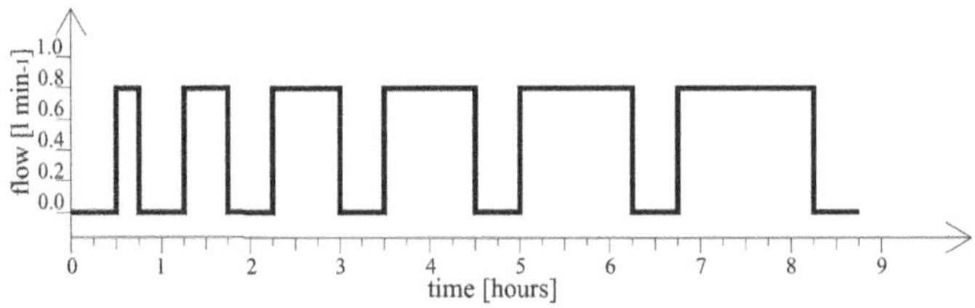

Figure1 Illustration of the adopted stagnation/flow sequence.

The length of the flow time was fixed to 30 minutes, in order to wash the internal surface of the cell and remove the layers of corrosion scales eventually formed during stagnation.

During each stagnation episode the electrochemical potential value of the corroding metal surface was acquired on-line using a voltmeter (Agilent, model # 34970A) and stored on computer using a specific software (Agilent Benchlink Data Logger). Immediately after each stagnation episode, 250 ml of water were sampled and stored for subsequent analyses: 50 ml of sample were used to measure the value of temperature and pH, whereas the remaining 200 ml were used for metal detection. The latter was effectuated with an atomic absorption spectrometer (Varian, model SpectrAA) Sixteen series of tests were performed, varying water composition as shown in Table 1. All analyses were performed according to international analytical methods.

Table 1 Summarizing and explaining table of tests carried out

ID Test	Anion	Concentration [mg l^{-1}]
B_1	HCO_3^-	25
B_2	HCO_3^-	50
B_3	HCO_3^-	100
B_4	HCO_3^-	200
B_5	HCO_3^-	400
S_1	SO_4^{-2}	2
S_2	SO_4^{-2}	4
S_3	SO_4^{-2}	8
P_1	PO_4^{-3}	0.5
P_2	PO_4^{-3}	1
P_3	PO_4^{-3}	2
C_1	Cl^-	2
C_2	Cl^-	4
C_3	Cl^-	8
C_4	Cl^-	16
C_5	Cl^-	32

Mathematical model

Parallel to experiments, a mathematical model was developed to better understand the corrosion phenomenon. It can represent a valid tool to predict the copper concentration in drinking water knowing the system geometry, the water characteristic and the stagnation times.

The main chemical and physical processes and the correspondent equations, written in the simplest form and relating to only copper, are hereafter listed:

- Corrosion process:

$$O_2 + 2H_2O + 2Cu \leftrightarrow 4OH^- + 2Cu^{++} \tag{4}$$

- Chemical complexation process.

$$Cu^{++} + rL \leftrightarrow Cu(L)_{r\,(aq)}^{\;2-r} \tag{5}$$

- Chemical precipitation and dissolution process:

$$\alpha Cu^{++} + rR^- \leftrightarrow Cu_\alpha(R)_{r\,(s)} \tag{6}$$

- Diffusion molecular process:

$$\frac{\partial[Cu^{++}]}{\partial t} = D_{Cu} \cdot \frac{\partial^2[Cu^{++}]}{\partial x^2} \tag{7}$$

in which L and R are, respectively, a chemical ligand and radical, and D_{Cu} is the copper molecular coefficient.

Equations (4), (5) and (6), are considered for all the chemical species dissolved in water. The model was written considering the presence of bicarbonates and implemented for the experimental cell. To validate it, a rate of copper release was found, utilising the experimental value for E_{corr} and total copper concentration.

Results and discussion

The corrosion potential during the stagnation/flow sequence shows a characteristic and reproducible trend whatever the composition of the stagnating water. As an example, in Figure 2, the E_{corr} curves are represented for waters containing different bicarbonates levels. At the beginning of each stagnation episode, in fact, E_{corr} quickly decreases, reaching a minimum, which is more pronounced for lower water salinity. After that, it tends to slightly increase. The values measured at the beginning of stagnation or after a certain time are not a function of the considered stagnation episode, but are only dependent on the type and the concentration of dissolved anions. Particularly, an increase of anion concentration is responsible for a shift of the experimental curves towards less negative corrosion potential.

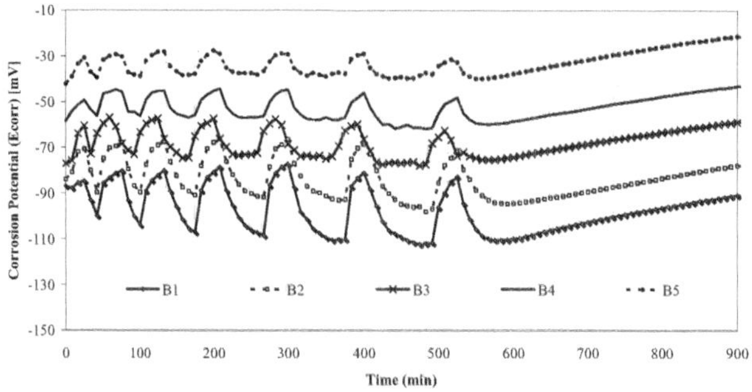

Figure 2 Corrosion potential varying bicarbonates concentration.

Data concerning copper release for different stagnation times and different bicarbonates concentration are summarized in Figure 3. As can be immediately observed, for an assigned stagnation time, copper release decreases, increasing bicarbonate concentration. This can be attributed, essentially to the effect of passivation processes taking place on the surface of the experimental cell. This trend, in fact, is likely to be associated with malachite $\{Cu_2(OH)_2(CO_3)_2\}$ and tenorite $\{CuO\}$ formation, that tend to precipitate protecting the metal surface and reducing copper release (Shin and Kim, 2004).

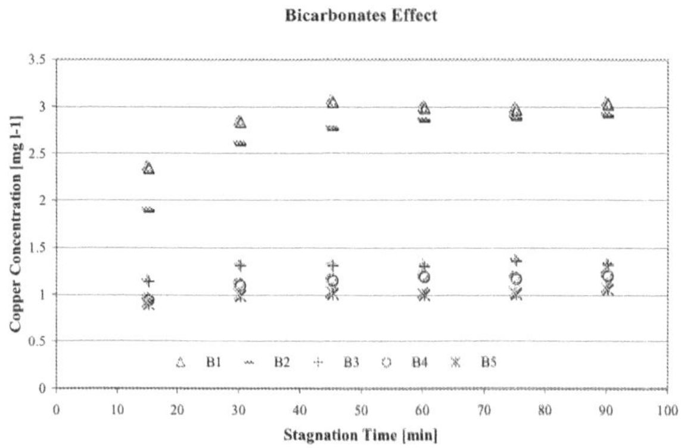

Figure 3 Copper concentration varying stagnation time and bicarbonate concentration.

The layers represent an obstacle to oxygen diffusion, and therefore inhibit copper release. Copper concentration becomes, then, only a function of the dynamic equilibrium established between precipitation and dissolution of tenorite and malachite, and does not increase anymore with increasing stagnation time. Experimental data show that the formation of the protective layer is reached in about one hour for low bicarbonates values (25–50 mg l L^{-1}) and in about half time for higher values (> 100 mg·L^{-1}).The effect of sulphates and phosphates on copper release is analogous. For each stagnation time, in fact, copper concentration decreases with increasing sulphates and phosphates concentration. A different behaviour is noted for chlorides (Figure 4), they seem to contribute to the corrosion process neither favourably nor unfavourably. It is not possible, in fact, to find a clear relationship between copper and chloride concentration in the stagnating water. For each concentration of sulphates, phosphates and chlorides, instead, metal release increases by increasing the length of the stagnation period, reaching a plateau for a value of stagnation time depending on the dissolved anion and its concentration. Within the limits of tested conditions, the stagnation time in which the copper concentration reaches a plateau are around 30 minutes in presence of sulphates, around 1 hour in presence of phosphates and around 1 hour and 20 minutes in presence of chlorides. The difference existing among these times can be attributed to different properties of the insoluble compounds produced by the reaction between copper ions and the other dissolved species (Shin and Kim, 2004).

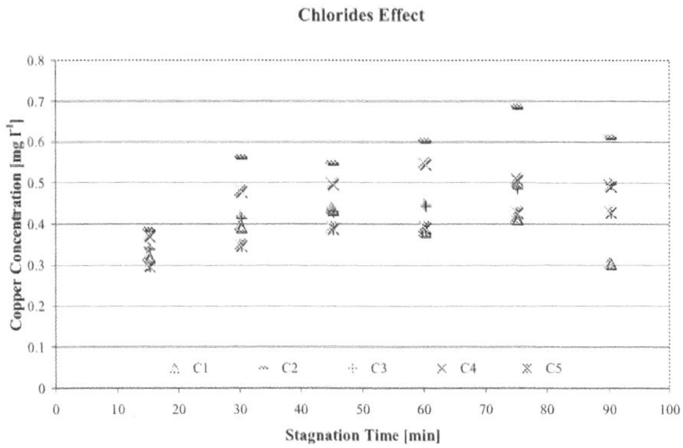

Figure 4 Copper concentration varying stagnation time and chlorides concentration.

Relating the data concerning the value of the corrosion potential at the end of each stagnation episode (E_{corr}) and the corresponding copper concentration in the stagnation water (Figure 5), it can be noted in a semi-logarithmic scale, for waters containing bicarbonates, sulphates and phosphates, the existence of linear interpolating curves, described by (3), for which the correlation coefficients (R^2) are not less than 0.81.

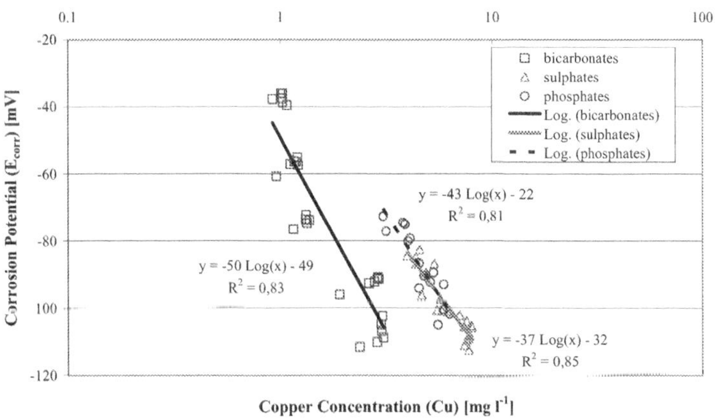

Figure 5 Relationship between E_{corr} and Log(Cu) in presence of different anions.

It is not possible to find, instead, an analogous relationship for water containing chlorides, and it confirms the absence of a clear dependence of chlorides on copper release as is previously hypothesized (Figure 6).

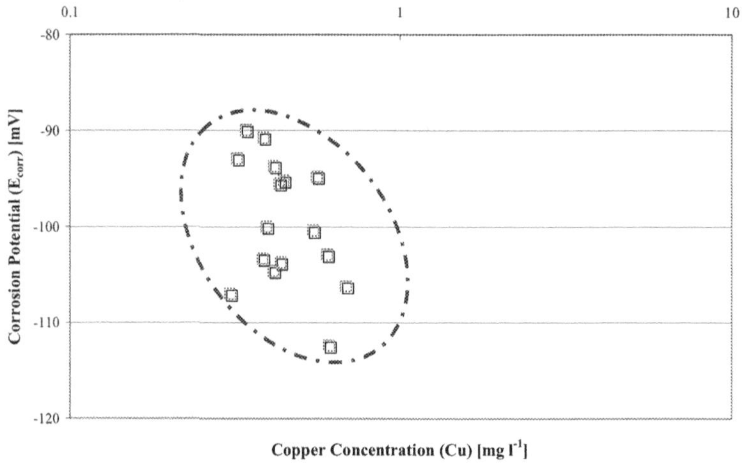

Figure 6 Relationship between E_{corr} and Log(Cu) in presence of chlorides.

Conclusions

The principal results of the presented research study can be summarized as follows:

- As consequence of electrochemical corrosion processes occurring in copper pipe, metal ions are released by tube internal surface and are responsible for changing drinking water characteristics.
- The amount of copper dissolved in water is proportional to stagnation time length, and depending on it, the final copper concentration can be higher than admissible limit for human health.
- The copper levels in drinking water are affected by water composition, in fact, depending on both type and concentration of chemical dissolved species, copper ions form different insoluble compounds which, precipitating on metal surface, protect it by corrosion processes in different ways.
- The corrosion potential measured on copper surface during the stagnation period supplies measurement of concentration of copper dissolved in water. For this reason, it can represents a surrogate parameter for copper concentration monitoring during water stagnation, reducing costs and time required for water quality control.

References

Broo, A.E., Berghult, B. and Hedberg, T. (1997). copper corrosion in drinking water distribution system – the influence of water quality. *Corrosion Science*, **39**, 1119–1132.

Broo, A.E., Berghult, B. and Hedberg, T. (1999). Drinking water distribution – the effect of the natural organic matter (NO<) on the corrosion of iron and copper. *Water Sci. Technol.*, **40**, 17–24.

Edwards, M., Hidmi, L., Gladwell, D. (2002). Phosfate inhibition of soluble copper corrosion by-product release. *Corrosion Science*, **44**, 1057–1071.

Hoekstra, E.J., Pedroni, V., Passerella, R., Trincherini, P.R., Eisenreich, S.J. (2004). *Elements in Tap Water. Part 3. Effect of sample volume and stagnation time on the concentration of the element.* IES, Italy

Kirmeyer, G., Murphy, B., Sandvig, A., Korshin, G.V., Shaha, B., Fabbricino, M. and Burlingame, G. (2004) *Post-Optimization Lead and Copper Control Monitoring Strategies.* IWA Publishing London, UK.

Lagos, G.E., Maggi, L. C., Peters, D., Reveco, F. (1999). Model for estimation of human exposure to copper in drinking water. *Science of the Total Environmental*, **239**, 49–70.

Lytle, D.A. and Schock, M.R. (2002). Impact of stagnation time on metal dissolution from plumbing materials in drinking water. *J. Water Supply Res. Technol.—AQUA*, **49**, 243–257.

Merkel, T.H., Groß, H.-J., Werner, W., Dahlke, T., Reicherter, S., Beuchle, G., Eberle, S.H. (2002). Copper corrosion by-product release in long-term stagnation experiments. *Water Research*, **36**, 1547–1555.

Pehkonen, S.O., Palit, A., Zhang, X. (2002). Effect of specific water quality parameters on copper corrosion. *Corrosion*, **58**, 156–165.

Sidhu, K.S., Nash, D.F., McBride, D.E. (1995). Need to revise the national drinking water regulation for copper. *Regulatory Toxicology and Pharmacology*, **22**, 95–100.

Shin, J.J. and Kim, J.G. (2004). Copper corrosion in potable water distribution systems: influence of copper products on the corrosion behaviour. *Materials Letters*, **58**, 2002–2006.

Xiao, W., Taylor, J., Hong, S., Shi, B., (2004). Simulation of copper release in pipe distribution system and copper rule compliance. *Florida Water Resources Journal*, **9**, 14–20.

Zietz, B.P., Dieter, H.H., Lakomek, M., Shneide, H., Kebler-Gaedtke, B. and Hartmut, D. (2003). Epidemiologic investigation on chronic copper toxicity to children exposed via the public drinking water supply. *Science of the Total Environment*, **302**, 127–144.

Wastewater biotechnology

A fundamental study of the biological and electrochemical processes in microbial fuel cells

Stefano Freguia and Jürg Keller

Advanced Wastewater Management Centre (AWMC), University of Queensland, St. Lucia, Brisbane 4072, Australia (E-mail: *stefano@awmc.uq.edu.au*)

Abstract Microbial fuel cells are emerging as a novel technology for the removal of wastewater COD with concomitant energy recovery as electricity. Their numerous advantages over anaerobic treatment have sprung a great interest in this area. While researchers have placed great efforts in understanding the microbiology of the process and in designing anodes and cathodes with increased power output, there is still lack of fundamental understanding of the biological and electrochemical processes that drive the electron transfer process. This study describes two techniques that can provide valuable information on (i) the bio-electrochemical reactions occurring at the anode and (ii) the electrochemical behavior of microbial fuel cells. In the first experiment, a microbial fuel cell was operated as a batch, with acetate as electron donor. The combination of online gas analysis, online current monitoring and offline liquid sample analysis provided a powerful tool to close COD and carbon balances, and revealed that carbon dioxide is produced stoichiometrically and comprises most of the gas production at the anode. In the second experiment, electrochemical monitoring allowed the quantification of activation overpotentials and Ohmic losses, revealing that the latter are important and of similar value as the anodic overpotentials.

Keywords Carbon balance; COD balance; microbial fuel cell; off-gas analysis; overpotential

Introduction

Microbial fuel cells (MFCs) are a novel technology that can achieve the removal of organic carbon compounds from wastewater with simultaneous production of electricity.

This technology is capable of harvesting the chemical energy contained in organic substrates and converting it directly to electricity in a fuel cell setup. A basic schematic of the working principle of a microbial fuel cell is shown in Figure 1.

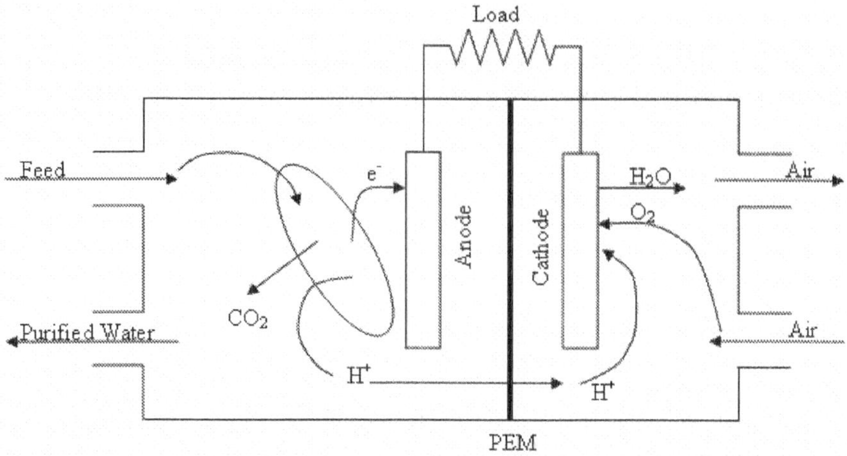

Figure 1 Schematic of a continuous microbial fuel cell with aeration at the cathode.

The oxidation and reduction half reactions occur in two separate compartments, each containing an electrode, called anode and cathode, respectively. The two electrodes are connected through an external circuit, where a resistor or electrical load can harvest the electrical energy generated by the process. The reactions are of electrochemical type, thus occurring at the interface between the liquid and the electrode surface and involving direct transfer of electrons. If the organic substrate is represented as glucose and the electron acceptor at the cathode is oxygen, then the two half reactions are those shown in equations (1) and (2).

$$C_6H_{12}O_6 + 6H_2O \rightarrow 6CO_2 + 24H^+ + 24e^- \tag{1}$$

$$6O_2 + 24H^+ + 24e^- \rightarrow 12H_2O \tag{2}$$

The two compartments are separated by a proton or cation exchange membrane (generally abbreviated as PEM), which transfers protons or other cations between anode and cathode compartments, and thus guarantees that the electrical circuit is closed.

Although the redox reaction written above is thermodynamically spontaneous (the electromotive force is 1.1 V at pH 7) its activation energy is high and it would not happen without a suitable catalyst. At the anode the catalysis is carried out by a biofilm growing on the electrode surface. The bacteria that grow on the anode can transfer electrons through their outer membrane, directly or through the use of redox active metabolites, which act as electron shuttles. Several studies have identified many species of electrochemically active bacteria. The most intensely studied ones include *Geobacter metallireducens* (Bond *et al.*, 2002), *Geobacter sulfurreducens* (Bond and Lovley, 2003), *Rhodoferax serrireducens* (Chauduri and Lovley, 2003), *Shewanella putrefaciens* (Kim *et al.*, 2002), *Pseudomona aeruginosa* (Rabaey *et al.*, 2004). The bacteria use a fraction of the energy liberated by the reaction for their own growth and maintenance.

The steps that determine current generation in a microbial fuel cell are: (i) mass transfer of the organic substrate from the bulk solution to the anode surface, (ii) substrate metabolism by the bacteria, (iii) electron transfer from microbial cells to anode, (iv) resistance of the external circuit, (v) proton transfer through the membrane, (vi) oxygen reduction at the cathode.

A single cell can typically produce a potential of 0.5–0.7 V when oxygen or hexacyanoferrate are used as electron acceptors at the cathode. The current and power productions depend on the anode and cathode surface areas. The maximum power that has been obtained to date is 6.0 W/m² anode area (Schröder *et al.*, 2003). As power production is rather small, MFCs cannot compete with chemical fuel cells as power producers, but they have a few potential applications, in particular as novel wastewater treatment technology: wastewater chemical oxygen demand (COD) can be thoroughly removed, while producing a "CO_2 neutral" power which can cover the electricity requirements of a wastewater treatment plant. Other niche applications include: (i) electricity production in remote areas: MFCs can produce power from organic waste, which is usually readily available; (ii) bacterial batteries: MFCs produce environmentally friendly power that could be used for small appliances; (iii) nitrogen and sulfur removal: by a tight control of the anodic potential, conditions can be found to selectively reduce nitrate to nitrogen at the cathode or oxidise sulfide to elemental sulfur at the anode.

Wastewater treatment by microbial fuel cells combines the benefits of both aerobic and anaerobic treatment. In fact, MFCs can achieve organics removal similar to aerobic processes, even from dilute influents. Like anaerobic processes, microbial fuel cell technology can produce enough power to cover the electricity requirement of a wastewater treatment plant. The biomass growth yield is comparable to that of anaerobic processes, which leads to lower sludge production than aerobic treatment. Like in anaerobic processes, aeration is not necessary to provide oxygen to the biomass, thus saving approximately 0.5 kWh per cubic meter of treated wastewater. Oxygen needs to be provided to the cathode, but this can be done by passive aeration using air cathodes.

Microbial fuel cells are an alternative to anaerobic processes as a technology for combined wastewater treatment and energy production. MFCs have a few advantages over methane generation followed by combustion in a thermal cycle: (i) MFCs can produce power and treat wastewater in one stage, whereas anaerobic processes require two stages, i.e. anaerobic wastewater treatment and power production in a gas turbine cycle or chemical fuel cell; (ii) MFCs have been shown to have energetic efficiencies of up to 65% at ideal conditions, whereas methane combustion following anaerobic digestion is limited to around 35–40%; (iii) unlike anaerobic treatment, microbial fuel cells operate efficiently at low temperature and low strength waste; (iv) as the substrate is converted directly to CO_2, there is no need for gas treatment. In anaerobic digestion, the methane produced contains traces of H_2S. The gas stream may need to be purified if it is used in a turbine (due to environmental regulations) or fed to a chemical fuel cell (sulfide poisons the catalysts).

Microbial fuel cells are at an early stage of research and, as such, they have some bottlenecks, which demand further research in the area. Firstly, activation overpotentials are responsible for significant losses of cell voltage, and therefore of energy efficiency and power output: at the anode, the typically high overpotential is due to a complex mechanism of the reactions involved in the electron transfer from the substrate to the electrode; at the cathode, the overpotential is significant if oxygen is the electron acceptor, due to the slow kinetics of the oxygen reduction at the electrode. Just like in chemical fuel cells, catalysts such as platinum are normally required to reduce these cathodic losses. A second bottleneck is the fact that only a monolayer of bacteria (except when mediators or nanowires – see Roguera *et al.* (2005) – are used) attached to the anode can be involved in the electron transfer. This limits the amount of electrochemically active biomass that can be present in the anodic compartment, therefore limiting the maximum loading rate and power output. A third problem of MFCs is that the presence of electron acceptors in the wastewater fed to the cell can capture electrons from COD that would otherwise be oxidised at the anode.

Most of the research done in the field so far has focused on the identification of bacterial species involved in the electron transfer process and on the optimization of anode and cathode configurations in order to increase the specific power output. Not much attention has been placed on the fundamentals of the biological and electrochemical processes at the basis of microbial fuel cells. The aim of this study was to deepen the understanding of the biological reactions occurring at the anode of microbial fuel cells, as well as determining the limiting steps to current production, by quantifying electrode overpotentials and Ohmic losses. The outcome of this study would then provide researchers with important knowledge to better address the bottlenecks of this technology.

Materials and methods

Reactor setup

A two-chamber microbial fuel cell built in rectangular geometry was used for the experiments. Figure 2 shows a sketch of the reactor. The volume of the anodic and cathodic compartments was 480 cm^3, of which 44 cm^3 of headspace. The liquid portion of the compartments was filled with granular graphite (El Carb 100, Graphite Sales, Inc., USA). The granules were 2–6 mm in diameter and had a porosity of 45%, leaving a liquid volume of 242 mL in the compartment. The granular matrix served as biofilm support (anode side) and conducting matrix, conveying electrons to a graphite rod. The anodic and cathodic rods were connected through an external variable resistance (maximum 100 Ω) to close the circuit. Two Ag/AgCl reference electrodes (Ref 201, Radiometer Analytical) were placed in the anodic and cathodic compartments of each cell and connected to the respective electrodes, for measurements of half cell potentials. The PEM was an Ultrex cation exchange membrane (Membranes International, USA).

The anodic solution was recirculated at fast rate (approximately 200 mL/min) in order to maintain well-mixed conditions and to avoid clogging of the granular matrix.

The electron acceptor used at the cathode was either potassium hexacyanoferrate or oxygen. The former was supplied as a 150 mM solution which was recirculated around the cathode, the latter was supplied by sparging air through a buffered solution which was also recirculated.

Electrochemical monitoring

The anode and cathode of each cell were connected to an Agilent 34970A data acquisition unit. Every 20 seconds, a full scan was performed and the voltage data were stored. The current was deduced from the external resistance, using Ohm's law. The Ag/AgCl reference electrodes were also connected to the instrument. The reading of the voltages between reference electrodes and operating electrodes was a measurement of anodic and cathodic half-cell potentials.

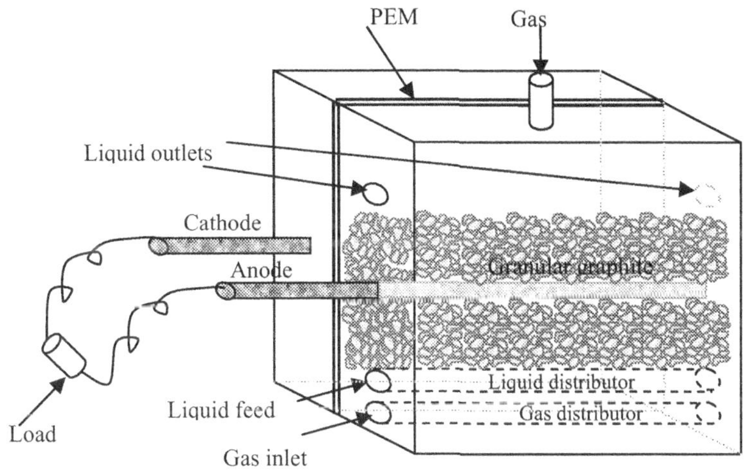

Figure 2 Schematic of microbial fuel cell in rectangular geometry.

Measurement of gas production to close COD and C balances

The microbial fuel cell was connected to a Titration and Off-Gas Analysis (TOGA) sensor, developed by Pratt *et al.* (2003). The sensor uses a mass spectrometer for online monitoring of gas production. Helium was used as a carrier and was sparged into the anode to strip carbon dioxide, methane and hydrogen produced in the anaerobic anode. The gases are then detected in the mass spectrometer. The mass spectrometer data were integrated with current data from the above mentioned data acquisition unit and with offline liquid sample analysis for COD, in order to close COD and carbon balances and thus provide information on bacterial growth yields. The pH was kept constant at 7.0 via pH control.

The anode was sparged overnight with helium to strip all the CO_2 from the liquid solution, while the reactor was run continuously on acetate at a loading rate of 1.0 kg $_{COD}$/m^3 d and external resistance of 20 Ω. After stopping the feed the reactor was kept under starvation until the current production levelled off to a base signal, which can be attributed to endogenous COD production. Subsequently, the reactor was spiked with 15 mL of a solution containing 1.034 g/L sodium acetate, 0.161 g/L NH_4Cl and trace elements. This injection corresponded to an addition of 12.10 mg COD in the reactor. The resistor was kept at 20 Ω throughout the cycle, which lasted about 90 minutes and was repeated six times.

Determination of limiting step to current production

The microbial fuel cell was run continuously with a feed rate of 0.5 litre/day, giving a hydraulic retention time (HRT) of 10.5 hours. The influent was made up of a medium containing 6.0 g/L Na_2HPO_4, 3.0 g/L KH_2PO_4, 0.1 g/L NH_4Cl, 0.5 g/L NaCl, 15 mg/L $CaCl_2 \cdot 2H_2O$ and trace elements. The substrate was acetate, fed at a concentration corresponding to 0.5 g_{COD}/L.

The reactor was operated at open circuit for at least three hours in order to load the anode solution with acetate and prevent any COD mass transfer limitations. The circuit was then closed and the resistance progressively reduced from 100 Ω to 10 Ω. At each resistance the electrochemical parameters (voltage, current and half cell potentials) reached a pseudo steady state in approximately 20 minutes. At that point the data were recorded and the resistance lowered. This method generated a so-called polarization curve for the fuel cell (cell voltage as a function of current). The anodic and cathodic overpotentials were calculated as difference between the values of the respective half-cell potentials at closed and open circuit. The two overpotentials were then subtracted to the total voltage loss of the cell to calculate the Ohmic losses. The knowledge of the relative importance of anodic overpotential, cathodic overpotential and Ohmic loss sheds light on the limiting steps to the electron transfer process. This experiment was carried out with both hexacyanoferrate and oxygen as electron acceptors.

Results and discussion

Gas production and COD/carbon balances

Immediately following injection of acetate, the current shows a rapid increase, followed by a flat phase and then a gradual decline. The CO_2 signal follows the current closely, with some delay at the end of the cycles, which is due to the mass transfer of CO_2 from the liquid phase to the gas phase. This is shown graphically in Figure 3, which also shows that amounts of hydrogen and methane were detected in the off-gas. Over a cycle, approximately 85% of the gas production was CO_2, with the remainder equally divided between CH_4 and H_2. Methane probably comes from methanogenesis occurring in zones of the biofilm where electrical contact is poor. Hydrogen may be produced by the bacteria to transport electrons from inside the cell to the electrode.

The following reactions (Equations 3–6) were proposed to describe the anodic process.

$$CH_3COO^- + (2-3Y)H_2O + 0.4Y\,NH_4^+ \rightarrow 2(1-Y)CO_2 + (7-8Y)H^+ + \\ + (8-8.4Y)e^- + 2Y\,CH_{1.8}O_{0.5}N_{0.2} \tag{3}$$

$$CH_3COO^- + H^+ \rightarrow CH_4 + CO_2 \tag{4}$$

$$CH_3COO^- + H^+ \rightarrow CH_3COOH \tag{5}$$

$$HCO_3^- + H^+ \rightarrow CO_2 + H_2O \tag{6}$$

Equation (3) is the main bioreaction and it assumes that acetate is oxidised directly to CO_2 with production of electrons and protons. The coefficient Y is the molar growth yield for the biomass, in C-mol biomass/C-mol acetate. In Equation (4) the biomass growth term was neglected because methanogens normally have very small yields and methanogenesis is a secondary reaction in this system. The data obtained from the TOGA, in conjunction with electrochemical measurements and offline liquid samples for COD were used to validate these reactions and to estimate the bacterial growth yield in the short term batches.

Based on the reactions shown above, no COD is destroyed in the anodic reactions. The COD consumed in the liquid phase must be converted into electrons (current), methane or hydrogen in the off-gas, or biomass growth, according to Equation (7):

$$COD_{consumed} = COD_{current} + COD_{CH_4} + COD_{H_2} + COD_{growth} \tag{7}$$

From the COD data averaged over the six cycles, $COD_{consumed}$=0.391 ± 0.05 mmol O_2. From current data, $COD_{current}$=0.253 ± 0.023 mmol O_2. From TOGA data, COD_{CH4}= 0.052 ± 0.008 mmol O_2 and COD_{H2}= 0.016 ± 0.005 mmol O_2. If the balance is entirely attributed to growth, the growth term becomes: COD_{growth}= 0.070 ± 0.041 mmol O_2. This gives a biomass growth yield of 0.173 ± 0.088 $g_{COD,\ biomass}/g_{COD,\ removed}$. Figure 4a represents a pie chart with the breakdown of the COD consumption.

The carbon balance is expressed by Equation (8): the carbon addition as acetate is equal to the sum of the carbon found in the off-gas as CO_2 or CH_4, the carbon accumulating in the solution as bicarbonate, and the carbon taken up for biomass growth.

$$C_{acetate} = C_{CO_2,offgas} + C_{CH_4,offgas} + C_{CO_2,accumulation} + C_{biomass} \quad (8)$$

The carbon accumulation in the liquid phase is practically negligible since the solution was completely stripped of residual CO_2 and the pH was held constant. From the CO_2 balance expressed by Equation (9), the carbon accumulation is calculated at –0.014 ± 0.017 mmol CO_2.

$$CTR = CO_{2,current} + CO_{2,methanogenesis} + CO_{2,accumulation} \quad (9)$$

In Equation (9) CTR is the carbon dioxide transfer rate, i.e. the CO_2 molar flow rate measured in the off-gas. CTR was measured at 0.322 ± 0.025 mmol CO_2 in these batch tests.

In Equation (8) $C_{CO2,\ off\text{-}gas}$ and $C_{CH4,\ off\text{-}gas}$ are easily derived from the off-gas analysis. $C_{biomass}$ is calculated from COD_{growth} assuming the standard formula for the biomass used in Equation (3). With this method the carbon balance closes with an error of 2.2%. Figure 4b shows the pie chart of the carbon consumption breakdown.

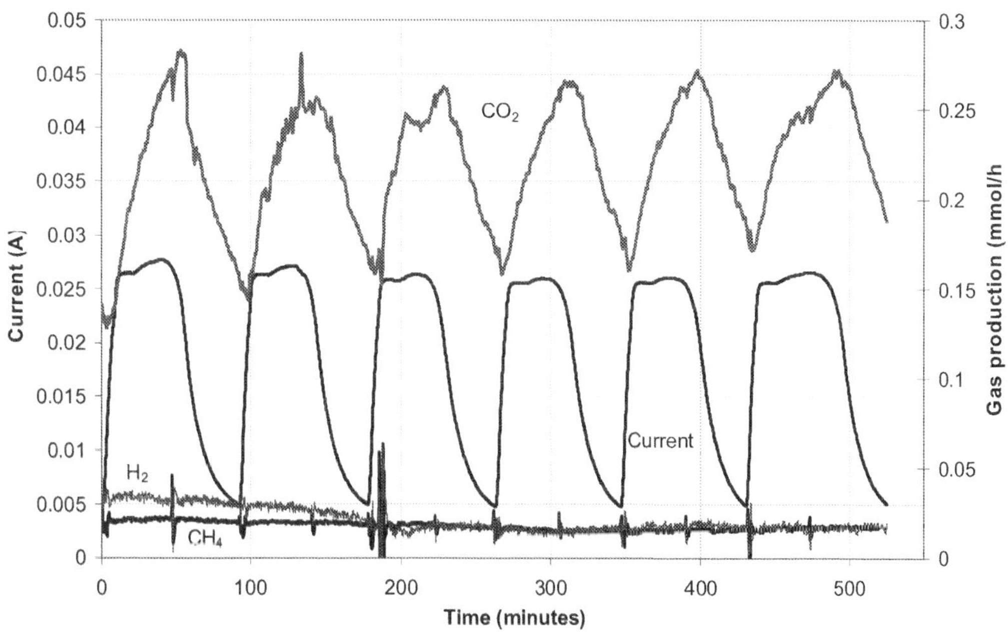

Figure 3 Profiles of current and CO_2 production during TOGA batch tests.

This analysis shows that the combination of off-gas analysis, off-line liquid analysis and online electrochemical monitoring is able to provide a set of independent measurements that closes COD and carbon balances. The results prove that the model described by Equations (3), (4), (5) and (6) is an accurate description of the overall biological and chemical reactions occurring at the anode, when acetate is the sole electron donor.

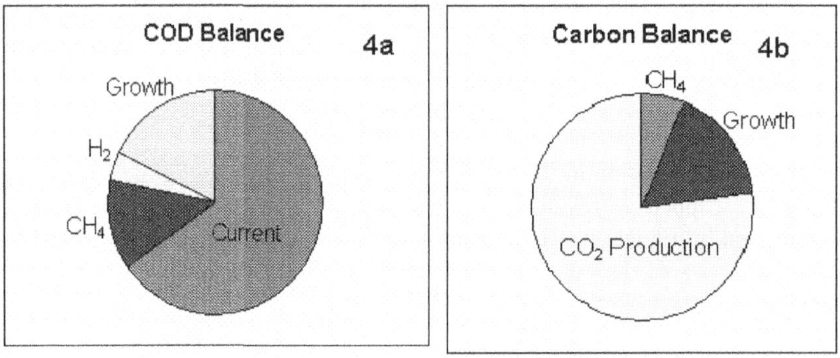

Figure 4 Pie charts representing the breakdown of COD balance (a) and carbon balance (b).

Determination of limiting steps to current production

The electrochemical measurements showed that the cathodic overpotential is by far the limiting step to current production when the electron acceptor is oxygen discharging on plain graphite. As shown in Figure 5b, the cathodic loss is up to ten times higher than the anodic or Ohmic loss. Because oxygen is the only electron acceptor that can be used in a large-scale application, the development of effective low-cost catalysts is necessary before microbial fuel cells can be commercialized. Successful results have been obtained so far with platinum (Oh *et al.*, 2004), iron porphyrines (Zhao *et al.*, 2005) and a ferric salt blended in the cathode (Park & Zeikus, 2002).

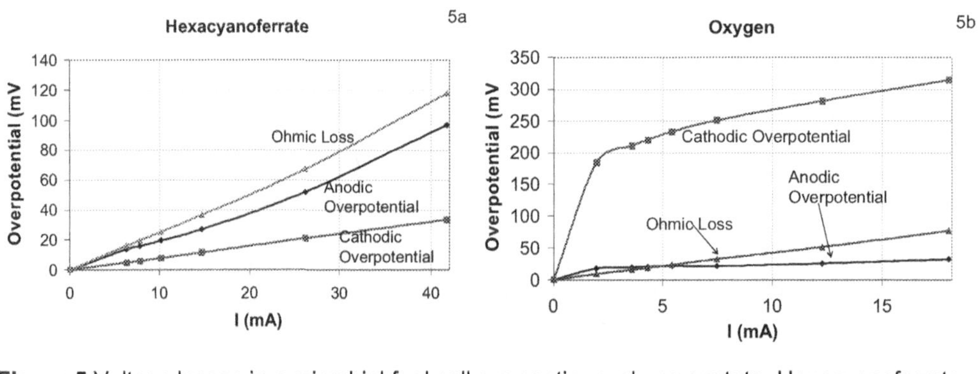

Figure 5 Voltage losses in a microbial fuel cell run continuously on acetate. Hexacyanoferrate (a) and oxygen (b) as electron acceptors.

In Figure 5a it can be seen that hexacyanoferrate has a low cathodic overpotential, making it suitable for laboratory studies focused on the anodic processes. Even though the anodic reaction is slow and limited by the surface area of the anode, the main loss is due to the internal Ohmic resistance. This includes the resistance of the PEM to ion transport and the diffusion of ions through the electrolyte solution. From the electrochemical data obtained in this study, the internal resistance was estimated at 2.65 ± 0.22 Ω for the cell running on hexacyanoferrate, and 4.39 ±0.16 Ω for the cell running on oxygen. The difference shows that the resistance depends on the electrolyte, therefore it cannot entirely be attributed to the PEM.

These results show that even if the anodic overpotential is reduced through the use of better configurations or materials, the performance of MFCs will not greatly improve until the internal resistance is also diminished. This task demands the development of cells with low distances between the electrodes and the PEM, or cells with increased PEM surface. An alternative is to eliminate the PEM completely (Liu and Logan, 2004) and allow direct cation transport via the electrolyte. However, the direct contact of the anodic and cathodic liquids leads to substantial losses in the efficiency of the electron transfer due to the competition from the electron acceptor directly.

Conclusions

This study describes two techniques that can be used to deepen the understanding of the biological and electrochemical processes occurring in microbial fuel cells.

When acetate is the sole substrate, CO_2 comprises about 85% of the total gas production, with the balance made up of methane and hydrogen.

The integration of off-gas analysis, online current monitoring and off-line liquid analysis creates a powerful tool that can be used to close COD and carbon balances, as well as estimating missing parameters, such as the growth yield. In this case the yield was estimated with this method to be 0.173 ± 0.088 $g_{COD,\ biomass}/g_{COD\ removed}$.

Electrochemical measurements can provide quick information on the limiting steps to current production in microbial fuel cells. The technique described shows that the cathode is strongly limiting if oxygen is used as electron acceptor on plain graphite granules. Ohmic loss is always higher than anodic loss. Attention needs therefore to be placed on reduction of both losses in order to attain high performing microbial fuel cells.

References

Bond, D. R., D. E. Holmes, L. M. Tender and D. R. Lovley (2002). Electrode-reducing microorganisms that harvest energy from marine sediments. *Science* **295**(5554): 483–485.

Bond, D. R. and D. R. Lovley (2003). Electricity production by *Geobacter sulfurreducens* attached to electrodes. *Applied and Environmental Microbiology* **69**(3): 1548–1555.

Chaudhuri, S. K. and D. R. Lovley (2003). Electricity generation by direct oxidation of glucose in mediatorless microbial fuel cells. *Nature Biotechnology* **21**(10): 1229–1232.

Kim, H. J., H. S. Park, M. S. Hyun, I. S. Chang, M. Kim and B. H. Kim (2002). A mediator-less microbial fuel cell using a metal reducing bacterium, *Shewanella putrefaciense*. *Enzyme and Microbial Technology* **30**(2): 145–152.

Liu, H. and B. E. Logan (2004). Electricity generation using an air-cathode single chamber microbial fuel cell in the presence and absence of a proton exchange membrane. *Environmental Science & Technology* **38**(14): 4040–4046.

Oh, S., B. Min and B. E. Logan (2004). Cathode performance as a factor in electricity generation in microbial fuel cells. *Environmental Science and Technology* **38**(18): 4900–4904.

Park, D. H. and J. G. Zeikus (2002). Impact of electrode composition on electricity generation in a single-compartment fuel cell using *Shewanella putrefaciens*. *Applied Microbiology and Biotechnology* **59**(1): 58–61.

Pratt, S., Z. G. Yuan, D. Gapes, M. Dorigo, R. J. Zeng and J. Keller (2003). Development of a novel titration and off-gas analysis (TOGA) sensor for study of biological processes in wastewater treatment systems. *Biotechnology and Bioengineering* **81**(4): 482–495.

Rabaey, K., N. Boon, S. D. Siciliano, M. Verhaege and W. Verstraete (2004). Biofuel cells select for microbial consortia that self-mediate electron transfer. *Applied and Environmental Microbiology* **70**(9): 5373–5382.

Rabaey, K. and W. Verstraete (2005). Microbial fuel cells: novel biotechnology for energy generation. *Trends in Biotechnology* **23**(6): 291–298.

Reguera, G., K. D. McCarthy, T. Mehta, J. S. Nicoll, M. T. Tuominen and D. R. Lovley (2005). Extracellular electron transfer via microbial nanowires. *Nature* **435**(7045): 1098–1101.

Schröder, U., J. Niessen and F. Scholz (2003). A generation of microbial fuel cells with current outputs boosted by more than one order of magnitude. *Angewandte Chemie – International Edition* **42**(25): 2880–2883.

Zhao, F., F. Harnisch, U. Schroder, F. Scholz, P. Bogdanoff and I. Herrmann (2005). Application of pyrolysed iron(II) phthalocyanine and CoTMPP based oxygen reduction catalysts as cathode materials in microbial fuel cells. *Electrochemistry Communications* **7**(12): 1405–1410.

Competition between phosphorus-accumulating organisms (PAOs) and glycogen-accumulating organisms (GAOs) in a mixed culture

C.M. Lopez-Vazquez*,*, C.M. Hooijmans*, D. Brdjanovic**, H. Gijzen* and M.C.M. van Loosdrecht*****

* Department of Environmental Resources, UNESCO-IHE Institute for Water Education, P.O. Box 3015, 2601 DA Delft, The Netherlands (E-mail: *c.lopezvazquez@unesco-ihe.org*; *t.hooijmans@unesco-ihe.org*; *h.gijzen@unesco-ihe.org*)

** Department of Municipal Urban Infrastructure, UNESCO-IHE Institute for Water Education. P.O. Box 3015, 2601 DA Delft, The Netherlands
E-mail: *d.brdjanovic@unesco-ihe.org*)

*** Department of Biotechnology, Delft University of Technology, Julianalaan 67, 2628 BC Delft, The Netherlands (E-mail: *m.c.m.vanloosdrecht@tnw.tudelft.nl*)

Abstract A PAO/GAO competition was created for the first time by inoculating a lab-scale SBR with a PAO and a GAO enriched culture in similar biomass loadings (on MLVSS basis). Two different SBRs were operated under same conditions (10 d SRT, 7.0 pH, 20 °C) and substrate characteristics but with different influent phosphorus concentrations to enrich a PAOs (SBR1) and a GAOs (SBR2) culture (15 and 2.2 mgPO_4^{3-}-P/L, respectively). The PAO/GAO SBR (SBR3) operated under the same (non-P limiting conditions) as SBR1. The activity of the PAOs/GAOs in SBR3 was followed for an experimental phase of 6 SRTs. At the end of the experiment, GAOs were able to totally dominate the PAOs/GAOs competition.

Keywords Enhanced biological phosphorus removal (EBPR); mixed culture; phosphorus accumulating organisms (PAOs); glycogen accumulating organisms (GAOs)

Introduction

Enhanced biological phosphorus removal (EBPR) has become in the last two decades a well-established process widely applied in several activated sludge wastewater treatment plants. Simple system configuration, economy, high efficiency and lower sludge production and sludge handling costs compared to other P-removal processes are some factors that have positively influenced its popularity.

The process requires a simple basic configuration: an anaerobic and an aerobic tank in series. Directing the influent to the anaerobic tank (and thus its organic load) and recycling the biomass between the two zones promote the enrichment of the treatment system with phosphorus accumulating organisms (PAOs). Under anaerobic conditions PAOs take up readily biodegradable organic matter present in the influent (mostly volatile fatty acids, VFA, like acetate and propionate) and store it as polyhydroxyalkanoates (PHA) such as polyhydroxybutyrate (PHB) and, in lower amounts, polyhydroxyvalerate (PHV). The energy needed to transport the VFA through the cell membrane and store them as PHA comes partially from the hydrolysis of glycogen and mainly from the degradation of intracellular stored polyphosphate (poly-P). The energy needed for anaerobic cellular maintenance is also produced from poly-P degradation. As a result of poly-P degradation, orthophosphate is released to the bulk liquid. Under the aerobic conditions the previously stored PHA is used as carbon and energy source for the production of biomass, synthesis of glycogen, aerobic maintenance requirements and orthophosphate uptake. Higher amounts of orthophosphate are taken up to recover the poly-P intracellular level. Phosphorus removal can be achieved by means of wasting or handling to sidestream facilities part of sludge taken at the end of the aerobic phase when the biomass contains a high level of poly-P.

Despite that the process configuration creates an environment to favor PAOs enrichment, different studies either at lab- or full-scale treatment plants have reported that another type of microorganisms, identified as glycogen accumulating organisms (GAOs), are able to take up VFA in the anaerobic phase (Chech and Hartman, 1993; Liu *et al.*, 1996) and even to compete for substrate with PAOs (Filipe *et al.*, 2001a; Saunders *et al.*, 2003; Oehmen *et al.*, 2005).

GAOs show a similar metabolism as PAOs (Filipe *et al.*, 2001a; Zeng *et al.*, 2003). They are able to take up VFA under anaerobic conditions and store them as PHA but without anaerobic P release and subsequent aerobic P uptake. Since GAOs do not store poly-P, the energy required for the transport and storage of VFA and anaerobic maintenance comes totally from the hydrolysis of the intracellular stored glycogen. In the aerobic phase, GAOs oxidize the anaerobic stored PHA to provide the carbon and energy necessary for biomass growth and glycogen replenishment.

A significant presence of GAOs in full-scale plants can not only increase the carbon requirements but also, when VFA are limiting, lead to deteriorated EBPR performance (Saunders *et al.*, 2003; Thomas *et al.*, 2003). Some cases, GAOs may furthermore totally outcompete PAOs and cause complete failure of the EBPR process (Satoh *et al.*, 1994).

Although the PAOs/GAOs competition has recently attracted the attention of scientists and engineers the counteraction of these two microorganisms remains unclear. In addition, no a direct and designed competition between PAOs and GAOs has been created and analyzed.

In this study, for the first time, a defined mixed culture of PAOs and GAOs has been studied under controlled laboratory conditions using synthetic wastewater and inoculums from full-scale wastewater treatment plants. The main aim has been to create a competition between these two microorganisms by means of inoculating a SBR with enriched PAOs and GAOs cultures in separate reactors. The initial inoculum of the lab-scale PAOs/GAOs SBR comprised half of the biomass transferred from an enriched PAO culture and another half from an enriched GAO culture on a MLVSS basis. The activity of the reactor was followed for 60 days (6 SRT) to allow sufficient time to study whether and which group of microorganisms was able to dominate the PAOs/GAOs competition.

Materials and methods

Continuous operation of the PAOs and GAOs Sequencing Batch Reactors (SBR1 and SBR2)

The study was carried out in three physically identical laboratory scale reactors (2.5L) at 20°C. The reactors were operated and controlled automatically in a sequential mode (SBR). The pH was maintained at pH 7.0±0.1 using 0.3M HCl and 0.2M NaOH. Activated sludge from domestic wastewater treatment plants was used as inoculum for the first two reactors (Haarlem Waarderpolder for SBR1 and Hoek van Holland WWTP for SBR2). The SBRs were operated in cycles of 6 hours consisting of a 2.25 h anaerobic, 2.25 h aerobic and 1.5 h settling phase. Nitrogen gas was supplied during the anaerobic phase at a flow rate of 30 L/h to remove the oxygen that may be present in the liquid and the headspace of the SBRs. Air was provided during the aerobic stage at a flow rate of 60 L/h. In the first 5 minutes of the anaerobic period nitrogen gas was bubbled to remove all oxygen remaining from the previous cycle. Following the first 5 minutes of the cycle 1.25 L of synthetic medium was fed to the reactor over a period of 5 minutes. 250 mL of mixed liquour were removed on a daily basis (62.5 mL per cycle) from the SBRs, resulting in a sludge retention time (SRT) of 10 d. The SRT was corrected weekly due to loss of biomass through the effluent. At the end of the settling period 1.19 L of supernatant was pumped out from the reactors, in this way the SBRs were operated at an overall hydraulic retention time (HRT) of 12 h. Mixing was provided at 500 rpm except during settling and decant phases.

Inoculation of PAOs/GAOs reactor (SBR3)

The mixed culture of PAOs and GAOs was developed by mixing sludge loadings of a PAO (SBR1) and a GAO (SBR2) enriched cultures in a third PAOs/GAOs reactor (SBR3). The PAOs/GAOs SBR was set up after the PAOs and the GAO SBRs reached steady-state conditions (operated for more than 90 days). The sludge mixing ratio was determined on the basis of the mixed liquor volatile suspended solids concentration (MLVSS) of the first two SBRs. The main aim was to start the SBR3 with an estimated 50% PAOs and 50% GAOs enriched population. After its inoculation, the performance of the PAOs/GAOs SBR was followed for 60 days (6 times its nominal SRT) at the same operating conditions as of SBR1 and SBR2. Due to technical problems, on day 14 the sludge was removed from the PAOs/GAOs SBR and stored at 4°C for 20 days, then the experiment was extended for 20 days (up to 80 days) to follow the competition for the same initially proposed period.

Synthetic wastewater

The main difference between the operating conditions of the first two SBRs was the phosphorus content in the synthetic wastewater supplied to each reactor. Since the first step in the experimental phase was to enrich a PAO culture in SBR1 and a GAO culture in SBR2 the synthetic phosphorus concentration in the synthetic wastewater of the SBR2 was limited to 2.2 mgPO_4^{3-}-P/L (Liu *et al.*, 1997) meanwhile medium for the PAO SBR (SBR1) contained 15 mgPO_4^{3-}-P/L (Smolders *et al.*, 1994). The synthetic medium of the PAOs/GAOs SBR had 15 mgPO_4^{3-}-P/L to avoid phosphorus limitation. In addition, such a composition of the influent of SBR3 was chosen to evaluate whether PAOs could outcompete GAOs under favorable operating conditions for PAOs despite their similar initial sludge loadings (in MLVSS weight). Besides the different phosphorus concentration, the synthetic wastewater for all reactors was prepared as described in Smolders *et al.* (1994). Synthetic wastewater was autoclaved, prior to use, at 110°C for 1 h.

Analysis

Orthophosphate (PO_4^{3-}-P) was determined by the ascorbic acid method, ammonia and nitrate were measured spectrophotometrically, all analysis in accordance with Standard Methods (APHA, 1998). Mixed liquor suspended solids (MLSS) and mixed liquor volatile suspended solids (MLVSS) determination was performed as described in Brdjanovic *et al.* (1997). PHA content of the washed (KP-buffer solution) and freeze-dried biomass was determined by extraction, hydrolyzation and esterification in a mixture of hydrochloric acid, 1-propanol, and dichloroetane at 100°C. After cooling, the organic phase was extracted and with water to remove free acids. PHA was analyzed by gas chromatography. Benzoic acid was used as internal standard throughout the procedure. For glycogen determination 4.5 mL of sample was added to 0.5 mL of 6 M HCl. The samples were digested at 100°C for 5 hours. After cooling, the samples were centrifuged and the glucose concentration was measured with high-pressure liquid chromatography (HPLC).

Results and discussion

Performance of the GAO and PAO SBRs

The GAO and PAO SBRs were operated for more than 90 days. After both SBRs reached steady-state conditions, cycle measurements were carried out on each SBR to determine the biomass activity. A typical cycle profile of a SBR enriched with PAOs is presented in Figure 1a that exhibits anaerobic acetate consumption, significant anaerobic P-release and full aerobic P-uptake. Meanwhile, Figure 1b shows typical cycle profiles for acetate (HAc), orthophosphate (PO_4^{3-}-P) and ammonium (NH^{4+}-N) for the GAO culture in the SBR2 where acetate is fully consumed in the anaerobic phase and anaerobic P-release is negligible. Since nitrification was inhibited due to ATU addition, the slight consumption of ammonium was due to requirements for biomass growth. Nitrate was absent in both reactors during the whole experimental phase (data not shown).

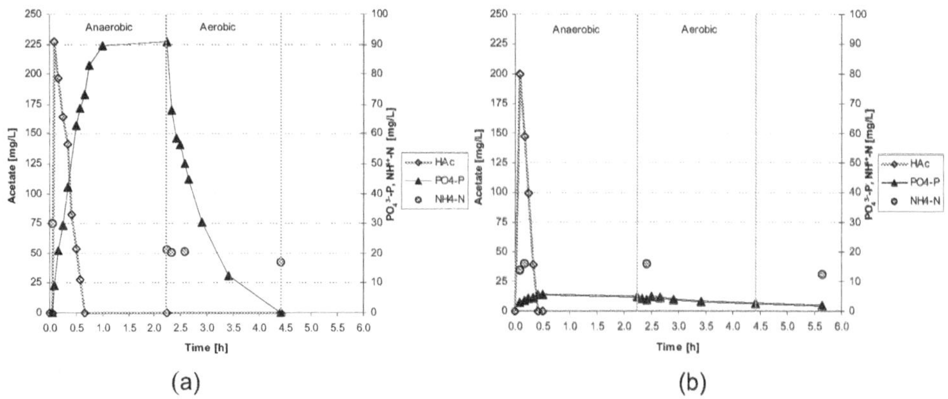

Figure 1 Cycle profiles of HAc (◊), PO_4^{3-}-P (▲) and NH^{4+}-N (○) for a) SBR1 enriched with PAOs and, b) SBR2 enriched with GAOs.

In the PAO SBR, the maximum acetate uptake rate was 0.260 mgHAc/mgVSS-h which is higher than most of the observed values reported in the literature (Table 1) but lower than the rate (0.50 mgHAc/mgVSS-h) reported by Smolders *et al.* (1994). The P/HAc ratio measured in SBR1 was 0.40 mgPO_4^{3-}-P/mgHAc, lower than the value for PAOs (0.50 mgPO_4^{3-}-P/mgHAc) determined by Smolders *et al.* (1994) but similar to the ratio (0.38 mgPO_4^{3-}-P/mgHAc) measured by Brdjanovic *et al.* (1997). The low P/HAc ratio compared to the theoretical value at the same conditions, could be caused by the presence of GAOs in the PAOs SBR. The average MLSS and MLVSS concentration in PAOs SBR were 2980 and 2105 mg/L, respectively. The average MLVSS/MLSS ratio was 0.71. The anaerobic P/HAc ratio and total aerobic P-uptake show that SBR1 was enriched with PAOs.

The maximum acetate uptake rate for GAOs SBR was 0.248 mgHAc/mgVSS-h, which is higher than the values (0.196–0.220 mgHAc/mgVSS-h) reported by Zeng *et al.* (2003), but lower than the rate (0.294 mgHAc/mgVSS-h) observed by Filipe *et al.* (2001b) for a GAOs culture (see Table 1). The anaerobic P/HAc ratio (P-released/Acetate uptake ratio) was 0.014 mgPO_4^{3-}-P/mgHAc which is insignificant compared with the theoretical value for PAOs (0.50 mgPO_4^{3-}-P/mgHAc) observed by Smolders *et al.* (1994) under similar operating conditions. This demonstrates that PAOs were present in the GAO SBR but in a practically negligible concentration due to the low phosphorus concentration in the influent (P-limiting conditions). The average MLSS and MLVSS concentration in the GAO SBR were around 3190 and 2879 mg/L, respectively. The average MLVSS/MLSS ratio was 0.90. Based on the activity of the reactor 2, GAOs were the dominant microorganisms in SBR2.

In general, the maximum acetate uptake rates measured in the GAO and PAO cultures were similar between these two types of microorganisms at pH 7.0 and 20°C. In addition, both measured values are within the range reported by previous studies (Table 1).

Table 1 Comparison of the maximum acetate uptake rates of PAO and GAO at 20°C reported by various researchers (in mgHAc/mgVSS-h).

Microorganism	Reference	Acetate uptake rate
PAO	Smolders *et al.* (1994)	0.500
	Filipe *et al.* (2001b)	0.230
	Kuba *et al.* (1996)	0.230
	Brdjanovic *et al.* (1997)	0.199
	This study	0.260
GAO	Filipe *et al.* (2001a)	0.294
	Zeng *et al.* (2003)	0.196-0.220
	This study	0.248

Start-up of the PAOs/GAOs mixed culture

The SBR3 was inoculated after the GAOs and PAOs SBRs reached steady-state conditions (estimation based on steady-state values of both reactors). The sludge transferred from the first two SBRs was determined based on the average MLVSS concentration in the reactors 5 days before inoculation. Average measured MLVSS concentrations were 2105 and 2879 mg/L in the PAO and GAO SBRs, respectively. The sludge was transferred at the end of the cycle of the two first SBRs, just after the supernatant was discharged. In that moment, PAO and GAO reactors had only 50% of their total volume. In this way, the sludge was stirred and 530 mL of mixed liquor from the PAO SBR and 720 mL from the GAO SBR were transferred to SBR3 (PAO/GAO SBR). The initial MLSS and MLVSS concentrations in the PAO/GAO SBR were 1859 and 1538 mg/L, respectively, with a MLVSS/MLSS ratio of 0.83. Figure 2 shows the first cycle measurement of the PAOs/GAOs SBR (SBR3).

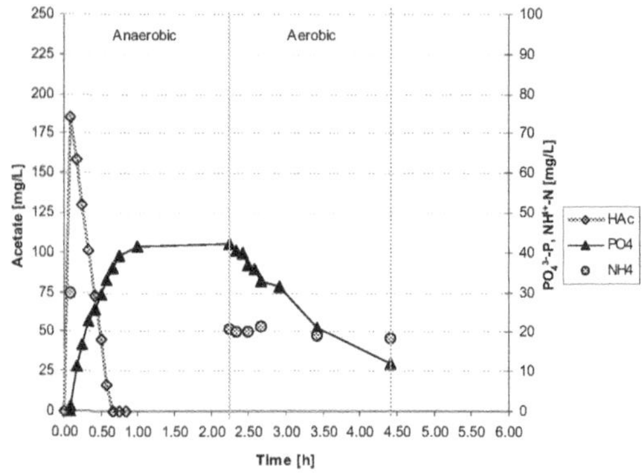

Figure 2 Profile of HAc (◊), PO_4^{3-}-P (▲) and NH^{4+}-N (○) for the first cycle measurement of the PAOs/GAOs mixed culture in SBR3 after its inoculation.

The acetate uptake rate observed in the first cycle of the PAO/GAO SBR (0.252 mgHAc/mgVSS-h) is within the range of the acetate uptake rates of GAOs and PAOs cultures (See Table 1). The P/HAc ratio measured in the first cycle of the PAO/GAO reactor was 0.19 mgPO_4^{3-}-P/mgHAc (Figure 2) which is slightly lower than the expected P/HAc value (0.20 mgPO_4^{3-}-P/mgHAc). Since similar sludge (on a MLVSS basis) were transferred from the GAO and PAO SBRs to the PAO/GAO reactor, the expected P/HAc value was calculated taking into account the average of the observed P/HAc ratios in the GAO (0.014 mgPO_4^{3-}-P/mgHAc) and PAO reactors (0.40 mgPO_4^{3-}-P/mgHAc). Nevertheless, the P/HAc ratio measured in PAO/GAO SBR confirms that, according to the transferred sludge, approximately 50% of the total biomass in this reactor was comprised of PAOs. Additionally to the P/HAc ratio, the observed acetate uptake of PAO/GAO SBR, which is within the rates measured in the GAO and PAO SBRs, confirms that PAO and GAO were present in the PAO/GAO reactor in similar concentrations. After its inoculation, the performance of the PAO/GAO SBR was followed for 60 days (six times its nominal SRT) to study the competition of the microorganisms. The observed net P-released/Acetate consumed (P/HAc ratio) and MLSS, MLVSS and MLVSS/MLSS ratio during the 6 SRT in the PAO/GAO mixed culture are shown in Figure 3a,b respectively.

The activity of the PAO/GAO SBR was analyzed based on the fact that, for PAO cultures, the anaerobic P-released is directly coupled with the acetate consumed by these microorganisms (Smolders *et al.*, 1994). If the operating conditions (pH, SRT, temperature and influent concentrations) remain constant during the whole experimental phase, then the P/HAc ratio can be used as an indicator of the activity of PAOs. Additionally, it can be assumed that, under the previously described initial operating conditions and characteristics of the PAO/GAO SBR (similar initial populations of PAOs and GAOs) any change on the activity of PAOs was result of the activity of GAOs.

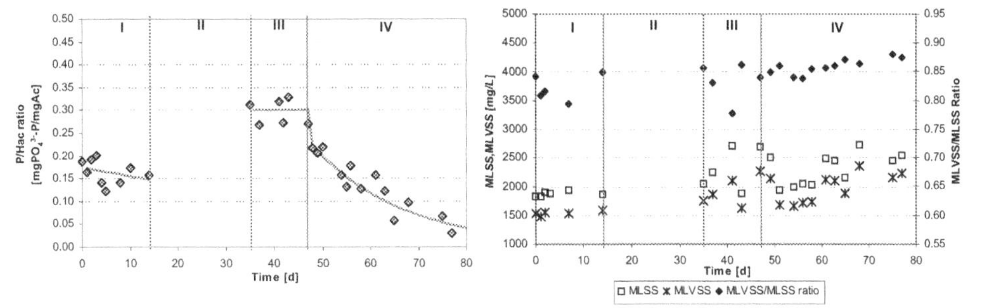

(a) (b)

Figure 3 (a) P-released/Acetate consumed (P/HAc ratio), and (b) MLSS, MLVSS and MLVSS/MLSS ratio during the operation of the PAO/GAO SBR.

Table 2. P/HAc ratio in the stages observed during the PAOs/GAOs SBR operation.

Stage	I	II	III	IV
Days	0-14	14-35	35-47	47-80
P/HAc ratio [mgPO$_4^{3-}$-P/mg HAc]	0.15±0.04	-	0.30±0.03	From 0.22 to 0.03
Comment	Slightly decrease of PAO activity	Technical failure, biomass stored at 4°C	Stable PAO activity	PAOs activity decreased

According to the previous paragraph, the activity of PAOs in the PAO/GAO SBR followed three different trends during the whole experiment. The gray lines on Figure 3a show the trends of the P/HAc ratio. In this way, 3 different stages can be identified. In addition, due to technical problems, an extra stage (Stage II) can be considered (see Figure 3a and Table 2). The P/HAc ratio during the four phases is described in Table 2. In the first 14 days (Stage I) an average P/HAc ratio of 0.15 ± 0.04 mgPO$_4^{3-}$-P/mgHAc was observed which is slightly lower than the P/HAc ratio (0.19 mgPO$_4^{3-}$-P/mgHAc) measured at the beginning of the test. This shows that the activity of PAOs decreased in the first 14 days of the experiment respect to their initial observed activity. However, due to a technical failure (Stage II) from day 14 to 35 the biomass was removed from the SBR and stored in the fridge at 4°C. After the problem was solved, the biomass was returned to the SBR and due to unclear reasons, in the Stage III, from day 35 to 47, the P/HAc ratio increased and reached an average P/HAc ratio of 0.30 ± 0.03 mgPO$_4^{3-}$-P/mgHAc. This ratio shows that PAOs increased significantly their activity in the Stage III (two times the measured P/HAc ratio on Stage I) and dominated the competition. Likely, PAOs presented a higher activity in this phase because, at lower temperatures, the PAOs metabolism is less inhibited than GAOs metabolism. In this way, after the biomass was stored at 4°C for 2 weeks in the fridge, PAOs acclimatized faster than GAOs when biomass was returned to the PAOs/GAOs SBR. From day 47–80, the PAOs activity decreased considerably from 0.22 to 0.03 mgPO$_4^{3-}$-P/mgHAc (Stage IV). This indicates that GAOs increased their activity and PAOs were outcompeted by GAOs in the PAOs/GAOs SBR system. There are not clear reasons to explain the sudden deterioration of PAOs activity. However, excluding the Stage III where some extra factors could influence, in the Stages I and IV always the activity of PAOs tended to decrease.

Similar trends observed in the Figure 3a can also be seen in Figure 3b on the basis of the MLSS and MLVSS concentrations and the MLVSS/MLSS ratio. In the first 14 days of the experiment (Stage I), the PAOs/GAOs SBR showed stable and constant MLSS and MLVSS concentrations (1857 ± 43 mg/L and 1535 ± 37 mg/L, respectively) but the MLVSS/MLSS ratio increased from an initial average of 0.82 ± 0.02 to 0.85. From day 14 to 35 (Stage II), the SBR was stopped due to the previous mentioned technical problem. In the Stage III, after the biomass was returned to the reactor average MLSS and MLVSS increased around 25% (2312 ± 375 mg/L and 1920 ± 260 mg/L). In the same period (Stage III), the average MLVSS/MLSS ratio remained at 0.85. In the last part of the experiment (Stage IV), the average MLSS concentration presented a similar value (2300 ± 275 mg/L) as in Stage III but the average MLVSS concentration increased slightly (3%) up to 1978 ± 251 mg/L. Consequently, the MLVSS/MLSS ratio increased from 0.85 up to 0.87, indicating that the inorganic content within the biomass decreased. In general, from day 0 to 80, the MLSS and MLVSS increased considerably (nearly 40%) as well as the MLVSS/MLSS ratio (5%).

The decrease of the P/HAc ratio in the Stage I and IV (Figure 3a) shows that, in the anaerobic phase, PAOs were not able to take up acetate as fast as GAOs and, as a consequence, PAOs produced less PHA. Thus, under aerobic conditions, PAOs had less PHA available than GAOs to cover their aerobic metabolism requirements (glycogen production, poly-P accumulation, biomass growth and maintenance). The increase of the MLVSS/MLSS ratio in the same periods (Figure 3b) indicates that the biomass present in the PAOs/GAOs SBR accumulated less inorganic matter (presumably less poly-P). The latter corroborates that PAOs, in the aerobic phase, had less PHA available. Since PAOs had less PHA available, PAOs not only accumulated less poly-P but also PAOs produced less biomass and glycogen and lost the PAOs/GAOs competition. At the end of the experimental phase (day 80), GAOs were the dominant microorganism in the PAOs/GAOs SBR.

Zeng *et al.* (2003) compared the net biomass growth per acetate consumed of both microorganisms (0.29 gPAO/gAc and 0.30 gGAO/gAc), at similar operating conditions as in the present study, and found similar yields. This shows that there are not significant differences on their aerobic metabolisms and means that neither PAOs nor GAOs had any advantage on the aerobic phase. Thus, the PAOs/GAOs competition only takes place in the anaerobic phase. Considering that PAOs and GAOs only competed for substrate under anaerobic conditions, the higher observed acetate uptake of PAOs (see Table 1) should have been a strong kinetic advantage to totally outcompete GAOs in the PAO/GAO SBR. However, due to not clear reasons the activity of PAOs always tended to decrease.

Different studies (see Table 1) have reported that GAOs are able to take up substrate faster than PAOs. Since the PAOs/GAOs competition is highly influenced by the substrate uptake rates, GAOs should lead always the competition in the EBPR process either at lab- or full-scale treatment plants and the removal of phosphorus should be usually an unstable process. However, this is not the case. As a consequence, how different operating conditions and parameters (such as temperature, dynamic conditions, type of substrate, pH, presence and concentration of both microorganisms in the origin of the inoculum and the configuration of the treatment systems) can influence the metabolism of PAOs and GAOs may play a significant role to define the usual dominance of PAOs and thus the popularity of EBPR process in full-scale treatment plants.

On the other hand, coupling the effect on the metabolism of PAOs and GAOs of the previous mentioned operating conditions and parameters (i.e. temperature and pH) with the use of experimentally validated metabolic models successfully applied to describe the activity of these microorganisms (Smolders *et al.*, 1994; Zeng *et al.*, 2003) may provide a useful tool to study and get a better understanding of the effect of different conditions and scenarios on the PAOs/GAOs competition.

The present experiment may be repeated under similar operating conditions (avoiding any technical problem) to evaluate if the PAOs/GAOs competition follows the same trend. In addition, the influence of the actual initial PAOs/GAOs ratios supported with the use of molecular techniques must be used to get a better understanding of the competition.

Conclusions

For the first time, a designed PAO/GAO competition was created by inoculating a lab-scale SBR with a PAOs and a GAOs enriched cultures in equal biomass loadings. After a 6 SRTs experimental phase and despite that the PAOs culture showed a higher acetate uptake rate than the GAOs culture, they did not outcompete GAOs. On the opposite, the activity observed in the PAO/GAO SBR demonstrated that GAOs were able to finally dominate the competition.

References

APHA, (1998) *Standard methods for the examination of water and wastewater*, 20th edn. American Public Health Association/American Water Works Association/Water Environment Federation, Washington, D.C.

Brdjanovic, D., van Loosdrecht, M.C.M., Hooijmans, C.M., Alaerts, G.J., Heijnen, J.J. (1997) Temperature effects on physiology of biological phosphorus removal. *ASCE Journal of Environmental Engineering*, **123** (2):144–154.

Chech, J.S., Hartman, P. (1993) Competition between phosphate and polysaccharide accumulating bacteria in enhanced biological phosphorus removal systems. *Water Research*, **27**: 1219–1225.

Filipe, C.D.M., Daigger, G.T., Grady, Jr C.P.L. (2001a) A metabolic model for acetate uptake under anaerobic conditions by GAOs: stoichiometry, kinetics and effect of pH. *Biotechnology and Bioengineering* **76**(1): 17-31.

Filipe, C.D.M., Daigger, G.T., Grady, C.P.L. Jr (2001b) pH as a key factor in the competition between GAOs and PAOs. *Water Environment Federation*, **73**(2): 223–232.

Liu, W.T., Mino, T., Nakamura, K, Matsuo, T. (1996) Glycogen accumulating population and its anaerobic substrate uptake in anaerobic-aerobic activated sludge without biological phosphorus removal. *Water Research*, **30**(1): 75–82.

Liu, W.T., Nakamura, K., Matsuo, T., Mino, T. (1997) Internal energy-based competition between polyphosphate- and glycogen-accumulating bacteria in biological phosphorus removal reactors-effect of P/C feeding ratio. *Water Research*, 31(6):1430–1438.

Oehmen, A, Yuan, Z, Blackall, LL, Keller, J (2005) Comparison of acetate and propionate uptake by PAOs and GAOs. *Biotech Bioengng* **91**(2): 162–168.

Satoh, H., Mino, T., Matsuo, T. (1994) Deterioration of enhanced biological phosphorus removal by the domination of microorganisms without polyphosphate accumulation. *Water Science and Technology*, **30**(6): 203–211.

Saunders, A.M., Oehmen, A., Blackall, L.L., Yuan, Z., Keller, J. (2003) The effect of GAOs (glycogen accumulating organisms) on anaerobic carbon requirements in full-scale Australian EBPR (enhanced biological phosphorus removal) plants. *Water Science and Technology*, **47**(11): 37–43.

Smolders, G.J.F., van Der Meij, J., van Loosdrecht, M.C.M., Heijnen, J.J. (1994) Model of the anaerobic metabolism of the biological phosphorus removal processes: stoichiometry and pH influence. *Biotechnology and Bioengineering*, **43**: 461–470.

Thomas, M., Wright, P., Blackall, L.L., Urbain, V., Keller, J. (2003) Optimization of Noosa BNR plant to improve performance and reduce operating costs. *Water Science and Technology*, **47**(12): 141–148.

Zeng, R.J., van Loosdrecht, M.C.M., Yuan, Z., Keller, J. (2003) Metabolic model for glycogen-accumulating organisms in anaerobic/aerobic activated sludge systems. *Biotechnology and Bioengineering*, **81**(1): 92–105.

The pH shock of seed sludge for enhanced biohydrogen production

Zhen-Peng Zhang*, Kuan-Yeow Show*, Joo-Hwa Tay* and David Tee Liang**

*School of Civil and Environmental Engineering, Nanyang Technological Univ., Nanyang Ave., Singapore 639798 (E-mail: *zpzhang@pmail.ntu.edu.sg*; *ckyshow@ntu.edu.sg*; *jhtay@ntu.edu.sg*)

** Institute of Environmental Science and Engineering, Nanyang Technological Univ. Nanyang Drive, Singapore 637723 (E-mail: dtliang@ntu.edu.sg)

Abstract Hydrogen consumption that resulted from hydrogenotrophic methanogenesis and homoacetogenesis might occur when a reactor was maintained continuously for long term operation. The effect of acid and base incubation on hydrogen production was investigated in batch tests after seed sludge was incubated for 0.5 h, 2 h, 6 h, 12 h and 24 h at pH of 3 and 10, respectively. The results showed that the activity of methanogens was inhibited completely by acid and base incubation. A short period of acid incubation (0.5-2 h) and base incubation (2-6 h) exhibited a short lag phase and significantly enhanced the production of hydrogen. The highest hydrogen production potentials of 44.5 mL and 43.1 mL were observed in the tests with acid incubation for 0.5 h and base incubation for 2 h, respectively, while the highest specific hydrogen production rates of 10.7 mL H2/g VSS·h and 12.4 mL H_2/g VSS·h were obtained in the tests with acid incubation for 2 h and base incubation for 6 h, respectively. It was speculated that the enhancement of hydrogen production might be derived from inhibited activities of methanogens or other non-spore hydrogenotrophic bacteria by acid and base incubation. These finding imply that the performance of microbial cultures in hydrogen production could be enhanced by acid or base incubation with an appropriate period.

Keywords Acid/base incubation; biohydrogen production; homoacetogenesis; hydrogen consumption; methanogenesis

Introduction

Hydrogen production by biological processes has attracted worldwide attention due to its potential to become an inexhaustible, low-cost and renewable source of clean energy (Levin *et al.* 2004). Anaerobic fermentation route is a promising biological system for hydrogen production owing to the fact that hydrogen can be produced continuously at high rate from renewable organic compounds as compared to other biological processes, i.e., biophotolysis and photodecomposition from water and organic compounds, respectively (Das and Veziroglu 2001).

Stable and sustainable production of hydrogen is pivotal for a continuously operating biohydrogen production system. However, hydrogen consumption resulted from hydrogenotrophic bacteria such as methanogens and homoacetogens in a mixed acidogenic culture has been observed (Hussy *et al.* 2003; Hwang *et al.* 2004). To maintain hydrogen producers as the dominant microorganisms, methanogens can be inhibited by low pH

(Sparling *et al.* 1997; Fang and Liu 2002; Kim *et al.* 2004). A pH range between 4.0-6.3 was suggested as an optimal operating condition, which is beyond the range for most methanogenic bacteria to function (Lay *et al.* 1997). However, the activity of hydrogenotrophic methanogens was reported at pH 5.0 and 4.5 in a semi-continuous reactor (Hwang *et al.* 2004; Kim *et al.* 2004). In our studies, little hydrogen gas was produced in anaerobic fluidized bed reactors operated at a pH range of 4.5-5.5 and hydraulic retention time (HRT) of 2 h (data not shown). These results indicated that the hydrogen would be consumed to form methane via reduction of carbon dioxide as shown in Eq. (1):

$$4H_2 + CO_2 \rightarrow CH_4 + 2H_2O \quad (1)$$

Accordingly, it is likely that pH condition as low as 4.5 may not completely inhibit the activity of hydrogen utilizing methanogens in a long-term hydrogen-producing reactor if other operating conditions are favourable for the growth of hydrogenotrophic methanogens. On the other hand, there is the potential for other competing reactions that consume hydrogen when using a mixed microflora. Hussy *et al.* (2003) found hydrogen consumption resulted from homoacetogens and propionate formers were apparent at both pH 5.2 and pH 4.5 in a continuous reactor following Equations (2) and (3), respectively.

$$2CO_2 + 4H_2 \rightarrow CH_3COOH + 2H_2O \quad (2)$$

$$C_6H_{12}O_6 + 2H_2 \rightarrow CH_3CH_2COOH + 2H_2O \quad (3)$$

Owing to resistance to heat and harsh chemicals, *Clostridium*, a high efficient hydrogen production species could be obtained by heat treatment or acid-base enrichment from natural inocula (Chen *et al.* 2002; Oh *et al.* 2003), thus it was possible to prevent hydrogen consumption that caused by hydrogenotrophic non-spore formers if heat treatment and acid-base enrichment were applied to the acclimated hydrogen-producing sludge in a continuous hydrogen-producing system. Only one study reported so far, however, elucidated sustainable hydrogen production could be achieved by using a combination of pre- and repeated heat treatments of the biomass in anaerobic bioreactors, and also indicated a long lag phase after repeated heat treatments might have prevented good performance of hydrogen-producing bioreactor (Sung *et al.* 2002).

The present study was undertaken to investigate the impact of acid and base incubation periods for acclimated hydrogen-producing sludge on shortening lag phase and enhancing hydrogen production. The mechanisms of hydrogen consumption and hydrogen enhancement were also proposed.

Materials and methods

Inoculum

The seed sludge was obtained from an anaerobic fluidized bed reactor operated at pH of 5.00±0.05 for hydrogen production for 60 days in the laboratory. Acclimated sludge was washed three times to remove the accumulated metabolites and residual glucose. The washing procedure was composed of a centrifugation step (4000 rpm, 10 min), followed by removing the supernatant and adding a same volume of distilled water and re-suspended sludge pellet using Vertex mixer (2500 rpm, 2.5 min).

Procedure

Batch tests were conducted by using 320 mL serum vials with a working volume of 200 mL. Base on a study by Chen *et al.* (2002), pH 3 and pH 10 were chosen as the incubation conditions for seed sludge in this study. The pH of a 20 mL seed sludge was adjusted to 3 and 10 using 2 M hydrochloric acid and 2 M sodium hydroxide, respectively, and then incubated (Shaking incubator, DK-S1001, Korea) at 37 °C for 0.5 h, 2 h, 6 h, 12 h and 24 h, respectively. After the acid and base incubation, each vial was added with 2.0 g glucose, and 1 mL nutrient stock solution containing the following nutrients in mg/L: 200 NH_4Cl, 200 peptone, 200 KH_2PO_4, 30 $FeCl_36H_2O$, 20 $MgCl_2·6H_2O$, 10 $MnCl_2·4H_2O$, 5 $CoCl_2·6H_2O$, 5 $Cu(NO_3)_2·3H_2O$, 5 $ZnSO_4·7H_2O$, 6 $CaCl_2·2H_2O$, and 5 $NiSO_4·6H_2O$. The final volume of culture was controlled at 200 mL by adding distilled water. Initial pH of each vial was adjusted to 5.00 and immediately capped with butyl rubber stopper. Nitrogen gas was flushed through the medium for 30 second to maintain anaerobic conditions. The vials were placed in a reciprocal water bath shaker (Firstek, B701, UK) at 37 °C and at a horizontal rotational speed of 150 rpm. Biogas production was recorded at an interval of 1 h using water displacement method, and hydrogen production was calculated using hydrogen percentage in the headspace and biogas production in each corresponding interval.

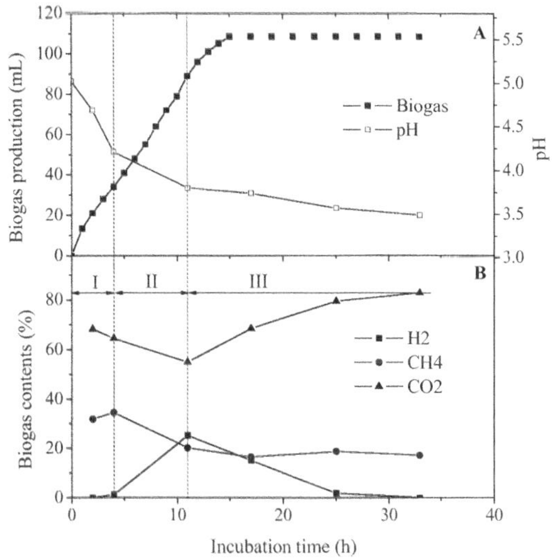

Figure 1 Variations in biogas production, biogas contents and pH in control test (Stage I: hydrogenotrophic methanogenesis; Stage II: H_2 accumulation; Stage III: homoacetogenesis).

Analytical methods

The biogas composition was analyzed by a Micro-Gas Chromatograph (Varian 4900, USA) equipped with a thermal conductivity detector, Molsieve 5A Plot column and Propac Q column for the analysis of biogas composition.

Aqueous samples were filtered through a 0.45 μm membrane before analysis. Volatile fatty acids (VFAs) and alcohol were determined by a gas chromatograph (Agilent 6890, USA) equipped with a mass selective detector, and a 30m×0.25 μm HP-FFAP fused-silica capillary column. Glucose concentration was determined by the phenol-sulphuric acid method (Dubois *et al.* 1956). Measurements of suspended solids and volatile suspended solids (VSS) were performed according to the Standard Methods (APHA 1998). The pH valves of culture were measured by a pH meter (Oakton-Ion 6, USA).

Data analysis

The modified Gompertz equation (Equation (4)) was used to describe the progress of cumulative hydrogen production obtained from a batch experiment (Lay *et al.* 1999).

$$H(t) = P \cdot \exp\left\{ -\exp\left[\frac{R_m \cdot e}{P}(\lambda - t) + 1 \right] \right\} \quad (4)$$

in which $H(t)$ is the cumulative hydrogen production (mL) during the incubation period, t (h); P is hydrogen production potential (mL); R_m is the maximum hydrogen production rate (mL/h); e, constant, 2.718282…; and λ is lag-phase time (h). Parameters (P, R_m and λ) were estimated using the solver function in Microsoft Excel 2003 (Microsoft, Inc.) as described by Van Ginkel and his co-workers (2001). R_s, the specific hydrogen production rate (SHPR, mL/g VSS·h) was calculated by dividing the R_m by the sludge weight (g VSS).

Table 1 Evaluated parameters in modified Gompertz equation

Incubation (condition/time)	Hydrogen production potential, P (mL)	Specific hydrogen production Rate, R_s (mL/g VSS·h)	Lag time, λ (h)	Correlation coefficient, R (-)
pH3/0.5h	44.50	8.14	1.72	1.00
pH3/2h	42.01	10.72	0.40	0.99
pH3/6h	28.81	9.42	0.81	0.98
pH3/12h	20.80	7.81	1.37	0.98
pH3/24h	20.91	6.64	1.45	0.98
pH10/0.5h	26.00	8.56	1.08	0.99
pH10/2h	43.90	11.97	–0.60	0.97
pH10/6h	41.96	12.36	–0.26	0.98
pH10/12h	13.71	2.29	2.39	1.00
pH10/24h	5.28	2.09	11.34	1.00

Results and discussion

Biohydrogen production

Before production of biogas, lag times ranging from 0 to 12 h were observed in all the tests (Figures 1A, 2A, and 3A). Acid incubation time did not exert a significant effect on lag time, and a production of biogas in all the tests with acid incubation was observed after a short lag phase of 0.4 ~1.7 h. However, lag times of 2.4 h and 11.3 h were observed in base enrichment times of 12 h and 24 h, respectively, which were much longer than that of short term (0.5~6 h) base incubations (Table 1). In the blank test, biogas production was accumulated to 108.5 mL after a 15 h cultivation, but biogas contents varied significantly meanwhile. Biogas was composed of 31.7% methane, 68.3% carbon dioxide and no hydrogen at 2 h (Figure 1B). The percentage of hydrogen concentration increased to 1.23% at 4 h associated with 34.34% methane and 64.4% carbon dioxide, and then reached a maximum of 25.1% at 11 h, but it declined to 15.1% and 1.9% at 17 h and at 25 h, respectively. The initial increase of hydrogen content could be resulted from inhibition of the activity of methanogens as a result of a drop of pH due to the accumulation of VFAs. Obviously, hydrogen-consuming methanogens remained active during the period of 0-4 h (Stage I), but they were suppressed thereafter. Because of 5-volume in hydrogen and carbon dioxide instead of 1-volume methane as products when this conversion ceased completely, accordingly an increase of biogas production rate during a period of 4-11 h (stage II) was observed. Based on biogas production and hydrogen content as shown in Figure 1, hydrogen production of 11.12 mL was obtained in the blank test.

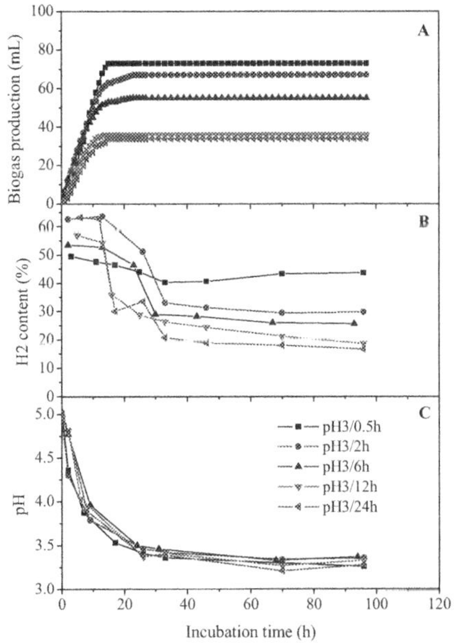

Figure 2 Variations in biogas production (A), hydrogen content (B) and pH (C) after acid incubation.

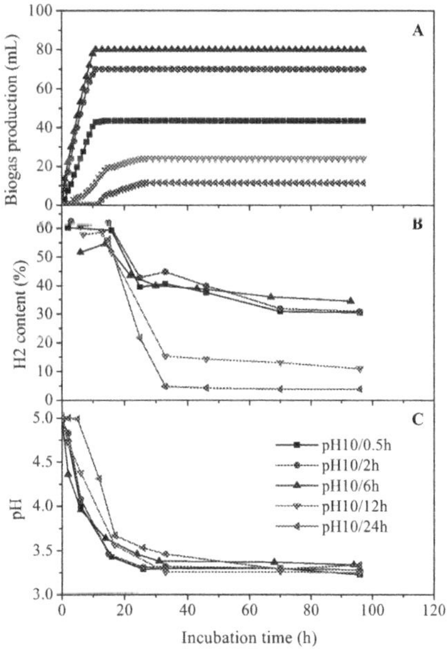

Figure 3 Variations in biogas production (A), H_2 content (B) and pH (C) after base incubation.

The biogas produced in all the tests with acid and base incubation was composed entirely of hydrogen and carbon dioxide, whereas no methane was detected, which was mainly attributed to the inhibition of methane formers by acid and base incubation as compared to the blank test. The maximum of hydrogen concentration of 47–63% in the tests with acid and base incubation was detected at the initial stage of cultivation, and kept approximately consistent during the period of biogas production. After that, however, hydrogen concentration in the headspace decreased to 4–44% before the culture pH declined to about 3.30 (Figures 2C and 3C). It was possible that there was a balance in biogas production and consumption, but a variation in biogas contents might occur during this period. These results indicated that hydrogen consumption in the headspace occurred to a certain extent in a pH range of 3.30-3.70, which might result from the facts that a drop of pH probably induced a metabolic shift of bacteria from hydrogen production to utilization around pH 3.70, whereas the hydrogen utilization was completely inhibited when the culture pH fell to around 3.30. Incubation time was negatively correlated with the amount of hydrogen production in the tests with acid incubation. The highest amount of hydrogen production was 44.53 mL in the tests with acid incubation 0.5 h. In the tests with base incubation, the highest and second highest hydrogen productions of 43.11 mL and 41.61 mL were observed in the test with incubation time of 2 h and 6 h, respectively, followed by 26.08 mL (0.5 h), 13.39 mL (12 h) and 5.27 mL (24 h). These observational values of accumulative hydrogen production and lag phase were quick consistent with the estimated hydrogen production potential (P) and lag time (λ) using modified Gompertz equation, as evidenced by high correlation coefficients R ranged of 0.97–1 (Table 1). These indicated that the modified Gompertz equation could be used to estimate the hydrogen production potential (P), hydrogen production rate (R_m) and lag time (λ).

Clearly, the results summarised in Table 1 shows that acid and base incubation increased hydrogen production and specific hydrogen production rate in comparison with the blank test. The range of SHPRs was 6.64–10.72 mL H_2/g VSS·h for the acid incubation, and the highest value of 10.72 was obtained in the test with acid incubation 2 h. The effect of base incubation periods on SHPR seemed to be more significant. The highest and lowest values of 12.36 mL H_2/g VSS·h and 2.09 mL H_2/g VSS·h were obtained in base incubation for 6 h and 24 h, respectively. Table 1 also reveals that the incubation period for seed sludge exposure to acid and base had a significant effect on hydrogen production and duration of lag phase. It appears that the period of acid and base incubation resulted in substantial differences in performance of hydrogen-producing bacteria. It has been known that *Clostridium* forms spores when unfavourable conditions are encountered, and the spores need time (observed lag phase, λ) to germinate and become vegetative cells when favourable conditions return. It has been established that the hydrogen-producing performance of *Clostridium* is dependent on the amount and activity of vegetative cells. The results indicated that acid and base incubation for seed sludge had a significant and immediate effect on germination and activity of hydrogen-producing spore formers, and an inhibitory effect on germination of spore or activity of vegetative cells might be expected from the long term acid and base incubation. Therefore, as results indicated, an appropriate incubation period of acid (0.5–2 h) and base (2–6 h) has been shown to accelerate germination of spore formers or shorten the lag phase and enhance activity of hydrogen-producing bacteria.

Table 2 VFAs and alcohol concentrations in all the tests at 17h

Metabolites (mM)	Blank	pH3 0.5h	pH3 2h	pH3 6h	pH3 12h	pH3 24h	pH10 0.5h	pH10 2h	pH10 6h	pH10 12h	pH10 24h
Ethanol	6.98	6.79	7.57	5.12	5.61	7.73	7.26	5.40	5.65	6.98	5.02
Acetate	13.37	7.65	6.86	5.66	6.39	6.01	6.18	15.71	7.08	10.41	5.72
Propionate	2.86	2.97	3.27	3.33	3.03	2.36	1.54	2.68	2.55	2.07	1.93
n-butyrate	8.36	4.98	4.90	4.60	4.13	3.48	2.87	6.62	6.91	2.96	3.28
Valerate	1.19	1.30	1.33	1.19	0.96	0.83	0.74	1.28	1.33	0.75	0.70
Caproate	1.22	0.75	0.77	0.67	0.58	0.58	0.54	0.64	0.72	0.55	0.54

Acetate, butyrate and ethanol were predominant aqueous metabolites in all the tests, and their concentrations increased with the biogas production. These VFAs and ethanol were listed in Table 2 when biogas production ceased. Propionate was present in all the tests with a low concentration of 1.54–3.33 mM. Relatively stable concentrations of these VFAs and ethanol were observed after biogas production ceased in all the tests, except for the blank test as shown in Figure 4.

Hydrogen consumption

A decrease in hydrogen content from 25.12% at 11 h to 1.85% at 25 h was observed in the blank test. This might be due to homoacetogenesis. Although the concentration of glucose (3.75 g/L) was still high when biogas production ceased, however, it has been known that the Wood-Ljungdahl pathway enables growth on H_2 and CO_2 according to Equation (2) (Muller 2003). This reaction for homoacetogenesis is favourable thermodynamically because the overall free energy change ($\Delta G_o' = -95$ kJ/mol) could allow for the synthesis of 1 to 2 mol of

ATP. Therefore, hydrogen gas remaining in the headspace was reduced to a very low level associated with an apparent increase of acetate production in the blank test (Figure 4).

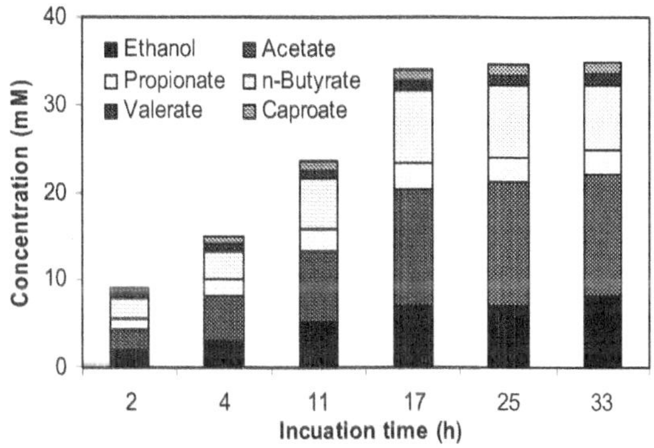

Figure 4 VFAs and alcohol concentration variation over time in the blank test.

Acid and base treatments were effective methods to inhibit methanogenesis, but hydrogen consumption was still observed in these tests. However, its mechanisms of hydrogen consumption appeared to be different with the blank test. The hydrogen-consuming acetogens that resulted in hydrogen consumption at a pH of around 3.8 in the blank test were likely non-spore formers, but these non-spore formers had been killed by acid and base incubation, thus no apparent hydrogen consumption occurred at a pH of around 3.8 in these tests with acid and base incubation. It was probably that the hydrogen consumption in the tests with acid and base incubation resulted from the metabolic shift of clostridial bacteria induced by culture pH drop, otherwise a substantial drop in hydrogen concentration would be expected through the whole cultivation, but this was not observed. Therefore, hydrogen consumption might be prevented by controlling operational pH above 3.80.

Conclusions

In summary, the results obtained in the present study demonstrated that activity of methane formers was inhibited completely by acid and base incubation, and a short period of acid incubation (0.5–2 h) and base incubation (2–6 h) exhibited a short lag phase and significantly enhanced the production of hydrogen. The highest hydrogen production potentials of 44.5 mL and 43.1 mL were observed in the tests with acid incubation for 0.5 h and base incubation for 2 h, respectively, while the highest specific hydrogen production rates of 10.7 mL H_2/g VSS·h and 12.4 mL H_2/g VSS·h were obtained in the tests with acid incubation for 2 h and base incubation for 6 h, respectively. The hydrogen consumption that resulted from hydrogenotrophic methanogenesis and homoacetogenesis occurred in the blank test. It was speculated that the enhancement of hydrogen production might be derived from inhibited activities of methanogens or other non-spore hydrogenotrophic bacteria by acid and base incubation, whereas the variations in hydrogen production potential after acid and base incubation were probably attributed to influence of incubation period on the activity of hydrogen-producing bacteria.

References

APHA (1998) *Standard methods for the examination of water and wastewater*. American Public Health Association.

Chen, C. C., Lin, C. Y., Lin, M. C. (2002) Acid–base enrichment enhances anaerobic hydrogen production process. *Applied Microbiology and Biotechnology* **58**(2): 224–228.

Das, D., Veziroglu, T. N. (2001) Hydrogen production by biological processes: a survey of literature. *International Journal of Hydrogen Energy* **26**(1): 13–28.

Dubois, M., Gilles, K. A., Hamilton, J. K., Rebers, P. A., Smith, F. (1956) Colorimetric method for determination of sugars and related substrates. *Analytical Chemistry* **28**(3): 350–356.

Fang, H. H. P., Liu, H. (2002) Effect of pH on hydrogen production from glucose by a mixed culture. *Bioresource Technology* **82**(1): 87–93.

Hussy, I., Hawkes, F. R., Dinsdale, R., Hawkes, D. L. (2003) Continuous fermentative hydrogen production from a wheat starch co-product by mixed microflora. *Biotechnology and Bioengineering* **84**(6): 619–626.

Hwang, M. H., Jang, N. J., Hyun, S. H., Kim, I. S. (2004) Anaerobic bio-hydrogen production from ethanol fermentation: the role of pH. *Journal of Biotechnology* **111**(3): 297–309.

Kim, I. S., Hwang, M. H., Jang, N. J., Hyun, S. H., Lee, S. T. (2004) Effect of low pH on the activity of hydrogen utilizing methanogen in bio-hydrogen process. *International Journal of Hydrogen Energy* **29**(11): 1133–1140.

Lay, J.-J., Lee, Y.-J., Noike, T. (1999) Feasibility of biological hydrogen production from organic fraction of municipal solid waste. *Water Research* **33**(11): 2579–2586.

Lay, J. J., Li, Y. Y., Noike, T., Endo, J., Ishimoto, S. (1997) Analysis of environmental factors affecting methane production from high-solids organic waste. *Water Science and Technology* **36**(6–7): 493–500.

Levin, D. B., Pitt, L., Love, M. (2004) Biohydrogen production: prospects and limitations to practical application. *International Journal of Hydrogen Energy* **29**(2): 173–185.

Muller, V. (2003) Energy conservation in acetogenic bacteria. *Appl. Environ. Microbiol.* **69**(11): 6345–6353.

Oh, S. E., Van Ginkel, S., Logan, B. E. (2003) The relative effectiveness of pH control and heat treatment for enhancing biohydrogen gas production. *Environmental Science and Technology* **37**(22): 5186–5190.

Sparling, R., Risbey, D., Poggi-Varaldo, H. M. (1997) Hydrogen production from inhibited anaerobic composters. *International Journal of Hydrogen Energy* **22**(6): 563–566.

Sung, S., Raskin, L., Duangmanee, T., Padmasiri, S., Simmons, J. J. (2002). Hydrogen production by anaerobic communities exposed to repeated heat treatment. In CD-ROM Proceedings of Water Environment Federation 75th Annual Conference & Exposition, Chicago, IL.

Van Ginkel, S., Sung, S. W., Lay, J. J. (2001) Biohydrogen production as a function of pH and substrate concentration. *Environmental Science & Technology*. **35**(24): 4726–4730.

On-line evaluation of competition behaviour between PAOs and GAOs in EBPR SBRs

Y. W. Cui*, C. Y. Peng, S. Y. Wang*, Y. Z. Peng* and L. Ye***

*College of Energy and Environmental Engineering, Beijing University of Technology, 100 Pingleyuan, Chaoyang District, Beijing P.R. China 100022 (E-mail: *envirotto@yahoo.com.cn*)
**Public Utility Board, Environment Building, 40 Scotts Road, Singapore 228231
(E-mail: *prissycn@hotmail.com*)

Abstract To maintain perfect performance of enhanced biological phosphorus removal (EBPR), monitoring competition behaviour between phosphorus-accumulating organisms (PAOs) and glucose-accumulating organisms (GAOs) has been brought into sharp focus. In this paper, information provided by pH and ORP in the EBPR process has been deeply investigated to develop a method that could monitor real-time population size shift of PAOs and GAOs. Three EBPR SBR systems developed to different selective advantages to PAOs or to GAOs by means of feeding glucose or P/C to the EBPR system were applied to treat real sewage. The response of pH and ORP at the steady states was evaluated in the three A/O SBRs. The variations in pH are highly dependent on the competition between PAOs and GAOs in anaerobic phase. When providing GAOs with a selective advantage over PAOs, pH increased despite of phosphorus release, while pH decrease with phosphorus release when PAOs outcompete GAOs. Anaerobic phosphorus release could be rationed according to variations of ORP in aerobic phase. On-line monitoring pH and ORP provides a highly valuable avenue to real-time feedback of population structure of GAOs and PAOs.

Keywords On-line evaluation; PAOs; GAOs; EBPR

Introduction

Enhanced biological phosphorus removal (EBPR) is widely accepted as one of the most economical and sustainable processes for the removal of phosphate from wastewater. This process utilizes the capability of phosphorus-accumulating organisms (PAOs) and requires alternating anaerobic–aerobic conditions to enrich the organisms responsible for phosphorus removal (Barnard, 1975). PAOs take up volatile fatty acids (VFAs) anaerobically and convert them to poly-β-hydroxyalkanoate (PHA), with energy obtained from the release of intracellularly stored polyphosphate (polyP) and simultaneous conversion of glycogen to PHA. Aerobically, PAOs accumulate polyP in excess of the anaerobic P release to provide net P removal from wastewater. They are commonly described to have abilities to derive energy from hydrolysis of polyP to orthophosphate for the uptake and conversion of VFAs into either PHA (Comeau *et al.*, 1986; Satoh *et al.*, 1992) or glycogen (Nakamura *et al.*, 1991).

In recent years, poor performance or complete failure of EBPR processes have been reported. In addition to conventional process effect factors (Smolders *et al.*, 1994, 1995), glycogen-accumulating organisms (GAOs) as a potential competitor to PAOs in EBPR systems have attracted considerable interest (Fukase *et al.*, 1984; Randall *et al.*, 1994; Cech *et al.*, 1990, 1993; Mino *et al.*, 1994; Carucci *et al.*, 1994; Matsuo, 1994; Liu *et al.*, 1994, 1997; Jeon and Park, 2000). In many of these cases, significant populations of GAOs are often found. GAOs show similar traits to PAOs except they produce a different PHA and lack the EBPR ability or polyp accumulation by PAOs (Cech *et al.*, 1993; Mastsuo *et al.*, 1994; Liu *et al.*, 1994, 1996). GAOs can compete effectively with PAOs under anaerobic–aerobic conditions. GAOs, however, do not store polyP and use glycogen as their sole energy source for VFA uptake. GAOs are clearly detrimental to the phosphorus removal performance since they consume the limited VFAs available in the influent, but make no contribution to P removal, thereby possibly causing instability or even failure of EBPR. The stability of EBPR process is highly dependent on maintenance of growth of PAOs in the majority. Therefore, the fundamental understanding of microbial ecology structure is essential before the upset of performance in EBPR process.

To create conditions that are favourable for PAOs and disadvantageous for GAOs is a highly relevant research question at present. However, it was not clear what caused the large differences in the abundance of GAOs in EBPR systems. A long sludge residence time (Fukase *et al.*, 1984), a high ratio of anaerobic/aerobic hydraulic retention time (Matsuo *et al.*, 1994), pH (Filipe *et al.*, 2001), a limited P source or a low P/C ratio in the feed (Liu *et al.*, 1994, 1997), and substrate feed (Randall *et al.*, 1994; Cech *et al.*, 1990, 1993; Mino *et al.*, 1994; Carucci *et al.*, 1994; Jeon *et al.*, 2000) were suspected as the causative agents. Although there are conflicted conclusions about the carbon source effect on ecology structure of EBPR, more researches implied possible means of providing PAOs with a selective advantage over GAOs is through the variation of the carbon source fed to the EBPR system. Due to the distinctive difference between PAOs and GAOs in their ability to utilise different carbon source, influencing competition ability of PAOs and GAOs through selective carbon source fed to the EBPR system is a sound means. Acetate and glucose were often chosen as carbon source for enrichment of the PAO and GAO phenotype since PAOs can utilize acetate directly while not for glucose (Cech *et al.*, 1990, 1993; Mino *et al.*, 1994).

The success of activated sludge plants for EBPR mainly depends on the ecology community. Without understanding the microbial community structure and function in EBPR process, a reliable performance of EBPR may prove difficult. Therefore, how to monitor competition behaviour between PAOs and GAOs has been brought to a sharp focus in order to immediately take actions to maintain perfect performance of EBPR. On-line monitoring of pH and ORP has been proved to be a practical and useful technique for process control of nitrogen removal treatment systems (Charpentier *et al.*, 1998; Plisson-Saune *et al.*, 1996; Yu *et al.*, 1997; Andreottola *et al.*, 2001). Since process of phosphorus removal was related to the variations of VFAs and phosphate and the oxidation-reduction, pH and ORP was considered as a valuable tool to provide information about the size of PAOs and GAOs in systems. A correlation has been found between phosphate release and ORP (Shapiro *et al.*, 1965; Koch and Oldham, 1985). However, information provided by pH and ORP measurements in EBPR processes has not been deeply investigated, especially when PAOs or GAOs were dominating in the biological culture. In both cases, information provided by pH and ORP could be rated to size shifts of PAOs and GAOs.

This research aimed at investigating pH and OPR variation characteristics in EBPR SBR systems where PAOs or GAOs were in the majority in the biological culture. More emphasis was put on to develop a method that could realize on-line evaluation of competition behaviour between PAOs and GAOs.

Materials and methods

Reactor setup and operation

Three laboratory-scale sequencing batch reactors (SBRs) with working volume of 10 L were seeded with sludge from the lab-scale MUCT process which purified domestic wastewater with good performance (Mean phosphorus and nitrogen removal efficiency is 93% and 74%, respectively). The reactors were operated under A/O conditions for nutrient removal in parallel in a temperature-controlled room at 18–22°C. The reactors were equipped with ORP and pH electrodes. A data acquisition program was used to continuously save and graph on a personal computer the measured data. Supplementary external carbon source, menthol, was added when anoxic phase began to obtain complete nitrogen removal. Initially, three reactors are operated in parallel with the same mixed liquor suspended solids (MLSS) at level of 2500–3000mg/L. The effective SRT (sludge age) is about 10 days.

Wastewater feed

The feed wastewater was collected from the sewer of Beijing University of Technology, Beijing. The influent characteristics are listed in Table1. To avoid the effect of influent fluctuations, the wastewater was collected at the same time every day. According to experimental design, glucose and sodium acetate was added into raw sewage to simulate three selective cases, i.e. domestic-fed SBR, glucose-mainly-fed SBR and acetate-mainly-fed SBR.

Table 1 Influent characterization

Parameters	pH	COD (mg/L)	BOD (mg/L)	TN (mg/L)	NH_4^+-N (mg/L)	NO_2^--N (mg /L)	NO_3^--N (mg/L)
Range	7.2–7.7	190–325	102–180	51–95	32–85	0.05–0.30	0.05–0.84

Experiment design

Seeded with the same biomass, three SBRs received different wastewater feed, which is domestic waste (SBR1), domestic waste and glucose (SBR2), and domestic waste and acetate (SBR3), respectively. Three SBRs were operated under constant conditions during about one month to achieve a steady state. When the steady state was achieved the SBRs were extensively sampled to characterize the phosphorus metabolism and dynamic variation of pH and ORP. After that, three SBRs were fed the same domestic waste stopping any glucose and acetate.

Table 2 Operational conditions tested

	Stage A		Stage B	
	Feed	P/C	Feed	P/C
SBR1	domestic	0.08~0.15	domestic	0.08~0.18
SBR2	domestic and glucose	0.027~0.062	domestic	0.08~0.15
SBR3	domestic and acetate	0.027~0.058	domestic	0.1~0.18

Analytical methods

Samples withdrawn from the liquid medium during processing were centrifuged at 3000rpm for 5 minutes to remove microorganism from liquid medium. Most routine chemical analyses (i.e. PO_4^{2-}-P, TN, NH_4^+-N, NO_3^--N, NO_2^--N, COD) were conducted according to the standard methods. ORP, DO and pH was measured by WTW inoLab Oxi level2 hand-held instruments.

Results and Discussions

Effect of wastewater composition on EBPR

Table 3 summarised the mean performance of three SBR at stage A. During the course of operation, effective COD reduction and phosphate release at the end of anaerobic stage and phosphate uptake at the end of aerobic stage were obtained. The same mixed population show the different EBPR potential when fed with different substrate. In SBR1, fed by domestic wastewater, over 90% phosphorus removal efficiency was found. Due to adding glucose and acetate into the influent, more amounts of anaerobic phosphate release and aerobic phosphate uptake were found in SBR2 and SBR3. However, only 85% phosphate removal was achieved in SBR2 with glucose as main carbon, while over 90% phosphate removal in SBR3 with acetate as main carbon. In comparison with SBR1, glucose addition is clearly detrimental to the phosphorus removal performance even lead to lower EBPR performance while acetate addition enhanced EBPR ability.

Another outstanding observation is the different values of specific anaerobic COD consumed related to P release in three systems. Elevated anaerobic organic consumption and the carbon requirements for phosphorus release were found in SBR2 fed by glucose-containing domestic. The occurrence may be explained by the postulation that increases of glucose in influent possibly stimulated the growth of GAOs, which is proved by Cech and Hartman (1993). Therefore, the different phosphorus removal capability and metabolic behaviour was owed to the dynamic shift of EBPR population structure. Liu et al. (1997) has proved that the P/C feeding ratio strongly affected the dominance of PAOs and GAOs. The high ratio (20/100) seemed to favour the growth of PAOs while low ratio (2/100) resulted in a shift of the dominant population to GAOs. Even so, PAOs and GAOs remained as two major populations in reactors. In SBR2, a low P/C ratio of 0.027–0.062 was applied couple with glucose as main carbon. Although with low P/C ratio of 0.027–0.058 as SBR2, SBR3 give good EBPR performance and activity during feeding acetate as main carbon, which need more research to explore whether low P/C ratio stimulates the growth of GAOs. Therefore, three systems with PAOs as dominant species and GAOs as dominant species were established by feeding different carbon and controlling P/C ratio in influent. Evidenced by experimental results, PAOs was dominating in SBR1 and GAOs seemed to be developed in SBR2. This conclusion has been further proved by the experimental results at stage B, where all SBRs were fed by the same domestic wastewater. After about 20 days when selective conditions were applied to achieve the selection of microbial structure, three SBRs showed the different phosphorus removal performance. Performance in SBR2 was kept deteriorating to about 50%. Over 90% phosphorus removal was obtained in SBR1 and 85% for SBR3. These results implied that despite of feeding by acetate as main carbon low P/C still results in the growth of GAOs in SBR3.

Table 3 Dynamics of phosphorus removal in three SBRs at stage A

	SBR1	SBR2	SBR3
Anaerobic P release(mg/L)	7.00~15.00	12.00~22.00	15.00~24.00
Aerobic P uptake(mg/L)	11.00~25.00	22.00~30.00	25.00~35.00
Mean P removal efficiency (%)	90	85	95
Specific anaerobic COD consumed relate P release (mg/L COD/(mg/L P •g/L MLSS))	4.92	6.89	3.57

On-line monitor at stage A

During the steady states, SBR2 and SBR3 were monitored based on tracing variations of pH and ORP in this process, showed in Fig.1 and Fig.2. Fig.1a and Fig.2a showed the typical dynamics of pollutants in a cycle containing two consecutive stages: anaerobic (A) and aerobic (O) stage. The corresponding pH and DO variation was depicted in Fig.1b. With glucose or acetate as main carbon, in anaerobic stages, phosphate release is companied by decrease of COD. In the aerobic period, ammonia was oxidised to nitrite and nitrate. In anaerobic stage, for SBR2 system, pH profile displays a down-up mode, which contributed to an obvious breakpoint (point a in Figure 1b). A speedy decrease was detected in pH curve during the first hour or so. Then, a subsequent increase occurred in the residual time during anaerobic stage. Glucose decomposition is possiblye responsible for the fast decrease of pH. Organic acids (such as formate, acetate, and lacate) were produced from the degradation of glucose, which results in a prompt decrease in pH curve. Meanwhile, organic acids generated by glucose decomposition were utilized by PAOs, which is one possible reason that good phosphorus removal performance can be maintained in SBR2. Due to no or a little fatty acid formation, there a brief decrease in pH profile in SBR3 fed by acetate as main carbon (Figure 2b). In anaerobic stage, fatty acid was uptaken companied by phosphate release, which cause the variation of pH. Therefore, an increase of pH will be found when the amount of acid uptake is below the amount of phosphate release (Figure 1b and 2b). Since PAOs and non-PAOs all can utilise fatty acid in anaerobic stage, the amount of acid, size of PAOs, and size of non-PAOs determined pH variation. Although in this case, both addition of glucose and acetate caused the increase of pH, relative size of PAOs and GAOs in EBPR ecology could not estimated due to the lack of ration to the exact amount of acid in influent.

Another breakpoint also appeared on the curve when concentration of ammonium was close to zero in aerobic period. Increases of pH during aerobic phase are due to phosphorus uptake. However, due to accumulation of CO_2, pH decreases, which cause that maximum pH value is not obtained when phosphate is exhausted.

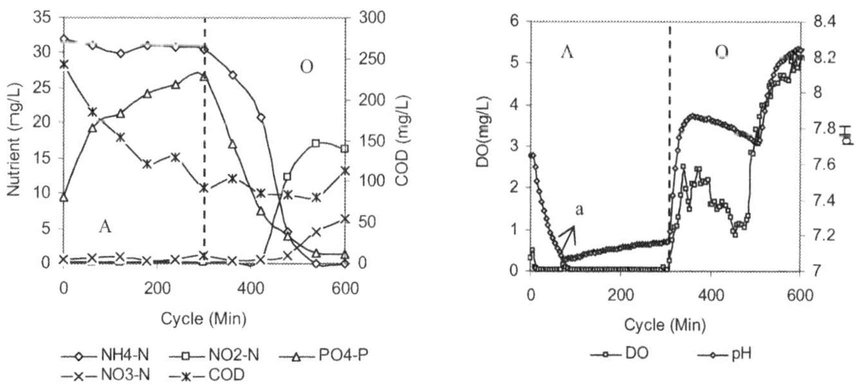

Figure1 (a) Cycle dynamics in SBR2; (b) Variation of pH and DO in SBR2.

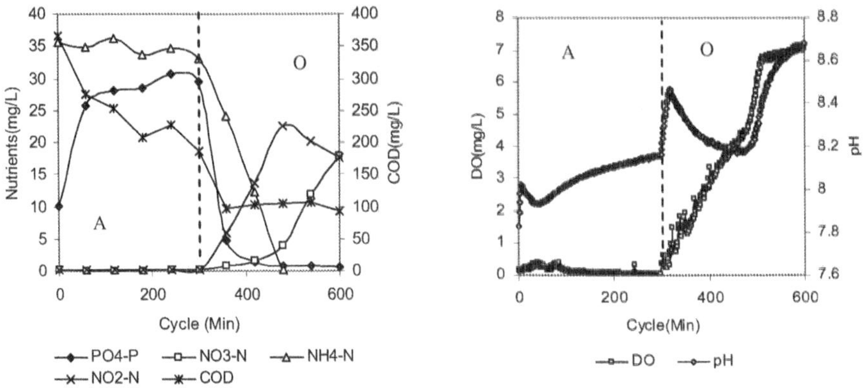

Figure 2 (a) Cycle dynamics in SBR2; (b) Variation of pH and DO in SBR2.

In this case, with feeding by different fatty acid concentration or composition in influent, variation of pH could not provide valuable information to estate the relative size of PAOs and GAOs in two EBPR systems.

On-line monitor at stage B

The same domestic wastewater was applied into the three SBRs to investigate variation of pH and ORP. Due to the same feed, amount of fatty acid was thought to be constant. So, variation of pH is directly related to the size of PAOs and non-PAOs.

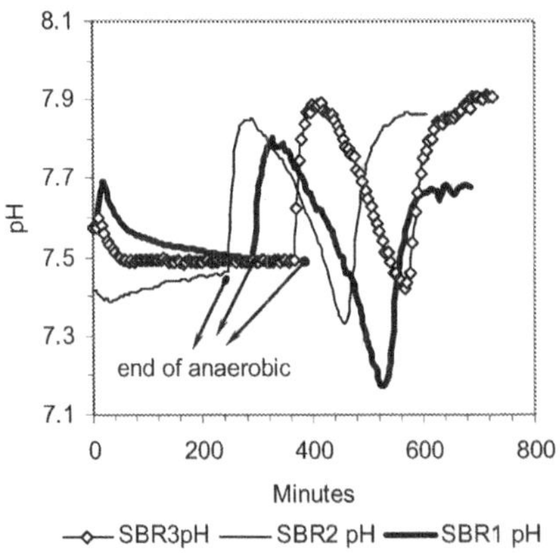

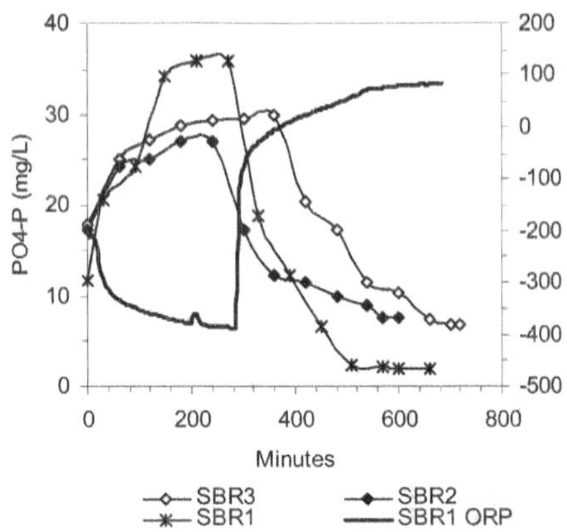

Figure 3 pH profile in one cycle of three SBRs Fig.4 P dynamics of three SBRs and OPR fed by same wastewater profile of SBR1.

Through controlling P/C in influent and carbon addition, three different ecology structures were established. When feeding with the same wastewater, EBPR capability was distinguished. According to the EBPR performance, PAOs is clearly dominant in SBR1, GAOs is dominant species in SBR2, and many PAOs and GAOs coexisted in SBR3. Due to the different microbial community structure, pH profile in anaerobic stage distinguished from each other. A sharp decrease of pH was found in SBR1 while SBR2 showed an increased trend. A two-stage curve of pH, a steeped decrease due to phosphate release and a subsequent even stage due to balance of acetate uptake and phosphate release is obvious in

SBR3. Although acetate is taken up by PAOs with the correspondent proton, pH decreases due to phosphorus release. On the same way, GAOs also take up acetate with the correspondent proton. However, as polyphosphate is not involved in GAOs metabolism, phosphorus is not released and pH increases drastically. Hence, pH variation in anaerobic stage is related to the relative ratio of PAOs and GAOs population.

In three SBRs, the correlation of OPR variation with phosphate dynamics is extremely similar. Fig.4 showed the variation of ORP with phosphate concentration. The rate of ORP decrease displayed a perfect relativity with the phosphorus release rate. Almost linear relations can be obtained between OPR and concentration of phosphate in anaerobic. Thus, anaerobic phosphate release can be forecasted by evaluating the variation of ORP.

Conclusions

Three systems with PAOs as dominant species or GAOs as dominant species were successfully established by feeding glucose as main carbon and controlling P/C ratio in influent, which providing with a selective advantage over PAOs or GAOs. In anaerobic stage, fatty acid was uptaken companied by phosphate release, which cause the variation of pH. A two-stage curve of pH, a steeped decrease due to glucose decomposition and a subsequent increase due to more amount of acetate uptake than phosphate release is obvious in SBR feeding glucose as main carbon. In this study, both addition of glucose and acetate caused the increase of pH in anaerobic stage, which means variation of pH in anaerobic stage could not provide valuable information to estate the relative size of PAOs and GAOs in two EBPR systems with different wastewater. Although acetate is taken up by PAOs with the correspondent proton, pH decreases due to phosphorus release. On the same way, GAOs also take up acetate with the correspondent proton. However, as polyphosphate is not involved in GAOs metabolism, phosphorus is not released and pH increases drastically. When same wastewater was applied to different system, an increase trend of pH profile in anaerobic stage was found in a GAOs-dominant ecology while a decrease of pH in a PAOs-dominant ecology. The rate of ORP decrease displayed a perfect relativity with the phosphorus release rate. Almost linear relations can be obtained between OPR and concentration of phosphate in anaerobic. Anaerobic phosphate release can be forecasted by evaluating the variation of ORP. Thus, anaerobic pH profile jointly with ORP could be used to detect relative size of PAOs and GAOs the SBR in case the same wastewater was applied.

The experiments established a method of on-line evaluation of competition behaviour between PAOs and GAOs in EBPR SBRs, which allows taking actions to control bacterial groups and treatment performance. The presented method is valuable to offer a solution to realize real-time monitor of bacterial population of PAOs and GAOs. It is simple, inexpensive and applicable.

Acknowledgements

This work was financially supported by the project under key international cooperative programs of NSFC (NO.50521140075) and the program of NSFC (NO.20377003)

References

Andreottola, G., Foladori, P. and Ragazzi, M. (2001) On-line control of a SBR system for nitrogen removal from industrial wastewater. *Water Science and Technology*, **43** (3), 93–100.

Barnard J. L. (1975) Biological nutrient removal without the addition of chemicals. *Water Res.* **9**, 485–490.

Carucci, A., Lindrea, K., Majone, M. and Ramadori, R. (1994) Dynamics of the anaerobic utilization of organic substrates in an anaerobic /aerobic sequencing batch reactor. *Wat. Sci. Technol.* **31**(20), 35–43.

Cech, J.S. and Hartman, P.(1990) Glucose induced breakdown of enhanced biological phosphate removal. *Environ. Technol.* **11**, 651–656.

Cech, J.S. and Hartman, P. (1993) Competition between polyphosphate and polysaccharide accumulating bacterial in enhanced biological phosphate remobal system. *Wat. Res.* **27**(7), 1219–1225.

Charpentier, J., Martin, G., Wacheux,H and Gilles, P. (1998).ORP regulation and Activated sludge: 15Years of Experience. *Wat. Sci. Tech.*, **38**(3), 197–208.

Comeau, Y., Hall, K. J., Hancock, R. E. W. and Oldham W. K. (1986) Biochemical model for enhanced biologi-cal phosphorus removal. *Water Res.* **20**, 1511–1521.

Filipe, C.D.M., Daigger, G.T. and Grady, C.P.L. Jr (2001) pH as a key factor in the competition between glycogen-accumulating organisms and phosphorus accumulating organism. *Water Environment Research* **73**(2), 223–1232.

Fukase T., Shibata M. and Miyaji. Y. (1984) The role of an anaerobic stage on biological phosphorus removal. *Wat. Sci. Technol.* **17**, 69–80.

Jeon C.O. and Park J.M. (2000) Enhanced biological phosphorus removal in a sequencing batch reactor supplied with glucose as a sole carbon source. *Wat. Res.* **34**(7), 2160–2170.

Koch, F.A. and Oldham, W.K. (1985).Oxidiation-reduction potential-a tool for monitoring, control and optimization for biological nutrient removal systems. *Wat. Sci. Technol.* **17**, 259–281.

Liu W.T., Mino T., Nakamura K. and Matsuo T. (1994) Role of glycogen in acetate uptake and polyhydroxyalkanoate synthesis in anaerobic-aerobic activated sludge with a minimal polyphosphate content. *J. Ferment. Biotechnol.* **77**, 535–540.

Liu W.T., Mino T., Nakamura K. and Matsuo T. (1996) Glycogen accumulating population and its ananerobic substrate uptake abilities in anaerobic-aerobic activated sludge without biological phosphorus removal. *Wat. Res.* **30**, 75–87.

Liu W.T., Mino T., Nakamura K., Matsuo T. and Mino T. (1997) Internal energy-based competition between polyphosphate- and glycogen-accumulating bacteria in biological phosphorus removal reactors—effect of P/C feeding ratio. *Wat. Res.* **31**(6), 1430–1438.

Mastsuo, Y. (1994) Effect of the anaerobic solids retention time on enhanced biological phosphorus removal. *Wat. Sci. Technol.* **30**(6), 193–202.

Mino, T., Satoh, H. and Matsuo, T. (1994) Metabolisms of different bacterial populations in enhanced biological phosphate removal process. *Wat. Sci. Technol.* **29**(7), 67–70.

Nakamura, K., Masuda, K. and Mikami, E. (1991) Isolation of new type of polyphosphate accumulation bacterium and its phosphate removal characteristics. *J. Ferment. Bioeng.* **71**, 258–263.

Plisson-Saune, S. *et al.* (1996) Real-time control of nitrogen removal using three ORP bending points: signification, control strategy and results. *Wat. Sci. Technol.* **33**(1), 275–280.

Randall, A.A., Benefield, L.D., and Hill, W.E. (1994) The effect of fermentation products on enhanced biologicial phosphorus removal, polyphosphate storage, and microbial population dynamics. *Wat. Sci. Technol.* **30**(6), 213–219.

Satoh, H., Mino, T. and Matsuo, T. (1992) Uptake of organic substrate and accumulation of polyhydroxyalkanoates linked with glycolysis of intracellular carbohydrates under anaerobic conditions in the biological excess phosphate removal process. *Wat. Sci. Technol.* **26**, 933–942.

Shapiro, J., Levin, G.V. and Zea, G. (1965). Anoxically induced release of phosphate in wastewatertreatment. *J. Wat. Pollut. Control Fed.*, **39**, 1810–1818.

Smolders, G.J.F., van Loosdrecht, M.C.M. and Heijnen, J.J. (1994) Model of the anaerobic metabolism of the biological phosphorus removal process; stoichiometry and pH influence. *Biotech. Bioeng.* 43, 461–470.

Smolders, G.J.F., Klop, J.M., van Loosdrecht, M.C.M. and Heijnen, J.J. (1995) A metabolic model of the biological phosphorus removal process; effect of the sludge retention time. *Biotech. Bioeng.* **48**, 222–233.

Yu, R.F. *et al.* (1997) Monitoring and control using on-line ORP on the continuous-flow activated sludge batch reactor system. *Wat. Sci. Technol.* **35**(1), 57–66.

Comparison of free ammonia and free nitrous acid inhibition on the anabolic and catabolic activities of enriched *Nitrosomonas* and *Nitrobacter* cultures

Vel Murugan Vadivelu, Jürg Keller and Zhiguo Yuan*

The Advanced Wastewater Management Centre, The University of Queensland, St Lucia, Queensland 4072, Australia (E-mail: *zhiguo@awmc.uq.edu.au*)

Abstract The inhibitory effect of free ammonia (FA) and free nitrous acid (FNA) on the metabolisms of *Nitrosomonas* and *Nitrobacter* was investigated using a method that allows the decoupling of the growth and energy generation processes. Lab-scale sequencing batch reactors (SBRs) were operated for the enrichment of *Nitrosomonas* and *Nitrobacter*. Fluorescent In-Situ Hybridization (FISH) analysis showed that the reactors were 82% and 73% enriched with *Nitrosomonas* and *Nitrobacter*, respectively. Batch tests were performed to measure the oxygen consumption rate by the enriched cultures at various FA and FNA levels, in the presence or absence of inorganic carbon (CO_2, HCO_3^- and CO_3^{2-}). FA up to 16.0 mgNH$_3$-N L^{-1} was not found to have any inhibitory effect on either the catabolic or anabolic processes of the *Nitrosomonas* culture, but both these processes were inhibited by FNA. While an FNA level of 0.40–0.63 mgHNO$_2$-N L^{-1} inhibited the energy production capability of *Nitrosomonas* by 50%, the growth process of the culture was completely inhibited by an FNA concentration of 0.40 mgHNO$_2$-N L^{-1}. Both FA and FNA were found to have strong inhibition on the anabolic processes of *Nitrobacter*, but with limited inhibitory effect on the catabolism of this culture. The biosynthesis of *Nitrobacter* was totally inhibited at an FA level of 6.0 mgNH$_3$-N L^{-1} (or above) or an FNA level of 0.02 mgHNO$_2$-N L^{-1} (or above). At the same level of FA, the energy production capability of *Nitrobacter* was only inhibited by 12%, whereas an FNA level of up to 0.024 mgHNO$_2$-N L^{-1} did not show any inhibition on the energy production of *Nitrobacter*.

Keywords Anabolism; catabolism; free ammonia inhibition; *Nitrosomonas*; *Nitrobacter*

Introduction

Nitrification is regarded as a two-step process where the reduced compound of nitrogen, ammonia is oxidized to nitrite (by ammonia oxidizer) and followed by the nitrite oxidation to nitrate (by nitrite oxidizer). Generally, nitrate is the main product of nitrification, and as such ammonia oxidation is typically considered the rate limiting step (Gil and Choi 2001). However, incomplete nitrification with various levels of nitrite accumulation has also been widely reported in literature (Hanaki *et al.* 1990, Yang and Alleman 1992). Traditionally, nitrite build up is considered undesirable in biological wastewater treatment (BWT) systems. In recent years, however, several processes have been developed which removes nitrogen via nitritation (ammonia to nitrite) and denitritation (nitrite to di-nitrogen gas). These included the well-known SHARON (Single reactor High Activity Ammonia Removal Over Nitrite) (Hellinga *et al.* 1999) and also a novel sequencing batch reactor (SBR) system reported in Lai *et al.* (2004). In comparison to full nitrification and denitrification, removing nitrogen via this so-called nitrite pathway reduces the oxygen and carbon demand by 25% and 40%, respectively (Fux *et al.* 2003). All these processes rely on the elimination or inhibition of nitrite oxidizing bacteria so that nitrite accumulates as the end product of nitrification.

Over the past few decades considerable research has been conducted for the identification of inhibiting factors involved in the inhibition of NOB. Factors such as pH, temperature, and the concentrations of dissolved oxygen, CO_2 and heavy metal were all found to influence on nitrite build-up. However, the major causes are believed to be the different inhibitory effects of free ammonia (FA) and free nitrous acid (FNA) (Painter 1970, Anthonisen *et al.* 1976, Mauret *et al.* 1996). However, limited information has been reported in literature on the mechanism involved in the FA and FNA inhibition. This study aims to provide further experimental information on the inhibitory effects of FA and FNA on the activities of ammonia oxidizing bacteria (AOB) and nitrite oxidizing bacteria (NOB). Different from most previous studies, the inhibitory effects of FA and FNA on the energy generation and growth processes are assessed separately with the aim to gain further insight into the inhibition mechanisms, and on this basis to provide support to the design and operation of nitrogen removal via nitrite.

Materials and Methods

Enrichment of NOB

An SBR was operated to selectively grow an enriched culture of NOB. The reactor had a working volume of 8 L and was fed with nitrite (synthetic wastewater with 1000 mg NO_2^--N L^{-1}) as the sole energy source and bicarbonate as the sole carbon source. The SBR was operated with a cycle time of 6 hours and a hydraulic retention time (HRT) of 1 day. The sludge age was 13 days. Each cycle consisted of a 270 minutes aerobic feeding, and a 20 minutes aerobic reaction period, followed by a 60 minutes settling and a 10 minutes decanting period. The reactor was operated in a temperature controlled room with temperature between 22 ± 1 °C. The dissolved oxygen (DO) concentration of the culture was maintained within the range of 2.75–3.25 mg L^{-1} using an ON/OFF controller. While not controlled, the pH in the reactor varied in a small range between 7.2–7.4, due to the strong buffer in the wastewater. Long-term monitoring of the reactor performance revealed that complete nitrite oxidation was achieved, with the nitrite level in effluent almost undetectable. Cycle studies showed that there was only a low level (2 mgN L^{-1}) accumulation of nitrite during the course of nitrite feeding, and the accumulated nitrite was oxidized almost immediately after feeding stopped (data not shown). Fluorescence *in situ* hybridization (FISH) analysis showed that 73% of the microbial population of the culture was *Nitrobacter*, while *Nitrospira* was undetected. The remaining 27% of bacterial community is believed to be heterotrophs surviving by using the lysate of *Nitrobacter*. Details of the operation and performance of the reactor can be found in Vadivelu *et al.* (submitted–c).

Enrichment of AOB

The SBR had a working volume of 11 L and was fed with synthetic wastewater containing 1000 mg NH_4^+-N L^{-1} as the sole energy source and bicarbonate as the sole carbon source. The cycle time of the reactor varied between 8–12 hours determined dynamically by an on-line control system. Each cycle consisted of five distinct aerobic periods with automatically controlled lengths followed by a 30 minutes settling and a 10 minutes decanting period. The synthetic feed of 1L was fed in 10 minutes at the beginning of each aerobic period, which initiated the oxidation of ammonia and caused pH to drop. The ammonium to bicarbonate ratio (molar) in the feed was 1:1, which is typical in sludge digestion liquor. As such, pH dropped sharply as soon as bicarbonate was completely consumed (corresponding to the oxidation of 50% of the ammonium added). Once the pH of the SBR decreased to a threshold value of 6.4, the next feeding period was started. At the end of the fifth aerobic period, 2 L of biomass was removed, giving rise to a sludge age of 1.7 days by also considering the impact of solids in effluent. The average hydraulic retention time was 1 day. The temperature in the reactor was controlled at 30 ± 1 °C using an aquarium heating apparatus, to mimic the temperature in a typical nitritation reactor treating sludge digestion liquor. The dissolved oxygen (DO) concentration was maintained within the range of 2.75–3.25 mg L^{-1} using an ON/OFF controller. Long-term performance of the SBR showed that ammonium oxidation completely stopped at nitrite with negligible nitrate production. The average ammonium and nitrite concentrations in effluent were around 500 mgN L^{-1}. FISH analysis revealed that 82% of the bacterial population bound to NEU probe specifically designed for *Nitrosomonas europaea*, while no other bacteria bound to other AOB and NOB probes applied. The remaining fraction of the bacterial community is believed to be heterotrophs, which likely grew utilizing the cell lysate of the *Nitrosomonas* culture as the energy and carbon sources. Detailed explanation of the reactor operation and performance data given in Vadivelu *et al.* (submitted–a).

Experiments to determine the inhibitory effects of free ammonia and free nitrous acid

Batch tests were carried out on the Titration and Off-Gas Analysis (TOGA) sensor (Pratt *et al.* 2003) to determine the inhibitory effects of FA and FNA on the metabolisms of the enriched AOB and NOB cultures.

The TOGA sensor is a reactor-based instrument designed for the study of biological processes. It contains a quadrupole mass spectrometer to measure the concentrations of various gaseous components in the off-gas from the sealed reactor (O_2 and CO_2 in this case). The measured off-gas composition along with the known flow rate and composition of the feeding gas (specific gas mixtures with known compositions are used in experiments) allows the calculation of the transfer rates of gaseous components of interest (the O_2 transfer rate–OTR and the CO_2 transfer rate–CTR in this study). The TOGA sensor also consists of control systems to control the dissolved oxygen (DO) and pH in the reactor at pre-selected set-points through manipulating the flow rate of the gas stream containing oxygen and an acid/base titration system, respectively. More details of the TOGA sensor principle and operation can be found in Pratt *et al.* (2003).

Each batch test with the TOGA sensor lasted a few hours. The reactor was filled with 1.3 L of mixed liquor taken from the parent SBR. Various amounts of ammonium and nitrite (see Results and discussion) were then added into the reactor to initiate the biomass activities and/or inhibitory effects. DO was controlled at around 3 mg L^{-1}, the levels similar to those in the parent SBR. The OTR and CTR were recorded during the entire experiment. Ammonium, nitrite and nitrate were measured offline.

Each batch test consisted of two phases, with and without the presence of inorganic carbon in the liquid phase. Two different gas mixtures were used in the two phases. In the first (growth) phase, the reactor was supplied with a total gas flow of 580 ml/min, which comprised of 31% (all gas concentrations are vol%) reaction gases and 69% carrier gas (helium). The reaction gas consisted of 50% oxygen, 0.5% carbon dioxide, 3% nitrogen and 46.5% argon. At the end of the growth phase, the reaction gas stream was stopped and the carrier gas alone was passed through the TOGA reactor to strip dissolved inorganic carbon (CO_2, bicarbonate and carbonate) from the reactor. The second phase in each experiment was started after the completion of CO_2 stripping, which was confirmed by a very low intensity off-gas CO_2 signal (below the detection limit, 0.8 ppm). The gas mixture used in this phase is similar to that used in the growth phase but without CO_2 therefore growth of AOB and NOB is not expected to occur.

Four sets of experiments were carried out to investigate the inhibitory effect of FA and FNA on the metabolism of *Nitrosomonas* and *Nitrobacter* cultures, respectively. Each set consisted of four to seven repeated tests.

Model used for data interpretation

The following model (Pirt 1982) was used to assist the evaluation of substrate utilization rate due to the growth and maintenance processes of the AOB and NOB cultures:

$$q = \frac{\mu}{Y_G} + m^0 + m'(1 - k\mu) \tag{1}$$

where q is the substrate (e.g. ammonia/nitrite and oxygen) utilization rate, Y_G is the true bacterial growth yield on the substrate concerned. The term $m^0 + m'(1 - k\mu)$ models the substrate utilization rate due to maintenance processes, where m^0 is a constant maintenance term (independent of μ), and $m'(1 - k\mu)$ characterizes the maintenance energy consumption rate dependent on the growth rate. Parameter k has been suggested to be $1/\mu_{max}$ (the inverse of the maximum specific growth rate) (Pirt 1982).

According to Equation 1, the difference in the substrate utilization rate (Δq) when a bacterium grows at a specific growth rate of μ and when there is no growth is:

$$\Delta q = \left(\frac{1}{Y_G} - \frac{m'}{\mu_{max}} \right) \mu \tag{2}$$

Results and Discussion

Oxygen uptake rate by *Nitrosomonas* and *Nitrobacter* in the presence and absence of CO_2

Bacteria spend energy on both maintenance and growth processes. In the presence of both the energy (ammonia/nitrite) and carbon (bicarbonate) sources (in the growth phase of the experiments) it was expected that the nitrifiers used the consumed O_2 ($OUR_{withCO2}$) to generate energy for both maintenance and growth processes. In the absence of externally supplied CO_2, the cultures were unlikely to grow due to the lack of carbon source. Hence, $\mu_{withoutCO2} = 0$. Thus, in the absence of externally supplied CO_2, the oxygen consumed ($OUR_{withoutCO2}$) and the energy generated from ammonia/nitrite oxidation were likely used to fulfill only the non-growth related processes of the cultures. With the assumption that $\mu_{withoutCO2} = 0$, the following equation holds true (referring to Equation 2):

$$\Delta OUR = OUR_{withCO2} - OUR_{withoutCO2} = \left(\frac{1}{Y_O} - \frac{m'}{\mu_{max}} \right) \mu \tag{3}$$

where Y_O is the true biomass growth yield with respect to oxygen, μ is specific growth rate of *Nitrosomonas/Nitrobacter* cultures when CO_2 is in excess, which is dependent on the ammonium/ammonia concentration. Other variables are as explained in the previous section.

In the above analysis, the OUR caused by heterotrophic bacteria present in the sludge oxidizing cell lysate was not considered. Also neglected was the growth of nitrifiers in the absence of externally supplied CO_2, utilizing the CO_2 produced by heterotrophs through oxidizing cell lysate. Justification of these assumptions can be found in Vadivelu *et al.* (submitted-a) and Vadivelu *et al.* (submitted–c).

The specific OUR of the cultures were calculated by dividing the measured OUR by the biomass concentration present in the reactor during each test. The specific OUR profiles determined are plotted as function of FA and FNA concentrations in Figure 1. In the figure, each specific OUR is expressed as a fraction of the maximum specific OUR measured in the presence of CO_2 and absence of inhibition. $OUR_{withCO2}$ and $OUR_{withoutCO2}$ indicate the oxygen consumption of the nitrifiers in the presence and absence of CO_2 respectively. ΔOUR (which is proportional to μ) is calculated from the difference between $OUR_{withCO2}$ and $OUR_{withoutCO2}$, and presented in the same figure.

Inhibitory effects of FNA on the anabolic and catabolic processes of *Nitrosomonas* and *Nitrobacter*

The $OUR_{withCO2}$, $OUR_{withoutCO2}$ and ΔOUR shown in Figure 1(c) clearly indicate that both the energy generation and growth processes of the *Nitrosomonas* culture were inhibited by FNA. However the degree of inhibition on these two processes is different. While the FNA inhibition on both processes initiated at similar levels (0.08–0.10 $mgHNO_2\text{-}N\ L^{-1}$), the biosynthesis process seemed to have ceased completely at an FNA concentration of 0.4 $mgHNO_2\text{-}N\ L^{-1}$, at which the energy generation process was only inhibited by 50%. The study therefore shows that FNA has a more severe inhibitory effect on the growth capability than on the energy generation capacity of the *Nitrosomonas* culture. The inhibition of FNA on AOB activity has been observed widely in literature. For example, Fux *et al.* (2003) in their study on the inhibitory effect of FNA on an AOB culture grown under conditions similar to those used in this study, observed that the ammonia oxidation rate decreased by 20–25% when the FNA concentration was increased to 0.17 $mgHNO_2\text{-}N\ L^{-1}$ from below 0.06 $mgHNO_2\text{-}N\ L^{-1}$, which agrees well with the 20% drop observed in this study for the same FNA concentration in the presence of CO_2. A further interesting observation on Figure 1c is that the increase of FNA from 0.40 to 0.63 $mgHNO_2\text{-}N\ L^{-1}$ did not cause further decrease in OUR. A similar observation was made in van Hulle *et al.* (2004), where the activities (also measured as OUR in the presence of CO_2) of AOB in their SHARON reactor treating sludge digestion liquor were inhibited by 60% when the FNA concentration was increased to above 0.7 $mgHNO_2\text{-}N\ L^{-1}$. However, in all these studies the inhibitory effects of FNA on the anabolic and catabolic processes were not assessed separately. As revealed in this study, the FNA inhibition on the biosynthesis processes of *Nitrosomonas* is much stronger than that on the energy production. These inhibitory effects could be underestimated by simply measuring $OUR_{withCO2}$, as done in most previous studies.

The dependency of $OUR_{withCO2}$ and $OUR_{withoutCO2}$ of *Nitrobacter* on the FNA concentration is depicted in Figure 1(d). Interestingly the data showed that $OUR_{withoutCO2}$ is independent of the FNA concentration up to 0.024 $mgHNO_2\text{-}N\ L^{-1}$, although $OUR_{withCO2}$ reached its minimum (85% of the maximum value) at around 0.02 $mgHNO_2\text{-}N\ L^{-1}$. This suggests that the energy generation processes of *Nitrobacter* are not inhibited by FNA up to 0.024 $mgHNO_2\text{-}N\ L^{-1}$. On the other hand, ΔOUR (which is proportional to specific growth rate) decreased with the increase in FNA concentration. It appears that the growth process started being inhibited when HNO_2 exceeded 0.7 μM (0.01 $mgN\ L^{-1}$), and ceased to occur when the HNO_2 level reached 1.4 μM (0.02 $mgN\ L^{-1}$). In an additional experiment reported in Vadivelu *et al.* (submitted-b), it was showed that the energy production capability of *Nitrobacter* was not inhibited at an FNA concentration of 0.050 $mgHNO_2\text{-}N\ L^{-1}$. This inhibition threshold is considerably lower than those reported in literature. For example, Prakasam and Loehr (1972) reported an HNO_2 inhibition threshold value of 0.07 mgN L^{-1}, and Anthonisen *et al.* (1976) concluded that nitrite oxidizing microorganisms would be inhibited at an HNO_2 concentration of 0.22 mgN L^{-1}. It needs to be recognized, however that those inhibitory studies did not distinguish between growth and maintenance processes. It is likely that previous studies assessing the inhibitory effects of nitrite/FNA on *Nitrobacter* by simply comparing the overall substrate utilization rates have seriously under-estimated the inhibitory effects.

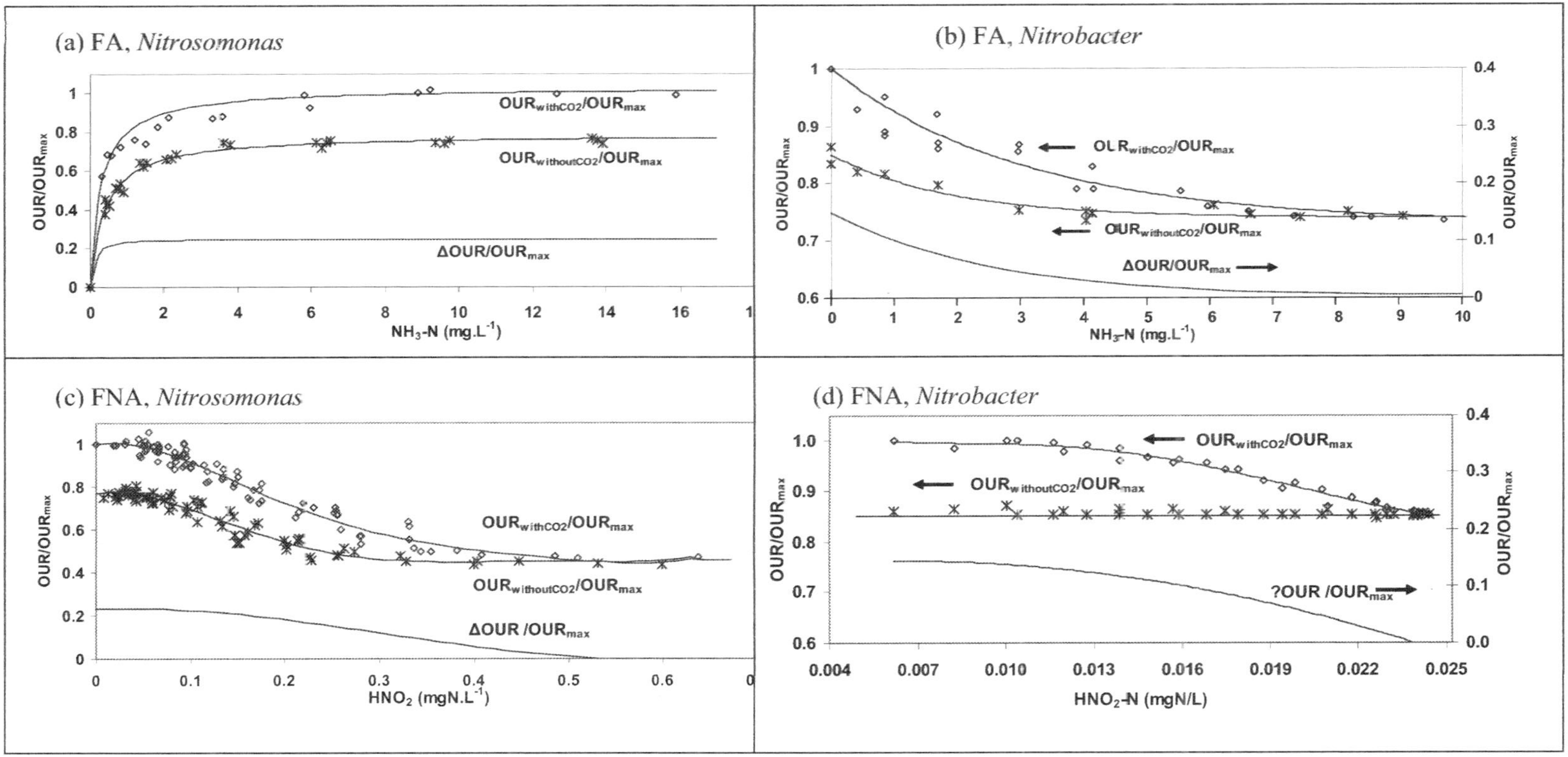

Figure 1 $OUR_{withCO2}/OUR_{max}$ and $OUR_{withoutCO2}/OUR_{max}$ as a function of the FA and FNA concentrations. $\Delta OUR/OUR_{max}$, calculated as the difference between $OUR_{withCO2}/OUR_{max}$ and $OUR_{withoutCO2}/OUR_{max}$, is proportional to the specific growth rate (μ) of the enriched *Nitrosomonas* and *Nitrobacter* cultures. (a) and (b) illustrate the FA inhibition and (c) and (d) demonstrate the FNA toxicity on *Nitrosomonas* and *Nitrobacter* cultures, respectively.

Inhibitory effects of FA on the anabolic and catabolic processes of *Nitrosomonas* and *Nitrobacter*

The specific OUR profiles determined from the FA inhibition studies on *Nitrosomonas* are plotted as a function of FA concentration in Figure 1(a). The OUR was clearly limited by the ammonia concentration when the latter was below 4 mgNH$_3$-N L^{-1}. Both OUR$_{withCO2}$ and OUR$_{withoutCO2}$ reached their maximum levels at around 4 mgNH$_3$-N L^{-1}. A further increase in ammonia (up to 16 mgNH$_3$-N L^{-1}) did not result in an increase in the ammonia oxidation rate. On the other hand, OUR$_{withCO2}$ and OUR$_{withoutCO2}$ did not show any tendency to decrease with the increase of the ammonia concentration, suggesting that the level of FA used in this study (up to 16 mgNH$_3$-N L^{-1}) did not inhibit the ammonia oxidation rate. This further means that the energy generation capability of the *Nitrosomonas* culture was not inhibited by a FA concentration of up to 16 mgNH$_3$-N L^{-1}. The result is in line with previous reports. For example, Hellinga *et al.* (1999) and van Hulle *et al.* (2004) found that AOB cultured in SHARON reactors were not inhibited by FA at a concentration of up to 70 - 300 mgNH$_3$-N L^{-1}. Knowing that ΔOUR is proportional to μ (Equation 3), Figure 1(a) clearly shows that an FA concentration of up to 16.0 mgNH$_3$-N L^{-1} does not inhibit the anabolic processes of the *Nitrosomonas* culture (μ does not decrease with increase of NH$_3$ concentration within this range).

On the other hand, Figure 1(b) clearly shows that the addition of FA immediately decreased both OUR$_{withCO2}$ and OUR$_{withoutCO2}$ of *Nitrobacter*. OUR$_{withCO2}$ reached its lowest level (75% of maximum OUR$_{withCO2}$) when FA was above 6 mgNH$_3$-N L^{-1}. Whereas, OUR$_{withoutCO2}$ reached the same minimum value at a lower concentration of ammonia (4 mgNH$_3$-N L^{-1}). The OUR$_{withoutCO2}$ data which are not related to the growth processes of *Nitrobacter* (as described above), show that the energy generation processes of *Nitrobacter* are inhibited by FA. However this inhibitory effect is limited. The results indicate that at an FA concentration of 4.0 mgNH$_3$-N L^{-1}, the energy production ability of *Nitrobacter* was only inhibited by 12%. The increase of FA concentration from 4.0 to 10.0 mgNH$_3$-N L^{-1} (the highest level used in this study) did not further reduce the energy generation capability by *Nitrobacter*. The results presented in an earlier work by Fux *et al.* (2003) agree well with the above findings. In their study, it was discovered that the nitrite oxidation rate of the biomass from a SBR system treating dewatering liquor decreased by only 10% when the FA concentration was at 24 mgNH$_3$-N L^{-1} to 80 mgNH$_3$-N L^{-1}. The ΔOUR suggests that FA has a strong inhibitory effect on the anabolic processes of *Nitrobacter*. The inhibition on growth processes (ΔOUR) initiated as soon as FA was added and completely ceased at around 6.0 mgNH$_3$-N L^{-1}. Upon this concentration, ΔOUR becomes zero (OUR$_{withCO2}$ and OUR$_{withoutCO2}$ reached an identical value), which implies that the presence of CO_2 indeed did not cause an increase in the respiration rate of the *Nitrobacter*. Considering that OUR$_{withoutCO2}$ was measured in the absence of CO_2 (and hence no growth occurred), it is very likely that growth did not occur even in the presence of CO_2.

Summary

The results presented above on the FA and FNA inhibition on the metabolism of *Nitrosomonas* and *Nitrobacter* are summarized in Table 1. Clearly, *Nitrosomonas* has a much higher level of tolerance to FA and FNA in comparison to *Nitrobacter*, which may have contributed significantly to the elimination of NOB from systems treating high nitrogen containing wastewater through partial nitrification.

While the experimental data obtained in this study strongly support that FA and FNA have a more severe inhibitory effect on the growth capability than on the energy generation capacity of *Nitrosomonas* and *Nitrobacter*, the method does not provides additional evidence for the fundamental mechanisms involved. It is believed that the toxicity of FA and FNA could be due to the selective inhibition of some key steps in the carbon transformation performed for biosynthesis. However, other reasons may also be possible; thus further studies need to be carried out to investigate the mechanisms behind the different inhibitory effect of FA and FNA on the energy generation and growth processes of *Nitrosomonas* and *Nitrobacter*.

Table 1 Summary of FNA and FA inhibition on *Nitrobacter* and *Nitrosomonas*

	Nitrobactor		*Nitrosomonas*	
	Anabolism	**Catabolism**	**Anabolism**	**Catabolism**
FNA	Completely stopped at 0.02 $mgHNO_{2}$-N L^{-1}	No inhibition up to 0.024 $mgHNO_{2}$-N L^{-1}	Totally inhibited at 0.40 $mgHNO_{2}$-N L^{-1}	50% inhibition at 0.40-0.63 $mgHNO_{2}$-N L^{-1}
FA	Completely inhibited at above 6.0 $mgNH_{3}$-N L^{-1}	Inhibited by 12% at 6.0 –10.0 $mgNH_{3}$-N L^{-1}	No inhibition at up to 16.0 $mgNH_{3}$-N L^{-1}	No inhibition at up to 16.0 $mgNH_{3}$-N L^{-1}

Conclusions

The effect of FA and FNA on the energy generation and growth processes of *Nitrosomonas* and *Nitrobacter* was assessed separately using enriched cultures. The following conclusions are drawn:

- Both the FA and FNA inhibit *Nitrobacter*. Both were found to have stronger inhibition on the growth capability than on the energy generation capacity of *Nitrobacter*. The biosynthesis of *Nitrobacter* was totally ceased at above 6.0 mgN $L^{-1}$and 0.02 mg.N L^{-1} of FA and FNA, respectively. At the same level of FA, the energy production capability of *Nitrobacter* was only inhibited by 12%, whereas FNA up to 0.024 $mgHNO_{2}$-N L^{-1} did not show any inhibition on the energy production of *Nitrobacter*.
- Neither the anabolic nor the catabolic processes of the *Nitrosomonas* culture were inhibited by FA at a concentration of up to 16.0 $mgNH_{3}$-N L^{-1}. FNA inhibits both anabolic and catabolic processes of *Nitrosomonas*. While the growth processes are completely stopped at around 0.40 $mgHNO_{2}$-N L^{-1}, only 50% inhibition was found on the energy generation processes at up to 0.63 $mgHNO_{2}$-N L^{-1}.
- The conventional methods of assessing the inhibitory effects of FA and FNA on nitrifiers (and possibly other bacteria as well), through simply comparing the overall substrate utilizing rates under inhibitory and non-inhibitory conditions may produce misleading results. It is evident that the inhibitory effect of FA on the growth rate of *Nitrobacter* and the inhibitory effect of FNA on the growth rate of both *Nitrobacter* and *Nitrosomonas* are considerably more significant than the overall OUR ($OUR_{withCO2}$) data may have suggested. It is recommended that the catabolic and anabolic processes should be investigated separately.

Acknowledgement

The Australian Research Council is acknowledged for funding this research through Project DP0342658. The first author also thanks the Universiti Sains Malaysia, Malaysia, for providing scholarships.

References

Anthonisen, A. C., R. C. Loehr, T. B. S. Prakasam, and E. G. Srinath. (1976). Inhibition of nitrication by ammonia and nitrous acid. *Journal Water Pollution Control Federation* **48**: 835–852.

Fux, C., K. Lange, A. Faessler, P. Huber, B. Grueniger, and H. Siegrist. (2003). Nitrogen removal from digester supernatant via nitrite – SBR or SHARON? *Water Science and Technology* **48**(8): 9–18.

Gil, K. I. and E. S. Choi. (2001). Modelling of inhibition of nitrite oxidation in biological nitritation processes by free ammonia. *Biotechnology Letters* **23**: 2021–2026.

Hanaki, K., C. Wantawin, and S. Ohgaki. (1990). Nitrification at low-levels of dissolved-oxygen with and without organic loading in a suspended-growth reactor. *Water Research* **24**: 297–302.

Hellinga, C., M. C. M. Van Loosdrecht, and J. J. Heijnen. (1999). Model based design of a novel process for nitrogen removal from concentrated flows. *Mathematical and Computer Modelling of Dynamical Systems* **5**: 351–371.

Lai, E., S. Senkpiel, D. Solley, and J. Keller. (2004). Nitrogen removal of high strength wastewater via nitritation/denitritation using a sequencing batch reactor. *Water Science and Technology* **50**(10): 27–33.

Mauret, M., E. Paul, E. PuechCostes, M. T. Maurette, and P. Baptiste. (1996). Application of experimental research methodology to the study of nitrification in mixed culture. *Water Science and Technology* **34**(1–2): 245–252.

Painter, H. A. (1970). A Review of Literature on Inorganic Nitrogen Metabolism in Microorganisms. *Water Research* **4**: 393–.

Pirt, S. J. (1982). Maintenance energy – a general-model for energy-limited and energy-sufficient growth. *Archives of Microbiology* **133**: 300–302.

Prakasam, T. B. and R. C. Loehr. (1972). Microbial Nitrification and Denitrification in Concentrated Wastes. *Water Research* **6**: 859–869.

Pratt, S., Z. Yuan, D. Gapes, M. Dorigo, R. J. Zeng, and J. and Keller. (2003). Development of a novel titration and off-gas analysis (TOGA) sensor for study of biological processes in wastewater treatment systems. *Biotechnology and Bioengineering* **81**: 482–495.

Vadivelu, V. M., J. Keller, and Z. Yuan. (Submitted–a). Stoichiometric and kinetic characterisation of *Nitrosomonas* in mixed culture by decoupling the growth and energy generation processes. *Journal of Biotechnology*.

Vadivelu, V. M., Z. Yuan, C. Fux, and J. Keller. (Submitted–b). The inhibitory effects of free nitrous acid on the energy generation and growth processes of an enriched *Nitrobacter* culture. *Environmental Science and Technology*.

Vadivelu, V. M., Z. Yuan, C. Fux, and J. Keller. (Submitted–c). Stoichiometric and kinetic characterisation of *Nitrobacter* in mixed culture by decoupling the growth and energy generation processes. *Biotechnology and Bioengineering*.

van Hulle, S. W. H., E.I.P. Volcke, J. Lopez Teruel, B. Donckels, M. C. M. v. Loosdrecht, and P. A. Vanrolleghem. (2004). Influence of temperature and pH on the kinetics of the SHARON nitritation process. In *4th IWA World Water Conference, Marrakech, Morocco*.

Yang, L., and J. E. Alleman. (1992). Investigation of Batchwise Nitrite Buildup by an Enriched Nitrification Culture. *Water Science and Technology* **26**(5–6): 997–1005.

Magnetic molecules for the selective removal of aqueous contaminants

I. Pascual*, J.R. Duffield*, N.M. Ratcliffe*, G.R. Elder and S. Bushart*****

* Faculty of Applied Sciences, University of the West of England, Frenchay Campus, Coldharbour Lane, Bristol BS16 1QY, UK (E-mail: *ilona.pascual@uwe.ac.uk*)
** Bradtec Decon Technologies Ltd, The Wheel House, Bond's Mill Estate, Stonehouse, Gloucestershire GL10 3RF, UK (E-mail: *grelder@globalnet.co.uk*)
*** Electric Power Research Institute, 3420 Hillview Avenue, Palo Alto, California, 94304, USA (E-mail: *Sbushart@epri.com*)

Abstract The aim of this project is to develop a magnetic molecule with specific chelating groups on its surface to selectively separate and extract specific metal ions from aqueous waste streams. Magnetoferritin was chosen as the base unit for the magnetic molecule technology. Initially a broad spectrum chelating agent DTPA was chosen to create a model ferritin-DTPA molecule using native horse spleen ferritin. DTPA was successfully attached to native ferritin and used for separation experiments. These tests indicated that a maximum of over 1000 chelating agents could be bound to the surface of a ferritin. Magnetoferritin was also chelated with approximately 700 DTPA molecules. Calix[4]arene-crown-6 and its derivatives in the 1,3-alternate conformation exhibit outstanding efficiency and selectivity for caesium. A representative, 25,27-bis[(4-carboxybutyl)oxy]calix[4]arene-crown-6 was synthesised and an average of 10 of this molecules were attached to the ferritin. The ability of this molecule to complex caesium was demonstrated. Radioactive measurements indicated that 1g of Ferritin-Calixcrown could remove 0.36GBq of radioactive caesium.

Keywords Calixarene; magnetic molecule; selective separation; specific chelating agents

Introduction

A significant amount (~16%) of the world's electricity requirement is supplied by nuclear power stations. One problem associated with nuclear power stations is radioactive waste water. In reactor liquid cooling systems and effluents radioactive species may exist in very low molar concentrations (typically about 10^{-15} to 10^{-12} moles per litre) while other harmless dissolved species are present in much greater amounts. In nuclear applications, it is desirable to selectively remove only the radioactive species while leaving the harmless species in the water. The separated radioactive waste requires careful containment and disposal therefore the volume of radioactive waste must be rigorously minimized for safety and economic reasons.

There are many known techniques for removing dissolved impurities from water including evaporation, ion exchange and reverse osmosis. Although these techniques produce pure water, they are not capable of selectively removing certain target impurities while leaving all other species dissolved in the solution. Therefore there is a need for the development of a new separation technology.

The aim of this project is to develop a magnetic molecule with specific chelating groups on its surface to selectively separate and extract specific metal ions from aqueous waste streams.

The operational parameters required for the new separation technology are high selectivity and satisfactory absorption capacity for the target contaminant as well as fast reaction kinetics. The particles should be easily separated from the treated liquid and should produce minimum amount of secondary waste.

Highly selective molecules exist for a wide range of nuclides. A considerable amount of work has been done over the last fifty years and the data now available forms the basis for the selection of specific chelating groups (Pulpoka and Vicens 2004; O'Boyle *et al.* 1997). It is not possible to use the chelating molecules alone as these could not subsequently be removed from solution more easily than the original contaminants. The chelating molecule must be attached to a discreet particle for its removal from solution. The use of a small particle gives a much higher surface area per volume of material, possibly increasing the number of chelating agents attached to the particles. Using small particles also reduces the volume of secondary waste. However, small particles can be difficult to remove from solution by standard barrier filtration due to 'blinding' of the filter membrane and the necessary use of high pressures. An alternative to barrier filtration is magnetic filtration. In this technology the liquid is passed through an open steel mesh while subjected to a strong magnetic field. Any magnetically susceptible particles are trapped on the steel mesh surface. Magnetic filtration is a well-proven industrial technology which is used, amongst other applications, to remove para and ferro magnetic iron oxide impurities from 25% clay slurries to produce high quality white clay used in paper manufacture.

Ferritin is an iron storage protein found in animals, plants and bacteria (Harrison and Arosio 1996). The ferritin molecule is composed of polypeptide chains arranged in a highly symmetrical way to give a hollow shell with an outer diameter of about 12nm and internal diameter of about 8nm as illustrated in figure 1.

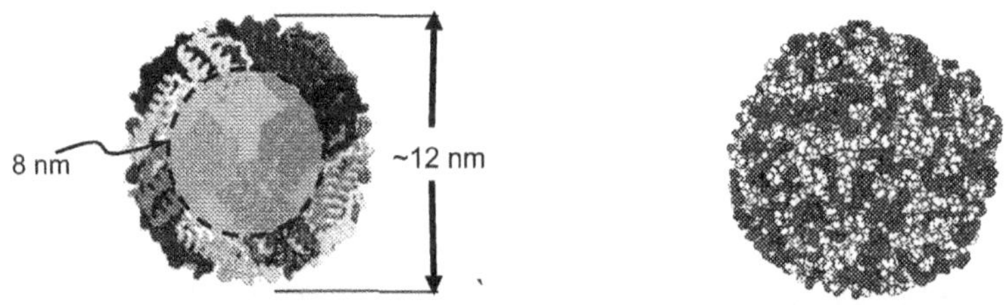

Figure 1 Graphic representation of ferritin (adapted from Chasteen and Harrison 1999), and surface functional groups on a ferritin molecule (adapted from: Ha,Y. et al 1999).

The iron is stored within the supramolecular cage of the ferritin as an antiferromagnetic oxide (Theil *et al.* 2000). The iron core can be removed by reductive dissolution and the cavity of the ferritin can be used as a confined reaction environment for the synthesis of a ferromagnetic mineral such as magnetite (Meldrum *et al.* 1992; Dickson *et al.* 1997). The material produced in this way is called magnetoferritin. Magnetoferritin is the base unit for the magnetic molecule technology. The advantages of this molecule are three fold. First, the magnetic susceptibility is uniform and sufficient to respond to magnetic filtration. Second, the size is uniform and extremely small. Finally, magnetoferritin possesses surface functional groups which make it possible to couple chelating agents to its surface. Figure 1 shows the free amino (blue) and carboxyl (red) groups on the surface of a ferritin/magnetoferritin molecule.

Specific chelating agents exist or can be synthesised for a wide range of nuclides. Calix[*n*]arenes are macrocyclic compounds made up of *n* phenol units linked by *n* methylene groups. The most commonly used calix[4]arene is illustrated in figure 2. Calix[4]arenes can be easily and selectively modified both at the lower and upper rim leading to molecules with selective host-guest properties (Otsuka and Shinkai, 1996; van Loon *et al.* 1990). Calixcrowns are molecules in which the phenolic oxygens of the calixarenes are bridged by a polyetherchain (Asfari *et al.* 1995). The cavity size of the calixcrowns can be modified by either changing the number of phenol units or by the variation of the polyetherchain. The selectivities of the calixcrowns are a function of the cavity size. For example the calix[4]crown-5 shows a 12000 fold selectivity for K^+ over Na^+ while the calix[4]crown-6 shows a 33000 fold selectivity for Cs^+ over Na^+ (Ungaro *et al.* 1994).

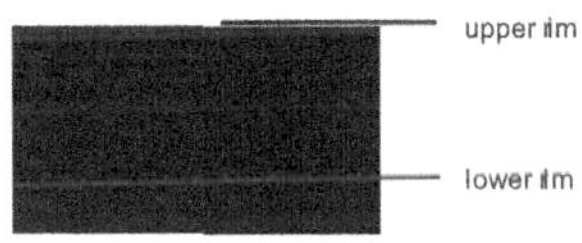

Figure **2** Calix[4]arene

Methods

Preliminary measurements

Calcium ion separation

Initially a broad spectrum chelating agent diethylenetriaminepentaacetic acid (DTPA) was chosen to create a model ferritin-DTPA molecule using native horse spleen ferritin.
The synthesis of Ferritin-DTPA included the chemical modification of DTPA molecules and their consequent attachment to ferritin. The synthetic method used by Krejcarek and Tucker (Krejcarek and Tucker 1977) was adopted. Briefly, the DTPA was modified in two steps: First, all the carboxyl groups were protected by the formation of pentatriethylammonium DTPA (PA-DTPA). Second, one of the groups was activated by its convertion to carboxycarbonic anhydride (CCA-DTPA). Both PA-DTPA and CCA-DTPA were characterised by Nuclear Magnetic Resonance Spectroscopy (NMR) and Infrared Spectroscopy (IR). The produced CCA-DTPA was then attached to the ferritin. The produced Ferritin-DTPA solution was dialysed against 0.1moldm^{-3} buffer (acetate or HEPES) solution to remove the by-products and the non-conjugated DTPA.

The dialysed Ferritin-DTPA solution was used to separate calcium ions (Ca^{2+}) from sodium ions (Na^+). 5ml aliquots of the solution were transferred to dialysis bags and dialysed against 10^{-4}–10^{-6} moldm^{-3} calcium chloride ($CaCl_2$) in either acetate or HEPES buffer. As a control, native ferritin was added to 0.1 moldm^{-3} sodium bicarbonate solution and dialysed in the same way as the test solutions. The amount of chelated Ca^{2+} was measured using Inductively Coupled Plasma Atomic Emission Spectrometry (ICP-AES).

Cobalt ion separation

In the next set of experiments the reaction conditions were modified in order to try to maximise the number of DTPA molecules bound to the ferritin molecules. Changes of the separation conditions were also applied to assess the stability of the synthesised molecules. The freshly prepared Ferritin-DTPA molecules were used to separate cobalt ions (Co^{2+}) from solutions in the pH range of 5–7. These solutions as well as all the calibration solutions were prepared using deionised water (not buffer). The pH was adjusted by either hydrochloric acid (HCl) or sodium hydroxide (NaOH). The amount of chelated Co^{2+} was measured by ICP-AES.

Testing the magnetoferritin

Magnetoferritin-DTPA molecules were synthesised the same way as the model Ferritin-DTPA molecules and used to separate Ca^{2+} and also Co^{2+}. The amount of chelated ions was measured using ICP-AES.

Caesium specific chelating agents

Synthesis of 25,27-bis[(4-carboxybutyl)oxy]calix[4]arene-crown-6

The next phase of development required the identification and addition of a caesium specific chelating agent to achieve specific separations. A possible caesium specific chelating agent 25,27-bis[(4-carboxybutyl)oxy]calix[4]arene-crown-6 (calixcrown-6) was synthesised. The synthetic route used by Nechifor *et al.* (Nechifor *et al.* 1996) was adopted and is illustrated in figure 3.

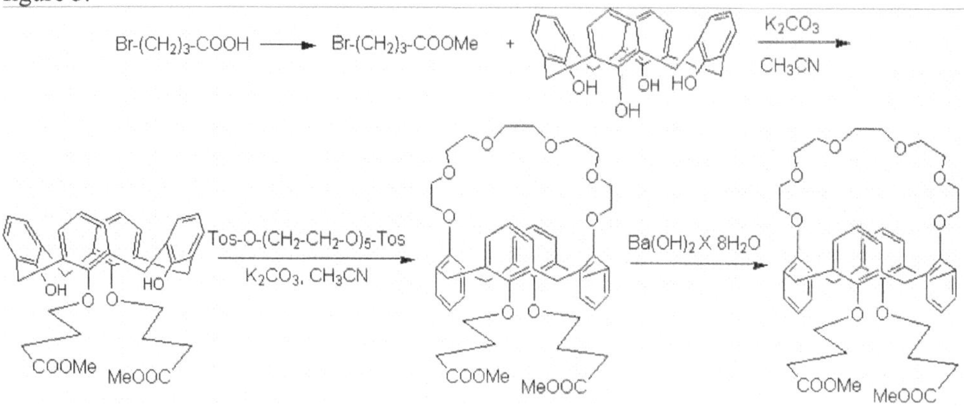

Figure 3 Synthesis of 25,27-bis[(4-carboxybutyl)oxy]calix[4]arene-crown-6

The ability of the synthesised calixcrown-6 to separate Cs^+ was assessed by proton NMR and a solvent extraction method. First, NMR of the freshly synthesised calixcrown-6 was taken in deuterated chloroform. To this solution cesium picrate was added in 1:1 molar ratio. The solution was left for half an hour before another NMR was taken. The two NMRs were compared. For the solvent extraction, the synthesised calixcrown-6 was dissolved in distilled ethyl acetate. A known amount of radioactive Cs^+ and caesium picrate was dissolved in deionised water. The two solutions were mixed for half an hour. After mixing, 0.5 ml was extracted from each of the aqueous and organic layers, and the radioactive caesium was counted in a gamma well counter.

Attachment of calix-crown-6 to ferritin

Possible pathways to couple the calixcrown-6 to the ferritin were investigated. Possibilities included using carbodiimide as coupling agent, producing mixed carbonic anhydrides and by formation of acyl chlorides as intermediates. Since the best coupling was achieved by using carbodiimide, only this method is summarised. Native ferritin was dissolved in 20ml of

0.1moldm^{-3} sodium bicarbonate solution and left at room temperature. Calixcrown-6 was dissolve in 1ml of anhydrous DMSO and 2 equivalent of N,N'-dicyclohexylcarbodiimide and 1-hydroxybenzotriazole hydrate was added. The two solutions were mixed and left at room temperature overnight. Next day the solution was dialysed against deionised water for 24 hours to discard any unreacted reagents.

Separation of caesium ion

Ferritin-Calixarene molecule was used for Cs^+ separation. The complexation was achieved by dialysis against non-radioactive and radioactive Cs^+ solutions. In the non-radioactive tests 5ml of Ferritin-Calixarene was placed in dialysis bags and dialysed against 10^{-4} and 10^{-5} moldm^{-3} CsCl solution for 4 hours. Native ferritin solution was subjected to the same treatment for control. The amount of Cs^+ within the dialysis bags was determined by ICP-AES.

In the second set of experiments 2.5 ml of each of the test and blank solutions were placed in dialysis bags and dialysed against 100 ml of radioactive Cs-137 solution (151 Bq/ml). 0.5 ml of the outer caesium solution was removed and the loss of caesium was measured in a gamma well counter. To establish the absorption behaviour under these conditions a test was also performed with a known absorber, IRN-77 proton exchanger.

Results and discussion

Preliminary measurements

DTPA is readily available in pure form, it is economic and it can be bound to the surface of a protein *via* the formation of amide bond. Although DTPA is a broad spectrum chelating agent it can separate bivalent cations from monovalent cations.

Several analytical methods have been investigated to obtain direct evidence of DTPA binding to ferritin. These included surface plasmon resonance (IAsys), MALDI-TOF mass spectrometry, electrophoresis and ion exchange chromatography. To date, while considerable efforts have been expended in this regard, direct evidence for ferritin-DTPA coupling remains elusive. This is because all of these methods look for a change in some property (*e.g.* surface charge, molecular mass) of a large and complex protein molecule following its interaction with a relatively small ligand. As a result of the above difficulties, coupling of DTPA to ferritin has continued to be assessed by the ability of the bound molecule to separate and concentrate Ca^+ across a dialysis membrane.

The Ca^{2+} concentration inside the dialysis sac following equilibrium was measured by ICP-AES. Since the amount of ferritin inside the dialysis bag was known it was possible to estimate the average number of DTPA molecules bound to each ferritin molecule. The calculated molar ratios of DTPA to ferritin are summarised in Table 1.

Table 1 Calculated Ferritin DTPA molar ratio

Buffer solution pH	Acetate 6	Acetate 7.4	HEPES 7.4	Magferritin 7.4
Dialysis sac contents and system	Ferritin-DTPA molar ratio			
Ferritin-DTPA dialysed against 10^{-6}M $CaCl_2$	38	13	15	12
Ferritin-DTPA dialysed against 10^{-5}M $CaCl_2$	6	28	38	37
Ferritin-DTPA dialysed against 10^{-4}M $CaCl_2$	46	117	100	90

The results indicate that a maximum of over one hundred DTPA molecules were attached to each ferritin molecule. It also suggests that pH and buffer have some affect on the ability of ferritin-DTPA to bind Ca^{2+}.

Reaction conditions have been optimised during Co^{+} separation experiments. Table 2 shows the calculated ferritin DTPA molar ratios. The results indicate that ferritin has been functionalised with a maximum of over 1200 DTPA molecules per mole of ferritin. This separation has been carried out over a range of pH values from 5 to 9. The table also indicates that the number of attached particles is pH dependent.

Table 2 Calculated Ferritin DTPA molar ratio

Type of ferritin pH	Ferritin 5	Ferritin 7	Ferritin 9	Magferritin 5	Magferritin 7	Magferritin 9
Dialysis sac contents and system	Ferritin-DTPA molar ratio					
Ferritin-DTPA dialysed against 10^{-5}M $CoCl_2$	248	150	66	363	331	247
Ferritin-DTPA dialysed against 10^{-4}M $CoCl_2$	1288	1270	1364	467	737	319

Magnetoferritin-DTPA molecules were also synthesised and used for Ca^{+} and Co^{+} separation. Table 1 and 2 show that magnetoferritin can also be functionalised by DTPA and used for ion complexation. The number of DTPA attached to the magnetoferritin is less then the number attached to the ferritin and it is also pH dependent.

Caesium-specific chelating agents

Calix[4]arene-crown-6 and its derivatives in the 1,3-alternate conformation exhibit outstanding efficiency and selectivity for caesium. Consideration was given to the attachment of a calix[4]arene-crown-6 based molecule to the ferritin. Examples exist in the literature of synthesising two pendant alkyl chains to the basic molecule (Arena *et al.* 1996). Such modified calix[4]arene-crown-6 have been coupled to both membrane (Duhart *et al.* 2001) and particle surfaces and have continued to exhibit selectivity for caesium. By modification of this synthesis it is possible to produce alkyl chains with a carboxylic end groups suitable for attachment to the ferritin in the already demonstrated manner as with DTPA. Hence, 25,27-bis[(4-carboxybutyl)oxy]calix[4]arene-crown-6 (calixcrown-6) was synthesised.

The ability of the calixcrown-6 to separate Cs^+ before attaching it to the protein was first demonstrated by NMR. The NMR spectra taken after complexation showed that all the peaks corresponding to the caliccrown-6 structure shifted to higher δ values, compared to before separation. This clearly indicated the presence of Cs^+ in the structure. The chelation of Cs^+ was further assessed by radioactive measurements. These studies showed that most of the radioactivity was transferred within half an hour from the aqueous to the organic phase. The short time required was also indication of the fast reaction kinetics.

The attachment of the calixcrown-6 was complicated by the fact that it is soluble in organic solvent while ferritin is water soluble and it becomes unstable if in contact with big volume of organic solvent. To overcome this problem DMSO was used as solvent in very small volumes.

The attachment was assessed by separation experiments. The non-radioactive measurements showed that it was possible to attach an average of 10 calixcrown-6 molecules to the ferritin. This number is much smaller then the number of attached DTPA which was expected due to its bigger size and different solubility. However, this number can be improved on. Radioactive measurements indicated that 1g of Ferritin-Calixcrown can remove 0.36GBq of radioactive caesium. For comparison the IRN-77 absorber could separate much less radioactive caesium during the same time interval.

Conclusions

A new selective separation technology has been proposed based on the use of selective chelating agents bound to the surface of magnetic molecule for subsequent removal by magnetic filtration. Magnetoferritin has been identified as a suitable magnetic moiety. A broad spectrum chelating agent DTPA was successfully attached to native ferritin and magnetoferritin and used for separation experiments. These tests indicated that a maximum of over 1000 chelating agents could be bound to the surface of a ferritin and approximately 700 to the magnetoferritin.

A specific caesium selective chelating agent 25,27-bis[(4-carboxybutyl)oxy]calix[4]arene-crown-6 was synthesised and its ability to separate Cs^+ was demonstrated. The molecule was successfully attached to the ferritin protein and used for radioactive separation.

The choice of the calixarene molecule, as a base for specific chelation, has an additional benefit. A great deal of work has been done and is still ongoing in order to use the calixarene molecule as a basic unit from which to build selective a family of chelating agents. The work covers transition metals, lead, mercury and cadmium, lanthanides, actinides alkali and alkaline metals. By establishing a single method for attachment of the calixarene to the ferritin the ion selective options can be increased by substituting in the appropriate modified calixarene.

These results indicate that the Ferritin-Calixcrown molecule has the potential for industrial applications. The most immediate area for application is the radioactive waste effluents in the commercial nuclear power plant. However, magnetic molecules could also be considered for use in ground water clean up. Ground water has many naturally occurring and beneficial salts and organic matter which is desirable to maintain. Any harmful contaminants could be removed by channelling the ground water from large areas through a restricted channel and by applying the magnetic molecule technology.

References

Arena, G., Casnati, A., Contino, A., Mirone, A., Sciotto, D. and Ungaro, R. (1996). Synthesis of new calixcrowns and their anchoring to silica gel for the selective separation of Cs^+ and K^+.*Chemistry Communications*, 2277–2278.

Asfari, Z., Wenger, S. and Vicens, J. (1995). Calixcrowns and related molecules. *Pure and Applied Chemistry*, **67**, 1037–1043.

Chasteen, N.D. and Harrison, P.M. (1999). *J. Structural Biology* **126**, 182–194.

Dickson, D. P. E., Walton, S. A., Mann, S. and Wong, K. (1997). Properties of magnetoferritin: a novel boimagnetic nanoparticle. *NanoStructured Materials*, **9**, 595–598.

Duhart, A., Dozol, J. F., Rouquette, H. and Deratani, A. (2001). Selective removal of cesium from model waste solution using a solid membrane composed of an unsymmetrical calix[4]arene-biscrown-6 bonded to an immobilized polysiloxane backbone *Journal of Membrane Science*, **185**, 145–155.

Ha, Y. *et al.* (1999). *J. Biol. Inorg. Chem.* **4**, 243–256.

Harrison, P. M. and Arosio, P. (1996). The ferritins: molecular properties, iron storage function and cellular regulation. *Biochimica et Biophysica Acta*, **1275**, 161–203.

Krejcarek, G. E. and Tucker, K. L. (1977). Covalent attachment of chelating groups to macromolecules. *Biochemical and Biophysical Research Communications*, **77**, 581–585.

Meldrum, F. C., Heywood, B. R. and Mann, S. (1992). Magnetoferritin: in vitro synthesis of a novel magnetic protein. *Science*, **257**, 522–524.

Nechifor, A. M., Philipse, A. P., de Jong, F., van Duynhoven, J. P. M., Egberink, R. J. M. and Reinhoudt, D. N. (1996). Preparation and properties of organic dispersions of monodisperse silica receptor colloids grafted with calixarene derivatives or alkyl chains. *Langmuir*, **12**, 3844–3854.

O'Boyle, N. C., Nicholson, G. P., Piper, T. J., Taylor, D. M., Williams, D. R. and Williams, G. (1997). A review of plutonium(IV) selective ligands. *Applied Radiation and Isotopes*, **48**, 183–200.

Otsuka, H. and Shinkai, S. (1996). Stereochemical control of calixarenes useful as rigid and conformationally diversiform platforms for molecular design. *Supramolecular Science*, **3**, 189–205.

Pulpoka, B. and Vicens, J. (2004). 1,3-Alternate calix[4]arene: the sophisticated conformer of calix[4]arene. A review. *Collect. Czech. Chem. Commun.*, **69**, 1251–1281.

Theil, E. C., Takagi, H., Small, G. W., He, L., Tipton, A. R. and Danger, D. (2000). The ferritin iron entry and exit problem. *Inoganica Chimica Acta*, **297**, 242–251.

Ungaro, R., Casnati, A., Ugozzoli, F., Pochini, A., Dozol, J. F., Hill, C. and Rouquette, H. (1994). 1,3-Dialkoxycalix[4]arene-crowns-6 in 1,3-alternate conformation: cesium selective ligands that exploit cation-arene interactions. *Angewandte Chemi – International Edition in English*, **33**, 1506–1509.

van Loon, J. D., Arduini, A., Coppi, L., Verboom, W., Pochini, A., Ungaro, R., Harkema, S. and Reinhoudt, D. N. (1990). Selective functionalization of calix[4]arenes at the upper rim. *Journal of Organic Chemistry*, **55**, 5639–5646.

Electrochemical oxidation assisted technology applied to industrial wastewater

Gabriela Roa-Morales*, Carlos Barrera-Díaz*, Javier Agujera Cotero*, Thelma B. Pavón Silva* and Bryan Bilyeu**

* Universidad Autónoma del Estado de México. Facultad de Química. Paseo Colón intersección Paseo Tollocan S/N. C.P. 50120, Toluca, Estado de México, México (E-mail: *gabyroamo@yahoo.com.mx*; *cbarrera@uaemex.mx*)

** Department of Materials Science and Engineering, University of North Texas, PO Box 305310, Denton, Texas 76203-5310, USA

Abstract The removal of organic pollutants from food industry wastewater by an aluminium electrocoagulation and electrocoagulation/H_2O_2 processes were evaluated using a pilot batch reactor. Under optimal conditions of pH 4 and 18.2 mA/m^2 current density, the electrochemical method yields a very effective removal of all organic pollution. Treatment reduced chemical oxygen demand (COD) by 88%, biochemical oxygen demand (BOD_5) by 96%, total solids by 82% and fecal coliforms by 99.9%. The effect of reducing the pH of the aqueous solution with sulphuric acid on the species formed and its effect on the sludge phase formation was determined. The wastewater quality was monitored using UV-Vis spectrometry and cyclic voltammetry to characterize initial and final pollutant levels. Finally, the morphology and elemental composition of the resulting sludge was characterized using scanning electron microscopy (SEM) and energy dispersion spectra (EDS).

Keywords Organic pollutants; food wastewater; coagulation; oxidation; decontamination

Introduction

The food industry produces wastewater with high concentrations of organic compounds and refractory organics that are not completely removed by physicochemical preliminary treatments (i.e. coagulation/flocculation and sedimentation) or biological processes (i.e. anaerobic and aerobic attack). These waste materials originate as process residues or as the reprocessed scrap. The high organic load of food industry wastewater means that the process represents a major environmental problem. With the establishment of more stringent regulations concerning public waste disposal, there has been a growing interest in the developing of new technologies and procedures for the decontamination of water (Keenan and Sainkov, 2000; Beltran de Heredia, 2004).

In recent years there has been great interest in the development of practical electrochemical methods, such as anodic oxidation and indirect electrooxidation, for the destruction of toxic and biorefractory organic pollutants for wastewater treatment, has been made (Brillas and Casado, 2002; Vlyssides *et al*., 2005).

Electrochemical methods have been used as chemical coagulation processes to remove color and cloudiness from turbid industrial wastewater. In this application the electrochemical process generated numerous flocculates, achieving high efficiency in clearing the wastewater (Riveiro *et al*., 2000; Can *et al*., 2003). The electrochemical treatment techniques have attracted a great deal of attention because of their versatility and environmental compatibility, which makes the treatments of liquids, gases and solids possible. In fact, the main reagent is the electron, which is a ‘‘clean reagent’’ (Rajeshwar *et al.*, 1994; Janssen and Koene, 2002). A recent paper indicates the main processes that takes place into a electrochemical cell: An electrochemical cell is used to provide aluminum or iron to the waste (from the anode). These coagulant reagents desestabilize the colloidal pollutants (or break the emulsion), similar to the case of a conventional coagulation process. The turbulence generated by the oxygen (produced as secondary reaction in the anode) and the hydrogen (produced due to water reduction in the cathode) evolution generates a soft mix that helps the desestabilized colloids to flocculate (to link together and to generate bigger particles): This process is often called electroflocculation. Finally, the pollutants are removed by sedimentation, filtration, or flotation from the waste. The bubbles formed in the oxygen and hydrogen evolution can help to increase the efficiency of the later process (electroflotation) (Cañizares *et al.*, 2005).

There are two novel processes to remove pollutants from wastewater when iron is used as anode in acidic conditions. The electro-Fenton, which is carried out with a conventional Fe anode and electrogenerated H_2O_2 in acidic conditions to form hydroxyl radicals, and peroxi-coagulation based on sacrificial Fe anode which is electrodissolved supplying stoichiometric amounts of Fe^{2+} to the wastewater. Fe^{2+} reacts with electrogenerated H_2O_2 and the excess of Fe^{3+} formed precipitates as $Fe(OH)_3$.The former processes have been successfully applied to the degradation of aniline, herbicides, and biorecalcitrant pollutants in industrial wastewater have been successfully probe (Brillas *et al.*, 2003a,b; Boye *et al.*, 2003; Barrera-Díaz, *et al.*, 2003a,b).

The used of Al as sacrificial electrode has been used mainly as an electrocoagulation technique in the removal of dyes (Can *et al.*, 2003), pollutants in surface waters (Jiang *et al.*, 2002) and clay in potable water (Holt *et al.*, 2002, 2004), among others. However, in all these works the main process that has been studied is the anodic reaction.

The aim of this work is to study the electrodissolution of the aluminum in the pollutant removal in the discharge of a wheat food industry. It is presented also the influence in the pollutant removal efficiency when different pH, H_2O_2 doses and aeration is provided in an electrochemical cell. It is proposed the main mechanisms involved in the anode and cathode electrodes, and the aqueous media chemistry involved. The wastewater is characterized prior and after the electrochemical treatment by UV-Vis and Cyclic Voltammetry. Finally, the sludge morphology and elemental composition is shown.

Methods

Wastewater treatment plant

The food industry wastewater treatment plant under study employs a biological system to remove organic pollutants. The system consists of screens, a homogenization tank, an anaerobic up flow reactor, an extended aeration aerobic tank, secondary clarifiers and finally a disinfection system. Treated water coming from this plant does not comply with environmental discharge limits. The biological treatment system can’t respond quickly to the large variation in quantity and physicochemical composition of the influent throughout the year (i.e. COD goes from 1000 to 22,000 mg L^{-1}). Therefore, new ways to treat this wastewater are needed. Electrochemical techniques offer advantages over biological systems, including adaptability.

Wastewater samples

Samples of wastewater were collected from the treatment plant in the system influent. Samples were collected in plastic containers and cooled to 4 °C, then transported to the laboratory for analysis and electrochemical treatment.

Electrochemical reactor

The reactor consisted of a rectangular vessel and two shafts supporting a series of aluminum plates. Four plates were connected as cathodes and four as anodes. All the electrodes were flat with a smooth surface. To promote turbulence and enhance mass transfer, an aeration system was added. The aeration system consisted of one air diffuser placed in the bottom part of the rectangular vessel. The volume of liquid treated each time was 1.5 dm^3. A dc source was used to supply the system with 1, 2, 3 and 4 A at 9 V.

Methods of analysis

The effect of the electrochemical treatment was determined by analysis of the chemical oxygen demand (COD) at different time intervals. However, once the optimal conditions were found the raw and treated wastewater were analyzed using the biochemical oxygen demand (BOD_5), chemical oxygen demand (COD), as well as by the solids, pH, fecal coliforms, color and turbidity, as indicated in APHA (Clesceri *et al.*, 1989).

Thermodynamic study

The existence of Al(III) complexes in aqueous solution has been reported (Smith and Martell, 1974; Baes and Mesmer, 1986). With this information the distribution diagrams of chemical species were calculated using the MEDUSA program (Puigdomenech, 1997).

Cyclic voltammetric measurements

Cyclic voltammetry of crude and treated wastewater was performed using a standard three-electrode cell. The waveforms were generated by a BAS-100W model potentiostat/galvanostat, which was controlled by BAS software. The carbon paste electrodes (CPE) were circular with a surface area of about 3.5 mm^2. The CPE was prepared from a 1:1 w/w ratio mixture of 99.99% purity single crystal graphite (Alfa AESAR) and nujol oil (Fluka). The paste was inserted into a PVC tube and compacted to eliminate trapped air, then a copper conductor was inserted before the paste set. The surface of the electrode was renovated after each potential scan (Roa-Morales *et al.*, 2003); the scan rate was 100 mV s^{-1}. The reference electrode was Ag/AgCl saturated with KCl and the counter electrode was a platinum wire.

UV-Vis spectrometry

UV-Vis spectra were obtained from samples of raw and treated wastewater using a double beam Perkin-Elmer 25 spectrophotometer. The scan rate was 960 nm s^{-1} within a 900–200 nm wavelength range. The samples were scanned in quartz cells with a 1 cm optical path.

Sludge characterization

The sludge produced by the electrochemical process was analyzed by scanning electron microscopy (SEM) and X-ray microanalysis. The analysis was performed on a Phillips XL-30 microscope to determine the composition and configuration of the structure. SEM provides images of surface feature with a resolution of fractions of μm; while energy disperse X-ray spectroscopy offers *in situ* chemical analysis of the bulk.

Results and discussion

Electrocoagulation

The electrocoagulation treatment of wastewater was investigated in laboratory experiments by modifying the initial aqueous media pH and the current density. Likewise, the effect of adding hydrogen peroxide (30% w) effectiveness of the electrocoagulation process was evaluated. All experiments were carried out at 18 ± 1 °C.

Table 1 lists the values of the initial conditions of the operating variables in each experiment. The evolution of the elimination of the organic matter was monitored en each experiment by measuring the chemical oxygen demand (COD).

Table 1 COD removals obtained at different electrolysis time, different initial wastewater pH and different current density applied

Electrolysis	pH				Current intensity, A			
time (min)	1	2	3	4	1	2	3	4
15	10	10	30	30	10	20	25	30
30	45	45	45	55	20	35	40	50
45	55	60	60	75	25	40	55	64
60	75	75	80	90	35	45	65	75

Experimental data in Table 1 indicates that in all cases the greater electrolysis time the better COD removal. However, there is a major impact on the COD removal from the pH and current density. In the case of the pH the optimum value is at 4. This effect can be explained due to the increase of other chemical species in aqueous solution as described in the electrocoagulation mechanisms section. The current density is directly related with the Al concentration in aqueous solution. Since this wastewater contains high loads of pollutants, it is need a large amount of reagent in the media to decontaminate the water.

Treatment comparison

Figure 1 shows the COD pollutant removal from wastewater, using different treatments. In the first case, extended aeration was introduced into the wastewater from the diffuser inside the electrochemical reactor. After 1 hour of providing aeration some oxidation of the organic compound occurs. Then, the electrocoagulation treatment using 4 A at an adjusted pH of 4 was applied to the raw wastewater. Under these conditions, a COD removal of 75% is achieved. A third treatment consisted in applying the electrocoagulation (at the same conditions as before) and providing air diffusion. In this case, an 85% COD removal is reached. The fourth treatment use the same conditions as the third one but 10 mL of H_2O_2 (%W) is added to the incoming wastewater. The fourth treatment provides the highest COD removal yield.

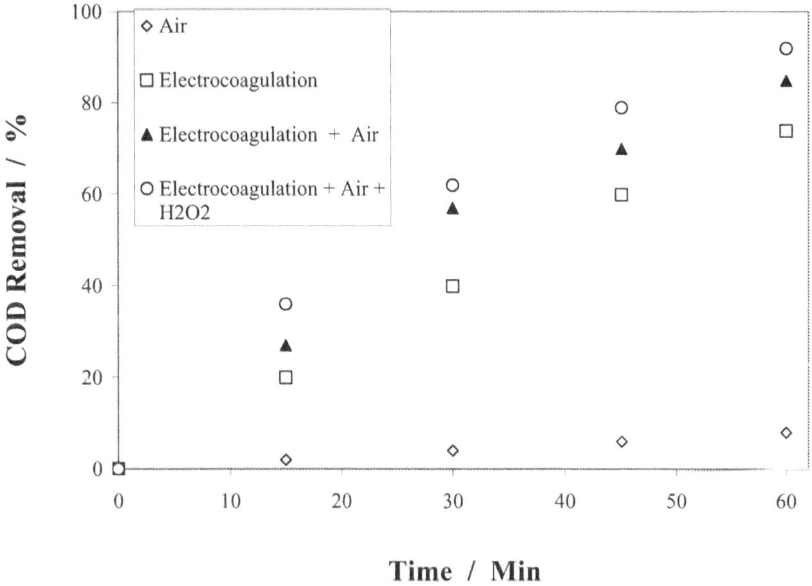

Figure 1 COD removal of different treatments applied to wastewater.

Electrocoagulation mechanisms

According to Picard *et al.* (2000), the chemical dissolution process corresponds to the oxidation of the aluminum sheets with the simultaneous reduction of water to form hydrogen, according to equation 1.

$$2Al + 6H_2O \rightarrow 2Al^{3+} + 3H_2 + 6OH^- \quad (1)$$

On the other hand, the electrochemical processes that occur on the anode and on the cathode surfaces are represented by equations 2-4. On the anode, aluminum dissolution and oxygen evolution compete. On the cathode, hydrogen evolution is the main reaction. As can be observed, in both the chemical and the electrochemical processes, hydrogen is generated.

$$Al \rightarrow Al^{3+} + 3e^- \quad (2)$$
$$2H_2O \rightarrow O_{2(g)} + 4H^+ + 4e^- \quad (3)$$
$$2H_2O + 2e^- \rightarrow H_{2(g)} + 2OH^- \quad (4)$$

As a result of the prior reactions, the formation of $Al(OH)_{3(s)}$ is expected.

Once the aluminum is dissolved (chemically or electrochemically), some species can be formed, depending on the pH of the solution and the presence of other chemical species.

Recent studies indicate that at dilute Al concentrations (5×10^{-3} M) the following chemical species can be found in aqueous solution: Al^{3+}, $Al(OH)^{2+}$, $Al(OH)_3$ $Al_{13}(OH)_4{}^{5+}$ and $Al_{13}(OH)_{32}{}^{7+}$. The distribution of the Al species depends on the pH of the aqueous solution. In the present study where the working pH is 4, the predominant species are Al^{3+} and $Al(OH)_3$, which are present in equal amounts. However, at pH 4.5, the predominant species is only $Al(OH)_3$, which can be soluble, insoluble or colloidal (Domínguez *et al.*, 1998; Pacheco-Hernández, 1999).

In actual experimental conditions, H_2SO_4 was used to adjust the pH to 4. The concentration of SO_4^{2-} affects the chemical equilibrium conditions of the process and new chemical species are formed. In Figure 4, the distribution of the chemical species in the actual solution with the sulfate ions as a function of pH is presented. Note that at pH 4 new insoluble species appear in the system, namely 98% $AlOHSO_4(s)$, 1% $Al(SO_4)_2^-$, and 1% $AlSO_4^{+1}$ and $Al(OH)_2^+$.

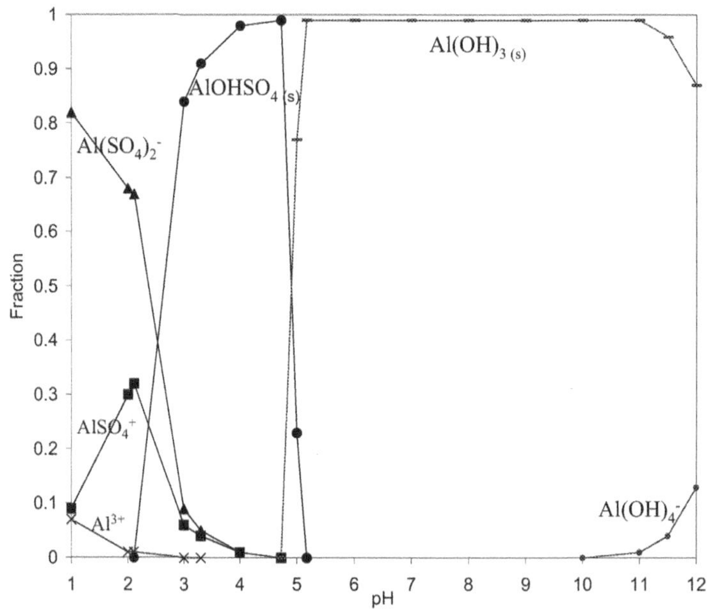

Figure 2 Distribution diagrams of Al(III) species in 0.06 M sulphates and 0.1 M of Na^+.

H_2O_2 effect

The reduction of H_2O_2 to form OH• by photoinduced electrons was reported in photoelectrochemical studies. In general, the irradiated TiO_2 has an ability to reduce H_2O_2 into OH• since the reduction potential of H_2O_2/OH• is about 0.87 V (vs NHE). Experimental observations indicate that he formation rate of OH• was increased by the addition of H_2O_2. (Hirakawa and Nosaka, 2002)

The H_2O_2 formation from the two-electron reduction of O_2 at a graphite electrode has been reported for (Do and Chen, 1994). This reaction in acidic medium can be written as:

$$O_2(g) + 2\,H^+ + 2\,e^- \rightarrow H_2O_2 \tag{5}$$

Gas diffusion cathodes may also be used to reduce oxygen to hydrogen peroxide in acidic solutions at rates which are appropriate for the needs of effluent treatment.
However, some reaction takes place as the initial reduction of H_2O_2 in the cathode (6)

$$H_2O_2 + e^- \rightarrow OH\bullet + OH^- \tag{6}$$

Organics are mainly destroyed by the action of OH• radicals produced in the aqueous media. The hydroxyl radical can react with organic matter as reaction (7) indicates

$$RH + OH\bullet \rightarrow R\bullet + H_2O \quad (7)$$

The hydroxyl radical has an oxidation potential of 2.8 V, just below fluorine 3.0; however, the ozone has 2.1 and the hydrogen peroxide only has 1.8 (Gandini *et al.*, 2000). Therefore, the main advantage of adding and electrogenerating H_2O_2 it to produce this hydroxyl radical which will react with the toxic and biorefractory organic pollutants present in wastewater. The organic radical can react with oxygen and hydroxyl radical to oxidation products, as reaction 8 and 9 show

$$R\bullet + O_2 \rightarrow \text{products} \quad (8)$$

$$R\bullet + OH\bullet \rightarrow \text{products} \quad (9)$$

Cyclic voltammetry

To obtain further information on the electrochemical processes occurring at the electrodes, a series of cyclic voltammetric experiments were performed using a CPE as the working electrode. Cyclic voltammetry results indicate that a chemically irreversible oxidation peak in the wastewater is detectable at potentials lower that those corresponding to oxygen evolution as shown in Figure 6. This peak corresponds to the direct electrochemical oxidation of pollutants present in wastewater. It is important to note that when cyclic voltammetry is applied after wastewater is electrochemically treated, the peak does not appear in the voltammogram indicating that pollutants in the solution have been oxidized. On the other hand, the voltammetric curves of reduction present similar behavior before and after the electrochemical treatment, possibly due to anions such as Cl^- and SO_4^{2-} in the wastewater. Thus, this voltammogram clearly indicate that there are processes attributable to direct oxidation of pollutants. These phenomena contribute to the destruction of organic matter present in the solution.

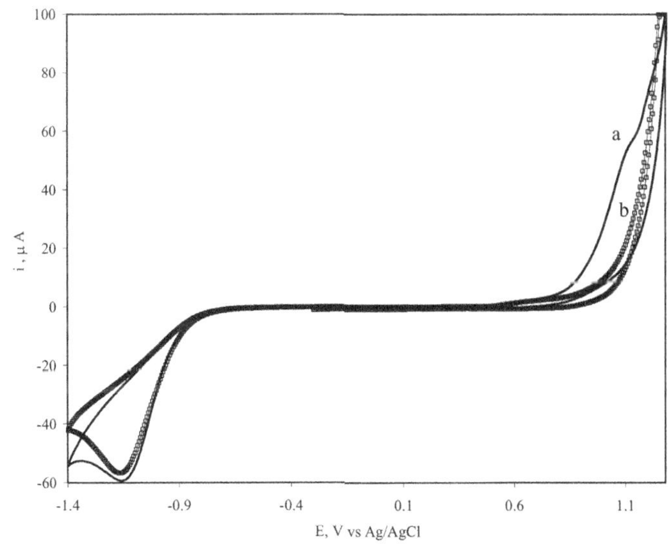

Figure 3 Cyclic voltammograms recorded at the CPE over the potential window from –1.4 to 1.2 V of the raw (a) and the electrolytic treated (b) wastewater at a scan rate of 0.1 Vs^1.

UV-Vis spectra

The UV-Vis spectra of the raw and electrochemically treated wastewater are shown in Figure 4. There are no peaks in the spectra corresponding to components of the raw and treated wastewater; it is only a continuous curve that has the higher absorbance in the zone of 300–400 nm. However, it is interesting to note that the absorbance of the curve decreases as electrochemical treatment time is increased.

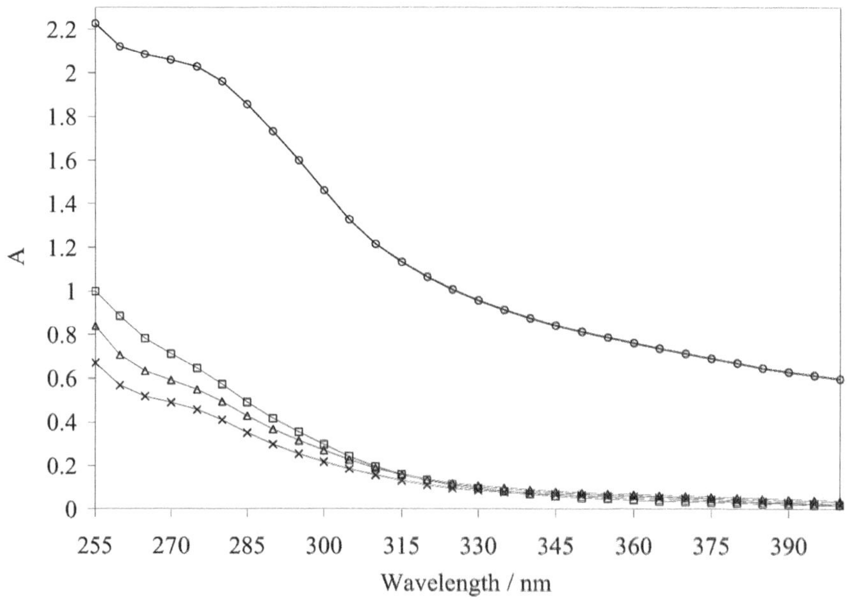

Figure 4 UV-Vis spectra of the raw wastewater (○) and treated water as a function of time, 20 min (□),40 min (Δ) and 60 min (×).

Scanning electron micrographs

To evaluate the structural features of the sludge that was generated by the electrochemical processes, a series of scanning electron microscopy experiments were performed. The micrograph in Figure 5 shows continuous flake-shaped aggregates with diameters of around 25 μm. This technique also allows elemental analysis of the sample by energy dispersion spectra, as shown in Figure 6. The peaks indicate that carbon, oxygen, aluminum, calcium and sulfur are present in the sludge. This information confirms that once the colloidal matter is destabilized, it can be separated from the wastewater and that the aluminum generated by the electrolysis forms complexes with the sulfur in the sludge.

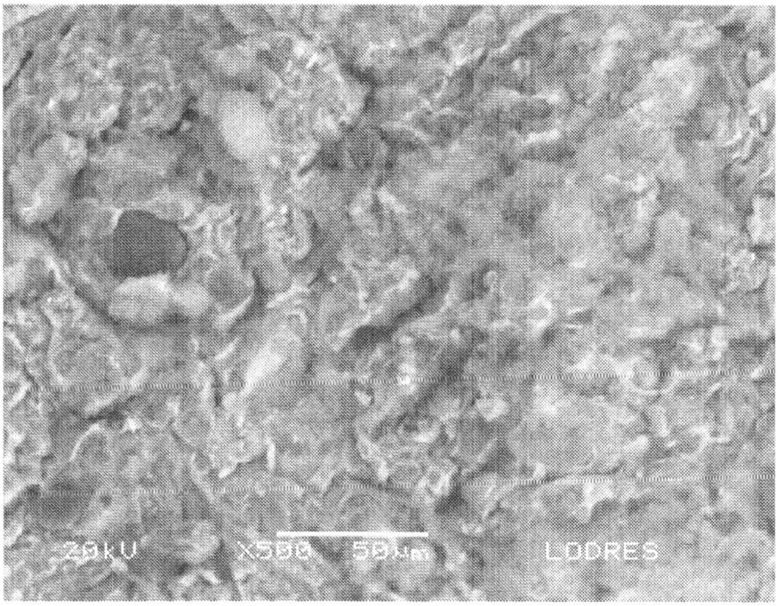

Figure 5 Micrograph of the sludge produced in the electrochemical process. The photograph was recorded at 500X and the marker is 50 μm.

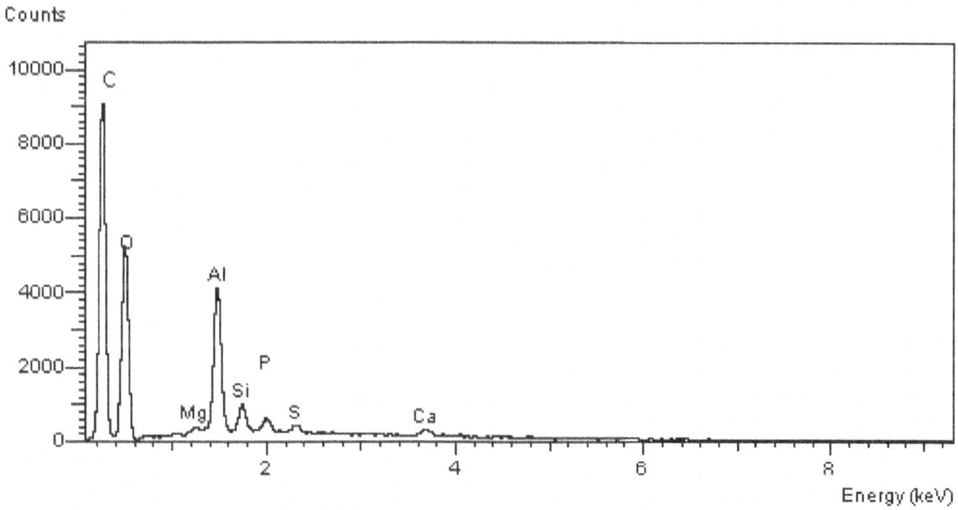

Figure 6. Energy dispersion spectra of the sludge produced in the electrochemical process showing the presence of the elements: C, O, Ca, Al and S.

Table 2 shows the characterization of the crude and treated wastewater using the optimal conditions found in this study. The BOD_5 and COD removal are 96 and 88% respectively. The color of the wastewater diminished by 33%; however the turbidity reduction is 97%. A very interesting effect is the disinfection by the electrochemical process, resulting in a 99.9% fecal coliforms reduction.

Table 2 Selected properties of raw and electrochemical treated wastewater. The electrolysis was applied to the wastewater for 40 minutes at pH 4 with a current density of 18.2 mA/m^2

Parameter	Raw wastewater	Treated wastewater
COD (mg/dm^3)	7489	910
BOD_5 (mg/dm^3)	3445	130
Color (Pt-Co U)	30	20
Turbidity (NTU)	1150	30
Chlorides (mg/dm^3)	110	77
pH	4.5	4.5
Fecal coliforms (MPN)	11,000	3
Helminth ovae	Not detected	Not detected
Total solids (mg/dm^3)	3804	892
Settable solids (mg/dm^3)	40	1

Conclusions

The electrochemical method presented in this study reduces the concentration of pollutants in wastewater. The use of predominant species in aqueous solution diagrams allowed the understanding of increasing pollutant removal from wastewater when pH is 4. Optimal conditions for the pollutant removal can be designed using this methodology. Cyclic voltammetry and UV-Vis confirms the improvement in the wastewater quality after the electrochemical treatment. Finally, scanning electron microscopy and elemental analysis shows the sludge morphology and the presence of C, O, Al, and S as its constituents.

Acknowledgements

We acknowledge the support given by the Universidad Autónoma del Estado de Mexico, specially the Facultad de Química (Project 2228/2006-01 and 22 /2004-2). To SEMARNAT CONACYT C01-0502 and to the SNI for supporting research.

References

Baes, C., Mesmer, R. (1986). *The Hydrolysis of Cations* (ed. Robert E. Kriegeer). Malabar, FL.

Barrera-Díaz, C.; Ureña-Nuñez, F.; Campos, E.; Palomar-Pardavé, M, Romero-Romo, M. (2003a). A combined electrochemical-irradiation treatment of highly colored and polluted industrial wastewater. *Radiat. Phys Chem.*, **67**, 657–663.

Barrera-Díaz, C., Palomar-Pardavé, M., Romero-Romo, M., Martínez, S. (2003b). Chemical and electrochemical considerations on the removal process of hexavalent chromium from aqueous media. *J. Appl. Electrochem.* **33**, 61–71.

Beltran de Heredia, J., Domínguez, J. R., López, R. (2004). Treatment of cork process wastewater by successive chemical-physical method. *J. Agric. Food Chem.*, **52**, 4501–4507.

Boye, B., Dieng, M., Brillas, E. (2003). Electrochemical degradation of 2,4,5-trichorophenoxyacetic acid in aqueous medium by peroxi-coagulation. Effect of pH and UV light. *Electrochimica Acta* **48**, 781–790.

Brillas, E., Casado, J. (2002). Aniline degradation by electro-fenton and peroxi-coagulation processes using a flow reactor for wastewater treatment. *Chemosphere* **47**, 241–248.

Brillas, E., Boye, B., Dieng, M. M. (2003a). Peroxi-coagulation and photoperoxi-coagulation treatments of the herbicide 4-Chlorophenoxyacetic acid in aqueous medium using an oxygen-diffusion cathode. *J. Electrochem. Soc.* **150**, E148–E154.

Brillas, E., Boye, B., Baños, M. A., Calpe, J. C., Garrido, J. A. (2003b). Electrochemical degradation of chorophenoxy and chlorobenzoic herbicides in acidic aqueous médium by the peroxi-coagulation method. *Chemosphere* **51**, 227–235.

Can, O. T.; Bayramoglu, M.; Kobya, M. (2003). Decolorization of reactive dye solutions by electrocoagulation using aluminum electrodes. *Ind. Eng. Chem. Res.* **42**, 3391–3396.

Cañizares, P., Carmona, M., Lobato, J., Martínez, F., Rodrigo M. A. (2005). Electrodissolution of aluminum electrodes in electrocoagulation processes. *Ind. Eng. Chem.* **44** (12), 4178–4185.

Do J. S., Chen C. P. (1994). In situ oxidative degradation of formaldehyde with hydrogen peroxide electrogenerated on the modified graphites. *J. Appl. Electrochem.* **24**, 936–942.

Domínguez, M. J.; Botello-Pozos, J. C.; López-Ortega, A.; Ramírez, M. T.; Sandoval-Flores, G.; Rojas-Hernández, A. (1998). Study of pillar precursos [Ga(III)-Al(III)-In(III)-Al(III), Zr(IV)] for hydrothermally stable pillared cáliz. *Catalysis Today*, **43**, 69–77.

Gandini D., Mahé E. Michaud P.A., Haenni W., Perret A., Comminellis Ch. (2000) Oxidation of carboxylic acids at boron-doped diamond electrodes for wastewater treatment. *J. Appl. Electrochem.*, **30** (12) 1345–1350.

Hirakawa T. and Nosaka Y. (2002). Properties of $O_2^{\bullet -}$ and OH• formed in TiO_2 aqueous suspensions by photocatalytic reaction and the influence of H_2O_2 and some ions *Langmuir*, **18**, 3247–3254.

Holt, P.K., Barton, G. W., Wark, M., Mitchell, C.A. (2002). A quantitative comparison between chemical dosing and electrocoagulation. *Colloids and Surfaces A: Physicochem. Eng. Aspects* **211**, 233–248.

Holt, P. K., Barton, G. W., Mitchell, C.A. (2004). Deciphering the science behind electrocoagulation to remove suspended clay particles from water. *Wat. Sci Technol.* **50**(12), 177–184.

Janssen, L. J. J., Koene, L. (2002). The role of electrochemistry and electrochemical technology in environmental protection. *J. Chem. Eng.* **85**, 127–136.

Jiang J., Graham N., Andre A. C., Kelsall G. H., Brandon N. (2002). Laboratory study of electro-coagulation–flotation for water treatment. *Wat. Res.*, **36**, 4064–4078.

Keenan, D., Sainkov A. (2000). Biological augmentation eliminates grease and oil in bakery wastewater. *Wat. Environ. Res.* **72**, 141–146.

Pacheco-Hernández, L. (1999). '*Estudio de especies de Al(III) – Ln(III)en solución acuosa*' MSc thesis, UAM-I: México.

Picard, T.; Cathalifaud-Feuillade, G.; Mazet, M.; Vandensteendam,C. (2000). Cathodic dissolution in the electrocoagulation process using aluminium electrodes. *J. Environ. Monit.*, **2**, 77–80.

Puigdomenech, I. (1997). *Hydrochemical equilibrium constants database (MEDUSA).* Royal Institute of Technology: Stockolm.

Rajeshwar, K., Ibañez, J. G., Swaine, G. M. (1994). Electrochemistry and Environment. *J. Appl. Electrochem.* **24**, 1077–1091.

Riveiro, A. B., Mateus, E. P., Ottosen, L. M., Bech-Nielsen, G. (2000). Electrodialytic removal of Cu, Cr, and As from chromated copper arsenate-treated timber waste. *Environ. Sci. Technol.* **34**, 784–788.

Roa-Morales, G., Galicia, L., Ramírez-Silva, M. T. (2003). Evidence of ternary Inclusión complexes formation using factorial design and determination of constant. *J. Inc. Phenom.* **46**, 139–145.

Smith, M. R., Martell A.E. (1974). *Critical Stability Constants*. Plenum Press, USA.

Vlyssides A., Barampouti E. M., Mai S. (2004). Degradation of methylparation in aqueous solution by electrochemical oxidation. *Environ. Sci. Technol.*, **38** (22), 6125–6131.

Keyword index